anatomy

The National Medical Series for Independent Study

anatomy

Ernest W. April, Ph.D.
Associate Professor of Anatomy and Cell Biology
College of Physicians & Surgeons
Columbia University
New York, New York

Illustrated by
Anne Erickson
and
Salvatore Montano

A WILEY MEDICAL PUBLICATION
JOHN WILEY & SONS
New York • Chichester • Brisbane • Toronto • Singapore

Harwal Publishing Company, Media, Pa.

Library of Congress Cataloging in Publication Data

April, Ernest W.
Anatomy.

(The National Medical series for independent study)
(A Wiley Medical publication)
Includes index.
1. Anatomy, Human—Outlines, syllabi, etc.
2. Anatomy, Human—Examinations, questions, etc. I. Title. II. Series. III. Series: Wiley Medical publication.
[DNLM: 1. Anatomy—Examination questions.
QS 18 A654a]
QM31.A67 1984 611'.002'02 84-6715
ISBN 0-471-09624-5

10 9 8 7 6 5 4 3

Contents

Preface

I. PURPOSE

A. OVERALL OBJECTIVES

1. **To organize** and **list** items and topics for easier comprehension, study, and review.
2. **To maximize** use of the limited time available to the student physician.
3. **To direct** the new student toward the most fruitful aspects of anatomic study by illustrating and emphasizing structural and functional detail pertinent to the skillful practice of up-to-date medicine.

B. INTERMEDIATE OBJECTIVES

1. **To present** pertinent **basic anatomy** with illustrations, diagrams, and tables; functional anatomic concepts; and **clinical notes** to provide the student physician with a framework upon which he or she may construct a working knowledge of human anatomy.
2. **To assist** the student physician in determining the relative importance of anatomic structures.
   - **a.** The treatment of the included subject material does not represent detailed discussions of every topic in anatomy, nor does it belabor the major points of anatomy, which are obvious and easily assimilated.
   - **b.** For the clinician in training some anatomic structures and concepts are more important than others.
   - **c.** Effort has been made to distinguish between the included *minutia* (clinically important detail) and the excluded *trivia* (inconsequential detail of interest only to anatomists and specialists).
3. **To provide** a **review** of human anatomy for:
   - **a. In-course examinations** in the professional health sciences curricula.
   - **b. Subsequent courses** in the medical curriculum.
   - **c.** Clinically oriented **licensure examinations:**
     - **(1)** The **NBME, Part I** (National Board Examinations).
     - **(2)** The new **FLEX I** (Federated Licensure Examination).
     - **(3)** The **ECFMG** (Examination Certifying Foreign Medical Graduates).
     - **(4)** Various medical **specialty board examinations.**

II. SUGGESTED METHOD OF USE

A. The subject presentation is **regional**, which is the method used in most medical teaching programs.

1. The **section order** is only one of a number of sequences that have proved workable, but one for which I have a preference.
2. Because the topics are extensively **cross-referenced**, the user of this outline and study guide need not proceed in any particular order.

B. To obtain maximal benefit from this outline and study guide:

1. Use in conjunction with:
   - **a.** A textbook of human anatomy.
   - **b.** An atlas of human anatomy.
2. Review from time to time the stated goals and learning objectives set forth before each section.

3. Use the pretest, section questions, and post-test.
   a. Taking the pretest will help the student to understand the direction and philosophy of this book.
   b. The questions at the end of each section and the post-test will not only indicate whether a basic knowledge of anatomy has been attained but also, because many of the questions are clinical in nature, enable the user to determine whether he or she has a working knowledge of human anatomy.

Ernest W. April

Acknowledgments

Because the function of a text or review book is to present concisely the basic information that forms an accepted body of knowledge, only the organization and presentation of those facts and concepts may be original. As such, I humbly acknowledge the numerous anatomic reference texts against which the material presented herein has been checked for accuracy. Even more humbly, I acknowledge those uncounted and anonymous deceased individuals as well as my numerous academic and clinical colleagues who have contributed over the years to my fund of knowledge and, finally, the student physicians against whom this anatomic knowledge has been honed. In addition, I am indebted to Dr. Timothy Chuter, a surgical colleague, for his thorough reading of the manuscript, his lively discussions, his contributions to the section on the extremities, and his assistance in writing much of the section on the head and neck.

The anatomic illustrations of Anne Erickson and Salvatore Montano confirm the importance of the visual aspects of anatomy. Much of their work has been based upon their own dissections.

Publisher's Note

The objective of the *National Medical Series* is to present an extraordinarily large amount of information in an easily retrievable form. The outline format was selected for this purpose of reducing to the essentials the medical information needed by today's student and practitioner.

While the concept of an outline format was well received by the authors and publisher, the difficulties inherent in working with this style were not initially apparent. That the series has been published and received enthusiastically is a tribute to the authors who worked long and diligently to produce books that are stylistically consistent and comprehensive in content.

The task of producing the *National Medical Series* required more than the efforts of the authors, however, and the missing elements have been supplied by highly competent and dedicated developmental editors and support staff. Editors, compositors, proofreaders, and layout and design staff have all polished the outline to a fine form. It is with deep appreciation that I thank all who have participated, in particular, the staff at Harwal—Debra L. Dreger, Jane Edwards, Gloria Hamilton, Jeanine Kosteski, Wieslawa B. Langenfeld, Keith LaSala, June A. Sangiorgio, Mary Ann C. Sheldon, and Jane Velker.

The Publisher

Introduction

Anatomy is one of seven basic science review books in a series entitled *The National Medical Series for Independent Study*. This series has been designed to provide students and house officers, as well as physicians, with a concise but comprehensive instrument for self-evaluation and review within the basic sciences. Although *Anatomy* would be most useful for students preparing for the National Board of Medical Examiners examinations (Part I, FLEX, and FMGEMS), it should also be useful for students studying for course examinations. These books are not intended to replace the standard basic science texts but, rather, to complement them.

The books in this series present the core content of each basic science area, using an outline format and featuring a total of 300 study questions. The questions are distributed throughout the book at the end of each chapter and in a pretest and posttest. In addition, each question is accompanied by the correct answer, a paragraph-length explanation of the correct answer, and specific reference to the outline points under which the information necessary to answer the question can be found.

We have chosen an outline format to allow maximal ease in retrieving information, assuming that the time available to the reader is limited. Considerable editorial time has been spent to ensure that the information required by all medical school curricula has been included and that each question parallels the format of the questions on the National Board examinations. We feel that the combination of the outline format and board-type study questions provides a unique teaching device.

We hope you will find this series interesting, relevant, and challenging. The authors, as well as the John Wiley and Harwal staffs, welcome your comments and suggestions.

Pretest

QUESTIONS

Directions: Each question below contains five suggested answers. Choose the **one best** response to each question.

1. Which of the following structures drains into the inferior meatus of the nose?

(A) Ethmoidal sinuses
(B) Frontal sinus
(C) Maxillary sinus
(D) Nasolacrimal duct
(E) Sphenoidal sinus

2. Hysterectomy (surgical removal of the uterus and ovaries) may result in injury to adjacent anatomic structures. One structure commonly injured is the

(A) external iliac artery
(B) rectum
(C) triangular ligament
(D) ureter
(E) urethra

3. A typical thoracic vertebra includes all of the following components EXCEPT

(A) a heart-shaped vertebral body
(B) inferior articular facets
(C) a neural canal
(D) superior costal facets
(E) transverse foramina

4. The tendon that can be seen and felt just posterior to the medial malleolus of the tibia during inversion of the foot is the tendon of which of the following muscles?

(A) Flexor hallucis longus
(B) Peroneus brevis
(C) Peroneus longus
(D) Tibialis anterior
(E) Tibialis posterior

5. Which of the following statements correctly describes the papillary muscles in the heart?

(A) They are rudimentary and have no major function
(B) They contract to close the atrioventricular valves during ventricular systole (contraction)
(C) They contract to open the atrioventricular valves during ventricular diastole (relaxation)
(D) They secure the chordae tendineae to the atrioventricular valve leaflets
(E) None of the above

6. All of the following circulatory changes normally occur immediately at birth EXCEPT

(A) decreased right atrial pressure
(B) increased blood flow through the lungs
(C) increased left atrial pressure
(D) reversal of flow through the ductus arteriosus
(E) reversal of flow through the foramen ovale

7. Which of the following statements best describes the teres major muscle?

(A) It contributes to the stability of the posterior part of the shoulder joint
(B) It divides the axillary artery into three parts
(C) It inserts on the humerus just distal to the infraspinatus muscle
(D) It is active in adduction of the shoulder joint
(E) It is innervated by the same nerve that supplies the deltoid muscle

8. Which of the following arteries frequently arises as a branch of the external iliac or inferior epigastric artery, instead of as a branch of the internal iliac artery?

(A) Internal pudendal
(B) Obturator
(C) Superior vesical
(D) Umbilical
(E) Uterine

9. Infection may spread from the nasal cavity to the meninges along the olfactory nerves. Olfactory fibers pass from the mucosa of the nasal cavity to the olfactory bulb via the

(A) anterior and posterior ethmoidal foramina
(B) cribriform plate of the ethmoid bone
(C) hiatus semilunaris
(D) nasociliary nerve
(E) sphenopalatine foramen

10. The muscles of the back receive motor innervation from

(A) dorsal primary rami
(B) dorsal roots
(C) posterior branches of the lateral perforating nerves
(D) ventral primary rami
(E) none of the above

11. Bilateral lumbar sympathectomy does not affect autonomic control of the descending colon because

(A) the descending colon receives its parasympathetic innervation from the pelvic splanchnics
(B) the descending colon receives its parasympathetic innervation from the vagus nerve
(C) the descending colon receives its sympathetic innervation from thoracic splanchnic nerves
(D) lumbar splanchnics innervate the pelvic viscera via the hypogastric nerve
(E) only presynaptic sympathetic fibers are severed

12. Femoral hernias pass deep (inferior) to the inguinal ligament. A segment of small intestine in this instance is in *direct* contact with the

(A) falx inguinalis
(B) lacunar ligament
(C) parietal peritoneum
(D) transversus abdominis muscle
(E) transversalis fascia

13. The tendon of which of the following muscles is involved when the tuberosity of the fifth metatarsal bone is avulsed (pulled off) in a sprain?

(A) Abductor digiti minimi
(B) Peroneus brevis
(C) Peroneus longus
(D) Tibialis anterior
(E) Tibialis posterior

14. Which of the following joints has only one degree of freedom (a one-arc joint)?

(A) Humeroulnar
(B) Metacarpophalangeal
(C) Radiocarpal
(D) Radiohumeral
(E) Sternoclavicular

15. Which of the following statements best characterizes the tunica dartos of the scrotum?

(A) It contains striated muscle fibers
(B) It is a continuation of the two layers of the superficial fascia
(C) It is involved in the cremaster reflex
(D) It is invested with adipose tissue
(E) It responds to cold temperature by lowering the scrotum away from the body

16. A characteristic of the intercostal neurovascular bundle that makes it particularly susceptible to injury from a fractured rib is that it lies

(A) behind the superior border of the rib
(B) beneath the inferior border of the rib
(C) between external and internal intercostal layers
(D) directly behind the midpoint of the rib
(E) halfway between two adjacent ribs

17. If cell bodies in the geniculate ganglion are damaged, you would expect the symptoms to be

(A) loss of sensation of pain from the face
(B) loss of sensation of touch from the anterior two-thirds of the tongue
(C) loss of taste from the anterior two-thirds of the tongue
(D) partial facial paralysis
(E) none of the above

18. Occlusion of the inferior mesenteric artery usually does not cause necrosis of the rectal mucosa because the

(A) arterial supply from the left colic artery compensates, via anastomoses across Sudeck's point
(B) inferior rectal artery, a major branch of the external iliac artery, also supplies the rectum
(C) major arterial supply to the rectum is from anastomotic connections with the superior mesenteric artery
(D) middle rectal artery, a major branch of the internal iliac artery, also supplies the rectum
(E) supply from the inferior mesenteric artery to the rectum is insignificant

19. All of the following structures pass deep to the flexor retinaculum EXCEPT the

(A) flexor digitorum profundus to the little finger
(B) flexor digitorum superficialis
(C) flexor pollicis longus
(D) median nerve
(E) ulnar artery

20. All of the following extrinsic extraocular muscles have a posterior site of attachment in the orbit EXCEPT the

(A) inferior oblique
(B) inferior rectus
(C) lateral rectus
(D) levator palpebrae superioris
(E) superior oblique

21. The anterior longitudinal ligament has which of the following characteristics?

(A) It anchors the emerging spinal nerves in place
(B) It limits the direction of nucleus pulposus extrusion during disk herniation
(C) It narrows anterior to the intervertebral disk
(D) It resists kyphosis
(E) None of the above

22. The ulnar nerve innervates which of the following muscles of the thumb?

(A) Abductor pollicis brevis
(B) Abductor pollicis longus
(C) Deep head of the flexor pollicis brevis
(D) Opponens pollicis
(E) Superficial head of the flexor pollicis brevis

23. The *median* umbilical fold is created by the

(A) falx inguinalis
(B) inferior epigastric arteries
(C) lateral borders of the rectus sheath
(D) obliterated umbilical arteries
(E) urachus

24. Which of the following muscles may be involved in flexion of the hip joint?

(A) Gluteus maximus
(B) Gluteus medius
(C) Gluteus minimus
(D) Piriformis
(E) Tensor fasciae latae

25. The cerebrospinal fluid enters the venous system

(A) at arachnoid granulations
(B) at the cisterna magna
(C) through subarachnoid veins
(D) via capillaries in the ependyma
(E) by none of the above routes

26. The external urethral sphincter functions to maintain urinary continence and

(A) has identical anatomic structure in both the male and female
(B) is a portion of the pelvic diaphragm
(C) is under involuntary control
(D) receives innervation by the pudendal nerve
(E) surrounds the prostatic portion of the urethra in the male

27. Intramuscular injections in the superomedial gluteal quadrant may injure which of the following nerves?

(A) Inferior gluteal
(B) Posterior femoral cutaneous
(C) Pudendal
(D) Sciatic
(E) Superior gluteal

28. The gallbladder, which stores and concentrates bile, is located

(A) between the right and caudate lobes of the liver
(B) between the right and quadrate lobes of the liver
(C) in the coronary ligament
(D) in the falciform ligament
(E) in the lesser omentum

29. The arterial supply to the maxillary and mandibular teeth comes from

(A) a single branch of the maxillary artery
(B) branches of the internal carotid artery
(C) branches of the maxillary and sublingual artery, respectively
(D) the maxillary and facial arteries
(E) separate branches of the maxillary artery

30. Which of the following portions of the gastrointestinal tract become predominantly secondarily retroperitoneal during development?

(A) Appendix
(B) First part of the duodenum
(C) Liver
(D) Sigmoid colon
(E) Uncinate process of the pancreas

Directions: Each question below contains four suggested answers of which **one or more** is correct. Choose the answer

A if **1, 2, and 3** are correct
B if **1 and 3** are correct
C if **2 and 4** are correct
D if **4** is correct
E if **1, 2, 3, and 4** are correct

31. The pectoralis minor muscle is a useful landmark structure that is characterized by which of the following statements?

(1) It is located at the level of the cords of the brachial plexus
(2) It is attached to the acromion process
(3) It divides the axillary artery into three portions
(4) It is innervated by the middle subscapular nerve

32. Structures that drain into the submandibular lymph nodes include the

(1) pinna of external ear
(2) nasal mucosa
(3) parotid gland
(4) teeth and gingivae

33. Correct statements describing the pelvic diaphragm include which of the following?

(1) It is comprised of the levator ani and coccygeus muscles and their fasciae
(2) It functions to suspend and support pelvic organs
(3) It does not share the same plane with the urogenital diaphragm
(4) It is innervated on its perineal surface by twigs from the pudendal nerve

34. Characteristics of external hemorrhoids, which develop in branches of the inferior hemorrhoidal vein, include which of the following?

(1) They appear in the rectum
(2) They are painful
(3) They are superior (proximal) to the pectinate line
(4) They lie underneath the mucosa

35. Cell bodies that are located in the superior cervical ganglion have which of the following functions?

(1) They prevent ptosis (drooping) of the eyelid
(2) They contribute to the greater superficial petrosal nerve
(3) They dilate the pupil
(4) They contract the ciliary muscle

36. Blood is returned to the left side of the heart via which of the following vessels?

(1) Anterior cardiac veins
(2) Thebesian veins
(3) Coronary sinus
(4) Pulmonary veins

37. Which of the following structures pass through the lesser sciatic foramen?

(1) Tendon of the obturator internus muscle
(2) Piriformis muscle
(3) Pudendal nerve
(4) External pudendal artery and vein

38. One annoying symptom of Bell's palsy that affects the facial nerve (CN VII) is hyperacusis, which is caused by involvement of the

(1) tympanic membrane
(2) tensor tympani muscle
(3) cochlear nerve
(4) stapedius muscle

39. Correct statements describing the lunate bone include which of the following?

(1) It may produce carpal tunnel syndrome if displayed anteriorly
(2) It provides an attachment for the flexor retinaculum
(3) It articulates with a fibrocartilaginous disk
(4) It is a component of the carpometacarpal joint

40. The perineum is supplied by which of the following nerves?

(1) Genitofemoral
(2) Inferior rectal
(3) Ilioinguinal
(4) Pudendal

41. To equalize air pressure on both sides of the tympanic membrane, which of the following muscles might contract?

(1) Salpingopharyngeus
(2) Tensor veli palatini
(3) Tensor tympani
(4) Levator veli palatini

42. Correct statements concerning the kidney include which of the following?

(1) Nephroptosis produces pain caused by traction on the renal vessels
(2) Destruction of the sympathetic ganglia associated with levels T10 to L2 produces diuresis secondary to renal vasoconstriction
(3) The perirenal fascia permits spread of infection along the ureters and between the kidneys and pelvic structures
(4) Major collateral circulation is provided to the kidney from supernumerary renal arteries when present

43. Respiratory mechanics involve coordinated activity of numerous muscles. Contraction of which of the following muscles contributes to forced expiration?

(1) Transverse abdominis 
(2) External oblique
(3) Rectus abdominis
(4) Internal oblique

44. Infection within the pterygopalatine fossa may track directly into which of the following cavities?

(1) Nasal cavity 
(2) Middle cranial fossa
(3) Oral cavity
(4) Orbital cavity

45. Reposition of the thumb (antagonistic to opponens action) is due to the

(1) abductor pollicis longus
(2) extensor pollicis brevi
(3) extensor pollicis longus
(4) adductor pollicis

46. The vestibule in the female perineum is characterized by which of the following statements?

(1) It receives both the urethra and the vagina
(2) It is bordered by the labia majora
(3) It receives drainage from the greater vestibular and paraurethral glands
(4) It receives sensory innervation from the nervi erigentes

47. When the tibial nerve is injured, some flexion may still be possible at the knee joint because of the actions of which of the following muscles?

(1) Gracilis
(2) Biceps femoris (long head)
(3) Biceps femoris (short head)
(4) Gastrocnemius

48. During a spinal tap, which of the following structures would be penetrated if the needle was inserted precisely in the midline?

(1) Dura mater
(2) Posterior longitudinal ligament
(3) Supraspinous ligament
(4) Ligamentum flavum

49. Myocardial infarction limited to the interventricular septum might be expected to produce problems, including

(1) aortic valvular insufficiency
(2) tricuspid valve regurgitation
(3) mitral valve regurgitation
(4) disturbances of cardiac impulse conduction

50. The superomedial boundary of the popliteal fossa is formed by which of the following muscles?

(1) Semimembranosus
(2) Long head of the biceps femoris
(3) Semitendinosus
(4) Short head of the biceps femoris

SUMMARY OF DIRECTIONS

A	B	C	D	E
1, 2, 3 only	1, 3 only	2, 4 only	4 only	All are correct

51. Correct statements concerning the superior laryngeal nerve include which of the following?

(1) It has an internal branch that pierces the thyrohyoid membrane to innervate the laryngeal mucosa
(2) It produces muscle contraction that lengthens the (true) vocal folds
(3) It provides the afferent limb of the cough reflex
(4) It innervates all of the laryngeal musculature by an external branch, except the cricothyroid muscle

52. Structures that transit the diaphragm via the esophageal hiatus include the

(1) azygos vein
(2) thoracic duct
(3) hemiazygos vein
(4) right vagus nerve

53. Pronounced dilation of a smooth muscular tube in the body generally produces pain. Sensory fibers for pain from the cervix travel to the spinal cord via

(1) a plexus around the uterine arteries
(2) the lateral pelvic plexus and inferior hypogastric plexus
(3) nervi erigentes
(4) white rami of L1 and L2

54. Structures that are both medial to the biceps tendon and deep to the bicipital aponeurosis include the

(1) brachial artery
(2) deep (profunda) brachial artery
(3) median nerve
(4) median cubital vein

55. A longitudinal incision just lateral to the medial border of the rectus sheath to gain access to the abdominal cavity will have which of the following results?

(1) It will spare the blood supply to the rectus muscle
(2) It will require division of the transversus abdominis muscle
(3) It will avoid paralysis of the rectus muscle
(4) It will be parallel to Langer's lines of cleavage

56. In moving the eye outward the lateral rectus as well as the superior and inferior oblique muscles are used. These muscles are innervated by which of the following nerves?

(1) Abducens nerve
(2) Inferior division of the oculomotor nerve
(3) Trochlear nerve
(4) Superior division of the oculomotor nerve

57. During full active extension of the knee joint, which of the following are tightened?

(1) Tibial collateral ligament
(2) Fibular collateral ligament
(3) Patella tendon
(4) Anterior cruciate ligament

58. If the motor root of the trigeminal nerve is injured, paralysis occurs in which of the following muscles?

(1) Posterior belly of the digastric
(2) Tensor tympani
(3) Buccinator
(4) Masseter

59. In angina pain of cardiac origin, the pain radiating across the precordium and perhaps down the arm to the wrist is mediated by increased activity in afferent fibers contained in the

(1) cervical cardiac nerves
(2) vagus nerves
(3) first four thoracic splanchnic nerves
(4) phrenic nerves

60. A gray ramus communicans, which extends between a sympathetic trunk ganglion and an anterior primary ramus of a spinal nerve, contains

(1) axons of a preganglionic neuron
(2) fibers that conduct general somatic afferent impulses
(3) myelinated fiber
(4) fibers that activate smooth muscle (errector pili) of the skin

ANSWERS AND EXPLANATIONS

1. The answer is D. *(Chapter 33 II B 3–4)* The nasolacrimal duct drains into the inferior meatus of the nose. The maxillary, frontal, and anterior ethmoidal sinuses drain into the middle meatus. The sphenoidal sinus drains into the superior meatus.

2. The answer is D. *[Chapter 22 VI D 2 i (2)]* The ureters, which course through the deep pelvis, lie in the transverse cardinal ligament beneath the uterine arteries. Ligation of the uterine arteries during hysterectomy must be accomplished with care so that the ureters are not injured.

3. The answer is E. *[Chapter 19 I C, D 1 c (2)]* Transverse foramina are characteristic of cervical vertebrae. These foramina are formed by a partial fusion of the transverse and costal processes of the cervical vertebrae. Except for the seventh cervical vertebra, the transverse foramina contain the vertebral artery as it passes toward the foramen magnum.

4. The answer is E. *(Chapter 25 I C 1 g; Chapter 26 II A 2 c)* The tibialis posterior tendon is palpable just posterior to the medial malleolus. Adjacent to this tendon is the tendon of the flexor digitorum longus and the posterior tibial artery. The posterior tibial pulse may be palpated here. The peroneus longus and brevis course posterior to the lateral malleolus, whereas the tibialis anterior is on the dorsum of the foot.

5. The answer is E. *[Chapter 13 VI B 3 b (2) (a)]* The papillary muscles attach the chordae tendineae of the atrioventricular valve cusps to the walls of the ventricular chambers. Contraction of the papillary muscles prevents eversion of the valve cusps as the ventricular chambers decrease in volume during the second (ejection) phase of systole.

6. The answer is E. *(Chapter 13 X C 1–3)* At birth major circulatory changes occur on expansion of the lungs. The resultant increased blood flow through the pulmonary vascular bed lowers the pressure within the right chambers of the heart. The pressure within the right atrium becomes less than that of the left atrium, thereby causing the valve of the foramen ovale to close. A brief reversal of flow through the ductus arteriosus occurs before this structure constricts as a result of elevated levels of prostaglandins in the blood circulation.

7. The answer is D. *(Chapter 6 Table 6-2)* The prime action of the teres major muscle is adduction of the humerus at the shoulder joint. Because this muscle inserts on the lesser tubercle of the humerus, it also acts as a medial rotator and, to a lesser degree, an extensor.

8. The answer is B. *[Chapter 15 X F 2 e; Chapter 20 IV C 4 c (3)]* The obturator artery, usually a branch of the internal iliac artery, may arise instead from the inferior epigastric artery or even from the external iliac artery. An aberrant obturator artery crosses the femoral ring, where it may become involved in a femoral hernia or complicate the repair of such a hernia.

9. The answer is B. *[Chapter 28 III C 2 b (1) (b) (iii); Chapter 29 V A 1 d; VI B 2 b]* The cribriform plate of the ethmoid bone provides the passageways for the olfactory nerves from the olfactory mucosa of the superior nasal meatus to the olfactory bulb of the brain. Infection may track along these nerves and thereby spread to the meninges. Fracture of the ethmoid bone may result in leaking of cerebrospinal fluid through the nose.

10. The answer is A. *(Chapter 4 III C 4 a, b; Chapter 19 III C 3 a, b)* The muscles of the back are innervated by dorsal primary rami of the spinal nerves. The ventral primary rami, giving off lateral and ventral perforating branches, innervate the lateral and ventral regions of the body wall. The dorsal roots are sensory.

11. The answer is A. *[Chapter 17 X D 8; XV A 2 b (4)(b)]* The control of peristalsis is primarily a function of parasympathetic innervation. The descending colon receives parasympathetic innervation from the pelvic splanchnic nerves, which arise from levels S2 to S4. Bilateral lumbar sympathectomy for the relief of intractable visceral pain, therefore, has little effect on the motility of the descending colon.

12. The answer is C. *(Chapter 15 X F 2 b, c)* A femoral hernia passes through the femoral ring, where it is surrounded by the inguinal ligament (anteriorly and superiorly), the lacunar ligament (medially), the pectineal ligament (posteriorly and inferiorly), and the femoral vein (laterally). The hernial sac is in contact with peritoneum, which is covered by attenuated transversalis fascia.

13. The answer is B. *(Chapter 25 III C 1 b (2); Table 25-1)* The peroneus brevis muscle, a plantar flexor and pronator of the foot, inserts into the tuberosity of the fifth metatarsal bone. The peroneus longus inserts onto the first metatarsal; the tibialis posterior, onto the navicular bone; the tibialis anterior, onto the first cuneiform bone and first metatarsal; and the abductor digiti minimi, onto the first phalanx of the fifth toe.

14. The answer is A. *(Chapter 7 III A 1 a, 2 a)* The humeroulnar joint has only one degree of freedom, which allows flexion-extension at the elbow joint. The humeroradial joint has two degrees of freedom, permitting rotation as well as flexion-extension at the elbow. The metacarpophalangeal and radiocarpal joints have two degrees of freedom, permitting abduction-adduction and flexion-extension. The sternoclavicular joint has two degrees of freedom, permitting circumduction.

15. The answer is B. *(Chapter 21 IV A 2)* The tunica dartos of the scrotum is formed by fusion of the superficial and deep layers of the superficial fascia. This layer contains smooth muscle that responds to cold by raising the scrotum toward the body.

16. The answer is B. *(Chapter 11 III D 3 d (2); V A 1, 2 b)* Each intercostal neurovascular bundle lies beneath the inferior border of a rib between the internal intercostal muscle and the innermost intercostal muscle, which is a division of the internal intercostal. Fracture of a rib may injure the nerve or blood vessels. If a fragment of rib also tears the pleura, hemothorax may result.

17. The answer is C. *(Chapter 29 V G 1 a)* The geniculate ganglion, the equivalent of a dorsal root ganglion, contains the cell bodies of the sensory neurons of the facial nerve (CN VII). Selective damage to these neurons results in loss of taste sensation from the anterior two-thirds of the tongue. The motor division of the facial nerve supplies the muscles of facial expression.

18. The answer is D. *(Chapter 17 XI C 1)* The rectum receives its blood supply from the superior rectal branch of the inferior mesenteric artery and the middle rectal branches of the internal iliac arteries. Because there usually are anastomoses between these two systems, occlusion of one or the other would not necessarily cause ischemic necrosis of the rectal mucosa.

19. The answer is E. *[Chapter 9 III A 2 a (6)]* At the wrist the ulnar artery lies between the tendons of the flexor carpi ulnaris muscle and the flexor digitorum superficialis muscle, where the ulnar pulse may be palpated. As the ulnar artery enters the hand, it lies superficial to the flexor retinaculum on the radial side of the pisiform bone.

20. The answer is A. *(Chapter 30 I C 5 b; Table 30-1)* The inferior oblique muscle of the eye originates from the maxilla at the anteromedial border of the orbit. All the other extraocular muscles, including the levator palpebrae superioris, originate from the apex of the orbit, which surrounds the optic foramen.

21. The answer is B. *(Chapter 19 I F 4)* The anterior longitudinal ligament provides support for the vertebral column and reinforces the anterior and lateral aspects of the intervertebral disks. Consequently, herniation of an intervertebral disk tends to be in a posterolateral direction; spinal nerves may become involved as they pass through the intervertebral foramina.

22. The answer is C. *(Chapter 10 VII C 1 b)* The thumb musculature is innervated by the ulnar, median, and radial nerves. The ulnar nerve innervates the deep head of the flexor pollicis brevis muscle as well as the adductor pollicis muscle. The opponens pollicis, abductor pollicis brevis, and superficial head of the flexor pollicis brevis are innervated by the median nerve. The abductor pollicis longus as well as the extensor pollicis longus and extensor pollicis brevis are innervated by the radial nerve.

23. The answer is E. *(Chapter 16 II E 2 d)* The median umbilical fold is formed by the peritoneum draping over the underlying urachus. If the urachus is patent to any extent, an incision carried across this structure may result in leakage of urine into the peritoneal cavity. The paired lateral umbilical folds are formed by the inferior epigastric arteries, whereas the obliterated umbilical arteries form the paired *medial* umbilical folds.

24. The answer is E. *[Chapter 23, II C 3 a (1)]* The tensor fasciae latae muscle, originating at the iliac crest and inserting into the iliotibial tract, is a flexor of the thigh at the hip joint. Because the iliotibial band passes anterior to the axis of rotation of the knee joint when the knee is extended, the tensor fasciae latae muscle also keeps the extended knee in the extended position.

25. The answer is A. *(Chapter 28 IV A 2 b (1); Chapter 29 Figure 29-9)* Cerebrospinal fluid is secreted by the choroid plexuses of the lateral ventricles, the third ventricle, and the fourth ventricle. It circulates through the ventricles and enters the subarachnoid space through the foramina of Luschka and Magendie in the roof of the fourth ventricle. The cerebrospinal fluid enters the superior sagittal sinus via arachnoid granulations.

26. The answer is D. *(Chapter 21 IV D 1 b (2); V D 4 b; Chapter 22 II C 1 b, 2 c)* The external urethral sphincter, which surrounds the membranous urethra, is a portion of the deep transverse perineal muscle and a component of the urogenital diaphragm. This sphincter is innervated by the perineal branch of the pudendal nerve and is under voluntary control.

27. The answer is E. *(Chapter 23 II E 1–3, F)* The superior gluteal neurovascular bundle lies in the superomedial quadrant of the gluteal region. The sciatic nerve and the inferior gluteal neurovascular bundle lie in the inferolateral and inferomedial gluteal quadrants, respectively. The safest location for intramuscular injection is in the superolateral gluteal quadrant.

28. The answer is B. *(Chapter 17 VI D 2)* The gallbladder develops in the ventral mesogastrium and lies between the right and quadrate lobes of the liver. Because the quadrate lobe is functionally a part of the left lobe, it is reasonable that the gallbladder be located between the two functional divisions of the liver.

29. The answer is E. *[Chapter 32, VIII A 2 a (4), c (2), (5)]* The blood supply to the maxillary teeth is from the posterosuperior alveolar and anterosuperior alveolar branches of the infraorbital branch of the (internal) maxillary artery. The blood supply to the mandibular teeth is from the inferior alveolar branch of the (internal) maxillary artery.

30. The answer is E. *(Chapter 16 VII B 2, 3)* The pancreas, including its uncinate process, become secondarily peritoneal. The liver is a peritoneal structure. It is suspended in the peritoneal cavity by the gastrohepatic ligament, hepatoduodenal ligament, falciform ligament, and coronary ligaments. The appendix and sigmoid colon are peritoneal structures supported by the mesoappendix and sigmoid mesocolon, respectively. The first part of the duodenum is peritoneal, supported by the hepatoduodenal ligament.

31. The answer is B (1, 3). *(Chapter 6 III C 2 a; Table 6-1)* As the pectoralis minor muscle courses toward its insertion on the coracoid process of the scapula, it divides the axillary artery into three regions. It is innervated by the medial pectoral nerve, a branch of the medial cord of the brachial plexus.

32. The answer is C (2, 4). *(Chapter 34 V D 1)* The lymphatic drainage of the anterior-inferior portion of the face, the nasal cavities, and the anterior portions of the oral cavity, including the anterior margin of the tongue, gingivae, and teeth, is through the submandibular lymph nodes to the deep cervical nodes. The lymphatic drainage of the external ear as well as the parotid gland and anterosuperior portion of the face is toward the superficial cervical lymph nodes.

33. The answer is A (1, 2, 3). *(Chapter 20 II A 1, 4 d)* The pelvic diaphragm, consisting of the levator ani and coccygeus muscles, is innervated by twigs from the sacral plexus. It functions to support the pelvic viscera and is reinforced anteriorly by the urogenital diaphragm.

34. The answer is C (2, 4). *(Chapter 17 XI C 2; XIII D 4 c; Chapter 21 III F 2 b)* Hemorrhoids are varicosities of the mucosal venous plexus. External hemorrhoids form as a result of enlargement of venous anastomotic connections between the middle and inferior rectal (hemorrhoidal) veins. External hemorrhoids are painful because the anal canal in which they form is innervated by rectal branches of the pudendal nerve.

35. The answer is B (1, 3). *(Chapter 34, VI D 2 b)* Postsynaptic sympathetic neurons that innervate the head are located in the superior cervical ganglion. Some of these neurons innervate the smooth muscle (of Müller) in the eyelid, which prevents ptosis. Other postganglionic neurons innervate the dilator pupillae muscle of the eye. Pupillary constriction is a function of the parasympathetic division of the oculomotor nerve (CN III). The greater superficial petrosal nerve is a parasympathetic portion of the facial nerve (CN VII).

36. The answer is C (2, 4). *(Chapter 13 IV B 3)* Blood is returned to the left side of the heart from the lungs via the pulmonary veins and from the innermost layers of the left atrial wall via the thebesian veins. The anterior cardiac veins, coronary sinus, superior vena cava, and inferior vena cava, as well as the thebesian veins, return blood to the right side of the heart.

37. The answer is B (1, 3). *[Chapter 20 I A 1 b (4)]* The pudendal nerve passes out of the greater sciatic foramen and through the lesser sciatic foramen on its way between the sacral plexus and the perineum. In addition, the tendon of the obturator internus muscle passes out of the deep pelvis via the lesser sciatic foramen on its way to the greater trochanter of the femur.

38. The answer is D (4). *[Chapter 32 III A 4 a, 7 d; B 3 c (5) (c)]* The stapedius muscle, innervated by the facial nerve (CN VII), functions to dampen loud sounds by limiting the movement of the stapes at the oval window. Nerve dysfunction with paralysis of the stapedius muscle produces hyperacusis. The tensor tympani muscle, which dampens the movement of the tympanic membrane, is innervated by the motor division of the trigeminal nerve (CN V).

39. The answer is B (1, 3). *(Chapter 8 II B 1 b)* The lunate bone, located in the proximal row of carpal bones between the scaphoid and triangular bones, participates in the radiocarpal (wrist) joint, where it articulates with the radius and the articular disk of the wrist. Anterior displacement of the lunate bone produces carpal tunnel syndrome by compressing the superficial and deep digital flexors as well as the median nerve.

40. The answer is E (all). *(Chapter 21 IV G; V F 1; Figures 21-8 and 21-13)* The principal innervation of the perineum is by the pudendal nerve. The perineal branches supply the urogenital triangle, and the inferior rectal branches supply the anal triangle. However, the ilioinguinal nerve and genital branch of the genitofemoral nerve supply the anterior portion of the labia majora or the scrotum. In addition, the posterior femoral cutaneous nerve innervates a posterolateral portion of the perineum.

41. The answer is C (2, 4). *[Chapter 33 III B 3 a (1) (c), (2) (c)]* The tensor veli palatini and the levator veli palatini arise in part from the lateral and medial sides of the cartilaginous portion of the tympanic tube. Tension in these muscles, such as raising the palate during swallowing, opens the auditory tube so that air pressure equalizes between the middle ear and the nasopharynx.

42. The answer is A (1, 2, 3). *(Chapter 18 I B 1–2, F 4 b, H 1 b)* The arteries to the kidney are end-arteries, each supplying a segment of renal parenchyma. Ligation of a supernumerary or polar renal artery results in ischemic necrosis of a portion of the kidney.

43. The answer is E (all). *(Chapter 11 VII D 2; Chapter 15 IV B 4)* The muscles of the abdominal wall (including the external oblique, internal oblique, transverse abdominis, and rectus abdominis) are antagonistic to the diaphragm and participate in forced expiration.

44. The answer is E (all). *(Chapter 31, II B 10 e; Chapter 32 VI D 3)* The branches of the nerves and vessels of the pterygopalatine fossa reach the nose, eye, and mouth through several foramina. The pterygopalatine fossa communicates with the middle cranial fossa via the foramen rotundum and the vidian canal, with the nasal cavity via the sphenopalatine foramen, with the orbital cavity via the inferior orbital fissure, and with the oral cavity via the palatine canal.

45. The answer is A (1, 2, 3). *(Chapter 10 IV D 2 e (2) (c); Table 10-2)* Reposition of the thumb is the result of contraction of the abductors and extensors. The adductor pollicis draws the thumb alongside the first digit.

46. The answer is B (1, 3). *(Chapter 21 V A 3)* The vestibule, bordered by the labia minora, is innervated by the perineal branches of the pudendal nerve. It receives the vagina and urethra as well as drainage from the paraurethral and vestibular glands.

47. The answer is B (1, 3). *(Chapter 24 Table 24-1)* The gracilis muscle, innervated by the obturator nerve, and the short head of the biceps femoris muscle, innervated by the common peroneal nerve, both cross the knee joint and may flex this joint, albeit weakly, if there is tibial nerve palsy. The long head of the biceps and the gastrocnemius muscle are both innervated by the tibial nerve.

48. The answer is B (1, 3). *(Chapter 19 I F; III E 1)* During a midline spinal tap, the supraspinous ligament, interspinous ligament, and dura mater are penetrated. The flaval ligaments lie to each side of the midline, extending between laminae of adjacent vertebrae. The posterior longitudinal ligament lies along the posterior surface of the vertebral bodies.

49. The answer is C (2, 4). *(Chapter 13 IV A 1–2; VI B 3 b; VII B 2)* An infarct within the interventricular septum can produce ventricular arrhythmias caused by involvement of the common atrioventricular conduction bundle (of His) or either the left or right bundle branch. In addition, because only the tricuspid (right atrioventricular) valve has septal papillary muscles, infarct of the septal region can lead to ischemic necrosis of the septal papillary muscles and avulsion of the chordae tendineae, with resultant tricuspid incompetence.

50. The answer is B (1, 3). *(Chapter 24 Figure 24-1)* The semimembranosus muscle and the tendon of the semitendinosus muscle together form the superomedial boundary of the popliteal fossa. The biceps femoris forms the superolateral boundary, the two heads of the gastrocnemius muscle form the inferior boundaries, and the popliteus muscle forms the floor.

51. The answer is A (1, 2, 3). *(Chapter 33 V C 2, E 3 c)* The large internal branch of the superior laryngeal nerve, a branch of the vagus nerve (CN X), innervates the laryngeal mucosa and provides the afferent limb of the cough reflex. The smaller external branch of this nerve innervates the cricothyroid muscle, which functions to lengthen the vocal folds. The other muscles of the larynx are innervated by the inferior (recurrent) laryngeal nerve.

52. The answer is D (4). *(Chapter 11 IV C 2 a, b; Chapter 14 II H 2 d)* The left and right vagus nerves, along with the esophagus, pass through the diaphragm via the esophageal hiatus. The azygos vein, the hemiazygos vein, and the thoracic duct pass through the aortic hiatus.

53. The answer is B (1, 3). *[Chapter 22 VI D 2 g (4) (b)]* Visceral afferent nerves from the cervix course along the uterine arteries to the lateral pelvic plexus and thence along the nervi erigentes to sacral segments S2 to S4 of the spinal cord. Consequently, pain of cervical dilation is referred to the midsacral dermatomes, which include the perineal region, posterior leg, and lateral aspect of the foot. Conversely, afferents from the body of the uterus course to the upper lumbar levels via the inferior hypogastric plexus and refer pain to the back, groin, and thigh.

54. The answer is B (1, 3). *[Chapter 7 IV A 3; Chapter 9 IV A 1 b (1)]* The brachial artery and median nerve are both deep to the bicipital aponeurosis and medial to the biceps tendon. The medial cubital vein is superficial to the bicipital aponeurosis. The deep brachial artery anastomoses with the recurrent radial artery, which is lateral to the biceps tendon.

55. The answer is B (1, 3). *(Chapter 15 IX A 1 a–c)* The blood supply to the rectus abdominis muscle is primarily from the superior epigastric artery, a branch of the internal thoracic artery, and the inferior epigastric artery, a branch of the external iliac artery. The innervation of the rectus abdominis muscle is provided by terminal branches of the spinal nerves, which enter the rectus sheath laterally. A longitudinal incision just lateral to the medial border of the rectus sheath avoids compromising the vascular and nerve supply to the muscle.

56. The answer is A (1, 2, 3). *(Chapter 30 Table 30-1)* The lateral rectus muscle, the principal abductor of the eye, is innervated by the abducens nerve (CN VI). The superior oblique muscle, an elevator and abductor of the eye, is innervated by the trochlear nerve (CN IV). The inferior oblique muscle, a depressor and abductor of the eye, and the inferior rectus muscles are innervated by the inferior division of the oculomotor nerve (CN III). The superior rectus and medial rectus muscles, as well as the levator palpebrae superioris, are all innervated by the superior division of the oculomotor nerve.

57. The answer is E (all). *[Chapter 24, II C 3 b (6), c (6), 4 b (2) (b)]* The lateral and medial collateral ligaments as well as the anterior cruciate ligament are tightened on full extension of the knee. Because extension is primarily a function of the quadriceps femoris muscle, the patellar tendon is also under considerable tension.

58. The answer is C (2, 4). *[Chapter 32 III B 3 c (5) (a)]* The trigeminal nerve provides motor innervation to those muscles derived from the first branchial arch. These include the muscles of mastication, of which the masseter is one, the tensor tympani, the tensor veli palatini, the mylohyoid, and the anterior belly of the digastric. The posterior belly of the digastric muscle and the buccinator muscle are both innervated by the facial nerve (CN VII).

59. The answer is B (1, 3). *(Chapter 13 VIII B; Chapter 14 II I 4 a)* Pain originating in the heart travels along visceral afferent nerve fibers, which take two pathways to reach the spinal cord. Some cardiac afferents run along the middle and inferior cardiac cervical (accelerator) nerves to reach the sympathetic chain and then travel inferiorly to the level of T1 or T2. Other cardiac afferents travel along the upper thoracic splanchnic nerves to reach the sympathetic chain. The cardiac afferent fibers leave the sympathetic chain through the white rami communicantes and travel along the spinal nerve to reach the corresponding level of the spinal cord. Pain of cardiac origin is referred to (appears as if orginating from) dermatomes T1 to T4.

60. The answer is D (4). *[Chapter 4 V D 1 b (2) (a)]* Each gray ramus communicans of the sympathetic chain is composed of unmyelinated postsynaptic sympathetic neurons, which reenter the spinal nerve and travel to the integument. These neurons innervate the sweat glands as well as smooth muscle associated with the hair follicles and the arterioles.

Part I
Introductory Overviews

1
Introduction

I. HUMAN ANATOMY, the oldest medical science, traces its origins to early Greek civilizations.

A. The Greek term for anatomy (*anatome*) means taking apart, as does the Latin term for dissection (*dissecare*).

B. From anatomy came the daughter medical sciences of pathology (morbid anatomy), physiology, neuroanatomy, histology and cytology (microscopic anatomy), embryology (developmental anatomy), and physical anthropology. Several of these, in turn, gave rise to other disciplines, such as biochemistry, microbiology, cell biology, and molecular biology.

II. GROSS (MACROSCOPIC) ANATOMY as a discipline concerns the structure and function of the human body.

A. SYSTEMATIC (SYSTEMIC) ANATOMY is organized according to the following systems:

1. The integument (Chap. 2)

2. The musculoskeletal system (Chap. 3)

3. The nervous system (Chap. 4)

4. The circulatory system (Chap. 5)

5. The viscera

6. The endocrine glands

B. REGIONAL ANATOMY is concerned with the systems found within a discrete portion of the body.

1. Back and extremities (Parts II, V, and VII)

2. Thorax (Part III)

3. Abdomen (Part IV)

4. Pelvis (Part VI)

5. Head and neck (Parts VIII and IX)

C. FUNCTIONAL ANATOMY concerns the correlations between structure and function. For example, parasympathetic innervation to the male genital organs mediates erection, and sympathetic innervation mediates ejaculation.

D. CLINICAL ANATOMY emphasizes structure and function as it relates to the practice of medicine and other health professions. For example, a displaced fracture of the humerus may injure the adjacent radial nerve, resulting in paralysis of the extensor muscles of the wrist and loss of sensation over most of the dorsum of the hand.

E. This book uses many of these approaches to cover:

1. The normal macroscopic structure of organs and systems.

2. Topographic relations between structures.

3. Developmental aspects as they apply to adult body structure.
4. The histologic structure as it elucidates function.
5. Neurologic considerations as they apply to normal function and loss of function.
6. Functional considerations as they apply to understanding structure.
7. Clinical considerations as they apply to functional disorders of structure.

III. ANATOMIC TERMINOLOGY

A. THE ANATOMIC POSITION

1. All structures are described and frequently named with reference to the anatomic position.
2. In the anatomic position the person is erect (or lying supine as if erect) with the arms by the sides, palms facing forward, and the legs together, feet directed forward (Fig. 1-1).

B. ANATOMIC PLANES (see Fig. 1-1)

1. The **midsagittal median plane** is vertical between the anterior midline and the posterior midline, dividing the body into left and right halves.
2. The **parasagittal (paramedian) planes** are parallel to the midsagittal plane.
3. The **coronal planes** are vertical and perpendicular to the midsagittal plane. The **midcoronal (frontal) plane** divides the body into anterior and posterior halves.
4. The **transverse (horizontal) planes** are mutually perpendicular to the midsagittal and coronal planes, dividing the body by cross sections.
5. The line formed by the intersection of any two mutually perpendicular planes defines an axis.
   a. The vertical axis is formed by the intersection of midsagittal and midcoronal planes.
   b. Anteroposterior axes are defined by the intersection of transverse and sagittal planes.
   c. Bilateral axes are defined by the intersection of coronal and transverse planes.

C. ANATOMIC ADJECTIVES are arranged as pairs of opposites (see Fig. 1-1).

1. **Anterior/posterior.**
   a. **Anterior (ventral)** is toward the front aspect of the body.
   b. **Posterior (dorsal)** is toward the back aspect of the body.
   c. **Palmar** is the ventral side of the hand.
   d. **Plantar** is the sole of the foot.
2. **Proximal/distal.**
   a. **Proximal** is close to the median or near the origin of a structure.
   b. **Distal** is away from the origin of a structure.
3. **External/internal.**
   a. **External (superficial)** is close to the surface of the body.
   b. **Internal (deep)** is close to the center of the body.
4. **Superior/inferior.**
   a. **Superior (cephalad, craniad, cephalic, rostral)** is toward the head.
   b. **Inferior (caudad, caudal)** is toward the tail or feet.
5. **Medial/lateral.**
   a. **Median** is in the midsagittal plane.
   b. **Medial** is toward the median.
   c. **Lateral** is away from the median.
6. **Central/peripheral.**
   a. **Central** is toward the center of mass of the body.
   b. **Peripheral** is away from the center of mass of the body.
7. **Prone/supine.**
   a. **Prone** is ventral surface down.
   b. **Supine** is ventral surface up.

D. ANATOMIC MOVEMENTS are usually described as pairs of opposites (Fig. 1-2).

1. **Flexion/extension** usually occurs in the midsagittal or a parasagittal plane.
   a. **Flexion** brings primitively ventral surfaces together (e.g., bending the arm at the elbow joint).

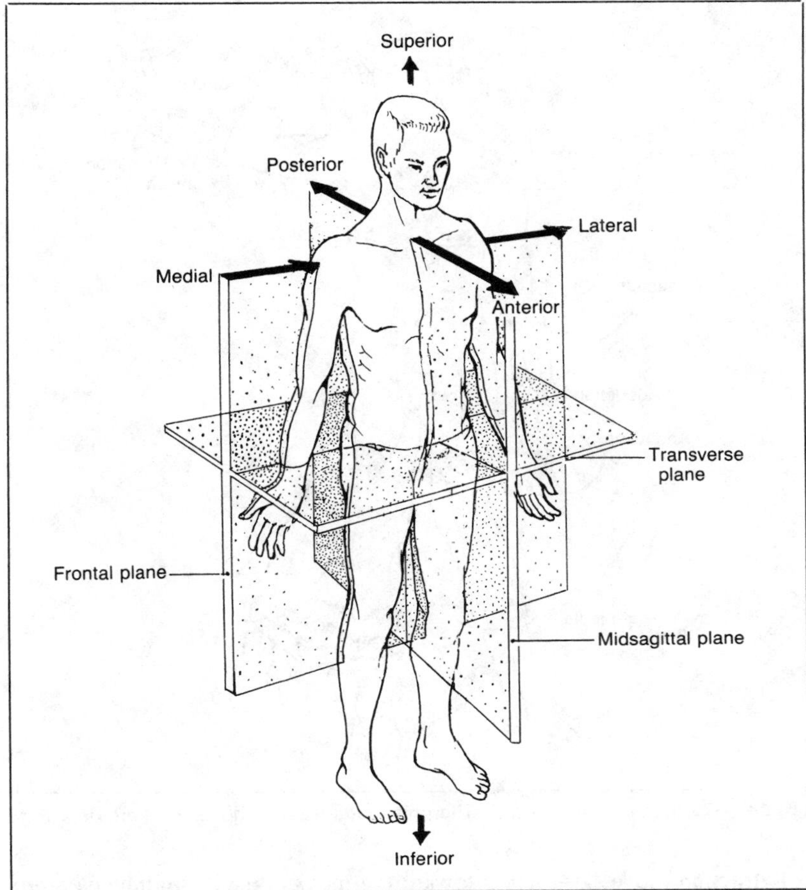

Figure 1-1. *Anatomic planes.* The body is in the anatomic position. The principal anatomic planes and some anatomic adjectives are indicated.

(1) Plantar flexion is downward flexion (flexion) of the foot at the ankle joint.
(2) Dorsiflexion is upward flexion (extension) of the foot at the ankle joint.
(3) Radial deviation (flexion) is abduction of the hand at the wrist joint.
(4) Ulnar deviation (flexion) is adduction of the hand at the wrist joint.

b. **Extension** is movement away from the ventral surface (e.g., straightening the leg at the knee joint).

2. **Abduction/adduction** usually occurs in the midcoronal plane.
   a. **Abduction** (lateral flexion) is movement away from the median, away from the middle finger, or away from the second toe (e.g., directing the eye laterally).
   b. **Adduction** is movement toward the median, toward the middle finger, or toward the second toe (e.g., bringing one leg adjacent to the other).

3. **Medial rotation/lateral rotation** usually occurs about a line described by the intersection of coronal and sagittal or parasagittal planes.
   a. **Medial** rotation is movement of a ventral surface toward the median (e.g., bringing a flexed arm across the chest).
   b. **Lateral** rotation is movement of a ventral surface away from the median (e.g., directing the head toward one side).

4. **Elevation/depression.**
   a. **Elevation** raises or moves a structure cephalad (e.g., shoulder shrug).
   b. **Depression** lowers or moves a structure caudally (e.g., directing the eye downward).

5. **Protraction/retraction**
   a. **Protraction** moves a structure anteriorly (e.g., jutting out the jaw).

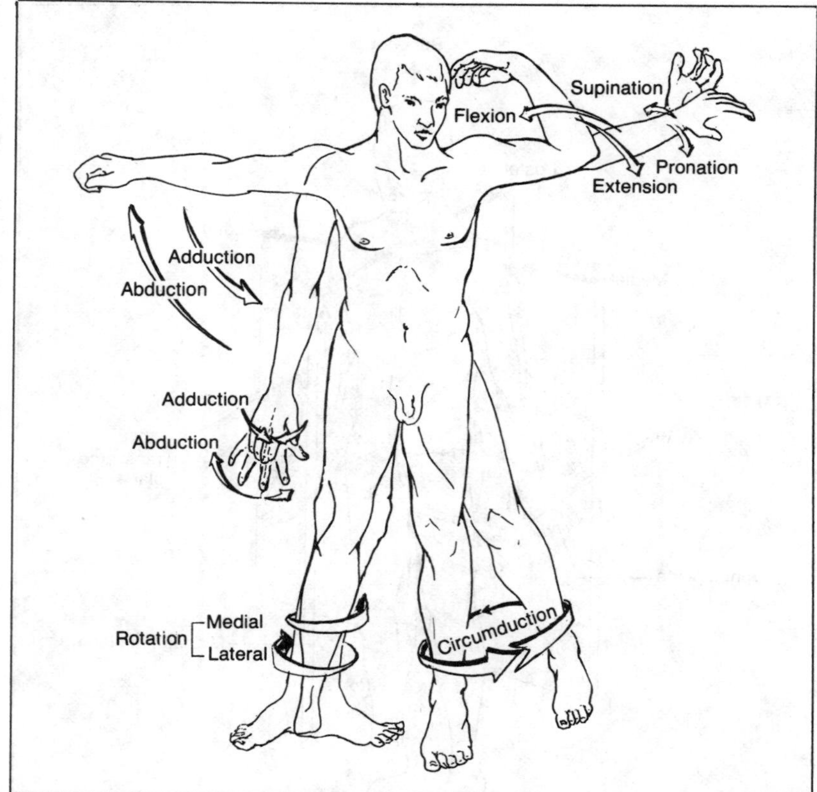

Figure 1-2. *Anatomic movements.* The principal anatomic movements are depicted.

b. **Retraction** moves a structure toward the median (e.g., withdrawing a protracted tongue into the oral cavity).

6. **Pronation/supination** refers to rotations of specific regions.
   a. **Pronation** of the arm, for example, is a medial rotation so that the palm faces posteriorly.
   b. **Supination** of the arm is a lateral rotation so that the palm faces anteriorly.
7. **Inversion/eversion** refers to rotations of specific regions.
   a. **Inversion** of the foot, for example, rotates the plantar surface inward.
   b. **Eversion** of the foot rotates the plantar surface laterally.
8. Other terms pertain to movement of particular structures.
   a. **Intorsion/extorsion** of the eye refers to rotation about an axis through the pupil.
   b. **Opposition/reposition** of the thumb is a uniquely human characteristic, which refers to rotation about a complex axis.
   c. **Circumduction** is a combined movement involving flexion/extension with abduction/adduction.

E. ANATOMIC VOCABULARY—THE LANGUAGE OF MEDICINE

1. Although the vocabulary of anatomy contains approximately 5000 terms, usually there is a rationale to a term that makes it easy to remember.
   a. To simplify a potentially complex terminology, the International Anatomical Nomenclature Committee has established the following guidelines for the naming of anatomic structures:
      (1) There shall be only one name for each structure, with no alternatives, and eponyms shall be discarded.
      (2) Names shall be in Latin whenever practical.
      (3) Terms shall be as short, simple, and informative as possible.
      (4) Spatially related structures shall have similar names whenever possible (e.g., femoral artery, femoral nerve, femoral vein, femoral ring, femoral canal).

(5) Differentiating adjectives shall be arranged as opposites (e.g., major, minor; medial, lateral).

b. Many structures are named in ancient or obsolete languages (e.g., *esophagus*, Greek; *ileum*, Latin; *liver*, Anglo-Saxon).

(1) Usually the name translates into a meaningful description [e.g., duodenum, *L.* 12 (finger breadths long)].

(2) Many of the original names have been transliterated into modern English (e.g., *arteria profunda brachii* becomes deep brachial artery).

c. Many structures have purely descriptive terms (vermiform appendix).

d. Many structures are named according to their relative position in the body (external intercostal muscle).

e. Many structures are named according to function (levator scapulae muscle).

f. Many structures, in addition to the accepted name listed in the *Nomina Anatomica*, have eponymic names associated with mythology (Achilles tendon), the first person to describe the structure (circle of Willis), or the first person to associate the structure with a malformation or disease state (Hunter's canal).

(1) Most anatomy books, in agreement with the International Anatomical Nomenclature Committee, do not acknowledge eponyms.

(2) Because clinical use of many eponyms persists, and because eponyms reflect the rich history of anatomy and medicine, the more common eponyms are acknowledged in this book.

2. Competence with the vocabulary of the body is only the first step in the mastery of anatomy.

a. Nomenclature aside, the study of anatomy is best approached with logic rather than rote memory.

(1) Comprehension of anatomy to the point of being able to use it as a medical tool requires integration of structure as it relates to function.

(a) An understanding of structure often leads to an understanding of function.

(b) An understanding of function often logically justifies the structure of a part.

(2) Understanding the development of a structure often clarifies complex relations (e.g., the innervation of the diaphragm).

b. An anatomic principle is very frequently the basis for the diagnosis or choice of treatment for a clinical problem (see study questions at the end of Part I).

c. Anatomy establishes the foundation for clinical observation, physical examination, and the interpretation of clinical signs and findings, as well as the basis for treatment.

2
The Integument and Mammary Glands

I. THE INTEGUMENT (SKIN)

A. SURFACE AREA. The integument may be considered the largest organ of the body with a surface area somewhat less than 2 m^2.

B. THE MULTIPLE FUNCTIONS of the skin include the following:

1. **Protection** by:
   a. Preventing fluid loss (the greatest problem in burn patients).
   b. Reducing abrasive trauma.
2. **Sensation** mediated by general sensory afferent (GSA) nerve endings (pain, touch, and temperature).
3. **Secretion** by:
   a. Sweat glands for temperature regulation mediated by general visceral efferent [GVE] (sympathetic) nerves.
   b. Mammary glands, which are modified sweat glands, under the primary control of endocrine hormones.

C. THE INTEGUMENT consists of two principal layers:

1. **The epidermis.**
   a. This is a superficial cellular layer of stratified epithelium.
   b. It is between 20 and 1400 μm thick, depending on location.
2. **The dermis.**
   a. This is an underlying layer of loose, irregularly arranged connective tissue.
   b. It is between 400 and 2500 μm thick, depending on location.
   c. It contains accessory structures such as hair follicles, sweat glands, the mammary glands, blood vessels, lymphatics, nerves, and special nerve endings.

D. THE CLEAVAGE LINES (of Langer) are important surgical considerations.

1. Although the connective tissue fibers in the dermis appear randomly oriented, there is a prevailing directionality in each area of the body which is denoted by the crease lines, or **Langer's lines**. The meshwork of collagen fibers (elastic fibers to some extent) provides overall mobility of the skin. The elasticity of the connective tissue fibers creates a slight tension. When the connective tissue fibers are severed, the skin retracts or gapes.
2. An incision *across* the prevailing directionality of the connective tissue fibers will retract considerably, producing a gaping wound and resulting in a prominent scar upon healing.
3. An incision made *parallel* to Langer's lines will sever fewer connective tissue fibers and therefore decrease the tendency to retract, thus resulting in a less unsightly scar.
4. Note the directions of Langer's lines on the thorax, the base of the neck, and the abdominal wall (Fig. 2-1). The primary surgical consideration is always adequate exposure; cosmetic considerations are, of course, secondary. On the female breast, Langer's lines tend to be circumferential to the nipple, an important consideration considering the frequency of tumor biopsy and cosmetic mammoplasty.

E. INNERVATION OF THE INTEGUMENT

1. In general, the skin is innervated in a segmental pattern. Most higher invertebrates and all chordates exhibit somatic segmentation (**soma, body**), or metamerism. As a first approximation, the body may be thought of as developing from a series of 42–46 identical **somites** (segments). Subsequently there is differential growth and development, so that segments may lose individual identity (such as in portions of the head) or develop extensively (such as the cervical and lumbosacral contributions to the extremities).
2. The adult body retains a segmented vertebral column. Between each two vertebrae a left and right mixed (afferent plus efferent) spinal nerve arises to supply a **dermatome** and a **myotome**. These are discrete portions of the skin and muscle of the body wall or limb that originated from that segment. The superficial branches of each spinal nerve terminate in the nerve endings of the corresponding dermatomes.
3. Adjacent dermatomes usually overlap somewhat. Understanding the dermatomal arrangement is essential to the conduct and interpretation of the physical examination of a patient.

F. SIGNIFICANCE OF THE INTEGUMENT

1. General medicine: manifestations of systemic disease such as vasoconstriction (cold and clammy skin), vasodilation (flushed skin), eruptions or rash, petechiae, ecchymoses, and edema.
2. General and plastic surgery: cosmetic incisions, skin grafts, loss of body fluids in severe burns.
3. Dermatology: skin disease.
4. Neurology: manifestations of neurologic dysfunction and disease.

II. FASCIAL LAYERS

A. SUPERFICIAL FASCIA (subcutaneous tissue) underlies the integument.

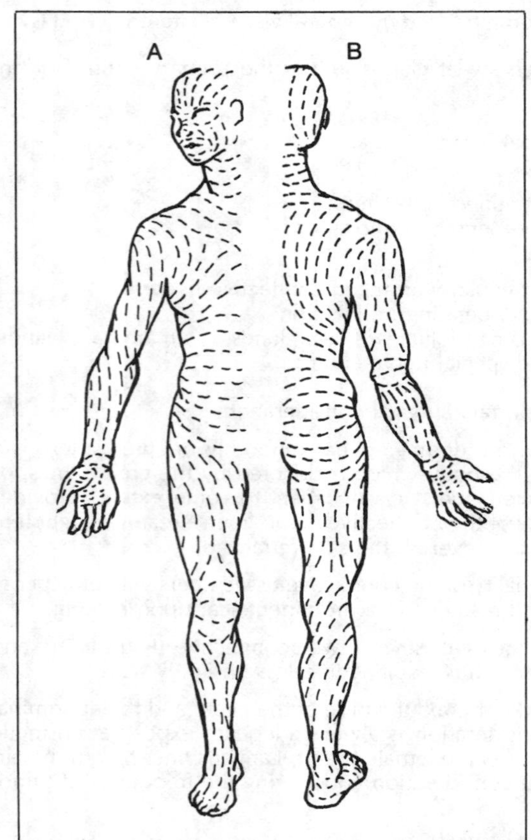

Figure 2-1. *The cleavage lines of the skin. A,* The anterior surface; *B,* the posterior surface.

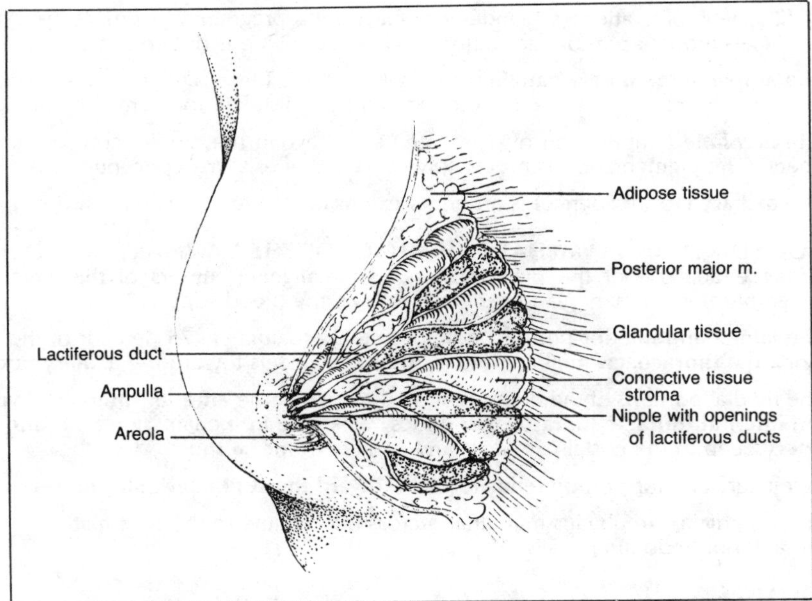

Figure 2-2. *The breast.* The right breast is partially dissected showing the secretory portion.

1. It is composed of loose, irregularly arranged connective tissue between the dermis and the deep (investing) fascia, and contains varying amounts of fat, depending on genotype, phenotype, and location.
2. This fascial layer serves as a loose packing material and matrix, through which course medium-sized (distributing) vessels and nerves.
3. It is divisible into a **superficial layer of the superficial fascia**, which is fatty, and a **deep layer of the superficial fascia**, which is membranous. These layers are especially obvious on the abdominal wall, where they are known as Camper's fascia and Scarpa's fascia, respectively. Only Scarpa's fascia will support sutures. In the perineum the superficial layer of the superficial fascia is referred to as Cruveilhier's fascia; the deep layer of the superficial fascia is known as Colles' fascia.

B. DEEP (INVESTING) FASCIA, a membranous layer that usually overlies muscle as epimysium, defines fascial planes between muscles.

1. With the aid of extracellular fluid, this tissue provides nearly frictionless surfaces for the motion of one muscle over another.
2. Unfortunately, these layers form potential pathways for infection or extravasation of fluids.

III. THE MAMMARY GLAND OR BREAST

A. STRUCTURAL CONSIDERATIONS (Fig. 2-2)

1. The breast consists of 15 to 20 pyramidal **lobes** of glandular tissue (which are only slightly developed in the male) contained entirely within and considered part of the superficial fascia.
   a. The lobes drain into separate **lactiferous ducts**, each of which opens onto the mammary papilla (nipple).
   b. The lobes are defined and interconnected by the **suspensory ligaments** (of Cooper), which are connective tissue septa.
      (1) The suspensory ligaments attach the breast to the skin and deep layer of the superficial fascia and determine the posture of each breast.
      (2) Tumors (benign or malignant) will often displace a suspensory ligament, causing retraction of the breast surface, which is usually a diagnostic sign.
   c. The glandular tissue of each lobe, surrounded by varying amounts of fat that determine the shape of each breast, is under hormonal control.
      (1) The monthly variation in the size of the breast is due to changes in estrogen titers.

(2) The proliferation of glandular tissue during pregnancy is due to estrogen and progesterone; secretory activation of glandular tissue is due to prolactin.

2. The **nipple** or **mammary papilla** receives 15 to 20 lactiferous ducts and is variable in size. (A large nipple may appear on a chest x-ray and be mistaken for a small pulmonary mass.)

3. The **areola**, a continuation of pigmented skin beyond the nipple, contains smooth-muscle fibers. (The pigmentation is irreversibly darkened after a first conception.)

4. The **axillary tail** (of Spence) is a normal extension of breast tissue toward or into the axilla.

B. BLOOD SUPPLY AND LYMPHATIC DRAINAGE OF THE MAMMARY GLAND are of especial importance because of the high incidence of malignant tumors of this organ. Lymphatic drainage of the mammary gland generally follows the blood supply.

1. The lateral and inferior portions, comprising approximately 75 percent of the breast, drain along the **thoracoacromial** and **lateral thoracic** vessels toward the **axillary nodes**.

2. The medial portions of the breast tend to drain along the **anterior intercostal vessels** toward the **internal thoracic (parasternal) nodes**, which lie in a chain along the **internal thoracic artery** deep to the costal cartilages and parallel to the sternum.

3. A smaller superior portion tends to drain toward the **supraclavicular nodes**.

4. The superficial lymphatics may drain across the midline to the contralateral breast or along the anterior abdominal wall.

C. ANOMALIES

1. Beyond an extensive normal range, the breasts frequently are overdeveloped or underdeveloped, both of which conditions are amenable to mammoplasty.

2. Accessory breasts and supernumerary nipples may develop along the "milk line" from the axilla to the groin.

3. Abnormal bilateral hypertrophy or hyperplasia of the male breasts (gynecomastia) is usually due to endocrine disorders, impaired liver function, or drugs.

3 The Musculoskeletal System

I. INTRODUCTION

A. The musculoskeletal system is composed of muscle and several types of connective tissue, including bone, cartilage, ligaments, and tendons.

B. This system provides the means of locomotion.

II. THE SKELETON

A. The adult skeleton consists of approximately 206 bones.

1. The **axial skeleton** consists of the bones of the head, the bones of the vertebral column, the ribs, and the sternum.

2. The **appendicular skeleton** consists of the bones of the extremities.

B. **BONE**, a calcified connective tissue, forms most of the adult skeleton.

1. The **skeleton** serves several basic functions.
a. Bone supports and protects certain internal organs.
b. Bones act as biomechanical **levers** on which muscles act to produce motion.
c. As a tissue, bone is in dynamic equilibrium with its bathing medium and serves as a reservoir of ions (Ca^{++}, PO_4^-, $CO_3^=$) in mineral homeostasis.
d. The bone marrow in the adult is the source of red blood cells, granular white blood cells, and platelets.

2. Bone is composed of living cells and an organic intercellular matrix with an inorganic component.
a. Connective tissue cells become **osteocytes** as the collagen meshwork that they secrete undergoes calcification and ossification.
b. The collagenous matrix provides tensile strength. If the mineral content of bone is removed by acid, the remaining collagenous meshwork is flexible but not particularly extensile.
c. The mineral content, a crystalline hydroxyapatite complex of calcium, provides shear strength and compressive strength. If the collagenous matrix is removed by incineration, the remaining inorganic matrix is very brittle.

3. There are several types of bone.
a. **Long bones**, such as those of the extremities (Fig. 3-1):
(1) Develop by replacement of hyaline cartilage.
(2) Usually provide the levers for movement.
(3) Are structurally distinct.
(a) The **diaphysis** (shaft) of long bones is composed of a thick collar of dense compact bone (cortical bone) beneath which is a thin layer of spongy trabecular bone adjacent to the marrow cavity.
(i) The microstructure of cortical bone is arranged for maximal strength.
(ii) Growth in thickness occurs by circumferential apposition of bone.
(b) The **metaphyses** (ends) of long bones are composed of a trabecular bony meshwork surrounded by a thinner collar of compact bone. The spicules are arranged along the lines of stress.

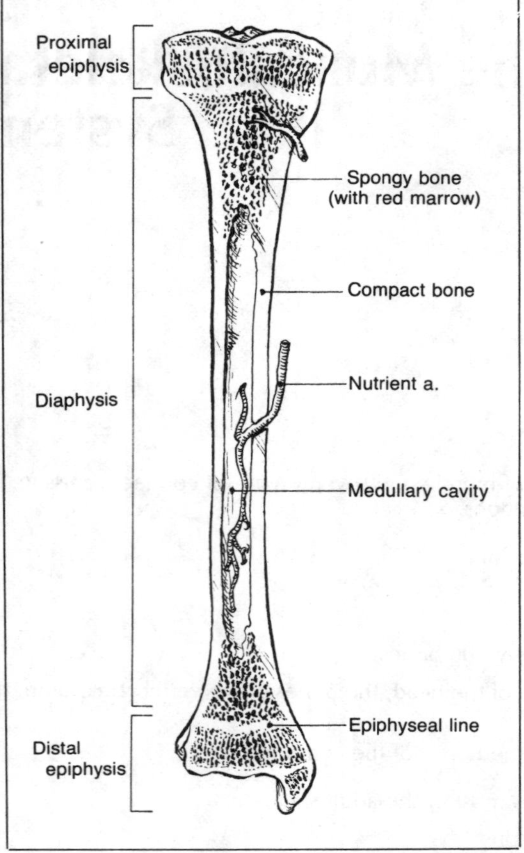

Figure 3-1. *The structure of a long bone.* The tibia is shown partially cut open to reveal the internal organization of the bone. The trabeculae of the spongy bone of the epiphyses align along the lines of force and transmit these forces to the compact bone of the shaft.

(c) The **epiphyses**, toward the ends of long bones, are separated from the metaphyses in a young person by cartilaginous growth plates, the **epiphyseal disks**.
 (i) Longitudinal growth occurs both proximally and distally from the epiphyseal plate by cartilage proliferation, calcification, and remodeling, until epiphyseal fusion occurs at about the time of puberty.
 (ii) The epiphyseal disks become ossified as longitudinal growth ceases.
 (iii) Unfused epiphyses may look like fractures radiographically.

(d) Each bone receives its blood supply through one or more nutrient foramina.

b. Flat (squamous) bones, such as ribs and the bones of the cranium:
 (1) Develop by replacement of connective tissue.
 (2) Generally serve protective or reinforcement functions.
 (3) Consist of two plates of compact bone separated by spongy bone (**diploe**) that bridges the marrow cavity.

c. Sesamoid bones:
 (1) Develop within tendons.
 (2) Serve to reduce attrition on the tendon.
 (3) Increase the lever arm of the muscle by moving the tendon away from the fulcrum.

4. Functional and clinical considerations.

a. As a living tissue, bone is in dynamic equilibrium with the body fluids and reacts to external forces.

b. These two factors interrelate in the process of bone remodeling and healing.
 (1) Bone remodeling.
  (a) The trabecular meshwork that comprises the internal structure of the metaphyses of bone and provides the directionality of the collagenous substructure of compact bony collar of the diaphysis develops along the lines of force (compression and stress) established by the mass of the body in response to gravitational force and voluntary exertion.
  (b) As these forces change over time (the result of a change in body weight, pregnancy, or exertion), bone undergoes a subtle remodeling to maximize intrinsic strength.

(c) The constant pull of a muscle at its sites of attachment produces ridges, crests, tubercles, and trochanters by the remodeling reaction.
(d) If bone is not stressed, due to illness, injury, or the weightlessness of outer space, calcium is rapidly resorbed.

(2) Fractures can be classified according to the degree of displacement (nondisplaced or displaced), whether there is compression of the bone fragments (comminution), and whether the skin is torn by displaced fragments (compound fracture).
(a) Fractures are usually visible radiographically. However, small fractures with no compression or displacement may not be visible radiographically until some bone resorption occurs, usually in about a week.
(b) Fracture of the cartilaginous epiphyseal plate in a young person may be difficult to detect unless there is compression or displacement. Healing of epiphyseal fractures in growing young persons can interfere with subsequent growth.
(c) Fragments of fractured long bones may jeopardize adjacent soft tissues, especially neurovascular bundles.
(d) In the head, many nerves and vessels pass through bony canals and foramina in flat bones. Fractures of these may compress or lacerate the nerve or vessels as well as compress or lacerate the brain.
(e) If there is a single major nutrient foramen involved, a fracture may isolate a significant portion of a long bone from its blood supply, with resultant avascular necrosis.

(3) **Bone healing.**
(a) Fracture of long bones may result in loss of integrity of the lever arm; fracture of flat bones results in loss of protective function.
(b) Fibroblasts peripheral to bone in the region of the fracture proliferate and secrete a collar of collagen about the fracture, the **callus**.
(c) The callus calcifies, providing an internal splint for the fracture.
(d) There is some bone resorption on either side of the fracture followed by unorganized proliferation of connective tissue across the fracture.
(e) The unorganized connective tissue in the fracture calcifies within 6 weeks, thereby healing the fracture.
(f) Over the next several months, remodeling occurs in the region of the fracture as well as in adjacent regions; at least in the young person, little evidence remains of the fracture.

C. CARTILAGE, a dense irregular connective tissue, forms a small portion of the skeleton.

1. Cartilage is formed by living cells and the intercellular matrix that they secrete. It is essentially avascular.

2. The composition of the intercellular matrix determines the type of cartilage.

a. **Hyaline cartilage:**
(1) Has an intercellular matrix especially rich in hyaluronic acid and mucopolysaccharides, which are natural lubricants.
(2) Forms the anterior portion of most ribs to complete the rib cage and provide the resiliency necessary for ventilation.
(3) Forms the articular cartilage in most joints.
(4) Provides the anlage for long bone development.

b. **Fibrocartilage:**
(1) Has an intercellular matrix rich in mucopolysaccharides and bundles of collagenous fibers.
(2) Has a very high osmotic pressure and a high water content as a result of a high concentration of mucopolysaccharides.
(3) Is an especially resilient and durable form of cartilage.
(4) Forms most symphyses, such as the pubic symphysis and the intervertebral joints, as well as certain joint disks, such as that of the temporomandibular joint.

c. **Elastic cartilage:**
(1) Has an intercellular matrix rich in mucopolysaccharides and bundles of elastic fibers, which provide a strong, yet flexible, support.
(2) Forms the skeletal structure of the external ear and the tip of the nose.

D. ARTICULATIONS (Fig. 3-2)

1. Most bones articulate with each other. There are three major types of articulations or joints.

a. **Synarthroses**, or fibrous joints, are barely movable or nonmovable, as in a **suture** (sagittal suture), **syndesmosis** (tibiofibular joint), and the **gomphosis** (tooth joint).

b. **Amphiarthoses**, or cartilaginous joints, such as the pubic symphysis and intervertebral disks, allow limited motion.

c. **Diarthroses**, or synovial joints, permit relatively free motion about at least one axis of rotation, as in the shoulder, knee, and temporomandibular joints.
(1) All diarthrodial joints have a number of common characteristics.
(a) There is a joint (synovial) capsule.
(b) The articular surfaces of the bones are covered by **articular cartilage**, which is usually hyaline cartilage but may occasionally be fibrous cartilage. Because articular cartilage is somewhat fluid, it can change shape so that mechanical forces are distributed over the largest possible surface area within the joint.
(c) The shapes of the two articular surfaces usually are reciprocally concave and convex.
(d) The diarthrodial joints are enclosed by a **synovial membrane**.
(i) The synovial cavity between the synovial membrane and the bone or cartilage is filled with **synovial fluid**.
(ii) Rich in hyaluronic acid, synovial fluid functions as a lubricant for the articular surfaces.
(e) Degeneration of articular cartilage—degenerative arthritis—results in loss of the smooth gliding surface and pain accompanying joint motion.
(2) Each joint acts as a **fulcrum** for a bony lever so that a motion about that fulcrum can be produced.
(3) The **motions** at all diarthroses can be reduced to **rotations** in one or more mutually perpendicular planes.
(a) All rotary movement takes place about an **axis of rotation**.
(b) The relative movements of bones at the common articulation are given pairs of description names that describe the motion occurring about a single axis of rotation:
(i) **Flexion-extension**.
(ii) **Abduction-adduction**, or sometimes, **elevation-depression**.
(iii) **Internal rotation-external rotation**.
(c) A combination of the three pairs of movements produces **circumduction**.
(d) Some diarthroses permit only one pair of motions, others permit two, and still others permit three pairs of motions.
(i) Joints can have one, two, or three axes of rotation.
(ii) There is **one degree of freedom** for each axis of rotation.

2. **Ligaments** are bands of dense, regularly arranged connective tissue that cross joints and frequently form an **articular capsule** about the joint.
a. On either side of a joint, ligaments are embedded in the bone by Sharpey's fibers so that they:
(1) Connect the bones.
(2) Reinforce the articulations.
(3) Contribute to joint stability.
b. Torn ligaments, sometimes referred to as sprains, present difficult clinical problems.
(1) Because ligaments are relatively avascular, unlike bone, tears heal slowly.
(2) When a ligament is detached from bone, its fibers do not grow back into the bone as extensively as before the injury; thus a healed ligament is usually weaker, predisposing to subsequent injury.
(3) Torn ligaments destablilize the joint and predispose to dislocation of the joint.
(a) Loss of joint stability is a sign of a torn ligament.
(b) Although ligaments usually cannot be visualized radiographically, displacement of bones at an articulation is a sign of torn ligaments.

III. MUSCLE

A. Muscle comprises approximately 48 percent of the body mass.

B. Muscles are composed of bundles of muscle cells (muscle fibers), which are contractile.

1. **Skeletal muscle** runs between two **points of attachment** on related bones or even on bones separated by a considerable distance.
a. **Tendons** are bundles or sheets (**aponeuroses**) of dense, regularly arranged connective tissue into which each end of a muscle inserts, and which, in turn, attach to the outer layer of the periosteum; a few tendons attach directly to bone through Sharpey's fibers.
(1) The more proximal attachment site of a muscle is often referred to as the **origin**; the more distal attachment site of a muscle, the **insertion**.
(2) Tendons may continue into the muscle as **septa**.
(3) Where a tendon is subjected to intense friction, a sesamoid bone may form in the tendon.

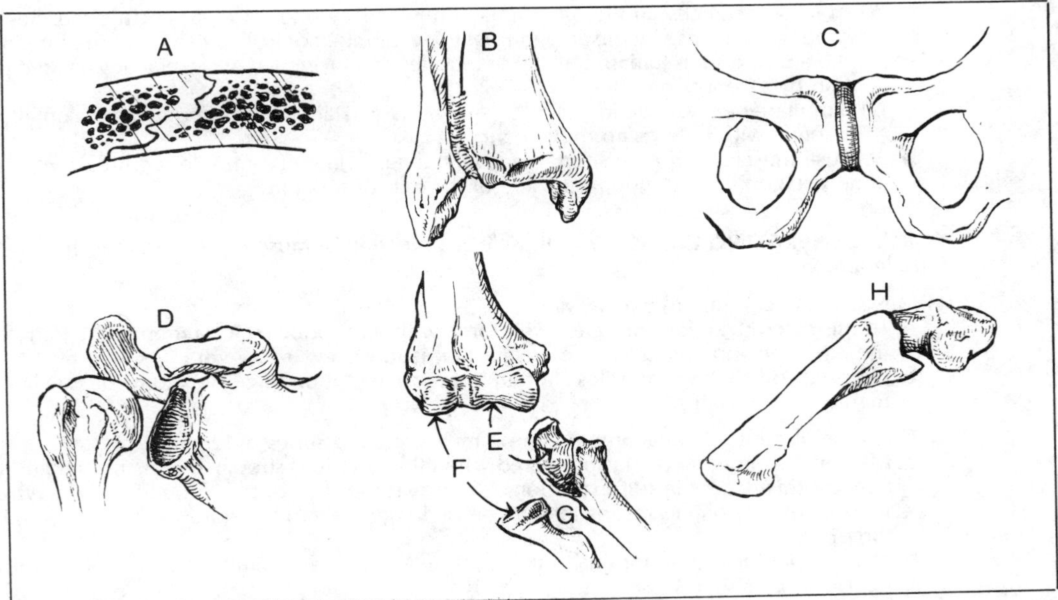

Figure 3-2. *Articulations. A,* A cranial suture; *B,* the tibiofibular syndesmosis; *C,* the pubic symphysis; *D,* the glenohumeral joint (ball-and-socket, 3° freedom); *E,* the humeroulnar joint (hinge, 1° freedom); *F,* the humeroradial joint (ball-and-pivot, 2° freedom); *G,* radioulnar joint (gliding pivot, 1° freedom); and *H,* the first carpometacarpal joint (saddle, 2° freedom).

(4) Muscle pulling at the sites of tendinous attachment to bones produces a remodeling reaction within the bones; thus attachment sites are indicated by ridges, crests, tubercles, and trochanters.

b. The line that best describes the mean direction of the muscle between the centers of any two such attachments is called the **line of action** of the muscle (see Fig. 3-3).

c. The **lever arm** of a muscle is a line drawn perpendicular to the line of action of that muscle through the axis of rotation (fulcrum) of the joint (see Fig. 3-3).

d. The arrangement of fascicles (groups of muscle fibers) within the muscle falls into several patterns.

(1) Fusiform: fascicles of muscle fibers lie parallel to the line of action along the long axis of the muscle (e.g., the sartorius muscle).

(2) Pennate: fascicles lie at an angle to the long axis of the muscle.

(a) Unipennate: all fascicles lie at the same angle on one side of the tendon (e.g., the flexor pollicis longus muscle).

(b) Bipennate: fascicles lie at an angle on either side of a tendinous septum (e.g., the soleus muscle).

(c) Multipennate: fascicles reach the tendinous septa from many directions (e.g., the deltoid muscle).

(3) Vectorial analysis demonstrates that the same amount of contraction produces slower movement but more force in a pennate muscle than in a fusiform muscle.

2. By shortening, muscles act on the bony levers to produce motion in one or both of the bones to which they attach.

a. All muscles exert *equal* and *opposite* tension at both attachments.

(1) Any muscle whose line of action crosses an unconstrained axis of rotation at a joint must produce movement at that joint.

(2) Movement at a joint is determined by the sum of the activity of all the muscles whose lines of action cross the axis of rotation.

(3) The bone that is least stabilized will move.

b. The strength of the muscle is a function of its cross-sectional area (4 kg/cm^2) and the length of its lever arm (mechanical advantage).

3. Regardless of the specific innervation, some muscles may act together as **synergists** to produce a specific motion; other muscles act together as **antagonists** to oppose this motion.

a. Synergistic muscles cross the same side of the axis of rotation; antagonistic muscles pass over the opposite side.

(1) In most instances, motion at a joint is initiated by one set of synergistic muscles and brought to a close by the antagonists. For example, controlled flexion of the forearm at the elbow joint is initiated by flexor muscles and brought to a close at any desired position by extensor muscles.

(2) Simultaneous contraction of both synergists and antagonists produces maximal joint stability with little or no movement.

b. Because muscles can only shorten actively, lengthening of a muscle requires contraction of an antagonist on the opposite side of the axis of rotation.

C. FORCE-GENERATING CAPACITY of muscle is a function of muscle stretch rather then overall muscle length.

1. Muscles may contract in two ways:

a. Isometric contraction: muscles exert force without producing a rotation at the joint (e.g., the elbow flexors trying to lift a weight that is too heavy to move).

b. Isotonic contraction: muscles shorten to produce motion (e.g., the elbow flexors lifting a manageable weight).

2. The force at a muscle attachment can be measured at a series of lengths.

a. Maximum isometric force is produced when the muscle is stretched to its rest length. (For a flexor this occurs in full extension; for an extensor this occurs in full flexion.) When a muscle reaches half its resting length, it is no longer able to generate significant contractile force.

b. The morphologic basis for this force-length relation lies in the ultrastructure of the muscle fibers.

(1) Maximum force of 4 kg/cm^2 occurs when there is just maximal overlap between myosin rods and actin filaments.

(a) At just-maximum overlap there can be a maximum number of force-generating cross-bridge interactions between these structural proteins.

(b) This occurs in maximally stretched muscle in situ.

(2) Shortening results in double overlap of contractile filaments in the center of the contractile unit.

(a) Double overlap of filaments interferes with cross-bridge interaction, resulting in loss of force-generating capacity.

(b) This occurs in maximally shortening muscle in situ, which is approximately one-half of the stretched length.

(3) Because maximal force is 4 kg/cm^2, the greater the cross-sectional area of a muscle, the greater the force-generating capacity.

D. THE MECHANICAL ADVANTAGE of a muscle is a function of the length of the lever arm.

1. In addition to the intrinsic force-length effect, as the joint is flexed the length of the lever arm of a muscle acting across that joint changes; thus, the mechanical advantage of the muscle either increases or decreases. This is shown in Figure 3-3.

a. For example, on full extension of the elbow, the lever arm of the flexor is relatively short because the joint angle is nearly 180° and the flexor muscle lies close to the joint. The mechanical advantage of the short lever arm is minimal.

b. As the elbow is flexed to 90°, the lever arm of the flexor is longer because the line of action of the flexor moves away from the joint. Here the mechanical advantage of the muscle is maximal.

c. With additional flexion of the elbow, the lever arm again becomes shorter and the mechanical advantage decreases.

2. Integrative summary.

a. Although the intrinsic force-generating capacity of the flexors is maximal in the fully extended arm, the lever arm of the flexor is minimal; hence, the muscle is strongest at the point at which its mechanical advantage is least.

b. As the arm flexes the force-generating capacity diminishes because of the force-length relation, but the mechanical advantage increases as the length of the lever arm increases; hence, as the muscle weakens its mechanical advantage increases. The net effect is maintenance of fairly constant strength.

c. Beyond 90° of flexion, the force-generating capacity of the flexor muscle and length of the lever arm both diminish; hence, the muscle becomes weaker and loses its mechanical advantage; the effect is a rapid loss of strength.

d. Finally, the speed of contraction is a function of both the weight of the load to be moved and the pennation.

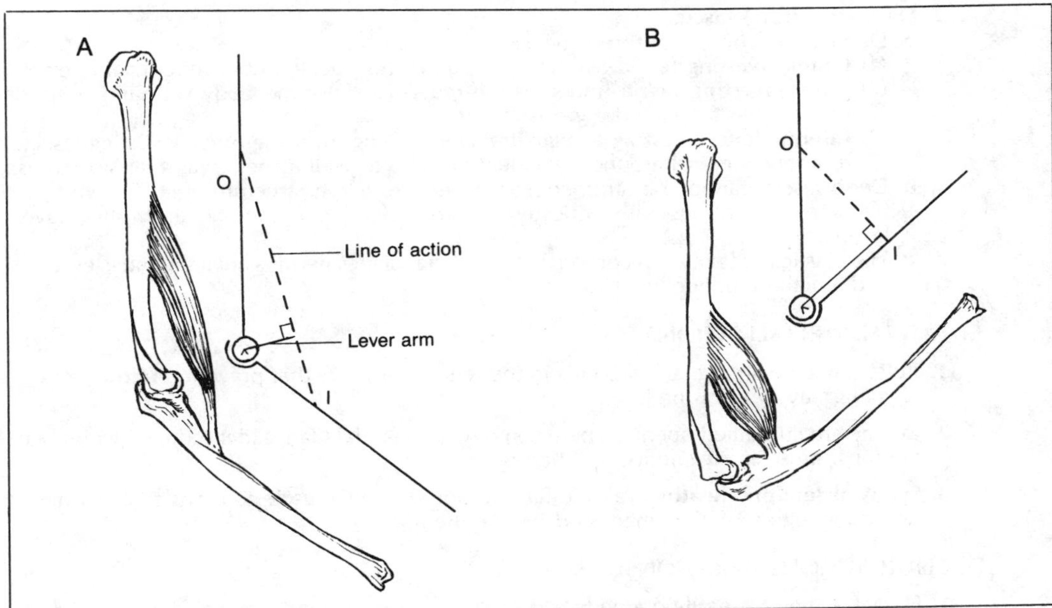

Figure 3-3. *Musculoskeletal action. A,* The stretched brachialis muscle, while intrinsically most powerful in the stretched position, has a short lever arm that provides little mechanical advantage. *B,* Upon isotonic contraction, while intrinsically less powerful in the shortened position, this muscle has a long lever arm that provides greater mechanical advantage.

E. NERVOUS CONTROL of musculoskeletal movement.

1. Physiologic recording demonstrates that electrical excitation passes along nerves to the flexor muscles. This is the basis for diagnostic **nerve conduction studies**.
2. Nerve electrical activity causes release of a neurotransmitter at the neuromuscular junctions, initiating electrical excitation along the muscle fiber and inducing contraction. This is the basis for diagnostic **electromyography**.
3. When gravity, friction, and inertia are overcome, the nerve becomes relatively silent, and flexion continues because of inertia.
4. To halt flexion, the nerve to the extensor muscle becomes active, and the antagonist contracts sufficiently to cease the flexion movement.

IV. SOMATIC FASCIA

A. FASCIA is loose, irregularly arranged connective tissue.

1. Composed of fibroblasts, collagen bundles, and some elastic fibers, fascia forms planes.
2. Fibroblasts secrete tropocollagen, which aggregates into liquid-crystalline collagen bundles.

B. FASCIAL SUBDIVISIONS

1. **Superficial fascia.**
   - **a.** Superficial fascia consists of two layers:
     - **(1)** The **superficial layer of superficial fascia** (of Camper):
       - **(a)** Is predominantly fatty—panniculus adiposus.
       - **(b)** Is of variable thickness and serves as insulation and padding.
       - **(c)** Contains the superficial arteries, veins, lymphatics, and nerves.
       - **(d)** Is particularly sensitive to estrogenic hormones.
     - **(2)** The **deep layer of superficial fascia** (of Scarpa):
       - **(a)** Is membranous and relatively thin.
       - **(b)** Holds sutures.
       - **(c)** Fuses with the deep fascia.
   - **b.** The superficial fascia is relatively mobile in most regions of the body; notable exceptions include the palms and soles.

2. Deep (investing) fascia.

a. Deep fascia consists of three layers:

(1) Outer investing fascia overlies the musculature beneath the superficial fascia.

(2) Inner investing fascia underlies the musculature of the body wall and supports the transversalis fascia and endopelvic fascia.

(3) Intermediate investing fasciae are septa arising from the outer investing fasciae that run between and around individual muscles as well as neurovascular structures.

b. Deep fascia cannot be stripped completely from the structures that it invests (i.e., it becomes continuous with periosteum, perimysium, perineurium, and other adventitial layers.

c. The elasticity of collagen comprising the fascial planes permits adjacent muscles to shorten and lengthen independently.

C. FASCIAL SPECIALIZATIONS

1. Retinacula are strong fascial bands in the regions of joints that prevent tendons from ''bow-stringing'' away from the joint.

2. Bursae are fluid-filled openings between or within fascial planes that reduce friction between tendons, muscles, ligaments, and bones.

3. Synovial tendon sheaths are fluid-filled tunnels about muscle tendons that permit a considerable degree of movement and reduce the friction.

D. CLINICAL CONSIDERATIONS

1. Fascial planes are easily opened:

a. By surgical blunt dissection.

b. By extravasation of fluid such as blood, urine, and pus.

2. Spread of infection across fascial planes is limited.

3. Infection may track along fascial planes; a classic example is the spread of tuberculosis of the lumbar vertebrae beneath the psoas fascia to present as an infection in the femoral triangle.

4
The Nervous System

I. INTRODUCTION

A. The nervous system is a complex organ system.

1. It provides a mechanism by which the organism can react to the ever-changing external and internal environments.

a. Sensory signals originating in the sensory receptors are monitored by, processed in, and transmitted through the nervous system.

b. These inputs may reach the conscious sphere or may be used at subconscious and reflex levels.

c. These actions are mediated through the somatic sensory system and the visceral sensory system.

2. The nervous system controls and integrates the activity of the various parts of the body (Fig. 4-1).

a. The neural messages modulating and regulating motor activity are processed in and conveyed through the nervous system to the muscles and glands.

b. These outputs may be voluntary, involuntary, or the result of reflex mechanisms.

c. These actions are mediated through the somatic motor system and the autonomic motor system.

3. The nervous system maintains the internal environment within narrow limits.

B. The nervous system is composed of **neurons** (nerve cells), which typically have two types of processes.

1. The **dendrites** are afferent processes that typically receive synaptic contact from the axons of other neurons.

2. The **axon** is the single efferent process through which each neuron communicates with other neurons and effectors (i.e., muscles and glands).

C. The bilaterally symmetric nervous system is subdivided anatomically into the **central nervous system** and the **peripheral nervous system**, and subdivided functionally into the **somatic nervous system** and the **autonomic nervous system** (see Fig. 4-1).

1. The **central nervous system (CNS):**

a. Is composed of the brain and spinal cord, which are encapsulated within the skull and vertebral column, respectively.

b. Is the center of perception and the site of integration of sensory information and initiation and coordination of motor activity.

2. The **peripheral nervous system (PNS):**

a. Includes the cranial nerves and spinal nerves that arise from the brain and spinal cord, respectively.

b. Conveys neural impulses:

(1) To the CNS as input from the sense organs and sensory receptors of the body.

(2) From the CNS as output to the muscles and glands of the body.

3. The **somatic nervous system:**

a. The **afferent** portion includes the neural structures of the CNS and PNS that are involved in conveying and processing conscious and unconscious sensory information.

b. The **efferent** portion includes the neural structures of the CNS and PNS that are involved in motor control of voluntary muscle.

4. The **autonomic nervous system (ANS):**
   a. Is composed of the neural structures of the CNS and PNS that are involved in motor activities influencing the involuntary (smooth) and cardiac musculature and glands of the viscera and skin.
   b. Has two divisions: the **sympathetic system** and the **parasympathetic system.**

II. THE CENTRAL NERVOUS SYSTEM

A. THE BRAIN

1. The brain is an enlarged, convoluted, and highly developed rostral portion of the CNS.
   a. The average adult human brain weighs about 1,400 g, approximately 2 percent of the total body weight.
   b. The brain can be divided into the **cerebrum, cerebellum**, and **brain stem.**
   c. The **gray matter** of the brain contains neuron cell bodies; the **white matter** consists of pathways (tracts) containing the axons.

2. The gelatinous brain is invested by a succession of three connective tissue membranes called **meninges** and is protected by an outer rigid capsule, the bony skull.
   a. The **pia mater:**
      (1) Is intimately attached to the brain.
      (2) Contains the blood supply to the brain and brain stem.
   b. The **arachnoid:**
      (1) Is a thin membranous layer external to the pia mater and connected to it by web-like trabeculations—hence its name.
      (2) Contains few, if any, vessels.
      (3) Contains the **subarachnoid space**, which:
         (a) Is located between the arachnoid and pia mater.
         (b) Surrounds the brain.
         (c) Is filled with **cerebrospinal fluid** (CSF).
         (d) May contain blood on hemorrhage of a cerebral artery.
   c. The **dura mater:**
      (1) Consists of two adherent fibrous membranes (hence its name) external to the arachnoid.
         (a) The outer dural layer is the periosteum of the cranial vault.
         (b) The inner dural layer is the true dura.
         (c) In certain locations, venous sinuses (draining blood from the brain) course between the two layers of dura mater.
      (2) Contains the **subdural space**, which:
         (a) Is a potential space located between the arachnoid and dura.
         (b) Does not contain CSF.
         (c) Is the location of a subdural hematoma, which is usually a low-pressure venous hemorrhage.
      (3) Contains the **epidural space**, which:
         (a) Is a potential space that lies between the inner and outer layers of the dura mater.
         (b) Contains the meningeal arteries.
         (c) Is the location of an epidural hematoma, which is usually a high-pressure arterial hemorrhage.
   d. The brain floats in CSF, which supports it and acts as a shock absorber when the head moves.

3. The brain functions in perception of sensory stimuli, in the integration and association of the stimuli with memory, and in neural activity resulting in the coordinated motor response to stimuli.
   a. Input to the brain is from the spinal cord as well as from cranial nerves.
   b. Output from the brain is through the brain stem and spinal cord as well as through cranial nerves.

B. THE SPINAL CORD

1. The cylindric spinal cord is located in the upper two-thirds of the **vertebral canal** of the bony vertebral column.
   a. A continuation of the brain stem, it extends from the foramen magnum at the base of the skull to its termination as the **conus medullaris**, usually located at the caudal level of the first lumbar vertebra.

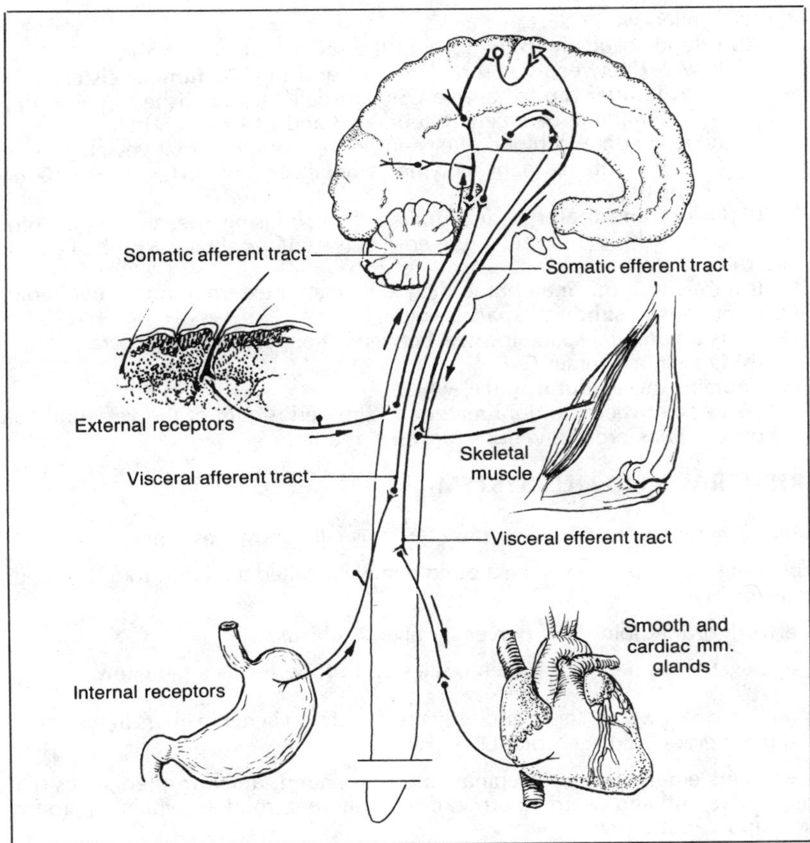

Figure 4-1. *Basic organization of the nervous system.* Somatic and visceral sensory input is schematized on one side while the somatic and autonomic motor output is schematized on the opposite side.

b. The non-neural **filum terminale** continues caudally as a filament from the conus medullaris to the coccyx.

c. The spinal cord is enlarged in those segments that innervate the extremities.

(1) The cervical (brachial) enlargement extends from spinal levels C5 to T1, the segments that innervate the upper extremities.

(2) The lumbosacral enlargement extends from spinal levels L3 to S2, the segments that innervate the lower extremities.

d. Because the vertebral column continues to grow after birth and the spinal cord grows very little in length, the adult spinal cord is much shorter than the bony vertebral column.

(1) The spinal nerves emerge from the vertebral column at levels below the spinal cord segments from which their corresponding rootlets leave the spinal cord.

(2) The lumbar and sacral nerves develop long roots that extend as the **cauda equina** (horse's tail) within the **lumbar cistern**.

2. The **spinal meninges.**

a. The **pia mater:**

(1) Is intimately attached to the spinal cord and its roots.

(2) Continues beyond the termination of the spinal cord as the filum terminale, which attaches to the coccyx.

(3) Contains the blood supply to the spinal cord.

b. The **arachnoid:**

(1) Is a thin membranous layer external to the pia mater and connected to it by web-like trabeculations—hence its name.

(2) Contains few, if any, vessels.

(3) Contains the **subarachnoid space**, which:

(a) Is located between the arachnoid and pia mater.

(b) Surrounds the spinal cord and its roots.

(c) Is filled with CSF.
(d) Extends caudally to the level of the second sacral vertebra.
(e) Is wider between vertebral levels L1 and S1—the **lumbar cistern**.
(i) A **lumbar tap** to sample CSF is usually accomplished by inserting a needle in the midline between vertebrae L3 and L4 or L4 and L5.
(ii) CSF contains blood after hemorrhage of a cerebral vessel.
(iii) CSF contains many polymorphonuclear leukocytes if there is bacterial meningitis.
(f) **Spinal anesthesia** is accomplished by infusing the anesthetic into the lumbar cistern through a midline needle between vertebrae L3 and L4 or L4 and L5.

c. The **dura mater:**
(1) Is a tough fibrous membrane (hence its name) external to the arachnoid.
(2) Contains the **subdural space**, which:
(a) Is a potential space located between the arachnoid and dura.
(b) Does not contain CSF.
(3) Contains the **epidural space**, which:
(a) Lies between the dura mater and the periosteum of the vertebral column.
(b) Contains profuse venous plexuses and fat.

III. THE PERIPHERAL NERVOUS SYSTEM

A. A bundle of nerve axons (fibers) in the PNS is usually known as a nerve.

1. Depending on its location, such a bundle may be called a rootlet, root, trunk, division, cord, ramus, or branch.

2. A network or interjoining of nerves is called a **plexus**.

3. An aggregation of nerve cells (cell bodies and processes) is a **ganglion**.

B. The spinal cord receives its input and projects its output via nerve fibers in the spinal rootlets and roots, spinal nerves, and their branches (Fig. 4-2).

1. Nerve fibers emerge from the spinal cord in a paired, uninterrupted series of dorsal (input, sensory, afferent) and ventral (output, motor, efferent) rootlets, which join to form 31 pairs of dorsal and ventral roots.

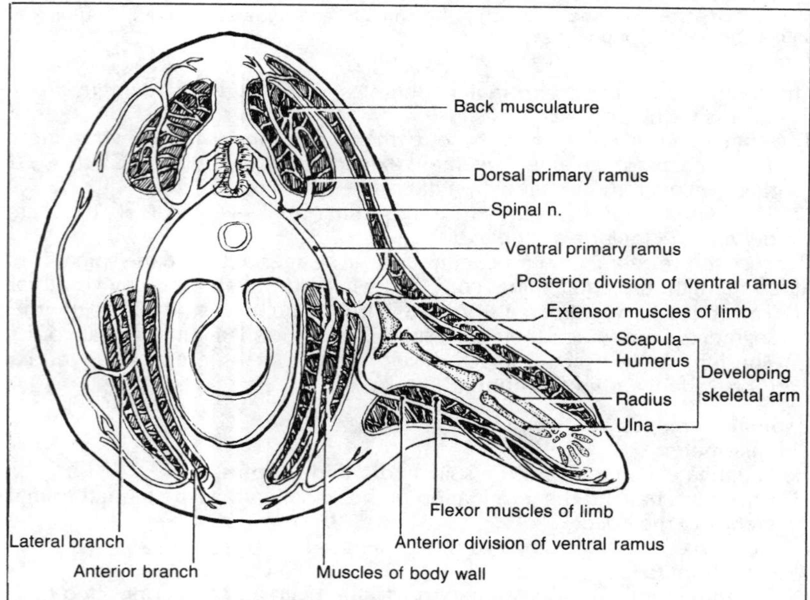

Figure 4-2. *The organization of the spinal nerve.* One side depicts the basic pattern of the spinal nerve with division into dorsal and ventral primary rami and the lateral branch of the ventral ramus. The other side depicts the formation of the brachial and lumbosacral plexuses with the lateral branch of the ventral ramus forming the posterior division while the continuation of the ventral ramus forms the anterior division.

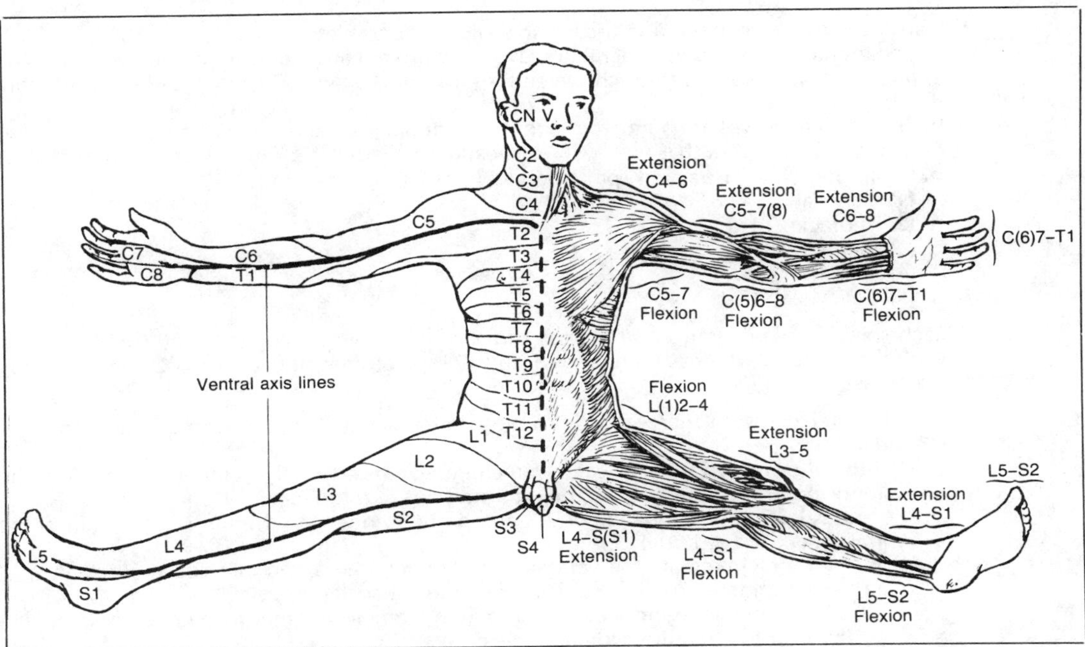

Figure 4-3. *The cutaneous and muscular nerve distributions in the primitive position.* The body is positioned in such a way that the primitive ventral surfaces are anterior. A typical arrangement of the dermatomes on the right side of the body depicts the dermatomal migration into the extremities during development. The innervating root numbers to the muscles that act across the joints of the upper and lower left extremities also show evidence of myotomal migration.

2. Spinal roots.

a. Dorsal (sensory) roots.

(1) The afferent fibers that comprise the dorsal root convey input from the sensory receptors in the body via the spinal nerves to the spinal cord.

(2) The cell bodies of the neurons lie in the dorsal root ganglion, which is located within the intervertebral foramen.

(3) Some fibers of the dorsal root of each spinal nerve supply the sensory innervation to a skin segment known as a **dermatome**, whereas other fibers provide nerve endings to deep structures (Fig. 4-3).

(4) There is usually no C1 or Co1 dermatome.

(5) Adjacent dermatomes overlap, and the loss of one dorsal root results in diminished sensation (not a complete loss) in that dermatome.

b. Ventral (motor) roots.

(1) The motor ventral roots, which comprise the ventral root, convey output from the spinal cord.

(2) The cell bodies lie in the ventral horn of the spinal cord gray matter and project axons which may:

(a) Innervate voluntary striated muscles (general somatic efferent).

(b) Synapse with neurons in peripheral ganglia which, in turn, innervate involuntary smooth muscles and glands (general visceral efferent).

(c) Innervate the striated muscle derived from the branchiomeric (gill) structures (special visceral efferent).

(3) The fibers of the ventral root of each spinal nerve supply the motor innervation to specific voluntary muscles known as **myotomes.**

(4) Usually several myotomes act over any joint (see Fig. 4-3).

(5) There is growing evidence that some visceral afferent fibers enter the spinal cord through the ventral root.

C. THE SPINAL NERVES

1. In the vicinity of an intervertebral foramen, a dorsal root and a ventral root meet to form a spinal nerve, which supplies the innervation of a segment of the body.

2. Spinal nerves are numbered in association with vertebral levels.
   a. The thoracic, lumbar, and sacral nerves are numbered after the vertebra just rostral to the intervertebral foramen through which they pass (e.g., nerve T4 emerges below vertebra T4).
   b. The cervical nerves are numbered after the vertebra just caudal (e.g., nerve C7 is rostral to vertebra C7), except that nerve C8 exits caudal to vertebra C7 and rostral to vertebra T1.
   c. In all, there are 31 pairs of spinal nerves:
      (1) Cervical: C1–C8.
      (2) Thoracic: T1–T12.
      (3) Lumbar: L1–L5.
      (4) Sacral: S1–S5.
      (5) Coccygeal: Co1.

3. Each spinal nerve contains both motor and sensory axons although C1 and Co1 have only ventral roots (i.e., whereas there are C1 and Co1 myotomes, there are no C1 and Co1 dermatomes).

4. The short mixed spinal nerve divides almost immediately into two primary rami and two secondary rami (see Fig. 4-2).
   a. The **dorsal primary ramus** of each spinal nerve arches dorsally to innervate the skin (the posterior portion of the dermatome) and muscles of the back.
   b. The **ventral primary ramus** continues anteriorly to innervate the muscle and skin of the ventral aspect of the body.
      (1) In the thoracic region, the ventral primary rami are termed intercostal nerves.
         (a) The 12th intercostal nerve is termed the subcostal nerve.
         (b) In the lumbar region the anterior primary rami form other named nerves (e.g., iliohypogastric and ilioinguinal nerves).
      (2) When the ventral primary ramus (e.g., intercostal nerve) reaches a point in line with the axilla, it gives off a **lateral cutaneous branch**, which penetrates the muscle to innervate the overlying skin.
         (a) The homologues of these branches form the **posterior divisions** of the brachial plexus and lumbosacral plexus.
         (b) The lateral branch of the ventral primary ramus then divides into anterior and posterior perforating branches, which innervate muscle and the lateral portion of the dermatome.
      (3) The ventral primary ramus continues anteriorly toward the midline after giving off the lateral cutaneous branch.
         (a) The homologues of this branch form the **anterior divisions** of the brachial plexus and lumbosacral plexus.
         (b) Just lateral to the midline the ventral ramus terminates as the **anterior cutaneous branch**, which innervates the anterior portion of the dermatome.
      (4) The cutaneous branches of the spinal nerves overlay in their distribution so that every region of the body is ensured innervation. The dermatomes also overlap in the anterior and posterior midline.
   c. The **meningeal ramus** is actually the first branch of the spinal nerve.
   d. The **rami communicantes.**
      (1) The **white ramus communicans:**
         (a) Connects the spinal cord and the sympathetic chain.
         (b) Conveys sympathetic preganglionic fibers to the paravertebral and prevertebral ganglia.
         (c) Conveys visceral afferent fibers to the spinal nerve.
      (2) The **gray ramus communicans:**
         (a) Connects the sympathetic chain and the spinal nerve.
         (b) Conveys sympathetic postsynaptic fibers from the paravertebral ganglia to the spinal nerve to reach the skin.

IV. THE SOMATIC NERVOUS SYSTEM

A. The somatic portions of the CNS and PNS are involved with conveying and processing conscious and unconscious sensory information and motor control of voluntary muscle.

B. AFFERENT NERVES (Fig. 4-4)

1. The afferent nerves convey sensory input from free nerve endings and special receptors to the CNS.

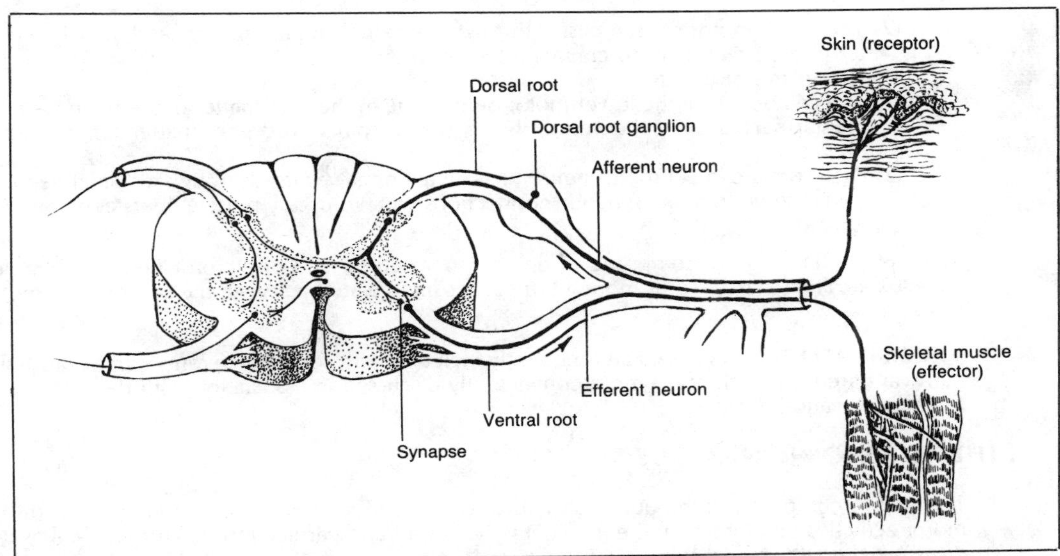

Figure 4-4. *The somatic reflex arc.* Somatic afferent neurons with cell bodies in the dorsal root ganglion convey sensory information from the periphery to the spinal cord. The information is passed through small internuncial neurons to the motor neuron. The efferent neurons with cell bodies located in the ventral horn of the spinal cord pass along the spinal nerve to contact effector organs.

2. Classification of afferent innervation:
   a. **General somatic afferent nerves** convey pain, temperature, touch, and proprioception from the head, body wall, and extremities to the CNS.
   b. **Special somatic afferent nerves** convey vision, hearing, and balance to the CNS.
   c. **General visceral afferent nerves** (frequently included in this classification for functional reasons) convey sensory information from the visceral organs to the CNS.
   d. **Special visceral afferent nerves** (frequently included in this classification for functional reasons) convey smell and taste sensations to the CNS.

3. One peripheral neuron and at least one central neuron are involved.
   a. The **peripheral neuron.**
      (1) The cell bodies of these pseudounipolar sensory neurons lie in the dorsal root ganglia of the spinal nerves or the cranial nerve equivalents.
      (2) The distal end of the axon either is a free nerve ending or makes contact with a specialized receptor; the proximal end of the axon enters the spinal cord through the dorsal rootlets.
      (3) The axons may synapse with a central neuron in the gray matter of the spinal cord or run cranially in the white matter to a region of gray matter in the brain stem at which synaptic contact is made.
   b. The **central neurons** lie in the gray matter of the spinal cord or brain stem and send axons to higher centers of the brain at which synaptic contact is made.

C. SOMATIC EFFERENT NERVES (see Fig. 4-4)

1. The basic role of the **somatic motor nerves** is to regulate the coordinated muscular activities associated with voluntary motion and the maintenance of posture.

2. Classification of somatic efferent innervation:
   a. **General somatic efferent nerves** supply the muscles of the head, body wall, and extremities, which arise from myotomes of the embryonic somites.
   b. **Special visceral efferent nerves** supply the muscles of the head and neck, which arise from branchiomeric (gill) structures.

3. Two sets of neurons are involved in producing somatic motor activity—one central and one peripheral.
   a. The **upper motor neurons.**
      (1) The cell bodies of the upper motor neurons lie in the gray matter of the motor areas of the cortex and various nuclei of the brain stem.

(2) The upper motor neuron passes through the cerebrum, brain stem, and the white matter of the spinal cord to contact a lower motor neuron.

b. The **lower motor neurons.**

(1) The cell bodies of the lower motor neurons lie in the gray matter of the brain stem for cranial nerves or in the ventral horns of the spinal cord gray matter for the spinal nerves.

(2) Each somatic lower motor neuron has an axon that courses through a cranial nerve or spinal nerve to make synaptic connections with voluntary muscle fibers at myoneural junctions.

4. Simple reflex arcs are composed of one sensory neuron and one motor neuron. Complex reflex arcs may have one or more interneurons intercalated between the sensory and motor neurons.

D. **VISCERAL AFFERENT NERVES** are frequently included with somatic afferents, but because the pathways are similar, they are more conveniently discussed in association with the visceral efferent innervation.

V. THE AUTONOMIC NERVOUS SYSTEM

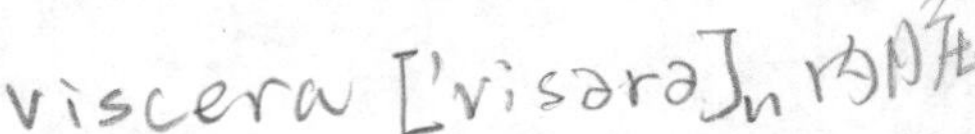

A. The ANS is composed of the neural structures of the CNS and PNS that are involved with the motor activities that influence the involuntary (smooth) and cardiac musculature as well as the glands of the viscera and skin.

B. The ANS is a visceral motor system, exclusively.

1. It is classified as **general visceral efferent** and supplies smooth muscles, cardiac muscle, and glands.

2. The ANS is often called the visceral or vegetative motor system because the effectors are associated with systems (e.g., cardiovascular, digestive, respiratory, perspiratory, vasodilatory) over which only minimal direct conscious control can be exerted.

C. Three sets of neurons are involved in producing motor activity—one central and two peripheral.

1. The **upper motor neuron.**

a. The cell bodies of the upper motor neurons are located in the gray matter of the autonomic areas of the brain.

b. The axons pass in the descending tracts at the brain stem and spinal cord.

2. The **preganglionic neuron.**

a. The preganglionic neuron, which may be seen as the equivalent of a lower motor neuron, originates in the gray matter of the brain stem or the lateral horn of the spinal cord gray matter.

b. Its axon courses through a cranial nerve or spinal nerve.

c. It terminates by synapsing with a postganglionic neuron located in a ganglion outside the central nervous system.

3. The **postganglionic neuron.**

a. The postganglionic neuron is unique to the ANS.

b. The cell body of the postganglionic neuron is located in an autonomic ganglion.

c. The postganglionic neuron has an axon that extends peripherally to terminate in endings associated with smooth muscles, cardiac muscle, or glands.

D. The ANS has two divisions.

1. The **sympathetic division** (Fig. 4-5).

a. As a very rough rule-of-thumb, the sympathetic system stimulates activities that are mobilized by the organism during emergency and stress situations—the so-called *fight, fright, and flight* responses.

b. The sympathetic system is also called the **thoracolumbar** or **adrenergic system**.

(1) Sympathetic **preganglionic** fibers originate from cell bodies located in spinal levels T1 through L2 and pass successively through:

(a) The ventral roots, referred to as the thoracolumbar outflow.

(b) The anterior primary rami of the spinal nerves.

(c) The myelinated **white rami communicantes**. Only spinal levels T1 through L2 have white rami communicantes.

(d) The **sympathetic chain** (composed of **paravertebral** ganglia and interconnecting fiber bundles), where they may:

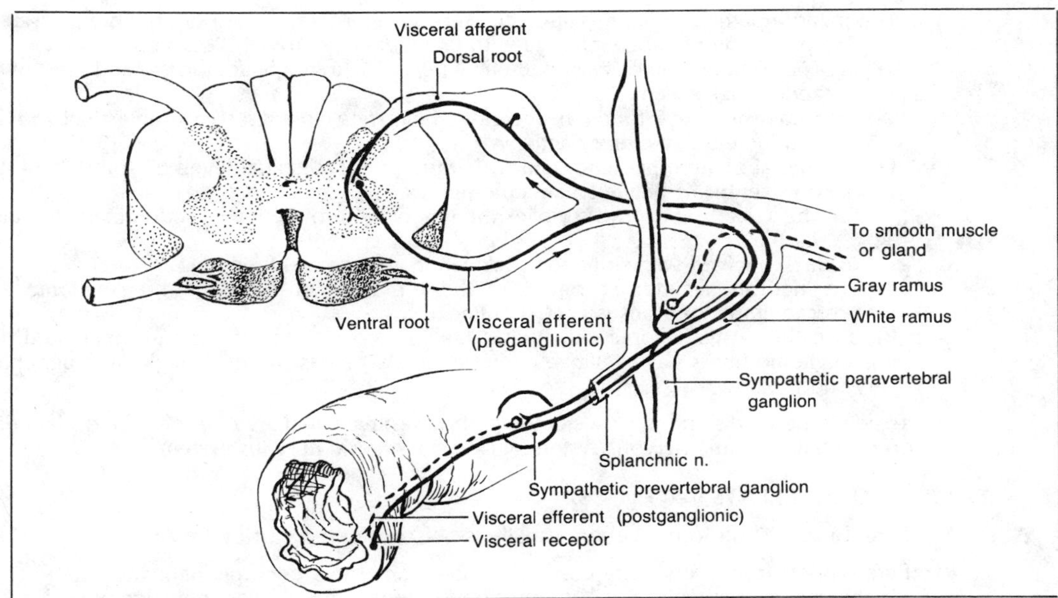

Figure 4-5. *Autonomic and visceral afferent neurons.* Presynaptic sympathetic neurons with cell bodies in the lateral horns of the spinal cord pass along the spinal cord and along a white ramus communicans to reach the sympathetic chain where they either synapse or pass along a splanchnic nerve to synapse in a prevertebral ganglion. The postganglion neurons contact the effector organs.

(i) Terminate by synapsing with a postganglionic neuron in the paravertebral ganglion associated with the spinal nerve level.

(ii) Course up or down the sympathetic chain to synapse with a postganglionic neuron located in a ganglion several levels away from that of entry.

(iii) Pass through the sympathetic chain without synapse to form a splanchnic nerve that leads to a prevertebral ganglion from the cervical to coccygeal regions.

(e) The neurosecretory transmitter released by the preganglionic sympathetic neurons is **acetylcholine**.

(2) The sympathetic **postganglionic** fibers from cells in the paravertebral ganglia pass either:

(a) Back to a spinal nerve along an unmyelinated **gray ramus communicans** from a paravertebral ganglion at *every spinal level* to terminate in the dermal sweat glands or the smooth muscles of blood vessels and hair (arrector pili) muscles of the body wall and extremities. Every spinal nerve receives a gray ramus from the sympathetic chain.

(b) To splanchnic nerves and perivascular plexuses leading to the visceral structures of the head, neck, thorax, abdomen, and pelvis.

(3) The neurotransmitter of the sympathetic postganglionic neuron is usually **norepinephrine**.

(4) The sympathetic postganglionic fibers from cells in the prevertebral ganglia course in the perivascular plexuses to innervate the abdominal and pelvic viscera.

(5) The cells of the **medulla** of the **adrenal gland** are specialized postganglionic sympathetic neurons.

(a) Preganglionic cholinergic fibers stimulate the adrenal chromaffin cells to release both norepinephrine and epinephrine into the circulatory system, which distributes these neurosecretions throughout the body with a time delay.

(b) In conjunction with the immediate norepinephrine release mediated by the sympathetic postganglionic fibers, the adrenal neurosecretions prolong the actions on the receptive tissues.

c. Because the paravertebral and prevertebral ganglia are located some distance from the organ innervated, the sympathetic preganglionic fibers are shorter, whereas the postganglion fibers are longer.

2. The **parasympathetic division.**

a. Following the same rough rule-of-thumb, the parasympathetic system stimulates those activities associated with conservation and restoration of body resources.

b. The parasympathetic system is also called the **craniosacral system** or **cholinergic system**.
  (1) The preganglionic fibers emerge with cranial nerves III, VII, IX, and X.
    (a) These supply the parasympathetic innervation to the head, thorax, and most of the abdominal viscera.
    (b) Parasympathetic function decreases heart rate, increases gastrointestinal activity, and increases secretory activity.
  (2) The sacral parasympathetic outflow is through sacral spinal levels S2 through S4, the **nervi erigentes** or **pelvic splanchnic nerves**.
    (a) The sacral spinal cord supplies the innervation to the lower abdominal and pelvic viscera.
    (b) It is involved with urination, defecation, and sexual function.
  (3) The neurotransmitter secretion at the terminals of both the preganglionic and postganglionic neurons is **acetylcholine**.

c. Because the postganglionic cell bodies are located close to the organ innervated, the preganglionic fibers have relatively long axons, whereas the postganglionic fibers have very short axons.

3. The neural networks and plexuses of the gastrointestinal canal are considered a distinct division of the autonomic nervous system, called the **enteric nervous system**.

E. VISCERAL AFFERENTS (see Fig. 4-5)

1. Afferent innervation to the viscera usually is not considered part of the ANS.

2. Afferent fibers from the viscera pass retrograde along the autonomic pathways.
  a. In the cervical region, visceral afferents travel along cervical splanchnic nerves such as the cardiac accelerator nerves, to reach the sympathetic chain, thence down the chain to the white rami communicantes of the upper thoracic levels to gain access to the spinal nerves and the upper thoracic levels of the spinal cord.
  b. In the thorax and abdomen they pass along splanchnic nerves to the sympathetic chain.
    (1) On reaching the sympathetic chain the afferents pass through the white rami communicantes to gain access to a spinal nerve.
    (2) If there is no white ramus communicans (above T1, below L2), the afferents course down or up the sympathetic chain until a white ramus is reached so that the spinal cord may be accessed.
  c. In the pelvic region there are two distinct afferent pathways.
    (1) From upper pelvic viscera, afferent neurons travel along sympathetic pathways to the lumbar splanchnic nerves, thence along white rami communicantes to the lumbar spinal nerves that bring the sensory information to the upper lumbar levels of the spinal cord.
    (2) From the lower pelvic viscera, afferent neurons travel along the parasympathetic nervi erigentes (pelvic splanchnic nerves) to reach midsacral (S2–S4) levels of the spinal cord.
  d. The visceral afferent pathways provide the anatomic basic for **referred pain**, whereby sensation from a visceral structure appears as if it originates from the somatic dermatome associated with the spinal level at which the visceral afferents enter the spinal cord.

VI. SUMMARY

A. AFFERENTS

1. **Somatic afferent neurons** take a direct pathway to the CNS from their specific receptors. The location, intensity, and quality of the stimulus reaches conscious awareness.

2. **Visceral afferent neurons** take rather indirect pathways to the CNS. Only noxious stimuli reach conscious awareness and are but poorly localized in somatic dermatomes by reference to their point of entry into the spinal cord.

B. EFFERENTS

1. All **primary peripheral motor neurons** have their cell bodies in the CNS. All secrete acetylcholine, which acts on similar receptors.

2. The **somatic peripheral neurons** act directly on striated muscle; the cholinergic autonomic primary motor neurons act on secondary motor neurons. The sympathetic and parasympathetic secondary motor neurons secrete different neurotransmitters, which mediate largely antagonistic effects at receptor organs.

5
The Circulatory System

I. INTRODUCTION

A. The circulatory system includes the heart and vessels (arteries, capillaries, veins) that conduct the blood through the body (see Fig. 5-1).

1. The **heart** is a muscular pump that propels blood through the circulatory system.
2. There are two circulatory loops.
   a. **Pulmonary circulation.** Blood is pumped from the right side of the heart through the lungs and returned to the left side of the heart.
   b. **Systemic circulation.** Blood is pumped from the left side of the heart through the tissues of the body and returned to the right side of the heart.
3. **Arteries** conduct blood from the heart to the capillary beds.
4. The **capillaries** provide for gaseous diffusion and exchange of nutrients and waste products.
5. **Veins** collect blood after its passage through capillary beds and return it to the heart.

B. The circulatory system also includes the lymphatic vessels, a set of channels that begins in the tissue spaces and returns excess tissue fluid to the bloodstream.

II. THE HEART

A. The heart is a set of two adjacent folded tubes of cardiac muscle (see Fig. 5-1).

1. The right side of the heart pumps blood from the venous system through the lung capillary beds and into the left side of the heart, which pumps blood through the systemic circulation.
2. Contraction of the muscular walls propels blood out of the chambers.
   a. Valves guarding the lumina maintain the blood flow in a fixed direction.
   b. The work done by the heart muscle is the product of the pressure and the volume of blood pumped.
3. The pumping rate of the heart is regulated by the autonomic nervous system, which controls an internal pacemaker (the sinoatrial node).
   a. An internal-impulse conducting system of modified muscle cells (atrioventricular node and bundle) coordinates the contraction in the ventricular chambers of the heart.
   b. Pathology in this impulse initiation and conduction system can lead to heart block or fibrillation (uncoordinated contractions).

B. The myocardium is supplied with blood by the **coronary circulation.**

1. The **right coronary artery** and **left coronary artery** begin in sinuses behind the right and left semilunar cusps of the aortic valve.
2. They distribute blood in large part to their own half of the heart, but anastomose with each other to some extent.
   a. Although anastomoses can be plentiful between different major branches of these arteries, they are rarely capable of much flow unless they are chronically hypertrophied.
   b. A thromboembolus (blood clot), which suddenly blocks flow in a major branch, results in ischemia and necrosis of the muscle served (an infarct).

c. Slow blockage, such as that associated with atherosclerosis, leads to a hypertrophy of the anastomoses and eventual substitution of an alternate blood supply.

3. Blood flow in the coronary arteries is maximal during diastole and minimal in systole.

C. FETAL CIRCULATION is different from circulation in the adult.

1. In the fetus oxygenation of the blood occurs in the placenta, and circulation through the lungs occurs to a much lesser degree than in postnatal life.
a. The right heart pumps very little blood to the collapsed fetal lungs.
b. The left heart pumps blood through the systemic circulation and the placenta.
c. Before birth a system of shunts operates to partially bypass the lungs and liver.
(1) The **foramen ovale** shunts blood from the right atrium to the left atrium, bypassing the pulmonary circulation.
(2) The **ductus arteriosus** shunts blood from the left pulmonary artery to the aorta, bypassing the pulmonary circulation.
(3) The **ductus venosus** shunts blood from the umbilical vein to the inferior vena cava, bypassing the liver.
d. These shunts close postnatally, and the adult circulatory pattern is established.

2. Failure of these shunts to close at birth represents a series of possible defects that may be surgically correctable.

III. THE ARTERIAL SYSTEM

A. CONDUCTING ARTERIES are the largest vessels of the body (Fig. 5-1).

1. The arterial system begins with the **aorta**, a single, large-diameter (3 cm) artery.

2. Wall structure is related to pressure.
a. The elastic tissue in the conducting arteries permits both stretch of the wall during the ejection phase of cardiac *systole* and a propulsive elastic recoil during ventricular *diastole*.
b. The aorta gives off smaller diameter distributing arteries.
(1) The higher the pressure, the greater the amount of elastic tissue relative to muscular tissue.
(2) As the arteries get smaller, intermittent pulsations (from cardiac systole) become a steady, continuous flow, and no pulsation is observed in the capillary bed.

3. Clinical considerations:
a. The elastic tissue in these vessels enables the pulse to be palpated.
b. Blood pressure is measured by auscultation of the bruit produced by the intermittent flow of blood through an artery.
(1) When sufficient pressure is applied by a pneumatic tourniquet to compress the artery completely, no bruit is heard.
(2) As pressure is released, an intermittent bruit is heard when systolic pressure just exceeds that applied by the tourniquet.
(3) As pressure is further released, the diastolic pressure equals that applied by the tourniquet, flow becomes continuous, and the bruit disappears.

B. DISTRIBUTING ARTERIES constitute the remainder of the named arteries of the body (see Fig. 5-1).

1. The distributing arteries progressively divide into smaller and smaller diameter arteries (although the total cross-sectional area increases).

2. Main arterial pathways:
a. Take the shortest possible course—for example, in the limbs they:
(1) Run on flexor surfaces.
(2) Usually do not pass directly through muscles, avoiding compression.
(3) Are somewhat extensible.
b. Branch in particular patterns such that:
(1) The angle of branching is related to hemodynamic factors and minimizes pull between parent stem branches.
(2) Branches of equal size (e.g., the common iliac arteries) make equal angles with the parent stem; the ideal angle is approximately 75°.
(3) With smaller branches, whose diameters barely affect the size of the parent stem (e.g., the intercostals), the angle can range from 70° to 90°.

3. Arterial **anastomoses** permit equalization of pressures and alternate channels of supply.

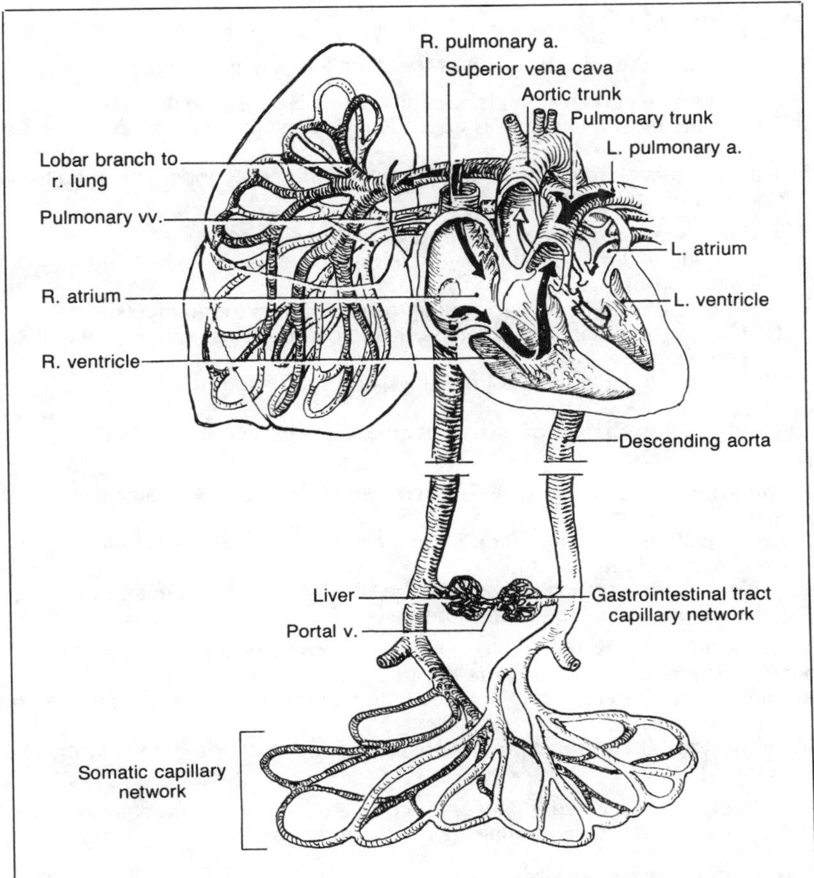

Figure 5-1. *A schematic representation of the circulatory system.* The pulmonary circulation is associated with the right side of the heart and the systemic circulation is associated with the left side of the heart.

- **a.** Abundant anastomoses occur in the regions of joints in which movement might temporarily occlude the main channel.
- **b.** In the brain, the circle of Willis serves to equalize the blood supply to brain.
- **c.** In the gastrointestinal tract, abundant anastomoses occur between some regions, whereas anastomoses are sparse between others.
- **d.** The surgeon must differentiate between *potential* and *functional* anastomoses and must be aware of variability, which might result in sparse or nonfunctional anastomoses where profuse functional anastomoses are expected.

4. End-arteries supply discrete regions of tissue that have no collateral supply.

- **a.** There are no direct anastomoses between end-arteries.
  - **(1)** End-arteries are found in the heart, kidneys, liver, brain, and organs of the gastrointestinal tract.
  - **(2)** A thrombosis or embolus lodged in an end-artery produces ischemia and necrosis (infarct) of the tissue supplied exclusively by that vessel.
- **b.** Potential anastomoses may not be functional at a crisis unless collateral channels develop with a slow-onset pathology.

C. ARTERIOLES, the terminations of the smallest arteries:

1. Are nearly as small as capillaries.

2. Are not merely conducting channels, but regulate the distribution of blood.

- **a.** The presence of muscles in the walls of the arterioles provides the basis for nervous regulation of the size of the vessel lumen.
- **b.** This muscular tissue runs in spiral patterns.

IV. THE CAPILLARY BEDS

A. CAPILLARIES are the exchange site of the circulatory system (Fig. 5-2).

1. The total cross-sectional area of the capillary bed is approximately 800 times that of the aorta.
   - a. The diameter of a capillary is about 5 μm, just sufficient for a red blood cell to squeeze through.
   - b. The velocity of circulation changes from 0.5 m/per second in the aorta to 0.5 mm/per second in the capillaries.
2. The walls are formed by a single layer of endothelial cells.
   - a. This endothelium and a thin layer of connective tissue fibers (the basal lamina) make up the diffusion barrier.
   - b. Gaseous diffusion occurs according to partial pressure gradients.
     - (1) Oxygen, released from the oxygenated blood, diffuses across the endothelium into the tissue spaces.
     - (2) Carbon dioxide diffuses from the tissue spaces into the blood.

B. In addition to an **exchange of gas** across the capillary endothelium, there is an **exchange of fluid**.

1. At the arterial end of the capillary bed, blood pressure is greater than the tissue osmotic pressure.
   - a. An ultrafiltrate of nutrient-rich plasma exudes from the arterial end of the capillaries into the tissue spaces.
   - b. This raises the osmotic pressure of the blood plasma remaining in the capillary bed and lowers the osmotic pressure of the tissue.
2. At the venous end of the capillary bed, the blood pressure is less and the plasma osmotic pressure is greater than the tissue osmotic pressure.
   - a. There is an osmotic uptake of tissue fluid that is high in metabolic waste products into the venous capillaries.
   - b. This restores the osmotic pressure of the tissue spaces and blood plasma in a constant process.
3. Density of a capillary bed in a tissue is correlated with the metabolic needs (e.g., endocrine glands are very vascular, cartilage is almost avascular).

C. CLINICAL CONSIDERATIONS

1. **Edema** is the collection of excess fluid in the tissue spaces.
   - a. When associated with trauma or inflammation edematous exudate is caused by increased permeability of the walls of the capillary bed whereby red blood cells are contained but plasma proteins leak out and raise the tissue osmotic pressure, thus interfering with the hydrostatic-osmotic pressure balance that normally results in fluid uptake at the venous end of the capillary bed.
   - b. When associated with elevated venous pressure, edematous transudate is caused by imbalance of the hydrostatic-osmotic pressures that normally result in fluid uptake at the venous end of the capillary bed.
2. **Hematoma** (black-and-blue mark associated with edema) results from a major loss of integrity of capillary walls or the walls of other blood vessels; red blood cells in addition to plasma leak into the tissue and cause discoloration.

D. SINUSOIDS substitute for capillaries in some organs, such as the liver, spleen, and red bone marrow.

1. Sinusoids often contain phagocytic cells.
2. Circulation is slowed in sinuses.

V. THE VENOUS SYSTEM

A. CAPILLARY BEDS drain into **venules**, which come together to form veins (see Fig. 5-2).

B. VEINS return blood to the heart (see Fig. 5-1).

1. The veins must carry the same volume of blood as the arteries, but at a lower pressure.
   - a. Compared to arteries, veins have:
     - (1) Large and somewhat more irregular lumina.

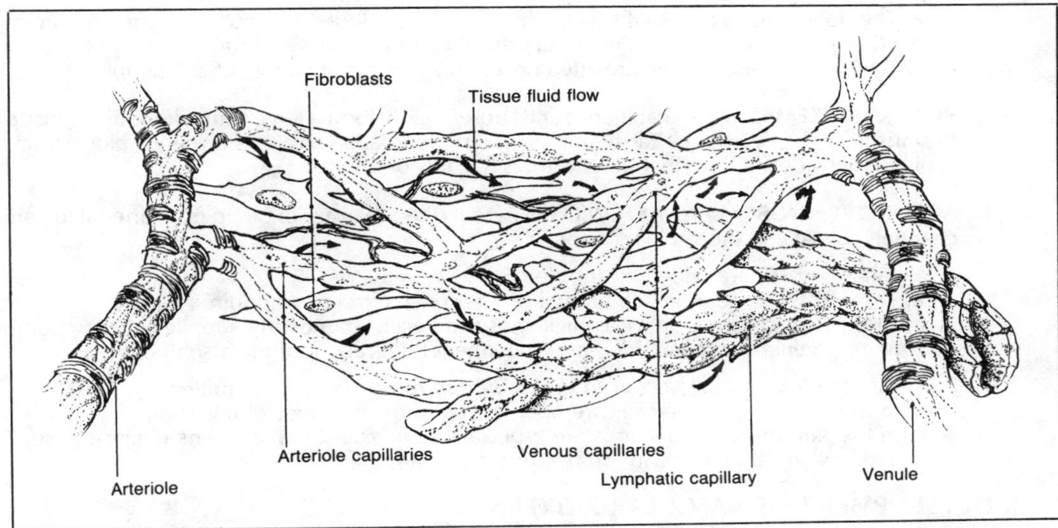

Figure 5-2. *The capillary bed.* The continuum from arteriole through the capillary bed to venule is depicted. *Arrows* indicate the flow of fluid. Blind lymphatic capillaries drain excess fluid from the tissue spaces.

(2) Thin walls.

b. Accordingly, veins are relatively compressible by external forces, a fact that aids in the flow of the venous blood.

c. When muscles are present in venous walls, they tend to be arranged in a loop near the point of drainage of the tributary to act as a "sluice gate" (e.g., the penile veins and the point at which the hepatic veins enter the vena cava).

d. The walls of larger veins have some elastic tissue to resist the pressure of right atrial systole.

e. Many veins contain **valves** that permit proximal flow only.

(1) Valves occur primarily in veins of limbs and movable viscera, but not in the cerebral veins.

(2) Valves are especially prevalent at junctions between tributaries and the larger vein.

2. Veins commonly run with arteries.

a. Venous patterns are far more variable than arterial patterns.

(1) Large veins are usually single.

(2) Medium-sized veins are frequently doubled (venae comitantes).

(3) In several regions, arterial patterns and venous patterns are quite separate and distinct (e.g., in the brain, liver, lungs, and penis).

b. The angulation of entry of tributaries is less related to hemodynamic factors than it is in arteries because venous pressure is lower.

c. The frequency of venous cross-anastomoses reflects retention of an embryonic condition in the adult (pedomorphism).

d. The "counter-current" pattern of arterial and venous patterns is a special arrangement related to transfer of water or heat, as in the kidney and the extremities, respectively.

3. Venous flow is dependent on the pressure gradients between the periphery and the right side of the heart, established by:

a. An arterial pressure of approximately 10 mm Hg transmitted through the capillary bed to the venous side.

b. The sucking bulb syringe action of the heart during right ventricular diastole.

c. The negative pressure, relative to atmospheric pressure, produced by the thoracic cage during inspiration.

d. The contractile activity of the muscles of the extremities, which "milks" the venous blood toward the heart.

(1) This action is aided in the limbs by the disposition of two sets of veins, deep and superficial.

(a) The muscles of the extremities, between which the deep veins lie, are surrounded by relatively inelastic fascial septa.

(b) Upon shortening, muscles become wider, thereby increasing the pressure on the deep veins and moving the blood in a direction guided by the internal valves.

(2) The movements of the segments of the limbs help move blood in the superficial veins.

e. The flow in the superficial veins, which can be occluded by application of mild pressure.
(1) A snug tourniquet occludes superficial veins and causes distal pooling of blood.
(2) The superifical veins provide convenient sites for obtaining blood samples.

C. PORTAL SYSTEMS begin in capillaries and subsequently form a large vein before breaking up into sinusoids. The hepatic portal system, for example, drains most of the venous blood from the capillary beds of the intestinal tract to the sinusoids of the liver.

D. ARTERIOVENOUS (AV) ANASTOMOSES permit direct transfer of blood from arterial to venous channels, bypassing the capillary bed.

1. AV anastomoses are widely distributed.
a. These vessels rise as a side branch of a terminal arteriole and join a venule.
b. Each AV anastomosis has a thick muscular wall, abundantly supplied with vasomotor nerves (general visceral efferent, sympathetic), thereby forming a sphincter.

2. AV anastomoses usually occur in organs whose functions are intermittent.
a. In the gut, AV anastomoses are open except during periods of digestion.
b. In the skin these connections are especially numerous in apical parts (fingers, nose, lips, ears), where they serve in temperature regulation.

VI. DEVELOPMENT OF VASCULAR PATTERNS

A. Although embryonic vessels develop before the initiation of circulation, and are therefore preadaptive, additional development is functionally regulated.

1. There is nearly complete remodeling of the embryonic vascular pattern (e.g., aortic arches, cardinal veins).
a. An increase in vascular pressure is followed by an increase in endothelial budding that forms new vessels to relieve the load.
b. An increase in vascular pressure, which exerts a circumferential stress on the walls, results in an increase in the diameter of the vessel and, because the amount of muscular and elastic tissue is a response to stress, an increase in wall thickness.

2. Incomplete remodeling results in vascular malformations.

B. The vascular supply to some organs, such as the kidney, is altered on migration within the body, whereas other organs, such as the gonads, carry their original vascular supply with them.

VII. THE LYMPHATIC SYSTEM

A. The lymphatic system is composed of:

1. An extensive network of extremely variable **lymphatic vessels.**

2. Lymph nodes, which serve as filters and a source of lymphocytes and plasma cells.

B. The system has evolved as a specialized mechanism to return to the bloodstream fluids that were not taken up by the blood capillaries.

1. The lymphatic vessels contain **lymph**.
a. Lymph is clear and colorless.
b. When lymph contains fat droplets, it is called **chyle**.
(1) These fatty substances are absorbed by the lymphatics from the gastrointestinal tract.
(2) Chyle is channeled to the blood for distribution to appropriate organs.

2. Proteins are constantly entering the lymph stream from the tissue spaces.
a. These proteins filter through lymph nodes on their way to the blood vascular system.
b. This association is of particular importance if the protein is "foreign."

C. STRUCTURE (see Fig. 5-2)

1. The **lymph capillaries** begin as cul-de-sacs that drain the tissue spaces.
a. The lymph capillaries are wider than blood capillaries.
b. They are irregular in diameter (from a few microns to 1 mm).
c. The lymphatic capillaries are numerous in the mucous membranes, serous surfaces, and dermis of the skin.
(1) The brain, spinal cord, and eyeball lack lymphatic capillaries; however, these structures have other channels for draining tissue fluids.

(2) Lymphatic capillaries are also absent in bone marrow, parenchyma of the spleen, and superficial fascia.

2. The larger **lymphatics** are formed by the convergence of lymph capillaries.
   a. Lymphatics have valves that give them a beaded appearance.
   b. They are more plentiful than veins, which they tend to accompany.
   c. The walls of the larger lymphatics become somewhat thicker as they acquire small amounts of smooth muscle and elastic tissue.
   d. Intercalated along the lymphatic vessels are the lymph nodes.

3. **Lymph ducts** are formed by the convergence of lymphatic vessels. The lymphatics from the lower extremities converge on lymph nodes located anteriorly and superficially in the uppermost part of the thigh.
   a. From the inguinal lymph nodes the main duct from each lower extremity enters the abdomen.
   b. After passing through the lumbar lymph nodes, the main duct becomes known as the **lumbar duct**, which enters the cisterna chyli.

4. The **cisterna chyli** is a dilated sac located between the diaphragmatic crura, opposite the first lumbar vertebra and behind the right side of the aorta.
   a. The cisterna chyli can vary in size and location.
   b. It contains some smooth muscle and is somewhat pulsatile.
   c. In addition to the lumbar lymphatics, the cisterna chyli also receives the large **common duct**, which conveys lymph from most of the intestinal tract.

5. The **thoracic duct** originates from the upper end of the cisterna chyli.
   a. The thoracic duct ascends upward on the anterior aspect of the vertebral column.
   b. It inclines slightly to the left and ends at the base of the neck, usually by entering the left brachiocephalic vein.
      (1) The thoracic duct thus receives the lymph from both legs, the pelvis, the abdomen, and the left side of the thorax.
      (2) In addition, it may drain the lymph from the left arm and from the left side of the head and neck, because the left jugular lymph duct and the left subclavian lymph duct may join the thoracic duct in the base of the neck.
         (a) The **left jugular duct** drains lymph from the left side of the head and neck, accompanying the left internal jugular vein in the neck.
         (b) The **left subclavian duct** drains the left axillary and subclavicular nodes of the left upper extremity and accompanies the left subclavian vein.
      (3) The thoracic viscera are drained in part by the **left bronchomediastinal lymph duct**, which frequently joins the thoracic duct.

6. The **right lymph duct** receives lymph from the right side of the head and neck, right upper extremity and the right side of the thorax.
   a. The channels are the:
      (1) **Right jugular lymph duct**, from the right side of the head and neck.
      (2) **Right subclavian lymph duct**, from the right axillary and subclavicular nodes.
      (3) **Right bronchomediastinal lymph duct**.
   b. The right lymph duct returns lymph to the great veins at the base of the neck on the right side.

7. There are valves in the thoracic duct and right lymphatic duct, including a particularly well-developed valve close to their termination before the lymph flows into the bloodstream.

D. **LYMPHATICOVENOUS COMMUNICATIONS** function when the lymphatic pressure is increased by blockage of the usual channels.

1. Direct lymphaticovenous connections have been demonstrated between:
   a. The thoracic duct and the hemiazygous vein.
   b. Abdominal lymphatic ducts and the inferior vena cava.

2. It is postulated that these connections are the result of enlargements of preexisting channels that normally convey little or no lymph.

E. **LYMPH FLOW**

1. The flow of lymph from a region is generally unidirectional toward the large veins at the base of the neck.
   a. In normal situations, the large volume of the lymph flowing through the thoracic duct is derived mainly from the liver and alimentary tract.

b. The progression and directionality results from:
   (1) Hydrostatic pressure from the volume of tissue fluids taken up by the lymphatic capillaries.
   (2) Mechanical forces:
      (a) The pressure resulting from the voluntary muscular activity on the valved lymph vessels (e.g., the flow of lymph in an immobile limb is almost negligible but becomes significant during muscular activity).
      (b) Respiratory movements: the alternation of pressures within the thorax during ventilation propels lymph in the valved lymphatics of the mediastinum.
      (c) The contractions of the muscles of the abdominal wall on expiration, coughing, and straining: these produce a positive abdominal pressure on the cisterna chyli that propels lymph.
      (d) The pulsations of adjacent blood vessels, which have a massaging effect on lymph vessels to aid the flow of lymph.
      (e) In some part of the body rhythmic flow may be aided by the contraction of the smooth muscles in the walls of the lymphatic vessels and the cisterna chyli.
   (3) Valves, which prevent backflow.

2. The average rate of lymph flow in the thoracic duct is from 1 to 2 ml/kg/hr.

F. CLINICAL CONSIDERATIONS

1. The **lymphatics** are the major route by which carcinoma metastasizes.
   a. Malignant cells may be trapped in lymph nodes where they proliferate.
   b. Malignant cells may be delivered to the venous system.
   c. Surgical removal of malignant tumors involves dissection of the major lymphatic vessels and nodes draining the involved region.

2. Injury to the thoracic duct may result in accumulation of fluid in thorax or pleural cavities, chylothorax. The diagnosis is made by the high lymphocyte count in a pleural aspirate.

3. Thoracic duct obstruction with backup of intestinal lymph to the kidney lymphatic capillaries may produce chyluria.

4. Lymphatic absorption of fluid in the peritoneal cavity is almost entirely through the peritoneal surface of the diaphragm.
   a. While the omentum has lymphatics, the absorption is minor compared to that of the diaphragmatic peritoneum.
   b. The rate of absorption from the peritoneal cavity is:
      (1) Exceedingly rapid; labeled protein injected intraperitoneally is detected in the thoracic duct in 10 to 15 minutes.
      (2) Normally about 1 L per day.
   c. This is the basis for peritoneal dialysis.

5. Lymph from liver contributes a considerable part of lymph flow of the thoracic duct.
   a. Hepatic lymph drains along the hepatic pedicle toward the hepatic nodes and thence to cisterna chyli.
   b. Ascites is the accumulation of extracellular fluid in the peritoneal cavity.
      (1) Ascitic fluid usually is a transudate from dilated subcapsular and hepatic hilar lymphatics.
      (2) It is related to increased capillary permeability, a hydrostatic-osmotic pressure imbalance, or altered lymphatic flow.
      (3) It is commonly, but not exclusively, associated with conditions that raise intrahepatic blood pressure (e.g., right-sided heart failure, cirrhosis).

6. The lymphatic drainage of the lungs is via the bronchomediastinal duct.
   a. Pulmonary edema is caused by either increased permeability or a hydrostatic-osmotic pressure imbalance in the pulmonary vascular bed with fluid accumulation in the tissue spaces and transudation of fluid into the alveoli.
   b. Pulmonary effusion (hydrothorax) can be caused by similar processes, but with transudation into the pleural cavity.

7. Lymph return from the trunk and extremities is to axillary and inguinal nodes.
   a. Lymph follows two sets of channels.
      (1) The superficial lymphatics accompany the superficial veins to the axillary and inguinal nodes. Infection may spread along superficial lymphatics causing fine striae.
      (2) The deep lymphatics accompany the deep veins to the axillary and inguinal nodes.

b. Lymphedema is the accumulation of tissue fluid as a result of lymphatic obstruction (e.g., elephantiasis caused by *Filaria bancrofti*).

8. Bacteria or antigens filtered by the lymph nodes may induce inflammation with swelling of the nodes.

9. Wound healing requires the regeneration of the lymphatics as well as the growth of blood capillaries.

Part I Introductory Overviews

STUDY QUESTIONS

Directions: Each question below contains five suggested answers. Choose the **one best** response to each question.

1. Circumduction is a movement that involves

(A) abduction/adduction and flexion/extension
(B) flexion/extension and medial rotation/lateral rotation
(C) medial rotation/lateral rotation and abduction/adduction
(D) one axis of rotation
(E) ball-and-socket joints only

2. The principal lymphatic drainage of the right breast is toward the

(A) right axillary nodes
(B) right internal mammary nodes
(C) right internal thoracic nodes
(D) right supraclavicular nodes
(E) thoracic duct

3. Which of the following statements about the lobes of the breast is true?

(A) They are defined by connective tissue septa, which determine the posture of the breast
(B) They are located between the deep layer of the superficial fascia and the deep fascia
(C) They contain fatty tissue, which is the source of breast size variation over the menstrual cycle
(D) They empty at the nipple via a common ampulla
(E) They number between 40 and 60

Directions: Each question below contains four suggested answers of which **one or more** is correct. Choose the answer

A if **1, 2, and 3** are correct
B if **1 and 3** are correct
C if **2 and 4** are correct
D if **4** is correct
E if **1, 2, 3, and 4** are correct

4. Correct statements concerning the epiphyseal disk in a child include which of the following?

(1) It may be mistaken for a fracture of the bone on x-ray
(2) It consists of fibrocartilage
(3) It can cause asymmetrical growth if fractured
(4) It is responsible for appositional (transverse) bone growth

5. Correct statements about torn ligaments include which of the following?

(1) They interrupt the connection between bone and muscle
(2) They can usually be visualized radiographically
(3) They, like bone, repair rapidly and strongly
(4) They destabilize the joint

6. The strength of a muscle is a function of the

(1) length of the lever arm
(2) cross-sectional area
(3) degree to which the muscle is stretched in situ
(4) length of the muscle in its fully stretched state in situ

7. The subarachnoid space between the arachnoid and pia meningeal layers contains which of the following substances?

(1) Blood upon hemorrhage of the middle meningeal artery
(2) Blood upon hemorrhage of a cerebral artery
(3) Blood upon hemorrhage of a cerebral vein
(4) Cerebrospinal fluid

SUMMARY OF DIRECTIONS

A	B	C	D	E
1, 2, 3 only	1, 3 only	2, 4 only	4 only	All are correct

8. Sympathetic postganglionic neurons, which join each spinal nerve via a gray ramus communicans, innervate

(1) arrector pili muscles
(2) smooth muscle of the peripheral vascular bed
(3) sweat glands and modified sweat glands
(4) skeletal muscle

9. A blood clot that develops in a peripheral vein will lodge in the vascular bed of the

(1) brain
(2) heart
(3) extremities
(4) lungs

10. Valves occur in which of the following vessels?

(1) Veins of the viscera
(2) Lymphatics
(3) Veins of the extremities
(4) Cerebral veins

ANSWERS AND EXPLANATIONS

1. The answer is A. (*Chapter 1 III D 8 c*). Circumduction is a combined movement involving both abduction/adduction and flexion/extension. It must, therefore, occur at joints with these two degrees of freedom.

2. The answer is A. [*Chapter 2 III B 1; Chapter 5 VII C 5 b (1)*]. Approximately 75 percent of the drainage of the breast is through the axillary nodes with a relatively small proportion draining through the internal thoracic (parasternal, internal mammary) nodes. Only the left axillary nodes drain into the thoracic duct.

3. The answer is A. (*Chapter 2, III A 1*). The 16 to 20 lobes of the breast are separated by connective tissue septa (Cooper's ligaments), which run between the epidermis and the deep layer of the superficial fascia. The connective tissue septa determine the posture of the breast.

4. The answer is B (1, 3). [*Chapter 3 II B 3 a (3) (c), 4 b (2) (b)*]. The epiphyseal plate in a child, composed of hyaline cartilage, is radiolucent on x-ray and thus appears as a fracture. The disk is the site of longitudinal bone growth, and a fracture of the disk, causing premature closure, results in asymmetrical growth.

5. The answer is 4 (D). (*Chapter 3 II D 2 b*). Ligaments connect bones and stabilize joints. Since ligaments are radiolucent, radiographic evidence of a ligamentous tear or sprain only occurs if a misalignment of bony structure accompanies such a tear. While ligaments heal by scar formation, they do not undergo the extensive remodeling seen in bone, and thus a healed ligament usually is not as strong.

6. The answer is A (1, 2, 3). (*Chapter 3 III B 2 b, C, D*). The strength of a muscle is a function of the cross-sectional area (the number of contractile units in parallel), the length of the lever arm (the mechanical advantage), and the degree to which the muscle is stretched in situ (the amount of interaction between contractile filaments). The relative length of a muscle in its fully stretched state (the number of contractile units in series) has no bearing on the strength.

7. The answer is C (2, 4). (*Chapter 4 II A 2*). The subarachnoid space contains cerebrospinal fluid (CSF), which bathes and protects the brain. The large distributing cerebral arteries lie in the subarachnoid space so that hemorrhage of one of these vessels will be evident by blood in the CSF. A middle meningeal hemorrhage produces an epidural hematoma, and a cerebral vein usually produces a subdural hematoma, neither of which has access to the subarachnoid space.

8. The answer is A (1, 2, 3). [*Chapter 4 V D 1 b (2)*]. Sympathetic preganglionic neurons leave the spinal cord in segments T1 through L2, gain access to the sympathetic chain via white rami communicantes, course up and down the chain, and synapse with a postsynaptic neuron. Those postsynaptic neurons that join a spinal nerve via gray rami communicantes innervate the arrector pili muscles, sweat glands, and the smooth muscle of the peripheral vascular bed.

9. The answer is D (4). (*Chapter 5 I A 2; II A 1; Figure 5-1*). A thromboembolus originating in the peripheral venous system will lodge in the vascular bed of the lung, producing a pulmonary embolus. Emboli that arise in the pulmonary veins or in the left side of the heart may lodge in the systemic vascular beds (i.e., the brain, heart, kidneys, and extremities).

10. The answer is A (1, 2, 3). (*Chapter 5 V B 1 e; VII C 7*). As a rule of thumb, the veins of the dependent portions of the body, including the extremities and viscera, have valves; the cerebral veins are valveless. The larger lymphatic vessels, lymph trunks, and the thoracic duct all have valves.

Part II
Upper Extremity

6
The Shoulder Region

I. INTRODUCTION TO THE UPPER EXTREMITY

A. BASIC PRINCIPLES

1. The basic pattern of the human upper extremity can be understood in terms of simulating the primitive position, i.e., by abducting the arm until it is horizontal, with the palm facing forward, so that the upper extremity approximates the pectoral fin of a fish.
   - **a.** The anterior side is the **primitively ventral surface**; the posterior side is the **primitively dorsal surface**. The flexor side corresponds to the primitively ventral surface; the extensor side corresponds to the primitively dorsal surface.
   - **b.** An imaginary line drawn through the long axis of the extremity will differentiate between a cephalad **preaxial border** and a caudal **postaxial border**. While there generally is overlap between sequential dermatomes, there is *no* overlap across the axial lines (see Fig. 4-3).
2. Flexing the elbow 90° and extending the wrist 90° from the fin-like position of the extremity approximates the amphibian/reptilian upper extremity.
3. In mammals, and in humans in particular, the amphibian/reptilian brachium has rotated 90° caudally, bringing the upper extremity alongside the body wall, so that in the erect posture the primitively ventral surface faces anteriorly.
4. The human upper and lower extremities are homologous, and the structural differences between them reflect different functions.
   - **a.** The upper extremity of the human, freed of its original role of support and propulsion, has developed high intrinsic mobility.
   - **b.** The function of the human upper extremity is to place its effector end, the hand, in a position to manipulate the environment. The hand can be positioned so that it touches every bodily part.
   - **c.** The hand has been developed for prehension, and only the human hand has a sophisticated two-point grasp between the thumb and index finger.
   - **d.** The hand is also a very sophisticated sensory organ.

B. The upper extremity can be divided into five regions.

1. The **shoulder**.
   - **a.** The **clavicle**.
   - **b.** The **scapula**.
2. The **arm** or **brachium**, containing the **humerus**.
3. The **forearm** or **antebrachium**.
   - **a.** The **radius**.
   - **b.** The **ulna**.
4. The **wrist**.
   - **a.** The **proximal** row of **carpal bones**.
   - **b.** The **distal** row of **carpal bones**.
5. The **hand**.
   - **a.** The **metacarpal bones**.
   - **b.** Three rows of **phalanges**.

C. The free portion of the upper extremity (the arm, forearm, wrist, and hand) is suspended from the axial skeleton by the pectoral girdle (the embedded portion of the upper extremity).

1. The only bony articulation between the pectoral girdle and the axial skeleton is through the clavicle at the sternoclavicular joint.

2. The attachment of the upper extremity to the thorax is primarily muscular.
   a. The muscles of the pectoral girdle provide support and stability, as well as produce movement.
   b. The pectoral girdle is very mobile at the expense of stability.

II. THE PECTORAL REGION

A. BONY LANDMARKS

1. The **clavicle**, an S-shaped strut between the scapula and sternum, is an easily observed and palpable subcutaneous bone.

2. The **scapula**, including:
   a. Spine.
   b. Acromion process.
   c. Coracoid process.
   d. Vertebral border.
   e. Inferior angle (apex).

3. The **humerus**, including:
   a. Head.
   b. Greater tuberosity.
   c. Intertubercular (bicipital) groove.
   d. Lesser tuberosity.

B. THE PECTORAL (shoulder) **GIRDLE** consists of the clavicles and the scapulae.

1. The **clavicle** (Fig. 6-1).
   a. The clavicle is probably the most commonly fractured bone in individuals below middle age.
      (1) During a fall onto an outstretched arm, forces transmitted through the clavicle to the sternum produce a shear within the S-shaped clavicle.
      (2) Due to the presence of major neurovascular structures between the clavicle and first rib, clavicular fracture can produce nerve damage, as well as severe, even fatal, internal bleeding.
   b. The clavicle articulates with the sternum at the **sternoclavicular joint**.
   c. It articulates with the scapula at the **acromioclavicular joint**.
   d. It is the first bone to begin ossification, during the fifth week of fetal development, but is the last bone to complete ossification, about the twenty-first year. It is the only long bone formed by intramembranous ossification.

2. The **scapula** (see Fig. 6-1).
   a. The scapula lies against the posterior aspect of the thoracic cage.
   b. It is a flat bone, shaped like an inverted triangle, with superior, vertebral, and lateral borders.
      (1) The subcutaneous, horizontal **spine** of the scapula is a reinforcing ridge along the dorsal surface, which divides the dorsal surface into supraspinous and infraspinous fossae.
      (2) The **acromion process** is the lateral expansion of the spine.
         (a) It articulates with the clavicle.
         (b) It provides an attachment for the trapezius and deltoid muscles.
      (3) The **coracoid process** projects anteriorly.
         (a) The remnant of a third bone of the pectoral girdle, it is palpable anteriorly, just medial to the head of the humerus.
         (b) It provides an attachment for several muscles and ligaments, one of which, the **coracoacromial ligament**, transmits tensile force from the coracoid process to the acromion process and spine of the scapula.
      (4) The **glenoid fossa** is the site of articulation with the humerus.
         (a) It faces laterally, but slightly anteriorly and slightly superiorly.
         (b) The fossa is very shallow, which predisposes this joint to dislocation.
         (c) The glenoid fossa is deepened slightly by a lip of fibrocartilage.
         (d) The supraglenoid and infraglenoid tubercles provide attachments for the long heads of the biceps and triceps muscles.

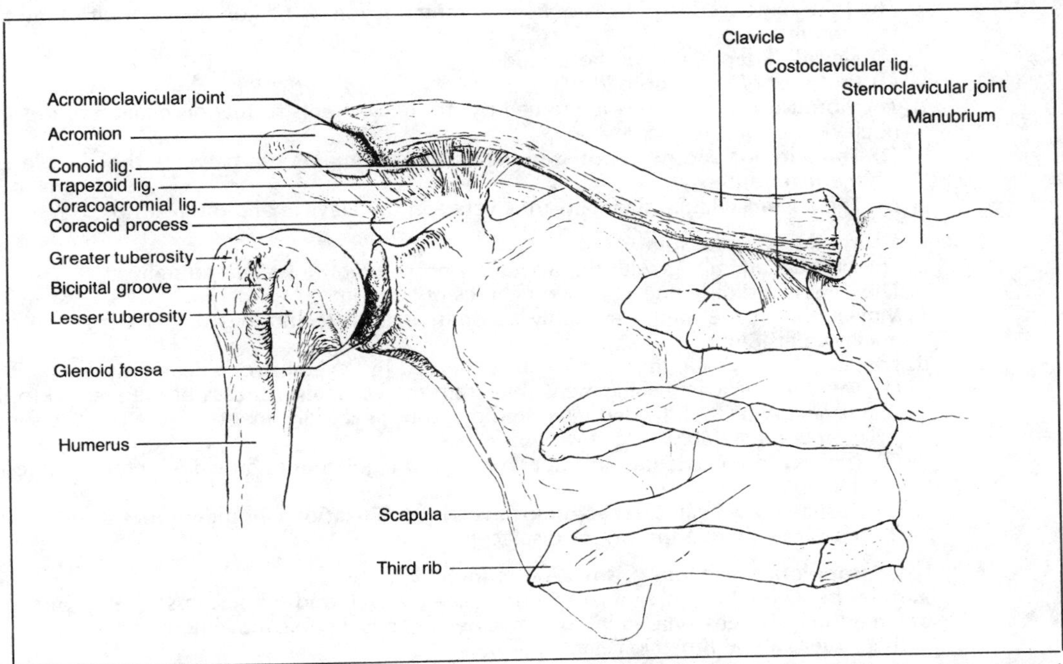

Figure 6-1. *The right pectoral girdle.* The right shoulder is shown in anterior view. The sternoclavicular and acromioclavicular joints are strongly reinforced by ligaments, while the glenohumeral joint is stabilized primarily by muscles.

c. Articulations:
   (1) **Acromioclavicular joint.**
   (2) **Glenohumeral (scapulohumeral) joint.**
   (3) **Scapulothoracic joint.**

C. THE HUMERUS is the long bone of the arm.

1. The proximal end of the humerus consists of a head and two tuberosities separated by an intertubercular groove (see Fig. 6.1).
   a. The **head** of the humerus articulates with the scapula at the glenohumeral (scapulohumeral) joint.
   b. The head is separated from the metaphysis by the **anatomical neck**, which marks the location of an epiphyseal plate that normally fuses between the nineteenth and twenty-first years.
   c. The **lesser tuberosity** lies on the anterior surface of the humerus, just distal to the anatomical neck, and provides an attachment for the subscapularis muscle.
   d. The **greater tuberosity** lies on the lateral surface of the humerus, just lateral to the anatomical neck, and provides attachments for the supraspinatus, infraspinatus, and teres minor muscles.
   e. The **intertubercular (bicipital) groove** lies between the greater and lesser tuberosities and contains the tendon of the long head of the biceps brachii muscle. It is bridged by the transverse humeral ligament, and the lateral and medial lips provide muscular attachments for the pectoralis major and teres major muscles, respectively, while the latissimus dorsi muscle inserts into the floor.
2. The **surgical neck**, the narrowest portion of the humerus, lies just below the tuberosities.
3. The **deltoid tubercle** lies on the lateral aspect of the midshaft.

D. THE ARTICULATIONS OF THE SHOULDER GIRDLE

1. The **sternoclavicular joint**.
   a. The clavicle articulates with the **manubrium** of the sternum proximally.
   b. The joint has a fibrocartilaginous articular disk that divides the joint so that there are two synovial capsules. It is attached to the superior edge of the articular surface of the clavicle and to the first rib.

c. The joint has two degrees of freedom. Circumduction is the combined result of the two movements.

(1) Elevation/depression of the shoulder.

(2) Protraction/retraction of the shoulder.

d. The fibrous joint capsule is reinforced by strong ligaments so that dislocation of the sternoclavicular joint is uncommon.

(1) The **anterior** and **posterior sternoclavicular ligaments** run between the clavicle and the manubrium.

(2) The **costoclavicular ligament** runs between the clavicle and the first rib.

2. The **acromioclavicular joint.**

a. The clavicle articulates with the **acromion process** of the scapula, distally.

b. The joint is a sliding one with two degrees of freedom.

c. Movements at the joint are mainly accommodations to motions that occur at the sternoclavicular joint.

d. The articular capsule, thin and lax, is reinforced by several strong ligaments.

(1) The **coracoclavicular ligament** runs from the **coracoid process** of the scapula to the clavicle and is subdivided into **conoid** and **trapezoid ligaments**. These provide superior-inferior stability.

(2) The **acromioclavicular ligament** bridges the joint and provides anterior-posterior stability.

(3) Despite these ligaments, **acromioclavicular subluxation (shoulder separation)** is common, usually due to upward displacement of the clavicle.

3. The **"scapulothoracic joint"** is not a true joint.

a. It lies between the subscapularis muscle (see Fig. 6-4) and the serratus anterior muscle.

b. It has three degrees of freedom, which allow for considerable movement of the scapula on the posterolateral thoracic cage.

4. Movement of the pectoral girdle is accomplished by muscles that run between the axial skeleton and the pectoral girdle (Figs. 6-2 and 6-3, Table 6-1).

a. Elevation/depression of the shoulder (10 cm) occurs about a bilateral axis through the sternoclavicular joint.

(1) Elevator muscles include the cervical portion of the trapezius, levator scapulae, and the rhomboids.

(2) Depressor muscles include the subclavius, pectoralis minor, and the thoracic portion of the trapezius.

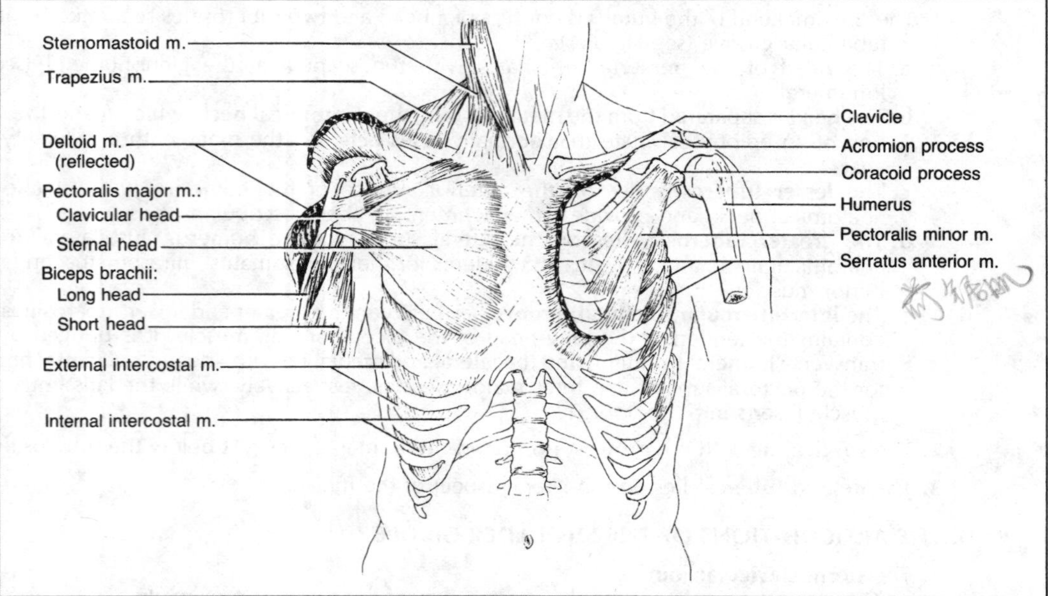

Figure 6-2. *The anterior pectoral musculature.* The right sternomastoid, trapezius, the clavicular and sternal heads of the pectoralis major, and the anterior portion of the deltoid, as well as the left pectoralis minor and serratus anterior are depicted.

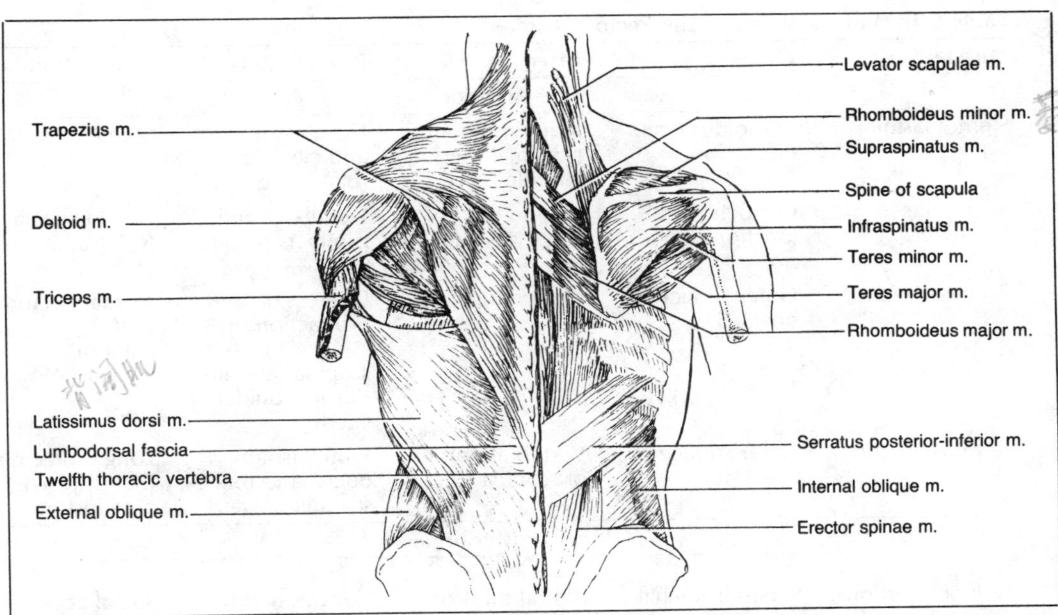

Figure 6-3. *The posterior pectoral musculature.* The left cervical and thoracic portions of the trapezius, the posterior and lateral portions of the deltoid, and the latissimus dorsi, as well as the right levator scapulae and major and minor rhomboids are depicted.

b. Protraction/retraction (15 cm) of the shoulder occurs about a vertical axis through the sternoclavicular joint.
   (1) Protractor muscles include the serratus anterior, pectoralis minor, and subclavius.
   (2) Retractor muscles include the rhomboids and the trapezius when both cervical and thoracic portions act synergistically.

c. Rotation of the shoulder occurs about an anteroposterior axis through the "scapulothoracic joint"; this axis is a result of the axes of the sternoclavicular and acromioclavicular joints.
   (1) Upward rotator muscles include both the cervical and thoracic portions of the trapezius and the serratus anterior.
   (2) Downward rotator muscles include the pectoralis minor, rhomboids, and levator scapulae.

d. Elevation of the arm between 60° and 170° involves upward rotation of the scapula.

5. The **glenohumeral (scapulohumeral) joint.**

a. As a ball-and-socket joint, it has three degrees of freedom.

b. Its lax fibrous capsule is reinforced by tough articular ligaments.
   (1) **Superior, middle,** and **inferior glenohumeral ligaments** run from the glenoid lip to the anatomical neck of the humerus.
   (2) The **coracohumeral ligament** between the coracoid process and the humerus supports the dead weight of the free portion of the upper extremity.

c. This rather unstable joint is dynamically reinforced by muscle action (see Figs. 6-2 and 6-3; Fig. 6-4).
   (1) The **rotator (musculotendinous) cuff** (supraspinatus, infraspinatus, subscapularis, and teres minor muscles) acts as a dynamic "ligament," keeping the head of the humerus pressed into the glenoid fossa.
   (2) The tendon of the long head of the biceps brachii muscle, restrained by the **transverse humeral ligament**, also forces the humeral head medially into the joint.

d. The **coracoacromial ligament** runs between the coracoid process and the acromion process.
   (1) Together with the acromion process, it forms the **coracoacromial arch**.
   (2) The coracoacromial arch buttresses the superior aspect of the glenohumeral joint to prevent superior displacement.
   (3) This ligament transmits tensile forces from the muscles that originate on the coracoid process to the acromion process and spine of the scapula.

e. Several bursae lie between various musculoskeletal components.

Table 6-1. Muscles Acting on the Pectoral Girdle

Muscle	Origin	Insertion	Primary Action	Innervation
		Anterior		
Sternomastoid	Mastoid process of cranium	Manubrium and proximal third of clavicle	Elevates sternum and clavicle; rotates the head	Spinal accessory n. (CN XI), C2–C3
Subclavius	Costochondral junction of first rib.	Middle third of clavicle	Depresses and protracts scapula	Nn. to subclavius (C5–C6, anterior)
Pectoralis minor	Outer surface of ribs 3–5	Coracoid process	Depression and protraction of scapula (elevates ribs if shoulder is fixed)	Medial pectoral n. (C8–T1, anterior)
Serratus anterior	Outer surface of ribs 1–9	Vertebral border of scapula	Protracts (abducts) and rotates scapula upward	Long thoracic n. (Bell's) [C5–C7, anterior]
		Posterior		
Trapezius:				
Upper portion	Superior nuchal line, ligamentum nuchae and spine of C7	Lateral third of clavicle and acromion process	Elevates and rotates scapula upward in elevation of arm	Spinal accessory n. (CN XI), C3–C4
Lower portion	Spines of T1–T12	Spine of scapula	Depresses and rotates scapula upward	Spinal accessory n. (CN XI), C2–C3
Levator scapulae	Transverse processes C1–C4.	Superior portion of vertebral border of scapula	Elevates scapula	Nn. to levator scapular (C3–C4)
Rhomboids:				
Minor	Lower part of ligamentum nuchae and spines of C7–T1	Proximal portion of spine of scapula	Retracts and elevates scapula	Dorsal scapula n. (C5, posterior)
Major	Spines of T2–T5	Vertebral border of scapula inferior to the spine	Retracts and elevates scapula	Dorsal scapula n. (C5, posterior)

(1) The **subacromial bursa** separates the acromion process and the supraspinatus muscle (see Fig. 6-4).

(a) This bursa is frequently the site of bursitis and supraspinatus tendinitis with calcium deposits.

(b) The pain associated with subacromial bursitis, mainly felt during the initial stages of abduction and forward flexion, severely limits mobility.

(2) The **subdeltoid bursa** separates the deltoid muscle from the head of the humerus and the insertions of the rotator cuff.

(3) The subacromial and subdeltoid bursae frequently communicate.

f. Despite the musculotendinous cuff and other stabilizing features, the extreme mobility of the glenohumeral joint together with the shallowness of the glenoid fossa result in a loss of stability which predisposes to dislocation.

(1) In **anterior dislocations**, the head of the humerus comes to lie inferior to the coracoid process. The axillary nerve is sometimes injured. Dislocation stretches the anterior capsule and may avulse the glenoid labrium, which predisposes to recurrence.

(2) In **posterior dislocations**, the humerus is displaced posteriorly.

g. Movement of the shoulder joint is accomplished by muscles that run between the axial skeleton and the humerus (see Table 6-1, Figs. 6-2 and 6-3), as well as by muscles that run between the pectoral girdle and the humerus (Table 6-2; see Fig. 6-4).

(1) Abduction/adduction of the arm occurs about an anteroposterior axis through the glenohumeral joint.

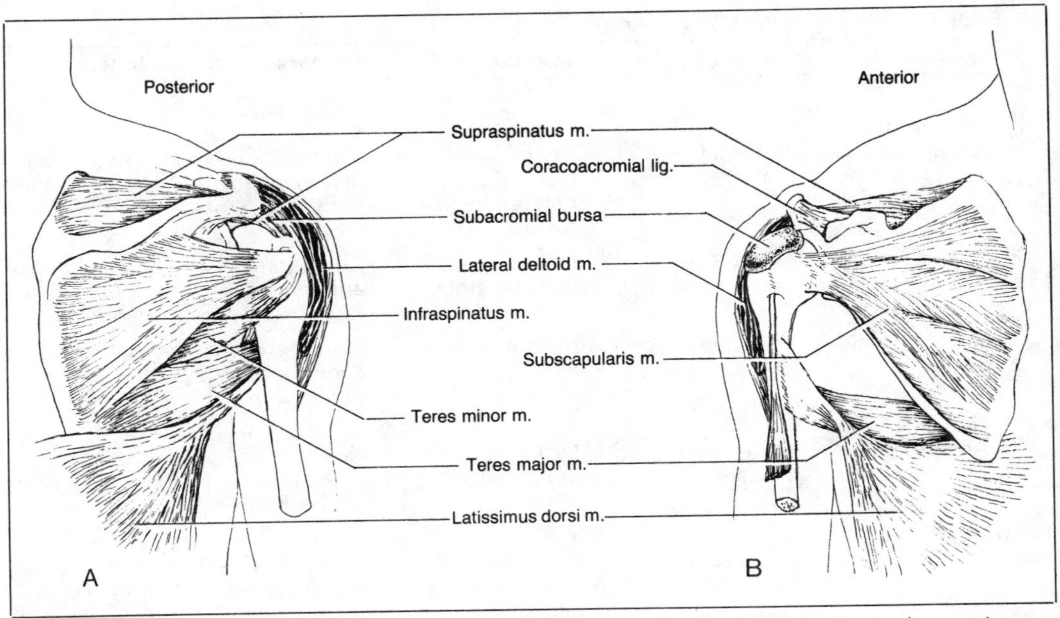

Figure 6-4. *A, The posterior muscles of the right rotator cuff.* The supraspinatus, infraspinatus, and teres minor are shown. *B, The anterior muscles of the right rotator cuff.* The supraspinatus and subscapularis are shown.

(a) Abductor muscles include the supraspinatus muscle (initial 15°) and the lateral portion of the deltoid muscle (10°–100°).

(b) Adductor muscles include the pectoralis major, latissimus dorsi, coracobrachialis, and the long head of the triceps brachii, as well as the anterior and posterior portions of the deltoid.

(2) Flexion/extension of the arm occurs about a bilateral axis through the glenohumeral joint.

(a) Flexor muscles include the anterior portion of the deltoid, the clavicular head of the pectoralis major, the coracobrachialis, and the biceps brachii.

(b) Extensor muscles include the posterior portion of the deltoid, the triceps brachii, the teres major, and the latissimus dorsi.

(3) Rotation of the arm occurs about a vertical axis through the glenohumeral joint.

(a) Lateral (external) rotator muscles include the posterior deltoid, infraspinatus, and teres minor.

(b) Medial (internal) rotator muscles include the anterior deltoid, teres major, subscapularis, latissimus dorsi, and the sternal head of the pectoralis major.

(4) In elevation of the arm, the first 60° occurs primarily at the glenohumeral joint.

E. CLINICAL CONSIDERATIONS

1. Fracture of the clavicle in the middle third (the most common fracture site) results in upward displacement of the proximal fragment, due to the pull of the sternomastoid muscle, and downward displacement of the distal fragment, due to the pull of the deltoid muscle and the effect of gravity.
2. Limitation of shoulder mobility may be due to rotator cuff tears, nerve or muscle paralysis, bursitis, tendonitis, joint disease, and pain.

III. THE AXILLA

A. The axilla is a space through which major neurovascular structures pass between the thorax and upper extremity.

B. It is shaped like a pyramid.

1. The apex is a triangular space limited by the first rib, the scapula, and the clavicle.
2. The anterior wall is the **anterior axillary fold**, which overlies the pectoralis major and pectoralis minor muscles.

Table 6-2. Muscles Acting on the Arm

Muscle	Origin	Insertion	Primary Action	Innervation
		Anterior		
Pectoralis major:				
Clavicular head	Proximal third of clavicle	Lateral lip of bicipital groove (intertubercular sulcus)	Flexes, adducts, and medially rotates arm	Lateral pectoral n. (C5–C7, anterior)
Sternal head	Sternum and costal cartilages 1–7	Lateral lip of bicipital groove of humerus	Adducts, flexes, and medially rotates arm	Medial pectoral n. (C8–T1, anterior)
Coracobrachialis	Coracoid process of scapula	Midhumerus	Flexes, adducts, and medially rotates arm	Musculocutaneous n. (C5–C7, anterior)
Deltoid:				
Anterior part	Lateral third of clavicle	Deltoid tuberosity of humerus	Flexes, adducts, and medially rotates arm	Axillary n. (C5–C6, posterior)
Biceps brachii:				
Long head	Supraglenoid tubercle	Radial tuberosity and antebrachial fascia	Flexes shoulder and stabilizes shoulder joint	Musculocutaneous n. (C5–C7, anterior)
Short head	Coracoid process	Radial tuberosity and antebranchial fascia	Flexes shoulder	
		Posterior		
Deltoid:				
Lateral part	Acromion process of scapula	Deltoid tuberosity of humerus	Abducts arm between 10° and 100°	Axillary n. (C5–C6, posterior)
Posterior part	Spine of scapula	Deltoid tuberosity of humerus	Extends, adducts, and laterally rotates arm	
Supraspinatus	Supraspinous fossa of scapula	Greater tubercle of humerus	Elevates arm first 15°. Stabilizes shoulder joint	Suprascapular n. (C5–C6, posterior)
Latissimus dorsi	Spinous processes of T6–L5, iliac crest, and ribs 10–12	Floor of bicipital groove of humerus	Extends, adducts, and medially rotates arm	Thoracodorsal n. (C6–C8, posterior)
Teres major	Inferior angle of scapula	Medial lip of bicipital groove of humerus	Extends, adducts, and medially rotates arm	Lower subscapular n. (C5–C6, posterior)
Triceps brachii:				
Long head	Infraglenoid tubercle	Olecranon process of ulnar	Extends and adducts arm	Radial n. (C6–C8, posterior)
Infraspinatus	Infraspinous fossa of scapula	Greater tubercle of humerus	Laterally rotates arm and stabilizes shoulder joint	Suprascapular n. (C5–C6, posterior)
Teres minor	Inferior angle of scapula	Greater tubercle of humerus	Laterally rotates arm and stabilizes shoulder joint	Axillary n. (C5–C6, posterior)
Subscapularis	Subscapular fossa	Lesser tubercle of humerus	Medially rotates arm and stabilizes shoulder joint	Upper subscapular n. (C5–C6, posterior)

3. The posterior wall is formed by the subscapularis muscle and the **posterior axillary fol**[...] which overlies the latissimus dorsi and teres major muscles.

4. The broad medial wall is the serratus anterior muscle overlying the thoracic cage.

5. The narrow lateral wall is the intertubercular (bicipital) groove.

6. The base is formed by the skin and fascia of the axillary fossa (armpit).

C. THE AXILLARY VASCULATURE (Fig. 6-5)

1. The **subclavian artery.**
   a. The **right subclavian artery** arises from the **brachiocephalic artery**.
   b. The **left subclavian artery** arises directly from the **aorta**.
   c. Each subclavian artery gives off several major distributing arteries in the base of the neck and shoulder.
      (1) The **vertebral artery** provides a primary supply to the brain.
      (2) The **internal thoracic (mammary) artery** provides a primary supply to the anterior thoracic and abdominal wall.
      (3) The **thyrocervical artery (trunk)** supplies portions of the neck and shoulder.
         (a) The **inferior thyroid artery** supplies the thyroid gland.
         (b) The **suprascapular artery**:
            (i) Passes through the **suprascapular notch** superior to the **transverse scapular ligament** to reach the dorsal aspect of the scapula.
            (ii) Anastomoses with the circumflex scapular artery and the dorsal scapular artery.
         (c) The **transverse cervical artery**:
            (i) Crosses the posterior triangle of the neck.
            (ii) Divides into a **superficial branch** and a **descending (dorsal) branch**, which anastomoses with the circumflex scapular and suprascapular arteries.
            (iii) More commonly the **superficial cervical artery** and the **dorsal scapular artery** arise separately, and there is no transverse cervical artery.
      (4) The **costocervical artery** passes back over the dome of the pleura to the neck of the first rib where it divides into deep cervical and superior intercostal arteries.

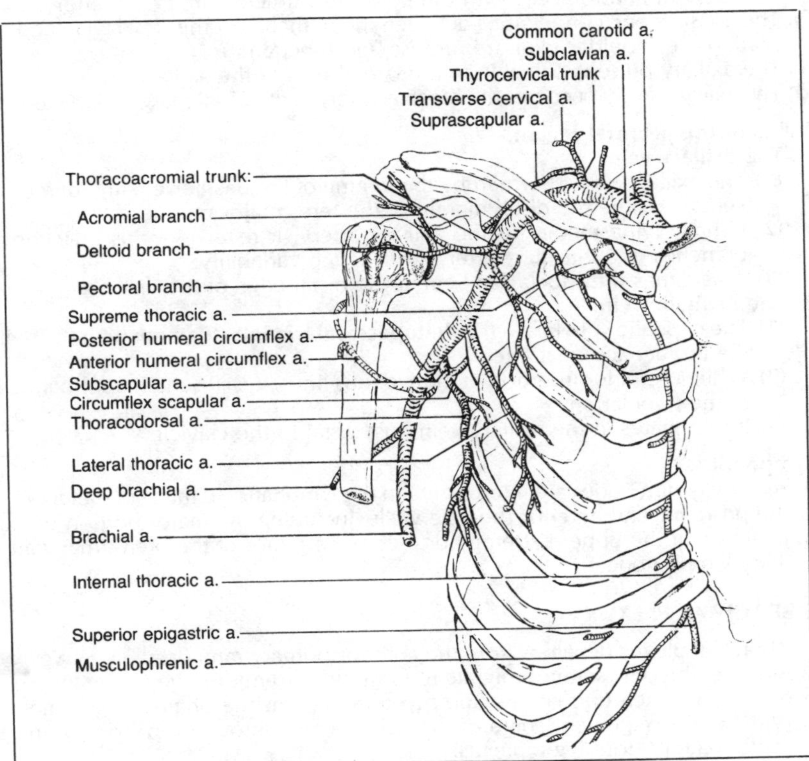

Figure 6-5. *The arterial supply to the right shoulder and proximal arm.*

The subclavian artery becomes the axillary artery as it passes beneath the clavicle and reaches the distal edge of the first rib.

The axillary artery.

a. The continuation of the subclavian artery, the axillary artery is divided into three parts by the tendon of the pectoralis minor muscle.
   (1) The upper part of the axillary artery gives off the **supreme (highest) thoracic** artery, which supplies the more superficial subclavicular portion of the chest.
   (2) The middle part of the axillary artery gives off two branches.
      (a) The **thoracoacromial artery (trunk)** gives off four or five branches.
         (i) **Pectoral branches** descend to the pectoralis major and minor muscles.
         (ii) The **acromial branch** crosses to the deltoid muscle.
         (iii) A **clavicular branch** runs medially to the subclavius muscle.
         (iv) A **deltoid branch** supplies the deltoid and pectoral major muscles and accompanies the cephalic vein in the deltopectoral triangle.
      (b) The **lateral (long) thoracic artery** courses along the thoracic wall, superficial to the serratus anterior muscle.
   (3) The lower part of the axillary artery gives off three branches.
      (a) The **subscapular artery**, the largest branch, bifurcates.
         (i) The **circumflex scapular artery** curves around the lateral border of the scapula and passes through the **triangular space** (teres major, teres minor, long head of triceps brachii) to reach the dorsum of the scapula where it anastomoses with the suprascapular and dorsal (descending) scapular arteries.
         (ii) The **thoracodorsal artery** lies on the anterior surface of the latissimus dorsi muscle.
      (b) The **posterior humeral circumflex artery** accompanies the axillary nerve.
         (i) This vessel passes posterior to the humerus, through the quadrangular space (teres major, teres minor, long head of triceps brachii, humerus).
         (ii) It anastomoses with the anterior humeral circumflex artery.
      (c) The **anterior humeral circumflex artery**.
         (i) This small branch passes beneath the coracobrachialis muscle, anterior to the humerus.
         (ii) It anastomoses with the posterior humeral circumflex artery.

b. The anastomotic connections between the branches of the subclavian and axillary arteries around the shoulder joint are profuse and functional.

c. The axillary pulse is palpable in the lateral wall of the axilla.

d. The axillary artery becomes the brachial artery at the distal margin of the teres major muscle.

3. Veins of the pectoral region.

a. The **axillary vein.**
   (1) The axillary vein is formed by the joining of the **basilic vein** and one or more **brachial veins** near the the distal margin of the teres major muscle.
   (2) Usually lying medial to the axillary artery, it receives veins that correspond to the branches of the axillary artery, with much variability.
   (3) It becomes the subclavian vein at the distal edge of the first rib.

b. The **cephalic vein.**
   (1) The cephalic vein lies in the **deltopectoral triangle**, together with the deltoid branch of the thoracoacromial artery.
   (2) While useful for insertion of intravenous lines, it is absent in approximately 10 percent of the population.
   (3) It terminates in the axillary vein, just distal to the clavicle.

4. Lymphatics.

a. Axillary lymph nodes are intercalated in the lymphatic drainage of the upper extremity and the posterior and anterior thoracic walls, including the major portion of each breast.

b. Infection of the upper extremity or metastatic tumor of the breast may cause swollen axillary lymph nodes.

D. THE BRACHIAL PLEXUS

1. The brachial plexus develops from the anterior primary rami of spinal nerves C4 through T1.
   a. As the embryonic somites migrate to form the extremities, they bring their own nerve supply, so that each dermatome and myotome retains the original segmental innervation.
   b. With somite migration, some of the nerves come into close proximity and fuse in a particular pattern, forming a plexus.

2. Subdivisions of the brachial plexus (Fig. 6-6).

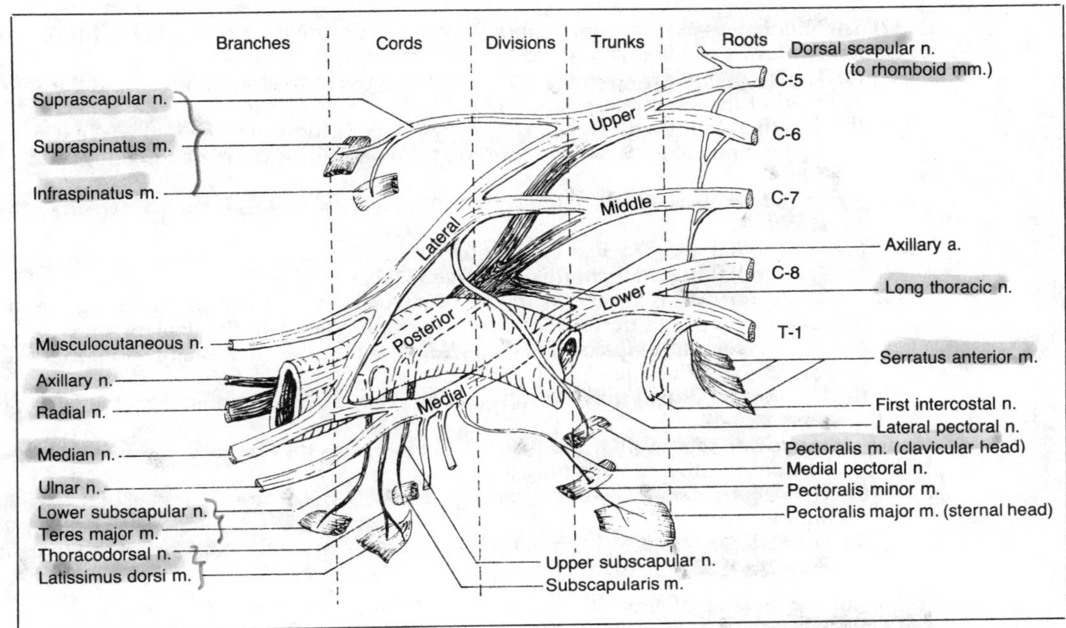

Figure 6-6. *The right brachial plexus.* The medial, lateral, and posterior cords are shown in relation to the axillary artery. The posterior division is *shaded*. The muscles innervated by branches of the roots and cords are indicated.

a. Roots represent the ventral primary rami of the spinal nerves, and each innervates a specific dermatome and myotome.

(1) Roots C3–C5 give branches to the **phrenic nerve**, which innervates the diaphragm.

(2) Roots C5–C7 give rise to two nerves:

(a) The **dorsal scapular nerve** (C5), which innervates the rhomboid and levator scapulae muscles.

(b) The **long thoracic nerve** (Bell's) [C5–C7], which innervates the serratus anterior muscle.

b. Trunks are formed by the joining of roots.

(1) The **upper trunk** is formed by the joining of roots C4–C6.

(a) The **suprascapular nerve** (C4–C6) passes through the suprascapular notch and beneath the transverse scapular ligament to innervate the supraspinatus and infraspinatus muscles.

(b) The **subclavius nerve** (C5–C6) innervates the subclavius muscle.

(2) The **middle trunk** is the continuation of root C7.

(3) The **lower trunk** is formed by the joining of roots C8 and T1.

c. Divisions are formed by bifurcation of the trunks.

(1) **Posterior divisions** of the trunks are equivalent to the lateral perforating branches of the spinal nerves. The divisions innervate the primitive dorsal musculature (i.e., the extensors).

(2) **Anterior divisions** of the trunks represent the continuations of the ventral primary rami. These divisions innervate the primitive ventral musculature (i.e., the flexors).

d. Cords are formed by the joining of either the anterior divisions of the trunks or the posterior divisions of the trunks. They lie beneath the pectoralis minor muscle and adjacent to the axillary artery.

(1) The **lateral cord** (C4–C7) is formed by the joining of the anterior divisions of the upper and middle trunks and is named for its position relative to the axillary artery.

(a) The **lateral pectoral nerve** (C5–C7) innervates the clavicular head of the pectoralis major muscle.

(b) The **musculocutaneous nerve** (C4–C6) innervates the coracobrachialis, biceps brachii, and brachialis muscles.

(c) Together with the medial cord, the lateral cord contributes to the **median nerve** (C5–C7), which innervates the numerous flexor and pronator muscles of the forearm, wrist, and hand.

(2) The **medial cord** is the continuation of the anterior division of the lower trunk and is named for its position relative to the axillary artery.
   (a) The **medial pectoral nerve** (C8–T1) innervates the pectoralis minor and the sternal head of the pectoralis major muscles.
   (b) Together with the lateral cord, the medial cord forms the **median nerve** (C5–C7), which innervates numerous flexor and pronator muscles of the forearm, wrist, and hand.
   (c) The **ulnar nerve** (C8–T1) innervates numerous flexor muscles of the forearm, wrist, and hand.
   (d) The **medial antebrachial cutaneous nerve** (C8–T1).
   (e) The **medial brachial cutaneous nerve** (T1).
(3) The **posterior cord** is formed by the fusion of the posterior divisions of the upper, middle, and lower trunks and named for its position relative to the axillary artery.
   (a) The **upper subscapular nerve** (C5–C6) innervates the major portion of the subscapularis muscle.
   (b) The **thoracodorsal (middle subscapular) nerve** (C6–C8) innervates the latissimus dorsi muscle.
   (c) The **lower subscapular nerve** (C5–C6) innervates the teres major and a small portion of the subscapularis muscle.
   (d) The **axillary nerve** (C5–C6) innervates the teres minor muscle and all of the deltoid muscle.
   (e) The **radial nerve** (C5–T1) innervates all of the numerous extensors of the forearm, wrist, and hand.

3. Lesions of the brachial plexus.

a. Injury to the roots will produce not only complete paralysis of the upper extremity but a winging of the scapula, due to paralysis of the long thoracic nerve, indicating root damage as opposed to plexus injury.

b. The most common cause of nerve root compression is cervical spondylosis. The patient first complains of pain in the arm or shoulder in a dermatomal distribution.

c. Injury to the cords produces distinct syndromes.

(1) **Upper trunk injury** (Erb-Duchenne paralysis) is the most common.
   (a) Violent downward displacement of the arm, such as results from being thrown from a horse or motorcycle, may tear the 5th and 6th roots or the upper trunk.
   (b) It involves paralysis of all nerves innervated by the anterior and posterior divisions of the upper trunk.
   (c) The loss of abduction, radial flexion, and external rotation produces an upper extremity that hangs by the side in internal rotation, the "waiter's tip" position.
   (d) Involvement of the lateral pectoral nerve produces inability to touch the opposite shoulder.
(2) **Lower trunk injury** (Klumpke's paralysis) is less common.
   (a) It is frequently caused by violent upward displacement of the arm, as in a difficult breach delivery or by dislocation of the shoulder, apical tumors of the lung, a cervical rib or scalene syndrome.
   (b) Paralysis of all nerves innervated by the anterior and posterior divisions of the lower trunk results.
   (c) There is a loss of ulnar flexion of the wrist and the use of many of the intrinsic muscles of the hand, so that unopposed extension of the hand produces a "claw-like" position.
   (d) Involvement of the medial pectoral nerve produces an inability to adduct the arm in the lowered position against resistance.
(3) **Posterior cord injuries.**
   (a) Thoracodorsal nerve injury produces paralysis of the latissimus dorsi muscle.
   (b) Axillary nerve injury results in deltoid paralysis (loss of shoulder abduction), loss of sensation over the deltoid muscle, and teres minor paralysis (weak external rotation).
   (c) Injury to the posterior cord ("**Saturday night palsy**," "**crutch palsy**") or to the radial nerve results in loss of the extensors of the arm, wrist, and hand.
(4) Injuries to the major distal nerves will be discussed subsequently.

d. **Scalene syndrome** can be caused by spasm of the anterior and medial scalene muscles or by a cervical rib.

(1) Such spasms can compress portions of the brachial plexus, most commonly the lower trunk, causing pain along the medial border of the arm and atrophy of some of the small muscles of the hand.
(2) Such spasms can also compress the subclavian artery, causing ischemia of the arm.

7
The Arm and Elbow Region

I. INTRODUCTION

A. The elbow region consists of the distal part of the **arm (brachium)** and the proximal part of the **forearm (antebrachium)**.

B. BONY LANDMARKS (see Fig. 6-1; Fig. 7-1)

1. The **humerus.**
- **a.** The head.
- **b.** The greater tuberosity.
- **c.** The bicipital (intertubercular) groove.
- **d.** The shaft.
- **e.** The lateral epicondyle.
- **f.** The medial epicondyle.

2. The **ulna.**
- **a.** The olecranon process.
- **b.** The shaft.

3. The **radius.**
- **a.** The head.
- **b.** The shaft.

II. BONES OF THE ELBOW REGION

A. THE DISTAL HUMERUS

1. Toward the distal end, the humerus flattens anteroposteriorly and broadens transversely into **medial** and **lateral supracondylar ridges**, which terminate in the **medial** and **lateral epicondyles**, all of which serve as muscle attachments.

2. The distal end of the humerus has two articulations with the forearm at the elbow joint (see Fig. 7-1).
- **a.** The **trochlea** is a medial pully-shaped articular surface which coapts the trochlear notch of the ulna, forming the **humeroulnar joint**.
  - **(1)** Anteriorly, proximal to the trochlea, the **coronoid fossa** receives the coronoid process of the ulna when the forearm is flexed.
  - **(2)** Posteriorly, proximal to the trochlea, the **olecranon fossa** receives the olecranon process when the forearm is extended.
- **b.** The **capitulum** is an anterolateral hemispherical articular surface that coapts the head of the radius, forming the **humeroradial joint**.

B. There are **two bones** in the **forearm**—the lateral **radius** and the medial **ulna**.

1. The **proximal ulna** (see Fig. 7-1).
- **a.** The proximal end of the ulna terminates in the thickened **olecranon process**.
  - **(1)** This process provides attachment for the triceps muscle.
  - **(2)** It is separated from the subcutaneous tissue by the **olecranon bursa**.
- **b.** The deep **trochlear (semilunar) notch**, located on the anterior surface, provides an articular surface for the humeroulnar joint.

c. The triangular **coronoid process**, just distal to the trochlear notch, provides the insertion for the brachialis muscle.
d. The **radial notch**, lateral to the coronoid process, provides an articular surface for the proximal radioulnar joint.
e. The ulna narrows as it extends distally.

2. The **proximal radius** (see Fig. 7-1).
   a. The proximal end of the radius consists of a cylindrical **head** that is concave on the top.
      (1) The top of the head provides an articular surface for the humeroradial joint.
      (2) The side of the head provides the articular surface for the radioulnar joint.
   b. Distal to the head is a short and narrow **neck** region.
   c. The **radial (bicipital) tuberosity**, on the anteromedial aspect of the neck, receives the tendon of the biceps brachii muscle.
   d. The radius broadens as it extends distally.

III. ARTICULATIONS AND MUSCLES.

The elbow region consists of articulations between the humerus, radius, and ulnar and the muscles that act across these joints to produce movement in the forearm.

A. There are **three articulations** at the elbow joint (see Fig. 7-1).

1. The **humeroulnar joint** is formed by the trochlea of the humerus and the trochlear notch of the ulna.
   a. This joint has one degree of freedom, permitting flexion/extension about a transverse axis through the trochlea.
   b. It is reinforced by the **ulnar (medial) collateral ligament**, a strong fan-shaped condensation of the fibrous joint capsule.
      (1) It is composed of anterior, intermediate, and posterior fiber bundles.
      (2) Sprain of this ligament will permit abnormal abduction at the elbow joint.
   c. The shape of the **trochlea** is variable among individuals and determines the angle (3° to 29°) that the extended forearm makes with the arm, the **"carrying angle."** The carrying angle is usually greater in women than men, the anatomical reason that women tend to bowl with a curve.

2. The **humeroradial joint** is formed by the capitulum of the humerus and the head of the radius.
   a. It has two degrees of freedom.
      (1) Flexion/extension occurs about an axis through the trochlea.
      (2) Pronation/supination (medial rotation/lateral rotation) occurs about an axis that passes through both centers of curvature of the proximal and distal radioulnar joints: the center of curvature of the proximal radioulnar joint is the head of the radius; that of the distal radioulnar joint is the head of the ulna.
   b. It is reinforced by the **radial (lateral) collateral ligament**, a strong fan-shaped condensation of the fibrous joint capsule.
      (1) It is composed of anterior, intermediate, and posterior fiber bundles.
      (2) Sprain of this ligament will permit abnormal adduction at the elbow joint.
      (3) "Tennis elbow" seems to involve inflammation of this ligament, the periosteum about its insertion, or the small underlying bursa, or even strain of the common extensor tendon.

3. The **proximal (superior) radioulnar joint** is formed by the side of the head of the radius and the radial notch of the ulna.
   a. It has one degree of freedom, permitting pronation/supination about an axis that passes through both the proximal and distal radioulnar joints.
   b. It is reinforced by two ligaments.
      (1) The **annular ligament** provides the major reinforcement.
         (a) Originating on the anterior lip of the radial notch of the ulna, it passes about the head and neck of the radius and inserts on the posterior lip of the radial notch of the ulna.
         (b) It permits rotation of the radius relative to the ulna.
      (2) The **interosseous membrane (septum).**
         (a) This is a broad sheet of strong connective tissue extending between the medial edge of the radius and the lateral edge of the ulna for nearly the entire length of the forearm.
         (b) Because the hand is attached primarily to the radius, and because the humeroulnar joint is the most stable joint of the elbow, force is transmitted to the radius from the hand and through the interosseus membrane to the ulnar and humerus.

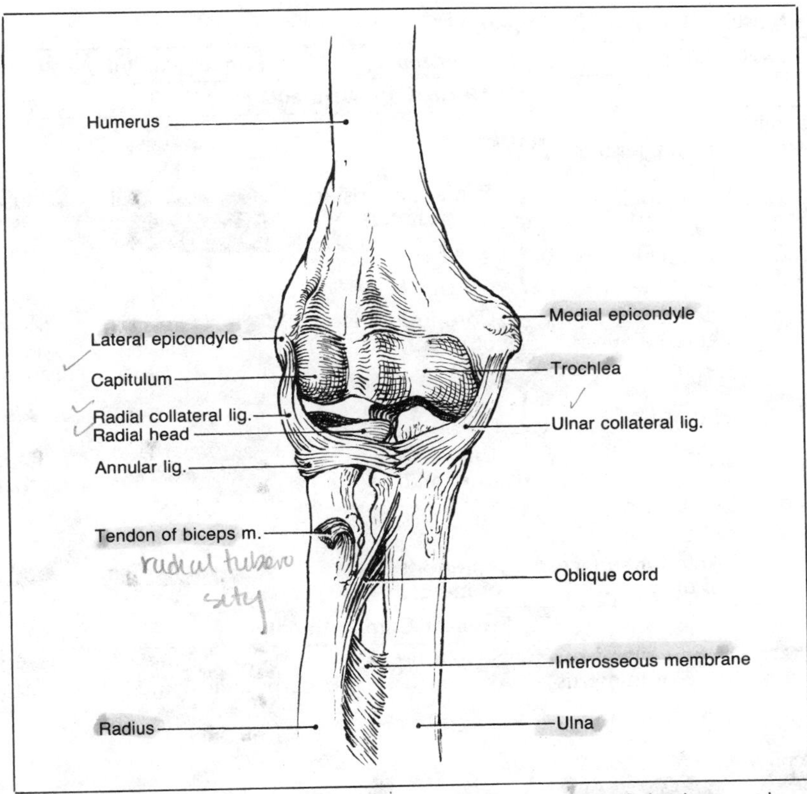

Figure 7-1. *The right elbow joint.* The elbow joint consists of the humeroulnar, humeroradial, and radioulnar joints and is reinforced by radial collateral and ulnar collateral ligaments, as well as by the annular ligament.

(c) The interosseous membrane also serves as attachment for the deep extrinsic flexor and extensor muscles of the hand.

(d) The **oblique cord** of the interosseous membrane runs inferolaterally between the ulna, just distal to the radial notch, and the radius, just distal to the radial tuberosity.

c. The proximal radioulnar joint functions in concert with the **distal (inferior) radioulnar joint** (see Fig. 9-1).

B. STABILITY OF THE ELBOW JOINT

1. In general, the elbow joint is very strong and stable. The coronoid process and the coaptation of the trochlea and the trochlear notch at the humeroulnar joint preclude dislocation except under extreme force or secondary to fracture.
2. The humeroradial joint cannot withstand excessive traction in children, however, because the radial head is undeveloped and can escape from the annular ligament, producing a "pulled elbow."
3. The elbow joint is particularly susceptible to impacting fracture, including:
   a. Radial head fracture, with or without fracture of the capitulum.
   b. Fracture of the coronoid process with posterior dislocation of the elbow joint.
   c. Fracture of the olecranon process or evulsion of the triceps brachii muscle insertion.

C. MUSCULATURE OF THE ARM (BRACHIUM)

1. The muscles of the arm produce movement at both the glenohumeral joint and the elbow joint.
2. The flexors lie in the **anterior (flexor) compartment** and correspond to the primitive ventral muscles; the extensors lie in the **posterior (extensor) compartment** and correspond to the primitive dorsal muscles.

Table 7-1. Muscles Acting on the Forearm

Muscle	Origin	Insertion	Primary Action	Innervation
		Flexor Compartment		
Biceps brachii:				
Long head	Supraglenoid tubercle of scapula	Radial tuberosity of the radius	Flexes and supinates forearm at the elbow joint	Musculocutaneous n. (C5–C6, anterior)
Short head	Coracoid process of scapula			
Brachialis	Distal half of anterior surface of humerus	Coronoid process of ulna	Flexes forearm	Musculocutaneous n. (C5–C6, anterior)
Pronator teres:				
Long head	Medial epicondyle of humerus	Lateral surface of midradius	Pronates and weakly flexes forearm	Median n. (C6–C7, anterior)
Short head	Distal ulna			
Pronator quadratus	Anterior surface of ulna	Anterior surface of radius	Pronates forearm	Median n. (C6–C7, anterior)
		Extensor Compartment		
Brachioradialis	Distal lateral surface of humerus	Styloid process of radius	Flexes, semipronates, and semisupinates forearm	Radial n. (C5–C6, posterior)
Triceps brachii:				
Long head	Infraglenoid tubercle of scapula	Olecranon process of ulna	Extends forearm at elbow joint	Radial n. (C7–C8, posterior)
Lateral head	Posterior surface of humerus			
Medial head	Posterior surface of humerus			
Anconeus	Posterior aspect of lateral epicondyle	Olecranon process	Extends forearm at elbow joint	Radial n. (C7–C8, posterior)
Supinator	Lateral epicondyle of humerus	Lateral aspect of midradius	Supinates forearm	Radial n. (C5–C7, posterior)

3. Movement of the forearm at the elbow joint is accomplished by muscles that run from the pectoral girdle or the arm to the forearm or the wrist (Table 7-1, Figs. 7-2 and 7-3).
   a. **Flexion/extension** (150°) occurs about an axis through the trochlea.
      (1) Flexor muscles include the biceps brachii (acting on the radius), brachialis (acting on the ulna), and brachioradialis when the forearm is semipronated. In addition, several of the wrist and hand flexor and extensor muscles that originate from the supracondylar ridges and epicondyles (see Table 9-1) serve as weak forearm flexors.
      (2) Extensor muscles include the triceps brachii, the anconeus, and the brachioradialis when the forearm is fully supinated and nearly fully extended.
   b. **Pronation/supination** (160°) occurs about an axis that passes through both the capitulum of the humerus and the distal ulna as well as through the interosseous membrane.
      (1) Pronator muscles include the pronator teres (long and short heads) and the pronator quadratus, which is wrapped around the distal end of the ulna and acts by unwinding.
      (2) Supinator muscles include the supinator, which wraps about the radius and acts by unwinding, and the biceps brachii, which inserts on the radial tuberosity. The strength of the biceps as a supinator accounts for why it is easier to drive a screw into wood with the right hand than with the left hand—even for left-handed persons.

4. **Clinical considerations.**
   a. Flexion of the forearm is mediated by the musculocutaneous, radial, and median nerves, thus a lesion of any *one* of these nerves will only weaken, not abolish, flexion.

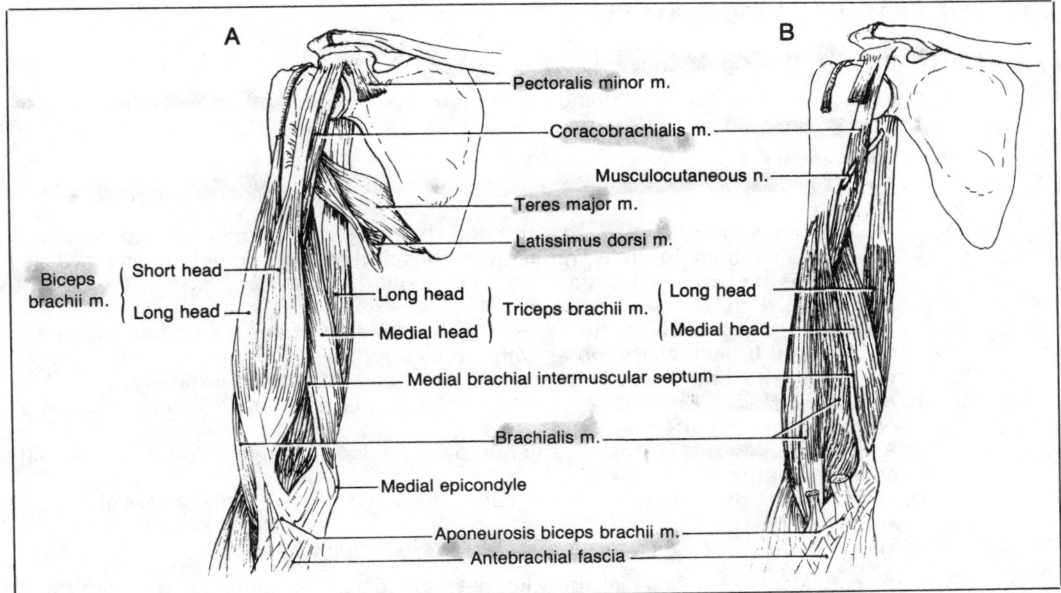

Figure 7-2. *A, The superficial muscles of the flexor compartment of the right arm. B, The deep muscles of the flexor compartment of the right arm.*

- **b.** Extension is mediated by the radial nerve.
- **c.** Pronation is a function of the median nerve, although weak semipronation from a supinated position is possible through the brachioradialis muscle.
- **d.** Supination is a function of the musculocutaneous and radial nerves, thus a lesion of only one of these nerves will only weaken supination.
- **e.** In fracture of the upper third of the radius, the supinator muscles draw the proximal fragment laterally, while the pronator muscles draw the distal fragment medially.
- **f.** In fracture of the middle radius, there is less displacement than in proximal fractures because the action of the supinator is opposed by the pronator teres; however, the pronator quadratus still draws the distal fragment medially.

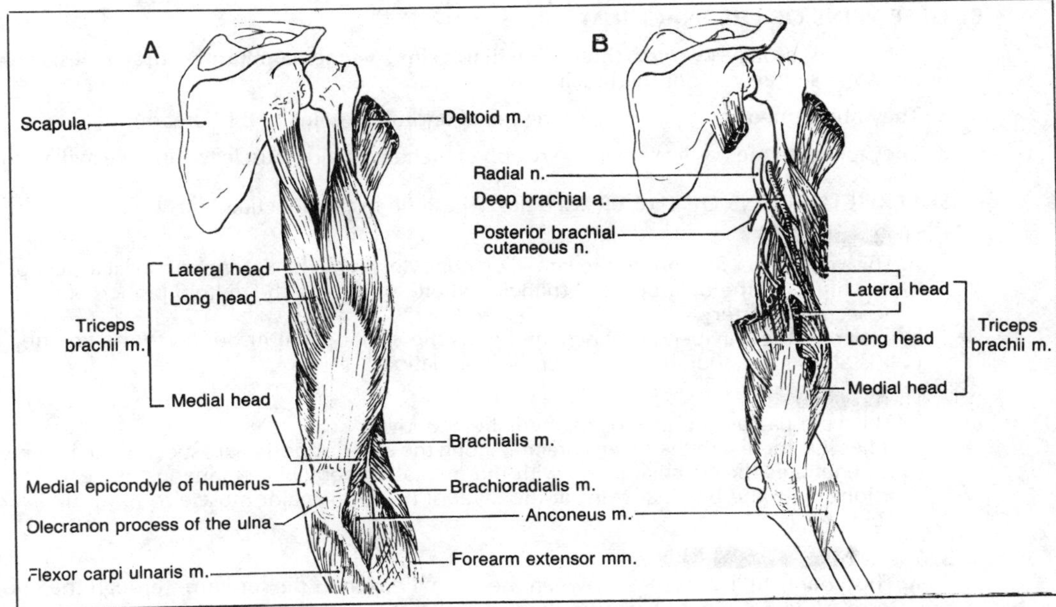

Figure 7-3. *A, The superficial layer of the extensor compartment of the right arm. B, The deep layer of the extensor compartment of the right arm.*

IV. THE BRACHIAL VASCULATURE

A. THE BRACHIAL ARTERY (Fig. 7-4)

1. The brachial artery, the continuation of the subclavian artery, begins at the distal edge of the teres major muscle.

2. Major branches:
   a. The **deep brachial (profunda brachii) artery** supplies the posterior compartment of the arm.
      (1) Together with the radial nerve, this deep branch passes posterior to and then lies adjacent to the humerus in the **radial (musculospiral) groove**. A midhumeral fracture may involve the two structures as they wind about the posterior aspect of the humerus in the radial groove.
      (2) A **recurrent branch** anastomoses with the posterior humeral circumflex artery.
      (3) A **lateral branch** anastomoses with the recurrent radial artery.
      (4) A **posterior branch** anastomoses with the recurrent interosseus artery.
   b. A **superior ulnar branch** passes posterior to the medial epicondyle to anastomose with the posterior ulnar recurrent artery.
   c. An **inferior ulnar branch** passes anterior to the medial epicondyle to anastomose with the anterior ulnar recurrent artery.
   d. The brachial artery bifurcates in the cubital fossa to form the **radial artery** and the **ulnar artery**.
   e. There is rich collateral circulation about the elbow joint.

3. In the cubital fossa, the brachial artery lies deep to the bicipital aponeurosis, superficial to the brachialis muscle, and medial to the biceps brachii tendon. It is here that a **brachial pulse** may be palpated.

4. The brachial artery may bifurcate anywhere in the brachium, resulting in a **superficial radial artery** or **superficial ulnar artery**.

5. Supracondylar fractures of the humerus may sever the brachial vessels and injure the median nerve.
   a. Traction on the triceps brachii muscle draws the proximal portion of the ulna posteriorly, while the brachialis muscle draws the distal humeral fragment anteriorly, jeopardizing the neurovascular bundle.
   b. Hemorrhage beneath the brachial fascia and antebrachial fascia may compress uninjured collateral blood vessels and thereby produce ischemia and atrophy of the forearm and hand musculature, **Volkmann's ischemic contracture**.

B. DEEP VEINS OF THE BRACHIUM

1. There are usually two or three **brachial veins (venae comitantes brachiales)**, which anastomose freely about the radial artery.

2. They are formed by the joining of the radial and ulnar veins of the forearm.

3. They join with the basilic vein in the region of the teres major muscle to form the axillary vein.

C. SUPERFICIAL VEINS OF THE BRACHIUM. These lie in the subcutaneous tissue.

1. The **cephalic vein.**
   a. This vein passes anterior to the lateral epicondyle, along the anterior preaxial aspect of the brachium, to the deltopectoral triangle, where it lies with the deltoid branch of the thoracoacromial artery.
   b. It pierces the clavipectoral fascia and joins the axillary vein just distal to the first rib.
   c. It is absent in about 10 percent of the population.

2. The **basilic vein.**
   a. This vein passes anterior to the medial epicondyle.
   b. It lies just medial to the biceps brachii, along the anterior postaxial aspect of the brachium.
   c. It penetrates the brachial fascia with the medial antebrachial cutaneous nerve.
   d. It joins with the brachial veins at the level of the teres major muscle to form the axillary vein.

3. The **median cubital vein.**
   a. This connecting vein runs between the cephalic vein in the forearm, through the cubital fossa, to join the basilic vein in the arm.
   b. It lies superficial to the bicipital aponeurosis.
   c. It is subject to extreme variation.

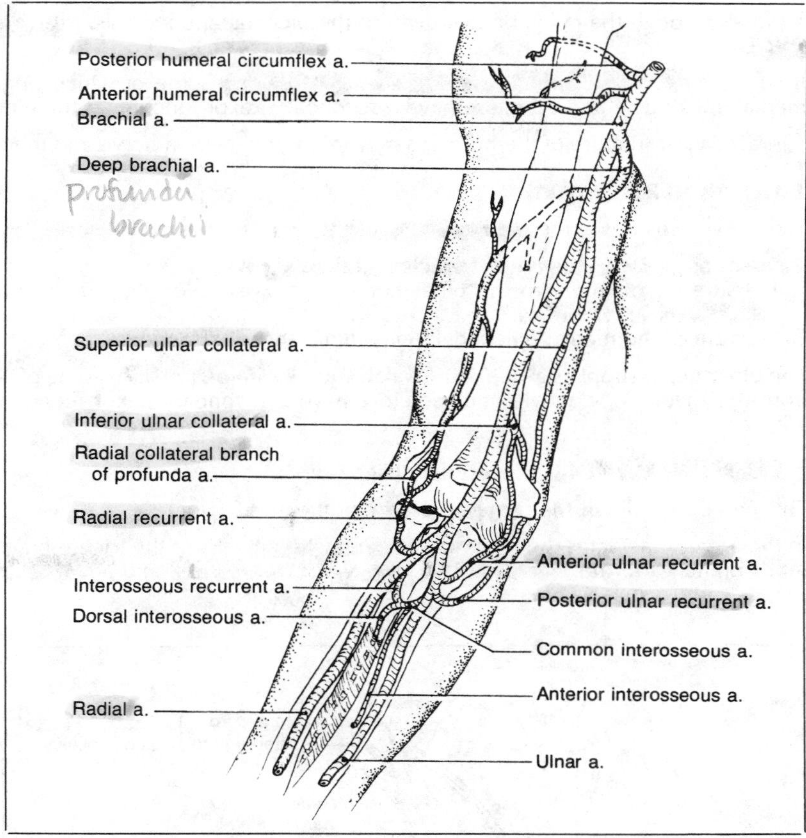

Figure 7-4. *The arteries of the arm and elbow region.*

D. LYMPHATIC DRAINAGE OF THE BRACHIUM

1. Deep lymphatics accompany the brachial vein and drain through the axillary lymph nodes.
2. Superficial lymphatics drain along the superficial veins.
   a. There are usually one or two supratrochlear lymph nodes in the distal brachium, just proximal to the medial epicondyle.
   b. The superficial drainage bypasses most of the axillary nodes, entering instead the subclavian nodes.

V. BRACHIAL INNERVATION

A. THE MUSCULOCUTANEOUS NERVE (C4–C6) [Fig. 7-5]

1. This nerve innervates the flexor muscles of the arm (brachium), including the coracobrachialis, biceps brachii, and brachialis muscles.
2. It continues into the forearm as the **lateral antebrachial cutaneous nerve**, which supplies the C6 dermatome along the radial side of the forearm.
3. Injury to the musculocutaneous nerve produces inability to strongly flex the forearm, loss of the **biceps tendon reflex**, and loss of sensation along the lateral aspect of the forearm. Some weak elbow flexion is possible despite injury because of the secondary flexor action of the brachioradialis (radial innervation) and the forearm muscles that originate on the medial epicondyle (medial and ulnar innervation).

B. THE MEDIAN NERVE (C6–C8)

1. This nerve runs down the anteromedial aspect of the arm and has no branches in the brachium.

2. It passes through the cubital fossa, deep to the bicipital aponeurosis, medial to the brachial artery.

3. It supplies flexor and pronator muscles in the forearm, some of which originate from the medial epicondyle and therefore have a secondary flexor function at the elbow joint.

4. Supracondylar fracture of the humerus may injure the median nerve and the brachial artery.

C. THE ULNAR NERVE (C8–T1)

1. This nerve runs down the medial aspect of the arm and has no branches in the brachium.

2. It passes posterior to the medial epicondyle in a groove.
   a. Pressure on or injury to the ulnar nerve as it passes along the ulnar groove produces "funny-bone" symptoms.
   b. Fracture of the medial epicondyle may injure the ulnar nerve.

3. The ulnar nerve supplies a few of the flexor muscles in the forearm, some of which originate from the medial epicondyle and therefore have a secondary flexor function at the elbow joint.

D. THE RADIAL NERVE (C4–T1, posterior) [Fig. 7-6].

1. This nerve runs down the posterior aspect of the arm.

2. In the arm it innervates the long, medial, and lateral heads of the triceps brachii muscle and sends branches to the brachioradialis, extensor carpi radialis, and anconeus muscles.

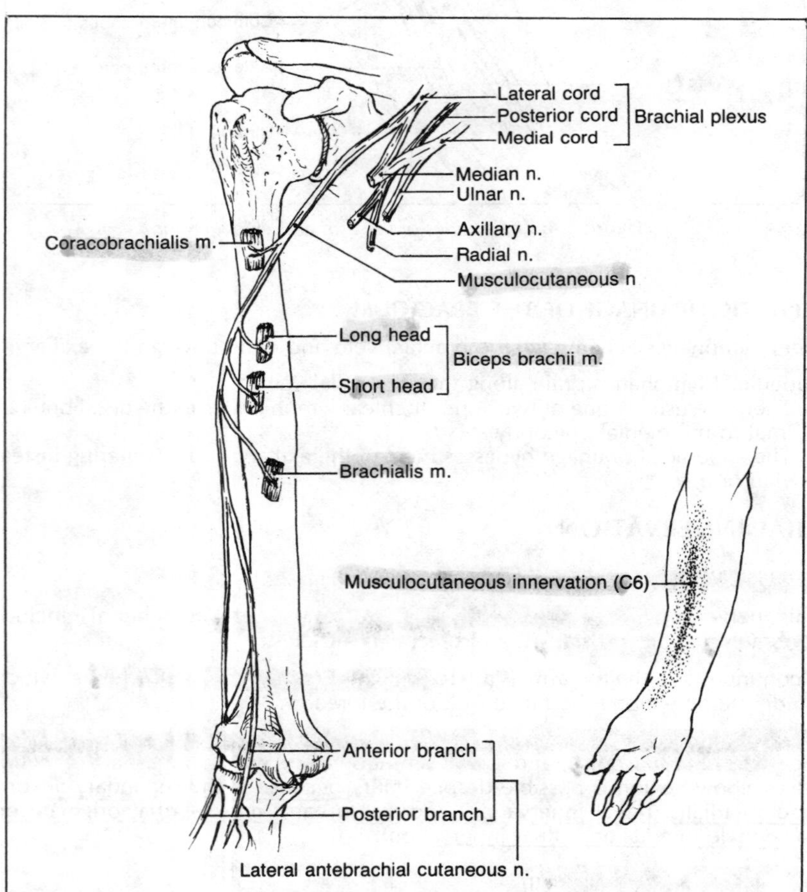

Figure 7-5. *The musculocutaneous nerve in the arm.* The muscles innervated by branches of the nerve are indicated. Also indicated is the region of dermatomal innervation by the lateral antebrachial cutaneous nerve in the forearm.

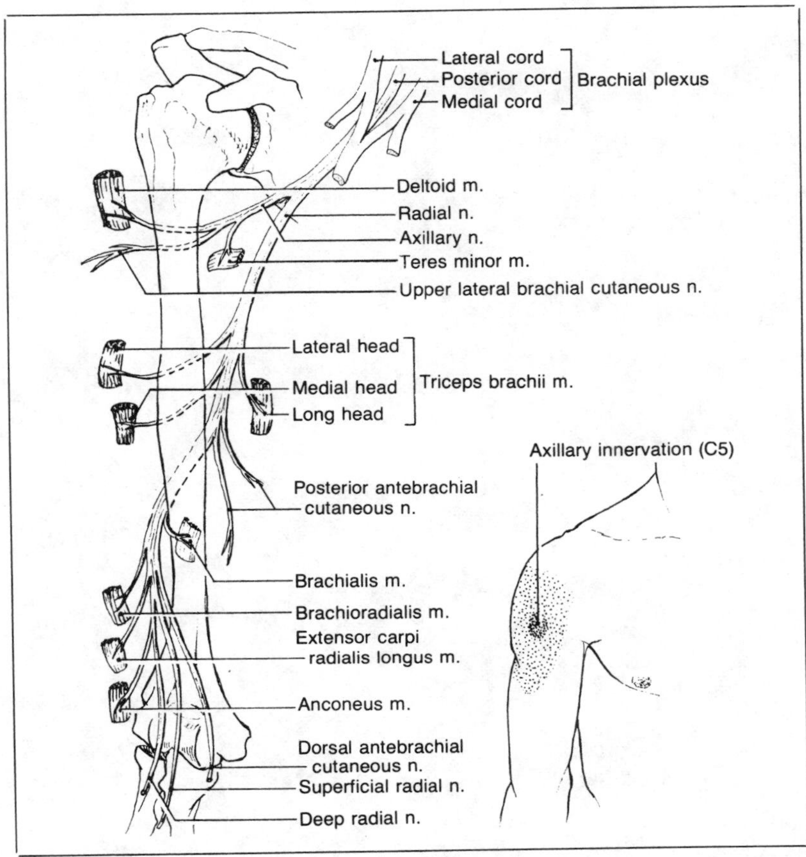

Figure 7-6. *The radial and axillary nerves in the arm.* The muscles innervated by branches of these nerves are indicated. Also indicated is the region of dermatomal innervation by the axillary nerve in the arm.

prof. brach

3. It lies posterior to the humerus in the musculospiral groove with the deep brachial artery; it then passes anterior to the lateral epicondyle, between the brachialis and brachioradialis muscles, as it enters the forearm.
   a. Midhumeral fracture may damage the radial nerve, causing paralysis of the extensor muscles of the wrist and hand.
   b. Because the branches of the radial nerve that innervate the triceps brachii muscle arise proximal to the musculospiral groove, extension of the forearm usually is not affected by midhumeral fracture, and some supination remains possible because of the biceps brachii muscle.
4. The radial nerve gives off the superficial **posterior antebrachial cutaneous nerve** and the lower lateral brachial cutaneous nerve as it winds around the humerus in the radial groove, deep to the long head of the triceps muscle.

E. THE AXILLARY NERVE (C5–C6) innervates the teres minor muscle and all of the deltoid muscle.

8
The Extensor Forearm and Posterior Wrist

I. INTRODUCTION

A. The primitively dorsal aspects of the forearm, wrist, and hand can be considered as a functional extensor unit.

B. BONY LANDMARKS (see Fig. 7-1; Fig. 8-1)

1. The **ulna**.
   a. The shaft.
   b. The head.
   c. The styloid process.
2. The **radius**.
   a. The shaft.
   b. The head.
   c. The styloid process.
   d. The dorsal tubercle.
3. The **carpal bones**.

II. BONES OF THE EXTENSOR FOREARM AND WRIST

A. The **radius** and **ulna** are the bones of the forearm.

1. The distal **ulna** (see Fig. 8-1).
   a. The ulna is strong at the elbow joint but becomes secondary at the wrist.
   b. At the distal end, it narrows and terminates as a slightly enlarged **head** and **styloid process**.
   c. Anterolaterally on the head, there is a surface that articulates with the ulnar notch of the radius, forming the **distal radioulnar joint**.
   d. The end of the ulna is separated from the triquetral bone by an **articular disk (triangular fibrocartilage)**.
      (1) The articular disk fills the gap between the ulna and variable portions of the triquetral and lunate bones.
      (2) This disk acts as a shock absorber, protecting the ulna and elbow joint from forces transmitted through the hand and wrist.
2. The distal **radius** (see Fig. 8-1).
   a. The radius is supplemental at the elbow joint but becomes primary at the wrist.
      (1) The radius articulates with the proximal row of carpal bones.
      (2) It receives most of the force transmitted from the hand and, as a consequence, is frequently fractured.
   b. The distal radial shaft widens transversely.
   c. Laterally, it terminates in the **styloid process**, to which the brachioradialis muscle attaches.
   d. Distally, the posterior surface is grooved by the tendons of the extensor carpi radialis and the extensor pollicis longus muscles. Between these grooves is a prominent ridge, the **dorsal (radial) tubercle**.
   e. The **ulnar notch**, located on the medial surface of the distal head, coapts the head of the ulna and forms the **distal radioulnar joint**.
   f. The base of the radius comprises the **carpal articular surface**, which contacts the **scaphoid bone** (laterally) and the **lunate bone** (medially) to form the **radiocarpal joint** (see Fig. 9-4).

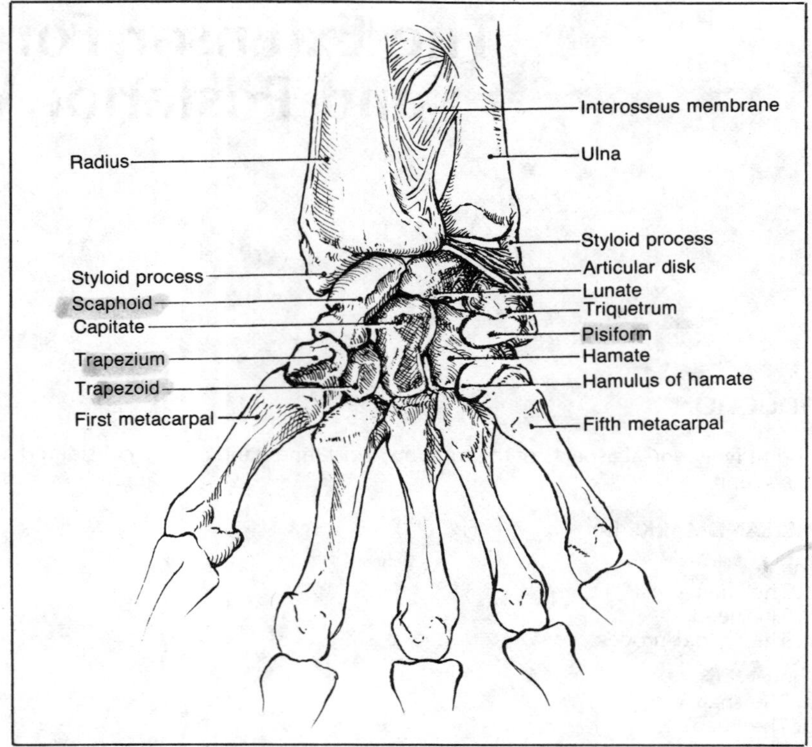

Figure 8-1. *The distal forearm and wrist.* Anterior view of the right forearm showing the radiocarpal, midcarpal, and carpometacarpal joints.

H
C
Td
Tm
Sc
Lun
Triqu

3. The **interosseus membrane (septum)** reinforces the radioulnar joint and functions as a shock absorber. Forces transmitted to the radius from the hand are moderated by the membrane in their transmission to the ulna.

4. Fracture of the distal radius (Colles' fracture) is only surpassed in frequency by fracture of the clavicle, ribs, and fingers.
   a. The distal radius, composed of large amounts of cancellous bone within a thin rim of cortex, is the weakest point.
   b. Anterior angulation of the proximal fragment may place the median nerve and radial artery in jeopardy.

B. **Eight carpal bones**, lying four in a row, comprise the skeleton of the wrist.
1. The **proximal carpal bones** (see Fig. 8-1).
   a. The **scaphoid (navicular) bone** is the most lateral.
      (1) It articulates with the radius proximally, with the lunate bone medially, and with the trapezium and trapezoid bones distally.
      (2) Because of its shape and location, which are responsible for the scaphoid transmitting forces from the abducted hand to the radius, the scaphoid is the carpal bone most prone to fracture.
         (a) Usually there is little displacement with fracture, and because the bony cortex of the scaphoid is very thin, fracture is easily missed on x-rays and can be misdiagnosed as a ligamentous sprain. If a second x-ray is obtained 10 days after fracture following the bone resorption stage of healing, the diagnosis will be apparent.
         (b) Because the blood supply enters the scaphoid bone distally, fracture may destroy the blood supply to the proximal fragment and cause avascular necrosis.
   b. The **lunate bone**.
      (1) The lunate bone articulates with the radius proximally, the scaphoid laterally, the triquetrum medially, and the capitate distally.
      (2) It also transmits forces from the hand to the radius.
      (3) It is not uncommonly dislocated anteriorly into the carpal tunnel, compressing the median nerve and causing "**carpal tunnel syndrome**."

(4) The lunate bone receives its major blood supply through the anterior and posterior radiocarpal ligaments, and dislocation of the lunate can thus result in avascular necrosis.

c. The **triquetral bone** (triquetrum) articulates with the articular disk proximally, the lunate bone laterally, the pisiform bone anteromedially, and the hamate bone distally.

d. The **pisiform bone** is the most medial of the proximal row of carpal bones.
 (1) It articulates only with the triquetral bone.
 (2) Developmentally, it is a sesamoid bone, embedded in the tendon of the flexor carpi ulnaris muscle. Thus, the proximal carpal row can be considered to consist of three true carpal bones.
 (3) It is the only proximal carpal bone with a muscular attachment.

2. The **distal carpal bones** (see Fig. 8-1).
 a. The **trapezium (greater multangular) bone**.
  (1) It is the most lateral of the distal row.
  (2) It articulates with the scaphoid proximally, the trapezoid medially, and the first and second metacarpals distally.
 b. The **trapezoid (lesser multangular) bone** articulates with the scaphoid proximally, the trapezium laterally, the capitate and the second metacarpal medially, and the first and second metacarpals distally.
 c. The **capitate bone** is the keystone of the **carpal arch**.
  (1) It articulates with the scaphoid and lunate bones proximally, the trapezoid laterally, the hamate medially, and the second, third, and fourth metacarpals distally.
  (2) It transmits forces from the second, third, and fourth fingers to the proximal row of carpal bones.
  (3) Lunate dislocation is accompanied by proximal displacement of the capitate, as a result of which the middle digit extends not much further than the second and fourth digits.
 d. The **hamate bone** is the most medial of the distal row of carpal bones.
  (1) It articulates with the triquetrum proximally, the capitate laterally, and the fourth and fifth metacarpals distally.
  (2) A hook-like process called the **hamulus**, which gives the hamate its name, projects toward the palm just distal to the pisiform bone.

III. THE TWO JOINTS OF THE WRIST

A. The **radiocarpal joint** is located between the radial head and the scaphoid and lunate bones (see Figs. 8-1 and 9-4).

1. The concave, ellipsoid articular surface of the distal end of the radius permits two degrees of freedom.
 a. Abduction (radial deviation)/adduction (ulnar deviation) occurs about an anteroposterior axis through the head of the capitate bone.
  (1) In abduction (15°), the scaphoid bone makes maximal contact with the radius, and the lunate contacts the articular disk.
  (2) In adduction (45°), the lunate bone makes maximal contact with the radius, and the triquetrum contacts the articular disk.
 b. Flexion/extension (170°) occurs about a transverse axis that passes between the lunate and capitate bones.
 c. Because there are two degrees of freedom, circumduction of the wrist is possible.

2. The radiocarpal joint is reinforced by several ligaments, but the high degree of mobility results in loss of strength.
 a. The **ulnar (medial) collateral ligament** connects the ulnar styloid process and the triquetrum.
 b. The **dorsal radiocarpal ligament** reinforces the dorsal side.
 c. The **radial (lateral) collateral ligament** connects the radial styloid process and the scaphoid bone.
 d. The **palmar radiocarpal ligament** reinforces the ventral side.
 e. There is also an **ulnocarpal ligament** on the palmar side.
 f. The **transverse carpal ligament**, from the pisiform bone and the hamulus medially and to the scaphoid and trapezium laterally, bridges and maintains the **carpal arch** and forms the **carpal tunnel.**

3. The radiocarpal joint is most stable in full flexion where abduction/adduction is not possible.

B. The **midcarpal joint** lies between the two rows of carpal bones (see Fig. 8-1).

1. Small amounts of gliding movements of accommodation occur in the midcarpal joint during abduction/adduction, flexion/extension, and as the hand is flattened or hollowed.

2. The intercarpal joints are reinforced by numerous **dorsal intercarpal ligaments** and a palmar **carpal radiate ligament**.

IV. MUSCULATURE OF THE ANTEBRACHIAL EXTENSOR COMPARTMENT

A. The antebrachial fascia (deep fascia of the forearm) envelops the musculature of the forearm.

1. Proximally, the **bicipital aponeurosis** of the biceps brachii muscle inserts into this fascial layer.

2. It gives origin to some of the more superficial fascicles of the flexor and extensor muscles of the forearm.

3. Septa penetrate between the individual muscles and also divide the arm into flexor and extensor compartments.

4. Posteriorly, the antebrachial fascia condenses at the wrist to form the **extensor retinaculum**.
   a. This retinaculum is subdivided into six compartments by attachment in several places to bone (see Fig. 9-4).
   b. The tendons of the extrinsic extensor muscles of the wrist, hand, and thumb pass through these compartments to gain access to the dorsum of the hand (see Fig. 8-3).
      (1) Beneath the extensor retinaculum, the extensor digitorum communis tendons are enclosed in synovial sheaths.
      (2) The sheaths of these tendons terminate just distal to the extensor retinaculum, and the tendons lie directly on the respective metacarpal bones.
      (3) The tendons proceed to the individual digits, but there are variable interconnections that tend to limit the individual movements of the fingers.
      (4) The extensor tendons insert into the **extensor aponeurosis** on the dorsum of the second through fifth digits.
   c. On the dorsum of the hand, the antebrachial fascia and the extensor tendons fuse to form the **deep dorsal fascia**.

5. The **dorsal subcutaneous space** of the hand lies between the skin and the deep dorsal fascia. The loose connective tissue of this space accounts for the mobility of the skin on the dorsum of the hand.

6. The **subaponeurotic space** lies between the deep dorsal fascia and the deep fascia covering the dorsal interosseous muscles and metacarpal bones.

B. The extensor muscles of the forearm are mostly "multijoint" muscles, acting across the elbow joint, the wrist joint, the carpal joints, and the metacarpophalangeal joints.

1. The extensor muscles of the wrist and digits originate on the preaxial (lateral) aspect of the distal arm and proximal forearm (the **extensor compartment**) and correspond to primitive dorsal musculature.
   a. The **superficial group** (Figs. 8-2A and 8-3, Table 8-1).
      (1) These extensor muscles of the wrist and hand originate from the lateral supracondylar ridge and the lateral epicondyle of the humerus, as well as from the proximal radius.
      (2) They consist of the brachioradialis (which is usually a forearm flexor), extensor carpi radialis longus and brevis, extensor digitorum communis, extensor digiti minimi, and extensor carpi ulnaris.
   b. The **deep group** (Fig. 8-2B; see Fig. 8-3 and Table 8-1).
      (1) These muscles originate from the midradius, the interosseous membrane, and the ulna.
      (2) They consist of the supinator (which supinates the forearm), abductor pollicis longus, extensor pollicis brevis, extensor pollicis longus, and extensor indicis proprius.

2. The action of each muscle at the radiocarpal joint is determined by the location of the tendon relative to the two axes of rotation.
   a. Flexion/extension occurs about a transverse axis (see L–M in Fig. 9-4).
      (1) Muscles that pass dorsal to the transverse axis of the radiocarpal joint are extensors of the wrist.
      (2) Muscles that pass ventral to this axis are flexors of the wrist.
      (3) The prime extensors of the wrist are the extensor carpi radialis longus and the extensor carpi ulnaris.
   b. Abduction/adduction occurs about an anteroposterior axis (see A–P in Fig. 9-4).
      (1) Muscles that pass on the radial side of the anteroposterior axis of the radiocarpal joint are abductors of the wrist.

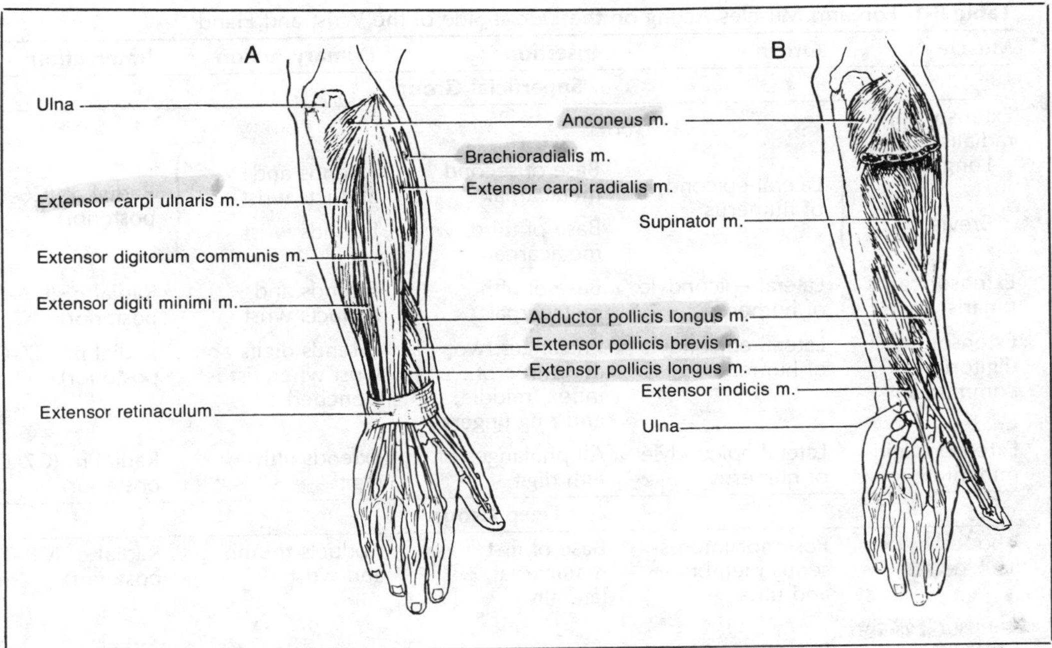

Figure 8-2. *A, The superficial layer of the extensor compartment of the right forearm. B, The deep layer of the extensor compartment of the right forearm.*

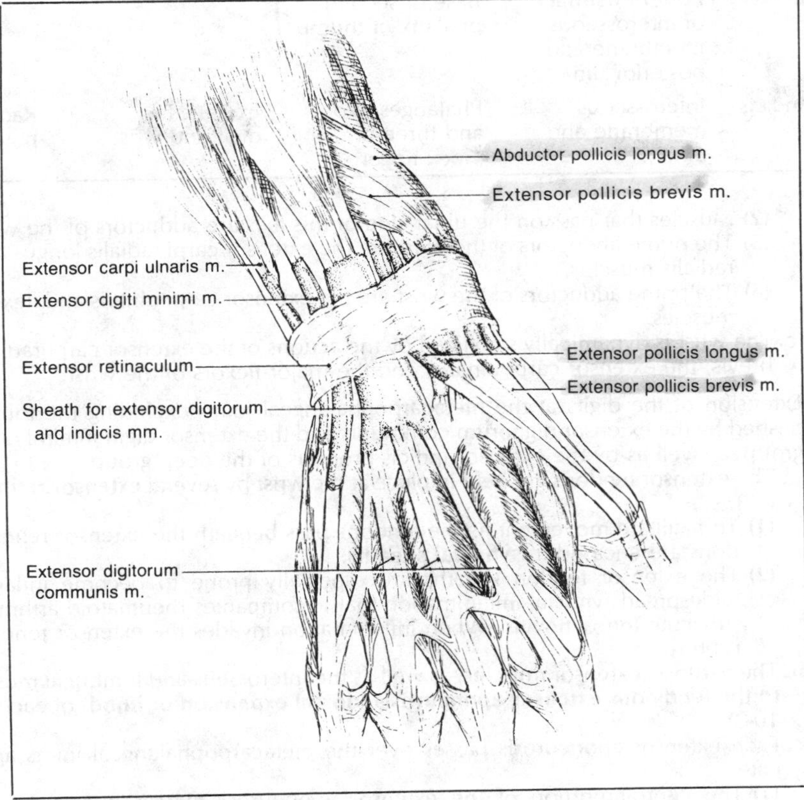

Figure 8-3. *The extrinsic extensor tendons of the dorsum of the right hand.*

Table 8-1. Forearm Muscles Acting on the Dorsal Side of the Wrist and Hand

Muscle	Origin	Insertion	Primary Action	Innervation
		Superficial Group		
Extensor carpi radialis:				
Longus	Lateral epicondyle of humerus	Base of second metacarpal	Extends and abducts wrist	Radial n. (C6–C7, posterior)
Brevis		Base of third metacarpal	Extends wrist	
Extensor carpi ulnaris	Lateral epicondyle of humerus	Base of fifth metacarpal	Extends and adducts wrist	Radial n. (C7–C8, posterior)
Extensor digitorum communis	Lateral epicondyle of humerus	Phalanges two and three of index, middle, and ring fingers	Extends digits and wrist when fist is clenched	Radial n. (C7–C8, posterior)
Extensor digiti minimi	Lateral epicondyle of humerus	All phalanges of fifth digit	Extends fifth digit	Radial n. (C7–C8, posterior)
		Deep Group		
Abductor pollicis longus	Posterior interosseous membrane and ulna	Base of first metacarpal, laterally	Abducts thumb and wrist	Radial n. (C8–T1, posterior)
Extensor pollicis:				
Brevis	Posterior midshaft of radius and interosseous membrane	Base of first phalanx of thumb	Extends thumb and abducts wrist	Radial n. (C8–T1, posterior)
Longus	Posterior surface of interosseous membrane and posterior ulna	Base of second phalanx of thumb		
Extensor indicis proprius	Interosseous membrane and ulna	Phalanges two and three of index finger	Extends first digit and wrist	Radial n. (C8–T1, posterior)

(2) Muscles that pass on the ulnar side of this axis are adductors of the wrist.

(3) The prime abductors of the wrist are the extensor carpi radialis longus and flexor carpi radialis muscles.

(4) The prime adductors of the wrist are the extensor carpi ulnaris and flexor carpi ulnaris muscles.

c. The wrist is dynamically stabilized by the actions of the extensor carpi radialis longus and brevis, the extensor carpi ulnaris, and the major flexors of the wrist.

3. Extension of the digits at the metacarpophalangeal joint (excluding the thumb) is accomplished by the extensor digitorum communis and the extensor digiti minimi of the superficial group, as well as by the extensor indicis proprius of the deep group.

a. The extensor tendons are held in place at the wrist by several **extensor retinacula** (see Fig. 8-3).

(1) To facilitate movement as the tendons pass beneath the extensor retinacula, the tendons are encased in **synovial sheaths.**

(2) The extensor tendon sheaths are especially prone to become inflamed from the widespread synovial inflammation that accompanies rheumatoid arthritis, resulting in subacute tenosynovitis. When inflammation invades the extensor tendons, they may rupture.

b. The extrinsic extensor muscles, as well as the interosseus and lumbrical muscles, insert into the tendinous **extensor aponeurosis (dorsal expansion** or **hood**) of each digit (see Fig. 10-3).

c. Each extensor aponeurosis passes over the metacarpophalangeal joints and then trifurcates.

(1) The central portion of the extensor aponeurosis passes over the proximal interphalangeal joint and inserts into the base of the middle phalanx.

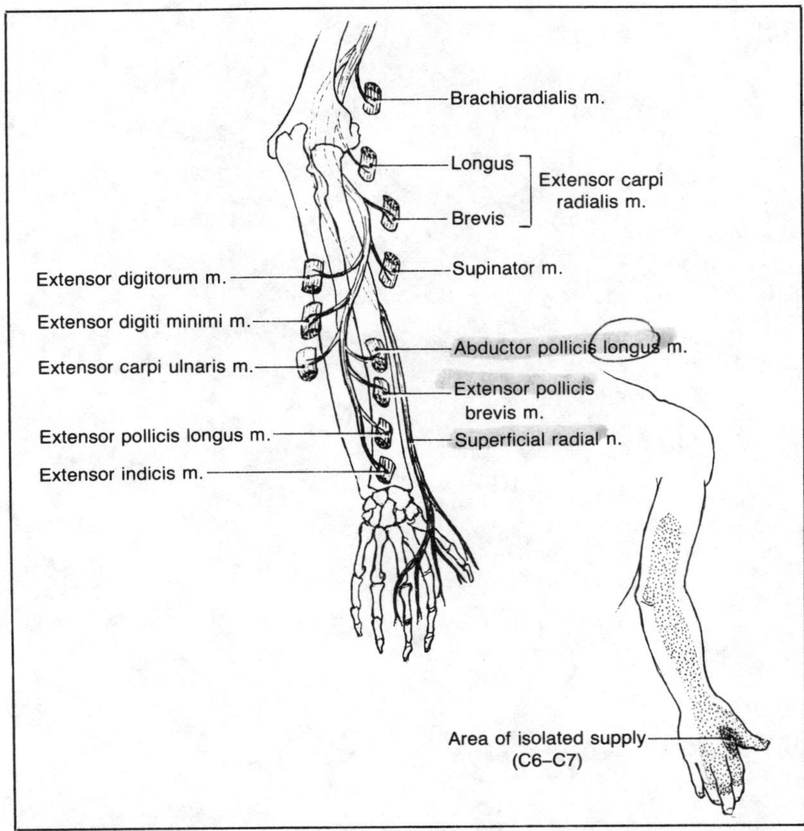

Figure 8-4. *The radial nerve in the forearm.* The muscles innervated by the branches of the radial nerve are indicated. Also depicted is the region of dermatomal innervation by the superficial branch of the radial nerve.

(2) The two lateral bands pass over the proximal interphalangeal joint and the distal interphalangeal joint to insert into the base of the distal phalanx.

d. The primary action of the extrinsic extensor muscles is to extend the metacarpophalangeal joints; secondarily, the extrinsic extensors assist with extension of the interphalangeal joints.

V. ANTEBRACHIAL VASCULATURE (see Chapter 9 III)

VI. INNERVATION OF THE ANTEBRACHIAL EXTENSOR COMPARTMENT

A. The radial nerve innervates all of the extensor muscles in the posterior compartment of the forearm. It passes out of the spiral groove into the anterior compartment of the arm where it gives off branches to the brachioradialis and the extensor carpi radialis muscles. It then divides into two branches (Fig. 8-4).

1. The **superficial radial nerve** usually arises at the level of the lateral epicondyle and supplies the lateral aspect of the dorsum of the wrist and hand, with a region of exclusivity between the thumb and index finger.
2. The **deep radial (posterior interosseous) nerve** is distributed to the muscles of the extensor compartment.

B. Radial nerve injuries.

1. Injury to the radial nerve above the epicondyles produces pronation of the hand, wristdrop with inability to extend the digits or thumb, and loss of sensation to the dorsum of the hand.
2. Injury to the radial nerve below the epicondyles produces wristdrop with inability to extend the digits or thumb but with no loss of sensation to the dorsum of the hand.

9
The Flexor Forearm and Anterior Wrist

I. INTRODUCTION

A. The primitively ventral aspects of the forearm, wrist, and hand can be considered a functional flexor unit.

B. Bony landmarks (see Chapter 8 I B).

C. The bones of the forearm (see Chapter 8 II A).

D. The bones of the wrist (see Chapter 8 II B).

II. THE MUSCULATURE OF THE ANTEBRACHIAL FLEXOR COMPARTMENT

A. THE ANTEBRACHIAL FASCIA (deep fascia of the forearm) envelops the musculature of the forearm.

1. Proximally, the **bicipital aponeurosis** of the biceps brachii muscle inserts into this fascial layer.
2. It gives origin to some of the more superficial fascicles of the flexor and extensor muscles of the forearm.
3. Septa penetrate between the individual muscles and also divide the arm into flexor and extensor compartments.
4. Anteriorly, at the wrist, the antebrachial fascia condenses to form the **flexor retinaculum**, which consists of two layers.
   - a. The more superficial **volar carpal ligament** separates the palmaris longus muscle from the underlying ulnar nerve and artery (see Fig. 9-4).
   - b. The deeper **transverse carpal ligament** (Fig. 9-1) bridges the carpal tunnel.
     - (1) The transverse carpal arch is formed by the nearly semicircular arrangement of the carpal bones.
     - (2) The transverse carpal ligament runs from the pisiform and hamulus to the scaphoid and trapezium and maintains the carpal arch.
     - (3) Beneath it run the median nerve, the flexor digitorum superficialis, the flexor digitorum profundus, and the flexor pollicis longus.
     - (4) It prevents the tendons of the extrinsic flexor muscles of the hand from bowstringing upon flexion.
     - (5) As the tendons of the extrinsic flexor muscles pass through the carpal tunnel, they are enclosed in synovial sheaths.
5. In the hand, the antebrachial fascia continues as the thickened **palmar aponeurosis**.

B. THE FLEXOR MUSCLES of the forearm are mostly ''multijoint'' muscles, acting across the elbow joint and the wrist joint, and in some instances the carpal joints, the metacarpal joints, and the interphalangeal joints.

1. The flexor muscles of the wrist and hand originate from the postaxial (medial) aspect of the distal arm and proximal forearm (the **flexor compartment**) and correspond to primitive ventral musculature.

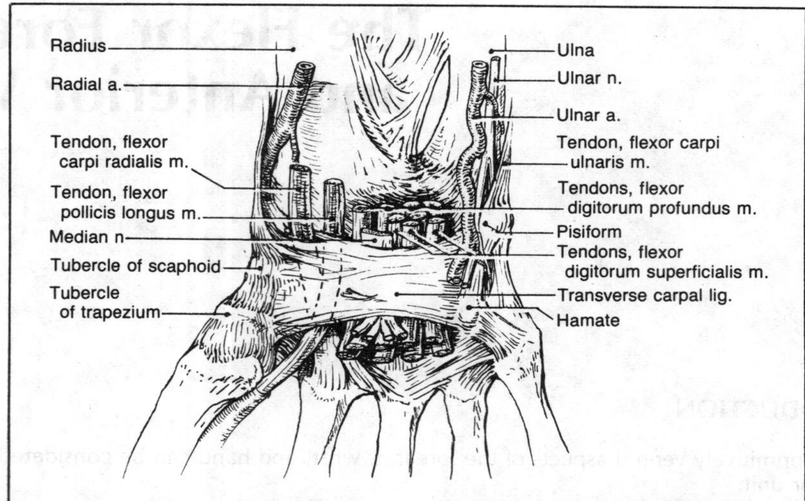

Figure 9-1. *The carpal tunnel.* The relations of the extrinsic flexor tendons of the hand as they pass beneath the transverse carpal ligament are indicated. Also, the relations of the median and ulnar nerves, as well as the radial and ulnar arteries, are indicated.

a. The **superficial group** (Fig. 9-2A, Table 9-1).

(1) The superficial muscles originate from the medial supracondylar ridge and medial epicondyle of the humerus, as well as from the anterior aspect of the proximal ulna.

(2) They include the pronator teres (a pronator of the forearm), flexor carpi radialis, palmaris longus (missing in 13 percent of forearms), and flexor carpi ulnaris.

(3) These muscles constitute the primary flexors of the wrist.

b. The **intermediate group** (see Fig. 9-2B and Table 9-1).

(1) The flexor digitorum superficialis, the only muscle of this group, originates from the medial epicondyle, proximal ulna, and proximal radius. The median nerve and ulnar artery pass between the humeroulnar and radial heads of this muscle.

(2) The flexor digitorum superficialis divides into four tendons, each of which divides to insert into the base of the middle phalanx of the second through fifth digits.

(3) This muscle is the primary flexor of the *proximal* interphalangeal joint.

(4) The tendons pass beneath the flexor retinaculum where they are enclosed in synovial sheaths.

c. The **deep group** (Fig. 9-3A; see Table 9-1).

(1) The deep muscles originate from the medial epicondyle of the humerus, the proximal ulna, the anterior surface of the interosseous membrane, and the anterior midradius.

(2) The deep group consists of the flexor digitorum profundus, flexor pollicis longus, and pronator quadratus.

(a) The flexor digitorum profundus passes through the split in the flexor digitorum superficialis tendons to insert into the base of the *distal* phalanx.

(b) The flexor digitorum profundus is the primary flexor of the distal interphalangeal joint.

(c) The flexor pollicis longus is the primary flexor of the distal phalanx of the thumb.

(d) The pronator quadratus is a pronator of the forearm.

(3) The tendons pass deep to the flexor retinaculum (see Fig. 9-3B).

2. The action of each muscle at the radiocarpal joint is determined by the location of the tendon relative to the two axes of rotation (Fig. 9-4).

a. Flexion/extension occurs about a transverse axis (L–M in Fig. 9-4).

(1) All of the anterior compartment muscles pass ventral to this axis and are, therefore, flexors of the wrist.

(2) All of the muscles of the extensor compartment pass dorsal to the transverse axis of the radiocarpal joint and are, therefore, extensors of the wrist.

(3) The prime flexors of the wrist are the flexor carpi radialis and flexor carpi ulnaris.

b. Abduction/adduction occurs about an anteroposterior axis (A–P in Fig. 9-4).

(1) Muscles that pass on the radial side of the anteroposterior axis of the radiocarpal joint are abductors of the wrist.

(2) Muscles that pass on the ulnar side of this axis are adductors of the wrist.

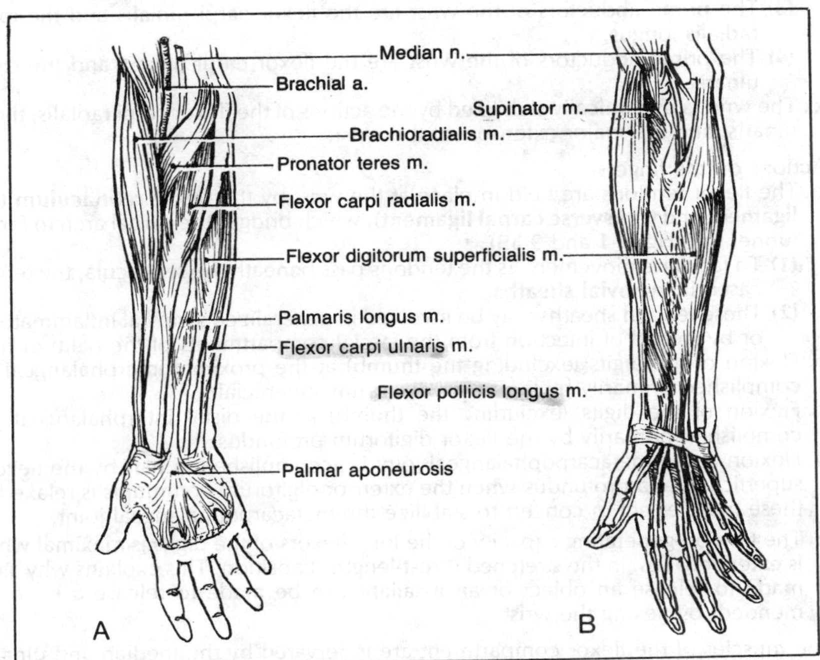

Figure 9-2. *A, Superficial muscles of the flexor compartment of the right forearm. B, Intermediate layer of muscles of the flexor compartment of the right forearm.*

Table 9-1. Forearm Muscles Acting on the Palmar Side of the Wrist and Hand

Muscle	Origin	Insertion	Action	Innervation
		Superficial Group		
Flexor carpi radialis	Medial epicondyle of humerus	Bases of second and third metacarpals	Flexes wrist, weakly flexes forearm	Median n. (C6–C7, anterior)
Palmaris longus	Medial epicondyle of humerus	Palmar aponeurosis	Flexes wrist	Median n. (C7–C8, anterior)
Flexor carpi ulnaris	Medial epicondyle of humerus	Pisiform bone and base of fifth metacarpal	Flexes wrist, weakly flexes forearm	Ulnar n. (C8–T1, anterior)
		Intermediate Group		
Flexor digitorum superficialis	Medial epicondyle of humerus	Base of second phalanx of second through fifth digits	Flexes proximal interphalangeal joint, metacarpophalangeal joint, and wrist	Median n. (C7–T1, anterior)
		Deep Group		
Flexor digitorum profundus	Anterior proximal ulna and interosseous membrane	Base of third phalanx of second through fifth digits	Flexes distal and proximal interphalangeal joint, metacarpophalangeal joint, and wrist	Heads for second and third digits: median n. (C7–T1, anterior). Heads for fourth and fifth digits: ulnar n. (C8–T1, anterior)
Flexor pollicis longus	Anterior midradius and interosseous membrane	Lateral aspect base of second phalanx of thumb	Flexes thumb, metacarpophalangeal joint, and wrist	Median n. (C7–T1, anterior)

(3) The prime abductors of the wrist are the flexor carpi radialis and the extensor carpi radialis longus.

(4) The prime adductors of the wrist are the flexor carpi ulnaris and the extensor carpi ulnaris.

c. The wrist is dynamically stabilized by the actions of the flexor carpi radialis, the flexor carpi ulnaris, and the major extensors of the wrist.

3. Actions on the fingers.

a. The flexor tendons are held in place at the wrist by the **flexor retinaculum (volar carpal ligament** and **transverse carpal ligament)**, which bridges the carpal arch to form the carpal tunnel (see Fig. 9-1 and 9-3B).

(1) To facilitate movement as the tendons pass beneath the retinacula, the tendons are encased in **synovial sheaths**.

(2) These tendon sheaths may be involved in generalized synovial inflammatory processes or by spread of infection from the fascial compartments of the palm or fingers.

b. Flexion of the digits (excluding the thumb) at the proximal interphalangeal joints is accomplished primarily by the flexor digitorum superficialis.

c. Flexion of the digits (excluding the thumb) at the distal interphalangeal joints is accomplished primarily by the flexor digitorum profundus.

d. Flexion at the metacarpophalangeal joint is accomplished in part by the flexor digitorum superficialis and profundus when the extensor digitorum communis is relaxed; otherwise, these muscles act in concert to stabilize the metacarpophalangeal joint.

e. The tension-generating capacity of the long flexors of the digits is maximal when the wrist is extended, i.e., in the stretched (''rest-length'') position. This explains why a child can be made to release an object or an assailant can be made to release a knife (not recommended) by flexing the wrist.

4. The muscles of the flexor compartment are innervated by the median and ulnar nerves.

III. ANTEBRACHIAL VASCULATURE

A. **THE RADIAL ARTERY AND ULNAR ARTERY** arise from the bifurcation of the **brachial artery** in the cubital fossa.

1. The **radial artery** (Fig. 9-5).

a. This major vessel crosses the biceps brachii tendon and descends along the anterior preaxial border of the forearm.

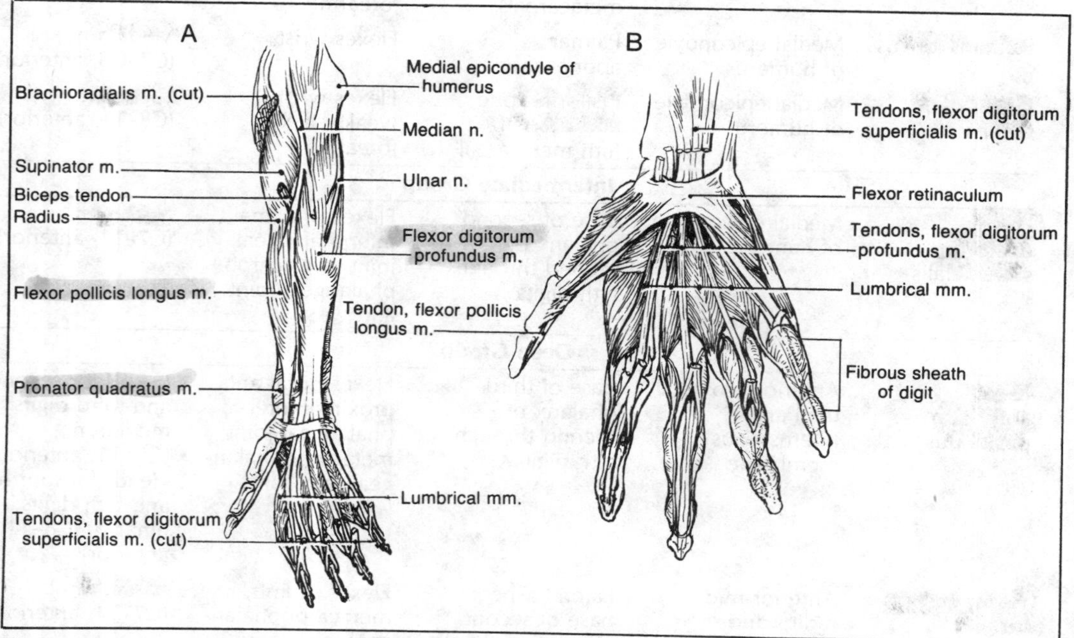

Figure 9-3. *A, Deep muscles of the flexor compartment of the right forearm. B, The extrinsic flexor tendons of the right hand.*

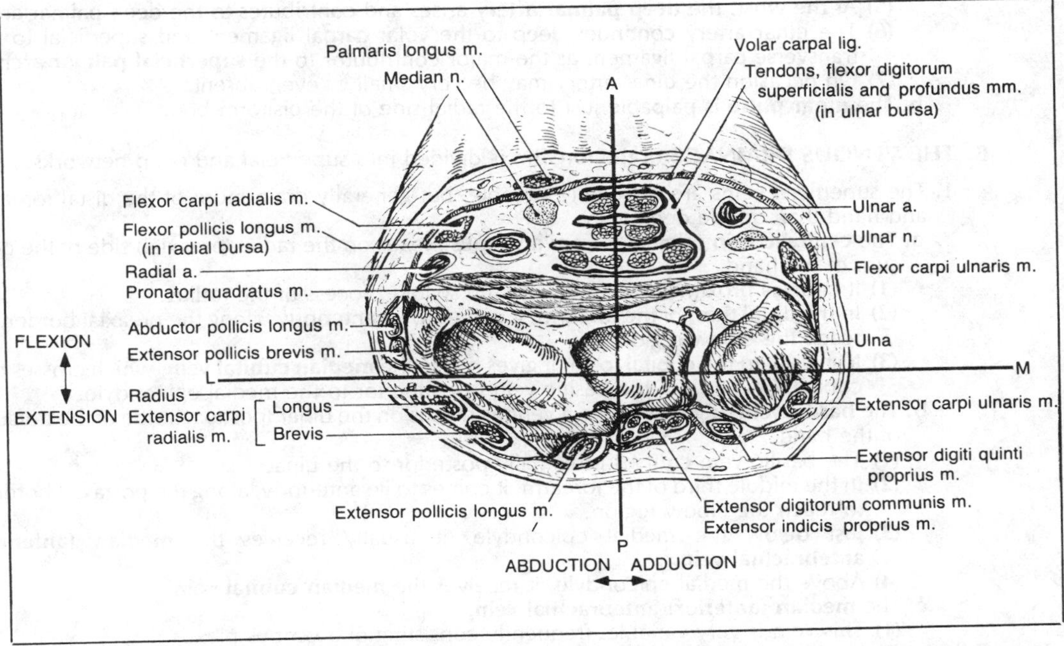

Figure 9-4. *The extrinsic flexor and extensor tendons of the wrist.* The right wrist is in the anatomical position. The transverse axis (L–M) and anteroposterior axis (A–P) of the radiocarpal joint are indicated. Muscles that lie to the lateral side of the A–P axis abduct, those to the medial side adduct. Muscles that lie to the palmar side of the L–M axis flex, those to the dorsal side extend.

(1) The **radial recurrent branch** passes anterior to the lateral epicondyle to anastomose with the radial (anterior) collateral of the deep brachial artery.
(2) Numerous muscular branches are given off.
(3) Distally, the radial recurrent branch contributes to the anterior and posterior carpal networks (rete).
(4) The **superficial palmar branch** provides the smaller radial contribution to the **superficial palmar arch**.
(5) The main stem of the radial artery continues deep to the volar (palmar) carpal ligament in the floor of the "anatomical snuff-box," toward the dorsal aspect of the hand. At the base of the first metacarpal, it passes between the two heads of the first dorsal interosseous muscle and becomes the major contributor to the **deep palmar arch**.

b. The **radial pulse** may be palpated at the wrist between the tendons of the brachioradialis and flexor carpi radialis muscles, as well as in the "anatomical snuff-box" between the tendons of the extensor pollicis longus and brevis muscles.

2. The **ulnar artery** (see Fig. 9-5).

a. This major vessel deviates toward the anterior postaxial border of the forearm.
(1) The **anterior ulnar recurrent branch** passes anterior to the medial epicondyle to anastomose with the inferior ulnar collateral of the brachial artery.
(2) The **posterior ulnar recurrent branch** passes posterior to the medial epicondyle to anastomose with the superior ulnar collateral of the brachial artery.
(3) The **common interosseous artery** bifurcates almost immediately.
(a) The **anterior interosseous artery** descends with the anterior interosseous nerve and passes through the interosseous membrane distally to anastomose with the posterior interosseous artery.
(b) The **posterior interosseous artery** passes through the interosseous membrane proximally.
(i) The **recurrent interosseous branch** passes behind the elbow to anastomose with the posterior branch of the deep brachial artery.
(ii) The posterior interosseous artery descends in the forearm and anastomoses with the anterior interosseous artery.
(4) Near the pisiform bone, the ulnar artery gives off contributions to the anterior and posterior carpal networks (rete).

(5) At the wrist, the **deep palmar artery** arises and contributes to the deep palmar arch.
(6) The ulnar artery continues deep to the volar carpal ligament and superficial to the transverse carpal ligament as the major contributor to the **superficial palmar arch**.
(7) On occasion the ulnar artery may be very small or even absent.

b. The **ulnar pulse** is palpable just to the radial side of the pisiform bone.

B. THE VENOUS DRAINAGE of the forearm is divided into superficial and deep networks.

1. The superficial veins are extremely variable, but generally drain more of the distal forearm and hand than do the deep veins.

a. The **cephalic vein** originates from a venous plexus on the radial (preaxial) side of the dorsum of the hand.
(1) It courses proximally, posterior to the styloid process of the radius.
(2) In the distal third of the forearm, it comes to lie anteriorly, along the preaxial border, to reach the elbow region.
(3) Just distal to the cubital fossa, it gives rise to the **median cubital vein**, which crosses the cubital fossa to join the basilic vein just superior to the medial epicondyle.

b. The **basilic vein** originates from a venous plexus on the ulnar (postaxial) side of the dorsum of the hand.
(1) The basilic vein courses proximally posterior to the ulna.
(2) In the middle third of the forearm, it comes to lie anteriorly, along the postaxial border, to reach the elbow region.
(3) Just below the medial epicondyle, it usually receives the **median (anterior) antebrachial vein**.
(4) Above the medial epicondyle, it receives the **median cubital vein**.

c. The **median (anterior) antebrachial vein**.
(1) This vein is very variable, frequently consisting of a venous plexus.
(2) It drains the palm of the hand and anterior aspect of the forearm.
(3) It usually drains into the basilic vein, inferior to the medial epicondyle, but may drain into the median cubital vein.

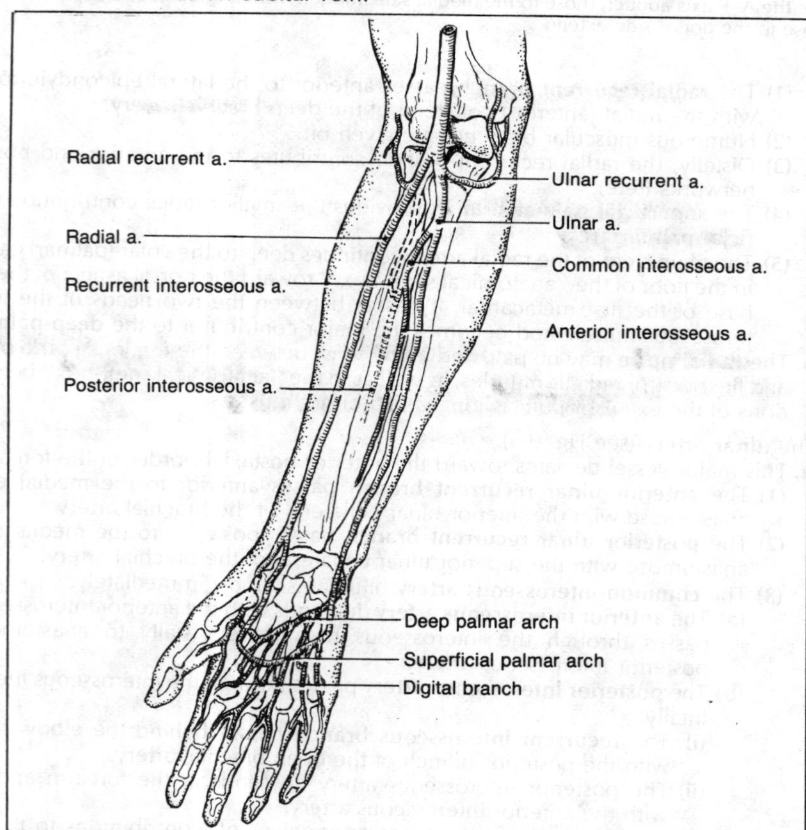

Figure 9-5. *The arterial supply of the forearm.*

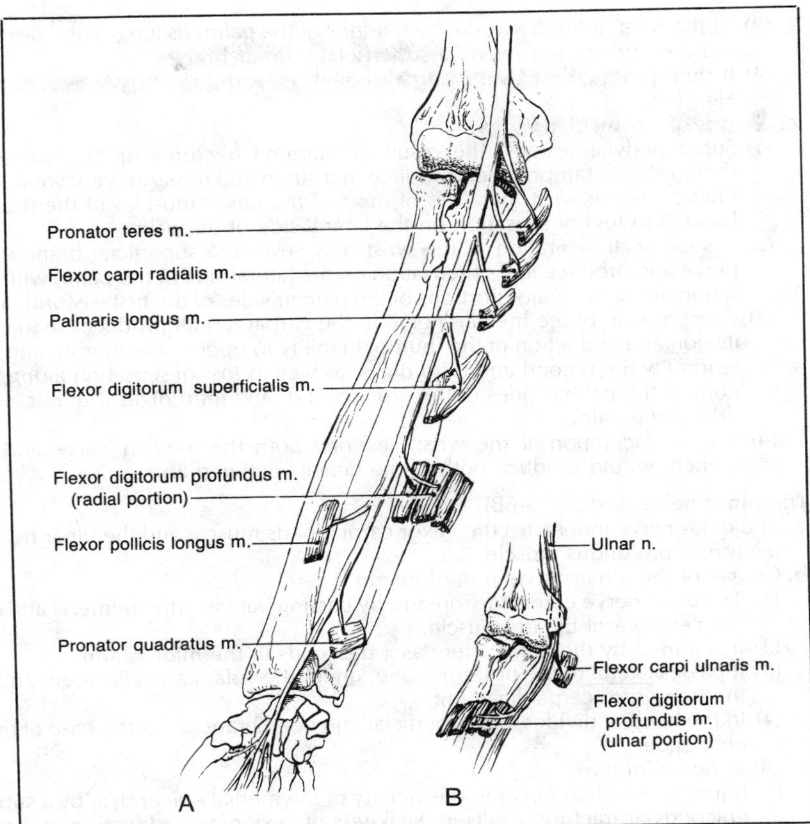

Figure 9-6. *A, The median nerve in the forearm.* The muscles innervated by branches of the median nerve are indicated. *B, The ulnar nerve in the forearm.* The muscles innervated by branches of the ulnar nerve are indicated.

2. The deep veins accompany the named arteries.

3. Phlebotomy or **venipuncture** is easily accomplished on the superficial veins of the cubital fossa.

a. A snug, but not tight, tourniquet will occlude the superficial venous return, causing distension of the superficial veins.

b. If a tourniquet is too tight, the arterial supply will be occluded, so that no venous return occurs.

C. THE LYMPHATIC DRAINAGE OF THE FOREARM, superficial and deep, accompanies the superficial and deep veins.

IV. INNERVATION OF THE ANTEBRACHIAL FLEXOR COMPARTMENT

A. The muscles of the flexor compartment are innervated by the median and ulnar nerves from the anterior divisions of the brachial plexus.

1. The **median nerve** (Fig. 9-6A).

a. The median nerve innervates the pronator teres, flexor carpi radialis, palmaris longus, flexor digitorum superficialis, radial half of the flexor digitorum profundus, flexor pollicis longus, and pronator quadratus muscles, i.e., all of the flexor and pronator muscles of the forearm *except* the flexor carpi ulnaris and the ulnar half of the flexor digitorum profundus.

b. Course of the median nerve in the forearm.

(1) In the cubital fossa, the median nerve lies medial to the tendon of the biceps brachii muscle and anterior to the brachialis muscle.

(2) It enters the forearm by passing between the ulnar and humeral heads of the pronator teres muscle, then between the humeroulnar and radial heads of the flexor digitorum superficialis along with the ulnar artery.

(3) At the wrist, it lies between the tendons of the palmaris longus and flexor carpi radialis muscles, where it gives off a **superficial palmar branch**.

(4) It then passes through the carpal tunnel beneath the transverse carpal ligament (see Fig. 9-1).

c. Lesions of the median nerve.

(1) Supracondylar lesions, the result of humeral fractures or brachial artery bleeding (hemorrhagic tamponade), produce in a supinated forearm, very weak flexion and abduction of the wrist, paralysis of most of the flexor muscles of the thenar side of the hand, and loss of sensation on the lateral side of the palm.

(2) A superficial laceration at the wrist may sever the superficial branch of the median nerve and produce loss of sensation on the lateral side of the palm, without loss of sensation along the major portions of the palmar sides of the first, second, and third digits.

(3) Compression of the median nerve in the carpal tunnel produces marked weakness of flexion and abduction of the thumb, inability to oppose the thumb, and inability to extend fully the second and third digits, as well as loss of sensation along the major portions of the palmar sides of the first, second, and third digits (but not along the lateral side of the palm).

(4) A deep laceration at the wrist, severing both the median nerve and its superficial branch, would produce both of the deficits outlined above.

2. The **ulnar nerve** (see Fig. 9-6B).

a. The ulnar nerve innervates the flexor carpi ulnaris muscle and the ulnar head of the flexor digitorum profundus muscle.

b. Course of the ulnar nerve in the forearm.

(1) The ulnar nerve enters the forearm by passing between the humeral and ulnar heads of the flexor carpi ulnaris muscle.

(2) It is joined by the ulnar artery as it descends in the midforearm.

(3) It passes lateral to the pisiform bone, under the volar carpal ligament and superficial to the transverse carpal ligament.

(4) In the hand, it divides into superficial and deep branches at the base of the hypothenar eminence.

c. Ulnar nerve injuries.

(1) Injury to the ulnar nerve in the vicinity of the medial epicondyle by a supracondylar or epicondylar fracture results in weakness of flexion and adduction of the wrist, as well as paralysis of the hypothenar muscles, most of the deep muscles of the hand, and some of the muscles of the thenar side of the hand.

(2) Injury to the ulnar nerve by fracture of the proximal ulna results only in paralysis of the hypothenar muscles, most of the deep muscles of the hand, and some of the muscles of the thenar side of the hand.

B. The median and ulnar nerves are not infrequently injured together at the level of the brachial plexus, the supracondylar region, or the wrist.

10
The Hand

I. INTRODUCTION

A. The hand has a unique combination of mobility, dexterity, and sensitivity.

1. The function of the upper extremity is to position the hand to interact effectively with the environment.
2. The function of the hand is dependent upon extrinsic and intrinsic mobility, muscle strength, and sensation.

B. Bony landmarks.

1. **Metacarpals.**
2. **Phalanges.**

II. THE SKELETON OF THE HAND

A. THE BONES OF THE WRIST (see Chapter 8 II B).

B. THE BONES OF THE HAND

1. Five **metacarpal bones** form the skeleton of the hand (see Fig. 8-1; Fig. 10-1).
   - **a.** Each metacarpal bone has a proximal **base**, a **shaft**, and a distal **head**.
   - **b.** The base of each metacarpal contacts the distal row of carpal bones to form the **carpometacarpal joint**.
   - **c.** The head of each metacarpal contacts a proximal phalanx at a **metacarpophalangeal** (MP) **joint**.
2. Three rows of **phalanges** comprise the skeletons of the second through fifth digits; the thumb has only two phalanges.
   - **a.** The proximal row of phalanges articulates with the middle row of phalanges at the **proximal interphalangeal** (PIP) **joints**.
   - **b.** The middle row of phalanges articulates with the distal row of phalanges at the **distal interphalangeal** (DIP) **joints**.

C. NUMEROUS JOINTS exist in the hand.

1. The **carpometacarpal joints** (see Fig. 10-1).
   - **a.** The second, third, fourth, and fifth carpometacarpal joints are condyloid, permitting two degrees of freedom.
     - **(1)** A slight amount of flexion/extension occurs, as well as some abduction/adduction.
     - **(2)** The combined circumduction at these joints allows the formation of the **transverse metacarpal arch** (the hollowing of the hand), which is essential for permitting the fingers to meet.
     - **(3)** Each joint is reinforced by carpometacarpal ligaments.
   - **b.** The first (thumb) carpometacarpal joint is a saddle-shaped joint with two degrees of freedom between the trapezium of the first carpal row and the first metacarpal.
     - **(1)** Flexion/extension (75°) of the thumb occurs about a complex axis.
       - **(a)** The plane of flexion/extension is approximately 60° to that of the hand.
       - **(b)** Flexion brings the thumb ventral to the plane of the hand toward the palm; extension brings the thumb back into the plane of the hand.

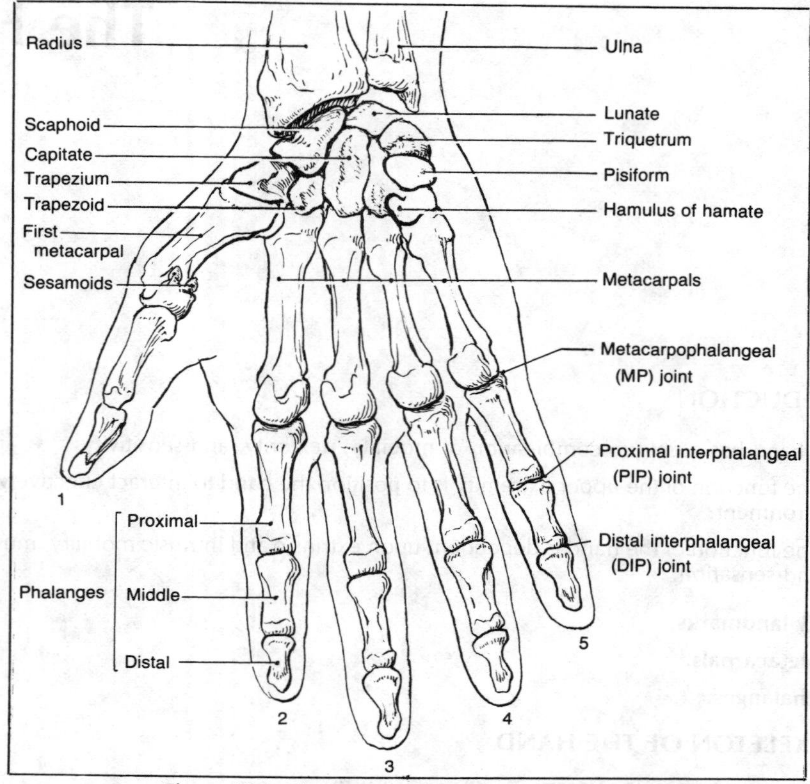

Figure 10-1. *The bones and joints of the right hand.*

(2) Abduction/adduction of the thumb occurs about an axis perpendicular to the plane of the hand.
  (a) Abduction (15°) is movement of the extended thumb away from the index finger.
  (b) Adduction (45°) brings the extended thumb against the index finger.

(3) Opposition is a form of circumduction which, because of the sellar shape of the joint surfaces, involves a rotational movement of the thumb. Thus, opposition is flexion accompanied by medial rotation; reposition is extension and abduction accompanied by lateral rotation.

2. The **metacarpophalangeal** (MP) **joints** (see Fig. 10-1) represent the knuckles.
  a. These joints are condyloid with two degrees of freedom.
    (1) Flexion/extension (100°) occurs about transverse axes.
    (2) Abduction/adduction (as much as 30° for the index finger) occurs about anteroposterior axes with reference to the middle finger.
    (3) Because there are two degrees of freedom at the metacarpophalangeal joint, circumduction (the combination of movements) is possible.
  b. The **metacarpophalangeal collateral ligaments** reinforce these joints. These ligaments are slack in the extended finger and taut in the flexed finger.
  c. The **deep transverse metacarpal ligament** interconnects the heads of the metacarpals.

3. The **interphalangeal joints** (see Fig. 10-1).
  a. These are hinge joints with one degree of freedom, permitting flexion/extension (90°).
    (1) The **proximal interphalangeal** (PIP) **joints** are between the proximal and middle phalanges.
    (2) The **distal interphalangeal** (DIP) **joints** are between the middle and distal phalanges.
  b. These joints are strongly reinforced by collateral ligaments, which are slacker when the finger is extended.

III. THE DORSAL HAND

A. THE DORSAL SUBCUTANEOUS SPACE

1. The skin of the dorsum of the hand is very mobile over the underlying fascia.

2. This layer is very vascular and contains a rich network of lymphatics that produce swelling of the dorsum of the hand upon infection of the thenar compartment, hypothenar compartment, or the fingers.

3. This space is limited by the **deep dorsal fascia**, which is formed by the fusion of the antebrachial fascia (which extends onto the dorsum of the hand) with the extensor tendons (which are unsheathed on the dorsum of the hand).

4. Posteriorly, the antebrachial fascia condenses at the wrist to form the **extensor retinaculum**.
   a. This retinaculum is subdivided into six compartments by attachment in several places to bone.
   b. The tendons of the extrinsic extensor muscles of the hand pass through these compartments to gain access to the dorsum of the hand (see Fig. 8-3).
      (1) The tendons are enclosed in synovial sheaths beneath the extensor retinaculum.
      (2) The sheaths of the extensor digitorum communis terminate just distal to the extensor retinaculum, and the tendons lie directly on the metacarpal bones.
      (3) The tendons proceed to the individual digits, but there are variable interconnections that tend to limit the individual movements of the fingers.

5. The **dorsal subaponeurotic space** lies between the extensor tendons and the deep fascia, covering dorsal interossei and metacarpals.
   a. Laceration of the knuckles and inoculation with pyogenic organisms (such as occurs upon the violent meeting of fist with teeth) may result in infection of the subaponeurotic space.
   b. Otherwise, this space is not frequently involved in hand infections.

B. THE EXTRINSIC EXTENSOR MUSCLES (see Chapter 8 IV B).

1. This group of muscles on the dorsum of the hand includes many of those of the extensor compartment of the forearm. It consists of the tendons of the extensor digitorum communis, extensor indicis proprius, extensor pollicis brevis, abductor pollicis longus, and extensor pollicis longus muscles (Table 10-1; see Fig. 8-3 and Table 10-2).
   a. The subcutaneous tendons of the extensor digitorum communis lie in sheaths as they pass beneath the extensor retinaculum.
   b. The extensor digitorum communis divides into four slips; each slip inserts into the extensor aponeurosis of one of the second through fifth digits (Figs. 10-2 and 10-3).

Table 10-1. Muscles Acting on the Second through Fifth Digits

Muscle	Origin	Insertion	Action	Innervation
		Extrinsic Muscles		
Extensor digitorum communis	Lateral epicondyle of humerus	Phalanges two and three of index, middle, and ring fingers	Extends digits and wrist when fist is clenched	Radial nerve (C7–C8, posterior)
Extensor digiti minimi	Common extensor tendon	All phalanges of fifth digit	Extends fifth digit	Radial nerve (C7–C8, posterior)
Extensor indicis proprius	Interosseous membrane and ulna	Phalanges two and three of index finger	Extends first digit and wrist	Radial nerve (C8–T1, posterior)
Flexor digitorum superficialis	Medial epicondyle of humerus	Base of second phalanx of each digit	Flexes proximal interphalangeal (PIP) joint, metacarpophalangeal (MP) joint, and wrist	Median nerve (C8–T1, anterior)
Flexor digitorum profundus	Anterior proximal ulna and interosseous membrane	Base of third phalanx of each digit	Flexes distal interphalangeal (DIP), PIP, and MP joints and wrist	Heads for second and third digits by median n. (C7–T1, anterior). Heads for fourth and fifth digits by ulnar n. (C7–T1, anterior)

Table 10-1. Continued

Muscle	Origin	Insertion	Action	Innervation
		Intrinsic Muscles		
Dorsal interossei (4)	Medial side of first metacarpal; both sides of the second, third, and fourth metacarpal; lateral side of fifth metacarpal	Tubercle of proximal phalanx and dorsal aponeurosis: laterally on second and third digits, medially on third and fourth digits	Abduct the second, third, and fourth digits from the midline of hand (third digit abducts to both sides). Flex MP joint and extend PIP and DIP joints	Ulnar nerve (C8–T1, anterior)
Palmar interossei (3)	Medial side of second and lateral side of fourth and fifth metacarpals	Tubercle of proximal phalanx and dorsal aponeurosis: medially on second digit, laterally on fourth and fifth digits	Adduct the second, fourth and fifth digits to the midline of hand. Flex MP joint and extend PIP and DIP joints	Ulnar nerve (C8–T1, anterior)
Lumbricales 1 and 2	Tendons of the flexor digitorum profundus in the deep palm	Lateral side of dorsal expansion of second and third digits	Flex MP joint and extend PIP and DIP joints	Median nerve (C8–T1, anterior)
Lumbricales 3 and 4	Tendons of the flexor digitorum profundus in the deep palm	Lateral side of dorsal expansion of fourth and fifth digits	Flex MP joint and extend PIP and DIP joints	Ulnar nerve (C8–T1, anterior)
Palmaris brevis	Medial border of palmar aponeurosis	Skin over hypothenar region	Corrugates palmar skin	Ulnar nerve (C8–T1, anterior)
Abductor digiti minimi	Pisiform bone	Ulnar side of base of fifth proximal phalanx	Abduct the fifth digit	Ulnar nerve (C8–T1, anterior)
Flexor digiti minimi brevis	Flexor retinaculum and hamulus	Ulnar side of base of fifth proximal phalanx	Flex fifth metacarpophalangeal joint	Ulnar nerve (C8–T1, anterior)
Opponens digiti minimi	Flexor retinaculum and hamulus	Ulnar side of fifth metacarpal	Flexion and opposition	Ulnar nerve (C8–T1, anterior)

(1) Each extensor aponeurosis passes over the metacarpophalangeal joint and then divides into a central and two lateral parts.

(a) The central portion passes over the proximal interphalangeal joint to insert into the base of the middle phalanx.

(b) The lateral bands pass over the proximal and distal interphalangeal joints to insert into the base of the distal phalanx.

(2) The primary action of the extrinsic extensor muscles is to extend the metacarpophalangeal joints. They also assist in the extension of the proximal and distal interphalangeal joints, as well as the wrist.

2. There are no intrinsic muscles on the dorsum of the hand.

3. The extrinsic extensor muscles are innervated by the radial nerve.

IV. THE PALMAR HAND

A. The **thenar eminence** consists of the intrinsic muscles of the thumb.

B. The **hypothenar eminence** consists of the intrinsic muscles of the fifth digit.

C. The **palm** lies between the thenar and hypothenar eminences.

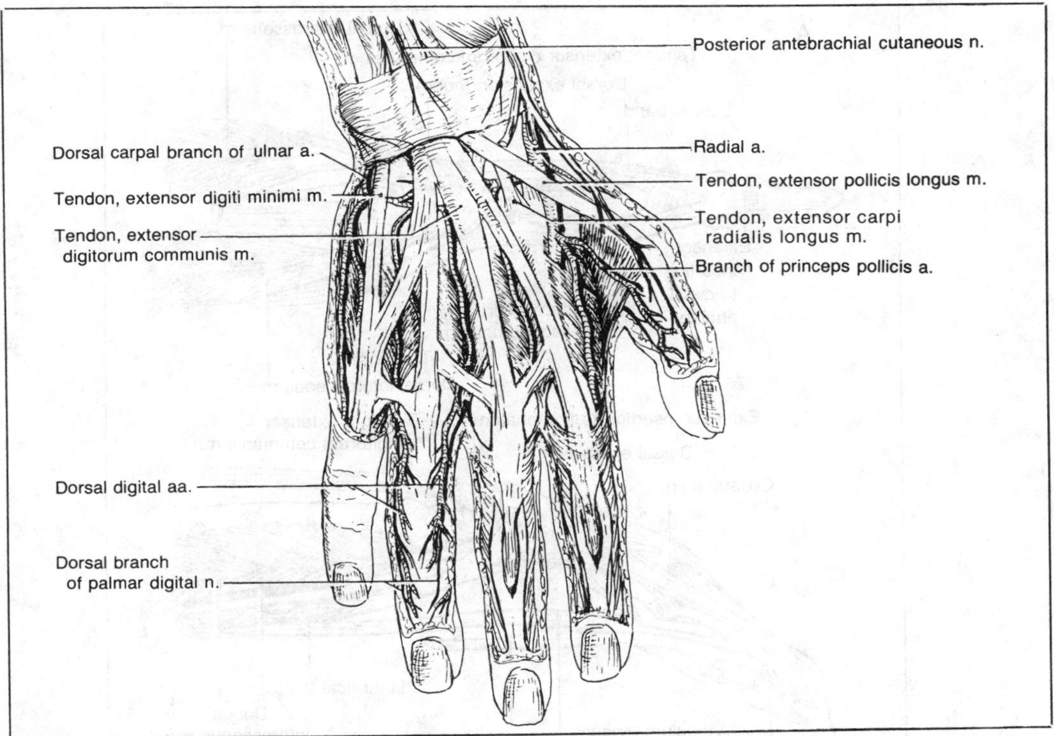

Figure 10-2. *The extrinsic extensor tendons of the dorsum of the right hand.*

1. The subcutaneous layer.

a. The **palmaris brevis** muscle and the **recurrent branch** of the median nerve lie superficial to the palmar aponeurosis.

b. Fibrous fasciculi connect the palmar skin to the palmar aponeurosis, precluding much independent movement of the palmar skin.

2. The **antebrachial fascia**, which formed the flexor retinaculum at the wrist, continues into the hand.

a. The deep fascial layer envelops the thenar muscles and hypothenar muscles, forming the **thenar compartment** and the **hypothenar compartment**.

(1) Beneath the thenar compartment, the deep fascia is attached along the length of the first metacarpal.

(2) Beneath the hypothenar compartment, it is attached along the length of the fifth metacarpal.

b. Between the thenar and hypothenar eminences, this deep fascial layer thickens as the **palmar aponeurosis**, beneath which lies the **palmar space**.

(1) Proximally, the palmar aponeurosis receives the insertion of the **palmaris longus** muscle, enabling that muscle to assist flexion of the hand.

(2) Distally, the palmar aponeurosis splits into four slips. Each slip has a superficial and deep part.

(a) The superficial parts insert into the skin at the base of the digits.

(b) The deep parts divide and pass along the sides of the digits to join the fibrous flexor sheaths and the transverse metacarpal ligaments.

(3) Medially, the palmar aponeurosis is attached along the length of the fifth metacarpal.

(4) Laterally, it is attached along the length of the first metacarpal.

(5) In the middle of the palm, it attaches along the length of the third metacarpal, thus subdividing the palmar space into a medial **midpalmar bursa** and a more lateral **thenar bursa**.

(6) In Dupuytren's contracture there is fibrosis and shortening of the palmar aponeurosis, more severe toward the ulnar side.

(a) Because the palmar aponeurosis inserts into the proximal and middle phalanges, shortening of the aponeurosis results in progressive flexion of the fourth and fifth digits.

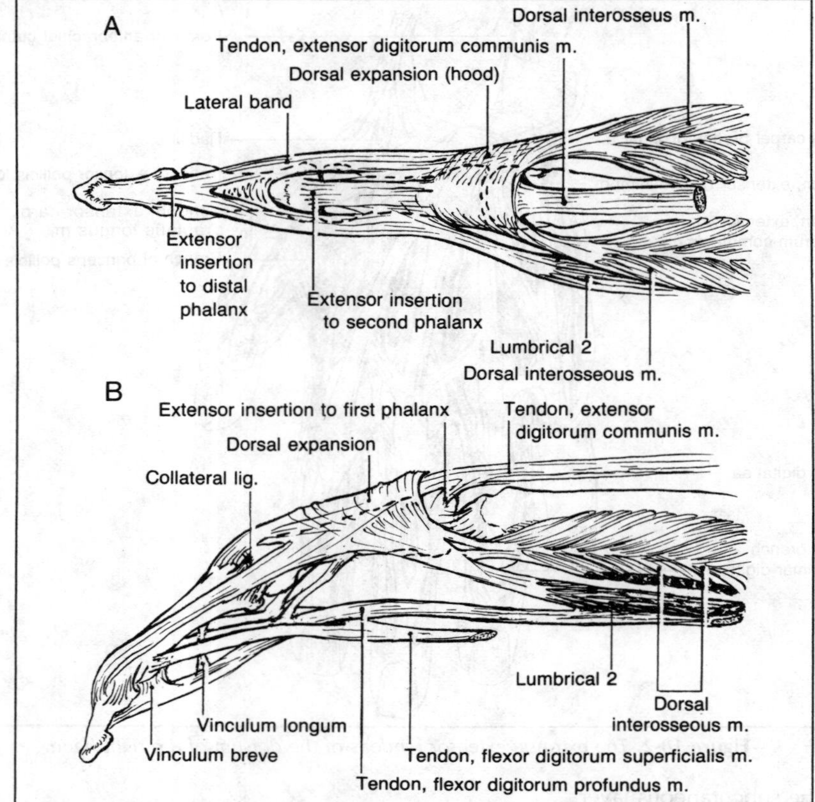

Figure 10-3. *The extensor aponeurosis of the third digit.* The extensor digitorum communis, interossei, and lumbricales all insert into the extensor aponeurosis (dorsal hood).

(b) A high correlation exists between Dupuytren's contracture and coronary artery disease, possibly as a result of vasospasm produced by the effect of referred pain upon the sympathetic innervation of the vasculature within the T1 component of the ulnar nerve distribution.

c. Deep to the palmar aponeurosis, septa extend around the ensheathed extrinsic flexor tendons of the digits.

(1) The **thenar space (bursa)** contains the flexor pollicis longus tendon and the extrinsic flexor tendons of the index finger.

(2) The **midpalmar space (bursa)** contains the extrinsic flexor tendons for the third, fourth, and fifth digits.

D. The palm of the hand can be divided functionally into two discrete muscular portions.

1. The **extrinsic flexor muscles** (see Chapter 9 II B).

a. This group of muscles, which includes many of those found in the flexor compartment of the forearm, consists of the palmaris longus, flexor digitorum superficialis, flexor digitorum profundus, and flexor pollicis longus (Table 10-2; see Tables 9-1 and 10-1).

b. The tendons of the extrinsic flexor muscles lie in the carpal tunnel beneath the transverse carpal ligament (see Figs. 9-1 and 9-3B).

c. The flexor digitorum superficialis muscle divides into four tendons.

(1) These tendons split near their insertions into the base of the middle phalanx of the second through fifth digits (see Fig. 10-3).

(2) The flexor digitorum superficialis primarily flexes the proximal interphalangeal joints but also flexes the metacarpophalangeal joints and wrist.

(3) The muscle is strongest when the wrist is extended.

d. The flexor digitorum profundus muscle divides into four tendons.

(1) These tendons pass through the split in the superficialis tendon to insert into the base of the distal phalanx of the second through fifth digits (see Fig. 10-3).

(2) The flexor digitorum profundus primarily flexes the distal interphalangeal joints but also flexes the proximal interphalangeal joints and wrist.
(3) The muscle is strongest when the wrist is extended.

e. **Vincula** attach the superficial and deep flexor tendons to the middle and distal phalanges, respectively (see Fig. 10-3).
(1) Each vinculum provides a route for the vascular supply to the tendons.
(2) Inflammatory swelling of the tendon sheaths may compress the vincular blood supply, causing ischemic necrosis of the tendons.

f. A fibrous **flexor sheath** extends along each digit.
(1) Fibers arch from both sides of the phalanges around the synovial sheaths forming **annular ligaments**, which prevent the tendons from bowing across the joints of a flexed digit.
(2) Over the joints, the fibrous sheaths are thinner and the fibers run obliquely, forming **cruciate ligaments**.

g. The extrinsic flexor muscles are innervated by the median and ulnar nerves.

h. The flexor tendon sheaths.
(1) The extrinsic flexor tendons are enclosed in synovial sheaths, which begin in the distal forearm and continue through the carpal tunnel.

Table 10-2. Muscles Acting on the Thumb

Muscle	Origin	Insertion	Primary Action	Innervation
		Extrinsic Muscles		
Abductor pollicis longus	Posterior interosseous membrane and ulna	Base of first metacarpal, laterally	Abducts thumb and wrist	Radial n. (C8–T1, posterior)
Extensor pollicis brevis	Posterior midshaft of radius and interosseous membrane	Base of first phalanx	Extends thumb and abducts wrist	Radial n. (C8–T1, posterior)
Extensor pollicis longus	Posterior surface of interosseous membrane and posterior ulna	Base of second phalanx	Extends thumb and abducts wrist	Radial n. (C8–T1, posterior)
Flexor pollicis longus	Anterior mid-radius and interosseous membrane	Lateral aspect base of first phalanx	Flexes thumb, MP joint, and wrist	Median n. (C7–T1, anterior)
		Intrinsic Muscles		
Abductor pollicis brevis	Anterior surface of trapezium, scaphoid	Lateral aspect of base of first phalanx	Abducts thumb	Median n. (C8–T1, anterior)
Opponens pollicis	Trapezium	Anterolateral surface of first metacarpal	Medially rotates (opposes) thumb	Median n. (C8–T1, anterior)
Flexor pollicis brevis:				
Superficial head	Transverse carpal ligament and trapezium	Lateral side of base of first phalanx	Flexes thumb	Median nerve (C8–T1, anterior)
Deep head	Lateral side of second metacarpal			Ulnar nerve (C8–T1, anterior)
Adductor pollicis:				
Oblique head	Anterior surface of capitate and second and third metacarpals	Medial side of base of first phalanx of thumb	Adducts thumb	Ulnar n. (C8–T1, anterior)
Transverse head	Distal half of third metacarpal			

(a) A common synovial sheath called the **ulnar bursa** usually surrounds the extrinsic digital flexor tendons as they pass through the carpal tunnel beneath the transverse carpal ligament.
(b) The ulnar bursa is continuous with the synovial sheath of the tendons along the length of the fifth digit.
(c) The more distal portions of the sheaths of the second, third, and fourth digits usually are not continuous, however, with the ulnar bursa; instead, they reform at the distal end of the metacarpals and continue to the proximal end of the distal phalanges.
(d) The synovial sheath of the flexor pollicis longus is continuous through the carpal tunnel, forming the **radial bursa**.

(2) The sheaths of the flexor tendons are somewhat variable.
(a) The radial bursa may or may not be continuous with the ulnar bursa.
(b) Occasionally an **intermediate bursa** invests the tendons of the flexor indicis. The intermediate bursa may or may not be continuous with the radial bursa, the ulnar bursa, or both.

(3) Infection may spread along the tendon sheaths.
(a) Infection of the flexor sheath of the thumb and of the fifth digit will track proximally into the radial and ulnar bursa, respectively.
(b) If the radial and ulnar bursae communicate, infection may track to the opposite side of the hand, "horseshoe abscess."
(c) Inflammatory swelling of the bursae beneath the transverse carpal ligament produces **carpal tunnel syndrome**, whereby compression of the median nerve results in thenar paralysis with entrapment of the extrinsic flexor tendons.

2. The **intrinsic muscles** (see Table 10-1).

a. The **dorsal interosseous muscles** (Fig. 10-4A).

(1) These consist of four bipennate muscles that originate from the first through fifth metacarpals.

(2) They insert into both the bases of proximal phalanges and the extensor aponeuroses (see Fig. 10-3).
(a) Abduction: The phalangeal insertions are so arranged as to abduct the second, fourth, and fifth digits from the midline of the hand and to abduct the third finger toward both the ulnar and radial sides of the hand.
(b) Flexion/extension: The insertions into the extensor aponeuroses are so arranged that they flex the metacarpophalangeal joints and extend the proximal and distal interphalangeal joints.

(3) The dorsal interossei are innervated by the ulnar nerve.

b. The **palmar interosseous muscles** (see Fig. 10-4B).

(1) These consist of three unipennate muscles that originate from the second, fourth, and fifth metacarpals.

(2) They insert into both the bases of the proximal phalanges and the extensor aponeuroses (see Fig. 10-3).

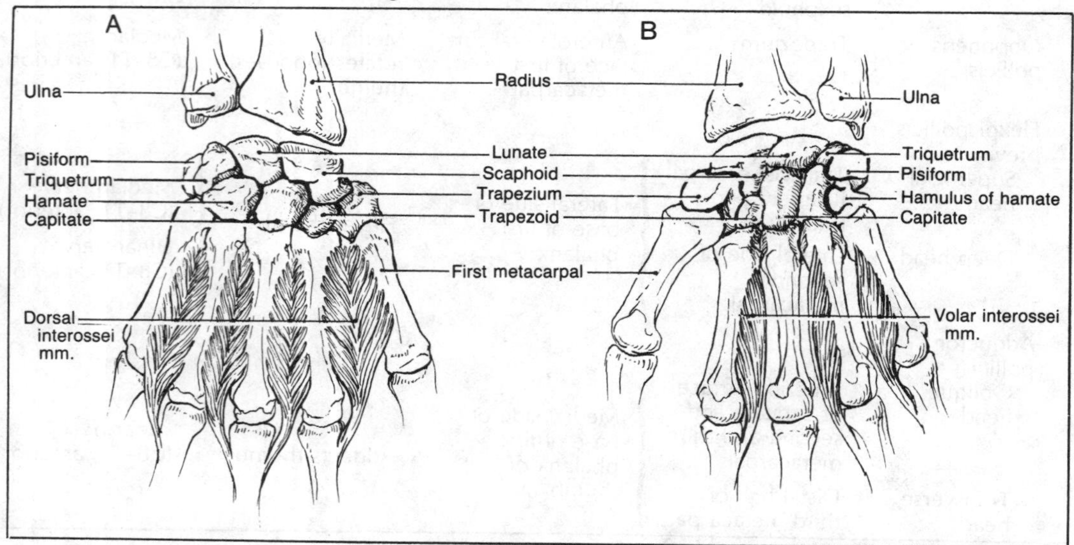

Figure 10-4. *A, The dorsal interossei of the right hand. B, The palmar interossei of the right hand.*

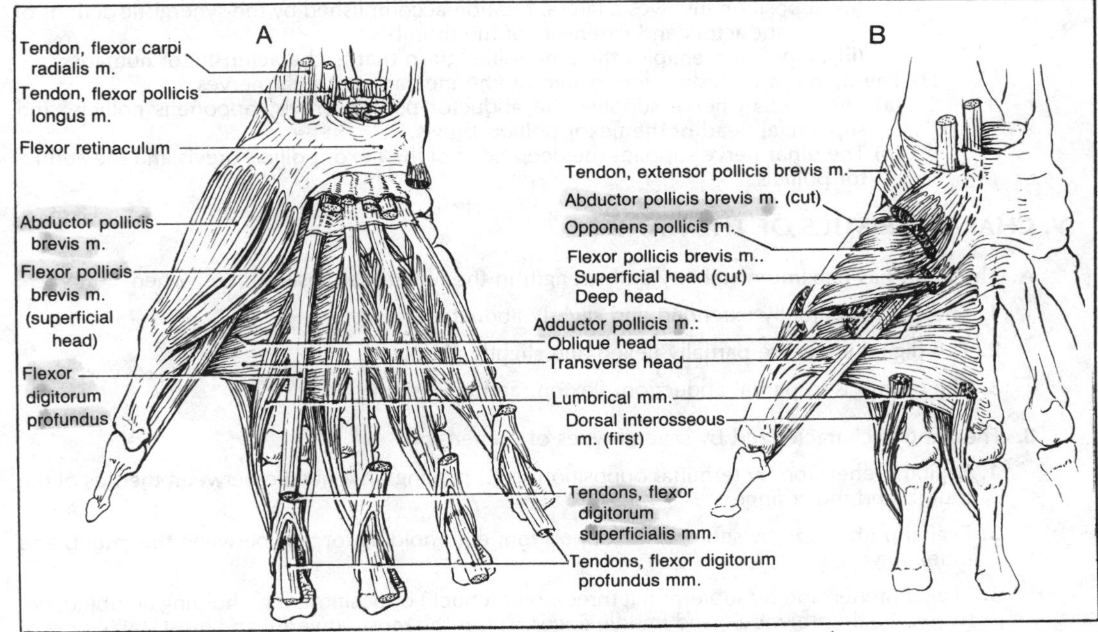

Figure 10-5. *A, The lumbrical muscles. B, The thenar muscles.*

(a) Adduction: The phalangeal insertions are so arranged as to adduct the second, fourth, and fifth digits toward the midline of the hand. The middle finger has no palmar interosseous muscle because the dorsal interossei give it full mobility.

(b) Flexion/extension: The insertions into the opposite sides of the extensor aponeuroses from the dorsal interossei are so arranged that they flex the metacarpophalangeal joint and extend the proximal and distal interphalangeal joints.

(3) The palmar interossei are innervated by the ulnar nerve.

c. The **lumbrical muscles** (Fig. 10-5).

(1) The lumbricales originate from the flexor digitorum profundus tendons deep in the hand.

(2) They insert into the radial side of the extensor aponeurosis of the second through fifth digits (see Fig. 10-3).

(3) They are stronger flexors of the metacarpophalangeal joint than the interossei, due to their longer lever arms. They also extend the proximal and distal interphalangeal joints.

(4) The first and second lumbrical muscles (to the second and third digits, respectively) are innervated by the median nerve; lumbricales 3 and 4 are innervated by the ulnar nerve.

d. The **hypothenar muscles** (see Table 10-1).

(1) Comprising the hypothenar group are the **abductor digiti minimi**, the **flexor digiti minimi brevis**, and the **opponens digiti minimi** (quinti).

(2) The names of these muscles describe their actions.

(3) The hypothenar muscles are all innervated by the ulnar nerve.

e. The **thenar muscles** (see Fig. 10-5 and Table 10-2).

(1) Comprising the thenar group are the **abductor pollicis brevis**, the **opponens pollicis**, the **flexor pollicis brevis**, which has a superficial portion and a deep portion, and the **adductor pollicis**, which has an oblique head and a transverse head.

(2) The actions of the thenar muscles are described by their names.

(a) Flexion/extension.

(i) Flexion of the thumb is accomplished by the flexor pollicis longus and brevis muscles.

(ii) Extension is accomplished by the extensor pollicis longus and brevis muscles.

(b) Abduction/adduction.

(i) Abduction is accomplished by the abductor pollicis longus and brevis muscles.

(ii) Adduction is accomplished by the adductor pollicis muscle.

(c) Opposition/reposition.

(i) Opposition involves a medial rotation at the first carpometacarpal joint accomplished by the opponens pollicis muscle and assisted by the flexor pollicis brevis muscle.

(ii) Reposition involves a lateral rotation accomplished by the synergistic actions of the abductors and extensors of the thumb.

(iii) Opposition enables the pincer-like grasp that is characteristic of humans.

(3) The thenar muscles are innervated by the median and ulnar nerves.

(a) The median nerve supplies the abductor pollicis brevis, opponens pollicis, and superficial head of the flexor pollicis brevis.

(b) The ulnar nerve supplies the deep head of the flexor pollicis brevis and the adductor pollicis.

V. CHARACTERISTICS OF THE HAND

A. The hand has maximum stability and strength in the functional position, i.e., when:

1. The wrist is partially extended and slightly abducted.
2. The digital joints are partially flexed and slightly abducted.
3. The thumb is in partial abduction, flexion, and opposition.

B. The hand is characterized by several types of prehension:

1. Digital prehension by terminal opposition, e.g., picking up a needle between the tips of the thumb and index finger.
2. Digital prehension by subterminal opposition, e.g., holding forceps between the thumb and index finger.
3. Digital prehension by subterminal three-point (chuck) opposition, e.g., holding an object between the thumb, index, and middle finger. This is the most powerful and most stable type of digital prehension.
4. Digital prehension by subterminal-lateral opposition, e.g., holding a card between the thumb and the side of the index finger.
5. Lateral digital prehension, e.g., holding an object between any two fingers.
6. Digital-palmar prehension, e.g., a grasp with opposition of the thumb, as in holding a bat.
7. Palmar prehension, e.g., a grasp with the thumb adducted into the plane of the hand.

VI. VASCULATURE OF THE HAND

A. THE ULNAR ARTERY continues into the hand at the medial side of the wrist (Fig. 10-6).

1. The **deep palmar branch** of the ulnar artery arises just distal to the lateral side of the pisiform bone, where the **ulnar pulse** is palpable.
   a. It passes through the hypothenar compartment.
   b. This branch contributes to the **deep palmar arch** by anastomosing with the terminal branch of the radial artery.
2. The **superficial palmar branch** is the major contributor to the superficial palmar arch.
   a. The **superficial palmar arch** lies just beneath the palmar aponeurosis.
   b. This arch is larger than the deep palmar arch.
   c. The **proper digital artery** to the fifth digit and the **common digital arteries** to the second, third, and fourth digits arise from the superficial palmar arch.
   d. Each common digital artery bifurcates into **phalangeal branches**, which run on the medial and lateral sides of adjacent digits.
   e. The superficial palmar arch anastomoses with the superficial palmar branch of the radial artery.

B. THE RADIAL ARTERY continues into the hand at the lateral side of the wrist (see Fig. 10-6).

1. The **superficial palmar branch** of the radial artery arises at the level of the end of the radius.
   a. It passes through the thenar compartment.
   b. It provides a minor contribution to the superficial palmar arch.
2. The **deep palmar branch** of the radial artery is the major contributor to the **deep palmar arch.**
   a. It courses around the base of the thumb in the "anatomical snuff-box," which is bounded by the extensor pollicis longus and the extensor pollicis brevis tendons and where a **radial pulse** may be palpated.
   b. It passes dorsal to the first metacarpal to gain access to the thenar space where it becomes the deep palmar arch.

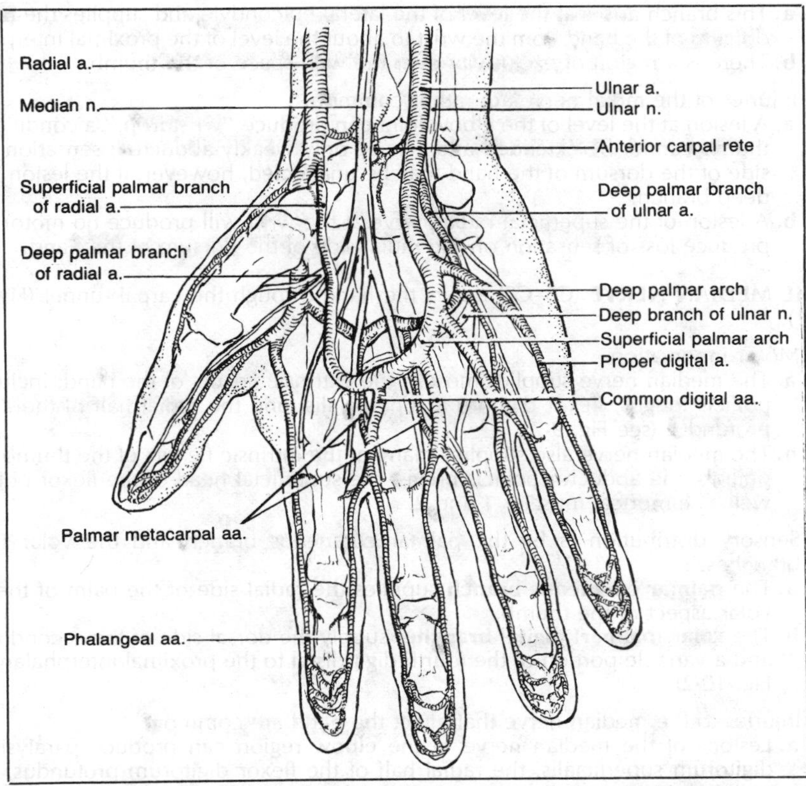

Figure 10-6. *The blood supply to the hand.* The radial and ulnar arteries form or contribute to the superficial and deep palmar arches.

c. Palmar metacarpal arteries arise from the deep arch and join the common digital arteries just proximal to the point where the latter bifurcate into **phalangeal arteries**.

C. THE VENOUS DRAINAGE OF THE HAND

1. The superficial veins provide a rich anastomotic network in the subcutaneous space of the dorsum of the hand. They coalesce laterally to form the **cephalic vein** and medially to form the **basilic vein**.
2. The deep veins, venae comitantes, accompany the radial and ulnar arteries as well as other branches.

D. THE LYMPHATICS OF THE HAND

1. **Superficial lymphatics.**
   - **a.** Most of the lymphatic drainage from the thenar compartment, hypothenar compartment, and digits is toward the dorsal subcutaneous space of the hand, explaining the extreme swelling of this region that accompanies infections of the digits or volar surface.
   - **b.** The lymphatics drain along the cephalic and basilic veins.
   - **c.** They eventually drain through the clavipectoral and supraclavicular nodes.
2. **Deep lymphatics.**
   - **a.** The thenar space, midpalmar space, and tendon sheaths drain through lymphatics that accompany the radial and ulnar vessels.
   - **b.** The deep lymphatics drain through the axillary nodes.

VII. INNERVATION OF THE HAND is provided by the radial, median, and ulnar nerves.

A. THE RADIAL NERVE (C4–T1) [see Fig. 8-4].

1. Motor innervation.
   - **a.** The radial nerve supplies the extrinsic extensors of the hand and thumb.
   - **b.** There are no intrinsic muscles on the dorsum of the hand.
2. Sensory distribution is by the **superficial branch**.

a. This branch arises at the level of the lateral epicondyle and supplies the radial side of the dorsum of the hand from the wrist to about the level of the proximal interphalangeal joint.
b. There is a region of exclusivity over the web space of the thumb.

3. Injuries of the radial nerve are rather common.
   a. A lesion at the level of the elbow joint can produce "wristdrop," a condition in which the thumb cannot be extended and can be only weakly abducted; sensation over the radial side of the dorsum of the hand may be unaffected, however, if the lesion affects only the deep branch.
   b. A lesion of the superficial radial nerve at the wrist will produce no motor deficit but will produce loss of sensation on the radial side of the dorsum of the hand.

B. THE MEDIAN NERVE (C5–C8) enters the hand through the carpal tunnel (Fig. 10-7; see Fig. 9-1).

1. Motor innervation.
   a. The median nerve supplies most of the extrinsic flexors of the hand, including the flexor pollicis longus, flexor digitorum superficialis, and the radial half of the flexor digitorum profundus (see Fig. 9-7).
   b. The median nerve also supplies many of the intrinsic flexors of the thumb (the opponens pollicis, the abductor pollicis brevis, the superficial head of the flexor pollicis brevis), as well as lumbrical muscles 1 and 2.

2. Sensory distribution is by the palmar cutaneous branch and the volar (palmar) digital branches.
   a. The **palmar cutaneous branch** supplies the radial side of the palm of the hand and the volar aspect of the thumb.
   b. The **volar (palmar) digital branches** supply the dorsal side of the second and third digits and a variable portion of the fourth digit distal to the proximal interphalangeal joints (see Fig. 10-2).

3. Injuries to the median nerve that affect the hand are common.
   a. Lesions of the median nerve in the elbow region can produce paralysis of the flexor digitorum superficialis, the radial half of the flexor digitorum profundus, and the flexor pollicis longus, as well as the muscles of the thenar compartment and lumbricales 1 and 2. In addition, they can cause sensory loss over the radial side of the palm and digits, lateral to the center line of the ring finger.
   b. Compression of the median nerve within the carpal tunnel due to inflammation or displacement of the lunate bone can produce paralysis of the muscles of the thenar compartment (recurrent branch), paralysis of lumbricales 1 and 2 (muscular twigs), and loss of sensation in the distal palm and digits (volar digital branches); but it will not produce a loss of sensation over the proximal palm (palmar cutaneous branch).
   c. Laceration of the **recurrent branch**, probably the most commonly injured nerve, produces loss of thenar muscles (weakness of flexion and loss of opposition) but no sensory deficit.

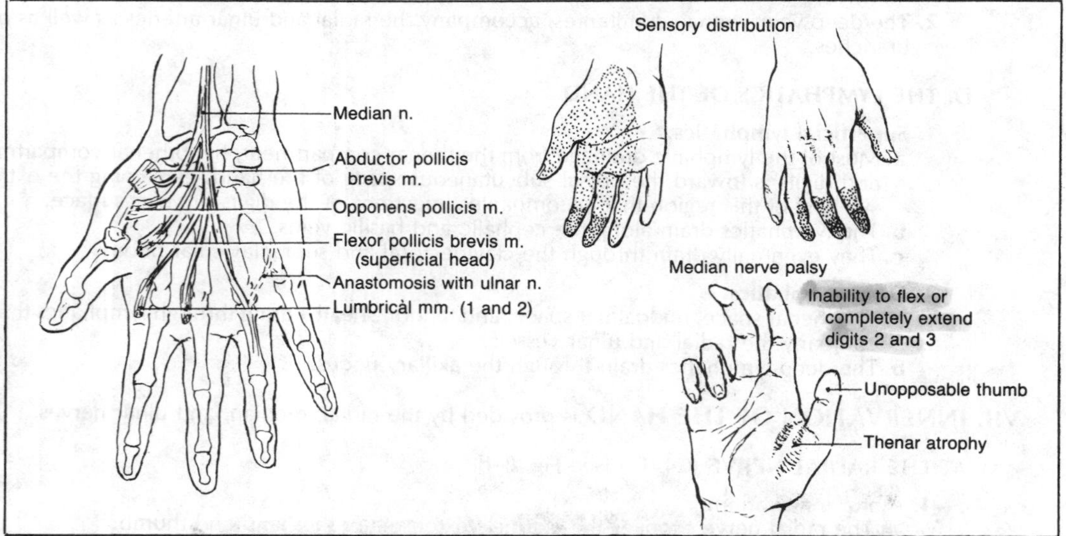

Figure 10-7. *The median nerve.* The muscles innervated by branches of the median nerve are indicated. Also indicated is the region of sensory distribution of the palmar cutaneous branch and the volar digital branches.

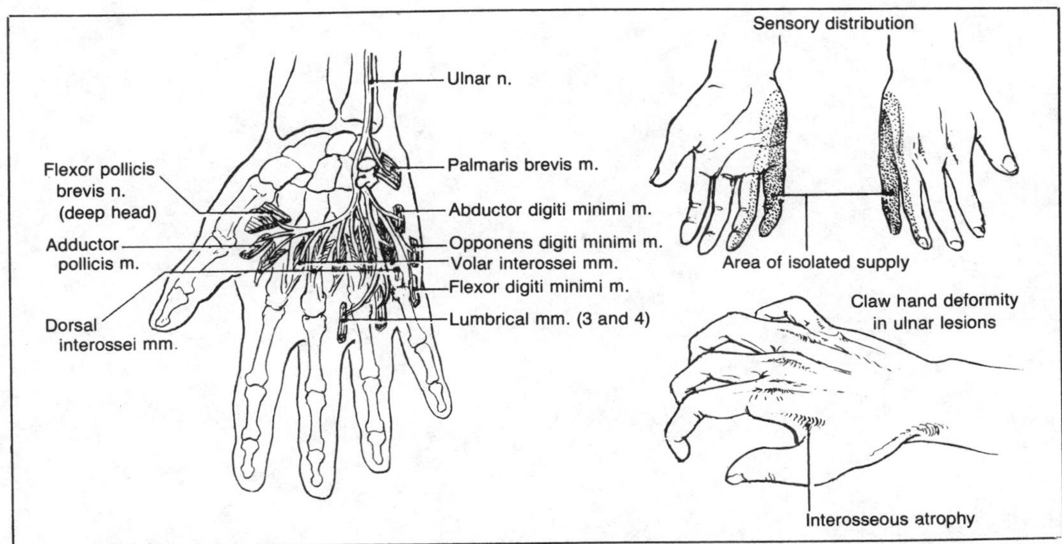

Figure 10-8. *The ulnar nerve.* The muscles innervated by branches of the ulnar nerve are indicated. Also indicated is the region of sensory distribution.

C. **THE ULNAR NERVE** (C8–T1) enters the hand with the ulnar artery (Fig. 10-8; see Fig. 9-1).

1. Motor innervation.
   a. The ulnar nerve supplies the ulnar half of the flexor digitorum profundus (see Fig. 9-3).
   b. The ulnar nerve also supplies most of the intrinsic muscles of the hand (the hypothenar muscles, the dorsal interossei, the palmar interossei, lumbricales 3 and 4, the adductor pollicis, and the deep head of the flexor pollicis brevis).

2. Sensory distribution is by the **dorsal branch**, the **dorsal digital branches**, and the **volar digital branches**.
   a. The **dorsal branch** and **dorsal digital branches** of the ulnar nerve supply the ulnar side of the dorsum of the hand, the ulnar side of the palm, the dorsum of the fifth digit, and a variable portion of the dorsum of the fourth digit.
   b. The **volar digital branches** supply the ulnar side of the palm, the fifth digit, and a variable portion of the fourth digit.

3. Injuries of the ulnar nerve are common.
   a. Lesions of the ulnar nerve in the region of the elbow will produce "ulnar claw hand."
      (1) Paralysis of the ulnar half of the flexor digitorum profundus, lumbricales 3 and 4, the dorsal and palmar interossei, and the hypothenar muscles results. When the metacarpophalangeal joints are extended in affected patients the paralyzed digits are semiflexed by the flexor digitorum superficialis.
      (2) Sensory deficit over the ulnar side of the dorsal and palmar aspects of the hand also is produced.
   b. Laceration of the ulnar nerve at the wrist leaves the innervation to the radial side of the digitorum profundus intact. Adduction of the thumb is lost, however, and a piece of paper cannot be held firmly between the thumb and lateral side of the palm.
   c. Coronary artery disease can refer pain to the T1 and T2 dermatomal components of the ulnar nerve.

D. Lesions of the radial, median, and ulnar nerves produce distinct deficits in the movements of the thumb.

1. Radial nerve palsy results in inability to extend and the ability to only weakly abduct the thumb, as well as loss of sensation in the dorsal web space.

2. Median nerve palsy results in inability to oppose the thumb, so that digital prehension by terminal opposition is lost.

3. Ulnar nerve palsy results in inability to adduct the thumb, so that digital prehension by subterminal opposition is lost.

4. With time, atrophy becomes obvious in the affected region of the hand.

Part II Upper Extremity

STUDY QUESTIONS

Directions: Each question below contains five suggested answers. Choose the **one best** response to each question.

1. Which of the following statements concerning the teres major muscle, a contributor to the posterior axillary fold, is true?

(A) It is a major contributor to the stability of the posterior aspect of the shoulder joint
(B) It divides the axillary artery into three parts
(C) It inserts on the humerus just distal to the infraspinatus insertion
(D) It is active in adduction of the glenohumeral joint
(E) It is innervated by the same nerve that innervates the deltoid muscle

Questions 2–6

A new medical clerk is asked to place an intravenous line in the right arm of a patient.

2. The clerk should know that the superficial veins of the arm are relatively consistent in their anatomic relationship and that all of the following statements concerning these veins are true EXCEPT

(A) at the level of the axilla, the basilic vein is joined by the cephalic vein to form the axillary vein
(B) the basilic vein runs along the medial aspect of the forearm
(C) the cephalic vein originates on the radial side of the dorsum of the hand
(D) the median cubital vein links the cephalic and basilic veins in the vicinity of the elbow joint
(E) the median cubital vein is separated from the brachial artery by the bicipital aponeurosis

3. In an attempt to distend the veins, the clerk applies a tight tourniquet approximately 2 inches proximal to the proposed site of venipuncture, but the veins do not distend when the patient's hand is vigorously flexed. The most probable reason for the failure of the veins to distend is

(A) that arterial flow to the forearm is obstructed
(B) that deep venous return has not been obstructed
(C) that superficial venous return has not been obstructed
(D) that the venae comitantes are abnormally absent
(E) none of the above

4. After several attempts, the veins become visible and a vessel in the cubital fossa is penetrated. A small amount of blood is aspirated into the line to ensure proper placement. The blood is bright red, which makes it likely that the needle is in the

(A) brachial artery
(B) deep brachial artery
(C) median cubital vein
(D) venae comitantes
(E) cephalic vein

5. If the clerk changed the site of venipuncture to a slightly more medial one, the patient might experience pain radiating down the ventral surface of the forearm and hand, including the thumb, index finger, and middle finger; the nerve penetrated in this case would most likely be

(A) a branch of the musculocutaneous nerve
(B) the median antebrachial cutaneous nerve
(C) the median nerve
(D) the radial nerve
(E) the ulnar nerve

6. Upon successful placement of the intravenous line, the clerk is relieved but fails to notice an approximately 1.5 ml air bubble in the line entering the vein. The region at highest risk for air embolus is

(A) the brain
(B) the hand
(C) the heart
(D) the lungs
(E) none of the above

(end of group question)

7. All of the following structures pass deep to the transverse carpal ligament EXCEPT the

(A) flexor digitorum superficialis tendon
(B) flexor digitorum profundus tendon
(C) flexor pollicis longus tendon
(D) median nerve
(E) ulnar artery

8. All of the following statements concerning the blood supply of the hand are true EXCEPT

(A) the arteries of the hand are accompanied by venae comitantes
(B) the blood supply reaches the long flexor tendons within the synovial sheaths through vincula
(C) the common digital arteries are branches of the deep palmar arch
(D) the deep and superficial palmar arches are supplied by the radial and ulnar arteries
(E) the dorsal carpal arch (rete) receives blood from the carpal branches of the radial, ulnar, and posterior interosseus arteries

Directions: Each question below contains four suggested answers of which **one or more** is correct. Choose the answer

A if **1, 2, and 3** are correct
B if **1 and 3** are correct
C if **2 and 4** are correct
D if **4** is correct
E if **1, 2, 3, and 4** are correct

9. With the upper limb adducted, hanging by the side and holding a heavy suitcase, support at the glenohumeral joint to prevent downward displacement is provided by the

(1) short head of the biceps brachii muscle
(2) coracohumeral ligament
(3) supraspinatus muscle
(4) coracoacromial ligament

10. Correct statements concerning the subacromial bursa include which of the following?

(1) It frequently communicates with the subdeltoid bursa
(2) It normally communicates with the synovial cavity of the glenohumeral joint
(3) It underlies the coracoacromial ligament
(4) It permits movement between the scapula and the thoracic wall

11. True statements concerning the pectoralis minor muscle include

(1) it attaches to the acromion process of the scapula
(2) it is an adductor and medial rotator of the humerus
(3) it is innervated by the middle subscapular (thoracodorsal) nerve
(4) it crosses the cords of the brachial plexus

12. Following complete severance of the musculocutaneous nerve, some weak flexion of the elbow is possible through contraction of the

(1) flexor carpi radialis
(2) brachioradialis
(3) flexor carpi ulnaris
(4) ulnar head of the pronator teres

13. Correct statements concerning the lunate bone include which of the following?

(1) It can compress the median nerve if displaced anteriorly
(2) It provides an attachment for the transverse carpal ligament
(3) It articulates with the fibrocartilage disk
(4) It is a component of the carpometacarpal joint

Questions 14–16

A tailor complains of difficulty picking up a needle and holding it while sewing. Physical examination reveals no sensory deficit. The patient is then tested for motor function of the second digit, as shown in the illustration below.

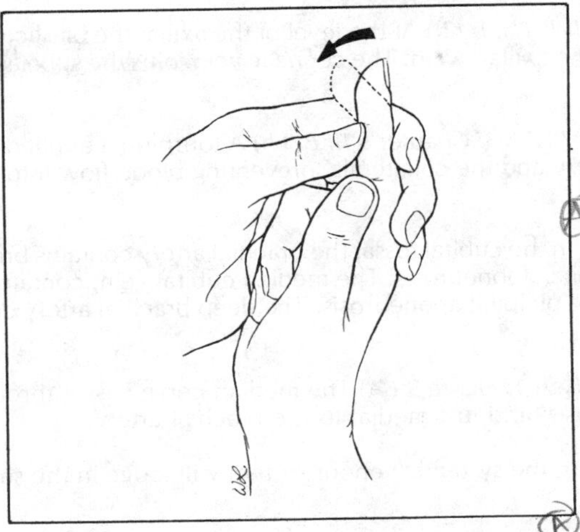

14. The muscles being tested in the illustration include the

(1) lumbrical 2
(2) flexor digitorum superficialis
(3) palmar interossei
(4) radial portion of the flexor digitorum profundus

15. If paralysis or paresis (weakness) of the terminal phalanx of the second digit were detected in the patient described, one might also expect to find

(1) paralysis or paresis of the fourth digit
(2) atrophy of the thenar eminence
(3) complete paralysis of the thumb
(4) weakness of pronation

16. Which of the following nerves supply muscles that produce antagonistic actions to that illustrated?

(1) Ulnar nerve
(2) Anterior interosseous nerve
(3) Median nerve
(4) Radial nerve

(end of group question)

17. In the distal portion of the upper extremity, a pulse may be palpated

(1) in the "anatomical snuff-box"
(2) at the radial side of the tendon of the flexor carpi radialis muscle
(3) between the tendons of the extensor pollicis longus and brevis muscles
(4) at the radial side of the pisiform bone

18. Reposition of the thumb is accomplished by the

(1) abductor pollicis longus
(2) extensor pollicis brevis
(3) extensor pollicis longus
(4) adductor pollicis

19. An infection of the synovial sheath within the fifth digit (ulnar bursa) may

(1) spread distally into the thumb via the radial bursa
(2) spread proximally through the carpal tunnel into the distal forearm
(3) result in ischemic necrosis of the flexor tendons to the fifth finger
(4) spread distally into adjacent digits

20. Laceration of the recurrent branch of the median nerve will paralyze which of the following muscles of the thumb?

(1) Superficial head of the flexor pollicis brevis
(2) Abductor pollicis longus
(3) Abductor pollicis brevis
(4) Deep head of the flexor pollicis brevis

ANSWERS AND EXPLANATIONS

1. The answer is D. (*Chapter 6 Table 6-2*) The teres major muscle, innervated by the lower subscapular nerve, inserts onto the medial lip of the bicipital groove of the humerus and thus is an adductor of the arm. The tendon of the pectoralis minor muscle, not the teres major muscle, divides the axillary artery into three parts.

2. The answer is A. [*Chapter 7 IV C; Chapter 9 III B 1 a, b (2)*] At the level of the axilla, the basilic vein and the brachial venae comitantes join to form the axillary vein. The cephalic vein joins the subclavian vein just distal to the clavicle.

3. The answer is A. (*Chapter 5 III A 3 b; Chapter 7 IV A 3; Chapter 9 III B 3 b*) A tourniquet applied too tightly to the arm will compress the brachial artery and the collaterals, preventing blood flow into the forearm and, as a consequence, venous return.

4. The answer is A. (*Chapter 7 IV A 3; Figure 7-4*) In the cubital fossa, the brachial artery contains bright red oxygenated blood and lies beneath the bicipital aponeurosis. The median cubital vein, containing darker deoxygenated blood, lies superficial to the bicipital aponeurosis. The deep brachial artery does not pass through the cubital fossa.

5. The answer is C. (*Chapter 7 V B 2; Chapter 9 IV A 1; Figure 9-6A*) The median nerve passes through the cubital fossa, deep to the bicipital aponeurosis, and just medial to the brachial artery.

6. The answer is D. (*Chapter 5 I A*) An embolus in the systemic venous system will lodge in the small vessels of the lungs.

7. The answer is E. (*Chapter 9 II A 4 a, b*) The ulnar nerve and ulnar artery pass deep to the volar carpal ligament but superficial to the transverse carpal ligament. The median nerve is the only major neurovascular structure that passes through the carpal tunnel.

8. The answer is C. (*Chapter 9 III A 1 a (3), 2 a (4); Chapter 10 IV D 1 e (1); VI A 1–2, B 1–2, C 2*) The deep palmar arch is supplied by the radial and ulnar arteries and gives off palmar metacarpal arteries. These arteries join the common digital arteries, which arise from the *superficial* palmar arch, before the latter bifurcate into phalangeal arteries.

9. The answer is A (1, 2, 3). (*Chapter 6 II D 4 b (2), c, d; Table 6-2*) The dead weight of the arm is supported primarily by the coracohumeral ligament. In addition, the supraspinatus muscle, coracobrachialis, the short head of the biceps brachii muscle, and the long head of the triceps brachii all provide dynamic support at the glenohumeral joint. The coracoacromial ligament prevents *superior* displacement of the humerus at the glenohumeral joint.

10. The answer is B (1, 3). (*Chapter 6 II D 4 e; Figure 6-4*) The subacromial and subdeltoid bursae may be separate, may communicate, or may form one large bursa. These bursae lie deep to the deltoid muscle, extend inferior to the coracoacromial ligament, and lie superior to the supraspinatus muscle. Only in dislocations of the glenohumeral joint, when the joint capsule is torn, will the synovial cavity of the glenohumeral joint communicate with the subacromial or subdeltoid bursa.

11. The answer is D (4). (*Chapter 6 III D 2 d; Table 6-1*) The pectoralis minor muscle crosses the cords of the brachial plexus as it attaches to the coracoid process of the scapula. It therefore has no direct action on the humerus. It is innervated by the medial pectoral nerve of the medial cord of the brachial plexus.

12. The answer is A (1, 2, 3). [*Chapter 7 III C 3 a (1)*] Aside from the brachialis and biceps brachii (both innervated by the musculocutaneous nerve), the brachioradialis (radial nerve), flexor carpi radialis (median nerve), and flexor carpi ulnaris (ulnar nerve) all take origin from the humerus and can weakly flex the elbow joint. The ulnar head of the pronator teres does not cross the humeroulnar joint.

13. The answer is B (1, 3). (*Chapter 8 II A 1 d, B 1 b; III A 1 a, 2 f*) The lunate bone articulates with the triangular fibrocartilage at the radiocarpal joint—much more so in abduction than in adduction. When displaced anteriorly, the lunate bone can impinge on the long flexor tendons of the digits and compress the median nerve, producing carpal tunnel syndrome.

14. The answer is D (4). (*Chapter 9 II B 3 b, c; Table 10-1*) The radial half of the flexor digitorum profundus flexes the terminal phalanx of the second and third digits. The flexor digitorum superficialis flexes the middle phalanx of the second through fifth digits. The lumbricales and interossei are extensors of the middle and distal phalanges.

15. The answer is C (2, 4). [*Chapter 9 IV A 1 a, 2 a; Chapter 10 VII B 3 a*] The radial half of the flexor digitorum profundus, as well as the flexor digitorum superficialis and flexor pollicis longus, are innervated by branches of the median nerve in the forearm. A lesion that affected these muscles would most likely also affect the pronators teres and quadratus, as well as some of the thenar muscles of the hand. Because the flexor digitorum profundus to the fourth and fifth digits is innervated by the ulnar nerve, rather than the median, paralysis of the fourth or fifth digit would not be expected to accompany paresis of the second digit; because some thumb muscles are innervated by the radial and ulnar nerves, rather than the median, injury of the median nerve would not cause complete paralysis of the thumb.

16. The answer is B (1, 3). (*Chapter 10 IV D 2 a–c; Table 10-1*) Extension of the terminal phalanges is accomplished primarily by the dorsal and palmar interossei, which are innervated by the ulnar nerve, as well as the lumbrical 2 innervated by the median nerve. The extensor digitorum communis and indicis proprius, which are innervated by the radial nerve, primarily extend the metacarpophalangeal joint.

17. The answer is E (all). (*Chapter 9 III A 1 b, 2 b; Chapter 10 VI B 2 a*) At the wrist, the radial pulse may be felt between the flexor carpi radialis and brachioradialis muscles, as well as in the "anatomical snuff-box" (i.e., between the tendons of the extensor pollicis longus and brevis). The ulnar pulse is felt on the radial side of the insertion of the flexor carpi ulnaris into the pisiform bone.

18. The answer is A (1, 2, 3). (*Chapter 10 II D 2 e (2); Table 10-2*) Reposition of the thumb, a combination of abduction and extension accompanied by lateral rotation, is accomplished primarily by the abductor pollicis longus and brevis muscles, as well as by the extensor pollicis longus and brevis muscles. Opposition of the thumb is accomplished by flexion accompanied by medial rotation.

19. The answer is A (1, 2, 3). (*Chapter 10 IV D 1 e (2), h*) An infection within the synovial sheath of the tendons of the fifth digit may compress the vincula and cause ischemic necrosis of the tendons. It also can spread along the synovial sheath through the carpal tunnel and into the distal forearm. If the ulnar and radial bursae communicate, infection may spread distally into the thumb. Infection of the fifth digit cannot, however, spread distally into the tendon sheaths of the second, third, and fourth digits.

20. The answer is B (1, 3). (*Chapter 10 IV D 2 e (3) (a); VII B 3 c; Table 10-1*) The recurrent branch of the median nerve in the palm innervates the muscles of the thenar compartment except the deep head of the flexor pollicis brevis, which is innervated by the ulnar nerve. The abductor pollicis longus is innervated by a branch of the median nerve in the forearm.

Part III
Thoracic Region

11
The Thoracic Cage

I. DEFINITIONS

A. THE THORAX is that region of the trunk delineated by the extent of the rib cage and diaphragm.

1. It communicates with:
   - **a.** The neck via the **superior thoracic aperture** or the **thoracic inlet.**
   - **b.** The abdomen via the **inferior thoracic aperture** or the **thoracic outlet**, with the **diaphragm** extending across and closing the thoracic outlet.
2. It includes the heart and lungs and neurovascular connections between the neck, upper extremities, and abdomen.

B. THE PECTORAL REGION is the portion of the anterior chest wall that provides the origin for the muscles that attach to the pectoral girdle and upper extremities.

II. SURFACE LANDMARKS

A. THE SPINES OF THE 12 THORACIC VERTEBRAE are palpable and are used to describe thoracic levels by number.

B. THE STERNUM may be palpated over its entire length. Particularly prominent are the:

1. **Jugular notch of the manubrium**, which approximates an anterior boundary between the thorax and the neck.
2. **Sternoclavicular joint** where the clavicle articulates with the sternum.
3. **Sternal angle (angle of Louis)**, which is the junction between the manubrium and the body of the sternum and is located at the articulations of the second ribs.
4. **Xiphoid process**, which, along with the costal margin, defines the **infrasternal angle** (notch).

C. THE 12 RIBS (except for the 1st, which underlies the clavicles) and the associated intercostal spaces are used to describe thoracic locations. The rib cage consists of:

1. Seven pairs of "true" ribs (1–7), which articulate with the vertebrae and the sternum.
2. Five pairs of "false" ribs (8–12), which articulate with the vertebrae and the immediately superjacent costal cartilage to form the anterior **costal margin**. The last two ribs (11–12), which articulate with vertebrae only, are sometimes called "floating" ribs.

D. THE PECTORAL GIRDLE consists of the clavicles and the scapulae.

1. The **clavicles** articulate with the manubrium at the **sternoclavicular joint** and with the scapulae at the **acromioclavicular joint**. The **midclavicular line**, an important landmark, passes vertically through the center of the clavicles and approximately through the nipple in men and variably medially to the nipple in women.
2. The **scapula** is the most prominent dorsal landmark of the thorax and has several palpable features.
   - **a.** The **acromion process** defines the superior surface of the shoulder.
   - **b.** The **coracoid process** represents the third bone of the pectoral girdle that fuses with the scapulae.
   - **c.** The **spine** of the scapula is obvious on the dorsal thoracic wall.

d. The **apex** (lowest point) of the scapula is usually located at the level of the 7th rib or 7th intercostal space.

E. THE MAMMARY GLAND is the most prominent superficial structure of the anterior thoracic wall. The **nipple** is located at the 4th intercostal space in men and variably lower in women.

F. Internal thoracic organs are not palpable, but during physical examination they may be outlined with reference to surface anatomy by percussion and auscultation.

III. THE THORACIC CAGE.

The primary function of the thoracic cage is respiratory, and thus deformities of or injuries to the thoracic cage may impair ventilation of the lungs. Its secondary function is protection of vital thoracic organs. The thoracic cage exhibits somatic segmentation.

A. The bony elements of the thoracic cage consist of the **thoracic vertebrae**, the **sternum**, and the **ribs** (Figs. 11-1 and 11-2).

B. THE THORACIC VERTEBRAE. Each of the 12 thoracic vertebrae has two main portions (see Fig. 19-3).

1. The **body** of each vertebra supports the mass of that portion of the person superior to it.
- **a.** It has facets for diarthrodial articulation with the heads of the immediately superior and inferior ribs (with some exceptions) near the junctions of the pedicles.
- **b.** It articulates with adjacent vertebral bodies through the **annulus fibrosus**, which is an amphiarthrosis (a fibrocartilaginous joint with no capsule). The enclosed gelatinous **nucleus pulposus** distributes pressure evenly throughout the joint.

2. The **neural arch** forms the **neural canal** and contains the spinal cord. It comprises:
- **a.** Paired **pedicles**, which arise from the posterior aspect of the vertebral body and have facets for diarthrodial intervertebral articulation.
  - **(1)** The **superior articular facets** face primarily posteriorly.
  - **(2)** The **inferior articular facets** face primarily anteriorly.
- **b.** Paired **laminae**, which continue from the pedicles and fuse posteriorly.
- **c.** A **spinous process**, which arises where the laminae meet.

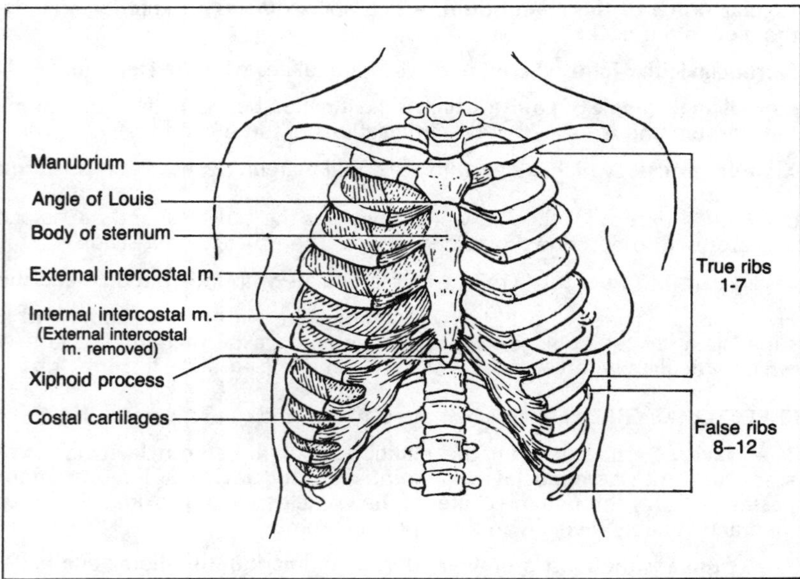

Figure 11-1. *The musculoskeletal thoracic cage.* The intrinsic thoracic musculature lies between the 12 pairs of ribs. The external intercostal muscles terminate at the costochondral junction, but the layer continues to the sternum as the external intercostal membrane. The internal intercostal layer is shown in the right 4th and 5th intercostal spaces.

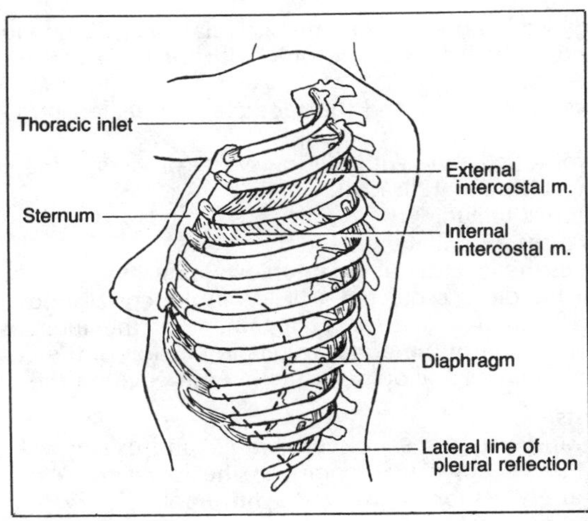

Figure 11-2. The differences in the directionality and extent of the external and internal intercostal muscles are shown in the 2nd and 3rd intercostal spaces, respectively. Also depicted is the normal end-expiratory position of the diaphragm.

d. Paired **transverse processes**, which arise at the junction of the pedicles and the laminae and have articular facets for the tubercles of ribs 1–10.

e. Paired **costal processes**, which form **ribs** in the thoracic region.

3. The **intervertebral canals** transmit the **spinal nerves** and are formed by:

a. The inferior border of the pedicle of the upper vertebra.

b. The superior border of the pedicle of the lower vertebra.

C. THE STERNUM. Having developed as a segmental structure, the adult sternum retains segmentation, which divides it into the **manubrium**, the **corpus**, and the **xiphoid**, and allows for movement during respiration (see Fig. 11-1).

1. The **manubrium** (handle):

a. Is anchored fairly firmly on each side by articulation with the clavicle at the sternoclavicular joint and the synchondrosis of the 1st rib (see Fig. 6-1).

b. Can be drawn upward slightly by the action of the **sternocleidomastoid** and **scalene** muscles during extreme inspiratory effort.

c. Articulates with the body of the sternum at the **sternal angle (angle of Louis)** or **manubriosternal synchondrosis**, which acts as a slight hinge allowing for anteroposterior expansion of the rib cage during inspiration.

d. Articulates with the superior facets of the second rib at the sternal angle.

2. The **corpus** or **body**:

a. Articulates with the inferior facets of the 2nd rib at the sternal angle, which is a useful landmark in counting ribs during physical examination.

b. Articulates with the 3rd–6th costal cartilages.

c. Is a preferred site for bone-marrow aspiration, since there is very little fat over this bone regardless of body build.

d. Is sensitive to nociceptive stimulation, and thus firm rubbing of the sternum with one's knuckles can be used to rouse a stuporous patient.

e. May be split (sternotomy) for surgical access to the middle mediastinum and superior mediastinum.

3. The **xiphoid process** (tip):

a. Articulates with the sternum at the **xiphisternal synchondrosis**, which usually fuses in the elderly.

b. Can be palpated in the epigastrium at the infrasternal angle.

D. THE RIBS. The 12 pairs of ribs form the main portion of the thoracic cage (see Fig. 11-2).

1. The true ribs (1–7) become progressively longer; the false ribs (8–12) become progressively shorter.

2. The ribs slope considerably downward posteriorly, before turning slightly upward anteriorly.

3. Each rib may be divided into several regions.

a. An expanded **head** usually articulates with its corresponding vertebral body and the immediately superjacent vertebral body via inferior and superior articular facets. Exceptions: the 1st, (sometimes 10th), 11th, and 12th ribs articulate only with their corresponding thoracic vertebrae.

b. The **neck** region, the portion between the head and the tubercle, is directed posterolaterally.

c. The **tubercle** of the rib has a facet for articulation with the transverse process of the corresponding vertebra (exceptions: the 11th and 12th ribs).

d. The **shaft** of the rib has several features, including:

(1) The **costal angle**, where the rib turns anterolaterally.

(2) The **costal groove**, which indicates the location of the **intercostal neurovascular bundle**. Although the neurovascular bundle is protected as it lies in the costal groove, a fractured rib may injure the contained structures. Enlargement of the intercostal arteries, such as occurs with coarctation of the aorta, results in erosion of the costal grooves, defects apparent on x-ray film as scalloping of the inferior edges of the ribs.

4. An anomalous 13th rib sometimes exists.

a. A **cervical rib** may arise from the transverse process of vertebra C7 and usually will articulate with the anterior third of the 1st rib; it may compress the lower trunk of the brachial plexus and the subclavian artery (**"thoracic outlet" syndrome**).

b. A **lumbar** ("gorilla") **rib** may arise from vertebra L1.

5. The **costal cartilage** represents the anterior portion of the rib that fails to ossify.

a. The costal cartilages articulate with the manubrium and the body of the sternum in the following manner:

(1) The 1st rib articulates with the manubrium below the sternoclavicular joint by synchondrosis (little or no movement is possible).

(2) The 2nd rib articulates with the sternum at the sternal angle, with one synovial joint on the manubrium and another on the body of the sternum.

(3) The 3rd to the 6th or 7th ribs articulate with the body of the sternum with synovial joints.

(4) The 7th or the 8th to 10th ribs articulate with the preceding costal cartilages with synovial joints.

(5) The 11th and 12th ribs (and 13th, if present) end in the musculature of the abdominal wall (floating ribs).

b. During *in*spiratory movements, the costal cartilages are bent upward and twisted. This flexion and torsion require energy which is subsequently released and utilized to assist expiratory effort (**extrinsic elastic recoil**).

c. Costal cartilages may calcify in old age; the loss of resiliency will affect respiration and predispose to fractures.

6. The axis of rotation of the ribs and degrees of freedom of movement.

a. Because there are two functional joints between ribs 1–10 and the corresponding vertebra (one at the body and another at the transverse process), movement can occur about only one **axis of rotation** drawn through the centers of the two joints (1 and 2 in Fig. 11-3). Movement about one axis of rotation is described as **one degree of freedom**.

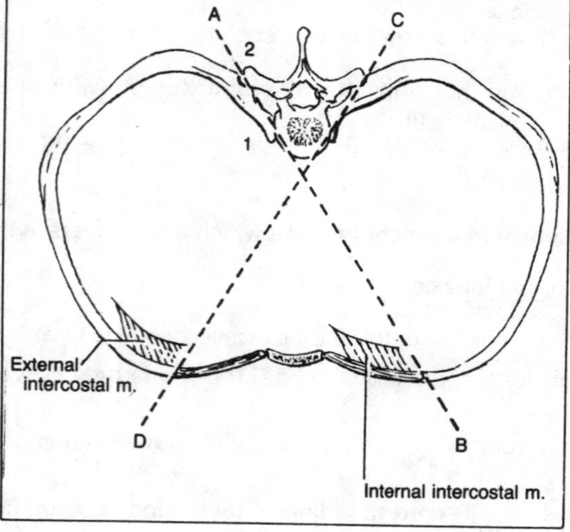

Figure 11-3. *Axis of rib rotation.* Rotation of the ribs occurs through two joints, one at the body of the vertebra (*1*) and the other at the transverse process (*2*), which define an axis of rotation (*AB*) that passes through the contralateral costochondral junction. Because the axes of the left and right ribs cross, the resultant movement of the shared arcs elevates the sternum.

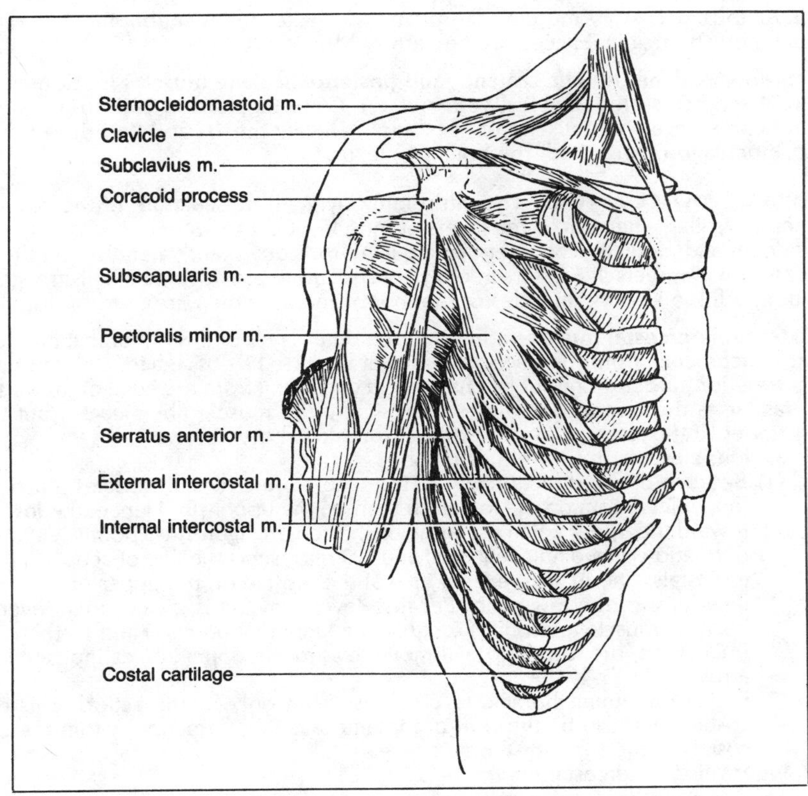

Figure 11-4. *Extrinsic thoracic musculature.* The pectoralis major muscle has been removed. In addition to their nonrespiratory actions, the pectoralis minor muscle elevates the ribs and the sternocleidomastoid muscle elevates the manubrium.

- **b.** The defined axis of rotation extends diagonally across the midline and through the contralateral costochondral junction.
- **c.** External intercostal muscles on one side of the body (from D to C in Fig. 11-3) plus any contralateral intercostal muscles that have the same fiber direction and are on the same side of the axis of rotation (from A to B in Fig. 11-3) will act synergistically to elevate the ribs.

IV. MUSCULATURE OF THE THORAX

A. EXTRINSIC MUSCULATURE. The extrinsic musculature of the thorax stabilizes and moves the pectoral girdle, upper limbs, and neck. Some of these muscles are used to a significant degree during extreme respiratory effort (Fig. 11-4).

1. Pectoralis major muscle (see Fig. 6-2).
- **a.** Attachments: from the proximal third of the clavicle (clavicular head) and from the sternum and costal cartilages of ribs 1–6 (sternal head) to the proximal humeral shaft.
- **b.** Action: accessory muscle of inspiration; elevates ribs 1–6 when the scapula is retracted and the humerus is extended posteriorly.
- **c.** Innervation: lateral pectoral nerve (clavicular head) and medial pectoral nerve (sternal head).

2. Pectoralis minor muscle.
- **a.** Attachments: from ribs 2–5 to the **coracoid process** of the scapula.
- **b.** Action: accessory muscle of inspiration; elevates ribs 2–5 when the scapula is retracted.
- **c.** Innervation: medial pectoral nerve.

3. Sternocleidomastoid muscle (a neck muscle).
- **a.** Attachments: from the mastoid process of the cranium to the manubrium and the proximal third of the clavicle.

b. Actions: accessory muscle of inspiration; elevates the manubrium.
c. Innervation: spinal accessory nerve (CN XI).

4. **Anterior scalene, middle scalene, and posterior scalene muscles** (neck muscles).
a. Attachments: from cervical vertebrae C1–C6 to the 1st and 2nd ribs.
b. Action: accessory muscles of inspiration; elevate the 1st and 2nd ribs.
c. Innervation: twigs from the brachial plexus.

B. INTRINSIC MUSCULATURE. The intrinsic (body wall) musculature of the thorax and the abdomen is similar. There are three layers of muscles. On the anterior body wall, the outer layer runs down and medially, the intermediate layer runs down and laterally, and the deepest layer tends to run transversely. This "three-ply construction" provides excellent strength through bias bonding. All the intrinsic muscles of the thorax and abdomen are used in respiratory effort.

1. **External intercostal muscles** constitute the outer layer of three thoracic musculature layers.
a. Attachments: from the inferior border of the first 11 ribs, diagonally and inferiorly to the superior border of the subjacent rib, and extending from the edge of the vertebral column as far as the costochondral junction, where the muscle fibers cease, but the layer continues to the sternum as the **external intercostal membrane.**
b. Action: elevates ribs from inspiration.
(1) Because of the directionality of the external intercostal muscles, each fascicle has a longer lever arm on the lower rib than on the upper rib. Hence, the lower rib is raised toward the upper when the muscles shorten. In Figure 11-5 points A and B are the axes of rotation of two adjacent ribs; line CD represents the line of action of the external intercostals; line AC is the lever arm of the muscle on the upper rib; and line BD is the lever arm of the lower rib. Because lever arm BD is greater than lever arm AC and because muscle tension (T) is equal at attachment points C and D, then (T) (BD) > (T) (AC). Thus, the lower rib will move toward the upper rib, as indicated by the heavy arrow.
(2) Each subsequent thoracic level is raised not only by the action of its external intercostals, but also by the sum of the amounts of contraction within the external intercostals of the superior thoracic levels.
c. Innervation: intercostal nerves 1–11.

2. The **internal intercostal muscles**, the intermediate layer of the thoracic musculature with a fiber direction generally opposite to that of the external intercostals, can be divided into two functionally distinct muscle groups.
a. The **internal inter*costal* portion** of the intermediate layer.
(1) Attachments: from the superior border of ribs 2–12, diagonally and superiorly to the inferior border of the superjacent rib, and extending from the costochondral junction as

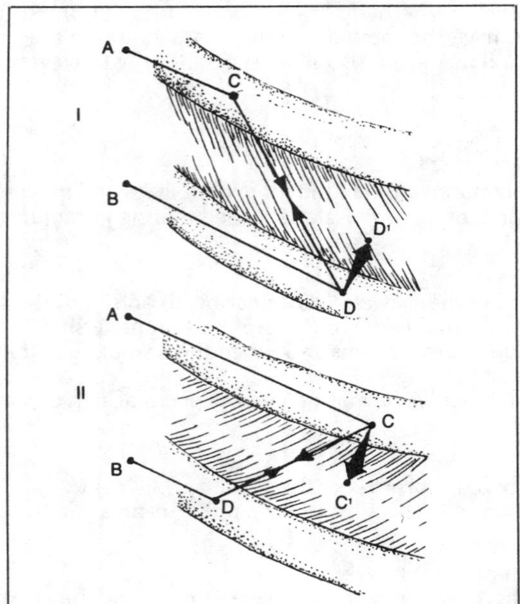

Figure 11-5. *Vectorial analysis of the intercostal muscle actions on the right side. I,* On contraction of the right external intercostal muscles, lever arm *BD* is longer than lever arm *AC*, resulting in elevation of the lower rib toward the upper rib. *II,* On contraction of the right internal intercostal muscles, lever arm *AC* is longer than lever arm *BD*, resulting in depression of the upper rib toward the lower rib.

far as the angle of the rib where the muscle fibers cease, but the layer continues to the vertebral column as the **internal intercostal membrane**.

(2) Action: depresses (lowers) the ribs for forced expiration.

(a) Because of the directionality of the internal intercostal muscles, each muscle fascicle has a longer lever arm on the upper rib than on the lower rib. Hence, the upper rib is drawn toward the lower rib. In Figure 11-5 points A and B are the axes of rotation of two adjacent ribs; line CD represents the line of action of the internal intercostals; line AC is the lever arm of the muscle on the upper rib; and line BD is the lever arm of the lower rib. Because lever arm BD is less than lever arm AC and because muscle tension (T) is equal at attachment points C and D, then (T) (BD) < (T) (AC). Thus, the upper rib will move toward the lower rib, as indicated by the heavy arrow.

(b) Each subsequent thoracic level is lowered, not only by the action of its internal intercostal muscle proper, but also by the sum of the amounts of contraction within this muscle layer at lower levels.

(3) Innervation: intercostal nerves 1–11.

b. The **internal inter*chondral* portion** of the intermediate layer.

(1) Attachments: runs between the cartilaginous portions of ribs 1–10 from the sternum to the costochondral junction with the same directionality as the external intercostal muscle on the contralateral side.

(2) Action: elevates ribs for inspiration. This paradox is explained by reference to Figure 11-4.

(a) The axis of rotation of each rib passes approximately through the costochondral junction. Because this inter*chondral* portion of the internal intercostal muscle lies on the side of the rib axis of rotation opposite the internal intercostal proper, action in a similar direction will be antagonistic to the internal intercostals.

(b) Because the internal inter*chondral* portion lies on the same side of the rib axis of rotation and has the same fiber directionality as the external intercostal of the opposite side of the chest, it will be synergistic with the external intercostals. Vectorial analysis illustrates these points.

(c) Innervation: intercostal nerves 1–11.

c. Innermost intercostals, frequently poorly developed, are functionally part of the internal intercostal layer. The **intercostal neurovascular bundle** artificially divides the intermediate muscle layer into internal and innermost intercostals (see Fig. 11-6B).

3. The **transverse thoracic (sternocostalis) muscles**, frequently poorly developed, comprise the innermost layer of the thoracic musculature.

a. Attachments: from the xiphoid process and the lower body of the sternum to ribs 2–6.

b. Action: lowers ribs in forced expiration.

c. Innervation: intercostal nerves 3–6.

C. THE DIAPHRAGM. The musculofibrous diaphragm separates the thoracic and abdominal cavities (see Fig. 18-4).

1. It is divided into several regions, reflecting its embryologic development.

a. The **costal portion** arises from the ribs that form the costal margin as far as the sternum; the **sternal portion** arises from the xiphoid process and lower sternum; and the **crura** (legs) of the **lumbar portion** arise from the bodies of vertebrae L1–L3.

b. The muscle fibers of the diaphragm arch upward and radially, converging into the **central tendon** of the diaphragm and forming a domed structure (see Fig. 11-2).

c. The right hemidiaphragm rises higher into the thoracic cavity than the left because of the liver, in the upper right quadrant of the abdomen, and the heart, toward the left in the thoracic cavity.

d. Medially, the left and right crura fuse over the aorta to form the **aortic hiatus**.

e. More laterally, the diaphragm arches from the vertebral bodies to the transverse process of L1, with the free border forming the **medial lumbocostal arch** (internal arcuate ligament), which bridges over the uppermost origins of the psoas major muscle and the sympathetic chain.

f. From the transverse process of L1, the diaphragm arches to the tip of the 12th rib, with the free border forming the **lateral lumbocostal arch** (external arcuate ligament); this arch bridges over the **quadratus lumborum muscle**, which stabilizes and depresses the 12th rib during expiration.

2. There are several **apertures through the diaphragm**.

a. The **aortic hiatus**, slightly to the left, contains the aorta, the azygos vein, and the thoracic duct.

b. The **esophageal hiatus**, located in the muscular portion on the right, contains the esophagus and the left and right vagus nerves; it is a frequent site of stomach herniation (hiatus or hiatal hernia).

c. The **vena caval foramen**, in the tendinous portion, contains the inferior vena cava and branches of the right phrenic nerve.

d. The splanchnic nerves and the hemiazygos vein pierce the crura.

3. **Development.**

a. The diaphragm develops in the cervical region from at least six primordia, some of them paired, which include contributions from the septum transversum in the cervical region, pleuroperitoneal membranes, and the somatic wall.

b. Failure of these primordia to fuse properly may result in anatomic defects through which herniation may occur.

(1) Incomplete closure of the pleuroperitoneal canal (usually on the left side) may cause herniation and passage of abdominal contents into the thoracic cavity (Bochdalek's hernia).

(2) A foramen of Morgagni, the result of incomplete closure of the sternocostal triangle, is another potential, although rare, site of herniation.

4. **Action.** Since the muscle fibers of the diaphragm are radially arranged and insert into the central tendon, contraction of the diaphragm pulls the dome inferiorly into the abdomen, thereby increasing the thoracic volume during inspiration.

5. **Innervation.** The motor and sensory innervation to the diaphragm is primarily by the phrenic nerves, which arise from levels C3–C5. However, there is some innervation from the lower thoracic and upper lumbar nerves to the crura. There is also some sensory innervation by intercostal nerves at the costal margin of the diaphragm.

a. Voluntary control of the diaphragm is subject to override by the respiratory center of the brain stem.

b. A phrenic nerve lesion may paralyze one hemidiaphragm.

(1) Paradoxical respiratory movements of a paralyzed hemidiaphragm result in poor oxygenation of blood in the corresponding lung.

(a) During inspiration, expansion of the thoracic cavity with decreased intrathoracic pressure draws a paralyzed hemidiaphragm upward into the thorax, so that there is little or no increase in thoracic volume on the affected side.

(b) During expiration, increased intrathoracic pressure draws a paralyzed hemidiaphragm inferiorly toward the abdomen, so that there is little decrease in the thoracic volume on the affected side.

(2) Paralysis of one hemidiaphragm is evident on x-ray examination.

c. Hiccups are recurrent spasms of the diaphragm. Phrenicectomy is sometimes undertaken as a last resort to relieve chronic hiccups.

6. Since the pericardial cavity and posterior mediastinum rest on the central tendon of the diaphragm, the diaphragm exerts a dynamic effect upon related structures. The position of the thoracic and abdominal organs varies with the respiratory cycle because of the changing thoracic volume. Conversely, postural changes alter the ventilatory effectiveness of the diaphragm.

V. VASCULATURE OF THE THORACIC WALL

A. The segmented skeleton of the thoracic cage is overlaid by segmentally innervated musculature (myotome) and skin (dermatome). On each side, each thoracic segment has a rib; a **neurovascular bundle** composed of a posterior intercostal artery and vein, as well as the spinal nerve; and three layers of intercostal musculature interposed between the ribs (Fig. 11-6A).

1. Posteriorly, the neurovascular bundle lies immediately behind the inferior edge of each rib in a slight depression, the **neurovascular (intercostal) groove** (Fig. 11-6B).

2. Laterally, the neurovascular bundle gives off collateral branches, which move inferiorly across the intercostal space to lie immediately above the next rib anteriorly.

a. Pleural taps usually are performed in the posterior or midaxillary line just above the superior edge of a rib.

b. Posterior or lateral fractures of a rib may tear the associated vascular structures. If the parietal pleura that lines the pleural cavity is also torn by the sharp edges of a fractured rib, there may be bleeding into the pleural cavity (hemothorax).

c. Intercostal nerve block is accomplished by injecting an anesthetic immediately beneath the inferior edge of a rib in the back.

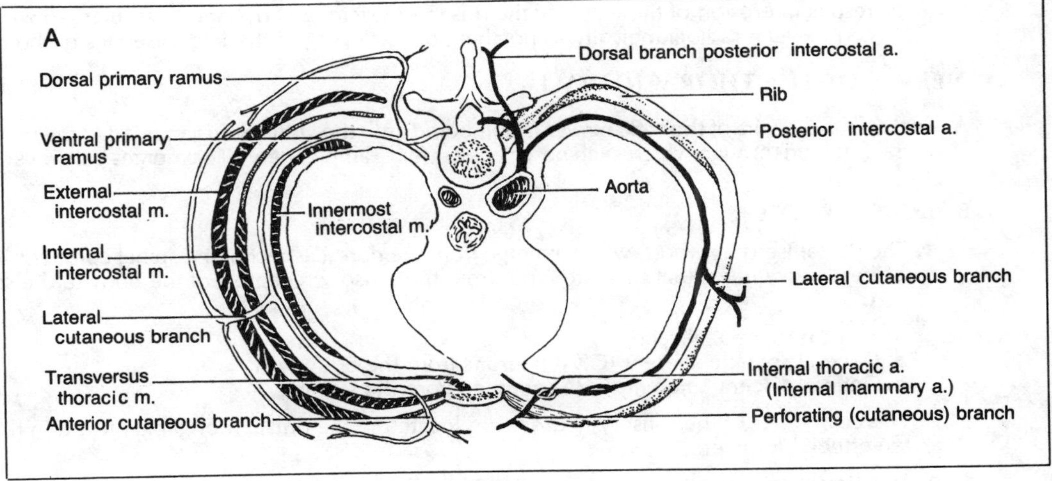

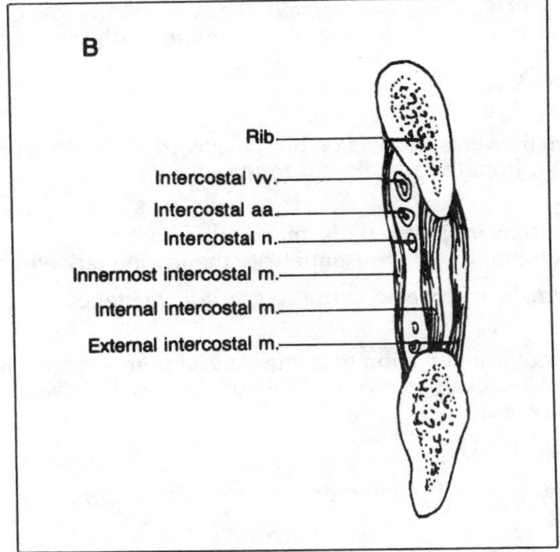

Figure 11-6. *Thoracic innervation and blood supply. A,* Branches of the right spinal nerve and left posterior intercostal artery. *B,* The relations of the intercostal neurovascular bundle to the rib and intrinsic thoracic musculature.

B. The blood supply to the thoracic wall is derived from several vessels (see Fig. 11-6A):

1. The **aorta**, which supplies the 3rd through 11th **posterior intercostal arteries** and the **subcostal artery**.
2. The **internal thoracic (mammary) artery** (a branch of the subclavian artery), which supplies:
   a. The 1st through 6th **anterior intercostal arteries**, which anastomose with the highest and posterior intercostal arteries.
   b. The **superior epigastric artery**, which anastomoses in the anterior abdominal wall with the **inferior epigastric artery** (a branch of the external iliac artery).
   c. The **musculophrenic artery**, which anastomoses with the 7th through 10th posterior intercostal arteries.
3. The **highest or superior intercostal artery** arises from the **costocervical trunk** (a branch of the subclavian artery) to supply the 1st and 2nd posterior intercostal arteries.
4. Pathways of **collateral circulation**:
   a. In the event of stenosis or coarctation of the aorta (a congenital or slow-onset narrowing to ultimate occlusion), anastomotic connections between the posterior and anterior intercostals, as well as between the superior and inferior epigastric arteries, provide important collateral pathways.
   b. Hypertrophy of the intercostal arteries, due to increased flow through collateral pathways,

results in erosion of the edges of the ribs in the vicinity of the neurovascular groove. This can be seen radiographically as notching or scalloping of the inferior edges of the ribs.

VI. NERVES OF THE THORACIC WALL

A. Each of the 12 pairs of **thoracic nerves** is numbered after the vertebra just craniad to the intervertebral foramen through which each nerve passes; for example, nerve T4 emerges below vertebra T4.

B. DORSAL ROOTS

1. The dorsal (sensory) roots, which emerge from the dorsal aspect of the spinal cord, consist of afferent nerve fibers that mediate input from the sensory receptors of the body to the spinal cord.
2. **Afferent** nerve classification.
   a. General somatic afferent (GSA) neurons from the body wall.
   b. General visceral afferent (GVA) neurons from the viscera.
3. The cell bodies of the sensory neurons are located in the dorsal root ganglion within the intervertebral foramen.
4. The dorsal root of each spinal nerve supplies a dermatome. Because most areas of skin receive some innervation from adjacent spinal nerves (dermatomes overlap), the loss of one dorsal root or spinal nerve to a dermatome results in possible diminution rather than complete loss of sensation.

C. VENTRAL ROOTS

1. The ventral (motor) roots emerge from the ventral aspect of the spinal cord; they consist of efferent nerve fibers that mediate output from the spinal cord to the effectors.
2. **Efferent** nerve classification.
   a. General somatic efferent (GSE) neurons innervate the somatic musculature.
   b. General visceral efferent (GVE) [autonomic] neurons innervate the involuntary effectors.
3. The cell bodies of the ventral root neurons are located in the ventral gray matter of the spinal cord (lateral and ventral horns).
4. The fibers of the ventral root supply motor innervation to a muscle group (myotome). Adjacent myotomes usually overlap, so that loss of a single ventral root usually results in weakness rather than paralysis of a group of muscles.

D. SPINAL NERVES

1. The dorsal and ventral roots generally join in the intervertebral foramen to form a mixed spinal nerve.
2. The spinal nerve is actually very short, dividing almost immediately into two major branches (rami) [see Fig. 11-6A].
   a. The **dorsal primary ramus** arches dorsally to innervate the posterior aspect of the dermatome and myotome (long muscles of the back).
   b. The **ventral primary ramus** continues ventrally as the **intercostal nerve**, which innervates the:
      (1) Parietal pleura.
      (2) Intercostal muscles.
      (3) Skin through the:
         (a) **Lateral cutaneous branch**, which penetrates the internal and external intercostal muscle layers at approximately the midaxillary line and subsequently bifurcates into **anterior** and **posterior branches** of the **lateral cutaneous nerve** to supply the lateral aspect of the dermatome.
         (b) **Anterior penetrating branch**, which penetrates the internal intercostal muscles and the external intercostal membrane just lateral to the sternum to supply the anterior aspect of the dermatome with some overlap across the midline.
   c. Injury to an intercostal nerve, producing paralysis of intercostal musculature, is evident as a sucking-in of the affected intercostal space upon inspiration and a bulging upon expiration.

VII. RESPIRATORY MECHANICS

A. MECHANICAL RESPIRATION (ventilation) can be divided into two categories—costal and diaphragmatic. Normally, ventilation is accomplished by a variable combination of costal and diaphragmatic contributions. In both types, air is moved into and out of the lungs by both muscular and mechanical forces.

B. DEFINITIONS AND CONSTRAINTS

1. Intrathoracic pressure (between lungs and thoracic wall) and intrapulmonic pressure (within the lungs) are measured or described relative to atmospheric pressure.
2. The second law of thermodynamics states that flow (of air) must always be from higher energy (pressure) to lower.
3. Boyle's law states that the product of pressure and volume is constant:

$$P_1V_1 = P_2V_2$$

4. Since inspiration increases thoracic volume with a concomitant decrease in the intrapulmonic pressure, air will flow into the lungs (if the glottis is open) until the pressures equalize. Conversely, since expiration decreases the thoracic volume with a concomitant increase in intrapulmonic pressure, air must flow out of the lungs (if the glottis is open) until the pressures equalize.

C. COSTAL VENTILATION

1. **Normal (relaxed) costal inspiration.**
   a. The external intercostals and the interchondral portion of the internal intercostals increase the transverse diameter of the thorax by moving the ribs upward and outward ("bucket-handle" effect, Fig. 11-7A).
   b. Concomitantly, since the axes of rotation of the ribs on each side cross behind the sternum, elevation of the ribs moves the sternum forward and upward, thereby increasing the anteroposterior diameter of the thorax ("pump-handle" effect, Fig. 11-7B). The pump-handle action changes the sternal angle from about 160° in inspiration to about 180° in expiration.
   c. Since the interior surface of the rib is concave, and since the rib rotates around its axis of rotation, there is a further increase in the transverse thoracic diameter when the ribs are elevated (Fig. 11-7C).
   d. Relaxed costal inspiration (along with the diaphragm) lowers the intrathoracic and intrapulmonary pressures from about −4 cm H_2O to between −5 and −10 cm H_2O, producing the **tidal volume** (0.5–0.6 L).

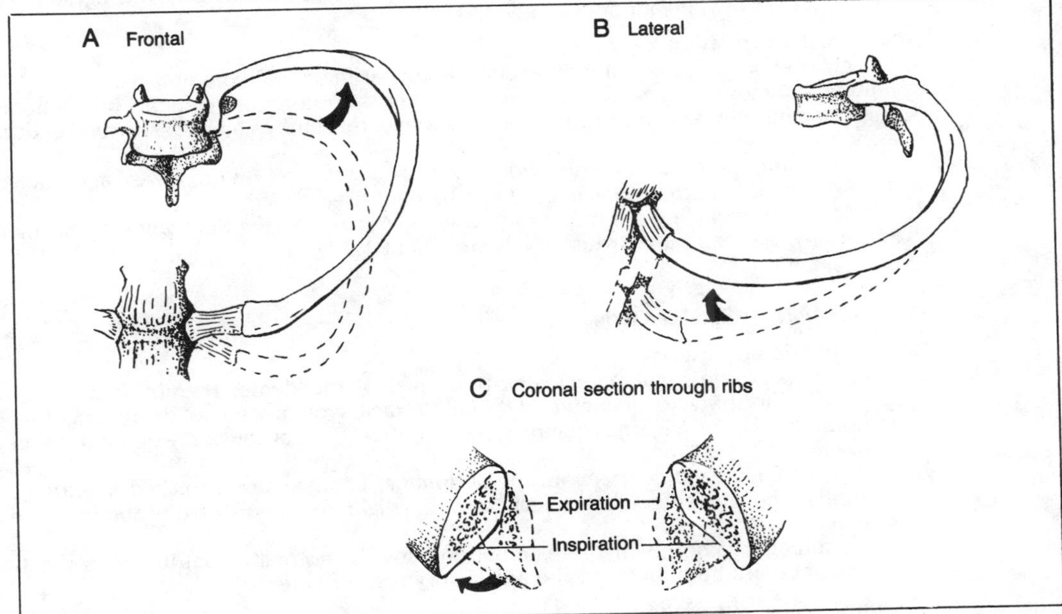

Figure 11-7. *Respiratory movements of the ribs. A,* Bucket handle motion of the left ribs whereby rotation about the axes produces elevation, which increases the transverse diameter of the thoracic cage. *B,* Pump handle motion of the ribs whereby rotation about the crossed axes produces anterior elevation, which raises the sternum and increases the anteroposterior diameter of the thoracic cage. *C,* Because of the ellipsoid cross-sectional shape of the ribs, rotation also slightly increases the diameters of the rib cage.

e. With a tidal volume of 0.5 L per breath and a respiratory rate of 14 breaths per minute, the lungs are ventilated with approximately 7.0 L of air per minute.

2. Forced costal inspiration.

a. In addition to the external intercostals, the pectoral muscles, which attach to the thoracic cage, assist with maximal elevation of the ribs to increase the transverse diameter of the thorax. The sternocleidomastoid and scalenes elevate the manubrium and 1st rib to increase the anteroposterior diameter of the thorax. (Recall the posture of a long-distance runner's neck and shoulders as he sprints for the finish line.)

b. Forced inspiration is required for ventilation in excess of tidal volume and contributes to the **inspiratory reserve volume** (about 3 L).

c. This mechanism comes into action when respiratory requirements exceed 70 to 100 L per minute.

3. Relaxed costal expiration.

a. No muscular effort is required to expel tidal volume.

b. Quiet tidal expiration is accomplished entirely by the elastic recoil of the costal cartilages and lungs.

(1) Extrinsic elastic recoil is provided by the costal cartilage.

(a) Elevation of the ribs during inspiration deforms and twists the costal cartilage. When the external intercostals relax, this stored energy is released, rolling the ribs back to their resting (end-expiratory) position. Gravity also assists this mechanism.

(b) The importance of this mechanism decreases with age because of calcification of the costal cartilages, which tends to immobilize the thoracic cage.

(2) Intrinsic elastic recoil or **pulmonary compliance** is provided by the lungs.

(a) The elastic fiber component of the interstitial tissue of the lung tends to cause the lung to shrink upon itself. Thus, the lung recoils after inspiratory stretching.

(b) In conjunction with the intrinsic elasticity of the lung, the tension between two surfaces (visceral pleura and parietal pleura) separated by a thin film of fluid produces a very high adhesive effect (surface tension). Thus, as the lungs deflate, they tend to pull the chest wall inward. Although this force, termed **intrathoracic pressure**, cannot be measured accurately, manometry shows that it may be between -3 cm H_20 and -5 cm H_20 (relative to atmospheric pressure) during relaxed tidal expiration, and may reach values from -5 cm H_20 to -10 cm H_20 during relaxed tidal inspiration.

(c) Diseases that reduce the compliance of the lung, such as emphysema, affect this aspect of ventilation.

4. Forced costal expiration.

a. Muscular effort is required for air exchange greater than tidal volume.

b. Internal intercostals decrease the transverse and anteroposterior diameters of the thoracic cage, and intrathoracic pressure becomes positive (reaching as high as 300 mm Hg during a sneeze).

c. The quadratus lumborum, running from the iliac crest to the 12th rib, lowers and stabilizes the 12th rib so that the thoracic cage can be depressed effectively.

d. In addition to expelling the inspiratory reserve volume and the tidal volume, this mechanism provides **expiratory reserve volume** (about 1.2 L).

D. DIAPHRAGMATIC VENTILATION

1. Diaphragmatic inspiration.

a. Contraction of the diaphragm lowers the level of the dome, thereby increasing the superoinferior thoracic diameter and the thoracic volume while decreasing the intrathoracic pressure. Vigorous diaphragmatic contraction may lower the dome by as much as 10 cm.

b. During diaphragmatic contraction, the abdominal contents are displaced inferiorly and anteriorly, the movement being facilitated by slight relaxation of the abdominal wall muscles.

c. The diaphragm, along with the external intercostals, normally contributes to the **tidal volume**, but also contributes to the **inspiratory reserve volume**.

2. "Diaphragmatic" expiration.

a. The abdominal musculature is antagonistic to the diaphragm. Since muscles accomplish work only when they contract, antagonistic muscles must contract to obtain work in the opposite direction.

b. Contraction of the abdominal muscles forces the abdominal viscera under the diaphragm

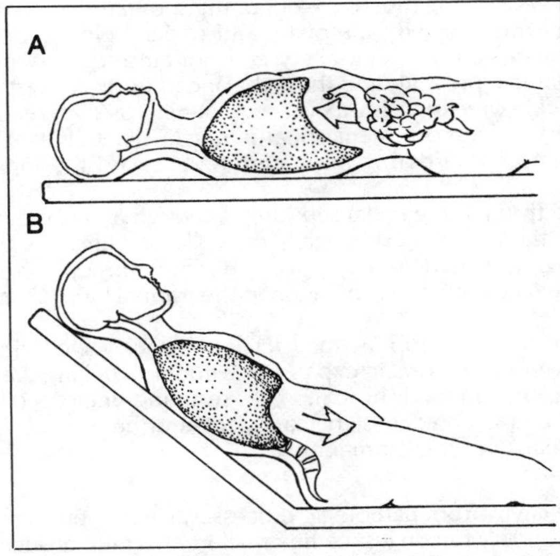

Figure 11-8. *Effects of posture on respiratory effort. A,* In a horizontal position, the weight of the abdominal viscera tends to stretch the diaphragm into the thoracic cavity. The patient must overcome this gravitational force on inspiration. *B,* A slight incline redistributes the abdominal viscera and facilitates inspiratory effort.

(much like a fluid piston), thereby stretching the diaphragm into the thoracic cavity and decreasing thoracic volume to increase intrathoracic pressure.

c. This mechanism contributes to **expiratory reserve volume**.

d. The abdominal musculature comes into action when the respiratory requirements exceed 40 to 60 L per minute.

3. When a patient is in the sitting or standing position, the force of gravity acting on the abdominal and thoracic viscera tends to lower the diaphragm. With a patient in the horizontal position, the force of gravity on the abdominal viscera tends to stretch the diaphragm into the thoracic cavity (Fig. 11-8A). The plus and minus effects of gravity on ventilation explain the rationale for propping up a bedridden person with respiratory difficulty (Fig. 11-8B).

4. The normal balance between costal and diaphragmatic ventilation depends upon sex, body type, age, profession, state of health, and clothing. Children and elderly persons tend to breathe diaphragmatically, as do horn players and singers; obese people and women in advanced stages of pregnancy cannot effectively contract the diaphragm.

E. SUMMARY OF APPROXIMATE VENTILATION CAPACITIES

1. **Tidal volume** = 0.5–0.6 L.
   a. Limited to normal quiet inspiration and expiration.
   b. Accomplished by the external intercostals and diaphragm; with *return* by elastic recoil of the thoracic cage and lungs.

2. **Inspiratory reserve volume** = 3 L.
   a. Begins at the end of normal inspiration.
   b. Accomplished by accessory muscles of inspiration in addition to the external intercostals and diaphragm; with *return* by elastic recoil of the thoracic cage and lungs (assisted as necessary by expiratory muscles).

3. **Expiratory reserve volume** = 1.2 L.
   a. Begins at end of normal expiration.
   b. Accomplished by *in*ternal intercostals, the quadratus lumborum, and abdominal muscles; with *return* accomplished by an expansionary elastic recoil of the thorax.

4. **Vital capacity** = 4.8 L (the sum of tidal volume and inspiratory and expiratory reserve volumes).

5. **Anatomic dead space** = 1.2 L (the nonrespiratory volume).

6. **Total lung capacity** = 6 L.

F. CLINICAL CONSIDERATIONS

1. **Pneumothorax.**
   a. Negative intrathoracic pressure (relative to atmospheric pressure) and surface tension normally hold the lung surface against the thoracic wall. Thus, the lung remains inflated

against its own intrinsic elasticity. A penetrating thoracic wound, the spontaneous rupture of a pulmonary bulla, a tear of abnormally fused pulmonary and parietal pleura, or the iatrogenic piercing of the pleural cavity—all of which allow entry of air into the pleural cavity—abolish the negative intrathoracic pressure and the unbalanced intrinsic elasticity results in collapse of the lung. A **sucking pneumothorax** is accompanied by hyperexpansion of the chest wall on the normal side and a mediastinal flutter (a slight shift of the mediastinal contents toward the normal side during inspiration and toward the injured side during expiration).

b. Because blood continues to perfuse through the collapsed lung, although at a somewhat reduced rate, and because there is little or no gas exchange in the collapsed lung, a significant portion of cardiac output is not oxygenated, and dyspnea and sometimes cyanosis result. Mediastinal flutter reduces the effectiveness of ventilation in the normal lung. Clearly, pneumothorax is an emergency.

c. In certain instances, the nature of the wound or rupture may act as a "check valve," drawing air into the pulmonary cavity with each thoracic expansion without expelling the air during thoracic compression (**tension pneumothorax**). The resulting increased intrathoracic pressure forces the mediastinal contents significantly toward the normal side, thereby interfering with the vital capacity of the normal lung.

2. Fluid in the pleural cavity.

a. Fluid may accumulate in the pleural cavity from pathologic processes, such as pulmonary effusion (due to inflammation or secondary to congestive heart failure) or trauma (such as torn vessels of the thoracic wall or lung). Such fluid accumulation reduces vital capacity of the lung and decreases the oxygenation-to-perfusion ratio.

b. As little as 500 ml of fluid may be seen on a thoracic x-ray film as blunting of the costodiaphragmatic recess.

3. Chest fractures.

a. Fracture of several ribs in two locations (usually anteriorly and posterolaterally) produces "flail chest." The structural integrity of the thoracic cage is diminished with the loss of effective lever arms for the thoracic musculature. As the diaphragm is lowered, the flail portion of the wall cannot withstand the decrease in intrathoracic pressure and moves inward; conversely, as the abdominal muscles contract in expiration, the flail portion moves outward (*paradoxical respiratory movement*). The result is reduced ventilation and a decreased oxygenation/perfusion ratio. A positive-pressure respirator will correct the oxygenation/perfusion ratio.

b. Fractured ribs may penetrate the thoracic wall or tear visceral pleura. Either circumstance can cause pneumothorax. Tearing of the associated blood vessels can result in hemothorax.

4. Respirators.

a. Negative-pressure devices, such as the Drinker ("iron lung" or tank) respirator, cyclically lower the extrathoracic and therefore the intrapulmonary pressure below atmospheric pressure, mimicking natural negative-pressure breathing.

b. Positive-pressure devices cyclically elevate the atmospheric pressure above normal and above intrapulmonary pressure so that air is forced into the lungs.

12
The Pleural Cavities and the Lungs

I. INTRODUCTION

A. The vertebrate body consists of a somatic wall surrounding the coelom, within which the hollow viscera are suspended by mesenteries. Subdivision of the common coelom produces paired pleural cavities, a single pericardial cavity, and the peritoneal (abdominopelvic) cavity.

B. THE THORACIC CAVITY contains:

1. The left and right **pleural cavities**, which are independent of each other, and which contain the **lungs**, each suspended by a mesentery.

2. The **mediastinum**, which separates the pleural cavities.
- **a.** The **anterior mediastinum** contains:
  - **(1)** The inferior portion of the **thymus gland.**
  - **(2)** The sternopericardial ligaments and connective tissue.
- **b.** The **middle mediastinum** contains:
  - **(1)** The **pericardial cavity** with its enclosed **heart**.
  - **(2)** The **phrenic nerves.**
  - **(3)** The **lung roots**.
- **c.** The **superior mediastinum** contains:
  - **(1)** The superior portion of the **thymus gland**.
  - **(2)** The **aortic arch** and associated great vessels.
  - **(3)** The **trachea** and **main stem bronchi**.
  - **(4)** The **esophagus.**
  - **(5)** The **superior vena cava** and its branches.
  - **(6)** The **autonomic nerves** to the heart and lungs.
- **d.** The **posterior mediastinum** contains:
  - **(1)** The **esophagus.**
  - **(2)** The **thoracic aorta**.
  - **(3)** The **thoracic duct**.
  - **(4)** The **inferior vena cava**.
  - **(5)** The **azygos** and **hemiazygos veins.**
  - **(6)** The **sympathetic chains.**
  - **(7)** The **vagus nerves (CN X)**.

II. THE PLEURAL CAVITIES

A. THE PARIETAL PLEURA

1. This layer of serous mesothelium lining the pleural cavities lies beneath the **endothoracic fascia**, which underlies the ribs, innermost intercostal muscles, and transverse thoracic muscle if present.

2. The parietal pleura is given several topographic names to describe its relation to adjacent structures:
- **a.** Costal pleura.
- **b.** Diaphragmatic pleura.
- **c.** Mediastinal pleura.
- **d.** Cervical pleura (cupula of the pleura).

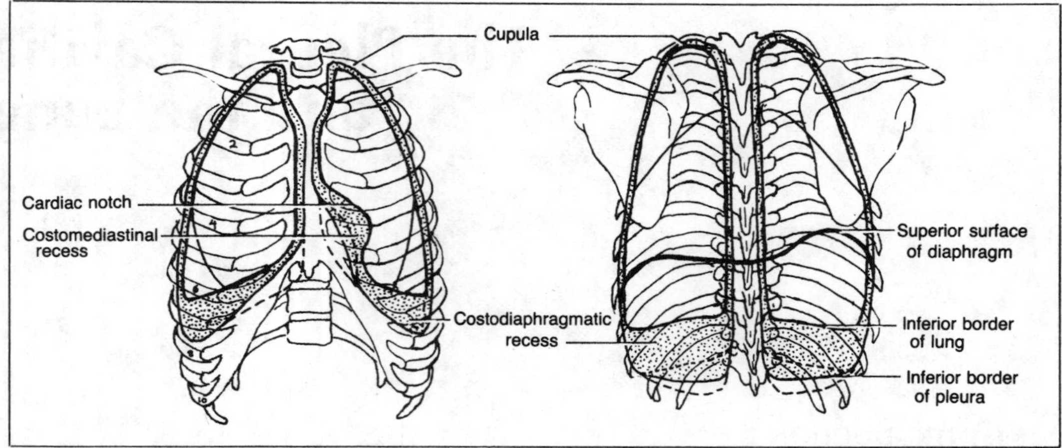

Figure 12-1. *Pleural cavities and lungs.* Because the lungs normally do not completely fill the pleural cavities, costodiaphragmatic and costomediastinal recesses occur. The pleural recesses provide potential space for lung expansion on inspiration. The lines of pleural reflection vary between the extremes indicated by the *dashed lines.*

3. Lines of pleural reflection (Fig. 12-1).

a. The **anterior border** of the **right** pleural cavity generally runs from the cupula (bounded by the base of the neck, clavicle, and trapezius muscle), through the right sternoclavicular joint, through the sternal angle in the midline, dropping in the midline to the level of the 6th costal cartilage or xiphoid process, where it begins to swing laterally to reach and then run just above the costal margin. The lower extent of this line of reflection is variable. A right subcostal incision carried into the sternocostal angle can nick the right pleural cavity, with resultant pneumothorax.

b. The **anterior border** of the **left** pleural cavity is similar, but it does not reach the midline; it deviates laterally at the level of the 4th rib to form the **cardiac notch** before diverging laterally at the level of the 6th or 7th costal cartilage.

c. The inferior line of reflection of both pleural cavities moves progressively superior to the costal margin, so that it is over the costochondral junction of the 8th rib and over the midshaft of the 11th rib.

d. The **inferior border** of both pleural cavities runs fairly horizontally across the 12th rib at midshaft to reach the vertebral column at about the 12th intervertebral level.

(1) The inferior borders of the pleural cavities are variable.

(2) A posterior abdominal incision inadvertently carried along the 12th rib runs a high probability of penetrating the pleural cavity and causing pneumothorax.

e. The **posterior border** of the pleural cavity runs vertically between the 12th intervertebral level and the cupula.

4. Pleural recesses.

a. The pleural cavities are larger than the pulmonary volumes (see Fig. 12-1).

b. The **pleural recesses** are regions of parietal pleura applied to parietal pleura at sites of pleural reflections where the lung fails to fill the pleural cavity. During deep inspiration the lung parenchyma expands into, but never completely fills, these recesses. The three pleural recesses are of considerable clinical importance. When air enters into or fluid collects within a pleural cavity, the pleural recesses become apparent on x-ray film.

(1) The **left** and **right costodiaphragmatic recesses** are those parts of the pleural cavities between the costal pleura and the diaphragmatic pleura.

(a) When a patient is standing or sitting, the most dependent (lowest) part of the costodiaphragmatic recess is usually where the pleura passes over the 11th rib posterolaterally.

(b) In the erect position, fluid in a pleural cavity accumulates in the costodiaphragmatic recess.

(c) Since the lung floats on pleural fluid, the fluid level can be determined by percussion.

(d) Blunting of the costodiaphragmatic recess will be visible radiographically with 400 to 500 ml of pleural fluid.

(2) The **costomediastinal recess** is that part of the pleural cavity on the left side between the mediastinal pleura and the costal pleura.

(a) The depth of the costomediastinal recess is variable.

(b) A needle inserted into the pericardial cavity or heart adjacent to the costomediastinal recess will avoid the pleural cavities and associated risk of pneumothorax.

5. **Pleural innervation.**
   a. The costal pleura is innervated by the **1st–12th intercostal nerves**.
   b. The peripheral diaphragmatic pleura is innervated by the lower **intercostal nerves**, and pain from irritation may be *referred* to (appear as if originating from) the lateral thoracic and anterior abdominal walls because lower intercostal nerves innervate those regions.
   c. The central diaphragmatic pleura and mediastinal pleura are innervated by the **phrenic nerves**; pain may be *referred* to the distribution of nerves C3–C5 (neck and shoulder).

6. Inflammation of the parietal pleura (pleurisy) is accompanied by pain (accentuated by respiratory movement), and a rasping sound (friction rub) may be heard on auscultation. If an effusion of fluid occurs, the friction rub disappears.

B. MESENTERIES (PULMONARY LIGAMENTS)

1. A mesentery (frequently termed ligament) is formed by two layers of back-to-back mesothelium and supports a visceral structure from the body wall.
2. The parietal pleura of the anterior and posterior walls of the thoracic cavity is reflected off the mediastinal wall toward the lungs to form the mesenteric **pulmonary ligament**.
3. The pulmonary ligament runs along the mediastinal surface of each lung from the **hilus** (the region where the vessels and air passages enter the lung) to the **base** (diaphragmatic surface).

C. THE VISCERAL PLEURA

1. The mesentery forming the pulmonary ligament diverges around the lungs to form the **visceral pleura (pulmonary pleura)**.
2. The parietal and visceral pleurae, through the pulmonary ligament, present a continuous surface. The potential space between these pleural layers is the **intrathoracic cavity**. Introduction of air, blood, or fluid into the intrathoracic cavity results in pneumothorax, hemothorax, or hydrothorax, respectively.
3. Normally, a thin film of extracellular fluid provides lubrication between the parietal and the visceral pleura, allowing the visceral surface to slide easily over the parietal surface. This thin fluid layer also provides tremendous surface tension between the parietal and the visceral pleural layers, which keeps the lung inflated. (As an example, move a microscope slide over another; they can be pulled apart easily. But place a drop of water between the slides and repeat the procedure. There is less sliding friction, but the slides cannot be pulled apart.)

D. FUNCTIONAL CONSIDERATIONS

1. The pleural cavities normally are filled *completely* by the lungs, with the exception of the costodiaphragmatic and costomediastinal recesses (see Fig. 12-1).
2. The pleural recesses are only *potential spaces*, since parietal pleura is apposed to parietal pleura (i.e., costal pleura to diaphragmatic pleura in the costodiaphragmatic recess or to mediastinal pleura in the costomediastinal recess).
3. The sizes of the pleural recesses vary with the respiratory cycle and with the degree of inspiratory and expiratory effort. As the thoracic cavities and lungs expand upon inspiration, a volume of lung tissue equivalent to the size of the inspiration moves into the pleural recesses. These changes in pulmonary volume between inspiration and expiration can be delineated by percussion.
4. Pneumothorax produces a hyperresonant percussive tone. Radiographically, the pleural cavity is markedly radiolucent, with the free edge of the collapsed lung visible.
5. Fluid from pathologic processes (pyothorax or hydrothorax) or traumatic events (hemothorax or chylothorax) can collect in the pleural recesses when the person is erect.
   a. Bleeding may result from injury to large or small thoracic vessels; chyle may issue from the thoracic duct; and the fluid may result from direct or indirect pathologic processes involving the lungs; for example, inflammation or heart failure.
   b. Fluid levels may be ascertained by x-ray or by the level at which the percussive tympanic tone changes abruptly to a dull tone.
6. Thoracentesis (pleural tap) usually is performed posterior to the midaxillary line with a patient in the sitting position. The fluid level is determined by percussion. At one or more intercostal spaces below the fluid level (but not below the 9th intercostal space), depending on the

amount of pleural fluid, a needle is inserted immediately above the superior margin of a rib to avoid injury to the intercostal neurovascular bundle.

7. Chest tubes for the reduction of pneumothorax uncomplicated by fluid accumulation usually are inserted into the pleural cavity anteriorly through the 2nd intercostal space immediately superior to the 3rd rib. If fluids are or might be present, the chest tube is usually inserted in the posterior axillary line through the 5th or 6th intercostal space. Chest tubes are ordinarily maintained at −10 cm H_2O (relative to atmospheric pressure), approximating normal intrathoracic pressure.

III. THE LUNGS

A. The **lungs** are covered by **visceral pleura**.

1. The developing lung buds can be thought of as expanding into the pleural cavities, pushing the primitive mesothelium of the mediastinal pleura ahead of them. In this manner, the lung buds are covered by mesothelium (visceral pleura), with a connection maintained to the parietal pleura (pulmonary ligament) along the mediastinum.
2. Although the right lung is shorter because the liver is on the right side, it is wider because the heart is offset to the left. The right lung has a slightly larger capacity than the left.

B. EXTERNAL STRUCTURE OF THE LUNGS

1. The lung surfaces are named to correspond to the related thoracic structures; for example, the diaphragmatic, costal, and mediastinal surfaces.
2. The **apex** lies in the **cupula**, which extends above the 1st rib into the base of the neck. Percussion and auscultation of the lung at the base of the neck and shoulder are performed routinely during thoracic examination.
3. The diaphragmatic surface defines the **base**.
4. The **radix** or **root** of the lung is on the mediastinal surface. Structures that enter or leave each lung at the **hilus** include:
   - **a.** The **main stem (primary) bronchus**.
   - **b.** The **pulmonary artery**.
   - **c.** The **pulmonary veins**.
   - **d.** The **bronchial lymphatics** and **hilar lymph nodes**.
5. The **pulmonary ligament**, the mesentery of the lung, is formed by embryonic evagination of the lung buds into the thoracic cavity.
   - **a.** It extends along the mediastinal surface from the hilus to the base.
   - **b.** It supports the lungs in the pleural cavity.
6. **The lobes.**
   - **a.** Fissures of the visceral pleura divide the lungs into lobes (Fig. 12-2, Table 12–1).
   - **b.** Each lobe is supplied by a **lobar (secondary) bronchus**, accompanied by the **lobar branch** of the pulmonary artery and a **lobar branch** of a pulmonary vein.
   - **c.** The **lingula** of the left lung, which corresponds to the middle lobe of the right lung, is formed by the more pronounced **cardiac incisure** in the left lung.
   - **d.** The upper lobes lie more anteriorly, and the lower lobes extend posteriorly.

Table 12-1. Lobes and Fissures of the Lungs

Right Lung (3 Lobes)		Left Lung (2 Lobes)	
Anteriorly:			
Upper lobe	Horizontal fissure	Upper lobe	Oblique fissure
Middle lobe		Lower lobe	
	Oblique fissure		
Lower lobe			
Posteriorly:			
Upper lobe	Oblique fissure		
Lower lobe			

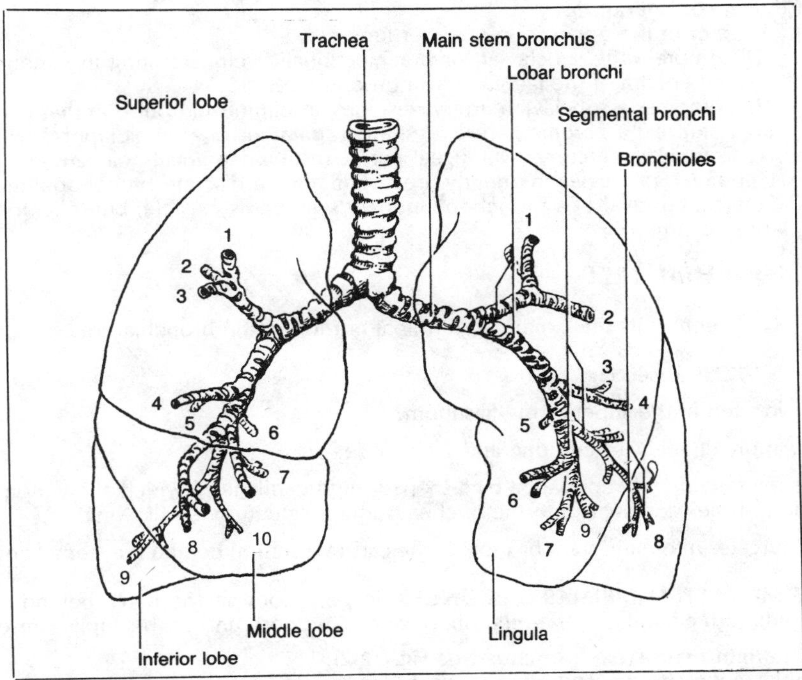

Figure 12-2. *The bronchial tree and lobes of the lung.* The right main stem bronchus divides into upper, middle, and lower lobar bronchi, which in turn give rise to a total of 10 segmental bronchi (*numbered*). The left main stem bronchus is more horizontal, divides into upper and lower lobar bronchi, and gives rise to 8 to 10 segmental bronchi (*numbered*).

7. **Bronchopulmonary segments.**
   - **a.** The bronchopulmonary segments, the functional units of the lungs, are characterized by:
     - **(1)** A central **segmental (tertiary) bronchus**.
     - **(2)** A central **segmental artery**.
     - **(3)** An **intersegmental septum** of connective tissue.
     - **(4) Intersegmental veins** forming a peripheral venous plexus.
   - **b.** There are 10 bronchopulmonary segments in the right lung and, depending upon the reference textbook selected, 8, 9, or even 10 described in the left lung (see Fig. 12-2, Table 12-2).

Table 12-2. Bronchopulmonary Segments of the Lungs

Right Lung	Left Lung	
Upper Lobe (3)	**Upper Lobe (4)**	
1. Apical	1. Apicoposterior	
2. Posterior		
3. Anterior	2. Anterior	
Middle Lobe (2)		
4. Medial	3. Superior lingular	
5. Lateral	4. Inferior lingular	
Lower Lobe (5)	**Lower Lobe (4–5)**	
6. Superior	5. Superior	
7. Basal medial	6. Basal medial	} or Basal anteromedial
8. Basal anterior	7. Basal anterior	}
9. Basal lateral	8. Basal lateral	
10. Basal posterior	9. Basal posterior	

8. Clinical considerations.

a. A tumor of the upper lobe of either lung may:

(1) Compress the subclavian or brachiocephalic vein, resulting in venous engorgement and edema of the face and arm on one side.

(2) Compress a subclavian artery, resulting in diminished pulse in that extremity.

(3) Compress a phrenic nerve, resulting in paralysis of a hemidiaphragm.

(4) Compress a recurrent laryngeal nerve, resulting in vocal hoarseness.

b. Pathologic processes frequently are confined to a discrete bronchopulmonary segment. Surgical removal of a bronchopulmonary segment is feasible, but lobectomy is currently more common.

IV. THE BRONCHIAL TREE

A. Each lung represents the expanded terminal portion of the bronchial tree.

B. THE TRACHEA (see Fig. 12-2)

1. Is located in the superior mediastinum.
2. Is approximately 12 cm long and 2 cm wide.
3. Is reinforced and kept patent by c-shaped rings (cartilages), which allow momentary expansion of the esophagus into the trachea during deglutition (swallowing).
4. Bifurcates into main stem bronchi at the **carina** (septum) behind the sternal angle at level T5.

C. THE MAIN STEM (PRIMARY) BRONCHI lie posteriorly in the hilus, behind the pulmonary artery (intermediate) and the pulmonary veins (anterior), and each supplies an entire lung.

1. **The right main stem bronchus** (see Fig. 12-2):
   - a. Is wider, shorter, and more vertical than the left main stem bronchus.
   - b. Is the most probable resting place for larger aspirated objects.
   - c. Divides external to the lung parenchyma into three lobar bronchi.
2. **The left main stem bronchus** (see Fig. 12-2):
   - a. Is narrower (because of smaller left lung volume), longer, and more horizontal than the right (because the heart is toward the left).
   - b. Divides external to the lung parenchyma into two lobar bronchi.

D. THE LOBAR (SECONDARY) BRONCHI (see Fig. 12-2)

1. Each bronchus supplies one lobe. There are three lobar bronchi in the right lung and two in the left lung.
2. The **right lower lobar bronchus** is most vertical, most nearly continues the direction of the trachea, and is larger in diameter than the left; it is therefore the most common resting place for small aspirated objects. Such foreign bodies may block the airway and cause atelectasis (local collapse), inflammation, or even abscess.
3. The lobar bronchi divide into segmental bronchi.

E. THE SEGMENTAL (TERTIARY) BRONCHI

1. Each segmental bronchus supplies a bronchopulmonary segment. It enters the center of the segment accompanied by a segmental branch of the pulmonary artery.
2. A segmental bronchus may branch 6 to 18 times to produce 50 to 70 respiratory bronchioles, which merge with alveolar sacs and terminate in the alveoli constituting the lung parenchyma.
3. Since the **superior segmental bronchi** of the left and right lower lobes are the most posterior when the body is supine, inhaled material (such as aspirated vomitus) easily enters the superior bronchopulmonary segments of the lower lobes. These segments thus are the ones most frequently involved in aspiration pneumonia (Mendelson's syndrome).

V. PULMONARY VASCULATURE

A. Two separate systems of arterial vessels supply the lungs.

1. **Pulmonary arteries** supply the respiratory (alveolar) parenchyma, carrying deoxygenated blood from the right ventricle. After transalveolar gaseous exchange, oxygenated blood returns to the left atrium via the **pulmonary veins**.

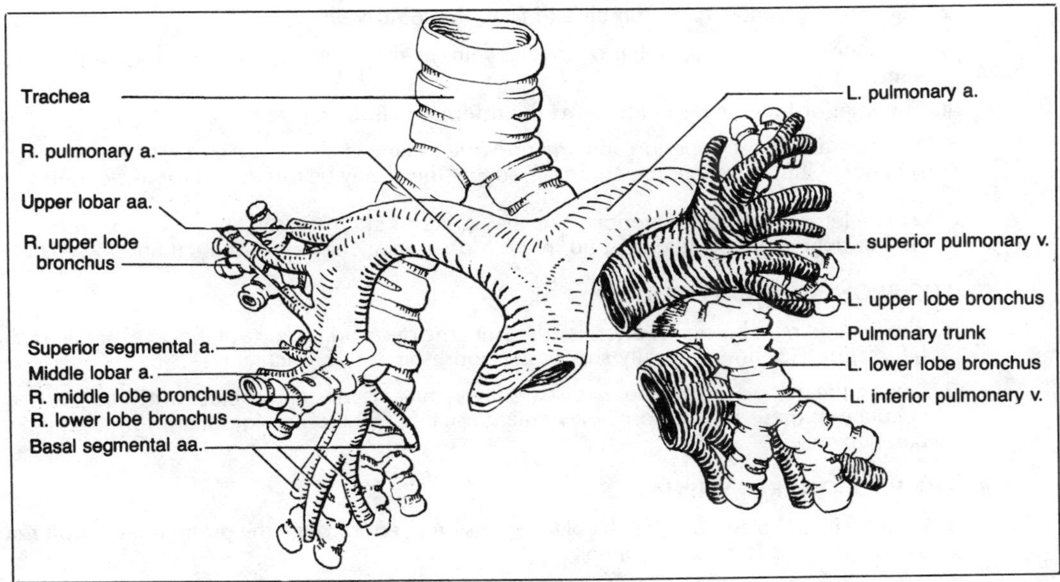

Figure 12-3. *The bronchial tree, right pulmonary arteries, and left pulmonary veins.*

2. **Bronchial arteries** supply the conducting (nonrespiratory) portions of the lung with oxygenated blood.
3. There are anastomotic connections between the capillary beds of the bronchial and the pulmonary arterial systems.

B. The **pulmonary trunk** arises from the **conus arteriosus** of the right ventricle of the heart; after approximately 5 cm it bifurcates into the left and right pulmonary arteries (Fig. 12-3).

C. THE PULMONARY ARTERIES (see Fig. 12-3)

1. **The right pulmonary artery:**
   a. Passes dorsal to the **ascending aorta** and the **superior vena cava.**
   b. Crosses the right main stem bronchus to lie superior to that bronchus.
   c. Lies posterior to the right superior pulmonary vein in the hilus.
2. **The left pulmonary artery:**
   a. Passes anterior to the arch of the **descending aorta**, where it is connected to the arch of the aorta by the **ligamentum arteriosum** (a remnant of the **ductus arteriosus** or left 6th aortic arch), around which courses the recurrent laryngeal branch of the left vagus nerve.
   b. Crosses the left main stem bronchus to lie superior to that bronchus.
   c. Lies posterior to the left superior pulmonary vein in the hilus.
3. The pulmonary arteries subdivide and follow the bronchial tree. Segmental arteries enter the center of each bronchopulmonary segment with a segmental bronchus. The segmental arteries continue to subdivide to supply the capillary plexuses about each alveolus.
4. A **pulmonary embolus** is a blocking body, such as a thrombus (thromboembolus) that is transported from one part of the systemic venous system (usually due to venous stasis or injury) or from the right side of the heart (usually due to valvular disease or myocardial infarct) to the pulmonary circulation, where it becomes impacted. An embolus results in local **ischemia**. If the ischemia is severe, the tissues may undergo **necrosis**. The area of necrosis is termed an **infarct**. Pulmonary embolism is one of the greatest single causes of death in elderly injured or postoperative patients. A large embolus in a main stem or lobar artery may result in neurogenic shock and death within minutes. A small embolus may produce only mild pain, pleurisy, and perhaps hemoptysis before it resolves with healing. Supposedly, less than one-half of all pulmonary emboli are correctly diagnosed.

D. THE PULMONARY VEINS (see Fig. 12-3)

1. Venous capillaries arise from the alveolar capillary plexuses and coalesce peripherally in the connective tissue septa of the bronchopulmonary segments as the **intersegmental veins**.

2. The intersegmental veins combine to form the **lobar veins**.
3. The superior and middle lobar or lingular veins usually unite to form the **superior pulmonary veins**.
4. The inferior lobar veins continue as the **inferior pulmonary veins**.
5. The left and right superior and inferior pulmonary veins generally enter the left atrium separately, but variations are common, so that there may be three, four, or five openings into the left atrium.
   a. The left superior and inferior veins may unite to enter the left atrium as one vein.
   b. The right superior, middle, and inferior lobar veins may enter the left atrium separately.

E. THE BRONCHIAL ARTERIES

1. The left and right bronchial arteries arise from the descending aorta at the level of the tracheal bifurcation (T5) and generally supply the nonrespiratory bronchial tree.
2. During the course of pulmonary disease, bronchial arteries anastomose with the pulmonary circulation; these anastomoses may enlarge significantly to provide blood to the respiratory parenchyma.

F. THE PULMONARY LYMPHATICS

1. Lymphatic drainage from the alveolar regions proceeds through the **pulmonary lymph nodes** associated with the lobar bronchi.
2. Subsequent drainage is toward **hilar** or **bronchopulmonary nodes** at the root of the lung.
3. Drainage is then toward the **tracheobronchial nodes**, along the main stem bronchi, and then to the bronchomediastinal or **tracheal nodes**, to unite with the **parasternal nodes**, which drain into the subclavian veins.
4. If any of these lymph nodes become blocked by metastatic growth, there may be backup through lymphatic connections to the contralateral lung or to the celiac nodes in the epigastric region of the abdomen.
5. Lymphatics provide pathways for metastatic spread of lung carcinoma.

VI. PULMONARY INNERVATION (Fig. 12-4)

A. General visceral afferent (GVA) innervation and general visceral efferent (GVE) innervation to the lungs are carried by branches from the thoracic sympathetic chain and the vagus nerve (CN X).

1. **Sympathetic nerves.**
   a. The left and right sympathetic chains provide short splanchnic branches to the **anterior** and **posterior pulmonary plexuses**, which are located on the anterior and posterior aspects, respectively, of the main stem bronchi on each side.
   b. Sympathetic GVE neurons mediate vasoconstrictive function in the pulmonary vascular bed and secretomotor activity to the bronchial glands.
2. **Parasympathetic nerves.**
   a. The left and right vagi in the mediastinum pass anterior to the main stem bronchi and in this vicinity give off twigs to the **anterior** and **posterior pulmonary plexuses**.
   b. Parasympathetic GVE neurons from the vagus innervate the bronchial smooth muscle. (Excessive stimulation produces the asthmatic syndrome.)
   c. The vagus nerves also carry respiratory reflex afferents (GVA).

B. REGULATION OF RESPIRATION

1. The lungs contain stretch receptors innervated by neurons that run in the vagus nerve. When the lung is stretched, these GVA fibers inhibit the respiratory center in the brain stem (end-inspiratory reflex). There is a similar GVA end-expiratory reflex.
2. The glossopharyngeal (CN IX) and vagus (CN X) nerves innervate the carotid and aortic bodies and sinuses, respectively. These small structures located in conjunction with the branchial arches monitor changes in the partial pressure of oxygen and hemodynamic pressure, respectively. A decrease in pO_2 will increase inspiration through reflex mechanisms in the brain stem. An increase in blood pressure will decrease respiration through similar reflex mechanisms.

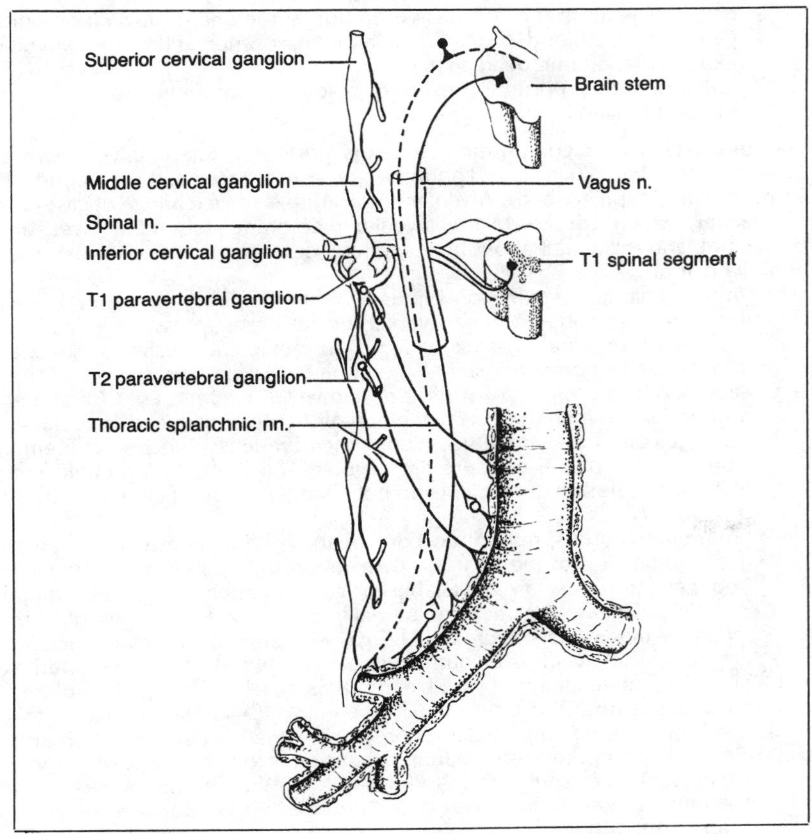

Figure 12-4. *Autonomic innervation of the bronchial tree.* The first, or preganglionic, sympathetic neurons lie in the thoracic spinal cord, send axons to the spinal nerves through the ventral roots, and leave the spinal nerves via white rami communicantes to gain access to the sympathetic chain. The second, or postganglionic, sympathetic neurons lie in the sympathetic chain and send axons along thoracic splanchnic nerves to reach the bronchial tree. The preganglionic parasympathetic neurons lie in the brain stem and send axons along the vagus nerve (CN X) to the bronchial tree. The postganglionic neurons lie in the walls of the bronchial tree. Reflex sensation from the bronchial tree travels along afferent neurons (*dashed lines*) in the vagus nerve to the brain.

VII. FUNCTIONAL ANATOMY OF THE RESPIRATORY SYSTEM

A. The lungs are the organs through which ambient oxygen passes into the blood and ultimately reaches the intracellular mitochondria for oxidative metabolism. Similarly, the lungs are the organs through which the products of oxidative metabolism—carbon dioxide and some water vapor—are expelled to the atmosphere.

B. Normal pulmonary function depends on **ventilation** and **alveolar respiration** or **gaseous exchange**.

1. **Ventilation**, the mechanical displacement of gases, is effected by the muscles of the thoracic cage and abdominal wall, intrinsic elastic recoil of the thoracic cage, and intrinsic elastic recoil of the lungs.
   a. Reduction of vital capacity (flail chest, diaphragmatic paralysis, pneumothorax, pleural effusion, loss of intrinsic elasticity) or obstruction of the airway (mucus, asthma, neoplasm) reduces ventilation, results in dyspnea, and may even produce cyanosis.
   b. Ventilation moves gases through a series of airways with constantly decreasing diameters. Functionally, the airway is divided into conducting and respiratory portions.
      (1) The conducting portion consists of the trachea and bronchi as far as the terminal bron-

chiole. The air in the conductive portion at the end of inspiration does not exchange gases with the blood; similarly, the air in this portion at the end of expiration cannot be expelled (anatomic dead space).

(2) The respiratory portion consists of respiratory bronchioles, alveolar ducts, alveolar sacs, and alveoli.

2. **Gaseous exchange** occurs in the respiratory portion, primarily alveoli, whose total surface area is more than 70 m^2, or about the surface area of a tennis court. The alveoli are surrounded by capillary beds. **Alveolar respiration** is the exchange of gases across the **blood-air barrier**, which consists of the alveolar membrane, basement layer, and capillary endothelium, and is brought about by the differences in partial pressure between venous blood and alveolar air.

a. Normal gas partial pressures of venous blood in the pulmonary arteries are p_VO_2 = 40 torr and p_VCO_2 = 45 torr. Because alveolar air has partial pressures of p_AO_2 = 100 torr and p_ACO_2 = 40 torr, and gas exchange is very efficient, freshly oxygenated blood in the pulmonary veins normally reaches the same partial pressures as alveolar air.

b. Gaseous exchange is normally efficient. However, excretion of CO_2 and absorption of O_2 may be reduced when the effective blood-air barrier is reduced (**alveolar-capillary block**).

(1) Gaseous diffusion across the alveolar membrane is a function of membrane thickness, surface area, partial pressure differentials, and the diffusion coefficients of the gases. Although diffusion coefficients do not change, the coefficient of CO_2 is about 20 times that for O_2.

(2) Pulmonary fibrosis, pulmonary edema, and hyaline membrane disease increase the *effective thickness* of the diffusion pathway, resulting in poor oxygenation. Emphysema reduces the *surface area* available for gas exchange and thereby reduces oxygen absorption. Finally, the alveolar *partial pressures* are functions of ventilation. As the pressure differentials decrease, the gas exchange becomes inefficient and the blood perfusing the alveoli is incompletely oxygenated—in effect, an arteriovenous shunt.

c. Ideally, alveolar ventilation ($\dot{V}_a$ in L per min) is matched volume for volume by pulmonary vascular perfusion ($\dot{Q}$ in L per min). Whereas $\dot{V}_a/\dot{Q}$ should equal 1.0, this is not usually attained through the lung because of regional variations in ventilation and perfusion.

(1) Because most of the reserve lung expansion is toward the base of the pleural cavities, there is greater ventilation in the upper lobes than in the lower. Similarly, because of the effect of gravity in the erect posture, there is greater perfusion in the lower lobes than in the upper.

(2) The $\dot{V}_a/\dot{Q}$ ratio for the lobes may run as high as 3.3 in the upper lobes and as low as 0.7 in the lower lobes. These differences affect the amount of gas exchange between the alveoli and the blood.

(3) When the reserve volumes are utilized, as in exercise, the unevenness of blood flow and ventilation is lessened.

13
The Pericardial Cavity and the Heart

I. THE MIDDLE MEDIASTINUM

A. The middle mediastinum is bounded by:

1. The left and right pleural cavities.

2. The diaphragm.

3. The anterior and posterior leaves of the fibrous pericardium.

4. The superior boundary of the left and right pulmonary arteries.

B. Its contents include:

1. The roots of the lungs.

2. The left and right phrenic nerves.

3. The pericardial cavity and heart.

4. The pulmonary trunk and left and right pulmonary arteries.

5. The left and right superior and inferior pulmonary veins.

6. The superior and inferior venae cavae.

7. The arch of the azygos vein.

8. The ascending aorta.

9. Lymph nodes.

II. THE PERICARDIAL CAVITY

A. The pericardial cavity (sac) contains the heart and the roots of the great vessels (Fig. 13-1A).

B. It lies behind the sternum and the 2nd through 6th costal cartilages on the right side. On the left side it extends inferolaterally beyond the costochondral junction of the 2nd through 4th ribs as far laterally as the midclavicular line.

C. Two layers of pericardium comprise the pericardial sac.

1. The **fibrous pericardium** is an outer layer of tough, dense, irregularly arranged connective tissue, which

a. Is attached to the central tendon of the diaphragm and to the sternal periosteum by numerous small sternopericardial ligaments.

b. Prevents transient overdistension of the heart but will stretch with time to accommodate chronic heart enlargement due to disease such as congestive heart failure.

c. Is continuous with the adventitia of the aorta, upper pulmonary trunk, superior and inferior venae cavae, and pulmonary veins.

2. The **serous pericardium** has two subdivisions.

a. The **parietal serous pericardium** is a serous mesothelium lining the pericardial cavity.

b. The **visceral serous pericardium (epicardium)** is a reflection of the mesothelial layer over the great vessels and surface of the heart to form the outermost layer of the heart wall. This layer can be thought of as being formed by invagination of the pericardial cavity by the developing heart.

c. The serous mesothelium secretes fluid, which

(1) Keeps the surface layers moist and slippery, allowing nearly frictionless beating of the heart within the pericardial cavity.

(2) Provides surface tension between the pericardial layers, with resultant expansion of the heart during inspiration. In conjunction with lowered inspiratory intrathoracic pressure, blood return to the right side of the heart is increased.

3. Reflection of the parietal pericardium around the great vessels to become the visceral pericardium results in the formation of sinuses.

a. The oblique sinus (Fig. 13-1B):

(1) Represents a cul-de-sac in the pericardial cavity posterior to the heart.

(2) Results from reflection of the parietal pericardium around the inferior vena cava, right inferior and superior pulmonary veins, superior vena cava, and left superior and inferior pulmonary veins. Although not so named, this reflection is equivalent to a mesentery and suspends the heart in the pericardial cavity.

b. The transverse sinus (see Fig. 13-1B):

(1) Is a transverse passage behind the ascending aorta and pulmonary trunk.

(2) Is separated from the oblique sinus by the reflection of the pericardium (equivalent to a supporting mesentery), which runs between the left and right superior pulmonary veins.

(3) Has clinical importance during cardiac surgery because a ligature may easily be passed through the transverse sinus and around the aorta and pulmonary trunk to control hemorrhage.

4. Clinical considerations.

a. Pericardial tamponade. An accumulation of fluid (blood or serous) in the pericardial sac will compress the heart and decrease diastolic capacity; this results in reduced cardiac output but an increased heart rate (indicated by weaker but rapid pulse) and increased venous pressure, with associated jugular vein distension, pulsating liver, edema, and dyspnea resulting from pleural effusion.

b. Pericardiocentesis and intracardiac injection.

(1) *Subcostal route.* A needle may be safely introduced into the pericardial sac via the sternocostal angle adjacent to the xiphoid process. The needle is inserted on the left side adjacent to the xiphoid process, angling upward at about 45° and to the left. The risk of iatrogenic pneumothorax is small because the pleural cavities are avoided, and any fluid may be effectively drained if the patient is propped up slightly. Also avoided is the risk of puncturing the left anterior descending or the right marginal artery.

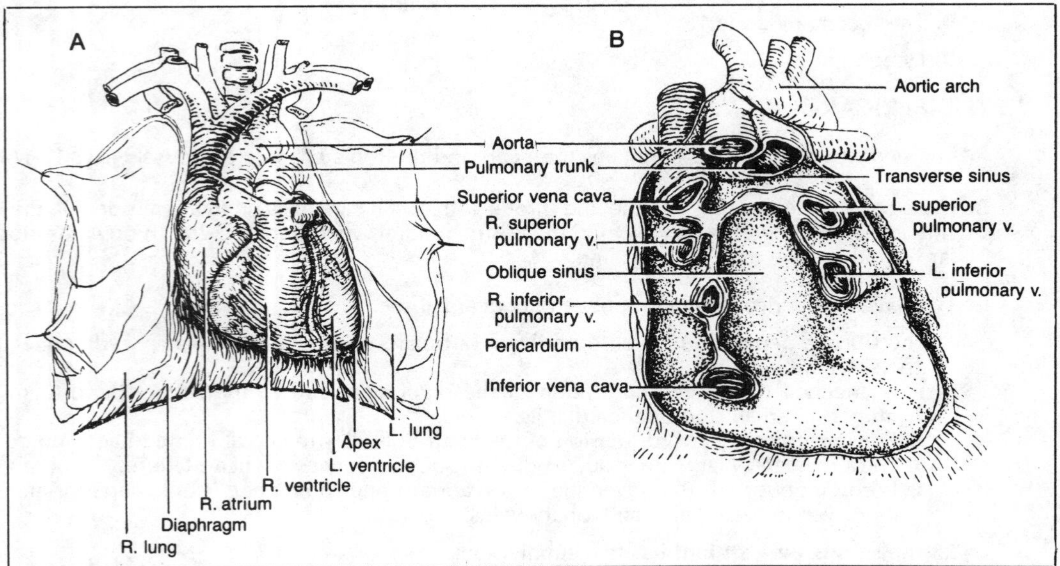

Figure 13-1. *A, The middle mediastinum.* The medial portions of the lungs have been retracted to show the location of the heart in the pericardial cavity. *B, The pericardial cavity.* With the heart removed, the line of pericardial reflection about the vessels that enter and leave the heart forms the oblique and transverse sinuses.

(2) *Parasternal route*. Because of the cardiac notch in the pleura on the left side, a needle may be introduced into the pericardial cavity or heart through the left intercostal space immediately adjacent to the sternum with little chance of inducing a pneumothorax. Since the internal thoracic artery generally passes about 1 cm or more lateral to the margin of the sternum, this artery is avoided if the edge of the sternum is located with the tip of the needle before plunging into the pericardial sac. This is usually the preferred route for intracardiac injection.

c. **Pericarditis.** Inflammation of the parietal serous pericardium results in pain referred to (appearing as if originating from) the precordium and the epigastrium. A friction rub may be heard on auscultation.

III. THE HEART

A. The heart is located in the pericardial sac and covered with visceral pericardium (see Fig. 13-1).

1. Normally it weighs about 250 g in women and 300 g in men.
2. The transverse diameter varies with inspiration and expiration, but normally measures 8 to 9 cm in transverse diameter on a radiograph at end-maximum inspiration in the standing position. Generally the transverse diameter should not be more than one-half the diameter of the chest.

B. EXTERNAL FEATURES OF THE HEART (Fig. 13-2)

1. Approximately one-third of the heart lies to the right of the midline and two-thirds to the left.
2. The **cardiac boundaries** are visible radiographically.
   a. The right border delineates the right atrium, superior vena cava, and inferior vena cava.
   b. The inferior border is delineated by the right ventricle.
   c. The **apex** is the tip of the left ventricle. An **apical pulse** or **point of maximal impulse** (PMI) can be palpated and sometimes visualized in thin men at the 5th intercostal space just beneath the nipple.
   d. The left (oblique) border is formed by the left ventricle.
   e. An upper border is indistinct radiographically.
3. **Surface structure of the heart.**
   a. **The coronary sulcus:**
      (1) Completely encircles the heart.
      (2) Divides the atria from the ventricles.
      (3) Has a nearly vertical orientation behind the sternum.
      (4) Contains the circumflex branch of the **left coronary artery, right coronary artery, coronary sinus**, and **small cardiac vein**, all embedded in fat.
   b. **The anterior interventricular sulcus:**
      (1) Marks the location of the **interventricular septum** separating the ventricles.
      (2) Runs diagonally along the sternocostal surface close to the left border of the heart.
      (3) Contains the **anterior interventricular (descending) artery** (a branch of the left coronary artery) and the **great cardiac vein.**
   c. **The posterior interventricular sulcus:**
      (1) Is the continuation of the anterior interventricular sulcus about the apex onto the diaphragmatic surface of the heart.
      (2) Delineates the interventricular septum.
      (3) Contains the **posterior interventricular (descending) artery** (usually a branch of the right coronary artery) and the **middle cardiac vein.**
4. **Surface characteristics of the chambers of the heart.**
   a. The heart is divided into right and left sides, each of which is composed of two chambers, an atrium and a ventricle.
   b. **The right atrium:**
      (1) Forms the right border of the heart, part of the base, and part of the sternocostal surface.
      (2) Gives off the auricular appendage.
      (3) Receives the superior vena cava, inferior vena cava, and coronary sinus.
      (4) Discharges into the right ventricle through the tricuspid valve.
   c. **The right ventricle:**
      (1) Forms the largest portion of the sternocostal surface and a small part of the diaphragmatic surface.
      (2) Projects to the left of the right atrium toward the apex.
      (3) Receives blood from the right atrium.

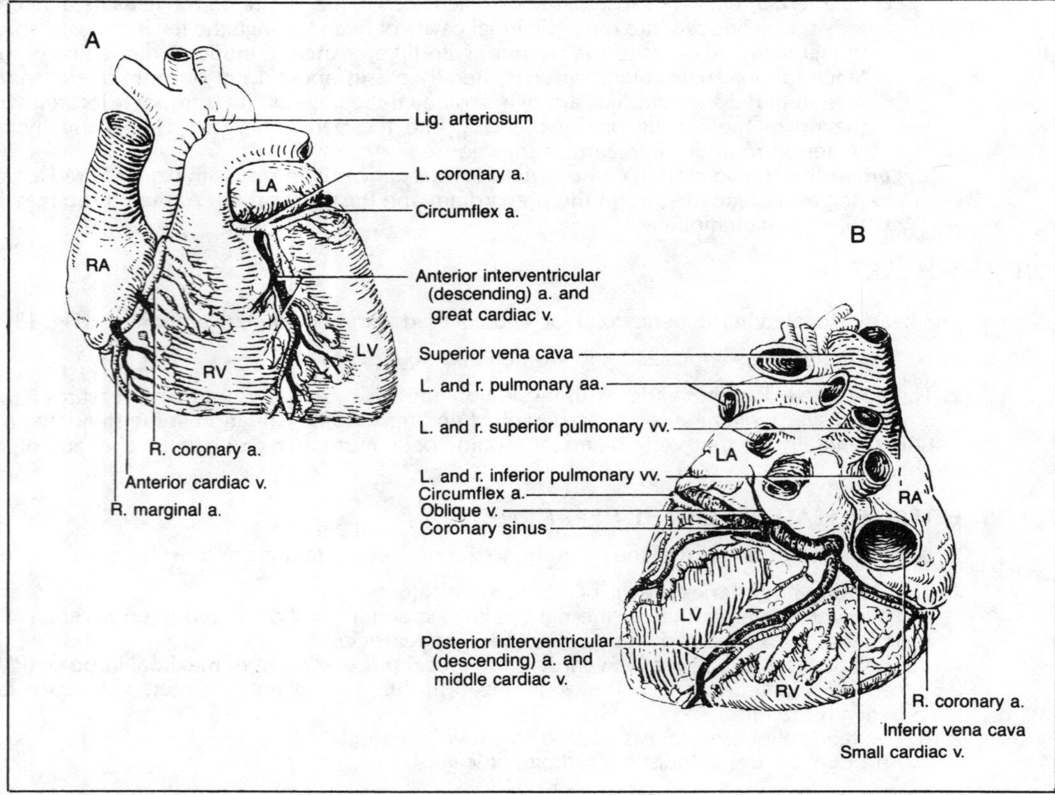

Figure 13-2. *A, Anterior aspect of the heart and coronary circulation. B, Posterior aspect of the heart and coronary circulation.*

(4) Discharges through the pulmonary semilunar valve into the pulmonary trunk just to the left of the aorta.

d. The left atrium:

(1) Forms the posterior surface (**base**) of the heart, lying entirely within the boundaries of the oblique sinus of the pericardial sac.

(2) Gives off the left auricular appendage.

(3) Receives two right pulmonary veins (sometimes three) and two left pulmonary veins (sometimes one).

(4) Discharges into the left ventricle through the mitral valve.

e. The left ventricle:

(1) Forms the left border, the **apex**, one-third of the sternocostal surface, and two-thirds of the diaphragmatic surface.

(2) Projects to the left of the left atrium.

(3) Receives blood from the left atrium.

(4) Ejects blood into the aortic vestibule and ascending aorta.

IV. THE CORONARY CIRCULATION

A. ARTERIAL CIRCULATION. The myocardium and epicardium (visceral pericardium) are supplied by two main arteries, the **left** and **right coronary arteries.**

1. The right coronary artery (see Fig. 13-2; Fig. 13-3):

a. Arises from the **right aortic sinus** (of Valsalva).

b. Courses in the coronary sulcus around the heart toward the right, beneath the right auricular appendage.

c. Has as major branches:

(1) The **nodal branch**, which courses toward the superior vena cava, supplying the right atrium and the sinoatrial (S-A) node.

(2) The **right marginal branch**, which passes along the inferior border toward the apex, supplying a portion of the right ventricle.

(3) The **posterior interventricular (descending) branch**, which gives off a branch to the atrioventricular (A-V) node and anastomoses with the anterior descending branch of the left coronary artery on the diaphragmatic surface in the vicinity of the apex, supplying the posterior third of the interventricular septum.

d. The right coronary artery generally supplies:

(1) The right atrium (including the **S-A node**).

(2) Superior parts of the right ventricle.

(3) The posterior third interventricular septum, including the **A-V node** and the **right branch of the A-V bundle** (of His). (Exception: In 10 to 15 percent of the population, this region is supplied by the left coronary artery.)

e. It usually continues a few centimeters beyond the posterior interventricular sulcus to anastomose with terminal twigs of the **circumflex branch** of the left coronary artery.

2. The left coronary artery (see Figs. 13-2 and 13-3):

a. Arises from the **left (left posterior) aortic sinus** (of Valsalva).

b. Runs for a short, variable distance between the pulmonary trunk and aorta before bifurcating into:

(1) The first branch, the **anterior interventricular (descending) artery**, which

(a) Enters the anterior interventricular sulcus.

(b) Generally supplies:

(i) The anterior aspects of the left and right ventricles.

(ii) The anterior two-thirds of the interventricular septum, via septal branches, including the **left branch of the A-V bundle** (of His), with some branches to the A-V node in about 40 percent of the population and total supply in 10 to 15 percent.

(c) Has as a major branch the **left marginal artery**, which courses along the left border of the heart to supply the left ventricle.

(d) Anastomoses with the posterior descending branch of the right coronary artery on the diaphragmatic surface of the heart in the vicinity of the apex.

(2) The **circumflex artery**, which

(a) Runs in the coronary sulcus toward the left border and around to the base of the heart.

(b) Generally supplies:

(i) The posterior aspect of the left atrium.

(ii) The superior portions of the left ventricle.

(c) Anastomoses with terminal twigs of the right coronary artery.

(d) Gives rise to the posterior descending artery in 10 to 15 percent of the population.

3. Variations in the coronary circulation.

a. Balanced coronary circulation:

(1) Is the coronary circulation described above.

(2) Is present in 60 to 65 percent of the population.

(3) Is sometimes termed "right dominant" because the right coronary artery gives off the posterior descending branch, but this term does not distinguish this condition from right preponderant circulation.

b. Left preponderant coronary circulation:

(1) Is the type of circulation in which the circumflex branch of the *left* coronary artery gives rise to the posterior descending branch, so that the entire interventricular septum, including the A-V node, is supplied by the left coronary artery.

(2) Reduces possibilities for development of effective collateral circulation in the septal region and therefore lowers the chances of survival after septal infarct.

(3) Is present in about 10 to 15 percent of the population.

(4) Is sometimes termed "left dominant" because the left coronary artery gives off the posterior descending branch.

c. Right preponderant coronary circulation:

(1) Is the type of circulation in which the right coronary artery, in addition to supplying the posterior descending artery, crosses the posterior interventricular septum to reach as far as the left marginal artery, so that a substantial portion of the diaphragmatic surface of the left ventricle is supplied by this extension of the right coronary artery.

(2) Is present in 20 to 25 percent of the population.

(3) Is sometimes called "right dominant" because the right coronary artery gives off the posterior descending branch, but this term does not distinguish this condition from balanced circulation.

d. Variations frequently occur in the manner in which the coronary arteries arise from the aorta or even, lethally, from the pulmonary trunk.

4. Functional considerations.

a. Normally, there are few anastomoses between left and right coronary arteries in the healthy young individual.

(1) In the normal healthy heart, most of the arteries are true end-arteries, which supply discrete volumes of myocardium with little overlap or few collaterals. Any anastomoses present are generally inadequate to maintain effective circulation in the event of sudden-onset occlusion of a major branch of the coronary circulation.

(2) With slow-onset occlusion (atherosclerotic coronary artery disease), collateral circulation develops.

(3) Obstruction in a coronary artery produces ischemia. Pain associated with ischemic cardiac tissue—**angina pectoris**—is *referred* to (appears as if originating from) the precordium, epigastrium, shoulder, and frequently, the left arm. Ischemia may result in necrosis and hemorrhage (an infarct), with severe and life-threatening damage to the myocardium. Even if a patient survives the myocardial insult, the thrombus that forms over the damaged endocardium may become dislodged and produce life-threatening emboli.

b. Because left ventricular systolic pressure and left ventricular transmural pressure are greater than or equal to aortic systolic pressure, blood flows through the coronary circulation of the *left* ventricle only during diastole. (Recall the second law of thermodynamics.) Thus, the heart has a low margin of safety with respect to the amount of perfusion loss before ischemia results.

B. VENOUS CIRCULATION (see Fig. 13-2)

1. The coronary sinus:

a. Receives most of the venous return from the epicardium and myocardium.

b. Is approximately 2 cm long.

c. Lies in the coronary sulcus, directed toward the diaphragmatic surface.

d. Opens into the right atrium between the opening of the inferior vena cava and the right atrioventricular valve.

e. Has four tributaries:

(1) The **great cardiac vein**, which

(a) Lies in the anterior interventricular groove.

(b) Drains the anterior portion of the interventricular septum and the anterior aspects of both ventricles in the vicinity of the septum.

(c) Accompanies the anterior descending (interventricular) branch of the left coronary artery.

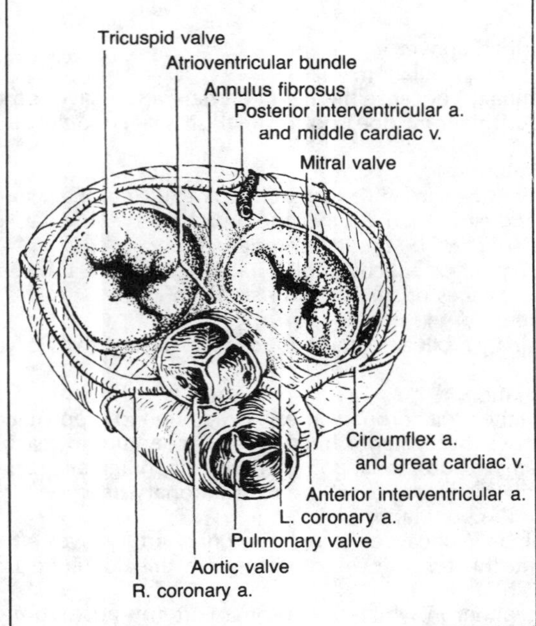

Figure 13-3. *The annulus fibrosus.* The left and right atria have been dissected away showing the fibrous ring that supports the atrioventricular and semilunar valves and showing the origins of the left and right coronary arteries.

(2) The **middle cardiac vein**, which
(a) Lies in the posterior interventricular groove.
(b) Drains the posterior portion of the interventricular septum and the posterior aspects of both ventricles in the vicinity of the septum.
(c) Accompanies the posterior descending (interventricular) branch of the right coronary artery.
(3) The **small (lesser) cardiac vein**, which
(a) Lies in the coronary sulcus to the right of the opening of the coronary sinus.
(b) Usually terminates as the **right marginal vein**, which occasionally opens separately into the right atrium.
(c) Drains the marginal aspect of the right ventricle.
(d) Accompanies the right coronary artery and its marginal branch.
(4) The **oblique vein** (Marshall's), which
(a) Courses superolaterally across the posterior aspect of the left atrium.
(b) Is the embryologic remnant of the left common cardinal vein and may occasionally develop into a persistent left superior vena cava whether or not the right vena cava is present.

2. **The anterior cardiac veins:**
a. Consist of one to several small veins that drain the anterior surface of the right ventricle.
b. Course across the coronary sulcus anterior to the right coronary artery.
c. Open separately into the right atrium.

3. **The least cardiac (thebesian) veins or venae cordis minimae:**
a. Drain the endocardium and innermost layers of myocardium.
b. Tend to empty directly into the atria, with fewer draining into the ventricles.

V. THE CARDIAC WALL

A. ENDOCARDIUM

1. The inner lining of the cardiac chambers is a thin, smooth endothelium.

2. Valve cusps are formed by a double layer of endocardium over a connective-tissue stroma.

B. MYOCARDIUM

1. This intermediate layer is composed of three layers of cardiac muscle, which
a. Are most definite and substantial in the ventricles.
b. Are oriented nearly perpendicular to each other.
c. Are arranged in a spiral manner that produces a wringing maneuver to reduce the volume of the chambers during systole.
d. Originate from the **annulus fibrosus** (fibrous ring).
e. Do not pass across the coronary sulcus.

2. The myocardium varies in thickness between chambers, depending upon the functional requirements of each chamber.

3. The myocardium contains excitable nodal tissue and the cardiac conducting system, consisting of:
a. The **S-A node**, which initiates atrial systole.
b. The **A-V node**, which is paced by the S-A node to initiate ventricular systole.
c. The **A-V bundle** (of His), which transmits electrical activity throughout the ventricles.

C. EPICARDIUM. The serous visceral pericardium and a variable amount of fatty tissue form the outermost layer of the heart.

D. THE FIBROUS RING (ANNULUS FIBROSUS) [see Fig. 13-3]

1. The annulus fibrosus is a layer of dense connective tissue, which
a. Is arranged in the A-V plane.
b. Forms the "skeleton" of the heart.
c. Demarcates the coronary sulcus.
d. Provides the site of origin for muscle layers of the atria and ventricles.
e. Surrounds and supports each valvular opening.

2. The connective-tissue stroma of the left and right A-V valves as well as the pulmonary and aortic valves is continuous with the fibrous ring.

3. Only the A-V bundle passes through the fibrous ring.

VI. THE CHAMBERS OF THE HEART

A. THE RIGHT ATRIUM (Fig. 13-4)

1. The right atrium has the thinnest walls of the four chambers.

2. The **crista terminalis** divides it into the sinus venarum and the atrium proper.

3. The sinus venarum:
- **a.** Is a smooth-walled region posterior and to the right of the crista terminalis.
- **b.** Represents the embryologic sinus venosus.
- **c.** Receives:
  - **(1)** The superior vena cava.
  - **(2)** The inferior vena cava, the valve of which (eustachian valve) is incompetent in the adult but in the fetus directs oxygenated blood from the inferior vena cava through the foramen ovale into the left atrium.
  - **(3)** The coronary sinus, the valve of which (thebesian valve) may reduce regurgitation of blood into the coronary sinus during atrial systole.
  - **(4)** The anterior cardiac veins.
  - **(5)** Many, but not all, of the thebesian veins.

4. The atrium proper:
- **a.** Comprises the anterior muscular portion of the atrium.
- **b.** Lies to the left of the **crista terminalis**, which
  - **(1)** Is a muscular ridge running anteriorly along the atrial wall from just medial to the opening of the superior vena cava to just medial to the inferior vena cava.
  - **(2)** Gives origin to the **pectinate muscles**, which run across the atrial walls.
- **c.** The atrium contains the **right auricular appendage**, which
  - **(1)** Has muscular walls.
  - **(2)** Is a potential site for the formation of thrombi that, if dislodged, can result in pulmonary embolism.
- **d.** The atrium proper empties into the right ventricle via the right A-V (tricuspid) valve.

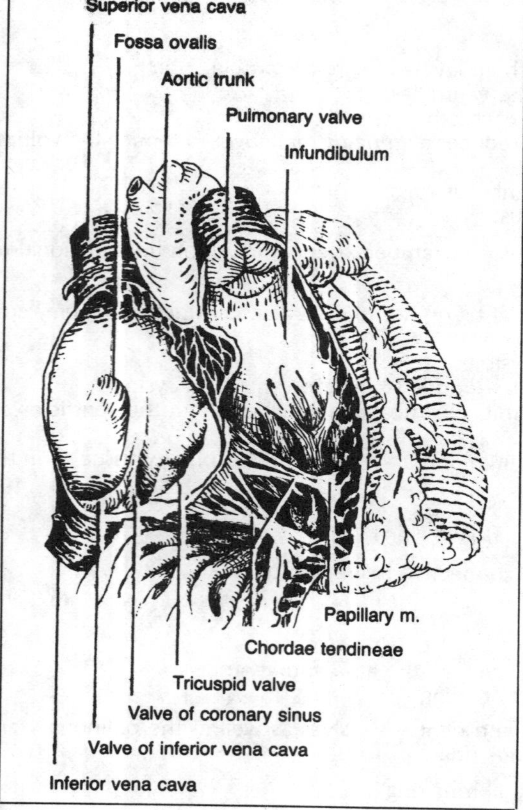

Figure 13-4. *Chambers of the right side of the heart.* The anterior heart wall has been dissected away revealing the right atrium, tricuspid valve, right ventricle, and pulmonic valve.

5. The atrial septal wall:
  a. Separates right and left atria.
  b. Contains the **fossa ovalis**, which
    (1) Is the embryologic remnant of the fetal **foramen ovale**, which connects the atria in the fetus and represents the **septum primum** (floor) and **septum secundum** (limbus or upper lateral edge).
    (2) May be probe-patent in 10 percent of the population.
    (3) Is the most common location of **atrial septal defects** (ASD).
      (a) Septum primum defects are rare and usually accompanied by endocardial cushion defects involving valvular structure.
      (b) Septum secundum defects (the typical patent foramen ovale) comprise 10 to 15 percent of all cardiac anomalies.
      (c) The left atrial pressure is normally slightly greater than the right atrial pressure, so that there is typically a **left-to-right shunt** through an ASD.
        (i) Usually there is no cyanosis associated with an ASD because oxygenated blood from the left side of the heart is shunted back to the right side.
        (ii) Atrial septal defects are usually compatible with a normal, healthy life unless extreme exercise, cardiac disease, or pulmonary disease alters normal chamber pressures, thereby inducing a **right-to-left shunt**, with resultant cyanosis.

B. THE RIGHT VENTRICLE (see Fig. 13-4)

1. To overcome resistance to flow in the pulmonary vascular bed, the right ventricle has a moderately thick wall.

2. It is separated into two regions, the **right ventricle proper** (for inflow) and the **infundibulum** (for outflow) by the **crista supraventricularis**, which lies between the right A-V valve and the pulmonary orifice.

3. The right ventricle proper:
  a. Has rough, muscular walls, with **trabeculae carneae** and anterior, posterior, and septal **papillary muscles.**
  b. Receives blood from the right atrium through the **right A-V (tricuspid) valve**, which
    (1) Has three valve **leaflets** or **cusps**: anterior, posterior, and septal; each cusp
      (a) Has a connective tissue core that is attached to the annulus fibrosus and covered with endothelium.
      (b) Is directed into the ventricle, with connective-tissue strands, **chordae tendineae,** running between the free edge and usually two papillary muscles.
      (c) Is associated with a large **anterior papillary muscle**, a smaller **posterior papillary muscle**, and usually several very small **septal papillary muscles.**
    (2) The tricuspid valve prevents regurgitation of blood into the atrium during ventricular systole.
      (a) During the isovolumic (pressure build-up) phase of ventricular systole, the papillary muscles tense to prevent eversion of the valve cusps.
      (b) Since the volume of the ventricle is reduced during the isotonic (blood ejection) phase of systole, shortening of the papillary muscles takes up any slack in the chordae tendineae to maintain the competence of the valvular closure.

4. The interventricular septum:
  a. Bulges into the right ventricle, giving the cavity a crescent shape in cross-section.
  b. Gives origin to the septal papillary muscle(s).
  c. Is mostly muscular but has a small membranous upper portion, which
    (1) Is composed of connective tissue continuous with the annulus fibrosus.
    (2) Is the usual site of **ventricular septal defects** (VSD). If small, a VSD may result in a left-to-right shunt; but in the presence of pulmonary stenosis, a VSD results in a **right-to-left shunt** producing cyanosis and the "blue-baby" syndrome. A VSD is usually the principal factor in **tetralogy of Fallot.**

5. The infundibulum:
  a. Is smooth-walled.
  b. Leads into the pulmonary orifice.
  c. Contains the pulmonary semilunar valve, which
    (1) Consists of three semilunar cusps or demilunes: left, right, and anterior; these cusps
      (a) Are attached peripherally to the annulus fibrosus.
      (b) Have their free edges strengthened with a rim of connective tissue and a **nodule** in the center of the free edges that completely closes the lumen.
      (c) Have their central free edges directed upward into the pulmonary trunk.

(2) The pulmonary semilunar valve prevents regurgitation of ejected blood from the pulmonary trunk into the ventricle during ventricular diastole.

6. The pulmonary trunk:

a. Starts at the level of the pulmonary valves.

b. Contains the **pulmonary sinuses** (of Valsalva), slight depressions in the walls of the pulmonary trunk behind each valve leaflet.

c. Lies in the pericardial cavity to the left of the aortic outflow pathway for the first 3 cm.

d. Bifurcates into the right and left pulmonary arteries at about 5 cm.

C. THE LEFT ATRIUM (Fig. 13-5)

1. The left atrium has slightly thicker walls than the right because of the increased effort required during atrial systole to overcome the elasticity of the extremely thick left ventricular walls.

2. This chamber receives a variable number of right and left pulmonary veins—usually two from each side, but frequently three on the right or one on the left.

3. It contains the **fossa ovalis** on the septal wall, the remnant of the embryonic **foramen ovale**.

4. It gives off the left **auricular appendage**, which

a. Has muscular walls.

b. Is a potential site for the formation of thrombi, which if dislodged can result in cerebral, systemic, or renal embolism.

5. The left atrium opens into the left ventricle via the left A-V valve.

D. THE LEFT VENTRICLE (see Fig. 13-5)

1. To overcome the vascular resistance of the systemic vascular bed, the wall of the left ventricle is three times as thick as the wall of the right ventricle and has well-developed trabeculae carneae.

2. The left ventricle is divided into the **left ventricle proper** and the **aortic vestibule**.

3. The left ventricle proper:

a. Has the most muscular wall of all the heart chambers.

b. Has a round cavity in cross-section because the interventricular septum bulges into the right ventricle.

c. Receives blood from the left atrium through the **left A-V (bicuspid, mitral) valve** (Fig. 13-6B), which

(1) Has a large **anterior cusp** and a smaller **posterior cusp** attached peripherally to the annulus fibrosus.

(2) Has a greater number of **chordae tendineae** than the right A-V valve. The chordae run from the free edges of the cusps to the papillary muscles.

(3) Has a large **anterior papillary muscle** and a smaller **posterior papillary muscle**.

(4) Prevents regurgitation of blood into the left atrium during ventricular systole in a manner similar to that described for the tricuspid valve [VI B 3 b (2)].

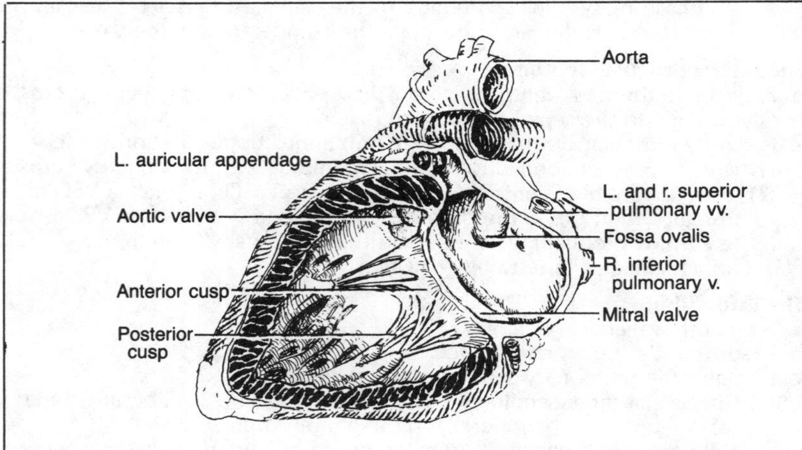

Figure 13-5. *Chambers of the left side of the heart.* The posterior heart wall has been dissected away revealing the left atrium, mitral valve, left ventricle, and aortic valve.

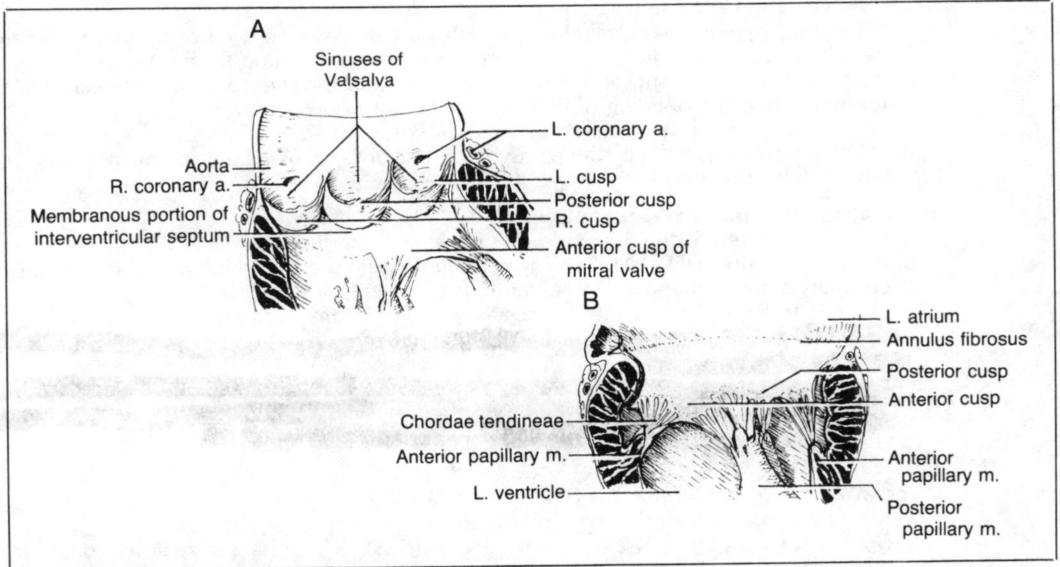

Figure 13-6. *Semilunar and atrioventricular valve structure. A,* The aorta has been opened between the left (posterior) and right semilunar valve cusps behind which are the ostia of the coronary arteries. *B,* The left ventricle has been opened showing the anterior and posterior valve cusps, the chordae tendineae, and the anterior and posterior papillary muscles of the mitral valve.

(a) Insufficiency of the mitral valve and transmission of the left ventricular systolic pressure to the left atrium and pulmonary vascular bed may lead to right heart failure (cor pulmonale).

(b) Mitral stenosis will manifest on auscultation as late diastolic murmur; mitral insufficiency will be heard as a low-pitched, late systolic blowing murmur.

4. The aortic vestibule:

a. Is located superior to and to the right of the mitral valve.

b. Leads into the **ascending aorta**.

c. Contains the **aortic valve**.

(1) This valve consists of three semilunar cusps or demilunes, right (anterior), left (left posterior), and posterior (right posterior), which have anatomic characteristics similar to those of the pulmonary semilunar valves (VI B 5 c).

(2) It prevents regurgitation of blood from the aorta into the ventricle during ventricular diastole.

(a) A common anomaly is a bicuspid aortic valve, which is especially prone to stenosis with age.

(b) Aortic insufficiency will manifest on auscultation as a diastolic murmur whereas aortic stenosis will tend to be heard as a high-pitched systolic murmur.

5. The ascending aorta:

a. Starts at the level of the aortic valve (Fig. 13-6A).

b. Contains the **aortic sinuses** (of Valsalva), which

(1) Are slight depressions in the walls of the aorta behind the valve cusps.

(2) Accommodate the volume of the open valve leaflets to reduce turbulent flow during ejection.

(3) Provide the origins of the **coronary arteries**.

(a) The **right coronary artery** normally originates from the **right aortic sinus**.

(b) The **left coronary artery** normally originates from the **left (left posterior) aortic sinus**.

c. The elastic properties of the aorta accommodate the ejected blood volume and maintain the range of diastolic arterial pressure.

VII. THE CONDUCTING SYSTEM OF THE HEART

A. Specialized cardiac muscle cells in certain regions of the myocardium have highly developed sensitivity and autorhythmicity.

1. **The sinoatrial (S-A) node** (Fig. 13-7):
   a. Is the autorhythmic pacemaker and initiates the contraction cycle with approximately 72 depolarizations per minute, which spread over both atria and to the A-V node.
   b. Is about 7 mm × 2 mm × 1 mm and is located in the myocardium between the crista terminalis and the opening of the superior vena cava.
   c. Usually is supplied by a branch of the right coronary artery.
   d. Is influenced principally by the parasympathetic division of the autonomic nervous system which slows the autorhythmicity.
2. **The atrioventricular (A-V) node** (see Fig. 13-7):
   a. Is not connected to the S-A node by specialized conducting fibers.
   b. Is located in the right atrial floor near the interatrial septum, medial to the ostium of the coronary sinus and above the septal cusp of the tricuspid valve.
   c. Usually (in 85 to 90 percent of cases) is supplied by the right coronary artery.
   d. Is autorhythmic, with approximately 40 depolarizations per minute, but is stimulated by atrial depolarizations.
   e. Has a slow conduction velocity (long latency), which allows atrial depolarization to spread over the entire right and left atria (0.09 second) before sending depolarization along the A-V bundle; thus the atria normally contract before the ventricles.
   f. Gives rise to the A-V bundle.
3. **The atrioventricular bundle (of His)** [see Fig. 13-7]:
   a. Arises from the A-V node and crosses the A-V ring (annulus fibrosus).
   b. Consists of modified muscle cells (Purkinje fibers), which have a rapid conduction rate (low latency), so that the A-V bundle begins to depolarize at about 0.16 second after the S-A node and activity spreads throughout the ventricles by 0.22 second.
   c. Descends along the posterior border of the membranous part of the interventricular septum to enter the muscular portion of the septum.
   d. Divides into **left** and **right branches (bundles** or **crura**); they descend into the interventricular septum and spread out into the walls of the ventricles, where they eventually become indistinguishable from the cardiac muscle fibers, which continue to conduct the impulse.

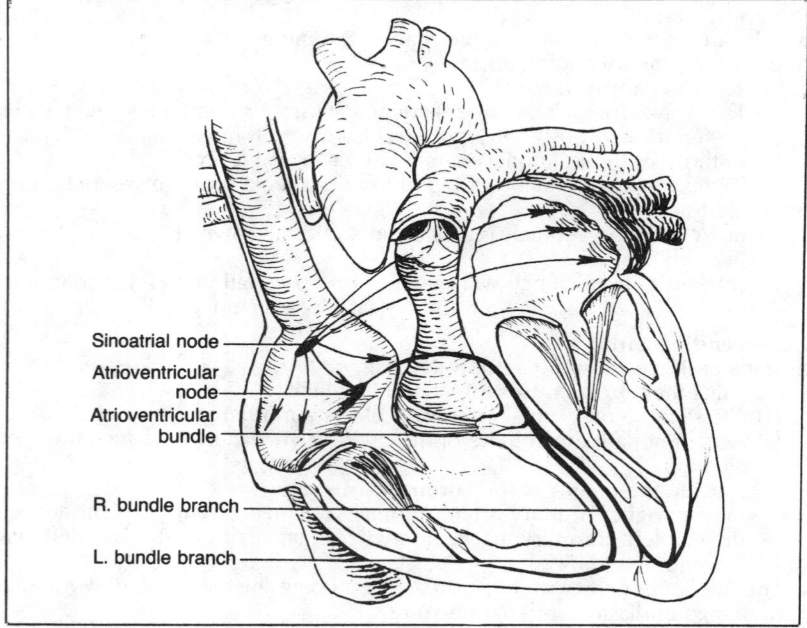

Figure 13-7. *The conducting system of the heart.* Electrical activity initiated in the S-A node spreads throughout the atrial myocardium within 0.1 second, producing atrial systole and exciting the A-V node. From the A-V node, electrical activity enters the ventricles by the A-V bundle, travels along left and right bundle branches, and spreads over the ventricular myocardium to produce ventricular systole within 0.22 second.

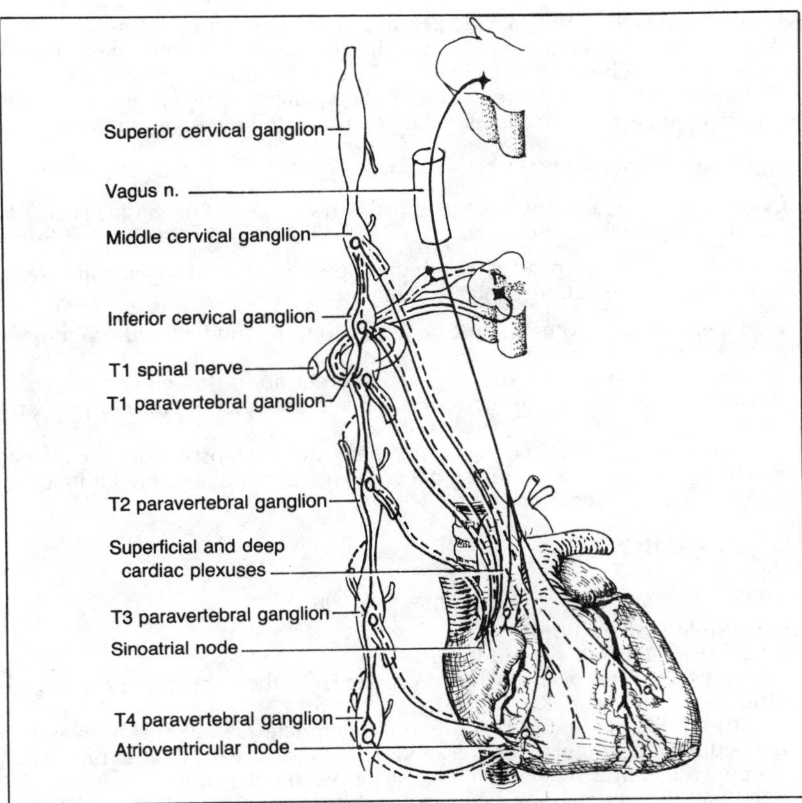

Figure 13-8. *Autonomic innervation of the heart.* The first, or preganglionic, sympathetic neurons in the thoracic spinal cord (levels T1–T4) send axons through the ventral roots, spinal nerves, and white rami communicantes to gain access to the sympathetic chain. The second, or postsynaptic, neurons lie in the sympathetic chain ganglia and send axons either through the cervical cardiac accelerator nerves or through the thoracic splanchnic nerves to the heart. The preganglionic parasympathetic neurons lie in the brain stem and send axons along the vagus nerve (CN X) to the heart. The short postganglionic neurons lie in the walls of the heart. Axons conveying painful sensation from the heart travel along the sympathetic pathways to the upper thoracic segments of the spinal cord.

B. CLINICAL CONSIDERATIONS

1. The contractile stimulus for any muscle is an electrical depolarization of its surface membrane. The heart has a large surface-to-volume ratio and consequently a large signal is generated that is detectable on the body surface by means of an electrocardiogram (ECG).
2. If a pathologic condition interrupts impulse propagation in the A-V bundle, heart block occurs, with asynchronous beating of the atria and ventricles. If depolarization is initiated outside of the nodal system, the ventricular muscle may then fibrillate, resulting in no blood being pumped.
3. **Bundles of Kent** are occasional abnormal muscle bridges between the atria and the ventricles, which may cause ventricular preexcitation by bypassing the A-V node (Wolff-Parkinson-White syndrome).

VIII. CARDIAC INNERVATION

A. The heart rate and ejection volume are controlled by the autonomic nervous system (Fig. 13-8).

1. *Parasympathetic division.* The **vagus nerve** sends fibers over the surface of the heart and to the nodal areas.
   a. These preganglionic fibers synapse with minute postganglionic fibers in the myocardium.
   b. Vagal activity slows the heart rate and reduces the stroke volume.

2. *Sympathetic division.* The **cardiac accelerator nerves** run in the neck to the heart from the superior, middle, and inferior ganglia of the cervical sympathetic chain. **Thoracic splanchnic nerves** run to the heart from ganglia T1–T3 of the thoracic sympathetic chain.
   a. These sympathetic fibers terminate in the vicinity of the S-A and A-V nodes.
   b. Sympathetic activity accelerates the heart rate and increases stroke volume.

B. AFFERENT SENSATION FROM THE HEART

1. Afferent nerves run along sympathetic pathways via both the cardiac accelerator nerves and the thoracic splanchnic nerves to reach levels T1–T3 of the spinal cord (see Fig. 13-8).
2. Pain originating in the heart is referred to (perceived as coming from) the arm, shoulder, precordium, and sometimes epigastric region.
3. The concept of referred pain must be understood to conduct and properly evaluate the results of a physical examination.
   a. Painful sensation mediated by visceral afferent fibers that enter the spinal cord at a particular level usually is referred to the somatic dermatome corresponding to that vertebral level.
   b. The mechanism of referred pain is not clear, but the most probable explanation involves a final common pathway in the spinal cord with cerebral interpretation toward the somatic rather than the visceral location.

IX. CARDIAC DYNAMICS

A. The cardiac cycle is approximately 0.8 second (Fig. 13-9).

1. **Atrial systole (contraction):**
   a. Is approximately 0.1 second long.
   b. Produces pressures of approximately 5 mm Hg in the right atrium and 11 mm Hg in the left atrium.
   c. Pumps a small amount of blood into the ventricles. (Since the atrioventricular valves are open during ventricular diastole and since ventricular filling is primarily passive, atrial systole results in a slight stretching of the ventricular muscle, thereby placing the heart muscle at the optimal point on the length-tension curve.) [Figs. 13-10 A and B]

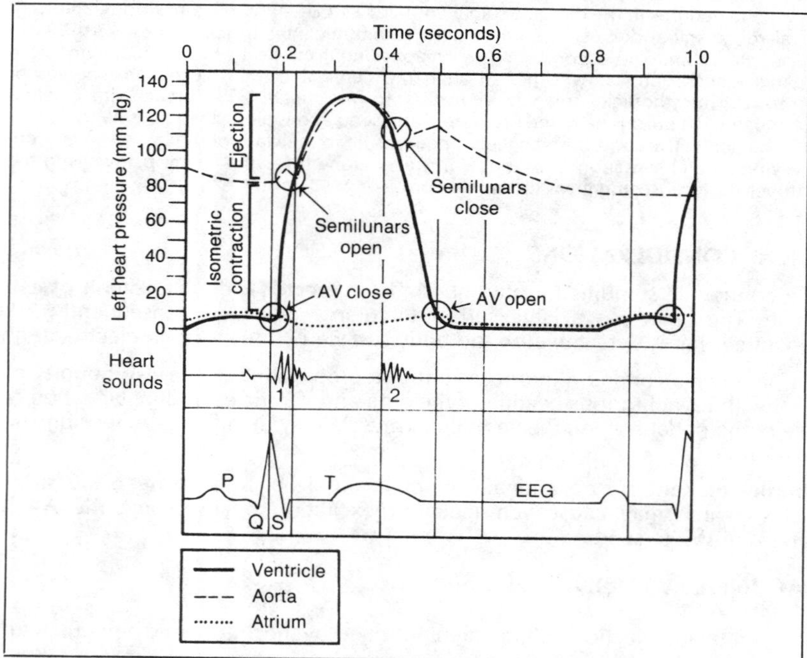

Figure 13-9. *Schematic diagram of the pressures developed during the cardiac cycle.* The atrioventricular valves close (S1, "lub") early in ventricular systole when ventricular pressures exceed atrial pressures. The semilunar valves close (S2 or "dup") early in ventricular diastole when arterial pressures exceed ventricular pressures.

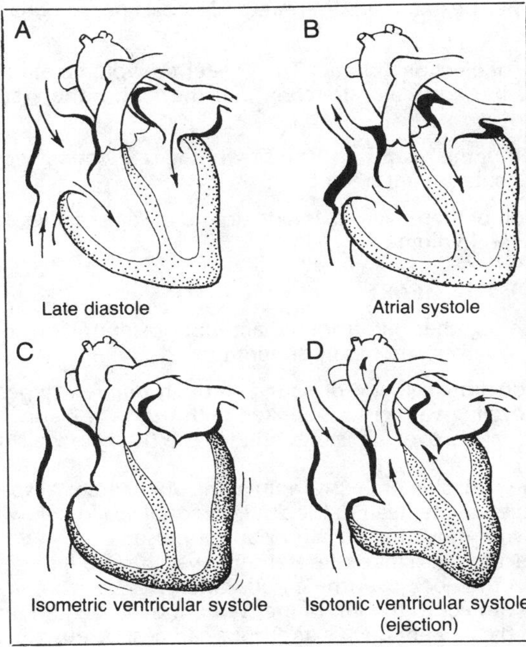

Figure 13-10. *The cardiac cycle.* The cardiac cycle normally requires 0.8 second. *A,* Late diastole takes 0.4 second to 0.8 second. *B,* Atrial systole takes 0.0 second to 0.1 second. *C,* Isometric ventricular systole takes 0.1 second to 0.2 second. *D,* Isotonic ventricular systole takes 0.2 second to 0.4 second.

2. Atrial diastole (relaxation):
   a. Begins at 0.2 second into the cardiac cycle and is approximately 0.7 second long.
   b. Coincides with pressures of – 5 mm Hg in the right atrium and 0 mm Hg in the left atrium.

3. Ventricular systole:
   a. Begins at approximately 0.1 second into the cardiac cycle and is approximately 0.3 second long.
   b. Produces pressures of approximately 23 mm Hg in the right ventricle and approximately 120 mm Hg in the left ventricle.
   c. Causes the atrioventricular valves to close when ventricular pressure exceeds atrial pressure, thus
      (1) Producing the S1 ("lub") heart sound.
      (2) Beginning a period of isometric contraction, during which pressure builds up (Fig. 13-10C).
   d. Ventricular systole causes the semilunar valves to open when ventricular pressure exceeds pulmonary diastolic pressure (10 mm Hg) and aortic diastolic pressure (80 mm Hg), thus
      (1) Beginning the period of isotonic contraction, during which blood is ejected into the outflow trunks (Fig. 13-10D).
      (2) Producing the maximum systolic pressure.

4. Ventricular diastole:
   a. Begins at approximately 0.4 second and has a 0.5-second duration.
   b. Results in approximately – 7 mm Hg pressure in the right ventricle and 0 mm Hg pressure in the left ventricle.
   c. Causes the semilunar valves to close when pulmonary (22 mm Hg) and aortic (120 mm Hg) pressures exceed end-systolic ventricular pressure producing:
      (1) The S2 ("dup") heart sound.
      (2) The dicrotic notch in the aortic pressure trace is probably due to elastic rebound when valve closure stops the momentary reverse flow of blood.
   d. Ventricular diastole coincides with rapid inflow of blood into the ventricular chambers caused by low intrathoracic pressure and the bulb-syringe effect (*vis a fronte*) of the ventricular walls (– 5 mm Hg right, 0 mm Hg left).

B. The duration of the cardiac cycle and the ejection volume are variable.

   1. The volume of blood ejected into the aorta during ventricular systole stretches that vessel and increases the elastic energy within its walls. This elastic pressure moves the blood from the aorta during diastole and maintains the range of diastolic pressure.

2. With an ejection volume of 60–70 ml per beat and a pulse rate of 80 beats per minute, the heart will pump 5 L per minute.
3. The systolic pressure can be a function of ejection volume. The greater the volume of blood ejected, the more the elastic walls of the aorta are stretched and the higher the systolic pressure.
4. The diastolic pressure can be a function of the heart rate. The slower the rate, the lower the aortic diastolic pressure falls before the subsequent heart beat.
5. Pulse pressure normally is a combination of these factors, in addition to other variables such as peripheral vascular resistance and blood volume.

C. PROJECTION OF SURFACE SOUNDS OF THE HEART

1. The four heart valves not only are close together, but from the anterior-posterior viewpoint, the semilunar valves, as well as the A-V valves, are nearly superimposed.
2. Because of the difference in the conduction velocity of sound through different types of tissues, the sound produced by each cycling valve is auscultated with maximal clarity and with minimal contribution from other valves over a distinct area of the thoracic wall (Fig. 13-11).
   a. **Tricuspid valve sounds** project to the right side of the sternum at the 5th intercostal space.
   b. **Mitral valve sounds** project to the apex of the heart at the 5th intercostal space below the left nipple—point of maximum outward movement (PMI) or apical pulse.
   c. **Aortic semilunar valve sounds** project to the right of the sternum over the 2nd intercostal space and also can be auscultated in the neck over the carotid artery.
   d. **Pulmonary semilunar valve sounds** project to the left of the sternum over the 2nd intercostal space. The murmur produced by a patent ductus arteriosus can best be auscultated just lateral to this point.
3. Inspiration lowers endothoracic pressure and increases venous return to the right side of the heart. This increased filling can result in closure of the pulmonary valve momentarily after the aortic valve, producing a "split" S2.
4. Valvular disorders produce murmurs because of turbulent blood flow.

X. FETAL AND EARLY POSTNATAL CIRCULATION

A. During fetal development, most of the blood passing through the heart must be shunted around the collapsed lungs (Fig. 13-12).

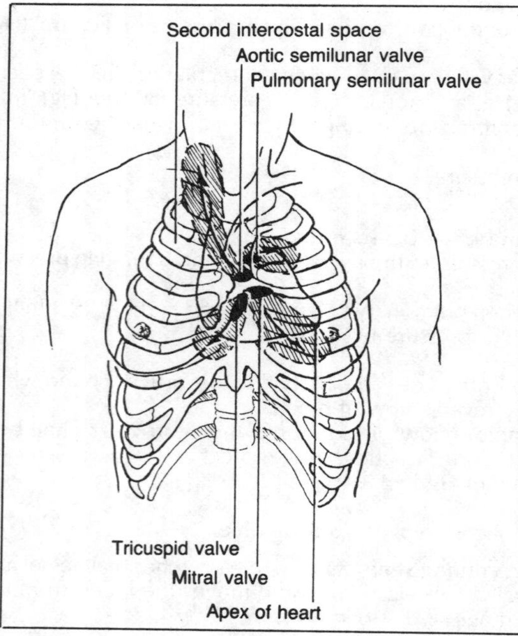

Figure 13-11. *Surface projections of the heart sounds.* The atrioventricular and semilunar valves lie in the plane of the coronary sulcus and are nearly vertical behind the sternum. The mitral valve sound projects to the 5th intercostal space on the left, and the tricuspid valve is heard most distinctly in a region just to the right of the sternum in the 4th or 5th intercostal space. The aortic valve projects to the 2nd intercostal space at the right of the sternum and radiates into the neck. The pulmonic valve projects to the 2nd intercostal space at the left of the sternum.

1. The **foramen ovale** in the developing interatrial septum remains patent and shunts a portion of the blood to the systemic side.

2. The **ductus arteriosus** (the 6th branchial arch on the left side) remains patent connecting the aorta (4th arch) and the left pulmonary artery. Blood entering the right ventricle flows into the aorta through this duct. Some blood does, however, flow through the lungs and is returned to the left atrium via the pulmonary veins.

B. FETAL BLOOD FLOW (see Fig. 13-12)

1. Fetal blood is charged with oxygen and nutriments in the placenta and returns to the fetus via the umbilical vein.
   a. Over one-half of the blood bypasses the liver through the ductus venosus and enters the inferior vena cava directly. The remaining blood from the umbilical vein drains through the portal vein and sinusoids of the liver before passing through the hepatic vein and joining the bypassed blood in the inferior vena cava.
   b. The blood flow from the umbilical vein is regulated by a muscular sphincter in the ductus venosus. When the sphincter relaxes, more blood passes through the ductus venosus; when the sphincter contracts, more blood is shunted to the portal vein and hepatic sinusoids.

2. After passing through the inferior vena cava, the blood enters the right atrium.
   a. Because the inferior vena cava also contains deoxygenated blood from the lower extremities, abdomen, and pelvis, the blood entering the right atrium is not quite as well oxygenated as that in the umbilical vein.
   b. The blood is guided by the valve of the inferior vena cava (eustachian valve) into the left atrium through the foramen ovale. The valve of the foramen ovale is located on the left side of the septum. The resistance to blood flow through the the collapsed lungs makes the systolic pressure greater in the right atrium than in the left atrium, causing the blood to flow through the foramen ovale into the left atrium.
   c. In the left atrium the well-oxygenated blood mixes with the small amount of deoxygenated blood returning via the pulmonary veins from the lungs. From the left atrium this blood flows successively through the left ventricle and ascending aorta.
   d. The coronary, carotid, and subclavian arteries are the first major arteries to receive blood from the ascending aorta. Hence, well-oxygenated and nutritional blood is directed to the heart, brain, head, neck, and upper extremities.

3. A small amount of blood entering the right atrium from the inferior vena cava is prevented from passing through the foramen ovale by the lower edge of the septum secundum, the so-called crista dividens, and thus remains in the right atrium.
   a. In the right atrium it mixes with the deoxygenated blood returning from the head and upper extremities by way of the superior vena cava and from the heart via the coronary sinus. This deoxygenated blood is directed through the tricuspid valve into the right ventricle and then into the pulmonary trunk.
   b. Because the resistance in the pulmonary vessels during fetal life is high, the main portion of this deoxygenated blood does not pass through the pulmonary arteries to the lungs. Rather, it is diverted from the left pulmonary artery through the ductus arteriosus into the descending aorta, where it mixes with well-oxygenated blood from the aortic arch. This admixture occurs *after* the coronary and carotid arteries have been given off.
   c. Most of the mixed blood in the descending aorta passes to the placenta for reoxygenation; the remainder circulates through the lower part of the body.

C. CHANGES DURING THE POSTNATAL PERIOD

1. The sudden changes occurring in the vascular system at birth are related to the cessation of placental flow and the commencement of pulmonary respiration.
   a. Occlusion and ligation of the placental circulation cause an immediate fall in blood pressure in the inferior vena cava and right atrium.
   b. As a result of the compression of the chest at birth, the amniotic fluid in the bronchial tree is expelled and is replaced by air; respiratory movements follow, and the lungs become aerated.
   c. Aeration of the lungs is associated with a drastic fall in pulmonary vascular resistance.
   d. The marked increase in pulmonary blood flow contributes to the drop in pressure in the right side of the heart and results in a rise in the blood pressure in the left atrium.

2. The differential blood pressure in the right atrium (lower) and the left atrium (higher) presses the valve of the **foramen ovale** (septum primum) against the atrial septum (septum secundum). Thus, the early functional closure of the foramen ovale is by pressure differential.

a. During the first days of life, however, the closure is reversible, and crying creates a right-to-left shunt in newborns accounting for cyanotic periods.
b. Constant apposition gradually leads to fusion of the two septa after several months, resulting in the formation of the **fossa ovalis**. In 10 to 15 percent of individuals, however, perfect anatomic closure may never be obtained.

3. The **ductus arteriosus** begins to close soon after birth, but during the first few postpartum days a left-to-right shunt is not unusual, and a murmur produced by a patent ductus may be auscultated.
   a. Because of the angle at which the ductus joins the aorta and the resultant Bernoulli effect, the amount of backflow during this period is less than might be expected.
   b. Initial closure of the ductus arteriosus is by muscular contraction mediated by bradykinin, a substance released from the lungs during their initial inflation. As closure occurs, the amount of blood flowing into the lungs and left atrium increases.
   c. Complete anatomic obliteration by proliferation of the intima takes from 1 to 3 months. In the adult, the obliterated ductus arteriosus is called the **ligamentum arteriosum**.
4. Closure of the umbilical arteries is accomplished by contraction of the smooth muscles in the wall of the vessels and is probably caused by thermal and mechanical stimuli and a change in oxygen tension.
   a. Functionally, the arteries are closed a few minutes after birth, but the actual obliteration of the lumen by fibrous proliferation may take from 2 to 3 months.
   b. The distal parts of the umbilical arteries then form the **medial umbilical ligaments**, while the proximal portions remain open as the superior vesical arteries.
5. Closure of the umbilical vein and ductus venosus occurs shortly after that of the umbilical arteries.
   a. After obliteration, the umbilical veins form the **ligamentum teres hepatis** in the lower margin of the falciform ligament.
   b. The ductus venosus, which courses from the ligamentum teres to the inferior vena cava, is also obliterated and forms the **ligamentum venosum**.

XI. CLINICAL CONSIDERATIONS

A. VALVULAR DEFECTS

1. A valve is **incompetent** or insufficient if it permits backflow. Valvular defects can be congenital or can result from an infarct in the vicinity of a papillary muscle, rupture of the ten-

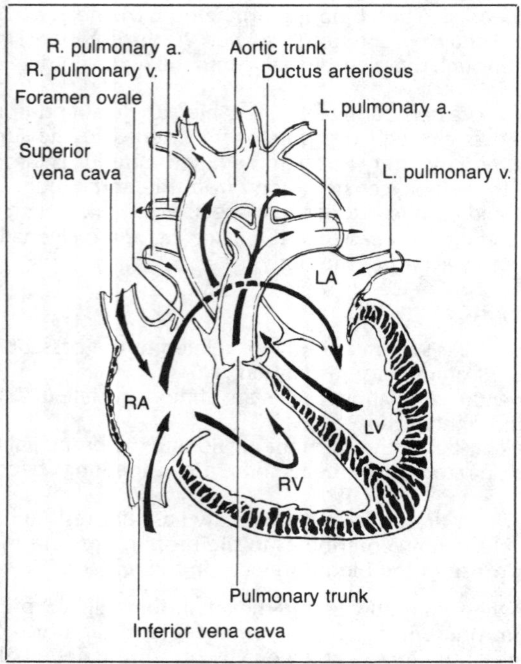

Figure 13-12. *Fetal circulation.* Oxygenated fetal blood, returning from the placenta through the inferior vena cava, passes predominantly through the foramen ovale into the left atrium, whence it passes through the left ventricle and supplies the vessels of the heart and head. Deoxygenated fetal blood, returning from the head through the superior vena cava, passes predominantly into the right ventricle and pulmonary trunk, whence most joins the aortic flow through the ductus arteriosus.

dinous cords, or endocardial inflammatory processes.

a. Tricuspid valve insufficiency leads to right-side heart failure in which right ventricular pressure is transmitted back to the venous system to produce distended neck veins, a pulsating liver, abdominal ascites, and edema.

b. Mitral valve insufficiency leads to left-side heart failure in which left ventricular pressure is transmitted back to the lungs to produce pulmonary edema and pleural effusion (hydrothorax). This condition may, in turn, induce right-side heart failure.

2. A valve is **stenotic** if it restricts forward flow; stenosis can be either congenital or secondary to endocardial inflammatory processes.

3. Both stenotic and incompetent valves produce murmurs that can be localized by auscultation.

4. As the volume of blood flow through the heart changes, the heart compensates by increasing or decreasing its work (pressure-volume product). Shunts and valvular stenosis or incompetence increase the need for cardiac compensation. However, a point is reached where decompensation occurs, with resultant cardiac failure.

5. Stenosis may be amenable to simple surgical correction, but insufficiency usually requires correction by means of a prosthetic valve.

B. CONGENITAL DEFECTS

1. Although the **ductus arteriosus** may remain patent throughout life with no major problems, it usually is ligated surgically when it is found to be open.

2. Septal defects.

a. A **patent foramen ovale** (one type of ASD) frequently is compatible with normal life and activity.

b. A VSD may result in **tetralogy of Fallot**, in which

(1) The VSD provides communication between ventricular chambers.

(2) The opening of the aorta overrides the VSD into the right ventricle.

(3) Distribution of hydrostatic pressure through the VSD from left ventricular systole results in hypertrophy of the right ventricular wall.

(4) Hypertrophy of the right ventricular wall in the vicinity of the infundibulum produces a functional pulmonary stenosis whereby myocardial contraction occludes the pulmonary outflow tract.

c. Septal defects with a **right-to-left shunt** of blood admix deoxygenated blood in the aorta, necessitating increased cardiac work to provide the same amount of oxygen to the body tissue, with the possibility of decompensation and failure.

3. In **dextrocardia** the heart is normal, but in mirror image (0.02 percent of the population).

a. A very high incidence of situs inversus viscerum (reversed rotation of the gastrointestinal tract) occurs with dextrocardia.

b. Dextrocardia may be associated with either a right-sided or left-sided aorta.

C. CIRCULATORY COLLAPSE (''SHOCK''). The heart can pump only the blood that flows into it. In shock there is venous pooling and the blood fails to return to the heart. Blood pressure drops and stroke volume decreases. The pulse becomes rapid but thready.

14
The Superior and Posterior Mediastinum

I. THE SUPERIOR MEDIASTINUM

A. The principal contents of the superior mediastinum are the thymus gland and the vascular structures running to and from the heart. The veins tend to be anterior and toward the right, whereas the arteries tend to be located behind the veins and to the left. The trachea and esophagus are close to the midline (Fig. 14-1).

B. THE THYMUS GLAND

1. The thymus gland lies anteriorly in the superior mediastinum and anterior mediastinum beneath the manubrium and upper body of the sternum and extends laterally beneath the upper four costal cartilages, sometimes as far as the hilum of the lung on either side. It rises slightly into the base of the neck under the sternohyoid and sternothyroid muscles. It lies anterior to the aorta, brachiocephalic vein, and fibrous pericardium.

2. It varies in size with age.
   - **a.** At birth, it weighs 10 to 15 g.
   - **b.** Before puberty, it grows to 30 to 40 g.
   - **c.** During late adult life, it regresses by involution and fatty atrophy to less than 10 g.

3. The thymus is a very important lymphoid organ associated with immunologic recognition.

4. Thymectomy is undertaken as a palliative measure for myasthenia gravis, an autoimmune disorder associated with the neuromuscular junctions.

C. THE BRACHIOCEPHALIC VEINS

1. The left and right brachiocephalic veins receive the drainage from the head, neck, and upper extremities.

2. The short **right brachiocephalic vein** and the longer, diagonal **left brachiocephalic vein** unite to form the **superior vena cava**, which continues the direction of the right brachiocephalic vein (see Fig. 14-1).

3. The superior vena cava receives the **azygos vein** before emptying into the right atrium.

4. **Anomalies.** The left brachiocephalic vein forms as an initial small shunt between the embryonic left and right anterior cardinal veins. As the shunt enlarges to become the left brachiocephalic vein, the left supracardinal vein normally regresses to form the oblique vein (Marshall's) of the left atrium. The left common cardinal vein, into which the supracardinal and subcardinal veins empty in the embryo, persists as the coronary sinus, which receives the oblique vein and empties into the right atrium. If this developmental sequence is reversed or modified, a single left superior vena cava can develop or, if the right vena cava persists, bilateral venae cavae may result.

D. THE AORTA

1. The aorta emerges from the pericardial sac between the superior vena cava and the pulmonary trunk.

2. The **ascending aorta** runs superiorly and toward the right before arching posteriorly as the **aortic arch** and then inferiorly as the **descending aorta** (see Fig. 14-4). The aortic arch produces the "aortic knob," a radiographic landmark on the left side of the mediastinum.

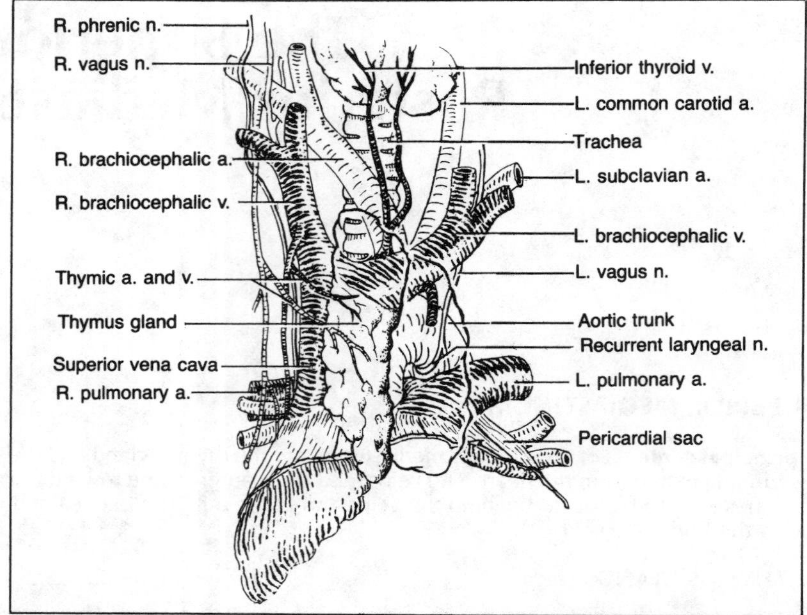

Figure 14-1. *The anterior mediastinum and the superior mediastinum.* The lower portion of the right half of the thymus gland is shown, and the remainder of the thymus is ghosted to show the major vessels of the anterior and superior mediastinum.

3. Branches of the ascending aorta and aortic arch include:
   a. The **left** and **right coronary arteries.**
   b. The **brachiocephalic artery** (on the right side only), which gives rise to the right common carotid and right subclavian arteries.
   c. The **left common carotid artery.**
   d. The **left subclavian artery**.
4. The branching pattern of the great vessels is somewhat variable, and the left common carotid artery sometimes arises from the brachiocephalic artery (especially in black people).
5. The definitive aortic arch is derived from the artery of the 4th left branchial arch of the embryo, whereas the right subclavian artery is derived from the artery of the 4th right branchial arch.

II. THE POSTERIOR MEDIASTINUM

A. LOCATION. By definition, the posterior mediastinum lies behind the pericardial sac; it is bounded by the mediastinal pleura on either side and by the diaphragm below. Superiorly, it is continuous with the superior mediastinum. It runs approximately from vertebral level T4 to T12.

B. THE DESCENDING AORTA

1. The descending aorta is located to the left within the mediastinum and bulges into the left pleural cavity, where it produces a distinctive radiopaque border on the right side of the mediastinum, superior to the heart shadow (see Fig. 14-4).
2. It is about 3 cm in diameter.
3. The main aortic branches in the thorax include:
   a. The **ligamentum arteriosum**, a remnant of the embryonic **ductus arteriosus**, which is derived from the artery of the 6th left branchial arch (the 6th arch degenerates completely on the right).
   b. Paired **intercostal arteries** at each vertebral level, which supply the thoracic wall and anastomose with the anterior intercostal branches of the internal thoracic (mammary) arteries.
   c. Paired **bronchial arteries**, which supply the bronchial tree.
   d. **Esophageal branches**, which supply the middle portion of the esophagus.

4. The descending aorta enters the abdominal cavity by passing through the **aortic hiatus**, between the crura of the diaphragm.

C. THE ESOPHAGUS

1. The esophagus lies anterior to the prevertebral fascia. High within the thorax, it lies in the midline behind the trachea, but inferior to the tracheal bifurcation it deviates slightly to the left so that it lies slightly to the left of the aorta. As it pierces the diaphragm through the **esophageal hiatus**, which is about 2½ cm to the left of the midline in the muscular portion of the diaphragm, it makes an abrupt turn to the left to enter the stomach (see Fig. 14-4).
2. The esophagus is 25 to 30 cm long in males and a few centimeters shorter in females.
3. Although normally it is flattened, the esophagus distends to allow passage of a bolus during deglutition (swallowing).
4. **Esophageal vasculature.**
   a. The **arterial supply** is derived from several sources.
      (1) The **inferior thyroid arteries** (branches of the subclavian arteries) supply the cervical portion of the esophagus.
      (2) The **esophageal branches** of the **aorta** supply the middle portion.
      (3) The **esophageal branches** of the **left gastric artery** (a branch of the celiac trunk) supply the lower thoracic and abdominal portions of the esophagus.
   b. The **venous return** is similar to the arterial supply.
      (1) The veins of the upper and lower portions parallel the supplying arteries and drain into the **inferior thyroid veins** and the **left gastric vein**, respectively.
      (2) The venous drainage of the middle portion is into the **azygos** and **hemiazygos veins.**
5. **Esophageal function.**
   a. The **esophagus** serves solely for the transport of food. It extends from the lower part of the laryngopharynx to the cardiac opening of the stomach.
      (1) Except during deglutition, the lumen of the esophagus is small and irregular, averaging 1 cm orally and 2 to 3 cm aborally.
      (2) The luminal shape is irregular because of tension within the inner (circular) layer of the muscularis externa, with resultant formation of longitudinal folds. Consequently, the esophagus is distensible and will accommodate anything that can pass through the pharynx.
   b. The esophagus is the most muscular segment of the alimentary tract. Its upper quarter is composed primarily of striated muscle, which is initially irregularly arranged but soon separates into inner circular and outer longitudinal layers.
      (1) Proceeding aborally, the amount of smooth muscle increases and the striated fibers disappear.
      (2) Between the longitudinal and circular muscle layers lie the parasympathetic postsynaptic neurons of the autonomic nervous system. The cell bodies of these neurons lie in the **myenteric plexus** (Auerbach's), receive presynaptic innervation from the vagus nerve, and innervate the esophageal musculature.
      (3) The **cricopharyngeal muscle** is the sphincter of the upper esophageal opening. It remains closed except during deglutition, eructation (belching), and emesis (vomiting).
         (a) A solid bolus entering the esophagus initiates an involuntary peristaltic wave, which reaches the cardiac sphincter in 5 to 6 seconds. The cardiac sphincter, normally closed, relaxes to allow the bolus to pass into the stomach.
         (b) A fluid bolus will outrun the peristaltic wave but will be stopped by the cardiac sphincter until the sphincter is inhibited by the approach of the peristaltic wave. Thus, only one sound will be heard when a solid bolus is swallowed, whereas there are two separate sounds with a fluid bolus. When a series of swallows is made in rapid succession ("chugging"), the cardiac sphincter may remain inhibited so that fluid passes directly into the stomach unassisted by peristaltic waves.
      (4) The **cardiac sphincter** is located at the cardioesophageal junction. The smooth muscle of this junction is continuous with that of the stomach, and the sphincter is not considered a true structural one. Nevertheless, it is a functional sphincter, intrinsic to the esophagus and not dependent upon external constriction by the crura of the esophageal hiatus of the diaphragm.
         (a) There is considerable evidence that the sphincteric barrier is, in part, due to valve-like folds of mucosa maintained by the tonic activity of the muscularis mucosae. This barrier, the diffuse muscular tonic activity of the muscularis externa, and the fairly constant tone within the muscle layers of the cardiac region of the stomach constitute the physiologic closing mechanism.

(b) The tension in the region of the sphincter is never very great, but it does maintain a broad zone of elevated intramural pressure at the aboral end of the esophagus. The pressure within the stomach is always greater than that in the midesophagus. Without the zone of increased transmural pressure due to esophageal tension, the gastric contents would reflux into the esophagus continuously. The cardiac sphincter can be opened by a pressure of 2 to 7 cm H_2O within the esophagus, but eructation will not occur until intragastric pressure exceeds 25 cm H_2O.

6. Clinical considerations.

a. Esophageal varices. The submucosa of the esophagus contains an extensive plexus of large blood vessels. Abundant anastomotic connections in the submucosa exist between the azygos and hemiazygos veins and the left gastric vein, which is a branch of the hepatic portal system. When portal hypertension occurs, this anastomotic route provides an important, but dangerous, shunt that can result in the development of esophageal varices.

(1) Varices are tortuous, dilated veins immediately beneath the mucosa that protrude into the esophageal lumen, usually in the distal third of the esophagus, where they are subject to mechanical trauma during deglutition, emesis, or the passage of instruments. Esophageal varices produce no symptoms until they rupture, whereupon massive hematemesis occurs.

(2) Among people who have advanced cirrhosis of the liver, over half the deaths result from rupture of an esophageal varix.

b. Occasionally, gastric contents **reflux** into the esophagus. The acidic peptic chyme burns the unprotected stratified squamous epithelium of the esophageal mucosa, producing the uncomfortable sensation commonly known as heartburn. Repeated reflux can result in **esophagitis** and the formation of **peptic ulcers** in the esophageal mucosa. Peptic ulceration of the esophagus can cause dysphagia (inability to swallow) and can be a serious source of hematemesis if erosion into a blood vessel occurs.

c. Achalasia. Dysfunction of the parasympathetic pathway at this level results in achalasia (cardiospasm), which is characterized by dysphagia associated with progressive dilatation and irregular tortuosity of the esophagus.

D. THE THORACIC DUCT

1. Essentially all the lymph from the body below the diaphragm returns to the systemic circulation through the thoracic duct at a flow rate of 2 to 4 L of chyle per day.

2. The duct originates in the **cisterna chyli**, which is located in the abdomen at vertebral level T12, between the crura of the diaphragm and posterior to the aorta (Fig. 14-2).

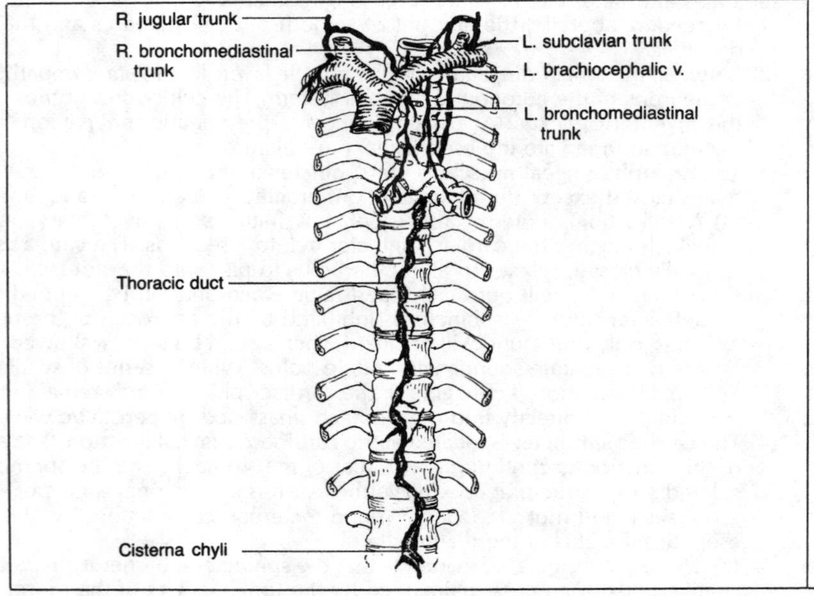

Figure 14-2. *The thoracic duct.* Lymph from the entire left side of the body as well as from the lower half of the right side is returned to the left brachiocephalic vein by the thoracic duct.

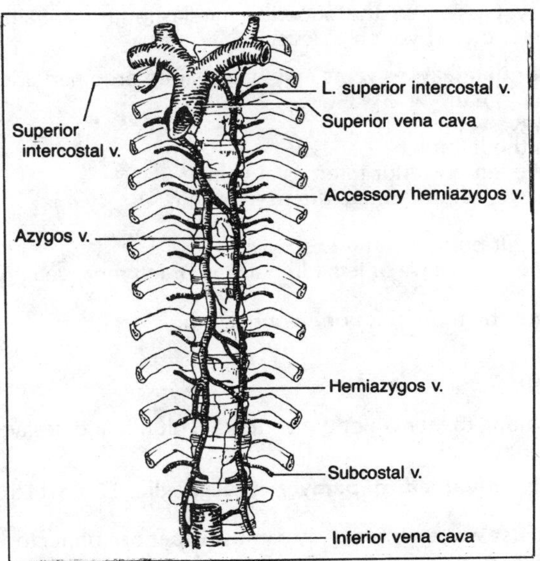

Figure 14-3. *The azygos system of veins.*

3. In the thorax the duct is located behind the esophagus and to the right of the aorta. It is very delicate and may occasionally be double or even triple.

4. In the superior mediastinum, the duct arches over the cupula of the left pleura to lie posterior to the left subclavian vein. It enters this vein at the angle formed by the juncture of the vein with the left internal jugular vein.

5. The thoracic duct receives branch trunks from:
   a. The intercostal nodes, draining the thoracic wall and breasts.
   b. The subclavian trunk, draining the left arm, left breast, and left side of the neck.
   c. The left jugular trunk, draining the left side of the head and neck.
   d. The left bronchomediastinal trunk, draining the left bronchial tree.

6. Chyle from a ruptured or severed thoracic duct may drain into the right or left pleural cavity (chylothorax).

E. THORACIC LYMPH NODES

1. **Preaortic nodes:**
   a. Lie anterior to the aorta.
   b. Drain the esophagus.
   c. Empty into the thoracic duct.

2. **Paraaortic nodes:**
   a. Lie alongside the aorta (intercostal nodes).
   b. Drain the posterior intercostal spaces (the anterior intercostal spaces are drained by parasternal nodes along the internal thoracic artery).
   c. Empty into the thoracic duct.

3. **Hilar nodes:**
   a. Are located around the root of the lung and within the lung parenchyma at the hilus.
   b. Drain the lung and peripheral bronchial tree.
   c. Empty into the bronchomediastinal nodes.

4. **Bronchomediastinal nodes:**
   a. Lie adjacent to the bronchi and trachea.
   b. Receive drainage from the hilar nodes.
   c. Usually drain into the brachiocephalic veins on each side, or occasionally, into the thoracic duct.

F. VENOUS DRAINAGE OF THE THORAX

1. **The azygos vein** (Fig. 14-3):
   a. Lies to the right of the esophagus on the vertebral column.
   b. Receives the right posterior intercostal veins, right bronchial veins, and hemiazygos veins.

c. Enters the superior vena cava by passing to the superior mediastinum beneath the mediastinal pleura and arching anteriorly at vertebral level T4.

2. The superior hemiazygos and **inferior hemiazygos veins** may be separate or joined and are subject to extensive variation (see Fig. 14-3).

a. The superior (accessory) **hemiazygos vein:**

(1) Receives bronchial veins from the left lung.

(2) Receives the upper three to five left posterior intercostal veins.

(3) Either joins the inferior hemiazgos vein or enters the azygos vein.

b. The (inferior) hemiazygos vein:

(1) Drains the lower eight or nine left posterior intercostal veins.

(2) May anastomose with the inferior vena cava or left adrenal vein or left renal vein in the abdomen.

(3) Joins the azygos vein, sometimes by more than one connection.

G. THE THORACIC SYMPATHETIC TRUNK

1. The sympathetic trunk is continuous along the length of the vertebral column and lies across the necks of the ribs (see Fig. 14-4).

2. It consists of a chain of interconnected **sympathetic** or **paravertebral ganglia**, 12 of which are in the thorax.

a. Each thoracic ganglion has five roots: white ramus, gray ramus, superior connector, inferior connector, and splanchnic nerve.

(1) White rami communicantes bring *pre*synaptic (myelinated) general visceral efferent (GVE) neurons from the spinal nerve to or through the sympathetic chain. Those neurons passing through the ganglia without synapses may go either up or down the chain to synapse at an adjacent level, travel in splanchnic nerves to synapse in prevertebral ganglia, or synapse in the ganglion at the level where they enter the chain.

(2) Gray rami communicantes return *post*synaptic (unmyelinated) GVE neurons to the spinal nerve for piloerection, sweating, and vasodilatory functions in the associated dermatome.

(3) The **superior** and **inferior connectors** carry presynaptic and postsynaptic neurons to adjacent or remote levels of the chain.

(4) Splanchnic nerves carry presynaptic neurons to prevertebral ganglia, such as the celiac or superior mesenteric ganglion, where they synapse. From the prevertebral ganglia, *post*synaptic neurons innervate visceral structures.

b. Ganglia T1–T3 contribute to the cervical chain since there are no white rami communicantes in the neck.

(1) The **superior, middle** and **inferior cervical ganglia** give off **cardiac (accelerator) nerves** to the heart, which are equivalent to splanchnic nerves.

(2) All presynaptic sympathetic neurons to the head arise from these upper thoracic levels.

c. Ganglia T1–T5 contribute **thoracic splanchnic nerves** to the heart and lungs.

d. Ganglia T5–T9 contribute to the **greater splanchnic nerve**, which innervates the upper gastrointestinal tract.

e. Ganglia T10–T11 contribute to the **lesser splanchnic nerve**, which innervates the midgut region.

f. Ganglion T12 contributes to the **least splanchnic nerve**, which innervates the gonads and the adrenal medulla.

H. THE VAGUS NERVE (CN X)

1. The vagus nerve leaves the cranium through the jugular foramen and passes through the neck and thorax into the abdomen to provide **parasympathetic** innervation to the cervical and thoracic viscera, foregut, and midgut.

2. The right vagus nerve (Fig. 14-4):

a. Gives off the **right recurrent (inferior) laryngeal nerve**, which arches inferior to and around the right subclavian artery, a remnant of the right 4th branchial arch (the right 5th and right 6th arches having degenerated completely) in the base of the neck.

b. Gives off twigs to the **cardiac plexus**.

c. Runs behind the right main stem bronchus and gives major contributions to the **anterior** and **posterior pulmonary plexuses**.

d. Passes along the right surface of the esophagus to form an **esophageal plexus**, which reforms as the **posterior vagal trunk** before passing through the esophageal hiatus of the diaphragm.

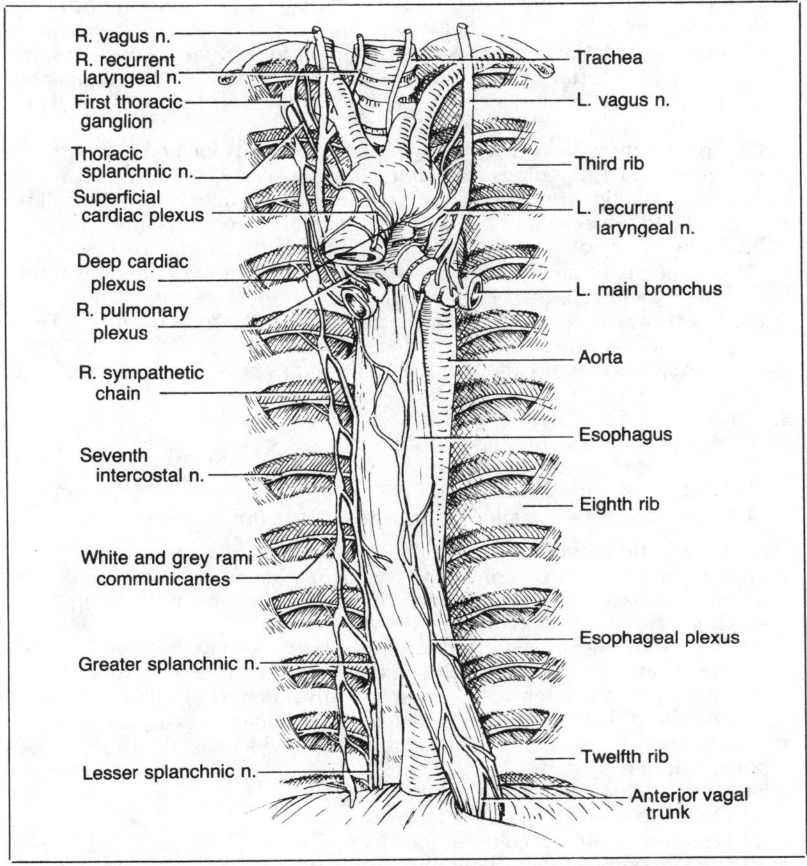

Figure 14-4. *The autonomic nerves of the thorax.* The sympathetic trunk and the vagus nerves bring sympathetic and parasympathetic innervation, respectively, to the thoracic viscera. Also shown are the aorta and esophagus.

3. **The left vagus nerve** (see Fig. 14-4):
   a. Gives off the **left recurrent (inferior) laryngeal nerve**, which courses around the ligamentum arteriosum (remnant of the 6th left branchial arch).
   b. Together with the recurrent laryngeal nerve gives off twigs to the **cardiac plexus**.
   c. Passes posterior to the root of the lung, where it gives off major contributions to the **posterior** and **anterior pulmonary plexuses**.
   d. Passes along the left side of the esophagus to form an **esophageal plexus**, which re-forms as the **anterior vagal trunk** just above the diaphragm.
4. Because of the rotation of the gut within the abdomen, the left and right vagi become the anterior and posterior vagal trunks, respectively (see Fig. 14-4).
5. The recurrent laryngeal nerves innervate the vocal musculature. Hilar or mediastinal tumors on the left side or an apical tumor on the right side, involving either of the respective nerves, will cause hoarseness, often the first sign of these tumors.

I. FUNCTIONS OF THE AUTONOMIC NERVOUS SYSTEM IN THE THORAX

1. The **cardiac plexus** is a region of intermingling sympathetic and parasympathetic (vagal) fibers located anterior to the bifurcation of the pulmonary trunk and posterior to the arch of the aorta. This plexus is continuous with another on the thoracic aorta as well as with the left and right **anterior** and **posterior pulmonary plexuses** associated with the roots of the lungs.
2. **Sympathetic innervation.**
   a. The preganglionic fibers convey sympathetic influences to the heart and lungs. These fibers:

(1) Have their cell bodies in the intermediolateral cell column of spinal cord levels T1 to T5.
(2) Pass through the ventral roots and ventral primary ramus of each spinal nerve T1 to T5.
(3) Leave intercostal nerves T1 to T5 via the white rami communicantes.
(4) Terminate in the three cervical and the upper few thoracic ganglia of the sympathetic chain.
(5) Apparently synapse with *post*ganglionic neurons located in the three cervical and the upper four (sometimes five) thoracic ganglia.

b. The *post*synaptic sympathetic fibers to the heart seem to have their cell bodies in the three cervical and upper four or five thoracic ganglia. However, some postsynaptic neurons may be located in small ganglia (Wrisberg's) within the cardiac plexus.
(1) The **cervical cardiac nerves** arise from the superior, middle, and inferior cervical ganglia and run deep in the neck to the cardiac plexus.
(2) The **thoracic** (splanchnic) **cardiac nerves** arise from T1 through T4 or T5 (variable on both sides) and run to the cardiac plexus.

c. The *post*synaptic sympathetic fibers to the lungs have their cell bodies mainly in thoracic ganglia T2 through T4.

d. Sympathetic stimulation results in:
(1) Increase in heart rate (tachycardia).
(2) Increase in stroke volume.
(3) Dilation of coronary arteries.
(4) Piloerection, perspiration, and dilation of peripheral vessels in the dermatomes.

3. Parasympathetic innervation.

a. The *pre*ganglionic fibers conveying parasympathetic influences to the heart, lungs, and abdominal viscera commence in the dorsal vagal nucleus of the brain and pass through the vagus nerve.
(1) The vagus and its recurrent laryngeal branches give off twigs to the cervical cardiac nerves and thereby reach the cardiac and pulmonary plexuses.
(2) These preganglionic fibers synapse with *post*synaptic fibers located in small ganglia (Wrisberg's) within the cardiac plexus, pulmonary plexus, and intrinsic cardiac ganglia.

b. The *post*ganglionic fibers arise in numerous small ganglia (Wrisberg's) located close to the point of innervation.

c. Parasympathetic stimulation results in:
(1) Decrease in heart rate (bradycardia).
(2) Decrease in stroke volume.
(3) Constriction of the bronchi and bronchioles.
(4) Increased bronchial secretion.

4. Visceral afferents.

a. Pain afferents.
(1) Pain afferents from the heart follow the sympathetic pathways (with the exception of the superior cervical cardiac nerve, which contains no afferents).
(2) All pain afferents from the heart, whether they travel along the cervical or thoracic cardiac nerves, pass through or along the sympathetic chain and leave the chain via the closest white ramus communicans to reach thoracic nerves T1 to T4.
(3) All visceral afferent neurons have their cell bodies in the dorsal root ganglia of the spinal nerve.
(4) The basis for *referred pain*.
(a) Painful stimuli mediated by visceral afferent neurons, which enter the spinal cord at a vertebral level, are interpreted by the brain as originating from (referred to) the somatic dermatome corresponding to that particular vertebral level.
(b) As such, painful stimuli originating from the heart will be referred to the T1 to T4 dermatomes, which include the medial aspect of the upper arm and anterior thoracic wall as far as the level of the nipple.

b. Reflex afferents.
(1) Afferents for the respiratory and cardiac reflex arcs travel exclusively along the vagal fibers.
(2) The cell bodies are located in the jugular ganglion of the vagus nerve.

Part III Thoracic Region

STUDY QUESTIONS

Directions: Each question below contains five suggested answers. Choose the **one best** response to each question.

1. With a patient in a sitting position, there is percussive dullness below the level of the 5th rib in the right midaxillary line. Thoracocentesis to remove the fluid is best accomplished here by inserting a needle

(A) adjacent to the sternum in the 2nd intercostal space
(B) adjacent to the sternum in the 5th intercostal space
(C) in the midaxillary line in the 5th intercostal space
(D) in the midaxillary line in the 10th intercostal space
(E) in none of the above spaces

2. Which of the following statements best characterizes bronchopulmonary segments, the functional units of the lungs?

(A) The bronchial tree and pulmonary vein are located centrally
(B) The pulmonary artery and pulmonary vein are distributed centrally
(C) The pulmonary artery and pulmonary vein are distributed peripherally
(D) The bronchopulmonary segments are supplied by a terminal bronchiole
(E) None of the above statements characterize bronchopulmonary segments

3. In a hospital emergency room, a 23-year-old man states that he "inhaled" a peanut. On bronchoscopy the peanut will most likely be located in the

(A) left lower lobar bronchus
(B) left main bronchus
(C) left superior segmental bronchus
(D) right lower lobar bronchus
(E) right superior segmental bronchus

4. Aspiration pneumonia is usually detectable by a distinct region of percussive dullness and a radiographic lightening of a particular bronchopulmonary segment. This bronchopulmonary segment is nearly always the

(A) left or right apical
(B) left or right superior
(C) left or right basal anterior
(D) left upper lingular
(E) left lower lingular

5. Cancer of the right lung may metastasize to all of the following lymph nodes EXCEPT the

(A) left tracheobronchial
(B) right axillary
(C) right bronchomediastinal
(D) right hilar
(E) right tracheobronchial

Questions 6–9

A 36-year-old male office worker reports to the clinic complaining of weakness and shortness of breath that become marked after moderate exercise. The patient complains of rapid, throbbing pulse on climbing two flights of stairs. Although the patient is not presently in pain, he admits to having occasional "heartburn" over the past year. Findings on physical examination include a weak and irregular pulse, a diastolic rumbling murmur attributed to the mitral valve, and rales (rattles) in the basal segments of the lungs.

6. The mitral valve is heard most distinctly on the

(A) left side, adjacent to the sternum in the 2nd intercostal space
(B) left side, adjacent to the sternum in the 5th intercostal space
(C) left side, in the midclavicular line in the 5th intercostal space
(D) right side, adjacent to the sternum in the 6th intercostal space
(E) right side, in the midclavicular line in the 6th intercostal space

7. Percussion reveals that the area of cardiac dullness extends two fingerbreadths to the right of the sternum. A chest x-ray confirms that the heart is enlarged, with specific right ventricular hypertrophy and right atrial hypertrophy. The preliminary diagnosis is mitral valve stenosis. A sequela of mitral valve stenosis is hypertrophy of the right ventricle and right atrium, which is caused by all of the following EXCEPT

(A) impeded blood flow to the left ventricle
(B) rise in left atrial systolic pressure
(C) rise in pressure in the pulmonary vascular bed
(D) rise in right ventricular systolic and diastolic pressures
(E) fall in left ventricular pressure

8. A complication of mitral stenosis is the formation of thrombi on the walls of the enlarged left atrium. A thrombus dislodged from the left atrium may produce all of the following conditions EXCEPT

(A) a cerebral infarct
(B) a myocardial infarct
(C) renal necrosis and kidney failure
(D) gangrene in a lower extremity
(E) a pulmonary embolus

9. The patient is scheduled for open heart surgery to correct the valvular defect. The heart is approached by resection of the 5th rib on the left side. Commissurotomy, the surgical reestablishment of the divisions between the mitral valve cusps, occasionally can be accomplished by insertion of a finger through the valve. Toward this end, the valve may be approached best through the wall of the

(A) left atrium between the left and right superior pulmonary veins
(B) left atrium between the left superior and inferior pulmonary veins
(C) left auricular appendage
(D) left ventricle on the posterior surface
(E) right atrium and foramen ovale

(end of group question)

Directions: Each question below contains four suggested answers of which **one or more** is correct. Choose the answer

A if **1, 2, and 3** are correct
B if **1 and 3** are correct
C if **2 and 4** are correct
D if **4** is correct
E if **1, 2, 3, and 4** are correct

10. During physical examination an elderly gentleman is observed to be breathing almost entirely with his diaphragm. However, there is no bulging of the intercostal space on expiration. One might expect that this respiratory pattern is the result of

(1) calcification of the costal cartilages
(2) paralysis of the thoracic musculature
(3) loss of extrinsic elastic recoil
(4) loss of intrinsic elastic recoil

11. Elevation of the rib cage during inspiration may be aided by the

(1) sternocleidomastoid muscle
(2) pectoralis minor muscle
(3) anterior, middle, and posterior scalene muscles
(4) pectoralis major muscle

12. The movement of the ribs during thoracic inspiration

(1) involves movement at the costovertebral joints
(2) involves movement at the sternomanubrial joint
(3) is produced by the crossing of the axes of rotation of the ribs
(4) increases the transverse thoracic diameter

13. Structures that are located in the anterior mediastinum include which of the following?

(1) Pericardial sac
(2) Trachea
(3) Heart
(4) Thymus

14. Two elderly women, both diagnosed as having pleurisy, were comparing their illnesses. One had lateral thoracic pain, whereas the other had neck and shoulder pain. Assuming the diagnoses to be correct, the explanation for these pain distributions is that the pleura is innervated by

(1) thoracic splanchnic nerves
(2) intercostal nerves
(3) vagus nerves
(4) phrenic nerves

15. During development, vessels that deliver blood to the left atrium include the

(1) pulmonary veins
(2) foramen ovale
(3) least cardiac veins (venae cordis minimae)
(4) ductus arteriosus

16. Myocardial infarction limited to the interventricular septum might be expected to produce which of the following problems?

(1) An aortic valve insufficiency
(2) A tricuspid valve incompetence
(3) A mitral valve incompetence
(4) Disturbance of cardiac impulse conduction

17. Circulatory changes that occur at birth normally include

(1) increased left atrial pressure
(2) increased blood flow through the lungs
(3) decreased right atrial pressure
(4) reversal of blood flow through the foramen ovale

18. Neurons that pass through the cardiac plexus include which of the following?

(1) Nonmyelinated postganglionic sympathetic fibers
(2) Visceral afferent fibers mediating reflexes
(3) Preganglionic parasympathetic fibers
(4) Visceral afferent fibers mediating pain

Directions: The group of questions below consists of lettered choices followed by several numbered items. For each numbered item select the **one** lettered choice with which it is **most** closely associated. Each lettered choice may be used once, more than once, or not at all.

Questions 19–24

For each situation listed below, select the artery that is most likely to be associated with it.

(A) Right coronary artery
(B) Circumflex branch of left coronary artery
(C) Both
(D) Neither

19. Is found in the coronary sulcus

20. Arises from the left anterior aortic sinus

21. Supplies the anterior portion of the interventricular septum

22. Supplies the atrioventricular node

23. Supplies the sinoatrial node

24. Terminates as the posterior interventricular (descending) artery in most people

ANSWERS AND EXPLANATIONS

1. The answer is E. (*Chapter 12 II D 6*) Thoracocentesis involves inserting a needle to withdraw fluid on which the lung floats and thereby reduces ventilation capacity. The needle is best inserted in the midaxillary line one or two ribs below the fluid level determined by percussion but not below the 9th intercostal space because of the proximity of the liver across the costodiaphragmatic recess.

2. The answer is E. (*Chapter 12 III B 7 a*) Bronchopulmonary segments are supplied by a segmental (tertiary) bronchus, which is distributed centrally along with a segmental branch of the pulmonary artery. The intersegmental veins lie in the connective tissue septa between the bronchopulmonary segments.

3. The answer is D. (*Chapter 12 IV D 2*) Aspirated objects usually drop into the right main bronchus because it is more vertical than the left main bronchus and thus more nearly in line with the trachea. Smaller objects tend to lodge in the right lower lobar bronchus, which more nearly continues the direction of the right main bronchus.

4. The answer is B. (*Chapter 12 IV E 3*) The left or right superior bronchopulmonary segment of the lower lobes are most dependent when a person is supine. Aspirated vomitus in the recumbent person drains into the nearly vertical superior segmental bronchi to produce a pneumonia and a region of percussive dullness just medial to the vertebral border of the scapula when the arm is elevated, as well as auscultatory rales.

5. The answer is B. (*Chapter 12 V F 2–4*) The lymphatic drainage from the right lungs does not pass through the axillary lymph nodes. If the bronchomediastinal lymph nodes become blocked by metastases, malignant cells may cross to the contralateral side.

6. The answer is C. (*Chapter 13 IX C 2 b*) The sound of mitral valve closure projects to the apex of the heart and is heard most distinctly in the left 5th intercostal space in the midclavicular line. In men this is usually just below the left nipple.

7. The answer is E. (*Chapter 13 XI A 1–4*) Mitral valve stenosis whereby blood flow to the left ventricle is impeded, results in elevated left atrial systolic pressure and consequent elevation of pulmonary pressure. This requires the right side of the heart to work more to overcome the pulmonary pressure, with consequent hypertrophy. The elevated pulmonary pressure produces pulmonary effusion and results in rales.

8. The answer is E. (*Chapter 13 VI A 4 c (2), C 4 b*) Thrombi dislodged from the left atrium can result in cerebral, coronary, renal, or systemic embolism. Thrombi dislodged from the right atrium produce pulmonary embolism.

9. The answer is C. (*Chapter 13 Figure 13-5*) The best approach to mitral commissurotomy is to gain access to the left atrium through the left auricular appendage and thence direct the finger through the mitral valve into the ventricle.

10. The answer is B (1, 3). (*Chapter 11 III D 5 b–c; VII D 4*) In the elderly frequently there is progressive calcification of the costal cartilages with resultant loss of thoracic cage elasticity that inhibits or even precludes thoracic respiratory movement. The person compensates by using the diaphragm for inspiration and the abdominal musculature for expiration. Paralysis of the thoracic musculature results in bulging of the intercostal space during expiration.

11. The answer is E (all). (*Chapter 11 IV A 1–4*) The pectoralis major and minor muscles as well as the scalene muscles attach to the ribs and can be used to elevate the ribs during exertional inspiration. The sternocleidomastoid muscle, attached to the manubrium, elevates the sternum and thereby increases the anteroposterior thoracic diameter during exertional inspiration.

12. The answer is E (all). (*Chapter 11 VII C 1 a–b*) The expiratory action of the external intercostal muscles and the interchondral portion of the internal intercostal muscles results in upward rotation of the ribs about an axis described by the costovertebral joints. Because the axes of rotation cross anteriorly, there is resultant upward movement of the sternum and a flexion at the sternomanubrial joint. The anteroposterior diameter of the thoracic cage is thereby increased.

13. The answer is D (4). (*Chapter 12 I B 2 a, c*) The anterior mediastinum contains the inferior portion of the thymus gland as well as sternopericardial ligaments and connective tissue. The superior portion of the thymus gland and the trachea are located in the superior mediastinum. The pericardial cavity and heart occupy the middle mediastinum.

14. The answer is C (2, 4). (*Chapter 12 II A 5 b c*) The parietal pleura is innervated by branches of the intercostal nerves, and parietal pleuritic pain is localized to the region of irritation (lateral thoracic and anterior abdominal walls). The diaphragmatic pleura is innervated in large part by the phrenic nerves, arising from C3–C5; thus, diaphragmatic pleuritic pain is referred to the midcervical dermatomes (neck and shoulder).

15. The answer is A (1, 2, 3). (*Chapter 13 IV B 3; VI C 2–3; X B 2–3*) The pulmonary veins, foramen ovale, and the least cardiac veins (venae cordis minimae) of the left atrial wall all deliver blood to the left atrium in the developing fetus. The ductus arteriosus shunts blood from the left pulmonary artery to the aorta.

16. The answer is C (2, 4). (*Chapter 13 VI B 3 a; VII B 2; XI A 1*) A myocardial infarction limited to the interventricular septum may produce tricuspid valve incompetence as a result of necrosis of the septal papillary muscles as well as irregularities in impulse conduction as a result of involvement of the atrioventricular bundle. The mitral valve, having no septal cusp or septal papillary muscles, is not directly affected nor is the aortic valve within the aortic outflow tract.

17. The answer is A (1, 2, 3). (*Chapter 13 X C 1–2*) At birth the lungs expand with a concomitant increase in the pulmonary blood flow, which both reduces the pressure in the right atrium and increases the pressure in the left atrium. This pressure reversal causes the foramen ovale to close, preventing a reverse flow of blood to the right atrium.

18. The answer is E (all). (*Chapter 13 Figure 13-8; Chapter 14 II I 1–4*) The cardiac plexus contains postganglionic sympathetic neurons from the sympathetic chain, preganglionic parasympathetic neurons from the vagus nerve, and afferent neurons for both pain and reflexes.

19-24. The answers are: 19-C, 20-D, 21-D, 22-A, 23-A, 24-A. (*Chapter 13, IV A 1–2*) Both the right coronary artery and the circumflex branch of the left coronary artery lie in the coronary sulcus. The left coronary artery arises from the left posterior aortic sinus, whereas the right coronary artery arises from the right aortic sinus. The right coronary artery usually supplies both the sinoatrial node and the atrioventricular node and usually terminates as the posterior interventricular (descending) artery. The anterior portion of the interventricular septum is normally supplied by the anterior interventricular (descending) branch of the left coronary artery.

Part IV
Abdominal Region

15 The Anterior Abdominal Wall

I. INTRODUCTION

A. The **abdomen** is the region of the trunk that is below the **diaphragm** and above the bones and ligaments of the **pelvic brim**.

B. SURFACE LANDMARKS OF THE ABDOMEN (Fig. 15-1A)

1. Landmarks provided by hard tissues include:
   a. The **xiphoid process** of the sternum.
   b. The **costal margin**.
   c. The **iliac crest** of the pelvis.
   d. The **anterior-superior iliac spine**.
   e. The **pubic tubercle**.
   f. The **inguinal ligament**, between the anterior-superior iliac spine and the pubic tubercle.
   g. The **pubic crest**.

2. Landmarks provided by soft tissues include:
   a. The **linea semilunaris**, just lateral to the rectus abdominis muscle.
   b. The **tendinous inscriptions** of the anterior sheath of the rectus abdominis muscle.
   c. The **linea alba** in the midline.
   d. The **umbilicus**.

C. REFERENCES TO QUADRANTS AND REGIONS are used to describe the location of anatomic structures within the abdomen, as well as to describe symptoms and the results of physical examinations pertaining to this area (see Fig. 15-1B).

1. The abdomen is divided by vertical and horizontal planes through the umbilicus into four **quadrants**:
   a. **Left** and **right upper quadrants.**
   b. **Left** and **right lower quadrants.**

2. The abdomen is divided into 9 **regions** by the horizontal **subcostal plane** (beneath the lowest extension of the thoracic cage) or the **transpyloric plane** (halfway between the jugular notch and the pubis), and the horizontal **intertubercular plane** (through the iliac tubercles), as well as the two vertical **midclavicular planes** or **midinguinal planes**, into:
   a. **Left** and **right hypochondriac regions**, which underly the rib cage.
   b. The **epigastric region.**
   c. **Left** and **right lateral** (posterolateral **lumbar** and anterolateral **flank**) **regions.**
   d. The **umbilical region.**
   e. **Left** and **right inguinal (iliac) regions**.
   f. The **pubic (hypogastric) region**.

D. NUMEROUS ABDOMINAL ORGANS AND STRUCTURES MAY BE PALPATED through the anterior abdominal wall (see Fig. 15-1B).

1. The **aorta** can be outlined both above and below the umbilicus from its pulse; the aortic pulse even may be seen in thin individuals. An abdominal aortic aneurysm can be discerned upon palpation.

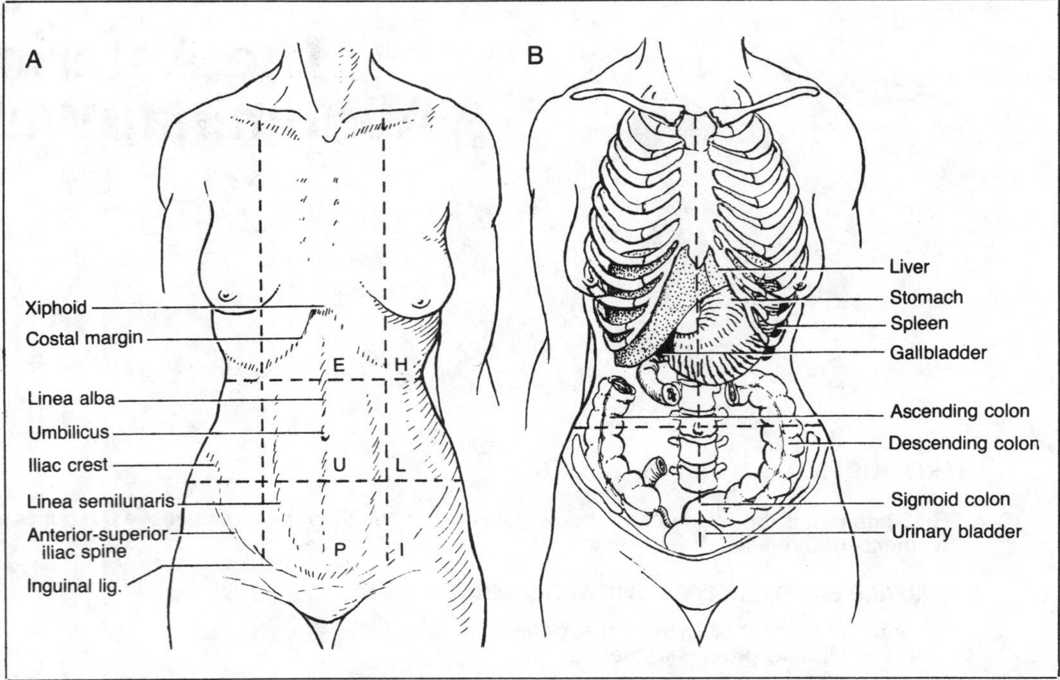

Figure 15-1. *A, The abdomen divided by regions.* Left and right hypochondriac (*H*), lateral (*L*), and inguinal (*I*) regions with medial epigastric (*E*), umbilical (*U*), and pubic (*P*) regions. Prominent landmarks and surface features are indicated. *B, The abdomen divided by quadrants.* Left and right upper and lower quadrants are depicted with the principal underlying structures.

2. The **liver** usually is palpable in the right upper quadrant, depending upon a person's build. A palpable liver may or may not indicate a pathologic condition. In persons with hepatomegaly, it may extend to the iliac crest.
3. The **gallbladder**, if it is enlarged or contains choleliths (gallstones), may be palpated in the right hypochondriac region at the junction of the linea semilunaris and the costal margin.
4. The **stomach** is palpable only if engorged.
5. The **spleen** normally is not palpable. When it is enlarged, however, it may be felt in the left hypochondriac region and may even be palpable in the lower quadrants.
6. The lower poles of the **kidneys** usually can be palpated in the lateral regions.
7. The **ascending colon** usually is palpable only if distended by gas or chyme.
8. The **descending colon** and **sigmoid colon**, containing solidifying feces, usually are palpable in the lower left quadrant.
9. The **urinary bladder**, when distended, is palpable in the pubic (hypogastric) region.
10. The **uterus** can be palpated bimanually in the pubic region, with the cervix being supported by the fingers of one hand.
11. The **ovaries**, if enlarged, may be palpated bimanually in the inguinal regions.

E. **SEVERAL ABDOMINAL ORGANS MAY BE PERCUSSED OR AUSCULTATED** through the abdominal wall.

1. A tympanic tone may be percussed from an empty stomach in the left upper quadrant in an upright patient and in the epigastric region in a supine patient.
2. Tympanic tones normally may be produced from the bowel; hypertympanic tones are elicited when the bowel is distended by gas.
3. Borborygmi (bowel sounds produced by peristalsis) normally are evident on auscultation; the frequency of borborygmi is an indication of intestinal motility.

4. The abdominal aorta and common iliac veins may be auscultated for pulse sounds and pathologic bruits.

II. THE INTEGUMENT

A. Review Chapter 2 I.

B. Beneath the **epidermis**, the connective tissue fibers of the **dermis** have a prevailing directionality.

1. The **cleavage lines (Langer's)** are of clinical importance. They define the prevailing arrangement of connective tissue fibers and are evident as crease lines in the skin.

2. Langer's lines are nearly horizontal along the anterior abdominal wall (see Fig. 2-1).

C. When connective tissue bundles are cut, they tend to retract because of their elastic properties (especially the collagen fibers).

1. An incision made across the prevailing direction of the connective tissue fibers will tend to gape, resulting in a prominent scar upon healing.

2. An incision made parallel to Langer's lines will sever fewer connective tissue bundles, will gape less, and will heal with a more cosmetic scar.

D. The dermis resists tearing, shearing, and localized pressure, but is flexible enough to permit some stretch. Skin that is stretched for long periods, as in pregnancy or in obese persons, can be damaged by rupture of the connective tissue fascicles, with subsequent scar formation called striae or "stretch marks." Striae appear perpendicular to Langer's lines.

III. THE FASCIAE OF THE ANTERIOR ABDOMINAL WALL

A. **THE SUPERFICIAL FASCIA** (subdermis or hypodermis) underlies the epidermis.

1. The **superficial layer of the superficial fascia (Camper's)** is predominantly an adipose layer, containing most of the fat of the subdermis. Camper's fascia continues:
   a. Over the chest as the superficial layer of the superficial thoracic fascia.
   b. Across the inguinal ligament to merge with the superficial fascia of the thigh.
   c. Over the pubis as the superficial layer **(Cruveilhier's)** of the superficial perineal fascia.

2. The **deep layer** of the **superficial fascia (Scarpa's)** is a fibrous layer that will hold sutures. Scarpa's fascia continues:
   a. Over the chest as the deep layer of superficial thoracic fascia.
   b. Into the upper thigh, where it is attached to the fascia lata just below and parallel to the inguinal ligament.
   c. Over the pubis as the deep layer **(Colles')** of the superficial perineal fascia.

3. Camper's and Scarpa's layers of the superficial fascia fuse over the penis and the scrotum.
   a. The reflection of superficial fascia over the penis, as the superficial fascia of the penis, has no apparent adipose contribution from Camper's fascia.
   b. The reflection over the scrotum, the **dartos layer**, contains smooth muscle.

4. The superficial fascia contains the superficial blood vessels, lymphatics, and nerves.

B. **THE DEEP FASCIA**

1. By definition, the deep fascia is the investing fascia of the musculature and aponeuroses. It cannot be separated easily from the underlying epimysium of muscle, perineurium of nerve, periosteum of bone, and perichondrium of cartilage.

2. The deep fascia continues:
   a. Over the spermatic cord as the **external spermatic fascia**.
   b. Over the pubis and perineal musculature as the deep perineal fascia **(Gallaudet's)**.
   c. Over the penis as the deep penile fascia (**Buck's**).

IV. MUSCULATURE OF THE ANTERIOR ABDOMINAL WALL

A. **THE INTRINSIC ABDOMINAL MUSCULATURE IS ARRANGED IN THREE LAYERS.** The outer layer runs superolaterally-inferomedially; the intermediate layer tends to run inferolaterally-superomedially; and the deepest layer tends to run transversely. As in the thorax, this three-ply bias bond construction provides maximum strength.

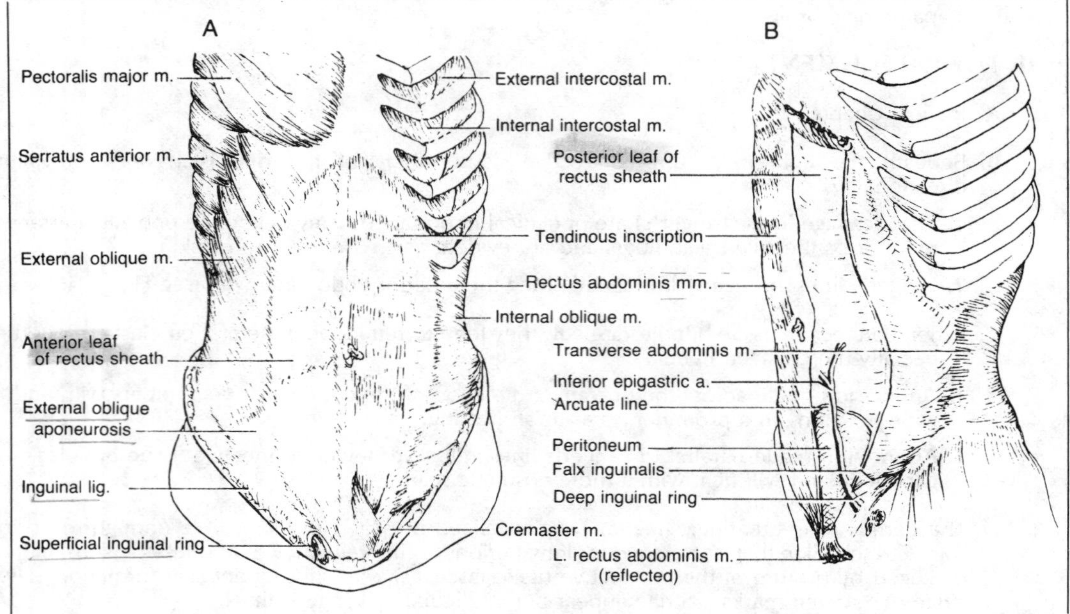

Figure 15-2. *The abdominal musculature. A,* The right external oblique and left internal oblique muscles are depicted. The anterior leaf of the rectus sheath is shown on the right side and dissected away on the left side to show the left rectus abdominis muscle. *B,* The left transverse abdominis and the right rectus abdominis muscles are depicted. The left rectus abdominis has been removed to show the posterior leaf of the rectus sheath and arcuate line.

1. **The external oblique muscle**, constituting the outer layer of the abdominal wall musculature, has fibers that run from superolateral origins to inferomedial insertions. The muscle layer becomes aponeurotic (broadly tendinous) at approximately the midclavicular line (Fig. 15-2).
   a. **Attachments.** From the inferior borders of ribs 5 to 12, this superficial muscle layer runs diagonally inferiorly and medially to insert into:
      (1) The iliac crest as far as the anterior-superior iliac spine.
      (2) The pubic tubercle and pubic crest, and the linea alba by decussating with fibers of the contralateral aponeurosis.
   b. The free posterior border of the external oblique muscle (between the rib cage and the iliac crest) forms one side of the **inferior lumbar trigone (Petit's)**. The iliac crest and the lateral border of the latissimus dorsi muscle form the remaining two sides of the **inferior lumbar trigone**.
   c. The lower margin of the external oblique aponeurosis (between the **anterior iliac spine** and the **pubic tubercle**) rolls under to form the **inguinal ligament (Poupart's)**.
   d. The **superficial inguinal ring**, although so termed, is actually a triangular defect in the aponeurosis of the external oblique muscle.
      (1) The aponeurosis of the external oblique muscle splits into two crura.
         (a) The **lateral crus** inserts onto the **pubic tubercle**, with some tendinous fibers reflected to the **superior pubic ramus** as the **lacunar ligament (Gimbernat's)**.
         (b) The **medial crus** inserts into the **pubic symphysis**.
         (c) **Intercrural fibers** strengthen the apex of the superficial inguinal ring.
      (2) The external inguinal ring transmits the **spermatic cord** (in the male) or the **round ligament of the uterus** (in the female).

2. **The internal oblique muscle** is the intermediate layer of the abdominal wall musculature. Fibers of the internal oblique muscle and aponeurosis originate inferolaterally and insert superomedially, usually running perpendicular to those of the external oblique. However, in the inguinal and hypogastric regions they run parallel to the fibers of the external oblique aponeurosis and provide muscular support for the abdominal wall in these regions. The intermediate muscle layer becomes aponeurotic at approximately the linea semilunaris (see Fig. 15-2).
   a. **Attachments.** From the thoracolumbar (thoracodorsal) fascia, transverse processes of the lumbar vertebrae, iliac crest, and inguinal ligament, this intermediate muscle layer inserts into:

(1) The inferior borders of ribs 12, 11, and 10.
(2) The linea alba, by decussating with fibers of the contralateral aponeurosis.
(3) The pubic symphysis.

b. In the male the spermatic cord passes through the internal oblique muscle from which the cremaster muscle is derived. Thus the **inguinal canal** passes diagonally through and is formed in part by the internal oblique muscle.

3. The transverse abdominal muscle constitutes the innermost layer of the abdominal wall musculature. Fibers of the transverse abdominal muscle and aponeurosis tend to run horizontally across the abdominal wall. At approximately the linea semilunaris, the muscle fibers cease and the layer becomes aponeurotic (see Fig. 15-2).

a. Attachments. From the inferior borders of costal cartilages 5 to 10, ribs 11 and 12, lumbodorsal fascia, and iliac crest, this innermost muscle layer inserts into:
(1) The linea alba, by decussating with fibers of the contralateral aponeurosis.
(2) The pubic crest, thereby reinforcing the inferomedial portion of the inguinal region.

b. The lower free edge of the transverse abdominal muscle and aponeurosis (**falx inguinalis**) is usually (in 95 percent of males) craniad to the inguinal canal and takes no part in forming the spermatic cord.

4. The rectus abdominis muscle (and the occasional **pyramidalis muscle**) represents the same layer as the occasional sternalis muscle in the thorax and the pubococcygeus muscle of the pelvis (see Fig. 15–3).

a. Attachments. From costal cartilages 5 to 7 and the xiphoid process, this innermost muscle layer inserts into the pubic crest.

b. The lateral margins define the **linea semilunaris (of Spigelius)**.

c. Medial margins define the **linea alba** in the midline.

d. There are three or four **tendinous inscriptions** between the anterior leaf of the rectus sheath and the rectus abdominis muscle craniad to the umbilicus. The rectus sheath is formed by the aponeuroses of the muscles of the lateral abdominal wall.

e. The **pyramidalis muscle** (frequently undeveloped, absent, or unilateral) lies anterior to the lower rectus abdominis muscle, between the linea alba and the pubic crest.

B. ACTIONS OF THE MUSCLES OF THE ANTERIOR ABDOMINAL WALL.

Depending upon the type of motion, the muscles of the abdominal wall can be antagonistic to or synergistic with each other.

1. Flexion of the vertebral column and pelvis by contraction of the external and internal oblique muscles, bilaterally, as well as the rectus abdominis muscle, the most powerful abdominal flexor.

2. Abduction of the trunk by ipsilateral contraction of the external and internal obliques, acting synergistically on one side.

3. Rotation of the trunk by contraction of the external oblique muscle on one side with the internal oblique of the contralateral side. The fibers of the external oblique on one side have the same prevailing direction as those of the internal oblique on the contralateral side.

4. Respiration. Abdominal expiration is accomplished by contraction of the abdominal musculature as a unit, which forces the abdominal viscera (as a fluid piston) against the inferior surface of the diaphragm and causes the diaphragm to stretch and elevate into the thoracic cavity, thereby decreasing thoracic volume.

5. Fixation of the abdominal wall.

a. Contraction of the abdominal musculature in concert with the thoracic musculature, and with the glottis closed, will result in increased thoracoabdominal pressure, the **Valsalva maneuver**.

b. Fixation of the thoracoabdominal wall is utilized in lifting heavy weights, coughing, sneezing, micturition, parturition, defecation, eructation, and emesis. In combination with weakness of the abdominal wall, herniation may result from excessive thoracoabdominal fixation.

C. THE RECTUS SHEATH INVESTS THE RECTUS ABDOMINIS MUSCLE (see Fig. 15-2).

1. The **rectus sheath** is formed by fusion of the aponeurosis of the internal oblique, external oblique, and transverse abdominis muscles.

2. Above the arcuate line (Douglas's).

a. The **anterior leaf** of the rectus sheath is formed by fusion of the external oblique aponeurosis *and* the anterior portion of the split internal oblique aponeurosis. Two or three **tendinous inscriptions** (remnants of segmentation) attach the anterior leaf of the rectus sheath to the rectus abdominis muscle and frequently are visible surface features superior to the umbilicus.

b. The **posterior leaf** of the rectus sheath is formed by fusion of the posterior portion of the split internal oblique aponeurosis *and* the transverse abdominis aponeurosis. There are no tendinous inscriptions on the posterior surface of the rectus sheath.

3. The arcuate line represents the inferior free edge of the posterior leaf of the rectus sheath.

a. One or two inches below the umbilical level, all of the aponeurotic layers of the lateral abdominal muscles pass *anterior* to the rectus abdominis muscle. The lower free edge of the posterior rectus sheath formed by this transition is the arcuate line.

b. If that portion of the posterior rectus sheath contributed by the internal oblique aponeurosis passes anterior to the rectus abdominis, superior (craniad) to that portion contributed by the transverse abdominis aponeurosis, there will be two arcuate lines.

c. The junction of the arcuate line with the linea semilunaris is the site of a **lateral ventral (Spigelian) hernia**.

4. Below the arcuate line.

a. The **anterior leaf** of the rectus sheath is formed by fusion of the aponeuroses of all three lateral abdominal muscles.

b. Since there is no posterior leaf of the sheath below the arcuate line, the rectus abdominis muscle lies against the transversalis fascia.

V. BLOOD SUPPLY TO THE ABDOMINAL WALL

A. THE BLOOD SUPPLY TO THE ABDOMINAL WALL IS DERIVED FROM SEVERAL SOURCES. Anastomoses between the sources provide an opportunity for collateral circulation.

B. THE EPIGASTRIC ARTERIAL SYSTEM (Fig. 15-3).

1. The superior epigastric artery:

a. Arises from the bifurcation of the **internal thoracic artery**.

b. Enters the rectus sheath beneath the costal margin.

c. Anastomoses with the inferior epigastric artery within the rectus abdominis muscle.

2. The inferior epigastric artery:

a. Arises from the **external iliac artery**.

b. As it passes obliquely upward from its origin, lies along the inferior and medial margins of the **deep inguinal ring**.

c. Enters the rectus sheath by passing superficial to the arcuate line.

d. Anastomoses with the superior epigastric artery within the rectus abdominis muscle.

3. Anastomoses in the epigastric arterial system can provide important shunts in conditions such as coarctation of the aorta to supply blood to the lower part of the body.

C. OTHER MAJOR ANTERIOR ABDOMINAL ARTERIES (see Fig. 15-3).

1. The musculophrenic artery:

a. Arises as the other bifurcating branch of the internal thoracic (internal mammary) artery.

b. Courses along the costal margin.

c. Supplies numerous branches to the anterior abdominal wall, as well as to the diaphragm.

2. The intercostal arteries 9 through 12:

a. Arise from the **thoracic aorta** and extend ventromedially beyond the costal margin into the abdominal wall.

b. Enter the rectus sheath laterally.

c. Anastomose with the superior and inferior epigastric arteries.

3. The superficial epigastric arteries:

a. Arise from the **external iliac artery**.

b. Supply the superficial inguinal and pubic regions.

c. Anastomose with lower intercostal arteries.

4. The deep circumflex iliac artery:

a. Arises from the **external iliac artery**.

b. Courses deep to the inguinal ligament toward the anterior-superior iliac spine.

c. Supplies the deep inguinal region.

d. Anastomoses with the lower intercostal arteries.

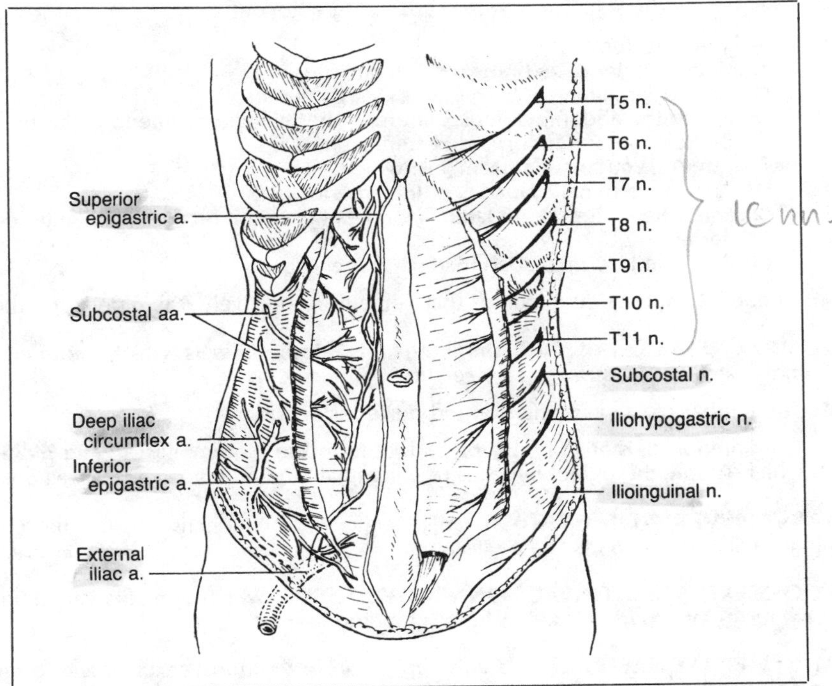

Figure 15-3. *The abdominal vasculature and innervation.* The vascular anastomoses between the superior epigastric, musculophrenic, intercostal, and inferior epigastric arteries are indicated. The 10th intercostal nerve supplies the umbilical region.

VI. VENOUS DRAINAGE OF THE ANTERIOR ABDOMINAL WALL

A. THE SUPERFICIAL VENOUS DRAINAGE of the anterior abdominal wall is toward the following veins:

1. The superior and inferior epigastric veins.

2. The axillary vein via the thoracic and thoracoepigastric veins.

3. The femoral vein via the superficial branches of the great saphenous vein.

4. The hepatic portal vein via the paraumbilical veins.
- **a.** Pathologic obstruction of the passage of blood from the gastrointestinal tract through the liver results in retrograde flow of blood to the systemic venous return of the body wall, via enlarged paraumbilical anastomoses, and is termed **caput medusae**.
- **b.** Although caput medusae may not be readily apparent in cases of portal hypertension, it nearly always is demonstrable in a photograph taken with infrared film.

B. THE DEEP DRAINAGE of the anterior wall is along veins that parallel the arterial supply previously noted.

C. ABDOMINAL LYMPHATIC VESSELS accompany the veins. Carcinoma of the breast may metastasize along the superficial lymphatics of the anterior abdominal wall.

VII. NERVES OF THE ANTERIOR ABDOMINAL WALL.

The **ventral primary rami** of the spinal nerves supply the dermatomes and myotomes of the abdominal wall (see Fig. 15-3).

A. INTERCOSTAL NERVES T7 THROUGH T12

1. They continue beyond the costal margin into the abdominal wall.

2. They lie between the internal oblique and the transverse abdominis muscles.

3. They penetrate the rectus sheath laterally to innervate the rectus abdominis muscle.

4. They supply the skin, muscles, and parietal peritoneum.

5. They branch to form:
   a. The **lateral cutaneous branches**, which:
      (1) Arise approximately at the anterior axillary line.
      (2) Pierce the abdominal musculature in the midaxillary line to reach the dermis.
      (3) Bifurcate into anterior and posterior branches.
   b. The **anterior cutaneous branches**, which:
      (1) Arise as terminal branches of the anterior rami.
      (2) Enter the rectus sheath laterally and exit from the rectus sheath anteriorly to reach the dermis.
      (3) Bifurcate into medial and lateral branches.

6. T4 and T10 supply dermatomes that contain, respectively, the nipple and the umbilicus.

B. **LUMBAR NERVES L1 AND L2** contribute to the **lumbar plexus**, which is formed by the anterior primary rami of T12 through L4 (see Fig. 18-6).

1. The lumbar plexus has anterior and posterior divisions.

2. The anterior divisions of T12, L1 and L2 form the **iliohypogastric** and **ilioinguinal nerves**, which supply the pubic (hypogastric), inguinal, and anterior perineal regions.

VIII. **THE TRANSVERSALIS FASCIA** is an inner layer of connective tissue that enwraps the abdominal cavity and supports the peritoneum.

A. **LOOSE EXTRAPERITONEAL CONNECTIVE TISSUE**, which contains varying amounts of fat, separates the transversalis fascia from the peritoneum.

B. **THE DEEP INGUINAL RING** is an interruption in the transversalis fascia formed by the embryonic extension of the **processus vaginalis** (peritonei) through the abdominal wall and subsequent passage of the testes through the transversalis fascia during the descent of the testes into the scrotum.

IX. CLINICAL CONSIDERATIONS RELATING TO THE ANTERIOR ABDOMINAL WALL

A. **ACCESS TO THE PERITONEAL CAVITY** through the anterior abdominal wall.

1. **Via the rectus sheath.**
   a. An incision into the anterior leaf of the rectus sheath medially has the advantage that the rectus abdominis muscle may be retracted laterally without compromising either its vascular or nerve supply; a second incision through the posterior leaf provides access to the peritoneal cavity. When the incisions in the posterior and anterior leaves are sutured, the undamaged rectus abdominis muscle will support the incisions well.
   b. If the rectus abdominis muscle is sectioned longitudinally, the nerves supplying the muscle medial to the incision will be severed, resulting in paralysis of the medial portion of the muscle and a weakened abdominal wall disposed to ventral herniation.
   c. If the rectus muscle is transsected, the arterial supply can be embarrassed, with possible resultant necrosis of the muscle and a weakened abdominal wall disposed to ventral herniation. Moreover, reanastomosis of a transsected muscle has been compared with trying to sew two paint brushes together.

2. **Via the linea alba.**
   a. The linea alba is a relatively avascular region, hence its name *alba*. A midline incision through this structure is relatively bloodless but may not heal well without leaving a large, unslightly roll of tissue.
   b. A midline incision may predispose to midline herniation, one of the most common postsurgical complications.

3. **Via the ventrolateral abdominal wall.**
   a. Although many transverse incisions can take advantage of Langer's cleavage lines, they also may transsect the nerve supply to the abdominal musculature unless appropriate precautions are taken.
   b. The muscle splitting approach (McBurney's) for appendectomy in the right lower quadrant utilizes Langer's lines for the superficial incision, then splits and retracts the external oblique muscle, splits and retracts the internal oblique muscle (locating and retracting the iliohypogastric nerve), splits and retracts the transverse abdominis muscle, and, finally, incises the peritoneum. Incisions in each layer close naturally with few, if any, sutures and

provide a strong repair. [Valsalva fixation (i.e., a cough or sneeze) tends to close a split muscle but to tear the sutures from a transsected muscle.]

B. ABDOMINAL HERNIAS

1. **Epigastric herniation** is a midline protrusion of fat or abdominal viscera through the linea alba (usually above the umbilicus). It is a frequent postoperative complication but rare otherwise.

2. **Umbilical herniation** is a protrusion of abdominal viscera through the umbilical ring.
   a. Congenital umbilical hernia, which has an embryologic basis, occurs in about 1 in 50 births.
      (1) It may be apparent at birth, omphalocele, or it may occur spontaneously in infants and children.
      (2) It may regress spontaneously.
   b. Acquired umbilical hernia usually occurs in adult life.
      (1) It is frequently the result of repeated Valsalva fixation, pregnancy, ascites, or obesity along with a weakened abdominal wall.
      (2) Repair of umbilical herniation may be complicated by portal hypertension, which causes the paraumbilical anastomoses to be greatly enlarged and bleed profusely if divided.
   c. Patent urachus, a congenital connection between the urinary bladder and the umbilicus, may result in seepage of urine from the umbilicus.

3. **Semilunar (spigelian) herniation** is a ventrolateral herniation that occurs most frequently at the junction of the arcuate line and the linea semilunaris but may occur along the linea semilunaris inferior to the arcuate line.

4. **Diaphragmatic herniation** involves upward displacement of abdominal viscera or fat into the thorax; it is more common on the left side because of the location of the esophageal hiatus.
   a. The **esophageal hiatus**, through which the esophagus passes into the abdomen, is a common site for herniation of the stomach, hiatus hernia.
   b. The diaphragm develops from at least six primordia, some of them paired, which include contributions from the septum transversum in the cervical region, pleuroperitoneal membranes, and somatic wall. Failure of these primordia to fuse properly results in defects through which herniation may occur.
      (1) Incomplete closure of the pleuroperitoneal canal (usually on the left side) may be a site of herniation of abdominal contents into the thoracic cavity (**Bochdalek's hernia**).
      (2) Incomplete closure of the sternocostal triangle (**foramen of Morgagni**) is another potential, although rare, cause of herniation.

X. THE INGUINAL REGION

A. INTRODUCTION

1. The inguinal region is best described with reference to descent of the testes.
   a. The **gonads** develop retroperitoneally from the urogenital ridge in the region of the kidneys. They migrate retroperitoneally within the transversalis fascia toward the scrotum during embryonic development, apparently guided by the **gubernaculum testis**, a ligamentous structure that runs between the lower pole of each gonad to each labial-scrotal fold.
      (1) By the third embryonic month, the gonads beneath the peritoneum have moved into the false pelvis.
         (a) In females, migration continues into the deep pelvis.
         (b) In males, a peritoneal pouch, the **processus vaginalis**, evaginates into the developing scrotum.
      (2) By the seventh month, the testes reach the abdominal end of the inguinal canal (always retroperitoneal).
      (3) Between the seventh and eighth months, the testes migrate through the inguinal canal *behind* the processus vaginalis (always retroperitoneal) into the scrotum to form the spermatic cord.
      (4) Before birth, the testes reach their definitive position in the scrotum behind the tunica vaginalis, the remnant of the processus vaginalis (always retroperitoneal).
      (5) The processus vaginalis may close before birth. However, it may remain patent and thereby predispose to congenital indirect hernia during the early years of life.
   b. The testes retain their original nerve and blood supply; this explains the course of the spermatic vessels and nerves.

c. The ovaries also migrate, but into the deep pelvis, also trailing their vessels and nerves.

2. Failure of testicular descent, **cryptorchidism**, occurs frequently.
   a. The testes may remain at the deep ring or within the inguinal canal.
   b. Cryptorchidism may correct spontaneously during the first year of life; if it persists, surgery is indicated. Failure to correct undescended testes before puberty results in sterility and, if bilateral, failure of the secondary sex characteristics to develop.

3. The inguinal region is a potential site of **abdominal herniation**.
   a. Passage of the testes through the ventral abdominal wall produces a weakened area which may predispose to herniation, with danger of strangulation, necrosis of the bowel loop, and embarrassment of the testicular blood supply.
   b. A persistent processus vaginalis (**funicular process**) may result in an **indirect congenital hernia**, in which a loop of gut passes into and may even continue through the patent canal into the scrotum.

B. THE INGUINAL CANAL passes through the anterior abdominal wall from the superficial inguinal ring to the deep inguinal ring.

1. The **superficial (external) inguinal ring** is a triangular interruption in the external oblique aponeurosis (Fig. 15-4A).
   a. It is formed by the splitting of the aponeurosis of the external oblique muscle into two crura.
      (1) The **lateral crus** inserts onto the **pubic tubercle**, with some fibers reflected to the **superior pubic ramus** as the **lacunar ligament** (Gimbernat's).
      (2) The **medial crus** inserts into the **pubis symphysis**.
      (3) **Intercrural fibers** strengthen the apex of the superficial inguinal ring.

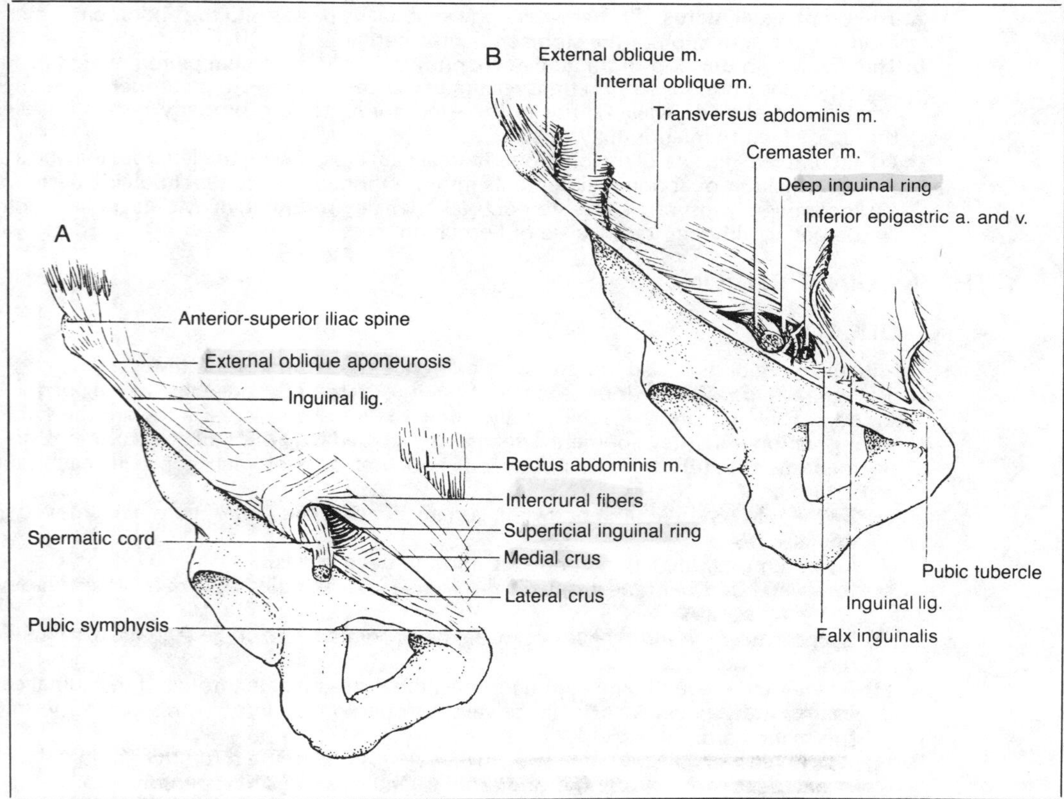

Figure 15-4. *A, The superficial inguinal ring.* The defect in the external oblique aponeurosis forms the superficial inguinal ring. The internal oblique muscle contributes the cremaster layer of the spermatic cord. *B, The deep inguinal ring.* The deep ring is the location of the fused processus vaginalis. The inguinal canal passes beneath the inferior free edge of the transverse abdominis muscle.

b. The superficial ring transmits the **spermatic cord** in males and the **round ligament of the uterus** in females.
c. The deep fascia (investing fascia of the external oblique muscle) continues over the spermatic cord as the **external spermatic fascia**.
d. The superficial inguinal ring may be palpated (but not normally entered) by inversion of the scrotum (preferably with the little finger) directed along the vas deferens toward the pubic tubercle.
e. At the point of exit of both **direct** and **indirect inguinal hernias**, a small hernia will bulge slightly at the superficial ring during Valsalva fixation, for example, a cough, whereas a large hernia will be palpable or even visually apparent.

2. The **deep (internal) inguinal ring** is a partial interruption in the transversalis fascia (Fig. 15-4B).
   a. It is covered completely by peritoneum (unless there is a patent funicular process).
   b. Some of the transversalis fascia is drawn into the spermatic cord during descent of the testes, forming the **internal spermatic fascia**.
   c. The ring may be the entry site of an **indirect inguinal hernia**.

3. The **walls of the inguinal canal** are formed by the muscular, aponeurotic, and fascial layers of the abdominal wall.
   a. The **anterior wall** of the inguinal canal is formed by the **external oblique aponeurosis** superior to the **intercrural fibers** and the fibers of the internal oblique muscle (Fig. 15-5).
      (1) The actual fibers of the external oblique muscle rarely extend as far as the inguinal canal.
      (2) Contraction of the internal oblique muscle tends to occlude and strengthen the inguinal canal.
      (3) Because the free edge of the transverse abdominis muscle lies superior to the deep ring, it seldom contributes (less than 5 percent of the time) to the anterior wall of the inguinal canal or to the cremaster muscle.
   b. The **superior wall** (roof) of the inguinal canal is formed by the **falx inguinalis**, which is the arcing free edge of the aponeurosis of the transverse abdominis muscle (see Fig. 15-4B).
      (1) The falx inguinalis (*falx, L. sickle*) runs from the lateral portion of the inguinal ligament to the rectus sheath and pubic ramus.
      (2) Many texts and atlas figures mistakenly and unfortunately refer to the falx inguinalis as the "conjoined tendon." The falx inguinalis actually is the lower, curving portion of the transverse abdominis aponeurosis. The conjoined tendon, on the other hand, results from the fusion (conjoining) of the aponeuroses of the internal oblique muscle and the transverse abdominis muscle as they form the anterior wall of the rectus sheath. Less than 5 percent of the time does a true conjoined tendon extend more than 1 cm lateral to the linea semilunaris. If it were not for this confusion, a true conjoined tendon would be regarded as the rarity it is.
   c. The **inferior wall** (floor) of the inguinal canal is formed by the **inguinal ligament** (Poupart's) and the **lateral crus** of the superficial ring.
      (1) The inguinal ligament runs from the anterior-superior iliac spine to the pubic tubercle (see Fig. 15-4B).
      (2) The floor also receives a contribution from the lacunar ligament which is formed by fibers reflected from the inguinal ligament posteriorly and runs superiorly to join the pectinate ligament along the pectinate line of the superior pubic ramus (see Fig. 15-4B).
   d. The **posterior wall** of the inguinal canal is formed by the **transversalis fascia** and the **interfoveolar ligament (Hesselbach's)**, a variable condensation of the transversalis fascia in the vicinity of the inferior epigastric vessels (see Fig. 15-4B).

C. THE SPERMATIC CORD

1. **Embryologic development** (see Fig. 15-5).
   a. The spermatic cord begins as an evagination of the peritoneal cavity through the anterior abdominal wall, the **processus vaginalis**. Normally, the processus vaginalis persists from the third to the ninth months.
   b. As the processus vaginalis and the testes descend through the abdominal wall in the inguinal region, contributions from some layers of the abdominal wall are incorporated into the spermatic cord. These contributions include:
      (1) A peritoneal remnant of the obliterated processus vaginalis or, if patent, the **funicular process**.
      (2) The **internal spermatic fascia**, derived from the transversalis fascia.
      (3) The **cremaster muscle**, derived from the internal oblique muscle, with contribution from the transverse abdominis muscle in less than 5 percent of individuals.

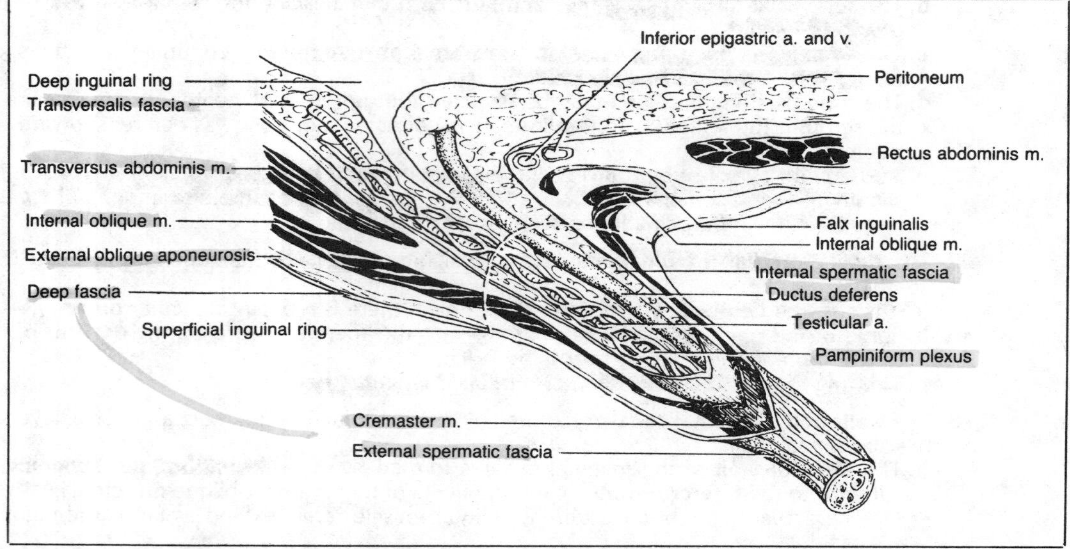

Figure 15-5. *The spermatic cord.* The deep fascia of the external oblique layer contributes the external spermatic fascia, the internal oblique layer contributes the cremaster muscle, the transverse abdominis layer usually makes no contribution, and the transversalis fascia contributes the internal spermatic fascia.

(4) The **external spermatic fascia**, derived from the deep (investing) fascia of the external oblique muscle, with no other contribution from the external oblique layer, since the aponeurosis splits to form the superficial inguinal ring.

(5) The **superficial fascia**, derived from Camper's and Scarpa's layers of the superficial fascia.

2. Because the testes remain attached to their original excretory ducts, vascular supply, and nerve supply, the contents of the spermatic cord include (see Fig. 15-5):

a. The **vas deferens**, the efferent duct of the testes.

b. The **testicular arteries**, each of which arises from the aorta just caudal to the renal arteries.

c. The **arteries of the vas deferens**, each of which arises from the internal iliac artery in the deep pelvis and is homologous to the uterine artery in females.

d. The **pampiniform plexus** of veins which coalesces into the **testicular veins**, which drain into the inferior vena cava (right) and the renal vein (left).

(1) This venous plexus provides a countercurrent heat exchange mechanism, which keeps the testes 1° to 3° F cooler than the core body temperature.

(2) Varicosities in the pampiniform plexus of veins occur on the left side in about 85 percent of cases.

e. The spermatic cord also contains autonomic nerves to and visceral afferent nerves from the testes, as well as the **genital branch** of the **genitofemoral nerve**, which innervates the cremaster muscle, and the sensory **ilioinguinal nerve**.

3. Only the **round ligament of the uterus**, which is a remnant of the gubernaculum, and a branch of the **ilioinguinal nerve** pass through the inguinal canal (Nuck's) in females.

D. THE NERVES OF THE INGUINAL REGION

1. These nerves arise from the **lumbar plexus** on each side.

a. They pass through the psoas major muscle and anterior to the quadratus lumborum muscle, posteriorly.

b. They course between the transverse abdominis and internal oblique muscles, laterally.

2. The nerves of the lumbar plexus include (see Fig. 18-6).

a. The iliohypogastric nerve.

(1) This nerve arises from the anterior divisions of T12 and L1.

(2) It is distributed to the skin of the hypogastric region just superior to the pubic crest and pubic symphysis.

(3) It usually can be observed penetrating the external oblique aponeurosis 1–2 cm superomedial to the superficial inguinal ring.

(4) It provides afferent (sensory) and one of the several efferent (motor) limbs of the **abdominal reflex**. Stroking the skin of the abdomen parallel to the inguinal ligament initiates a rippling of the rectus abdominis muscle or flank muscles of the abdomen.

b. The ilioinguinal nerve.

(1) It arises from the anterior division of L1.
(2) It accompanies the spermatic cord or round ligament of the uterus through the superficial inguinal ring.
(3) It is distributed to the medial aspect of the thigh, the base of the penis or mons, and the scrotum or labia majora.

c. The genitofemoral nerve.

(1) It arises from the anterior division of L1 and L2.
(2) It lies on the anterior surface of the psoas major muscle.
(3) Its **femoral branch**, which passes beneath the inguinal ligament, is distributed to the anterolateral aspect of the thigh and provides the afferent limb of the **cremaster reflex**, initiated by stroking of the medial thigh.
(4) Its **genital branch**, which lies lateral or posterolateral to the spermatic cord as it passes through the superficial inguinal ring, is distributed to the cremaster muscle and scrotum, and provides the efferent limb of the cremaster reflex whereby a testis is elevated within the scrotum.

E. THE INGUINAL TRIANGLE (Hesselbach's)

1. The **inguinal triangle** (Fig. 15-6) lies in the inferomedial inguinal region, where it is bounded by:
   a. The **linea semilunaris** (the lateral edge of the rectus sheath), medially.
   b. The **inguinal ligament**, inferolaterally.
   c. The **inferior epigastric artery**, laterally.
2. The inguinal triangle is an area of potential weakness and thus often the site of a **direct inguinal hernia**.
   a. The abdominal wall contributes little or no muscular support to the inguinal triangle.
   b. Anteriorly, the inguinal triangle is covered by the external oblique aponeurosis. It is weakened by a defect in the anterior abdominal wall, the superficial inguinal ring.
   c. Behind the superficial ring, the reinforcement of the inguinal triangle is provided mainly by the **falx inguinalis** and almost insignificantly by the **interfoveolar ligament**. The extent of support derived from the falx inguinalis varies among individuals, depending on the extent of its lateral insertion upon the superior pubic ramus.
   d. Toward the base of the inguinal triangle, support is provided only by the weak transversalis fascia and the presence of the spermatic cord within the superficial ring.
3. **Clinical considerations relating to inguinal hernias.**
   a. Fully 97 percent of all hernias in males and 50 percent in females occur in the inguinal region.
   b. Inguinal hernias occur superior to the inguinal ligament.

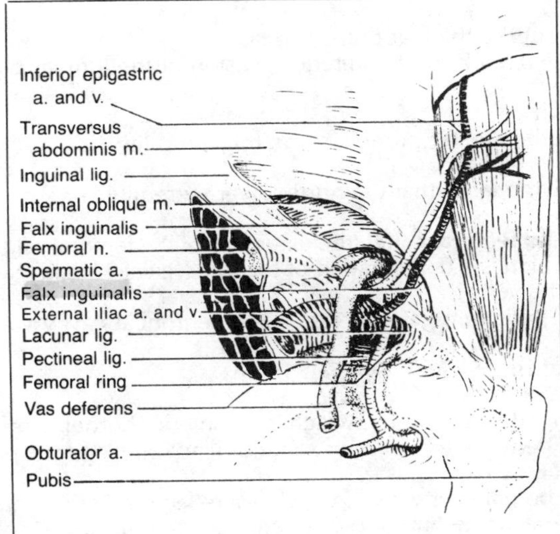

Figure 15-6. *The inguinal triangle.* The boundaries of the inguinal triangle (viewed from behind) are the lateral border of the rectus abdominis muscle, the inguinal ligament, and the inferior epigastric artery.

c. There are two types of inguinal herinas:

(1) Indirect inguinal hernias.

(a) Indirect herniation is with*in* the inguinal canal.

(i) The herniated viscus or fat lies with*in* the spermatic cord.

(ii) The hernia passes through the deep inguinal ring; therefore, it must pass *lateral* to the inferior epigastric artery to enter the inguinal canal; this places its origin, by definition, lateral to the inguinal triangle.

(iii) The hernia exits the abdominal wall through the superficial inguinal ring.

(iv) The hernia is directed by the spermatic cord toward the scrotum.

(v) The risk of strangulation and infarct is high because the hernia passes through the muscular inguinal canal.

(b) A **congenital** (funicular) indirect inguinal hernia occurs through a patent funicular process, usually during the first year or so of life.

(c) An **acquired** indirect inguinal hernia occurs through a weakened area behind the remnant of the fused processus vaginalis, usually in middle age or later.

(2) Direct inguinal hernia.

(a) Direct herniation through the inguinal triangle (Hesselbach's), pushes the peritoneum and transversalis fascia through the superficial inguinal ring.

(i) The herniated viscus or fat lies *adjacent to* (not within) the spermatic cord.

(ii) The hernia does *not* enter the deep inguinal ring, and does *not* pass through the inguinal canal; therefore, it passes medial to the inferior epigastric artery and, by definition, through the inguinal triangle.

(iii) The hernia exits the abdominal wall *directly* through the superficial inguinal ring.

(iv) Direct hernias seldom enter the scrotum.

(v) The risk of strangulation in direct inguinal herniation is low because passage is not through muscular layers or strong tendinous confines.

(b) Direct inguinal hernias are always acquired.

(i) They are most common in middle-aged males and seldom occur in females (although they have been seen more often in recent years).

(ii) Although these hernias are acquired, genetic factors predispose to the weakness or strength of the abdominal wall in the inguinal triangle (i.e., the lateral extent of the attachment of the falx inguinalis along the superior pubic ramus).

(3) All inguinal hernias, **in contrast to femoral hernias:**

(a) Lie superior to the inguinal ligament.

(b) Lie superior and medial to the pubic tubercle, to which the inguinal ligament and lateral crus of the superficial ring both attach.

F. THE FEMORAL TRIANGLE is bounded by the inguinal ligament, the sartorius, muscle, and the adductor longus muscle.

1. The area between the inguinal ligament and the superior pubic ramus is divided into two compartments (Fig. 15-7).

a. The **muscular compartment** contains:

(1) The **iliopsoas muscle**, a flexor of the thigh at the hip joint.

(2) The **femoral nerve**, which is derived from the anterior division of the lumbar plexus and innervates the anterior thigh.

b. The **vascular compartment** contains:

(1) The **femoral artery**, supplying the greater portion of the thigh and leg.

(2) The **femoral vein**.

(3) The **femoral ring**, containing lymphatics that drain the lower extremity.

(4) The **lacunar ligament**.

c. The femoral artery, femoral vein, and **femoral canal** (the extension of the femoral ring) are wrapped in the **femoral sheath**, which is a continuation of the transversalis fascia leaving the abdominal cavity along with the vessels.

(1) Lymphatics from the legs and perineum gain access to the peritoneal cavity via the femoral ring.

(2) The **boundaries of the femoral ring** include (see Fig. 15-7):

(a) The **femoral vein**, laterally.

(b) The **inguinal ligament** (Poupart's), anteriorly, which is the inferior border of the external oblique aponeurosis between the anterior-superior iliac spine and the pubic tubercle.

(c) The **lacunar ligament** (Gimbernat's), medially, which is a reflection of the inguinal ligament posteriorly to the pectinate line of the superior pubic ramus.

(d) The **pectineal ligament** (Cooper's), posteriorly, which is an extension of the lacunar ligament along the pectinate line of the superior pubic ramus.

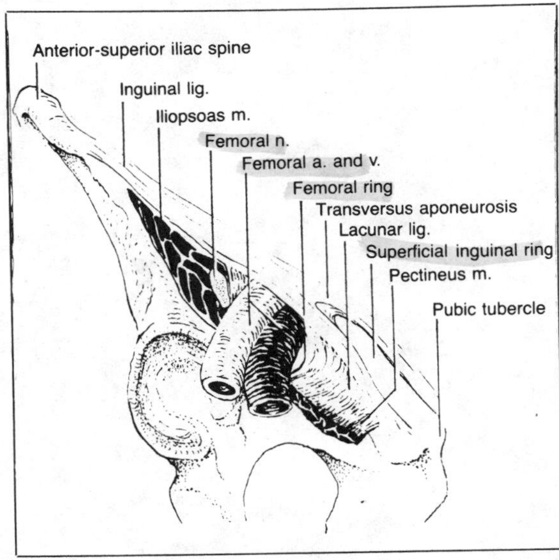

Figure 15-7. *The femoral ring.* The lateral muscular compartment contains the iliopsoas muscle and the femoral nerve. The medial vascular compartment contains the femoral vein, the femoral artery, and the femoral ring through which lymphatics pass.

2. Clinical considerations relating to femoral hernias.

a. Fully 34 percent of all hernias in females are femoral, but these hernias are rare in males.

b. Femoral hernias enter the femoral ring and femoral canal inferior to the inguinal ligament.

c. The femoral canal is located lateral to the lacunar ligament, anterior to the pectinate ligament, and medial to the femoral vein.

d. Because of the ligamentous boundaries, there is a high potential for strangulation and necrosis of a herniated loop of bowel.

e. These hernias may run medial to, lateral to, or even bifurcate to run on both sides of an **aberrant obturator artery**, which arises from the inferior epigastric artery (see Fig. 15-6). This situation can present a danger for the unaware during surgical repair of a femoral hernia.

f. Femoral hernias, **in contast to inguinal hernias:**

(1) Lie inferior to the inguinal ligament.

(2) Lie inferior and lateral to the pubic tubercle to which the inguinal ligament attaches.

(3) Usually lie in the thigh and present through the fossa ovalis of the great saphenous vein.

16
The Peritoneal Cavity

I. INTRODUCTION

A. THE ABDOMINOPELVIC CAVITY is the largest cavity of the body.

1. The **abdominal cavity** lies within the abdomen proper.
   - **a.** It extends from the inferior surface of the musculotendinous diaphragm to the brim of the minor (deep) pelvis.
   - **b.** It contains most of the gastrointestinal tract, the accessory digestive glands, the kidneys, and the major vessels.
2. The **pelvic cavity** lies within the true pelvis.
   - **a.** It extends from the brim of the minor (deep) pelvis to the pelvic diaphragm (levator ani muscle) and is about the size of a clenched fist.
   - **b.** It contains loops of small bowel, the sigmoid colon, the terminal portion of the gastrointestinal tract, and the organs of the urogenital system with associated adnexa.

B. THE PARIETAL PERITONEUM is a continuous layer of mesothelial serosa that lines the abdominopelvic cavity.

1. In males, the peritoneal cavity is completely enclosed.
2. In females, it is perforated by the ostia of the uterine tubes, which may carry infection from the external environment into the peritoneal cavity.

II. EARLY DIFFERENTIATION

A. THE EMBRYONIC DISK DEVELOPS during the third week. From the fourth to the eighth weeks, differential development produces tissues and organs.

B. THE PRIMITIVE GUT IS DIVIDED INTO THREE REGIONS. Differential growth of the **head fold, tail fold,** and **lateral folds** incorporates within the developing embryo a portion of the **upper pole** of the primitive **yolk sac** to form the **primitive gut** (Figs. 16-1 and 16-2).

1. **The foregut:**
   - **a.** Is formed by the head fold.
   - **b.** Is closed by the buccopharyngeal membrane while a blind tube and eventually opens through the stomodeum.
   - **c.** Is supplied by the **celiac artery**. CA
2. **The midgut:**
   - **a.** Is formed largely by lateral folds.
   - **b.** Retains a connection with the yolk sac by the **omphalomesenteric (vitelline) duct**, which normally ultimately degenerates but which may persist in whole or part, resulting in an **ileal diverticulum** (Meckel's).
   - **c.** Is supplied by the **superior mesenteric artery** in the midabdominal region. SMA
3. **The hindgut:**
   - **a.** Is formed by the tail fold.
   - **b.** Although a blind tube closed by the cloacal membrane, eventually opens through the proctodeum.
   - **c.** Is supplied by the **inferior mesenteric artery**.

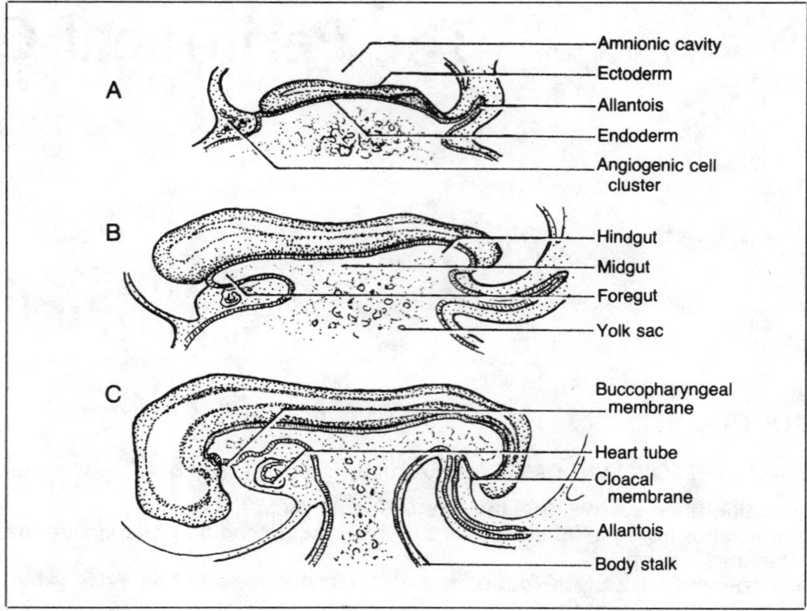

Figure 16-1. *Early differentiation of the gastrointestinal tract. A,* The germ disk at 19 days lies flat on the yolk mass. *B,* By 21 days head and tail folds form the early foregut and hindgut, respectively. *C,* By 24 days the head and tail folds have deepened and lateral folds have come together to form the midgut.

C. THE ENTERIC TUBE IS FORMED by the ventral fusion of the splanchnopleure (viscera primordium) as the two lateral folds meet with the head and tail folds at the umbilical stalk; the body wall is formed by the fusion of the somatopleure (body-wall primordium) [Fig. 16-3].

D. THE COELOMIC CAVITY (coelom) is the space enclosed between the splanchnopleure and the somatopleure (see Fig. 16-3).

1. The **visceral pleura** and **visceral peritoneum**, serous mesothelia, cover the outer surfaces of the splanchnopleure.

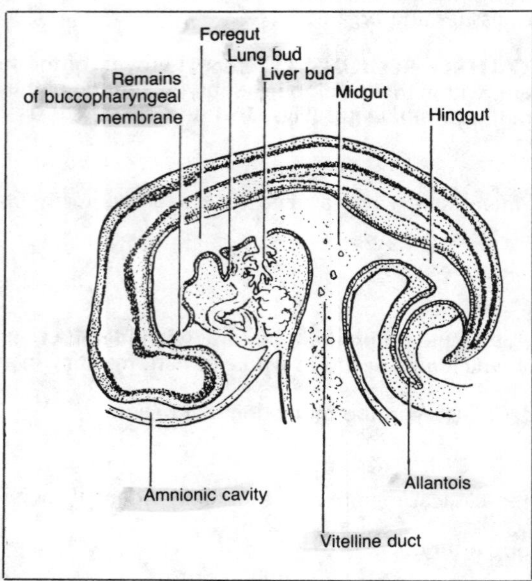

Figure 16-2. *Later differentiation of the gastrointestinal tract.* During the fourth week, flexion occurs with development of the body stalk. The midgut retains a connection with the yolk sac through the vitelline duct.

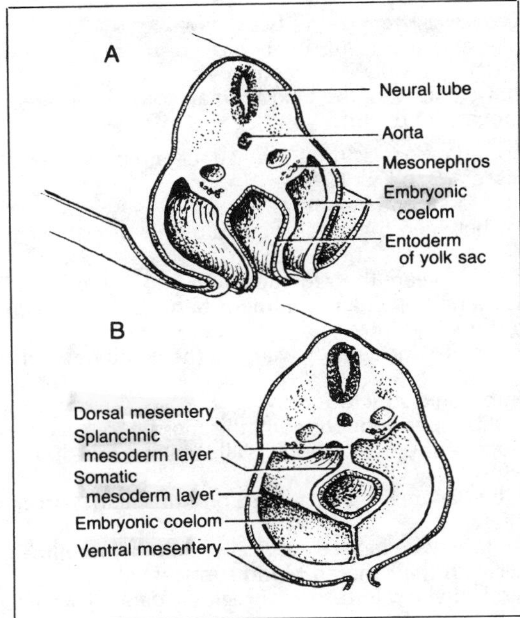

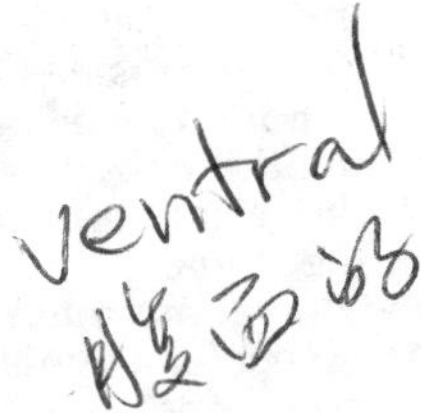

Figure 16-3. *Formation of the dorsal and ventral mesenteries. A,* About 21 days the lateral folds delineate the intraembryonic coelom and the midgut. *B,* By 28 days dorsal and ventral mesenteries become defined.

2. The **parietal pleura** and **parietal peritoneum**, serous mesothelia, cover the inner surfaces of the somatopleure.

E. **MESENTERIES** suspend the primitive gut within the coelom (see Fig. 16-3).

1. Reflections of parietal peritoneum to become visceral peritoneum constitute mesenteries which are, in effect, back-to-back layers of serous mesothelia.
2. The **ventral mesentery** largely degenerates, except for the cephalad and caudal portions. The liver develops *within* the cephalad portion of the ventral mesentery, dividing it into the lesser omentum and the falciform ligament.
   a. The **lesser omentum**, between the stomach and the liver, has two named portions (see Fig. 16-4).
      (1) The **gastrohepatic ligament** runs between the lesser curvature of the stomach and the liver.
      (2) The **hepatoduodenal ligament** runs between the superior duodenum and the liver.
         (a) This portion of the lesser omentum contains the hepatic artery, hepatic portal vein, and common bile duct.
         (b) The inferior free edge of the ventral mesentery forms the **epiploic foramen** (Winslow's).
   b. The **coronary ligaments** attach the liver to the diaphragm (see Fig. 17-7).
      (1) **Anterior** and **posterior coronary ligaments** provide continuity between the visceral peritoneum covering the liver and the parietal peritoneum lining the inferior surface of the diaphragm.
      (2) Nonapposition of the anterior and posterior leaves of the coronary ligaments result in an area between the liver and the diaphragm not covered by mesothelium, the **bare area**.
      (3) Lateral extensions of the anterior and posterior leaflets of the coronary ligaments form the **left** and **right triangular ligaments**.
   c. The **falciform ligament** (*falx, L. sickle*) attaches the liver to the ventral body wall.
      (1) The left and right leaflets of the anterior coronary ligament meet anteriorly to form the falciform ligament.
      (2) The falciform ligament extends between the ventral surface of the liver and the ventral abdominal wall as far inferiorly as the umbilicus.
      (3) It contains the **round ligament of the liver (ligamentum teres)**.
         (a) This ligament is the remnant of the obliterated **umbilical vein(s)**.
         (b) It may remain patent (rare in adults).
   d. The **median umbilical fold** is an inferior remnant of the ventral mesentery.
      (1) It runs from the apex of the urinary bladder to the umbilicus.

(2) It contains the **urachus**, which is the remnant of the fused allantoic stalk.

(a) A completely patent urachus (rare) will result in seepage of urine from the umbilicus, while isolated patency may produce a urachal cyst.

(b) Partial patency extending from the apex of the bladder may complicate surgery in the umbilical and pubic (hypogastric) regions.

3. The **dorsal mesentery** is the primary support of the gastrointestinal tract and provides passage for the blood vessels that supply the viscera (see Fig. 16-4).

a. The **dorsal mesogastrium** is divided into several ligaments.

(1) The **gastrophrenic ligament** runs between the greater curvature of the stomach and the diaphragmatic peritoneum.

(2) The **gastrosplenic ligament** runs between the stomach and the spleen, and the **lienorenal (splenorenal, phrenicolienal) ligament** continues between the spleen and the dorsal body wall superior to the left kidney.

(3) The **greater omentum** runs between the dorsal body wall and the greater curvature of the stomach.

b. The **dorsal mesoduodenum** is an embryonic structure only.

(1) Except for a minute portion along the superior duodenum, this mesentery becomes apposed and fixed to the peritoneum of the dorsal body wall, making the duodenum *secondarily retroperitoneal*.

(2) It contains the head and body of the pancreas, so that those portions of that organ also become *secondarily retroperitoneal*.

(3) The region of fusion (fascia of Toldt) may be incised surgically, thereby mobilizing the duodenum and pancreas together with their shared blood supplies.

c. The **dorsal mesentery** of the intestine is divided into several regions, each of which supports a segment of bowel.

(1) The **mesentery proper** supports most of the small intestine.

(a) This ligament runs from the dorsal body wall to the jejunum and ileum.

(b) It is fan-shaped, so that a root about 9 inches long supports about 20 feet of small bowel.

(c) The positions in which it supports the various loops of small bowel are extremely variable.

(2) The **ascending mesocolon** is an embryonic structure only.

(a) Except for a small portion along the inferior cecum, this mesentery becomes apposed and fixed to the peritoneum of the dorsal body wall, making the ascending colon *secondarily retroperitoneal*.

(b) The line of fusion may be incised surgically, thereby reestablishing the mesentery and mobilizing the ascending colon with its blood supply.

(3) The **transverse mesocolon** supports the transverse colon.

(a) It runs between the dorsal body wall and the transverse colon.

(b) It suspends the transverse colon in greatly variable positions (see Fig. 17-15).

(c) It fuses with the greater omentum to form the gastrocolic ligament (see Fig. 16-6).

(4) The **descending mesocolon** is an embryonic structure only.

(a) This mesentery becomes apposed and fixed to the peritoneum of the dorsal body wall, making the descending colon *secondarily retroperitoneal*.

(b) The line of fusion may be incised surgically, thereby reestablishing the mesentery and mobilizing the descending colon with its blood supply.

(5) The **sigmoid mesocolon** supports the sigmoid colon.

(a) It runs between the dorsal body wall and the sigmoid colon.

(b) It suspends the sigmoid colon in variable positions within the deep pelvis (see Fig. 17-16).

F. DEVELOPMENT OF THE LIVER (Fig. 16-4)

1. The liver appears by the third week as the **hepatic diverticulum** of the foregut.

a. The hepatic diverticulum, an outgrowth of entodermal epithelium, forms the bile duct and a portion of the liver parenchyma.

b. It develops within the **septum transversum** and the **ventral mesogastrium**, dividing the ventral mesogastrium into the **lesser omentum** and the **falciform ligament**.

c. The cystic duct and gallbladder differentiate as an outgrowth from the hepatic diverticulum.

2. By the tenth week, the liver has assumed a hematopoietic function.

3. During development, the liver grows more rapidly on the right side than on the left. This differential growth affects the position of adjacent abdominal organs and may be instrumental in the rotation of the stomach.

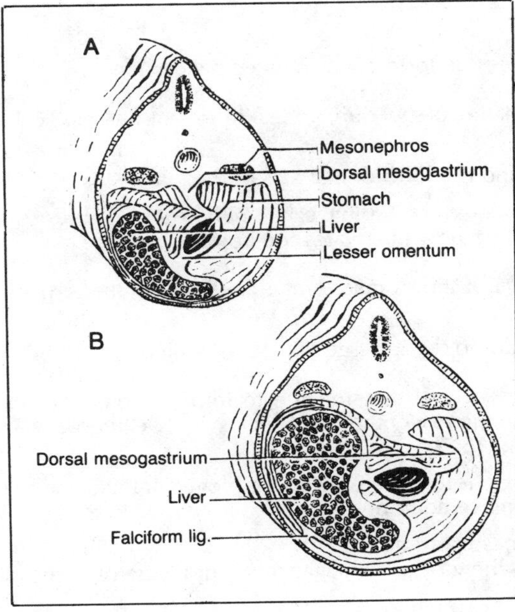

Figure 16-4. *Rotation of the stomach. A,* During the fifth week, the stomach rotates so that the primitively left side becomes the anterior surface. *B,* By the sixth week, the greater omentum begins to form, and the liver develops toward the right. The ventral mesentery caudal to the falciform ligament begins to degenerate.

4. The developing liver is an important metabolic center in the umbilical venous channels and represents 10 percent of the weight of the embryo.

G. DEVELOPMENT OF THE PANCREAS

1. During the third week, the pancreas develops from two duodenal diverticula.
   - a. The **dorsal pancreatic bud** develops in the dorsal mesogastrium craniad to the hepatic diverticulum.
   - b. The **ventral pancreatic bud** develops within the ventral mesogastrium either from or just caudal to the hepatic diverticulum.
2. During the fifth week, the ventral pancreatic anlage and the common bile duct migrate dorsally to the right.
   - a. The ducts of the dorsal and ventral pancreatic anlages fuse.
     - (1) **The primary pancreatic duct** (Wirsung's):
       - (a) Represents the distal portion of the dorsal pancreatic duct and entire ventral duct.
       - (b) Provides the principal drainage channel of the pancreas.
       - (c) Enters the descending duodenum at the **major duodenal papilla** (of Vater).
       - (d) Usually joins the bile duct to enter the duodenum by a common ostium but may enter the duodenum separately.
     - (2) **The secondary pancreatic duct** (Santorini's):
       - (a) Is formed by the proximal portion of the dorsal pancreatic duct.
       - (b) May or may not retain connection with the primary duct.
       - (c) Usually enters the descending duodenum independently as the **minor duodenal papilla**, just craniad to the papilla of Vater.
   - b. Extensive variation may occur within this developmental sequence.
     - (1) The embryonic ductal system persists in 10 percent of all adults (see Fig. 17-8).
     - (2) There may be variations at the ampulla of Vater such that the bile and pancreatic ducts may not join but enter the duodenum separately (see Fig. 17-11).
     - (3) An **annular pancreas** is a developmental anomaly.
       - (a) A portion of ventral pancreatic anlage may migrate to the left so that the pancreatic parenchyma fuses around the duodenum.
       - (b) The annular pancreas may produce stricture of the duodenum to the point of obstruction, necessitating surgical intervention.
     - (4) Nests of **ectopic pancreatic tissue** may occur.
       - (a) Ectopic pancreatic parenchyma may be found anywhere along the foregut and midgut.
       - (b) These nests are especially frequent in the gallbladder and are associated with Meckel's diverticula.

III. ROTATION OF THE FOREGUT

A. The definitive position of the gut in adults is due to rotation during development.

B. The gastric anlage develops as a dilatation of the foregut during the fifth week (see Figs. 16-4 and 16-5).

1. There is a 90° rotation to the right around the longitudinal axis of the gut.
2. The left side of the stomach becomes anterior; the left vagus nerve now innervates the anterior surface while the right vagus nerve now innervates the posterior surface.

C. During rotation, the primitive dorsal surface of the stomach undergoes accelerated growth to produce the greater curvature.

1. Differential growth pulls the mesogastrium to the left (see Fig. 16-4), which:
   a. Produces the greater omentum.
   b. Entraps a portion of the right coelom behind the stomach to form the **omental bursa (lesser sac)**, which communicates with the rest of the coelomic cavity (greater sac) via the **epiploic foramen** (Winslow's) [see Fig. 16-7].
2. Because of the concomitant growth of the intestines, the pylorus is moved craniad and to the right, and the stomach assumes the definitive adult position.
3. The **spleen** develops *within* the dorsal mesogastrium.
   a. The position of the spleen divides the dorsal mesogastrium into **gastrosplenic** and **lienorenal (phrenicolienal) ligaments**.
   b. The lienorenal ligament is continuous with the peritoneum of the posterior abdominal wall and contains the tail of the developing pancreas.
4. The dorsal mesogastrium continues to grow, doubling upon itself.
   a. The omental bursa (lesser sac) initially extends into the space between the folds of the dorsal mesogastrium.
   b. Fusion of the mesothelial leaflets of the dorsal mesogastrium on the lesser sac side forms the **greater omentum**.
   c. Fat develops in the greater omentum.
   d. The greater omentum is surprisingly mobile and may migrate into an area of inflammation.

D. During rotation, the **duodenum** assumes a C-shape.

1. It is moved posterior and to the right because of gastric rotation.
2. During the sixth week, it comes to lie against the dorsal body wall and becomes secondarily retroperitoneal as the dorsal mesoduodenum fuses with the peritoneum of the dorsal body wall.
3. Upon fusion of the dorsal mesoduodenum to the posterior body wall, the pancreas (with the exception of the terminal portion of the tail) becomes secondarily retroperitoneal.
4. During the second month, the duodenum passes through a solid stage and reopens somewhat later. Incomplete recanalization results in **pyloric** or **duodenal stenosis**, which must be surgically corrected.

IV. ROTATION OF THE MIDGUT

A. FIRST STAGE (Fig. 16-5A).

1. Rapid growth of the midgut during the fifth week results in formation of the **primary intestinal loop**, which is connected to the yolk sac via the **omphalomesenteric vitelline duct**.
   a. The cranial limb of the primary intestinal loop develops into the distal duodenum, the jejunum, and the ileum as far as the omphalomesenteric duct.
   b. The caudal limb of the primary intestinal loop develops into the terminal ileum, cecum and appendix, ascending colon, and most of the transverse colon.
   c. The omphalomesenteric duct, or its remnant in 3 percent of adults (Meckel's diverticulum), represents the boundary between the cranial and the caudal limbs of the primary intestinal loop.
2. A physiologic **umbilical herniation** of the primary intestinal loop normally occurs.
   a. Because of rapid hepatic and intestinal growth, the abdominal cavity becomes temporarily too small for the developing gut.
   b. The primary intestinal loop herniates into the umbilical stalk during the sixth week.

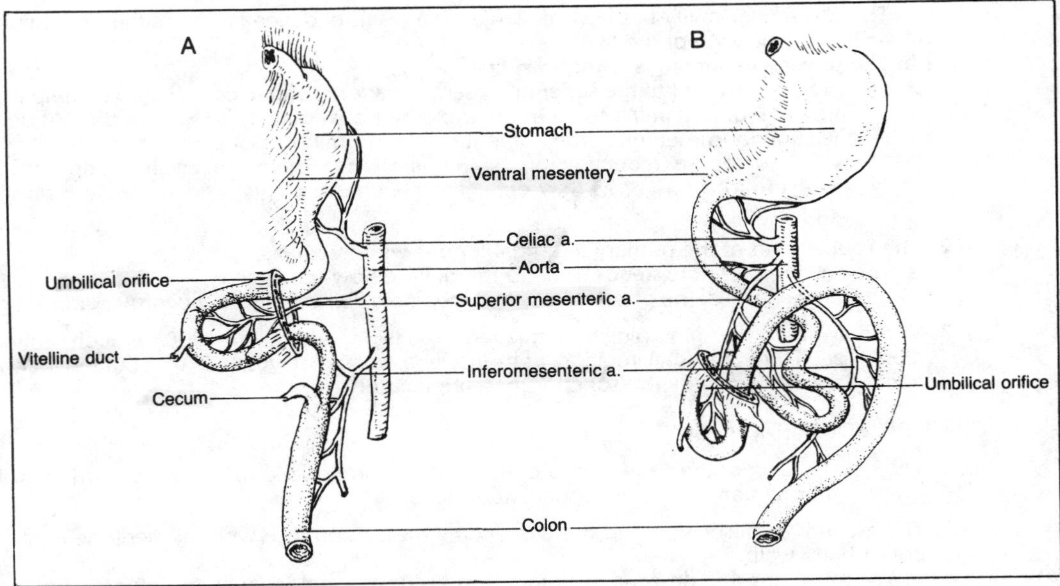

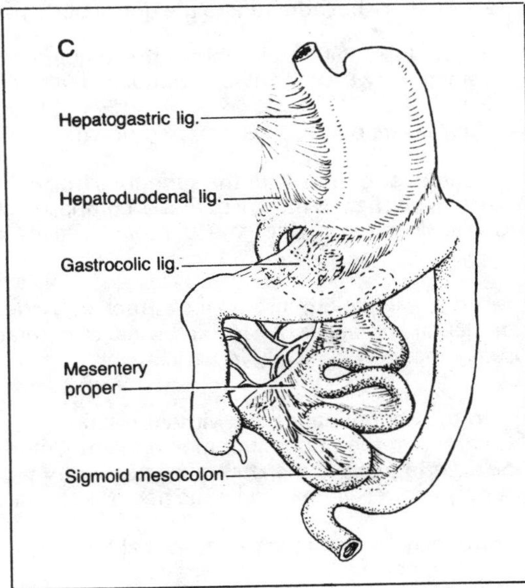

Figure 16-5. *Rotation of the gut. A, Stage I: umbilical herniation.* During the sixth week, the rapidly growing intestine herniates into the umbilical stalk. *B, Stage II: reduction of the umbilical herniation.* During the tenth week, the intraembryonic coils of intestine are rapidly withdrawn into the coelomic cavity. The cranial loop passes posterior to the caudal loop so that an effective rotation results. *C, Stage III: fixation of the gastrointestinal tract.* Once the intestines have returned to the body cavity, their mesenteries fuse with the parietal peritoneum in several regions, making those portions of the tract appear retroperitoneal, thus the term *secondarily retroperitoneal.*

3. The primary intestinal loop rotates 90° counterclockwise around an axis provided by the superior mesenteric artery, the artery of the midgut.
   a. The cranial limb comes to lie on the right and the caudal limb on the left within the umbilical stalk. This positioning is probably the result of extremely rapid growth of the cranial limb during this phase.
   b. The cranial limb forms secondary loops and coils within the umbilical stalk.
   c. A clockwise rotation at this stage produces **situs inversus viscerum** (with a high probability of visceral reversal in the thorax). Situs inversus occurs in about 1 in 10,000 births.
4. Rotation of the stomach and duodenal fixation to the dorsal body wall occur simultaneously with this stage of intestinal rotation.
5. Umbilical herniation reaches its maximum extent during the ninth week.

B. SECOND STAGE (Fig. 16-5B)

1. During the tenth week, the umbilical herniation is completely withdrawn.

a. The abdominal cavity has become larger as a result of degeneration of the mesonephros and slower growth of the liver.

b. The proximal jejunum is withdrawn first.

(1) It passes posterior to the superior mesenteric artery and the caudal limb of the primary intestinal loop, coming to lie in the upper left quadrant. This explains the position of the superior mesenteric artery anterior to the terminal duodenum in adults.

(2) With progressive reduction of the umbilical herniation, successive loops lie progressively to the right as they enter the abdominal cavity, and secondary intestinal loops form.

2. The caudal limb of the primary intestinal loop is withdrawn last.

a. The cecum, therefore, tends to lie to the right, below the liver.

b. The caudal limb of the primary intestinal loop now lies anterior to the former cranial limb.

3. Withdrawal of the physiologic herniation results in additional 180° rotation counterclockwise. Thus, the total rotation from first and second stages is 270° counterclockwise about an axis formed by the superior mesenteric artery.

C. THIRD STAGE (Fig. 16-5C)

1. The cecal region of the caudal limb of the primary intestinal loop moves inferiorly into the lower right quadrant, and the appendix develops.

2. The ascending mesocolon and the descending mesocolon fuse with the peritoneum of the dorsal body wall.

a. The ascending and descending colons thus become secondarily retroperitoneal, i.e., once supported in the peritoneal cavity by the mesentery but adherent to the dorsal body wall in adults.

b. The dorsal mesogastrium grows markedly and fuses with itself, obliterating that portion of the omental bursa that had been contained between the two redundant leaflets and becoming the greater omentum (Fig. 16-6).

(1) The greater omentum thus contains four layers of mesothelium, two of which are apposed and fused.

(2) The inferior surface of the greater omentum also fuses with the superior surface of the transverse mesocolon, forming the **gastrocolic ligament**, which thus is composed of six layers of mesothelium. The gastrocolic ligament divides the peritoneal cavity into supracolic and infracolic compartments.

D. ANOMALIES OF ROTATION predispose toward strangulation, intestinal obstruction, ischemic necrosis, and infarct. Although major developmental anomalies of gastrointestinal tract rotation occasionally are incidental findings, they usually manifest clinically in neonates.

1. Nonrotation of the gut.

a. The primary intestinal loop may return to the abdominal cavity without rotation.

b. The jejunum and ileum may lie on the right, and the duodenum may remain peritoneal. The small intestine may twist around the superior mesenteric artery (**volvulus**), obstructing the intestine or compressing its blood supply and resulting in ischemic necrosis and infarction of the small bowel.

c. The ascending colon and descending colon may lie on the left side, so that there is, in effect, no transverse colon.

(1) The descending colon may become fixed.

(2) Peritoneal bands or adhesions may develop because of partial fixation and lead to intestinal obstruction.

(3) The floating ascending colon may undergo volvulus.

2. Reversed rotation (situs inversus viscerum).

a. Situs inversus is rare, occurring in about 1 in 10,000 births.

b. The external appearance is normal, except that in males the right testis is lower. The visceral positions are a mirror image of normal.

c. This anomaly is caused by clockwise withdrawal of the primary intestinal loop into the abdominal cavity.

d. Situs inversus can seriously confuse the unaware.

(1) In one extensive study of situs inversus cases, 55 percent were recognized *prior* to surgery, 32 percent recognized *during* surgery, and 13 percent unrecognized until *after* surgery.

(2) In this same study, 14 percent of cases had a correct diagnosis and appropriate incision, 45 percent had an incorrect diagnosis, and 31 percent had an inappropriate incision; in 10 percent the abdomen was closed immediately because the surgeons became confused.

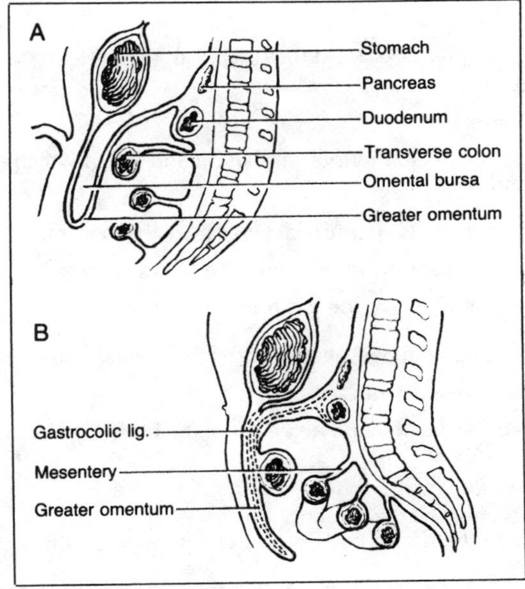

Figure 16-6. *Formation of ligaments. A,* The omental bursa (lesser sac) extends between the redundant leaflets of the greater omentum. *B,* Fusion of the redundant leaflets of the greater omentum and subsequent fusion with the transverse mesocolon form the gastrocolic ligament.

3. **Malrotation of the gut.**
   a. There are various stages or degrees of incomplete rotation.
   b. Depending on the degree of nonrotation and subsequent peritoneal fixation, volvulus or strangulation may occur.
   c. Nonfixation occurs most frequently at the cecum, and always to a variable extent in the region of the sigmoid colon, so that the normal length of the peritoneal portion of the sigmoid colon varies widely.
   d. Malrotation may produce **paraduodenal hernias**, which are readily detected on x-rays.
      (1) **Left paraduodenal hernia.** As the cranial limb of the primary intestinal loop withdraws from the umbilical herniation into the abdominal cavity, it may pass beneath the inferior mesenteric vein and thereby pass between the descending mesocolon and the parietal peritoneum. This passage prevents complete fusion of the descending mesocolon to the parietal peritoneum and forms an extensive left paraduodenal fossa containing loops of small bowel.
      (2) **Right paraduodenal hernia.** Similarly, in instances of nonrotation where the small intestine lies to the right, the ascending colon may move properly to the right and partially fixate to the parietal peritoneum, entrapping the small intestine beneath the ascending mesocolon and forming an extensive right paraduodenal fossa.

4. **Omphalocele and umbilical hernia.**
   a. Omphalocele is an anterior abdominal wall defect, covered only by peritoneum and perhaps a layer of amnion, through which viscera may protrude.
   b. An umbilical hernia results from incomplete reduction of the physiologic umbilical hernia during the 10th week, with retention of part of the primary intestinal loop within the umbilical cord.
      (1) This anomaly occurs in approximately 2 percent of infants.
      (2) A very small umbilical hernia may not be immediately recognized, and a loop of small intestine may be divided when the umbilical cord is sectioned and tied.

5. **Meckel's diverticulum** is a persistent remnant of the vitelline (omphalomesenteric) duct that is present in 3 percent of adults.
   a. It may remain attached to the umbilicus either as a fibrous cord or a vitelline fistula.
   b. Epithelium-lined cysts may persist along a fibrous connector.
   c. Vitelline vessels may persist with attachment at the umbilicus and anastomoses with somatic vessels of the anterior abdominal wall.
   d. Volvulus may occur around any persistent connection between the ileum and umbilicus, with possible obstruction or strangulation.
   e. Even if Meckel's diverticulum is not attached to the ventral body wall, there may be complications, for example:
      (1) Inflammation of the diverticulum.
      (2) Peptic ulcers of the ileum, because Meckel's diverticula frequently contain gastric mucosa that secretes acid into the normally alkaline ileum.

V. THE PERITONEUM

A. THE PARIETAL PERITONEUM is the innermost layer of the abdominal wall, completely covering the inside wall of the abdominal cavity.

1. It is composed of mesothelial serous membrane.

2. It is supported by extraperitoneal connective tissue (transversalis fascia and fat) in which are embedded the great vessels, kidneys, and ureters.

B. THE VISCERAL PERITONEUM is the reflection of the peritoneal layer over the viscera.

1. Histologically, it is composed of a mesothelial serous membrane, the tunica serosa.

2. The serosa comprises the external surface of much of the gastrointestinal tract.

C. MESENTERIES are continuations between the parietal peritoneum and the visceral peritoneum.

1. Mesenteries are formed by back-to-back layers of reflected peritoneum.

2. They provide pathways to the viscera for blood vessels and nerves from the body walls as well as supporting the ducts from the digestive glands.

3. **The primitive dorsal mesentery:**
   a. Persists throughout embryonic development and adult life.
   b. Suspends the entire abdominal gut from the dorsal body wall (even in regions where the gut becomes secondarily retroperitoneal).
   c. Has as named portions the **dorsal mesogastrium (gastrophrenic, gastrosplenic, phrenicolienal ligaments)**, and the **greater omentum,** the **mesentery proper**, the **mesoappendix**, the **transverse mesocolon,** and the **sigmoid mesocolon**.

4. **The primitive ventral mesentery:**
   a. Largely degenerates during development, but where persistent, it suspends the viscera from the ventral body wall.
   b. Has as named portions the **lesser omentum (gastrohepatic ligament** and **hepatoduodenal ligament)**, the **coronary ligament, left** and **right triangular ligaments**, the **falciform ligament**, and the **median umbilical fold**.

D. SMALL AMOUNTS OF SEROUS FLUID normally exude through the peritoneal membranes.

1. This serous fluid provides a lubricating film for the parietal and visceral surfaces, and facilitates free mobility of the viscera with minimal friction.

2. During inflammatory processes involving the peritoneum or other pathologic conditions, large quantities of fluid (ascites) may collect in the abdominal cavity.

E. THE OMENTAL BURSA (LESSER SAC) represents the primitive right coelomic cavity.

1. The omental bursa is isolated from the **greater sac** by rotation of the stomach as well as by elevation and fixation of the duodenum.

2. It lies posterior to the stomach and lesser omentum.

3. Access to the lesser sac is through the **epiploic foramen** (Winslow's), which is bounded by:
   a. The **caudate lobe** of the liver, superiorly.
   b. The parietal peritoneum over the **inferior vena cava**, posteriorly.
   c. The **superior duodenum**, inferiorly.
   d. The **hepatoduodenal ligament**, anteriorly, which contains:
      (1) The **common bile duct**.
      (2) The **hepatic artery.**
      (3) The **hepatic portal vein.**

4. The epiploic foramen is a potential site for intra-abdominal herniation. If a loop of bowel passes through the epiploic foramen and becomes incarcerated, none of the boundaries of the foramen can be safely incised. Instead, the bowel must be deflated with a needle, and the loop of gut can then be withdrawn easily.

VI. PERITONEAL RELATIONSHIPS

A. The definitive positions and attachments of the gastrointestinal tract are the result of rotation of the gut.

B. Certain regions of the gut and associated mesentery become applied to the peritoneum covering the dorsal body wall and appear to lose the supporting mesentery, i.e., they become **secondarily retroperitoneal**. Secondarily retroperitoneal portions of the gut may be surgically mobilized along with the supporting mesentery and its contained blood vessels and nerve supply by incising the fusion between the mesentery and the parietal peritoneum.

C. SUMMARY OF PERITONEAL AND SECONDARILY RETROPERITONEAL STRUCTURES

1. The **esophagus** is retroperitoneal, except for a small segment immediately after penetration of the diaphragm.
2. The **stomach** is suspended in the peritoneal cavity by the dorsal mesogastrium (gastrophrenic ligament, gastrosplenic ligament, phrenicolienal ligament, greater omentum) and ventral mesogastrium (gastrohepatic portion of the lesser omentum).
3. The **liver** is suspended in the peritoneal cavity by the ventral mesentery (lesser omentum, coronary, triangular, and falciform ligaments).
4. The **pancreas** is 2° retroperitoneal, except for a small portion of the tail which lies in the lienorenal ligament.
5. The **duodenum** is 2° retroperitoneal, except for a minute segment of the first portion, which is suspended by the hepatoduodenal ligament.
6. The **jejunum** and the **ileum** are suspended in the peritoneal cavity by the mesentery proper.
7. The **appendix** is suspended in the peritoneal cavity by the mesoappendix.
8. The **cecum** is usually considered 2° retroperitoneal, but the most inferior portion frequently is suspended by the ileocecal fold, an extension of the mesentery proper.
9. The **ascending colon** is 2° retroperitoneal (nonfixation is a developmental anomaly).
10. The **transverse colon** is suspended in the peritoneal cavity by the transverse mesocolon.
11. The **descending colon** is 2° retroperitoneal.
12. The **sigmoid colon** is suspended in the peritoneal cavity of the deep pelvis by the sigmoid mesocolon.
13. The **rectum** is retroperitoneal.

D. PERITONEAL RECESSES, SPACES, AND FOSSAE

1. Peritoneal reflections as mesenteries divide the abdominal cavity into two compartments and several blind fossae.
   a. The dorsal mesogastrium and lesser omentum separate the greater and lesser sacs.
   b. The gastrocolic ligament, the fusion of the greater omentum and transverse mesocolon, divides the greater sac into:
      (1) The **supracolic compartment**.
      (2) The **infracolic compartment**.
2. The peritoneal recesses, spaces, and fossae are potential sites of infection.
3. **Named recesses, spaces, and fossae** (Figs. 16-7 and 16-8) include:
   a. The **subphrenic (suprahepatic) recess**, located anterior and superior to the liver, beneath the diaphragm.
      (1) It is divided into **right** and **left subphrenic recesses** by the falciform ligament.
      (2) It is the second most frequently infected abdominal space.
         (a) Adhesions between hepatic and diaphragmatic peritoneum may result in abscess formation.
         (b) A pulmonary abscess may erode across the diaphragm, invading the abdominal cavity and involving the suprahepatic recess. This type of infection was a frequent cause of mortality before antibiotics and still is a serious condition.
   b. The **infrahepatic recess**, located inferior to the liver within the lesser sac.
   c. The **right subhepatic recess** and the **hepatorenal recess** form the **pouch of Morison**.
      (1) The pouch of Morison is located inferior and posterior to the liver in the supracolic compartment.
      (2) These spaces communicate with (see Fig. 16-8):
         (a) The **lesser sac** via the **epiploic foramen**.
         (b) The **right paracolic gutter** and, hence, the pelvic cavity.
         (c) The **subphrenic recess**.

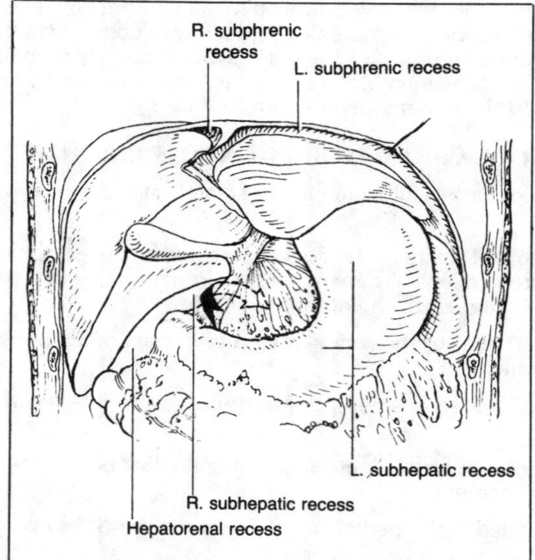

Figure 16-7. *Recesses of the supracolic compartment.* The named recesses are indicated. The *arrow* passes through the epiploic foramen into the omental bursa (lesser sac).

(3) When a person is supine, the hepatorenal recess becomes the lowest portion of the abdominal cavity and will collect fluid.

(4) The pouch of Morison is probably the most frequently infected abdominal space.

(a) If a patient has symptoms of an abdominal infection but no local abdominal signs, consider an abscess in the pouch of Morison and you will probably be correct.

(b) The primary causes of infection include appendicitis, a perforated duodenal or gastric ulcer, a perforated gallbladder or biliary tree, a liver abscess, and surgical procedures.

d. The **paraduodenal fossae.**

(1) The **superior, inferior, left**, and **right paraduodenal fossae** occur in association with the **terminal duodenum,** the **inferior mesenteric vein**, and the **left colic artery**.

(2) Herniation into a paraduodenal fossa may occur as a developmental anomaly.

(a) Malrotation of the gut during withdrawal of fetal umbilical herniation may result in formation of a left or right paraduodenal hernia.

(b) These hernias are usually pediatric problems and are very rare in adults.

e. Fossae of the cecal region (see Fig. 17-14).

(1) The **superior ileocecal fossa** is bounded by the **mesentery proper** and the **superior ileocecal (vascular) fold.**

(2) The **inferior ileocecal fossa** is bounded by the **mesoappendix** and the **inferior ileocecal** (bloodless) **fold**.

(3) The **retrocecal fossa** lies beneath the cecum, is variable in extent, is formed by adhesion of the cecum to the posterior parietal peritoneum, and frequently contains the appendix.

f. Two **paracolic recesses (gutters)** [see Fig. 16-8].

(1) The **right paracolic recess (gutter):**

(a) Lies lateral to the ascending colon.

(b) Communicates with the supracolic compartment, the pouch of Morison, and the pelvic cavity.

(c) Provides a route for spread of infection between the pelvis and the upper abdominal region.

(2) The **left paracolic recess (gutter):**

(a) Lies lateral to the descending colon.

(b) Communicates with the supracolic and infracolic compartments and the pelvic cavity.

(c) Seldom becomes infected.

g. The **infrasigmoid (parasigmoid) fossa**, which lies beneath the sigmoid colon, at the juncture of the sigmoid mesocolon and parietal peritoneum, is a potential site for intra-abdominal herniation of the sigmoid colon or small bowel.

h. The **abdominal wall fossae.**

(1) The **lateral inguinal fossae:**

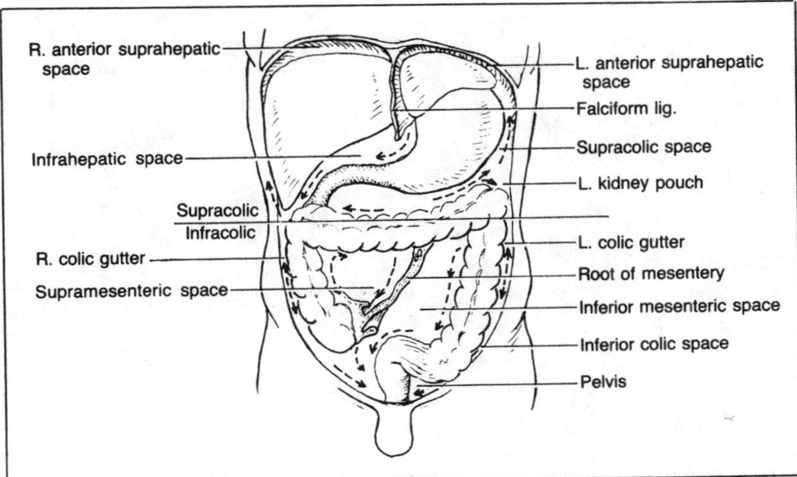

Figure 16-8. *Drainage pathways.* Infection can spread from one part of the peritoneal cavity to another via the recesses, spaces, and gutters.

(a) Lie lateral to the **lateral umbilical folds**, which are formed by the underlying inferior epigastric arteries.
(b) Are the sites where *in*direct inguinal herniation begins.

(2) The **medial inguinal fossae:**
(a) Lie medial to the lateral umbilical folds.
(b) Lie lateral to the **medial umbilical folds**, which are formed by the underlying obliterated umbilical arteries.
(c) Are the sites where direct inguinal herniation begins.

(3) The **supravesical fossae:**
(a) Lie medial to the medial umbilical folds and superior to the bladder.
(b) Lie lateral to the **median (middle) umbilical fold**, which is formed by the underlying urachus.
(c) Are the sites of supravesicular herniation.

E. CLINICAL CONSIDERATIONS

1. **Paracentesis.**
   a. Inflammatory exudate tends to collect in the subphrenic and hepatorenal recesses, as well as in the pelvic cavity, when a patient who has a peritoneal infection is supine.
   b. Diagnosis. The fluid level may be percussed along the lateral abdominal walls and seen on x-rays.
   c. Treatment includes puncture of the abdominal wall to withdraw fluid. A cannula inserted in the flank will pass through the skin, superficial fascia, deep fascia, aponeurosis of the external oblique muscle, internal oblique muscle, transverse abdominis muscle, transversalis fascia and extraperitoneal fat, and the parietal peritoneum.

2. **Peritonitis**, the "acute abdomen."
   a. Inflammation of the parietal and visceral peritoneum produces different symptoms.
      (1) The **parietal peritoneum** is innervated by somatic nerves of the anterior body wall. An inflamed parietal peritoneum is exquisitely sensitive to palpation or stretching in an area that is rather *discretely localized* on the abdominal wall. The abdominal musculature will tense—*guarding* or *splinting*—to produce a rigid abdomen and thereby minimize the pain.
      (2) The **visceral peritoneum** is innervated by visceral nerves, which travel along autonomic pathways. An inflamed visceral peritoneum results in diffuse, crampy or colicky abdominal pain, which may be *referred* to specific dermatomes on the abdominal wall.
   b. Infection may gain access to the peritoneal cavity by means of:
      (1) Traumatic or surgical penetration of the abdominal wall.
      (2) Perforation of the gastrointestinal tract.
      (3) Hematogenous or lymphatic spread.
      (4) The uterine tubes, which normally open into the peritoneal cavity.

17
The Gastrointestinal Tract

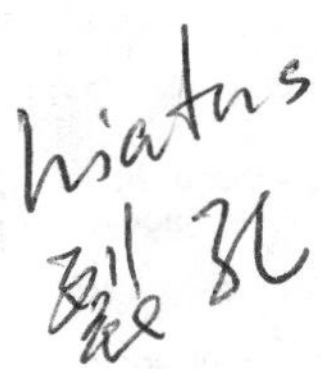

I. INTRODUCTION

A. The major portion of the gastrointestinal (GI) tract, as well as many of the associated digestive glands, is contained within the abdominopelvic cavity.

B. The principal functions of the GI tract are digestion and nutrient absorption.

C. The divisions, locations, and mesenteric attachments of the portions of the GI tract may be described best with reference to embryonic development.

1. The abdominopelvic portion of the GI tract is divided according to embryonic blood supply.
   - **a.** The **foregut** is that portion of the GI tract supplied largely by the celiac artery.
   - **b.** The **midgut** is that portion supplied by the superior mesenteric artery.
   - **c.** The **hindgut** is that portion supplied primarily by the inferior mesenteric artery.
   - **d.** Most of the venous drainage of the abdominal portion of the GI tract is through the liver.
2. The definitive locations of the portions of the GI tract are determined during development, as the gut elongates and rotates rapidly.
3. The mesenteric attachments and peritoneal fusions remaining subsequent to rotation provide definitive support for the GI tract.

D. Motor innervation of the GI tract is via the autonomic nervous system, the pathways of which are shared by visceral afferent nerves.

II. THE ESOPHAGUS

A. ESOPHAGEAL STRUCTURE

1. The structure of the esophagus subserves its principal function of deglutition (swallowing).
2. The esophagus extends from the lower part of the laryngopharynx to the cardiac opening of the stomach (Fig. 17-1).
3. It is approximately 25 to 30 cm long in males (in the erect position) and a few centimeters shorter in females.
4. High within the thorax, it lies in the midline behind the trachea, deviating slightly to the left after the tracheal bifurcation.
5. As it enters the abdominal cavity through the **esophageal hiatus,** it makes an abrupt turn to the left and enters the stomach.
   - **a.** The esophageal hiatus is slightly left of the midline in the muscular portion of the diaphragm.
   - **b.** The hiatus is a frequent site of herniation of the stomach into the thoracic cavity (hiatus hernia).

B. THE LUMEN of the esophagus is small and irregular except during deglutition. It averages 1 cm in diameter orally and 2 to 3 cm aborally.

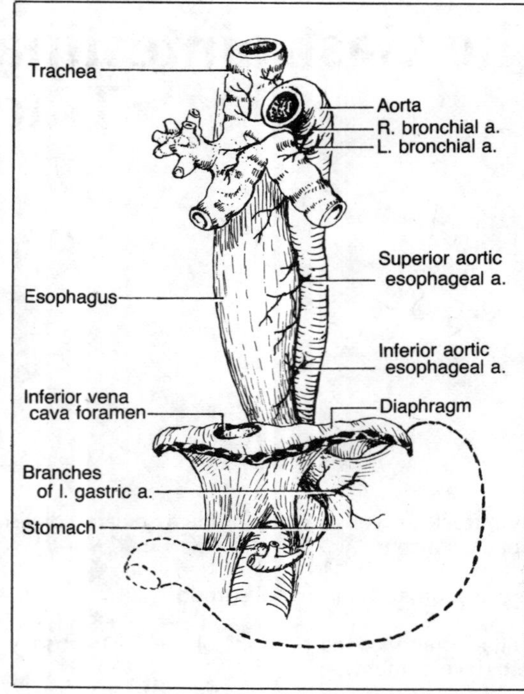

Figure 17-1. *The esophagus.* The esophagus passes through the esophageal hiatus of the diaphragm to gain access to the abdomen where it joins the stomach.

1. The shape of the lumen is irregular because tension within the inner (circular) layer of the muscularis externa causes formation of longitudinal folds.
2. Because of this folding, the esophagus is quite distensible, accommodating anything that passes the epiglottis.

C. MUSCULATURE. The esophagus is the most muscular segment of the alimentary tract. Its upper quarter is composed primarily of striated muscle that initially is arranged irregularly but soon separates into inner circular and outer longitudinal layers.

1. The cricopharyngeal muscle is the sphincter of the upper esophageal opening. It remains closed except during deglutition, eructation (belching), and emesis (vomiting).
2. Proceeding aborally, the quantity of smooth muscle increases and the striated fibers disappear.
3. The intra-abdominal portion of the esophageal musculature acts as a diffuse sphincter (the **cardiac sphincter**).

D. NERVES. The **left and right vagus nerves** form the **esophageal plexus**, which reunites into the vagal trunks.

1. Due to rotation of the stomach, the left and right vagus nerves change their relative positions to become **anterior** and **posterior vagal trunks**, respectively (see Fig. 14-4).
2. The parasympathetic presynaptic neurons of the vagus nerves synapse with the parasympathetic postsynaptic neurons of the autonomic nervous system, which are located between the longitudinal and circular muscle layers.
   a. The cell bodies of these parasympathetic neurons lie in the **myenteric plexuses** (Auerbach's) and receive presynaptic innervation from the vagus nerve.
   b. Dysfunction of the parasympathetic pathway at this level results in achalasia (cardiospasm), which is characterized by dysphagia (inability to swallow) and a progressively dilated, tortuous esophagus.

E. FUNCTIONAL ASPECTS OF THE ESOPHAGUS

1. A solid bolus entering the esophagus initiates an involuntary peristaltic wave, which reaches the cardiac sphincter in 5 to 6 seconds.

a. The cardiac sphincter, normally closed, relaxes with the approach of a peristaltic wave to allow the bolus to pass into the stomach.
b. A fluid bolus will outrun the peristaltic wave but will be stopped by the cardiac sphincter until those muscle fibers are inhibited by the approaching peristaltic wave.
(1) Only one sound is heard when a solid bolus is swallowed; there are two separate sounds with a fluid bolus.
(2) When a series of swallows is made in rapid succession, as in "chugging," the cardiac sphincter remains inhibited so that fluid passes directly into the stomach unassisted by peristalsis.

2. The smooth muscle of the cardioesophageal junction is continuous with that of the cardiac region of the stomach.
a. A true anatomic sphincter cannot be observed.
b. A true physiologic sphincter must exist because the pressure within the stomach is about 20 cm of water greater than that in the midesophagus. Without the esophageal zone of increased transmural pressure, gastric contents would reflux into the esophagus continuously.
c. Because of peristaltic inhibition, the cardiac sphincter is opened by a pressure of 2 to 7 cm of water within the esophagus, but eructation does not occur until intragastric pressure exceeds 25 cm of water.
d. This sphincteric mechanism is intrinsic to the esophagus and not dependent on external constriction by the diaphragmatic muscle around the esophageal hiatus.
(1) The muscular tension in the region of the sphincter is never very great, but it does maintain a broad zone of elevated intramural pressure at the aboral end of the esophagus.
(2) There is considerable evidence that the sphincteric barrier results, in part, from longitudinal mucosal folds maintained by the tonic activity in the muscularis mucosa.
(3) In addition, diffuse muscular tonic activity of the muscularis externa and the relatively constant tone within the muscle layers of the cardiac region of the stomach contribute to the sphincteric action.

F. CLINICAL CONSIDERATIONS

1. Esophagitis.
a. Occasionally, gastric contents reflux into the esophagus. The acidic peptic chyme burns and inflames the unprotected stratified squamous epithelium of the esophageal mucosa, producing esophagitis—the uncomfortable sensation commonly known as "heartburn."
b. Repeated reflux can result in peptic ulceration of the esophageal mucosa, which can cause dysphagia and serious bleeding.

2. Esophageal varices.
a. The submucosa of the esophagus contains an extensive plexus of large blood vessels. The esophageal veins drain into the azygos and hemiazygos veins.
b. The veins of the lower esophagus anastomose with the branches of the left gastric (coronary) vein of the stomach. These anastomoses provide an important but potentially dangerous shunt in the event of portal hypertension, and result in the development of esophageal varices.
(1) These tortuous and dilated veins directly beneath the mucosa protrude into the esophageal lumen, usually in the distal third of the esophagus, where they are subject to mechanical trauma during deglutition, emesis, or the passage of diagnostic instrumentation.
(2) Esophageal varices produce no symptoms until they rupture, causing massive hematemesis.
(3) Among patients with advanced cirrhosis of the liver, over half of the deaths result from rupture of an esophageal varix.

3. Hiatus hernia.
a. A disorder of the region of the gastroesophageal junction, herniation of the stomach through the esophageal hiatus produces a sac-like dilation above the diaphragm.
(1) In 90 percent of hiatus hernia cases the esophagus ends above the diaphragm, creating a bell-like dilation of that portion of the stomach superior to the diaphragm ("hourglass stomach" or sliding hiatus hernia).
(2) In the remaining 10 percent, the cardiac region of the stomach dissects alongside the esophagus, through a defect in the esophageal hiatus, to produce an intrathoracic sac (paraesophageal hernia).
b. Hiatus hernia may produce incompetence of the cardiac sphincter, resulting in regurgitative esophagitis, which may ultimately lead to peptic ulceration of the esophagus.

III. THE STOMACH

A. THE POSITION, SHAPE, AND SIZE of the stomach are quite variable (Fig. 17-2).

1. The stomach is located in the left hypochondriac and epigastric regions of the abdomen.
2. It is very distensible and can accommodate upwards of 2 liters.
3. A mobile, easily displaced organ, the stomach has no fixed position. Its shape varies among individuals and with degree of fullness.
   - **a.** In the empty state, the stomach is almost tubular or J-shaped, except for the bulge of the fundus; the stomach may be almost entirely under the rib cage.
   - **b.** After gorging, the stomach may pendulate below the intertubercular plane into the pelvis.

B. THE GREATER CURVATURE represents the primitive dorsal surface (Fig. 17-3). It receives ligamentous support from the primitive dorsal mesentery, which becomes:

1. The **greater omentum**, a redundant portion of the dorsal mesogastrium.
2. The **gastrophrenic ligament**, that portion of the dorsal mesogastrium between the fundic region and the dorsal body wall in the vicinity of the diaphragmatic crura.
3. The **gastrosplenic ligament**, that portion of the dorsal mesogastrium between the body of the stomach and the spleen.
4. The **lienorenal ligament**, a continuation of the dorsal mesogastrium between the spleen and the dorsal body wall in the vicinity of the left kidney.

C. THE LESSER CURVATURE represents the primitive ventral surface (see Fig. 17-3). It receives ligamentous support from the primitive ventral mesentery, which becomes the **gastrohepatic ligament** of the **lesser omentum**.

D. THE STOMACH IS DIVIDED INTO FOUR REGIONS (see Fig. 17-3).

1. The **cardia** is the region located in the vicinity of the esophagus.

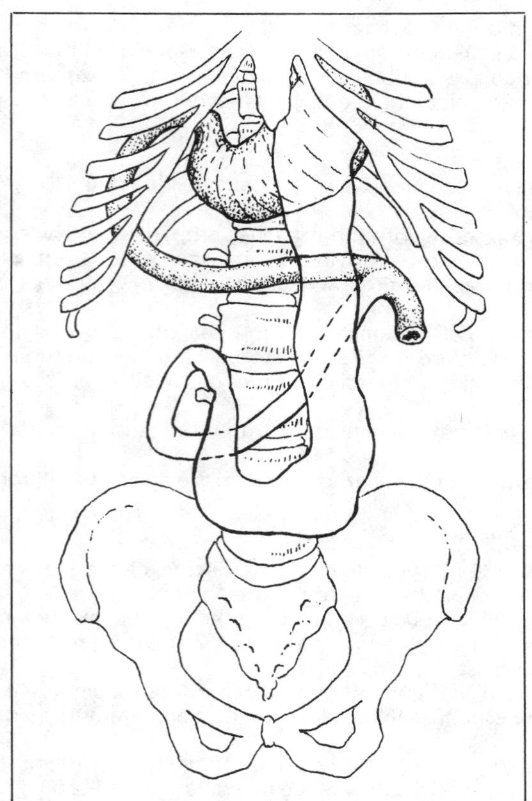

Figure 17-2. *The location of the stomach.* The location, size, and shape of the stomach is variable as indicated.

a. The stomach receives the esophagus at the cardiac opening, which lies at the juncture of the greater and lesser curvatures.
   (1) At this juncture, the stratified squamous epithelium of the esophagus changes abruptly to simple columnar epithelium.
   (2) This transition provides a line of demarcation between the relatively white esophageal mucosa and the pink gastric mucosa.

b. The **cardiac notch** is located on the lesser curvature between the **cardia** and the **fundus**.

2. The **fundus** is located above the level of the esophageal juncture.
   a. Air in the fundus, evident on x-ray film, is a useful radiographic landmark.
   b. A tympanic note may be percussed from the fundic air bubble.

3. The **corpus** (body) constitutes the major portion of the stomach.

4. The **pylorus** is subdivided into three regions:
   a. The **pyloric antrum** begins as a slight dilation at the **angular incisure** in the lesser curvature.
   b. The **pyloric canal** is 2 to 3 cm long.
   c. The **pyloric sphincter** is not demarcated from the circular muscle of the adjacent regions nor are there any discernible physiologic differences.
      (1) A pressure gradient of only 2 to 4 cm of water is sufficient to move chyme through the pylorus.
      (2) Unlike a true sphincter, the pylorus contracts synergistically with peristalsis, functioning as a unit with the pyloric end of the stomach.
      (3) The pylorus does not control the rate of stomach emptying but does control the size of the particles that enter the duodenum.
      (4) The pyloric sphincter may be abnormally thickened at birth (congenital hypertrophic pyloric stenosis), a condition requiring surgical correction.

E. **THE INTERNAL STRUCTURE** of the stomach (see Fig. 17-3) consists of gastric mucosa that is thrown into folds, rugae.

1. **Permanent rugae form the gastric canal or gastric gutter,** in the region of the lesser curvature which directs fluids toward the pylorus.
   a. The gastric canal is especially vulnerable to burns from accidental ingestion of caustic substances.
   b. Unlike the rugae in the rest of the stomach, those of the gastric canal do not diminish as the stomach fills.
   c. These rugae are usually evident on upper GI barium x-ray films.

2. **Gastric rugae in regions away from the gastric canal** are temporary high folds caused by tension in the muscularis mucosa which pleats the mucosa and underlying submucosa.
   a. These rugae run primarily in a longitudinal direction.
   b. They allow for stomach expansion, becoming flattened when the stomach is engorged.

3. The mucosa of the stomach is thrown into small permanent furrows, **gastric sulci**, that divide the surface into gastric or mamillated areas.
   a. Gastric areas do not flatten out with stomach distension.
   b. Within gastric areas, the surface of the distended stomach appears smooth.
   c. The gastric glands (pits) open onto the surface of the gastric areas.

F. **VASCULATURE** of the stomach (see Fig. 17-3)

1. The stomach receives blood from several sources.
   a. The **left gastric artery** supplies the right superior region of the stomach as well as the lower esophagus.
   b. The splenic artery gives off:
      (1) **Short gastric arteries** to the left superior region.
      (2) The **left gastroepiploic artery** to the left inferior region.
   c. The common hepatic artery gives off:
      (1) The **right gastric artery** to the right inferior region.
      (2) The gastroduodenal artery, which in turn gives off the **right gastroepiploic artery** to the right inferior region.

2. From a surgical standpoint, the blood supply to the stomach is so rich in extramural as well as intramural anastomoses that any of the branches along the stomach may be ligated with little risk of ischemia and necrosis.
   a. Anastomoses between gastric, esophageal, and splenic arteries occur constantly and abundantly.

b. Anastomoses between gastric and duodenal arteries are scant, resulting in the so-called "bloodless line" at the pyloroduodenal junction.

3. The veins of the stomach generally parallel the arterial supply but drain into the hepatic portal system.
   a. The anastomoses between the left gastric (coronary) vein and the esophageal veins are important portal-systemic anastomotic shunts.
   b. Anastomoses between the left gastroepiploic veins, or short gastric veins, and the esophageal veins may become enlarged if the splenic vein becomes occluded.

4. The routes of lymphatic drainage of the stomach generally follow the arteries and are so-named, although alternative names frequently are used.

5. The venous and lymphatic drainage of the stomach is such that malignant cancer at this site can spread to other organs or regions.
   a. Metastases to the liver may occur via either the portal vein or reverse flow along the lymphatics.
   b. Metastases as far as the pelvis may occur via retroperitoneal lymphatics.
   c. Metastases to any other part of the body may occur via the thoracic duct and circulatory system.

G. FUNCTIONAL ASPECTS

1. Trituration, formation of chyme, and acid enzymatic digestion, as well as serving as a reservoir, are the major functions of the stomach.
   a. Three muscle layers comprise the muscularis externa, but their predominant directions are difficult to characterize.
   b. The upper half of the stomach serves as a reservoir for ingested food and expands passively as it is filled.
      (1) Persistent tonic contracture in the region of the cardia and fundus aids in the establishment of the cardiac sphincter, and serves to deliver food to the lower, more motile regions of the stomach.
      (2) The gastric glands of both the fundus and the body contain cells that secrete hydrochloric acid; the mucosal glands of the cardia and pylorus are mostly devoid of acid-secreting cells.
         (a) In the upper regions of the stomach, boluses are bathed in gastric juice, which converts the surface of a bolus into a liquid mixture, chyme.
         (b) Pepsin, secreted by gastric glands, converts proteins to polypeptides at low pH.
   c. In the lower half of the stomach, the contents are thoroughly mixed with the secretions of the gastric glands.
      (1) The muscle layers increase in thickness from the cardia to the pylorus, and peristaltic activity increases in intensity along this gradient.
      (2) Peristaltic waves, which occur at about 20-second intervals, begin as ring-like contractions about midway in the body of the stomach and propagate toward the pylorus with increasing vigor.
      (3) Once formed, chyme is moved toward the pylorus.
   d. The factors controlling gastric motility are complex.
      (1) Gastric peristalsis is the result of an intrinsic rhythm within the enteric neurons.
         (a) Vagal activity accelerates gastric peristalsis.
         (b) Because gastric peristalsis is under the local endocrine control of enterogastrone, the introduction of acid chyme into the duodenum is inhibitory (the enterogastric reflex).
      (2) After an initial adjustment period, the stomach empties its contents at a relatively constant rate.

2. Some substances, such as alcohol, are absorbed directly through the gastric mucosa.

H. CLINICAL CONSIDERATIONS

1. **Gastritis.**
   a. Excessive vagal activity may produce gastritis, as may substances that irritate the gastric mucosa, such as aspirin and steroids.
   b. Chronic gastritis may lead to peptic ulceration.

2. **Peptic ulceration.**
   a. Peptic ulcers, due to a variety of causes, occur in the non-acid-secreting regions of the upper GI tract, such as the duodenum (primarily), the antral region of the stomach along the lesser curvature (less frequently), and the lower esophagus (rarely).

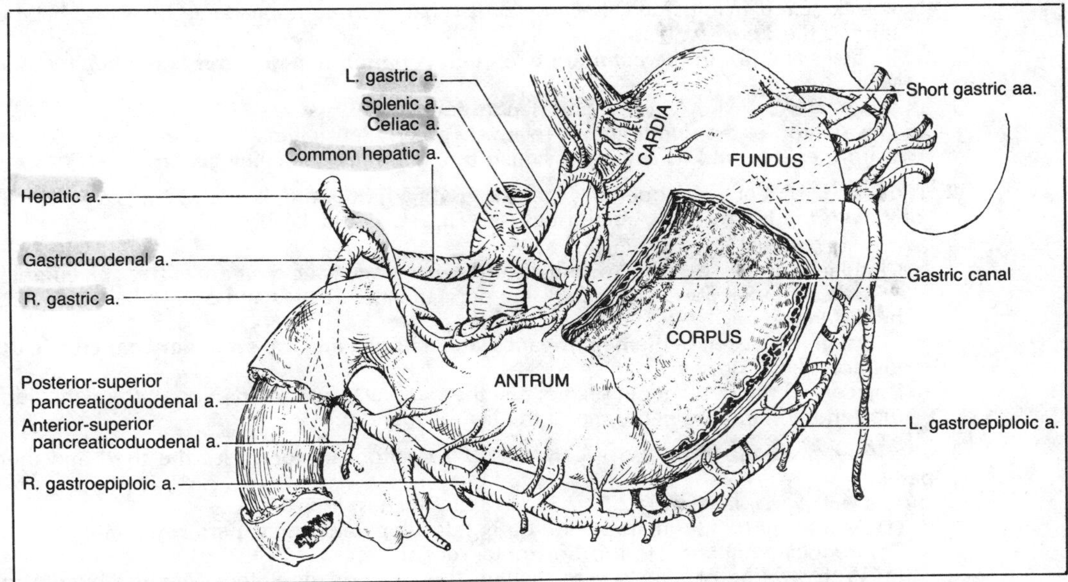

Figure 17-3. *The stomach.* The regions of the stomach are indicated. The window in the body of the stomach shows the rugae and the location of the gastric canal. The blood supply to the stomach is by the three branches of the celiac artery.

- **b.** Peptic ulcers may produce severe bleeding, obstruction (from edema or scarring), and perforation.
- **c.** Perforation into a subjacent blood vessel results in massive hematemesis or intra-abdominal bleeding with complicating peritonitis.
- **d.** Pain from peptic ulceration of the lower esophagus, stomach, or superior duodenum is referred to (appears as if originating from) the distributions of the 5th and 6th dermatomes, which include the epigastric and hypochondriac regions.
- **e.** Ulceration may be treated with drugs that block acid secretion.
- **f.** Selective vagotomy (section of the gastric branches of the anterior and posterior vagal trunks) reduces peptic secretion and may be combined with pyloroplasty.
- **g.** The gastric mucosa may be inspected with a fiberoptic gastroscope.

3. Gastrectomy.

- **a.** Surgical treatment of duodenal and gastric ulceration frequently involves resection of approximately one-third of the distal stomach and, because a collateral blood supply is inconsistent, the superior duodenum.
- **b.** Although the distal portion of the stomach produces little acid, it does produce most of the secretin, which induces acid secretion in the fundus and body. Resection of the lower portion of the stomach thus reduces acid secretion by the upper regions, and largely preserves the capacity of the stomach as a reservoir.

IV. THE DUODENUM

A. While the small intestine is arbitrarily divided into three parts, the morphologic and physiologic differences between the duodenum and jejunum, as well as between the jejunum and ileum, are slight. These differences occur by slow transitions over considerable distances. (See section IX for discussion of the jejunum and the ileum.)

B. The duodenum is the smallest portion of the small intestine. It is about 12 finger-breadths (*duodeni*, *L.* twelve) or 25 cm (10 inches) long.

C. The position and size of the duodenum are variable (see Fig. 17-2).

D. THE DUODENUM IS DIVIDED INTO FOUR PARTS (Fig. 17-4)

1. The first part, called the **superior duodenum**, is 3 to 5 cm long, with the following characteristics.

- **a.** It receives chyme through the pylorus.

b. It lacks circular folds and, because of its characteristic appearance on x-ray, has been termed the **duodenal cap**.
c. It passes anterior to the **common bile duct, common hepatic artery,** and **hepatic portal vein**.
d. It is peritoneal, supported by the **hepatoduodenal ligament**.
e. This section of the duodenum is relatively quiet. (Peristaltic activity becomes evident toward the jejunum.) Its activity is similar to a churning or milling action.

2. The second part of the duodenum, the **descending duodenum**, is 8 to 10 cm (3.5 to 4 inches) long, with the following characteristics (see Figs. 17-4 and 17-8).
   a. It is secondarily retroperitoneal.
   b. Circular folds (**plicae circulares**) make their appearance and increase in size and complexity toward the jejunum. These folds are permanent structures and cannot be flattened out by distension.
   c. This segment receives the **hepatopancreatic ampulla** and the **secondary pancreatic duct** (Santorini's), if present.
   d. Adjacent to the descending segment lie the transverse colon (anteriorly), the right kidney (posteriorly), and the right lobe of the liver (superiorly).

3. The **inferior duodenum** (also secondarily retroperitoneal) comprises the third and fourth parts.
   a. The **transverse portion** is 2.5 to 5 cm (1 to 2 inches) long.
      (1) As a result of intestinal rotation during development, the **superior mesenteric vessels** are located anterior to this horizontal section.
      (2) With wasting or after severe dieting, the superior mesenteric vascular bundle may compress the duodenum, producing obstruction.
   b. The **ascending portion** of the duodenum is 2.5 to 5 cm (1 to 2 inches) long.
      (1) This segment terminates at the duodenojejunal flexure.
      (2) The **suspensory ligament** (Treitz), a surgical landmark, holds the ascending portion in place.
         (a) The suspensory ligament of the duodenum runs from the fourth duodenal segment to the right crus of the diaphragm (see Fig. 17-4).
         (b) This ligament is composed of a tendon insinuated between slips of smooth muscle from the muscularis externa of the duodenum, and fascicles of striated muscle from the right crus of the diaphragm.
         (c) At the duodenojejunal flexure the small intestine becomes the peritoneal jejunum, which is supported by the **mesentery proper**.

E. ARTERIAL SUPPLY TO THE DUODENUM (see Fig. 17-4)

1. The blood supply to the duodenum is from the celiac artery via the right gastric, gastroduodenal, and superior pancreaticoduodenal arteries, as well as from the superior mesenteric artery via the inferior pancreaticoduodenal arteries.

2. The blood supply to the first (superior) part of the duodenum is sparse, sometimes to the point of being precarious, and exceedingly variable. Because of this, the superior duodenum is removed when the distal portion of the stomach is resected.
   a. The **supraduodenal artery**:
      (1) Is absent in 30 percent of individuals.
      (2) Usually arises from the gastroduodenal artery but may arise from the right gastric artery or the common hepatic artery.
   b. **Retroduodenal arteries**:
      (1) Are variable in number and location.
      (2) Usually arise from the gastroduodenal artery, right gastroepiploic artery, and superior pancreatic arteries, or any combination thereof.
   c. There are few, if any, extramural anastomotic connections between branches of the right gastric and gastroduodenal arteries.

3. A profuse blood flow to the second (descending) and third (inferior) portions of the duodenum is supplied by anastomosing arcades from the **superior pancreaticoduodenal** branches of the gastroduodenal artery and the **inferior pancreaticoduodenal** branches of the superior mesenteric artery.

F. FUNCTIONAL ASPECTS

1. While the first portion of the duodenum is mobile (suspended by the greater omentum and hepatoduodenal ligaments), the second, third, and fourth portions are secondarily retroperitoneal.

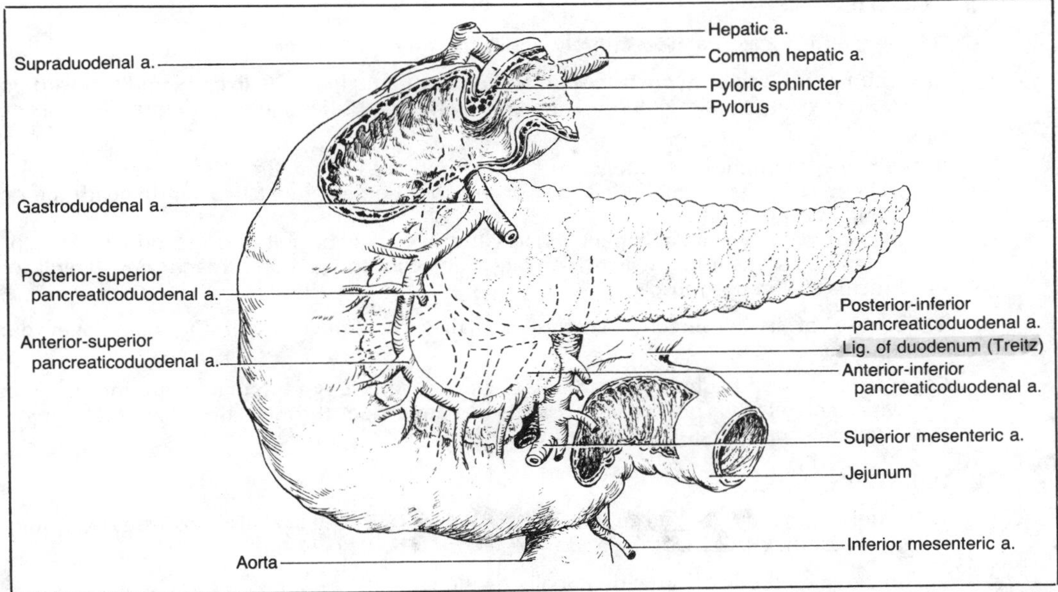

Figure 17-4. *The duodenum and pancreas.* The window in the superior portion of the duodenum shows the pyloric sphincter and beginnings of the plica circularis. The blood supply to the duodenum is by branches of the celiac and superior mesenteric arteries. The head of the pancreas is intimately associated with the duodenum, sharing a common blood supply.

2. Since the blood supply is from the medial side, an incision along the right edge of the descending duodenum mobilizes this viscus as well as the head of the pancreas and reestablishes the former dorsal mesentery without compromising the blood supply.
3. Since the venous and lymphatic channels may anastomose with those of the dorsal body wall, carcinoma in these sections frequently has a poor prognosis.
4. The small intestine has developed towards the function of nutrient absorption, accomplished through structural modifications that increase the surface area.
   a. The length of this viscus (5 to 8 m) provides a substantial surface area for absorption.
   b. The intestinal surface area is substantially increased by circular folds, or **plicae circulares**.
   c. Superimposed on the circular folds are surface projections of the mucosa (**intestinal villi**), which increase the surface area tremendously.
   d. The **microvilli**, histologic modifications of the mucosal cell luminal surfaces, increase the absorptive surface area by several magnitudes.

G. THE HEPATOPANCREATIC AMPULLA AND DUODENAL PAPILLA (Vater's) is located in the descending duodenum.

1. Usually the **common bile duct** and **primary pancreatic duct** join to form the **hepatopancreatic ampulla** (see Fig. 17-8).
   a. Where the hepatopancreatic ampulla enters the duodenum, a small protrusion occurs, the **duodenal papilla**.
   b. The papilla is surrounded by the hepatopancreatic sphincter (Oddi's).
2. Numerous variations are common (see Fig. 17-11).

V. THE LIVER

A. THE LIVER IS A COMPOUND TUBULAR GLAND

1. Morphologically, the liver appears to be one of the simplest organs; functionally, it is one of the most complex.
2. The liver, averaging 1200 to 1400 g in the female and 1400 to 1600 g in the male, is the largest gland in the body.
3. The liver secretes up to a liter of bile a day.

B. LOCATION

1. The liver is located almost entirely under the rib cage (see Fig. 15-1).
2. Occupying the **right hypochondriac** and **epigastric regions**, the liver extends upward under the rib cage as far as the 5th rib anteriorly on the right side, and the 5th intercostal space on the left.
3. Normally, the right lobe extends just beneath the costal margin.
   a. The liver in the normal adult may or may not be palpated 1 to 2 cm beneath the costal margin on the right.
   b. When palpable, as a patient inhales the liver will be felt to slide down beneath the examiner's fingertips, which are placed adjacent to the costal margin just lateral to the rectus abdominis muscle.
4. Because of its mesenteric attachments, the liver moves with the diaphragm during respiration.
5. Hepatic cysts and subphrenic abscesses can erode through the diaphragm into the pleural cavity; conversely, pulmonary abscesses can erode through the diaphragm into the subphrenic space and involve the liver.

C. LOBATION

1. In addition to the **right lobe** and the **left lobe** (Fig. 17-5), the liver also possesses two rudimentary lobes, the **quadrate lobe** and the **caudate lobe** (Fig. 17-6).
2. Functionally, the liver is nearly equally divided.
   a. Right and left lobes have separate biliary drainage and vascular supplies.
   b. The quadrate lobe, draining into the left hepatic duct and receiving blood supply from the left hepatic artery, is functionally part of the left lobe.
   c. The caudate lobe, receiving blood supply from both right and left hepatic arteries and secreting bile into both right and left hepatic ducts, is functionally part of both the right and left lobes.
3. The liver is covered by a thin fibrous capsule (Glisson's) that lies just beneath the visceral peritoneum.
   a. From the hepatic capsule, incomplete septa project inward into the hepatic stroma.
   b. Because of this arrangement of connective tissue and liver parenchyma, the surface of the liver appears mottled and the cut surface, granular.

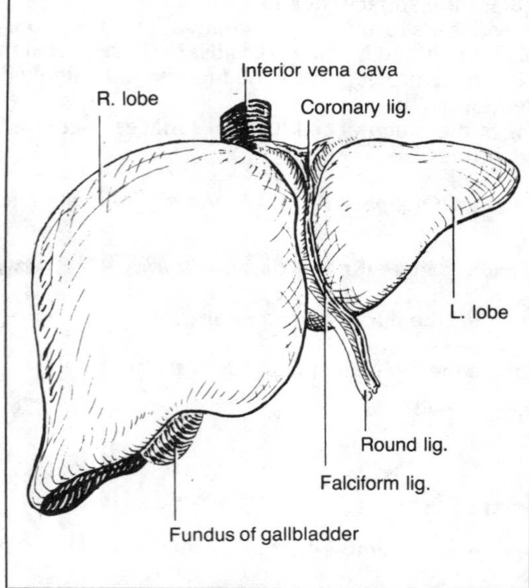

Figure 17-5. *The liver.* The anterior surface of the liver is depicted with the right and left lobes separated by the falciform ligament.

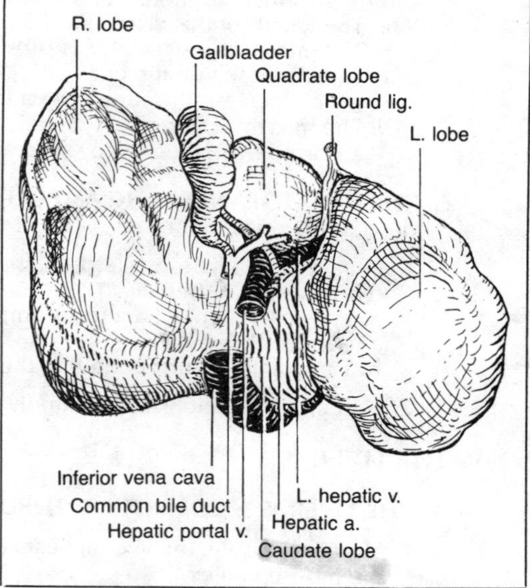

Figure 17-6. *The liver.* The inferior surface of the liver is shown with the left, right, quadrate, and caudate lobes as well as the gallbladder located between the left and quadrate lobes.

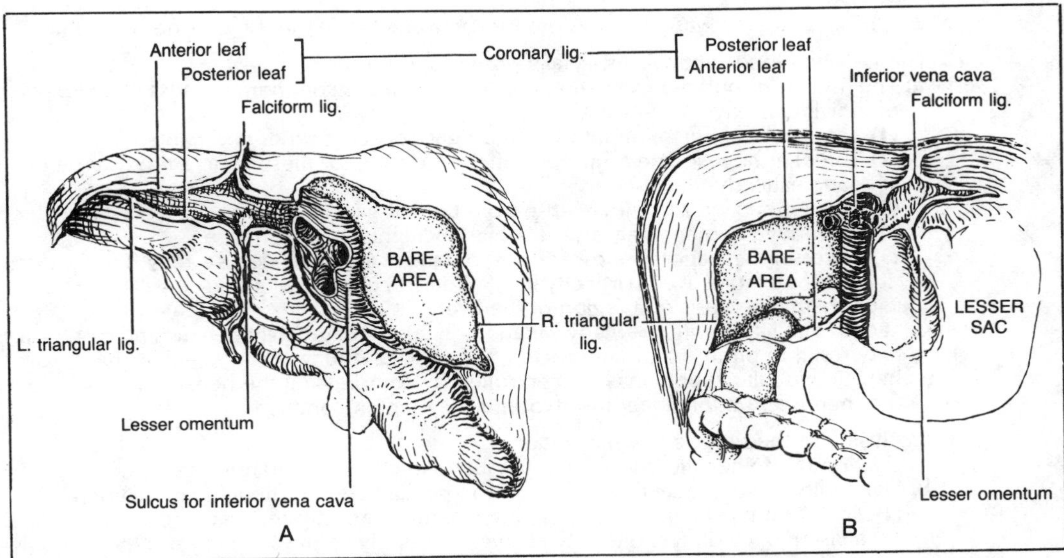

Figure 17-7. *The bare areas. A,* The bare area of the liver is bounded by the coronary and triangular ligaments. *B,* The bare area of the inferior surface of the diaphragm where the ligaments become parietal peritoneum is indicated.

c. Pain associated with acute hepatomegaly and transient venous congestion ("runner's stitch") is the result of sudden stretching of Glisson's capsule.

4. The liver is an extremely vascular structure, bleeding profusely when ruptured by blunt trauma or sectioned in surgery.

D. MESENTERIC ATTACHMENTS

1. The liver develops within the primitive ventral mesentery and divides the ventral mesogastrium into two ligaments (see Fig. 16-4).
   a. The **falciform ligament** is a continuation of the ventral mesentery, from the ventral surface of the liver to the midventral wall as far as the umbilicus (see Fig. 17-5).
      (1) This ligament contains the **round ligament (ligamentum teres)** of the liver, which is the remnant of the obliterated umbilical vein.
      (2) Patency may be reestablished in the round ligament by cannulation to measure portal venous pressure or to sample portal blood.
   b. The **coronary ligaments** are formed by reflections of ventral mesentery between the superior pole of the liver and the inferior surface of the diaphragm (Fig. 17-7).
      (1) The **bare area** is the region of the diaphragmatic and hepatic surfaces located between the leaflets of the anterior and posterior coronary ligaments.
      (2) The boundaries of the bare area, before returning to the falciform ligament in the anterior midline, include the falciform ligament, right anterior coronary ligament, right triangular ligament, right posterior coronary ligament, gastrohepatic ligament, left posterior coronary ligament, left triangular ligament, and left anterior coronary ligament.
      (3) The **inferior vena cava** transits the bare area.

2. The **lesser omentum**, running between the lesser curvature of the stomach and the liver, is comprised of the **hepatoduodenal ligament** and the **gastrohepatic ligament**.

E. BLOOD SUPPLY TO THE LIVER

1. The liver receives arterial blood from the hepatic artery and venous blood rich in nutrients from the hepatic portal vein (see Figs. 17-6 and 17-9).
   a. The **common hepatic artery**, one of the three usual branches of the celiac artery, becomes the **proper hepatic artery** once the gastroduodenal artery is given off.
   b. The proper hepatic artery ascends in the **hepatoduodenal ligament** (right edge of the **lesser omentum**) and divides into a **right hepatic artery** and a **left hepatic artery** (see Fig. 17-9).
      (1) The **right hepatic artery** supplies the right lobe and the right half of the caudate lobe.
      (2) The **left hepatic artery** supplies the left lobe, the quadrate lobe, and the left half of the caudate lobe.

(3) There are few vascular anastomoses between the left and right sides of the liver.

2. The hepatic blood supply is especially prone to variation.
   a. The usual textbook representation is actually rather uncommon, because 60 to 70 percent of individuals have some variation.
      (1) The level of bifurcation into left and right hepatic arteries is variable.
      (2) The right hepatic artery may arise from the superior mesenteric artery via the inferior pancreaticoduodenal artery.
      (3) The left hepatic artery may arise from the left gastric artery.
      (4) The above two anomalies have a high incidence of covariance.
      (5) Accessory hepatic arteries are not collateral arteries because they supply discrete regions of the liver parenchyma. Sudden occlusion of these vessels can result in ischemic necrosis of a region of the liver.
   b. Lack of familiarity with the many variations in course of the right hepatic and cystic arteries may result in profuse and unexpected hemorrhage during surgery. Hemorrhage from the hepatic vascular supply may be controlled by compressing the hepatic pedicle between the thumb and a forefinger inserted into the epiploic foramen.

3. Occlusion of the hepatic vascularization:
   a. Ligation of the hepatic portal vein may be tolerated up to 30 minutes.
   b. Depending on the location, ligation of the hepatic arterial supply may be tolerated.
      (1) The common hepatic artery may be permanently ligated because there are ample anastomotic connections between vessels supplying the stomach and those supplying the liver.
      (2) There is great risk with permanent ligation of the proper hepatic artery unless it gives off the right gastric artery.
      (3) The proper hepatic artery above the right gastric artery and the left and right hepatic arteries cannot be ligated without liver ischemia and necrosis.
         (a) "Accessory" and "replacing" arteries also cannot be ligated without resultant liver ischemia and necrosis.
         (b) Each artery entering the parenchyma of the liver supplies a discrete region of parenchyma, without major arterial anastomoses.
   c. If an occlusion develops over a prolonged period, adequate intrahepatic and extrahepatic collateral circulation will develop.

F. FUNCTIONAL ASPECTS

1. The liver is a primary metabolic center.

2. It has a uniform structure, consisting of anastomosing sheets of cells, which are separated by intervening venous sinusoids.

3. All nutrient-rich venous blood from the GI tract passes through the liver sinusoids.

4. Liver cells are supplied by blood from the hepatic portal vein and hepatic artery, both of which perfuse through the sinusoids.
   a. Hepatic cells are bathed in a mixture of venous blood from the alimentary canal and oxygenated blood from the aorta.
   b. Because of this uncommon vascular arrangement, a lower than usual oxygen content bathes the liver cells; this may explain, in part, the high susceptibility of the liver to damage and disease.
   c. The sinusoids drain into the central veins of the lobes, which unite to form **hepatic veins**; these, in turn, join the **inferior vena cava**.

5. Liver cells secrete up to a liter of bile per day into the small bile canaliculi. Among the constituents of bile are bile salts and bile pigments.
   a. **Bile pigments**, including bilirubin and biliverdin, are derived from the breakdown of hemoglobin in the spleen and liver, and produce the distinctive coloration of feces.
   b. The **bile salts**, formed from cholesterol, aid in digestion by emulsification of fats, thereby facilitating their absorption by the intestinal mucosa. Most of the bile salts are resorbed by the mucosa and returned to the liver by the portal veins for resecretion.

G. CLINICAL CONSIDERATIONS

1. Although the liver is a vital organ, portions of it may be removed surgically without irreparable damage to the body. The liver has great regenerative powers, which can be detrimental when uncontrolled, as in cirrhosis.

2. Icterus (jaundice) occurs either when the liver is overloaded beyond its capacity (excessive hemolysis of blood or liver damage) or when the biliary tree is obstructed. In both instances

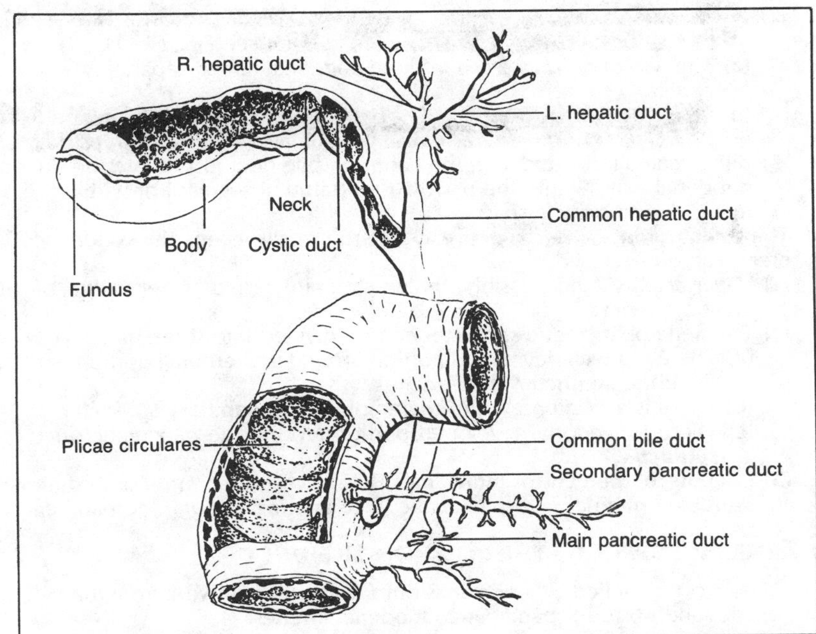

Figure 17-8. *The biliary tree, gallbladder, and descending duodenum.*

bile pigments back up in the blood, giving the skin a characteristic yellow coloration accompanied by itching.

VI. THE BILIARY APPARATUS (HEPATIC TREE AND GALLBLADDER)

A. THE BILE CANALICULI, which traverse the sheets of hepatocytes, unite to form the intrahepatic bile ductules. The latter lie in the portal triads with the branches of the hepatic portal veins and hepatic arteries.

1. The **bile ductules** converge into a consistent pattern to form the left and right hepatic ducts.
2. This pattern defines the segments of the liver.

B. THE HEPATIC TREE consists of **left** and **right hepatic ducts, common hepatic duct, gallbladder, cystic duct**, and **common bile duct** (Fig. 17-8).

1. **Right** and **left hepatic ducts** leave the liver and join to form the **common hepatic duct**.
   a. The **left hepatic duct** drains the left and quadrate lobes as well as the left half of the caudate lobe.
   b. The **right hepatic duct** drains the right lobe and the right half of the caudate lobe.
2. The **cystic duct** to the **gallbladder** arises from the **common hepatic duct**.
   a. The cystic duct is from 1 to 1.5 cm long.
   b. The spiral fold (valve of Heister) maintains the patency of the cystic duct.
   c. The course of the cystic duct is variable; rarely, it may be double.
3. The **common hepatic duct** and the **cystic duct** converge to form the **common bile duct**, which empties into the duodenum.
   a. The common bile duct, which is about 10 cm long and 1 cm in diameter, contributes to the **hepatic pedicle**.
      (1) The **hepatic pedicle** is located in the free edge of the **hepatoduodenal ligament**.
         (a) The **common bile duct** lies anteriorly to the right.
         (b) The **hepatic artery** lies anteriorly to the left.
         (c) The **hepatic portal vein** lies posteriorly.
      (2) The free inferior edge of the hepatoduodenal ligament forms the anterior boundary of the **epiploic foramen** (Winslow's).
   b. The common bile duct passes posterior to the superior and descending duodenum. It usually is embedded in pancreatic parenchyma but is easily dissected free.
   c. It enters the descending duodenum about 3 to 8 cm beyond the pyloric sphincter.
      (1) The intramural portion of the common bile duct is the narrowest part.

(2) The **hepatopancreatic ampulla** or **duodenal papilla** (Vater's) is formed by confluence of the common bile duct and pancreatic duct (see Fig. 17-11).
(a) The two ducts may open into the duodenal lumen separately, each having its own papilla.
(b) Blockage of the ampulla by a bile calculus, muscle spasm, or edema may precipitate pancreatitis, the result of bile reflux into and along the pancreatic duct.
(3) Biliary calculi may lodge in the common bile duct just before the duct penetrates the duodenal wall, within the narrowest intramural segment or within the ampullary segment, due to muscle spasm.

d. The **hepatopancreatic sphincter** (Oddi's) normally keeps the ostium of the hepatopancreatic ampulla closed.
(1) Vagal activity (and possibly cholecystokinin) results in relaxation of the hepatopancreatic sphincter.
(2) The hepatopancreatic sphincter can be divided into three parts:
(a) There is a well-developed portion around the terminal portion of the common bile duct: the **sphincter choledochus**.
(b) There is a portion around the ampulla: the **papillary sphincter**.
(c) There is a poorly developed portion around the pancreatic duct: the **pancreatic sphincter**.
(3) In about 65 percent of individuals, bile from the common bile duct is mixed in the **duodenal papilla**, with pancreatic juice arriving through the pancreatic duct.

C. THE ARTERIAL SUPPLY TO THE COMMON BILE DUCT

1. The bile duct is supplied by a rich anastomotic vascular network from the right hepatic, proper hepatic, and posterior pancreaticoduodenal arteries.

2. The location of the twigs is variable and inconstant.

3. These arteries course in the adventitia of the bile duct.

4. During manipulation of the common bile duct, great care must be taken to prevent isolation of the blood supply, as necrosis and infarct often result.

D. THE GALLBLADDER is 6 to 10 cm long, with a capacity of about 45 ml.

1. The gallbladder usually cannot be palpated unless it contains stones, whereupon it may be felt in the right upper quadrant at the angle between the costal margin and the rectus abdominis muscle.

2. Located between the right and quadrate lobes of the liver, the gallbladder lies in a fossa on the ventral surface of the liver.
a. It may be embedded within the liver parenchyma (rare); it may be adhered to the liver by peritoneum; it may be surrounded by peritoneum (floating); or it may have a short mesentery to the inferior surface of the liver.
b. Close to the superior duodenum and transverse colon, anteriorly, it may be adherent to either the duodenum or transverse colon.
(1) A large gallstone may erode through the adhered common walls into the bowel.
(2) To relieve common duct obstruction, the gallbladder may be anastomosed surgically to the duodenum.

3. The gallbladder is divided into three parts: the **fundus**, the **body**, and the **neck**. The neck may contain a pendulous pouch (**Hartmann's**).
a. During surgery, traction on Hartmann's pouch causes the cystic duct to stand out.
b. Hartmann's pouch may be adherent to the cystic duct, or even to the common bile duct, and may be included accidentally in a ligature passed around the cystic duct.

4. The **arterial supply** to the gallbladder.
a. Normally, the cystic artery arises as a single stem from the right hepatic artery within the **cystohepatic triangle** (Calot's) and bifurcates into **anterior** and **posterior cystic arteries** (Fig. 17-9).
b. Cystic arteries are quite variable; they may arise from right, left, or even proper hepatic arteries. The anterior and posterior branches may arise separately from any of the abovementioned arteries.

5. **Function** of the gallbladder.
a. The sphincter choledochus and papillary sphincter are normally closed so that bile, which is secreted continuously by the liver, refluxes through the cystic duct into the gallbladder.
b. The gallbladder stores and concentrates bile by mucosal absorption of electrolytes and water.

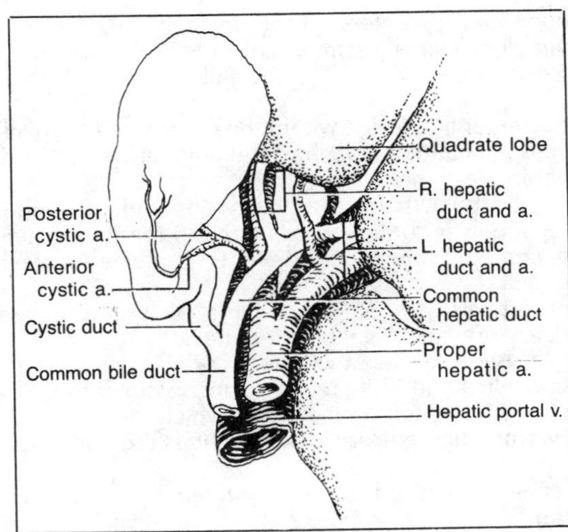

Figure 17-9. *The vascular supply to the liver and gallbladder.*

c. The ability of the gallbladder to concentrate bile forms the basis for radiographic visualization of that organ (cholangiography).

(1) When radiopaque, organic iodinated compounds are administered either orally or intravenously, they are excreted by the liver.

(2) As the bile is concentrated in the gallbladder, the concentration of the radiopaque dye increases so that the entire organ, and often the biliary tree, becomes radiopaque.

d. A spurt of fat through the pylorus into the duodenum causes release of cholecystokinin by the intestinal glands in the duodenum.

(1) The hormone cholecystokinin induces contraction of the gallbladder musculature, causing the fundus to rise and the body to narrow.

(2) The resulting rise in biliary pressure overcomes the sphincters of the lower biliary tree, in which the tone may have been decreased slightly by vagal action, and bile pours into the duodenum.

6. Clinical considerations.

a. The numerous variations in the biliary tree constitute a major hazard in surgery.

(1) Although procedures on the biliary tree are routine, complications in cholecystectomy (removal) and cholecystostomy (drainage) are surprisingly frequent.

(2) Inadvertent injury to the biliary tree, with leakage of bile into the peritoneal cavity, rapidly produces bile peritonitis.

(3) Obstruction of the duct system as the result of a stone, tumor, edema, or scar tissue produces pain and jaundice.

b. Variation in branches and relations of the hepatic artery within the triangle of Calot constitutes the major hazard in biliary surgery (see Fig. 17-9).

(1) The boundaries of the **triangle of Calot** include:

(a) The **common hepatic duct** on the left.

(b) The **cystic duct**, on the right.

(c) The **liver**, superiorly.

(2) The triangle of Calot contains most structures of importance in cholecystectomy:

(a) 90 percent of all **cystic arteries**.

(b) 82 percent of all **right hepatic arteries**.

(c) 95 percent of all **accessory right hepatic arteries**.

(d) 91 percent of all **accessory bile ducts**.

(3) Structures in the triangle of Calot and in the hepatoduodenal ligament should not be clamped or divided until dissected free and positively identified.

c. Overconcentration of the bile, perhaps due to biliary stasis, results in the dissolution of bile salts and bile pigments, leading to concretions (choleliths or gallstones).

(1) Gallstones may be composed purely of cholesterol, which may calcify and be visible in any ordinary radiograph; or purely of bile pigments, which appear as a filling defect in a cholangiogram; or mixed in concentric layers and calcified.

(2) In women, it is not uncommon to find several populations of stones; the stones of each population are approximately the same size and represent a different period of pregnancy.

(3) Gallstones are most common in overweight, multiparous women over the age of forty—the so-called "4-F individual" (fat, fertile, female, forty).

(4) Small stones pass through the ducts and sphincters without incident.
(5) Stones too large to enter the cystic duct irritate the mucosa (cholecystitis) and may result in ulceration, and eventual perforation of the wall of the gallbladder, with ensuing peritonitis.
(6) Stones of the appropriate size may enter and lodge within the duct system, especially at points of narrowing such as the sphincter choledochus, the point at which the duct enters the duodenal wall and within the duodenal papilla.
(a) If the stone lodges in the cystic duct, there is cholecystitis without jaundice, because the bile produced by the liver continues to drain into the duodenum.
(b) If the stone lodges in the common bile duct, there is jaundice as well as cholecystitis.
(c) Biliary obstruction is painful, especially when fatty chyme elicits the release of cholecystokinin, elevating biliary pressure (''gallbladder attack'').
(i) Pain associated with cholecystitis reaches the spinal cord, levels T5–T6, via visceral afferent neurons traveling with the greater splanchnic nerves.
(ii) Biliary pain is referred to the 5th and 6th dermatomes, which include the right upper quadrant of the abdomen and epigastrium, as well as the region of the right scapula.
(iii) When the peritoneum becomes involved, the pain becomes localized in the right hypochondriac region.
(d) Chronic cholecystitis and obstruction of the biliary tree are managed surgically.

d. Anomalies of the gallbladder include diverticula, duplication, and congenital absence, as well as ductal duplication, atresia, and stenosis.

VII. PANCREAS

A. DEVELOPMENT

1. The pancreas develops from two intestinal diverticula, one in the dorsal and the other in the ventral mesentery.

2. The developmental sequence accounts for two pancreatic ducts.

3. The location and secondarily retroperitoneal relation of the pancreas is explained by rotation of the gut.

B. STRUCTURE (Fig. 17-10)

1. The pancreas lies in the epigastric and left hypochondriac regions.

2. Except for the end of the tail, the pancreas is secondarily retroperitoneal.
a. The anterior surface is covered by the posterior wall of the omental bursa; the posterior surface is adhered to the peritoneum of the dorsal body wall by the fascia of Toldt.
b. Because of its deep position, the pancreas is difficult to palpate; only a large tumor is palpable and is, at this stage, nonoperable.

3. The **head** is located within the curve of the duodenum.
a. The intimate association between the pancreas and the duodenum results in shared vascular supplies.
b. The **uncinate process** of the head projects to the left, behind the **superior mesenteric vessels**.

4. The **body** lies posterior to the stomach.
a. The body of the pancreas has a variable blood supply.
b. A perforated gastric ulcer may result in adhesions between the stomach and pancreas, with subsequent erosion into pancreatic parenchyma, and resultant pancreatitis. Hemorrhagic pancreatitis may be rapidly fatal.

5. The **tail** contains the majority of the pancreatic islets—a major consideration in pancreatic resection.
a. The peritoneal portion of the tail projects to a variable extent into the lienorenal ligament and lies very close to the hilus of the spleen.
b. The splenic vessels are usually embedded in the parenchyma of the tail, providing the tail with an exceedingly variable blood supply.

C. THE PANCREATIC DUCTS (see Fig. 17-10)

1. The **primary pancreatic duct** (Wirsung's) results from the joining of the distal portion of the primitive dorsal duct to the proximal portion of the primitive ventral duct.

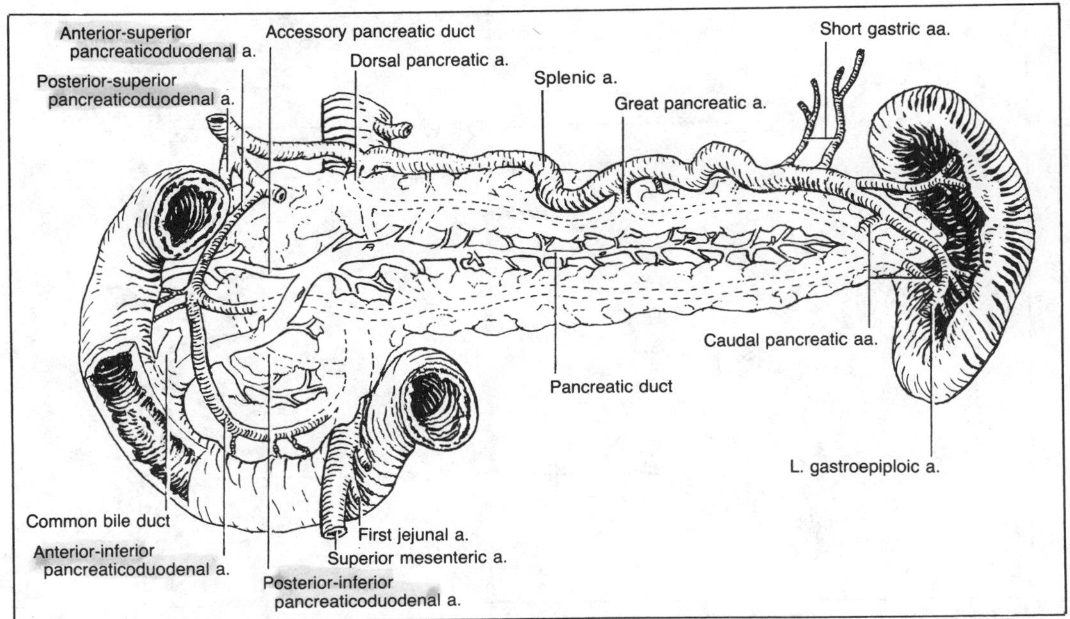

Figure 17-10. *The pancreas and spleen.* The primary and secondary (if present) pancreatic ducts empty into the descending portion of the duodenum. The pancreas receives its blood supply from branches of the celiac and superior mesenteric arteries. The spleen receives a profuse blood supply from the splenic branch of the celiac artery.

a. The duct courses along the entire pancreas.
b. It usually joins the common bile duct, which also develops in the ventral mesentery at the duodenal papilla (Vater's).

2. The **accessory (secondary) pancreatic duct** (Santorini's) represents the proximal portion of the primitive dorsal duct.
   a. This duct enters the duodenum separately at the **accessory duodenal papilla**.
   b. The distal end of the secondary duct may join the primary duct.

3. There is extensive variation in the ducts (Fig. 17-11).
   a. The embryonic pattern persists in 10 percent of individuals.
   b. The primary pancreatic duct and common bile duct may open into the duodenum separately, each with its own papilla.

D. THE ARTERIAL SUPPLY TO THE PANCREAS (see Fig. 17-10)

1. The **great pancreatic artery, caudal pancreatic artery**, and an occasional **dorsal pancreatic artery** all arise from the splenic artery.

2. The **superior pancreaticoduodenal artery**, arising from the gastroduodenal artery, bifurcates into anterior and posterior arcades, which supply the head of the pancreas and the duodenum.

3. The **inferior pancreaticoduodenal artery**, arising from the superior mesenteric artery, also bifurcates into anterior and posterior arcades, which supply the head of the pancreas and the duodenum.

E. PANCREATIC FUNCTION

1. Comprising the endocrine portion of the pancreas, the islets of Langerhans secrete insulin and glucagon into the circulatory system.

2. As a compound tubular acinar exocrine gland, the pancreas secretes 500 to 1200 ml of pancreatic juice per day into the duodenum.
   a. This colorless, viscous fluid is alkaline due to its high bicarbonate content. It maintains the basic pH of the duodenum.
   b. Pancreatic juice contains three classes of powerful digestive enzymes: trypsin, amylase, and lipase, all of which are activated at a high pH.

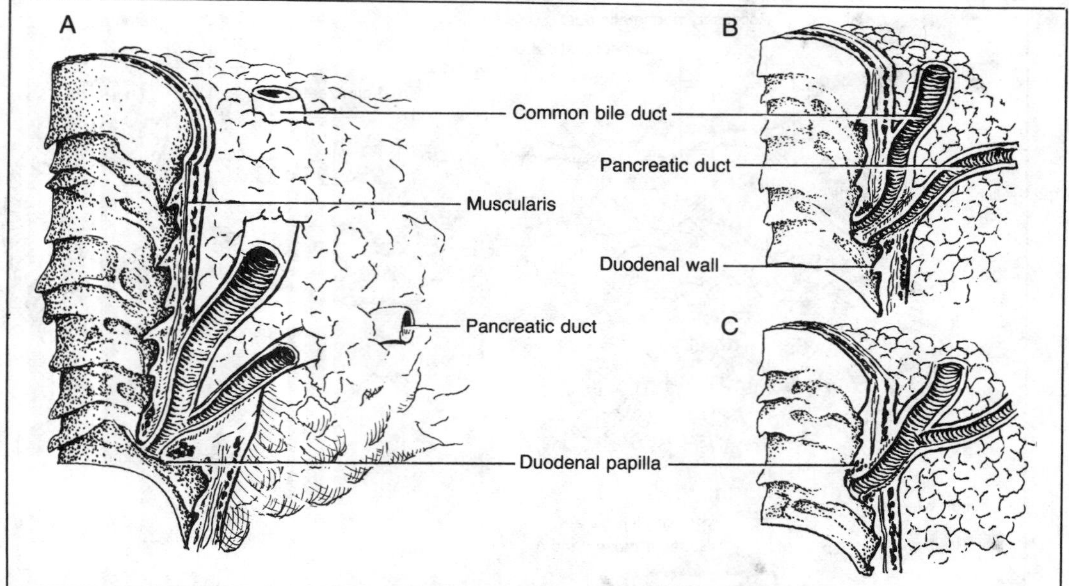

Figure 17-11. *The duodenal papilla. A,* The most usual configuration whereby the common bile duct and the primary pancreatic duct join to enter the duodenum through a hepatopancreatic duct. *B,* The common bile duct and primary pancreatic duct may enter the duodenum separately. *C,* A more proximal joining of the biliary and pancreatic ducts produces an unusually long hepatopancreatic duct.

c. The secretion of pancreatic juice is regulated by local chemical stimuli, as well as by parasympathetic innervation.

F. CLINICAL CONSIDERATIONS

1. Because the pancreatic and common bile ducts join and enter the duodenum via a common ostium in 65 percent of individuals, spasm of the hepatopancreatic sphincter or blockage of the duct at the duodenal papilla results in both reflux of bile into the pancreas and pancreatic stasis. This situation causes acute pancreatitis (chemical autolytic pancreatic necrosis).

2. Carcinoma of the pancreas is not uncommon.
   a. Due to the diffuse lymphatic drainage of the pancreas (in part due to its secondarily retroperitoneal nature), the prognosis for recovery is poor.
   b. Fully 80 percent of pancreatic carcinomas are located in the head region.
      (1) Jaundice with no indication of biliary pathology may result from a pancreatic tumor compressing the common bile duct—a lifesaving signal, as pancreatic tumors are frequently silent until it is too late for surgery.
      (2) Carcinoma frequently invades the splenic vein or the portal vein with symptoms of portal hypertension, such as esophageal varices.

3. Aside from the commonly occurring ectopic pancreatic tissue (possible in the stomach, small intestine, gallbladder, or spleen), the pancreas may completely surround the descending duodenum (annular pancreas), with constriction or even obstruction of this portion of the small bowel.

VIII. THE SPLEEN

A. STRUCTURE AND FUNCTION

1. Once thought to be the seat of bad humor ("...to vent one's spleen..."), the **spleen** is not a vital organ, although it is hematopoietic in the fetus and later functions in blood destruction. It has the capacity for blood and lymphocyte storage, and also participates in immunologic activity.

2. The size of the normal spleen varies within wide limits but is usually about 12 cm × 7 cm × 3 cm and weighs 150 to 200 g.

3. The spleen is an intensely vascular structure.
   a. It is composed of lymphoid tissue separated by blood-filled sinuses.

b. As a reticuloendothelial structure situated in the hepatic portal system, the spleen removes particulate matter and cellular residue from the blood.
c. Due to its vascularity and capacity to store blood, as well as its location in the portal system with a large splenic artery, the spleen hemorrhages profusely when ruptured, such as by a broken 11th rib.

4. The spleen is palpable only when splenomegaly occurs, which may be a sign of chronic infection (such as mononucleosis), blood dyscrasias, lipid-storage diseases, or lymphosarcoma.

5. One to ten accessory spleens in the dorsal mesogastrium are common. When splenectomy is indicated, all accessory splenic tissue must be located and excised.

B. RELATIONS

1. The spleen is located in the **left hypochondriac region** (see Fig. 15-1).

2. Arising within the dorsal mesogastrium, its supports include the **gastrosplenic ligament** and the **lienorenal ligament**.
a. Variable in position, it may prolapse into the pelvis, especially in splenomegaly.
b. Normally, the spleen is located anterior to ribs 11 and 12.
c. The lower pole generally lies between L1 and L2.
d. The spleen is adjacent to the stomach and superior to the left colic flexure where the transverse colon becomes the descending colon.

3. The spleen is supplied by the **splenic artery** and is drained by the **splenic vein**, a major branch of the hepatic portal vein (see Figs. 17-10 and 17-21).

IX. THE JEJUNUM AND ILEUM

A. STRUCTURE (Fig. 17-12)

1. The peritoneal section of small bowel averages 7 m (22 feet) and varies from 5 to 11 m (15 to 34 feet). (See Table 17-1 for distinguishing characteristics of the jejunum and ileum.)

2. The **jejunum** constitutes the upper three-fifths of the peritoneal portion of the small intestine.

3. The **ileum** constitutes the lower three-fifths of the peritoneal portion of the small intestine.

4. Because the distinctive characteristics of the jejunum gradually become those of the ileum, the one-fifth in common cannot be ascribed to either the ileum or jejunum alone.

5. The **mesentery proper** supports the jejunum and the ileum.
a. The radix of the mesentery proper is 15 to 20 cm (6 to 9 inches) long, fanning out to support more than 7 m of small bowel.
b. The mesentery proper passes obliquely across the posterior abdominal wall from upper left to lower right.

6. Distinguishing characteristics between proximal jejunum and distal ileum are summarized in Figure 17-12.
a. Although differences between the jejunum and ileum are subtle, it is relatively easy to distinguish between proximal jejunum and distal ileum when seen as a whole.
b. The size and number of arcades and arteriae rectae, as well as the relative quantity of mesenteric fat between these vessels, are the most obvious gross characteristics differentiating proximal jejunum from distal ileum.
c. On radiography, the distinction between jejunum and ileum is obvious from differences in the plicae circulares.
(1) The jejunal lumen has a feathery appearance because the plicae are tall, branched, and close together.
(2) The ileum appears denser because the lumen is narrower and the plicae are shorter and further apart.
(3) The jejunal loops usually are located in the left lateral region whereas the ileal loops lie in the pubic region. However, the small intestine is extremely mobile, and radiographic relationships may change from day to day.

B. PARADUODENAL RECESSES (FOSSAE) are formed by folds of peritoneum where the small intestine changes from secondarily retroperitoneal to peritoneal.

1. These recesses are potential sites for intra-abdominal herniation, especially during development.

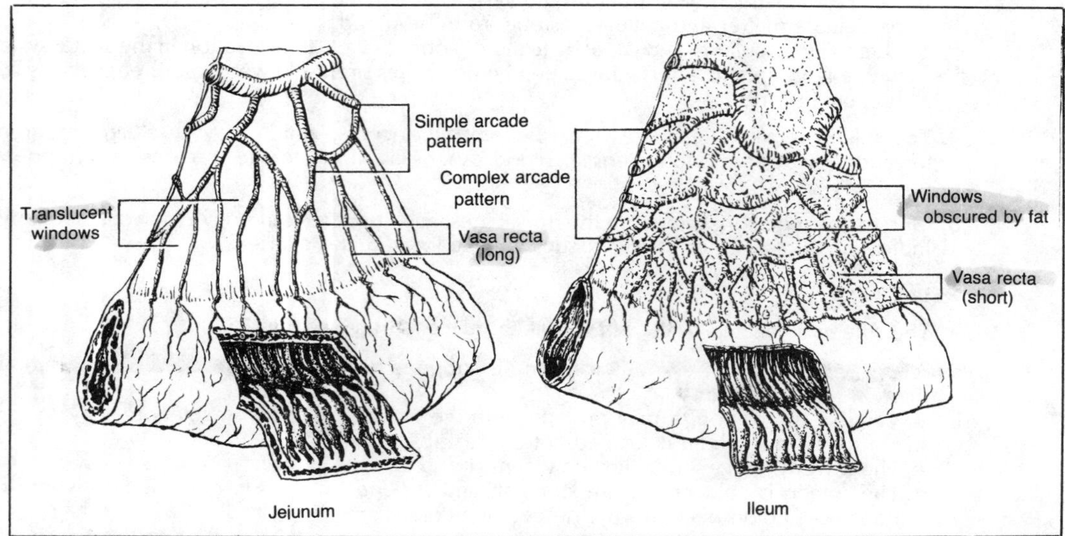

Figure 17-12. *The jejunum and ileum.* Several differences between the proximal jejunum and distal ileum are indicated.

Table 17-1. Distinguishing Characteristics of the Jejunum and Ileum

Characteristic	Jejunum	Ileum
Diameter	2 to 4 cm	2.5 to 3 cm
Wall	Thick, heavy	Thin, light
Vascularity	Greater	Less
Color	Deeper red	Paler pink
Arteriae rectae	Tall	Short
Arcades	Large loops, few	Short loops, many
Fat in mesentery	Less	More
Plicae circulares	Well developed	Rudimentary
Peyer's patches	Few	Many

2. The openings to the paraduodenal fossae are bounded by the duodenum, medially, and by the **inferior mesenteric vein** and the **left colic artery**, laterally.

C. THE ARTERIAL SUPPLY TO THE JEJUNUM AND ILEUM is from the **jejunal** and **ileal branches** of the **superior mesenteric artery** (Fig. 17-13).

1. Up to 20 jejunal and ileal arteries branch to form **arcades** within the mesentery proper.
   a. In contrast to those of the jejunum, the arcades of the ileum are smaller and more numerous.
   b. The arterial arcades provide abundant collateral circulation, especially important during peristaltic movements when some of the arteries may be constricted.
2. **Straight arteries (arteriae rectae)** run from arcades directly to the viscus without further anastomoses.
   a. In contrast to the jejunum, the arteriae rectae of the ileum are shorter and more numerous.
   b. The arteriae rectae give off anterior and posterior branches to the respective sides of the viscus.
   c. Arteriae rectae may anastomose with branches of adjacent arteriae rectae within the intestinal wall on the mesenteric side, but there are only nonanastomotic end-arteries on the antimesenteric side.
      (1) Intramural anastomoses on the mesenteric side are sufficient to sustain several centimeters of intestine in the event of embarrassment of the direct blood supply.

(2) By definition, an end-artery supplies a distinct region of tissue, with no anastomoses possible. If an end-artery to the antimesenteric side becomes occluded, local infarct results.

3. **Ileal branches** of the ileocolic artery provide the major supply to the terminal ileum.

D. THE MAJOR FUNCTION OF THE LYMPHATICS OF THE GUT is transportation of triglycerides into the circulation via the thoracic duct. Lymphatics of this region are of minor clinical importance because the jejunum and ileum are noted for a low incidence of carcinoma.

E. FUNCTIONAL ASPECTS OF THE SMALL INTESTINE

1. Intestinal motility propels chyme through the gut.
   a. **Segmentation** occurs at the rate of about 10 constrictions per minute and corresponds to the frequency of borborygmi (the sounds heard with a stethoscope applied to the abdominal wall).
      (1) Intestinal movements can be classified as segmental when columns of chyme are broken up into segments of equal size by contraction of the circular muscle.
      (2) Segmental contractions are the result of reflexes initiated by intestinal wall distension.
      (3) Subsequent contractions in adjacent regions divide segments of chyme in half, and the neighboring segments produced by the previous contraction unite to form new segments.
   b. **Pendular movements** are slight waves of contraction that pass bidirectionally and rapidly along the gut at a rate of 12 to 13 per minute.
      (1) Both muscular coats are involved in the production of pendular movements, which mix chyme with intestinal and pancreatic digestive juices as well as with bile.
      (2) These movements assist absorption by continuously bringing chyme into contact with the surface mucosa.
      (3) Pendular movements have an important massaging influence on the blood and lymphatic channels of the small intestine.
   c. **Peristalsis**, a unidirectional intestinal movement, occurs in the jejunum, ileum, and large intestine. While not very forceful in the jejunum, peristalsis becomes more pronounced in the ileum.
      (1) One form of peristaltic wave is a relatively slow (1 to 2 cm per second) advancing contraction that proceeds aborally over relatively short distances (4 to 5 cm).
      (2) The second form is **peristaltic rush**, which sweeps along the intestine without pause for great distances. This usually occurs after meals, especially the first of the day.
   d. Control of GI motility:
      (1) Segmental and pendular movements are probably enteric in origin.
         (a) Many of the nerve cells of the myenteric plexus seem to be intrinsic and involve local reflexes.
         (b) Enteric reflexes may explain the **Bayliss-Starling law of the gut**, whereby a bolus of chyme, exerting transverse pressure on the intestinal wall, results in increased muscular tone immediately oral to the bolus and relaxation in the adjacent aboral region. This reflex mechanism ensures unidirectional oral to aboral flow of intestinal contents.
      (2) Peristaltic contraction is influenced by autonomic neural control.
         (a) The small intestine is supplied by sympathetic (adrenergic) neurons from the celiac and mesenteric ganglia. Sympathetic activity inhibits peristalsis.
         (b) Parasympathetic (cholinergic) neurons lie in the myenteric plexus (Auerbach's) and are stimulated by the vagus nerve. Parasympathetic activity increases peristalsis, probably by facilitating the enteric neurons and the enteric reflexes.

2. The digestive functions of the small intestine require alkaline secretion (pancreatic juice) as well as the production of proteolytic enzymes.

3. The mucosa of the small intestine absorbs carbohydrates, triglycerides and fatty acids, amino acids, vitamins, and electrolytes—as well as water.

F. CLINICAL CONSIDERATIONS

1. Resection of up to one-third of the small intestine is compatible with normal life. Survival is possible with as little as 18 inches.

2. The stomach may be anastomosed to the jejunum (gastrojejunostomy) when the duodenum must be bypassed or resected (Whipple's procedure).

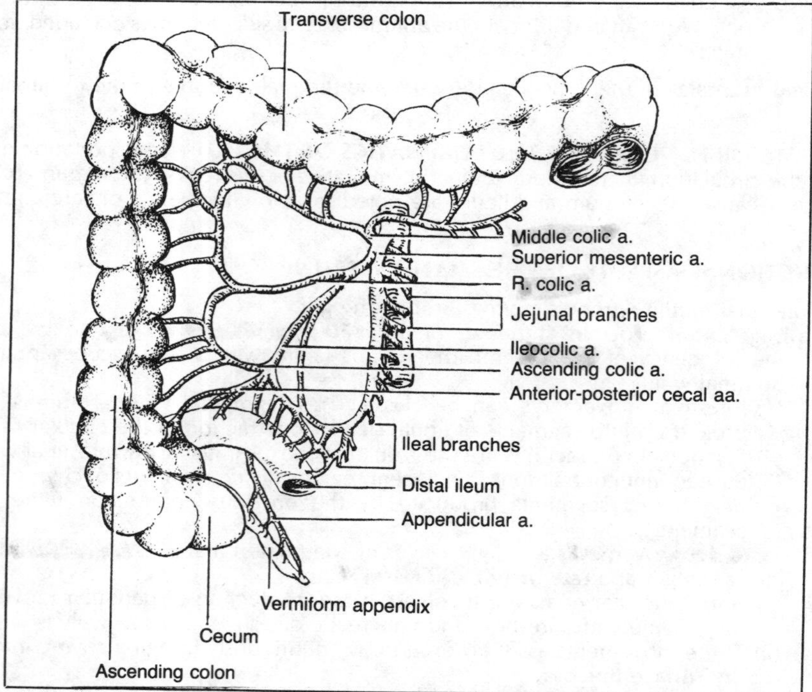

Figure 17-13. *The superior mesenteric arterial distribution.*

3. **Meckel's diverticulum.**
   a. Occasionally, a remnant of the omphalomesenteric (vitelline) duct of the embryo persists and becomes an **ileal diverticulum** (Meckel's) in the adult.
   b. Present in about 3 percent of the population, Meckel's diverticulum is located about 3 feet from the ileocecal junction (on the antimesenteric side) and is between 1 and 3 inches long.
   c. Meckel's diverticulum may contain gastric mucosa with oxyntic (acid-secreting) cells. Peptic ulceration of adjacent ileal mucosa is a frequent complication.
   d. Rarely, the omphalomesenteric duct remains a patent connection with the umbilicus so that chyme may be discharged to the exterior.
   e. This connection with the umbilicus also may persist with an obliterated lumen; however, the gut may revolve about this structure (volvulus) with resultant obstruction of the intestinal lumen, occlusion of blood vessels, and subsequent gangrene.
   f. An ileal diverticulum may become inflamed, with symptoms similar to those produced by an inflamed appendix, and require surgical intervention.
   g. During appendectomy, the terminal ileum usually is explored for a possible ileal diverticulum. If found, it is resected as a prophylactic measure.
4. Inflammation or trauma (even surgical manipulation) to abdominal or related organs, such as the kidneys or gonads, usually leads to inhibition of the small gut, paralytic ileus, or intestinal shut-down. The reemergence of borborygmi is anxiously awaited postoperatively as the sign of reestablished bowel function, an indication that feeding by mouth may be resumed.

X. THE COLON

A. FUNCTION AND STRUCTURE

1. The principal function of the colon is the conservation of fluid, a process that results in the formation of semisolid feces.
2. The mucosa of the colon is characterized by a mixture of absorptive cells, which extract water from the chyme, and mucous cells, which provide lubrication, thereby enabling the forming stool to be moved toward the rectum.
3. The large intestine is characterized by its capacity, its distensibility, the length of time that it retains its contents, and the special arrangement of musculature within the tunica muscularis. These properties are related to the role of the large intestine in the formation, transport, and evacuation of feces.

a. The large bowel is approximately 5 feet long (varying between 3 and 7 feet).
b. Fat-filled tags (**appendices epiploicae**) are scattered over its surface.
c. The outer longitudinal layer of the muscularis externa of the colon is incomplete, being restricted to three longitudinal bands, the **teniae coli**.
(1) These three strips of longitudinal muscle, about 1 cm wide, are most obvious on the cecum and ascending colon.
(2) Their position is somewhat variable along the transverse colon, becoming diffuse along the descending colon, rather indistinct on the sigmoid colon, and fusing to form a complete layer over both the appendix and rectum.
d. Muscle tone in the teniae coli results in sacculations (**haustra**) along the large intestine.
e. The ascending and descending colons are secondarily retroperitoneal.
f. The transverse and sigmoid segments, supported by the **transverse mesocolon** and **sigmoid mesocolon**, are considerably mobile, so that their blood supplies are at some risk of occlusion by volvulus.

4. The colon is divided according to its mesenteric attachments, or lack thereof, into ascending, transverse, descending, and sigmoid regions.

B. THE ASCENDING COLON

1. The **cecum**, a dilated sac, is the first part of the ascending colon (Fig. 17-14).
a. The cecum is located inferior to the ileocecal juncture.
b. The terminal part of the ileum enters the cecum posteromedially, with some variation.
(1) The ileum forms a conical projection into the cecum, the **ileal papilla**.
(2) The ileocecal ostium is bordered by two folds that form the **ileocecal valve**.
(a) The ileocecal valve is largely incompetent and of little mechanical importance; backflow of chyme is prevented by the general direction of peristalsis.
(b) Although this valve does impede regurgitation of cecal contents into the ileum, a barium enema that fills the colon completely penetrates the terminal ileum to a variable extent.
(3) The terminal ileum is particularly vulnerable to herniation of the ileum into the cecum (intussusception). The resultant functional obstruction necessitates immediate surgical repair.
c. **Mesenteric support** of the cecum.
(1) The cecal portion of the large intestine is peritoneal (supported by a **mesocecum** of variable dimensions) and is somewhat variable in size.
(2) The cecum is also secondarily retroperitoneal to a variable extent.
(a) Folds of peritoneum define three cecal fossae (see Fig. 17-14).
(i) The **superior ileocecal fossa** lies between the mesentery proper and the **superior ileocecal (vascular) fold** of mesentery, which contains the **anterior cecal artery**.
(ii) The **inferior ileocecal fossa** lies between the mesoappendix and the **inferior (bloodless) ileocecal fold**.
(iii) The **retrocecal fossa** is formed by incomplete fusion between the cecal (visceral) peritoneum and the parietal peritoneum of the posterior abdominal wall.
(b) The appendix is frequently coiled within the retrocecal fossa.
(3) The cecum may be mobilized with its blood supply by incision along its lateral juncture with the parietal peritoneum.
d. The teniae coli, which converge onto the appendix, are useful landmarks in locating the appendix at operation.
e. The **arterial supply** to the cecum is from the **ileocolic artery** (see Fig. 17-13).
(1) The **ileocolic artery** is the terminal branch of the superior mesenteric artery.
(2) At the juncture of the ileum and cecum, the ileocolic artery divides into numerous branches.
(a) The **ascending colic artery** supplies the basal portion of the ascending colon.
(b) **Anterior** and **posterior cecal arteries** supply the cecum.
(c) The **appendicular artery** runs in the mesoappendix to supply the appendix.
(d) **Ileal branches** of the ileocolic artery provide the major blood supply to the terminal ileum.
(i) There are weak anastomotic connections between the ileocolic artery and the ileal arcades of the superior mesenteric artery.
(ii) In cecal resection with removal of the ileocolic artery, it is customary to remove the terminal ileum to avoid the possibility of ischemic necrosis.

2. The **vermiform appendix**.
a. The appendix is a narrow, hollow, muscular viscus that arises from the posteromedial aspect of the cecum about 2 to 3 cm below the ileocecal orifice.

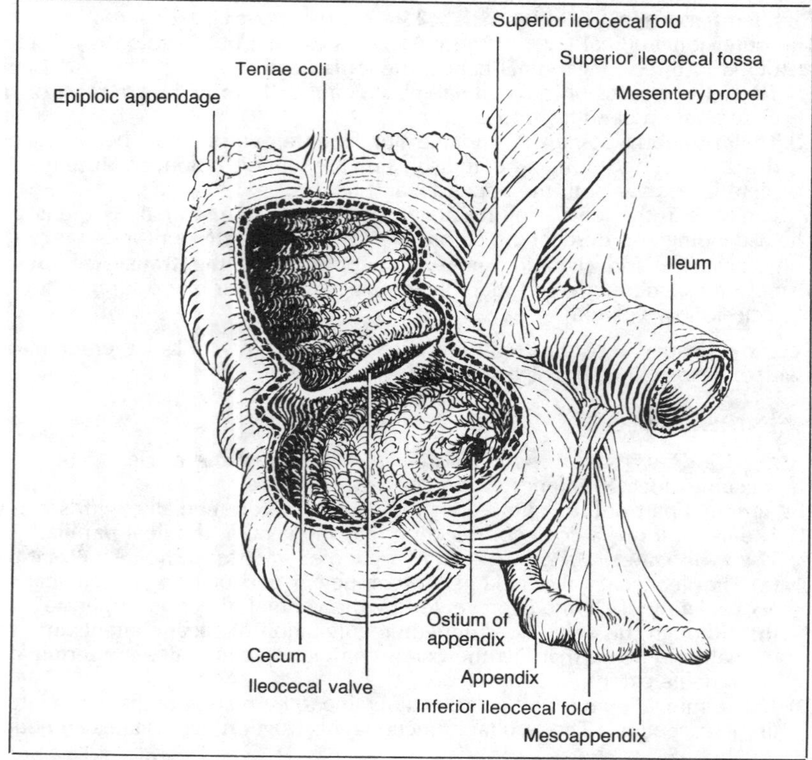

Figure 17-14. *The cecum and appendix.* The mesenteries and mesenteric folds supporting the cecum and appendix are shown as is the internal structure of the cecum.

b. The appendix represents the tip of the cecum, which fails to enlarge.
c. The appendix averages 9 cm (3.5 inches) in length, but may vary from 0.5 cm to more than 25 cm (0.25 to 10 inches).
d. A complete longitudinal muscular coat gives the appendix a smooth appearance.
e. Within the mucosa and submucosa there are large accumulations of lymphatic tissue.
f. The appendix is a peritoneal structure, suspended from the dorsal body wall by the **mesoappendix**.
  (1) The **appendicular artery**, a branch of the **ileocolic artery**, lies in the mesoappendix.
  (2) Because the appendix is supported by the **mesoappendix**, its position is extremely variable.
  (3) At operation, the appendix may be located in the retrocecal fossa (65%), in the iliac fossa (31%), in the right paracolic recess (2%), in the superior ileocolic fossa (1%), or in the inferior ileocolic fossa (1%).
g. The region of the appendix adjacent to the cecum has a thickened muscular coat and a slightly narrower lumen.
  (1) The luminal diameter varies with age.
    (a) The lumen tends to be rather wide (6 to 8 mm) in infants and youngsters.
    (b) It is often entirely obliterated by the time middle age is reached.
  (2) The lumen is dangerously narrow in adolescents and young adults. During this period, it may be occluded by a fecalith or even by edema and swollen lymphatic tissue associated with mild inflammation, explaining the high frequency of appendicitis in teenagers and young adults.
h. Clinical considerations.
  (1) The pathogenesis of appendicitis is poorly understood.
    (a) If the lumen is blocked, local bacteria multiply rapidly and their toxins produce inflammation, nausea, and fever.
    (b) The resultant distension of the appendix from gas or edema compromises the blood supply, which exacerbates the inflammatory process. Local ischemia with breakdown of the ischemic blood vessels produces intramural hemorrhage.
    (c) This situation is a surgical emergency. Delay results in gangrenous appendicitis, with danger of rupture and complicating peritonitis.

(d) Because the appendix is innervated by neurons of the lesser splanchnic nerve, initial colicky pain is referred to the 10th and 11th dermatomes, which include the periumbilical region.

(e) As the adjacent parietal peritoneum becomes inflamed, pain becomes exquisitely localized to the lower right quadrant, and the overlying musculature exhibits the reflex spasm of the acute abdomen (guarding, splinting). In *situs inversus*, pain may be on the left.

(2) The usual route for surgical access to the appendix is at McBurney's point, about 5 cm along a line from the anterior-superior iliac spine to the umbilicus. Surgeons frequently use McBurney's muscle-splitting incision for appendectomy.

3. The **ascending colon proper** (see Fig. 17-13).
   a. This section of the large bowel is 15 to 20 cm long (about 8 inches).
   b. The ascending colon is secondarily retroperitoneal and may be mobilized along with its mesenteric blood supply by incising along the line of peritoneal fusion.
      (1) Its relations are somewhat variable, depending on the degree to which the ascending colon is adherent to the peritoneum of the posterior abdominal wall.
      (2) Incomplete fixation can occur as a developmental anomaly.
   c. The ascending colon defines the **right paracolic gutter**, which connects the **supracolic** and **infracolic compartments**. Infection may track between the pelvic cavity and subhepatic recess along the right paracolic gutter (see Fig. 16-8).
   d. **Teniae coli** are remnants of the longitudinal muscle layer; contraction of these muscular strips forms the saccular **haustrations**.
   e. The ascending colon terminates at the **right colic (hepatic) flexure**, where the large intestine becomes peritoneal.
   f. The arterial supply to the ascending colon is via the **ileocolic** and **right colic arteries**.
      (1) The **ascending colic branch** of the **ileocolic artery** supplies the most inferior portion of the ascending colon.
      (2) The **right colic artery** is the most variable branch of the superior mesenteric artery.
         (a) The right colic artery may arise from the ileocolic artery or directly from the superior mesenteric artery.
         (b) This artery is completely absent in 18 percent of all individuals.
      (3) The **marginal artery** (Drummond's), an important anastomotic channel, lies in the mesentery adjacent to the colon (see Fig. 17-20).
         (a) The ascending colic branch of the ileocolic artery anastomoses with the descending colic branch of the right colic artery.
         (b) The descending and ascending branches of the right colic artery anastomose.
         (c) The ascending branch of the right colic artery anastomoses with the right branch of the middle colic artery.

C. THE TRANSVERSE COLON (see Fig. 17-13)

1. The second segment of large bowel begins at the **right colic (hepatic) flexure**, located anterior to the lower pole of the right kidney and the inferior segment of duodenum.

2. The transverse colon is a peritoneal structure suspended by the **transverse mesocolon** for 45 to 50 cm (18 to 20 inches).
   a. The location of the transverse mesocolon depends on the degree of fullness (Fig. 17-15).
   b. The **gastrocolic ligament** is formed by fusion of the superior surface of the transverse mesocolon to the inferior surface of the **greater omentum** (see Fig. 16-6).

3. The transverse colon is the most caudal portion of the GI tract to be supplied with parasympathetic innervation from the vagus nerve.

4. The blood supply to the transverse colon is via the **middle colic artery** (see Fig. 17-13).
   a. The middle colic artery is the first major branch of the superior mesenteric artery after the inferior pancreaticoduodenal artery.
   b. This artery lies within the transverse mesocolon, and its branches contribute to the **marginal artery** (see Fig. 17-20).
      (1) A right branch anastomoses with the ascending branch of the right colic artery or, if missing, the ascending branch of the ileocolic artery.
      (2) A left branch usually anastomoses with the ascending branch of the left colic artery from the inferior mesenteric artery. However, anastomoses may be poor or even absent in the vicinity of the splenic flexure.

D. THE DESCENDING COLON (Figs. 17-16 and 17-17)

1. This third segment of large bowel, the initial segment of the hindgut, begins at the **left colic (splenic) flexure** in the vicinity of the spleen and left kidney.

2. The descending colon is about 25 cm long (10 inches), terminating at the pelvic brim.
3. This segment is secondarily retroperitoneal but may be mobilized with its blood supply by incising along the line of peritoneal adhesion.
4. The descending colon defines the **left paracolic gutter** (see Fig. 16-8).
5. Because the bulk of feces decreases after absorption of water, the descending colon is smaller in diameter than the ascending colon.
6. The descending colon is especially prone to diverticulitis due to the more solid nature of its contents and to its function in fecal storage.
7. The blood supply to the descending colon is by the **left colic artery** off the **inferior mesenteric artery** (see Fig. 17-17).
   a. An ascending branch anastomoses inconsistently with the left branch of the middle colic artery.
   b. A descending branch anastomoses with the sigmoid arteries of the inferior mesenteric artery.
   c. These anastomoses contribute to the **marginal artery** (**Drummond's**) [see Fig. 17-20].
8. The innervation of the descending colon is by two pathways.
   a. The descending colon receives parasympathetic innervation from pelvic splanchnic nerves from S2–S4 and sympathetic innervation from the lumbar splanchnic nerves (L1–L2).
   b. The afferent nerves from the descending colon reach the spinal cord via lumbar splanchnic nerves (L1–L2), and those from the sigmoid colon and rectum reach the spinal cord via pelvic splanchnic nerves (S2–S4).
      (1) Pain originating within the descending colon is referred to the inguinal region and thigh.
      (2) Pain originating within the sigmoid colon and rectum is referred to the perineum and legs.

E. THE SIGMOID COLON (see Figs. 17-16 and 17-17)

1. The sigmoid colon begins where the colon becomes peritoneal, usually at the pelvic brim anterior to the left external iliac artery.
2. The length of the sigmoid colon ranges from 25 to 40 cm (10 to 15 inches).
3. The sigmoid colon is a peritoneal structure, suspended from the posterior abdominal wall by the **sigmoid mesocolon**.
   a. Its position is variable; a large loop may occupy the deep pelvis (see Fig. 17-16).
   b. The **intersigmoidal recess**, an occasional site for intra-abdominal herniation, is located at the juncture of the inferior surface of the sigmoid mesocolon and the dorsal body wall.
4. Anteriorly, the sigmoid colon is related to the urinary bladder in males and the posterior surface of the uterus and upper part of the vagina in females; posteriorly, it is related to the external iliac vessels and the sacrum.
5. The sigmoid colon becomes continuous with the **rectum** at the third sacral level.
6. The primary function of the sigmoid colon is the storage of feces.
7. The **blood supply** to the sigmoid colon is via the **sigmoid** and **rectosigmoid arteries** (see Fig. 17-17).
   a. There are from one to four sigmoid arteries; they are rich in arcades and supply the sigmoid colon.
      (1) They anastomose with the descending branch of the left colic artery, contributing to the **marginal artery** (see Fig. 17-20).
      (2) Arteriae rectae, penetrating incomplete longitudinal muscle layers, predispose the sigmoid colon to diverticulosis.
   b. The rectosigmoid artery supplies the terminal sigmoid colon and the superior rectum. It may or may not anastomose with the sigmoid and the superior rectal (hemorrhoidal) arteries.
   c. The region between the last sigmoid arcade and the rectosigmoid artery is known as the **critical point of Sudeck** because adequate anastomoses between these two arteries occur in only 52 percent of all individuals.
8. The **lymphatic drainage** of the terminal portion of the large bowel is of considerable importance due to the high frequency of colonic carcinoma.
   a. In general, the lymphatic drainage follows the arteries.
   b. There are regional nodes along the lines of drainage of the ileum, ascending colon, and

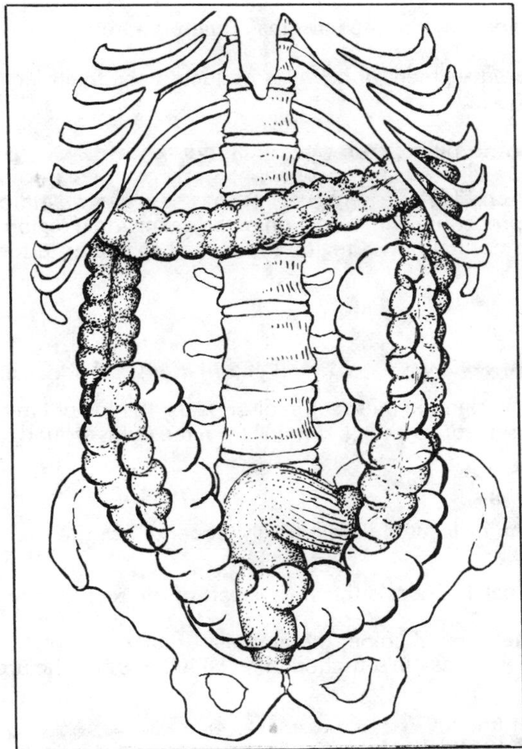

Figure 17-15. *Variation in the position of the transverse colon.*

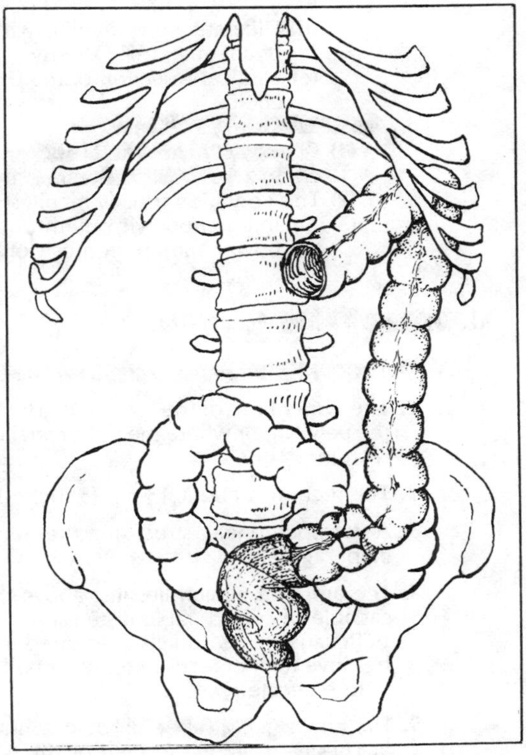

Figure 17-16. *Variation in the length and location of the sigmoid colon.*

transverse colon. The ileocolic, right colic, and middle colic nodes drain into the superior mesenteric nodes, and thence into the intestinal trunk, which empties into the thoracic duct.

c. The lymphatic channels of the descending colon are variable.

(1) The major lymph drainage from the descending and sigmoid colons is via the left colic nodes, which drain in turn into the inferior mesenteric nodes, the para-aortic nodes, and the thoracic duct.

(2) Lymph from the descending colon may drain to the celiac nodes via the lymphatic channels along the inferior mesenteric vein.

(3) Drainage also may be to the pelvic nodes via the lymphatics associated with the superior hemorrhoidal vessels.

(4) Thus, carcinoma of the descending colon, sigmoid colon, and rectum (the most frequent sites) may spread not only to the liver hematogenously, but to the lymph nodes of the abdominal and pelvic cavities.

9. Clinical considerations.

a. Diverticulosis.

(1) The sigmoid colon is the most common site of diverticula.

(2) Diverticula are small saccular protrusions of intestinal mucosa through the colonic wall along the margins of the teniae coli, usually adjacent to the penetrating blood vessels.

(3) Diverticula result from a combination of factors, such as an incomplete outer muscular layer of the large intestine and the penetration of the intestinal wall by arteries providing potential weak spots in the muscularis.

(a) The prevalence of diverticula increases with age. They are associated with increased intraluminal pressure brought about by straining at stool during chronic constipation.

(b) Diverticula may become occluded by fecaliths and may become inflamed (diverticulitis).

(c) An inflamed diverticulum may perforate, resulting in severe intra-abdominal bleeding and peritonitis.

(d) Ulcerative diverticulitis with bleeding into the intestinal lumen is a frequent cause of anemia in the elderly.
(e) Surgical resection of the diseased segment of bowel is frequently the treatment of choice.

b. Hirschsprung's disease.
(1) Congenital agenesis of the myenteric plexus in the terminal portion of the sigmoid colon precludes peristaltic activity and results in functional blockage with megacolon.
(2) This condition usually manifests as chronic constipation in neonates or young children.
(3) An abdominopelvic pull-through procedure with resection of the aganglionic region of the bowel and anastomosis of the normal bowel region to the anal sphincter restores function.

XI. THE RECTUM (Fig. 17-18)

A. The rectum is that part of the large intestine between the sigmoid colon and the anal canal.

1. The upper limit of the rectum is where the sigmoid mesocolon disappears; the lower limit is the pelvic floor where the puborectal muscle forms a sling, running from the pubis around the rectum.
2. The rectum is about 13 cm (5 inches) long.
3. A retroperitoneal structure, the rectum is located immediately superior to the pelvic diaphragm.
4. The **ampulla**, which lies just above the pelvic floor, is the widest part of the rectum and is capable of considerable distension.
   a. Because feces usually are stored in the sigmoid colon, the ampulla usually is empty.
   b. Movement of feces into the ampulla generates the sensation of rectal fullness and the urge to defecate.
5. The rectum pierces the pelvic diaphragm (the levator ani muscle) to become the anal canal.
   a. The **rectal sling** is formed by the **puborectal muscle**, which is the most medial portion of the **pubococcygeal muscle** of the **levator ani** group.

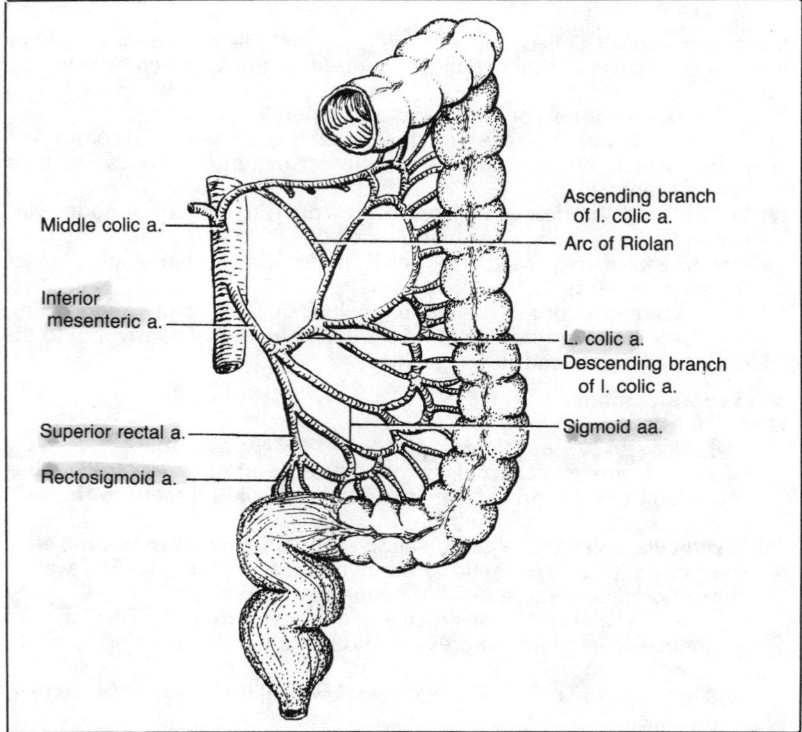

Figure 17-17. *The inferior mesenteric arterial distribution.* Anastomoses between the middle colic and left colic arteries via an arc of Riolan is inconsistent.

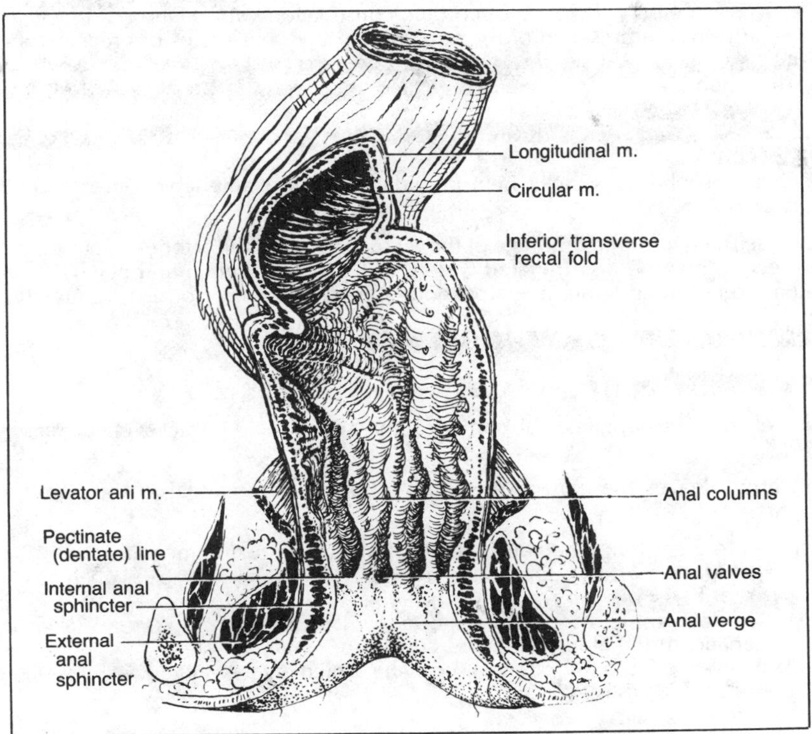

Figure 17-18. *The rectum and anal canal.*

(1) The puborectal muscle fibers loop posterior to the rectum, with some fibers attaching to the rectum.
(2) Tension in this muscle cants the rectum forward.
b. The rectal sling is the single most important factor in fecal continence.

B. Three transverse **rectal folds** (Houston's valves) are formed by the inner three layers of the intestinal wall within the rectum.

1. The **inferior rectal fold** projects from the left side about 2 cm (1 inch) above the anal canal and can be palpated digitally.
2. The **middle rectal fold** projects from the right about 2 cm (1 inch) above the inferior fold.
3. The **superior rectal fold** projects from the left about 2 cm (1 inch) above the middle fold.
4. The high incidence of rectal, colonic, and prostatic carcinoma makes digital and sigmoidoscopic examination important in the postmiddle-aged male.
   a. Most rectal cancers can be discovered by digital palpation.
   b. Carcinomatous spread from the rectum, posteriorly, may involve the sacral plexus, with pain distribution down the leg and over the perineum. Anterior spread may involve pelvic viscera, such as the prostate, bladder, uterus, or vagina.
5. Prolapse of the rectum through the anus is often related to damage of the levator ani muscle, usually the result of obstetric trauma.

C. The arterial supply, venous return, and lymphatic drainage of the rectum are of considerable importance.

1. The rectum receives branches from the inferior mesenteric artery via the **superior rectal (hemorrhoidal) artery**, from the internal iliac artery via **middle rectal (hemorrhoidal) arteries**, as well as from the internal pudendal arteries via **inferior rectal (hemorrhoidal) arteries**. There are abundant anastomotic connections in the submucosa between tributaries of these arteries (Fig. 17-19).
2. The **venous return** is along similarly named vessels that have especially rich anastomotic connections.

a. Portal hypertension (accompanying cirrhosis) or local compression of the inferior mesenteric artery (because of chronic constipation with dilation of the sigmoid colon or because of the presence of the fetus during the later stages of pregnancy) results in the shunting of blood from the rectum and anal canal to the systemic venous system.

b. This shunting results in variceal dilation of the submucosal anal and perianal venous plexuses (hemorrhoids).

c. Hemorrhoidal varices may rupture during the evacuation of feces, producing considerable blood loss.

d. Unlike esophageal varices, hemorrhoidal varices are amenable to treatment by ligature or cautery.

3. Because the **lymphatic drainage** of the rectum parallels the arteries, metastatic carcinoma of the rectum may be disseminated widely within the abdomen and pelvis, as well as the inguinal nodes. In addition, hematogenous spread to the liver occurs frequently.

XII. THE VASCULAR SUPPLY TO THE GI TRACT

A. THE CELIAC ARTERY (TRUNK, AXIS) [see Fig. 17-3]

1. The celiac artery supplies that portion of the abdominal GI tract derived from the primitive foregut.

2. This artery leaves the aorta in the midventral line just caudal to the aortic hiatus of the diaphragm.

3. Only 0.5 to 3 cm long, the celiac artery trifurcates almost immediately.

a. The **left gastric artery** supplies the right superior region of the stomach as well as the abdominal portion of the esophagus (see Fig. 17-3).

(1) The left gastric artery anastomoses with the right gastric artery (from the common hepatic artery) along the lesser curvature.

(2) It gives off the **esophageal branch**, which anastomoses with the midesophageal branches of the aorta.

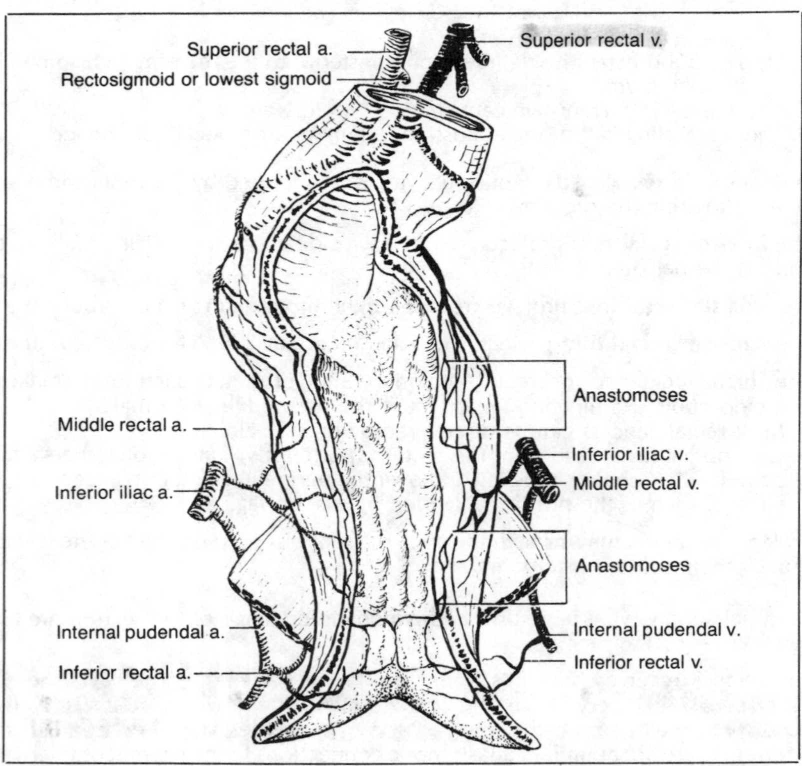

Figure 17-19. *The vasculature of the rectum and anal canal.* The arterial supply from the superior mesenteric, internal, and external iliac arteries is depicted on the right side. The parallel venous drainage is shown on the opposite side.

(3) This artery provides the pathway for the **posterior vagus nerve** to reach the **celiac plexus**.

(4) It may give rise to an aberrant **left hepatic artery**.

b. The **splenic artery** supplies the spleen and the entire left side of the stomach, as well as portions of the pancreas and greater omentum (see Figs. 17-3 and 17-10).

(1) **Short gastric arteries** supply the left superior portion of the stomach and anastomose with the arteries of the esophagus.

(2) The **left gastroepiploic artery** supplies the left inferior region of the stomach and portions of the greater omentum; it anastomoses with the right gastroepiploic artery along the greater curvature.

(3) Several **pancreatic arteries** supply the tail of the pancreas.

c. The **common hepatic artery** supplies the liver and gallbladder, as well as portions of the stomach, duodenum, and pancreas (see Figs. 17-3 and 17-9).

(1) The **right gastric artery** supplies the right inferior region of the stomach and anastomoses with the left gastric artery along the lesser curvature.

(2) The **gastroduodenal artery** gives off two major branches.

(a) The right gastroepiploic artery supplies the right inferior region of the stomach and a portion of the greater omentum, and anastomoses with the left gastroepiploic artery (from the splenic artery) along the greater curvature.

(b) The superior pancreaticoduodenal artery supplies the upper portion of the duodenum and the head of the pancreas.

(i) This artery divides into anterior-superior and posterior-superior branches.

(ii) It anastomoses with the anterior-inferior and posterior-inferior branches of the inferior pancreaticoduodenal arteries which, in turn, arise from the superior mesenteric artery.

(3) The **proper hepatic artery** is the continuation of the common hepatic artery.

(a) The **left hepatic artery** supplies the left lobe, the quadrate lobe, and the left half of the caudate lobe.

(b) The **right hepatic artery** supplies the right lobe, the right half of the caudate lobe, and the gallbladder.

(i) The **cystic artery** is usually a branch of the right hepatic artery.

(ii) The cystic artery bifurcates into the **anterior cystic artery** and the **posterior cystic artery**.

B. THE SUPERIOR MESENTERIC ARTERY (see Figs. 17-10 and 17-13)

1. The **superior mesenteric artery** supplies that portion of the GI tract derived from the primitive midgut.

2. This artery leaves the aorta in the midventral line a few centimeters caudal to the celiac artery. It gives off numerous branches.

a. The **inferior pancreaticoduodenal artery** divides into **anterior** and **posterior** branches, which anastomose with anterior and posterior branches of the superior pancreaticoduodenal artery and supply the descending and inferior portions of the duodenum, as well as the head of the pancreas.

b. **Jejunal** and **ileal branches** provide abundant collateral circulation to the peritoneal portion of the small bowel.

c. The **ileocolic artery** is the terminal branch of the superior mesenteric artery. Its branches include the **anterior** and **posterior cecal arteries**, the **appendicular artery**, the **ascending colic artery**, and the **ileal branches**.

d. The **right colic artery**, which is frequently missing, supplies the ascending colon with great variation.

e. The **middle colic artery** supplies the transverse colon.

3. Stenosis of the vascular supply may produce intestinal angina (colicky abdominal pain referred to the umbilical region).

4. Due to the pattern of blood supply, occlusion of the superior mesenteric artery by a thrombus results in infarction of the small intestine and of the large intestine as far as the vicinity of the splenic flexure.

C. THE INFERIOR MESENTERIC ARTERY (see Fig. 17-17)

1. The **inferior mesenteric artery** supplies that portion of the GI tract derived from the primitive hindgut.

a. The **left colic artery** supplies the descending colon; the ascending branch anastomoses with the left branch of the middle colic artery and the descending branch anastomoses with the sigmoid arteries.

b. The **sigmoidal arteries** supply the sigmoid colon and anastomose with the descending branch of the left colic artery.

c. The **rectosigmoid arteries** supply the terminal sigmoid colon and superior rectum, variably anastomosing with sigmoid and superior hemorrhoidal arteries at the critical point (Sudeck's).

d. The **superior hemorrhoidal (rectal) artery** supplies the upper rectum; this artery has poor anastomotic connections with sigmoid arteries.

2. Stenosis or occlusion of the inferior mesenteric artery results in ischemia of the descending colon and sigmoid colon.

D. THE MARGINAL ARTERY (Fig. 17-20)

1. Anastomoses between branches of the **superior mesenteric** and **inferior mesenteric arteries** result in the formation of a continuous arterial trunk, the **marginal artery** (Drummond's).

2. The marginal artery lies in the mesentery, close to the border of the large intestine, and runs from the ileocecal valve to the rectosigmoid junction.

3. There are several so-called "critical points" at which the marginal artery may not provide adequate collateral supply:

a. Anastomoses between the ileal branches of the superior mesenteric artery and the ileal branches of the ileocolic artery seldom provide adequate collateral circulation in the ileocecal region.

b. There may be few anastomotic connections between the left branch of the middle colic artery and the ascending branch of the left colic artery in the vicinity of the splenic flexure.

(1) Occlusion or ligature of one of these vessels may result in ischemic necrosis of a segment of large bowel.

(2) Riolan's arch occasionally provides a strong anastomotic connection between the inferior mesenteric artery and the left branch of the middle colic artery (see Fig. 17-17).

c. Anastomotic connection between the lowest sigmoid branch and the upper rectal branch of the superior rectal (hemorrhoidal) artery is sometimes by a very small vessel incapable of providing adequate collateral circulation in the vicinity of Sudeck's critical point.

XIII. THE HEPATIC PORTAL SYSTEM

A. The hepatic portal system begins as the venous capillaries of the GI tract and ends as the venous sinusoids in the liver; this system delivers nutrient-rich (sometimes toxic) blood to the liver.

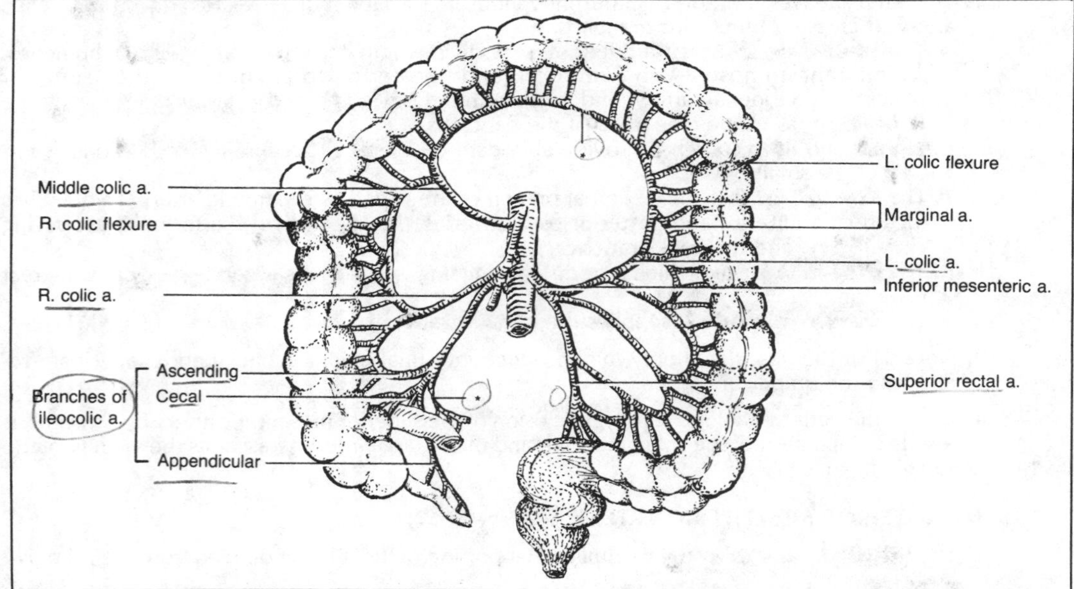

Figure 17-20. *The marginal artery.* Anastomoses occur between the branches of the superior mesenteric arteries as well as between the branches of the inferior mesenteric arteries. Inconsistent anastomoses occur in the regions of the ileocecal junction and the splenic flexure (*).

B. THE HEPATIC PORTAL VEIN

1. The hepatic portal vein lies in the **hepatoduodenal ligament** just posterior and slightly to the left of the common bile duct and the proper hepatic artery (see Fig. 17-9).

2. This vein receives three major veins and numerous smaller veins (Fig. 17-21).
   a. The **superior mesenteric vein** drains the small intestine and large intestine as far as the splenic flexure.
      (1) The superior mesenteric vein generally parallels and drains the regions supplied by the superior mesenteric artery.
      (2) It is the largest contributor to the hepatic portal vein.
   b. The **inferior mesenteric vein** drains the descending and sigmoid colons.
      (1) This vein generally parallels and drains the regions supplied by the inferior mesenteric artery.
      (2) It usually enters the superior mesenteric vein but may enter the splenic vein.
   c. The **splenic vein** drains the spleen, as well as portions of the stomach and pancreas.
      (1) The splenic vein generally parallels the splenic artery and may be embedded in pancreatic parenchyma.
      (2) This vein joins with the superior mesenteric vein to form the hepatic portal vein.
   d. Small veins corresponding to the named branches of the celiac artery drain directly into the hepatic portal vein.

C. HEPATIC VEINS

1. The hepatic sinusoids coalesce into interlobular veins, which come together to form the hepatic veins.

2. The hepatic veins may join before emptying into the inferior vena cava, or they may enter separately. There are usually three hepatic veins, draining the left, right, and quadrate lobes of the liver.

D. CLINICAL CONSIDERATIONS

1. **Portal hypertension.**
   a. Because the hepatic portal system has no valves, blood need not flow toward the liver.
   b. Liver disease (such as cirrhosis) or compression of a vein (as in pregnancy or constipation) results in blood shunting through the anastomotic connections in the systemic venous system.

2. **Esophageal varices.**
   a. The branches of the **left gastric vein** anastomose profusely with the **azygos**, as well as the **hemiazygos, veins**.
   b. With portal hypertension, increased shunting through the esophageal anastomoses results in **esophageal varices** within the mucosa of the esophagus.
      (1) A varix may rupture when a large bolus is swallowed, or upon emesis or the passage of diagnostic instrumentation.
      (2) Uncontrollable and frequently fatal hemorrhage results from variceal rupture.
      (3) Hematemesis associated with esophageal varices is the most common terminal event in alcohol-related liver disease.

3. **Caput medusae.**
   a. The **periumbilical veins** in the falciform ligament (and potentially even a recannulized **ligamentum teres**) anastomose abundantly with **superior** and **inferior epigastric veins** and other veins of the anterior abdominal wall.
   b. Although the grossly dilated veins of a caput medusae appear to be rare, this condition is always present in portal hypertension and can be demonstrated by infrared photography.
   c. These periumbilical anastomoses may bleed profusely during midline surgical approaches through the abdominal wall or during attempts to correct epigastric hernia.

4. **Hemorrhoids.**
   a. The **superior rectal vein** anastomoses profusely with the **middle rectal vein**, which is a branch of the **internal iliac vein**.
   b. In addition, the **middle rectal vein** anastomoses with the **inferior rectal vein**, which is a branch of the **internal pudendal vein**.
   c. Increased shunting through these anastomotic channels results in varices within the submucosa of the terminal rectum (**internal hemorrhoids**) and of the anal canal (**external hemorrhoids**).

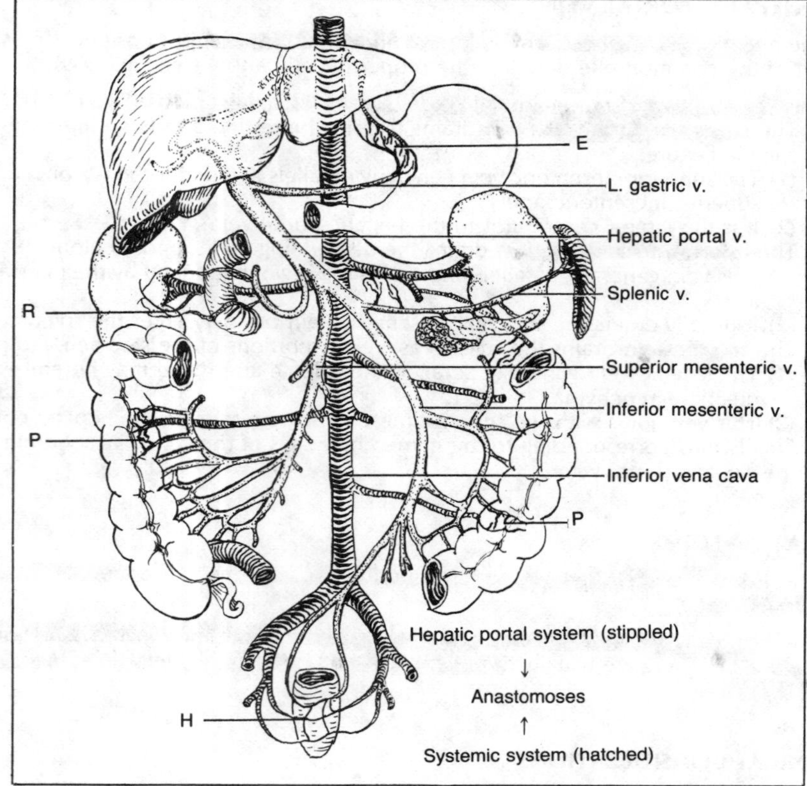

Figure 17-21. *The hepatic portal system.* Most of the blood returns from the gut through the liver. Anastomoses occur in several regions where varix development is possible. *E*, Esophageal veins; *H*, hemorrhoidal veins; *P*, veins of Retzius between secondarily retroperitoneal structures and the veins of the abdominal wall; and *R*, veins between the duodenum and spleen with renal veins.

5. **Veins of Retzius.**
   a. The veins of a secondarily retroperitoneal structure may anastomose with veins of the dorsal body wall, forming **veins of Retzius.**
   b. These varices usually are silent and produce few problems. However, they may bleed profusely when a secondarily retroperitoneal structure is mobilized surgically.
6. The life-threatening **sequelae of portal hypertension** are amenable to surgical correction to provide routes for visceral blood to bypass the obstructed liver.
   a. Surgical creation of a portacaval fistula produces a side-to-side anastomosis between the hepatic portal vein and the inferior vena cava, across the epiploic foramen.
   b. A splenorenal anastomosis is also a common procedure. Splenectomy is followed by end-to-side anastomosis of the splenic vein to the left renal vein.
   c. Although shunts reduce the dangers inherent in varices, toxic blood shunted around the obstructed liver has a long-term effect on the central nervous system, causing disorders such as the encephalopathy resultant from alcohol-related liver disease.

XIV. LYMPHATIC DRAINAGE OF THE GUT

A. THE LYMPHATIC CHANNELS generally follow the vascular system.

1. Lymph nodes are interposed in the lymphatic channels.
2. These nodes are of great importance due to lymphatic spread of carcinoma of the GI tract.

B. THE CELIAC NODES

1. **Celiac nodes** are located around the celiac axis and drain into the **intestinal lymph trunk**, which, in turn, drains into the **cisterna chyli**.

2. There are four drainage areas about the stomach.
   a. **Left gastric nodes.**
      (1) These are located along the lesser curvature of the stomach.
      (2) The left gastric nodes generally drain along left gastric vessels to the celiac nodes.
      (3) Because they communicate with esophageal lymphatics, the left gastric nodes provide a potential route for spread of carcinoma from the stomach to the esophagus.
   b. **Gastroepiploic nodes.**
      (1) This group is located along the inferior portion of the greater curvature, pyloric antrum, and pyloric canal.
      (2) The drainage is generally along the right gastroepiploic vessels toward the pyloric and celiac nodes.
      (3) This pathway gains significance from the high prevalence of gastric carcinoma (the most common carcinoma in the region), as well as the high prevalence of duodenal and pancreatic carcinoma.
   c. **Splenic nodes.**
      (1) These are located along the lateral portion of the greater curvature and fundus.
      (2) They generally drain along left gastroepiploic and splenic vessels to the celiac nodes.
   d. **Right gastric nodes.**
      (1) This group of nodes is located along the superior region of the pyloric antrum and the pyloric canal.
      (2) The drainage is along the right gastric vessels to the celiac and hepatic nodes.

C. THE SUPERIOR MESENTERIC NODES

1. Superior mesenteric nodes are located along the root of the superior mesenteric artery.
   a. These nodes receive drainage from the intermediate nodes, which are intercalated in lymph pathways along the named branches of the superior mesenteric artery.
      (1) **Mesenteric nodes** are associated with the small intestine.
      (2) **Ileocolic nodes** are associated with the cecum and appendix.
      (3) **Right colic nodes** are associated with the ascending colon.
      (4) **Middle colic nodes** are associated with the transverse colon.
   b. Superior mesenteric nodes receive drainage from the duodenum and pancreas.

2. There are numerous anastomotic connections with the celiac and inferior mesenteric nodes.

3. The superior mesenteric nodes and the celiac nodes give rise to the **intestinal trunk**, which drains into the **cisterna chyli**.

D. THE INFERIOR MESENTERIC NODES

1. Intermediate nodes intercalated in lymph pathways along the named vessels include the following:
   a. **Left colic nodes** are associated with the descending and sigmoid colon.
   b. **Superior rectal nodes** are associated with the rectum.
      (1) There are profuse anastomotic connections with:
         (a) **Middle rectal nodes**, which drain along the internal iliac vessels to the hypogastric nodes.
         (b) **Inferior rectal nodes**, which drain along the external iliac vessels to the inguinal and hypogastric nodes.
      (2) Carcinoma of the rectum may metastasize along abdominal as well as pelvic lymph channels, resulting in wide dissemination of the tumor.

2. **Inferior mesenteric nodes** drain into the superior mesenteric nodes. There is also considerable communication with the celiac nodes (from lymphatics, which follow the inferior mesenteric vein rather than the inferior mesenteric artery), as well as with the middle colic nodes.

E. THE CISTERNA CHYLI

1. The cisterna chyli receives the celiac and intestinal trunks and drains the **hypogastric nodes** from the systemic lymphatics of the lower body wall, lower extremities, and pelvis.

2. It drains into the **thoracic duct**, which empties into the left subclavian vein at the base of the neck.

XV. INNERVATION OF THE GI TRACT

A. THE AUTONOMIC NERVOUS SYSTEM, which innervates the GI tract, is subdivided into two divisions:

1. The **parasympathetic (craniosacral) division** (Fig. 17-22A):
   a. The **vagus nerve (cranial nerve X)** brings parasympathetic presynaptic neurons to the GI tract almost as far as the splenic flexure of the transverse colon.
      (1) The vagus reaches abdominal viscera via the esophageal plexus.
      (2) Left and right vagal plexuses consolidate into anterior and posterior vagal trunks as they pass through the diaphragm at the esophageal hiatus.
         (a) The **anterior vagal trunk** (left vagus nerve) courses along the anterior surface of the abdominal esophagus.
            (i) **Hepatic branches** course through the lesser omentum, to the **hepatic plexus** on the hepatic artery, to innervate the liver, gallbladder, and bile duct. Some fibers course retrograde along the hepatic artery to join the **celiac plexus** and innervate portions of the duodenum and the sphincter of Oddi.
            (ii) **Gastric branches** continue along the anterior surface of the stomach as far as the pylorus to innervate the smooth muscle and glands of the stomach.
            (iii) The **celiac branch** courses along the left gastric artery, contributing to and coursing within the **celiac plexus** and the **superior mesenteric plexus**. It supplies the small and large intestines approximately as far as the splenic flexure by coursing along the arteries.
         (b) The **posterior vagal trunk** (the right vagus nerve) courses along the posterior surface of the esophagus and stomach.
            (i) **Gastric branches** continue along the posterior surface of the stomach, not quite reaching the pylorus, to innervate smooth muscle and glands of the stomach.
            (ii) The **celiac branch** courses along the left gastric artery, contributing to and coursing through the **celiac plexus** and **superior mesenteric plexus**. It supplies the small and large intestines approximately as far as the splenic flexure by coursing along the arteries.
            (iii) The posterior vagal trunk does not appear to give rise to any hepatic branches.
      (3) **Functional considerations.**
         (a) The vagus nerve is composed of *pre*ganglionic (*pre*synaptic) neurons.
            (i) These neurons synapse with *post*ganglionic (*post*synaptic) neurons within the enteric plexuses.
            (ii) Parasympathetic presynaptic neurons utilize acetylcholine as the neurotransmitter (cholinergic transmission).
         (b) The *post*ganglionic neurons are located in the enteric plexuses within the walls of the gut.
            (i) These neurons are very short, intramural fibers.
            (ii) They synapse with effectors, such as smooth muscle cells and gland cells.
            (iii) Parasympathetic postsynaptic neurons are cholinergic.
         (c) The parasympathetic division functions primarily to promote digestion.
            (i) Vagal activity increases acid, alkaline, and enzymatic secretion from the various regions of the GI tract.
            (ii) It enhances peristaltic activity and decreases sphincteric tone.
      (4) **Clinical considerations.**
         (a) Selective vagotomy (section of the gastric branches of the anterior and posterior vagal trunks) reduces peptic secretion. This procedure is sometimes utilized as treatment for chronic gastric and duodenal ulcers.
         (b) Total vagotomy (section of the anterior and posterior vagi on the surface of the abdominal esophagus) has other sequelae, such as a decreased rate of stomach emptying, gallbladder dilation, biliary stasis with predisposition to stone formation, and dumping syndrome.
   b. The **nervi erigentes (sacral nerve, pelvic splanchnic nerves)** represent the sacral portion of the parasympathetic division of the autonomic nervous system.
      (1) The nervi erigentes arise from S2, S3, and S4 (see Fig. 17-22B) and innervate the lower gastrointestinal tract approximately between the splenic flexure and the pectinate line of the anal canal.
      (2) These nerves run in the adventitia of the rectum and contribute to the **lateral pelvic plexus**.
         (a) These fibers course retrograde along the rectum, sigmoid colon, and descending colon.
         (b) Other fibers course retrograde in the **hypogastric plexus** as far as the branches of the inferior mesenteric artery, which they follow to the descending bowel.
      (3) The *pre*ganglionic neurons of the pelvic splanchnic nerves synapse with *post*ganglionic fibers in enteric plexuses, utilizing cholinergic neurotransmission.
      (4) The *post*synaptic neurons have their cell bodies in enteric plexuses.
         (a) These *post*ganglionic neurons synapse with effectors, such as smooth muscle cells and secretory cells.

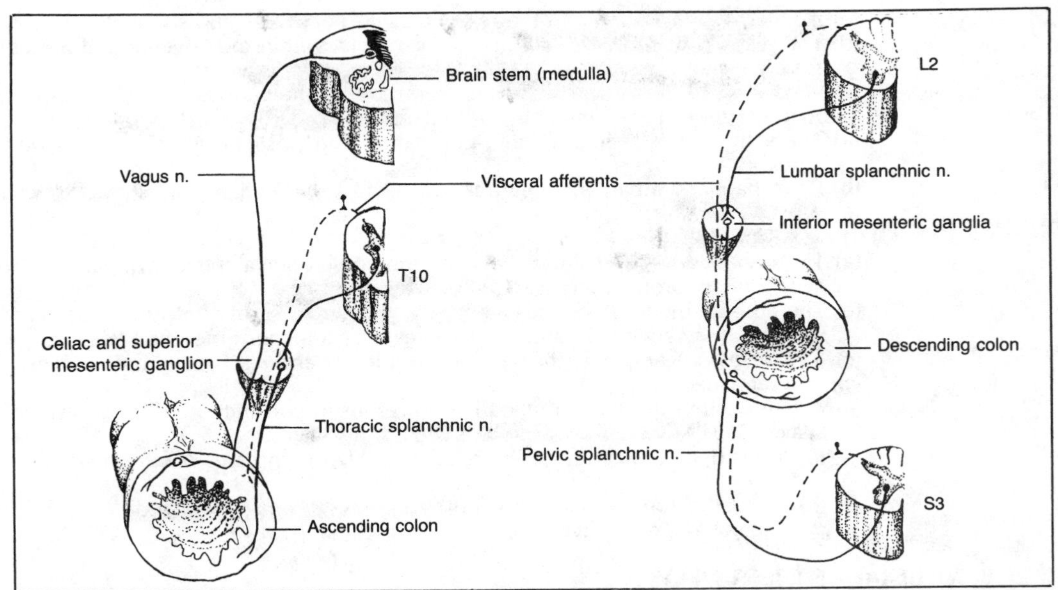

Figure 17-22. *Innervation of the gastrointestinal tract.* Preganglionic sympathetic neurons (*solid cell body*) in the thoracic spinal cord (levels T5–L2) send axons through the ventral roots, spinal nerves, white rami communicantes, the sympathetic chain, and splanchnic nerves to gain access to prevertebral ganglia. The postsynaptic sympathetic neurons (*open cell body*) of the prevertebral ganglia send their axons to the effectors in the gut walls. The cranial presynaptic parasympathetic neurons (*solid cell body*) in the brain send axons along the vagus nerve (CN X) to the enteric ganglia within the walls of the viscera as far as the splenic flexure. The sacral presynaptic parasympathetic neurons (*solid cell body*) in the spinal cord (levels S2–S4) send axons through the spinal nerves and nervi erigentes to distribute to the enteric ganglia of the lower colon. The short parasympathetic postganglionic neurons (*open*) of the enteric ganglia send axons to the effectors. Afferent axons (*dashed*) which convey painful stimuli from the gut, travel along the autonomic pathways to both thoracolumbar and pelvic segments of the spinal cord.

(b) The parasympathetic *post*synaptic neurons are also cholinergic.

(5) Clinical considerations.

(a) Aganglionosis (Hirschsprung's disease) is a genetic defect wherein the parasympathetic ganglia do not develop in the walls of the descending colon.

(b) In this condition, there is loss of initiation of propulsive function within the sigmoid colon—a functional blockage—with ensuing megacolon. With extreme stretching of the colon, ischemia, with abdominal signs, results. Toxic megacolon is usually treated surgically.

2. The **sympathetic (thoracolumbar) division** (see Fig. 17-22).

a. *Pre*ganglionic sympathetic fibers arise from levels T1 to L2, or sometimes, L3.

(1) The **greater splanchnic nerve** arises from T5 to T9 and synapses with postganglionic fibers in the **celiac ganglion**.

(2) The **lesser splanchnic nerve** arises from T10 and T11 and generally synapses with postganglionic fibers in the **superior mesenteric ganglion**.

(3) The **least splanchnic nerve** arises from T12, may be incorporated into the lesser splanchnic nerve, and synapses with postganglionic fibers in the **aorticorenal ganglion**.

(4) Lumbar splanchnic nerves arise from L1 to L2 (occasionally L3) and synapse with postganglionic fibers in small ganglia located in the **aortic plexus**.

(5) All sympathetic *pre*ganglionic neurons utilize acetylcholine as their neurotransmitter (cholinergic transmission).

b. *Post*ganglionic sympathetic neurons, which innervate the GI tract, lie in the prevertebral ganglia.

(1) The prevertebral ganglia are paired bilaterally and are variable in size—sometimes fused on either side to a variable extent.

(a) Celiac ganglia are located in the **celiac plexus** and supply sympathetic nerves to that portion of the viscera associated with the celiac artery.

(b) Superior mesenteric ganglia are located in the **celiac** or **superior mesenteric plexuses** and supply sympathetic nerves to that portion of the viscera associated with the superior mesenteric artery.

(c) **Aorticorenal ganglia** lie in the **celiac** or **aortic plexuses** and supply the kidneys.
(d) The **ganglia of the aortic plexus** (small ganglia within the adventitia of the aorta) supply viscera associated with the inferior mesenteric artery.

(2) The postganglionic sympathetic fibers contribute to the various named plexuses and follow the arteries to the effectors, usually the musculature of blood vessels.
(a) These fibers are generally long, compared with postganglionic parasympathetic fibers.
(b) Postsynaptic sympathetic neurons are adrenergic, utilizing noradrenalin as the neurotransmitter.

(3) The **adrenal medulla**.
(a) The cells of the adrenal medulla are embryonic equivalents of sympathetic postganglionic neurons; both are derived from neural crest tissue.
(b) The adrenal medulla secretes adrenaline and noradrenalin into the bloodstream, whereby they act on distant effector sites over a longer time period.
(c) It receives preganglionic innervation from lesser and least splanchnic nerves.

(4) **Functional aspects.**
(a) Activity within the sympathetic division results in constriction of blood vessels.
(b) Sympathectomy has no appreciable effect on the gut.
(i) It has been utilized to promote GI vasodilation in cases of extreme hypertension.
(ii) Sympathectomy has also been utilized to relieve intractable pain, because the visceral afferents travel along the sympathetic pathways.

B. VISCERAL AFFERENT NERVES

1. The visceral afferent nerves are not considered part of the autonomic nervous system, although they do share the pathways.

2. As a general rule, parasympathetic pathways carry reflex afferents, whereas sympathetic pathways convey pain afferents.
a. Afferent reflex neurons travel along with the vagus fibers to the brain (the gastric filling reflex), as well as along pelvic splanchnics to the spinal cord (the micturition reflex).
b. Pain and pressure afferents travel along the thoracic splanchnics (sympathetic), lumbar splanchnics (sympathetic), and pelvic splanchnics (parasympathetic—the exception to the rule) to the spinal cord.
(1) This arrangement explains the basis for **referred pain**.
(a) Nociceptive input from the viscera at any thoracic or lumbar level cannot be distinguished from that of somatic origin.
(b) The pain is referred to the somatic dermatome associated with the particular splanchnic nerve.
(2) Knowing approximate visceral innervation patterns is essential to understanding referred pain and making accurate diagnoses.

3. Because the pain fibers run along the sympathetic pathways, sympathectomy is undertaken to relieve visceral pain, as in untreatable carcinoma.

4. A summary of referred pain patterns is found in Table 17-2.

Table 17-2. Summary of Visceral Afferent Pathways

Organ	Afferent Pathway	Vertebral Level	Referred To
Heart	Middle inferior cardiac nerves Thoracic splanchnics	Cervical ganglia, thence T1-T2 T1-T4	Shoulder, postaxial arm, high thorax, mid-thorax postaxial arm
Esophagus Stomach Gallbladder Liver Bile duct Superior duodenum	Celiac plexus and great splanchnic nerve	T5-T9	Low thorax, epigastric region
Inferior duodenum Jejunum Ileum Appendix Ascending colon Transverse colon	Superior mesenteric plexus and lesser splanchnic nerve	T10-T11	Umbilical region
Kidneys High ureters Gonads	Aorticorenal plexus and least splanchnic nerve, upper lumbar splanchnic nerves	T12-L1	Lumbar and inguinal regions (unilaterally) (unilaterally)
Descending colon Sigmoid colon Midureter Uterus	Aortic plexus and lumbar splanchnics	L1-L2	Pubic and inguinal regions, anterior scrotum, or labia; thigh
Cervix Bladder Low ureters Rectum Superior anal canal	Pelvic plexus and pelvic splanchnics	S2-S4	Leg, foot, perineum

18
The Kidneys and Posterior Abdominal Wall

I. THE KIDNEYS

A. INTRODUCTION

1. The paired kidneys are located posteriorly in the abdominal cavity.
- **a.** The kidneys are retroperitoneal, embedded in a considerable amount of perinephric and paranephric fat.
- **b.** About the size of a clenched fist, each kidney normally weighs 150 ± 25 g in men and 135 ± 20 g in women.

2. Relations.
- **a.** Each kidney:
  - **(1)** Lies on the ventral surface of the quadratus lumborum muscle, just lateral to the psoas muscle and vertebral column.
  - **(2)** Is in contact posterosuperiorly with the inferior surface of the diaphragm.
  - **(3)** Is capped anteromedially by an adrenal gland.
- **b.** The right kidney:
  - **(1)** Is 2 to 8 cm (1 to 2 inches) lower than the left, due to the presence of the liver on the right side.
  - **(2)** Is usually related to the 12th rib, posteriorly.
  - **(3)** Is related to the liver, duodenum, and hepatic flexure of the colon, anteriorly.
- **c.** The left kidney:
  - **(1)** Is related to the 11th and 12th ribs, posteriorly.
  - **(2)** Is related to the pancreas, spleen, and splenic flexure of the colon, anteriorly.

B. THE RENAL (PERIRENAL) FASCIA (Gerota's; false capsule)

1. A discrete fascial layer surrounds each kidney (see Fig. 18-5).
- **a.** This layer seems to arise from the prevertebral fascia and passes anterior to the vertebral column.
- **b.** This fascial layer splits into two leaflets.
  - **(1)** Lateral and superior to each kidney the leaflets fuse, thereby partially enclosing each kidney with its associated adrenal gland.
  - **(2)** The renal fascia divides the fat associated with the kidney into two distinct regions.
    - **(a) Perirenal (perinephric) fat** is contained within the renal fascia.
    - **(b) Pararenal (paranephric) fat** is external to the renal fascia; it is found most abundantly posterolaterally.
  - **(3)** Although the renal fascia fuses around the renal vessels, it is open inferiorly along each ureter.
    - **(a)** Infection or renal abscess may be limited by the renal fascia and may track between the leaflets of this fascia into the pelvis.
    - **(b)** Air injected into the retroperitoneal tissue of the pelvis will rise within the renal fascia to surround the kidney (insufflation), enabling radiographic visualization of the kidneys and adrenal glands.

2. Mobility of the kidneys.
- **a.** Normally, kidneys do not exceed a 9-cm (3- to 4-inch) excursion when an individual changes position from supine to erect.
- **b.** The position of the kidneys is maintained primarily by deposits of paranephric and perinephric fat.

(1) This fat (similar to periorbital and ischiorectal fat) is very slow to be reabsorbed in wasting disease or acute starvation.

(2) Renal vessels provide some support, although it is uncertain how much.

c. **Nephroptosis** (floating kidney).

(1) This condition occurs frequently among truck drivers, horseback riders, and motorcyclists.

(2) An aberrant inferior polar artery may compress the ureter, with resultant hydronephrosis.

(3) The pain of nephroptosis is due, apparently, to traction on the renal vessels.

(a) Visceral afferent nerves from the kidneys reach the spinal cord via the least splanchnic nerve (T12) and the lumbar splanchnic nerves (L1–L2).

(b) Pain originating from the kidney or its vascular supply is referred to (appears as if originating from) the lumbar and inguinal regions.

C. **THE RENAL CAPSULE** is the true fibrous capsule of the kidney.

1. The capsule is stripped readily from a normal kidney but is adherent in certain pathologic conditions in which scarring has occurred.

2. The renal capsule provides a barrier against spread of infection.

D. **THE RENAL SINUS**, the cavity of the kidney (Fig. 18-1)

1. The sinus opens medially at the **hilus**.

2. Most structures enter and leave the kidney at the hilus.

a. The **renal artery** branches extensively within the sinus to supply **renal segments**.

b. The **renal vein** leaves the kidney at the renal hilus.

c. The **renal pelvis**, the funnel-shaped proximal end of the ureter, is continuous with the ureter proper distally and with the major calyces within the kidney.

d. There is a variable amount of fat between the various structures within the renal sinus.

E. **GENERAL STRUCTURE OF THE KIDNEY** (see Fig. 18-1)

1. The renal parenchyma is divisible into two parts.

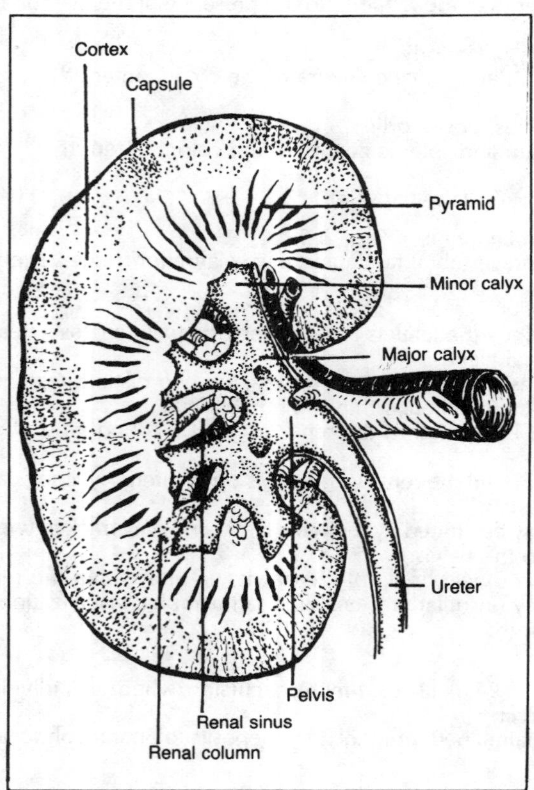

Figure 18-1. *The right kidney.* The division of the renal parenchyma into cortex and medulla is evident. Renal pyramids empty into calyces, which coalesce into the renal pelvis and ureter.

a. The **cortex:**
   (1) Is composed primarily of renal corpuscles, proximal convoluted tubules, and distal convoluted tubules.
   (2) Extends between the renal pyramids as **renal columns** (Bertini's).
b. The **medulla:**
   (1) Is formed by 9 to 14 **renal pyramids** (Malpighi's).
   (2) Consists primarily of straight tubules, Henle's loops, and collecting ducts.

2. The kidney is subdivided into lobes.
   a. Each **renal pyramid** with the overlying cortex and adjacent regions of renal columns forms the basic functional unit of the kidney (the lobe).
   b. Fusion of lobes (pyramids) at the apices forms 8 to 18 **renal papillae**.
   c. Each papilla opens into the cup-shaped end of a **minor calyx**, which tends to be arranged either anteriorly or posteriorly.
   d. Minor calyces fuse and join to form three to four **major calyces**.
   e. Each major calyx joins with others to form the **renal pelvis**.

F. **RENAL ARTERIES** (Fig. 18-2)

1. The blood supply to the kidneys is profuse, with the kidneys receiving approximately one-fifth of the cardiac output.

2. The left and right renal arteries arise from lateral aspects of the aorta, usually between L1 and L2 just below the origin of the superior mesenteric artery.

3. Left and right renal arteries give rise to:
   a. **Suprarenal arteries** to the adrenal glands.
   b. Numerous twigs to the ureters.
   c. Occasional **gonadal arteries** (15%).
   d. Occasional **inferior phrenic arteries** (10%).

4. Renal arteries divide into **segmental arteries** within the renal sinus.
   a. This division defines five renal segments:
      (1) Superior (apical).
      (2) Anterior-superior (upper).
      (3) Anterior-inferior (middle).
      (4) Inferior (lower).
      (5) Posterior.
   b. The distributions of the anterior-superior and anterior-inferior branches, as distinct from the posterior branch, divide the kidney into **anterior** and **posterior divisions**.
      (1) There are few anastomoses between the anterior segments and the posterior segment.
         (a) A longitudinal incision along the **avascular line** (Brödel's white line) produces minimal damage to the blood supply. This approach is utilized for removal of renal (staghorn) calculi.
         (b) The apical and lower segments have a more variable blood supply.
      (2) Ligation of a segmental artery results in necrosis of the entire segment.
   c. Each segmental artery gives rise to **interlobar arteries** (see Fig. 18-2).
      (1) Each interlobar artery enters a renal column.
      (2) Each gives off **arcuate arteries**, which run across the bases of the pyramids between cortex and medulla.
         (a) Arcuate arteries give off **interlobular arteries**, which course between the medullary rays.
         (b) Interlobular arteries supply the **afferent glomerular arterioles**.

5. Although the possibility of segmental resection for segmentally localized kidney disease is appealing, it seems yet to be of marginal practicality.

6. Vascular variations.
   a. Aberrant or supernumerary segmental arteries commonly (32%) arise from the renal artery or aorta.
      (1) Supernumerary renal arteries are derived from the fetal lobation pattern, with failure of the renal arterial segments to fuse.
      (2) About half of all aberrant segmental arteries are hilar (arising from the renal artery) and the other half, polar (arising from the aorta).
         (a) Polar arteries, arising directly from the aorta, tend to be larger than hilar arteries, which arise from the renal vessels.
         (b) Lack of awareness of arterial variations can result in overwhelming hemorrhage at surgery.
         (c) If torn, a polar artery usually separates where it arises from the aorta, leaving no stump to clamp and making control exasperatingly difficult.

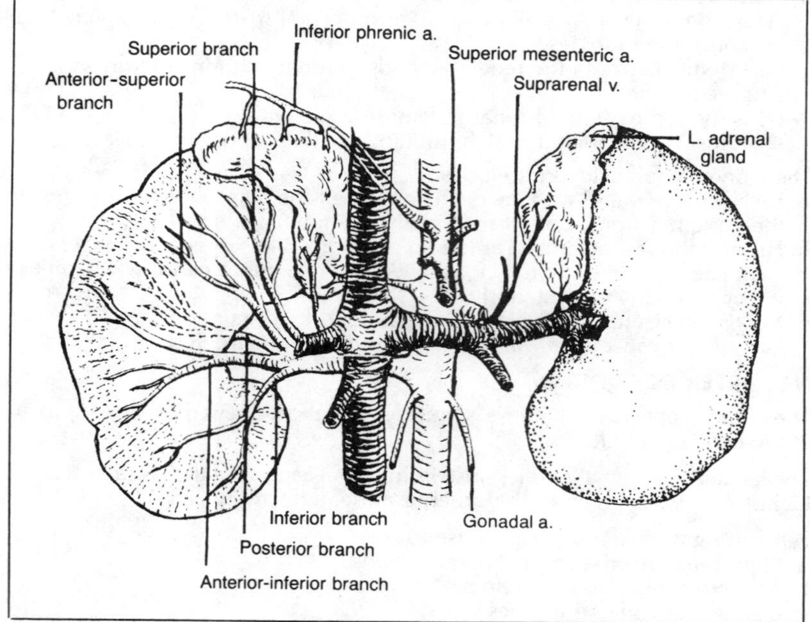

Figure 18-2. *The blood supply and venous drainage of the kidneys and adrenal glands.* The five segmental arteries of the right kidney are shown as are the three arteries supplying the right adrenal gland. The venous drainage of the left kidney and adrenal gland is depicted on the opposite side.

(d) Angiographic studies prior to renal surgery are usually advisable.

b. Supernumerary, or so-called accessory renal arteries, cannot be considered collaterals because each supplies one segment. Ligation almost always produces necrosis of a renal segment.

c. **Lower polar arteries** typically pass anterior to the ureter. Constriction of the ureter by an inferior polar artery produces hydronephrosis.

G. RENAL VEINS (see Fig. 18-2)

1. The **right renal vein**:

a. Enters the inferior vena cava at a lower point than the left renal vein.

b. Usually has no significant tributaries.

2. The **left renal vein**:

a. Is longer than the right renal vein.

b. Passes anterior to the aorta.

c. Receives the left gonadal vein, left suprarenal vein, and left inferior phrenic vein, and communicates with the azygos vein.

3. Variations.

a. Multiple renal veins are less common (14%) than supernumerary arteries (32%).

b. The most common variation is doubling of a renal vein (6%).

H. RENAL INNERVATION

1. Sympathetic nerves from T12 to L2 run in the least splanchnic nerve and lumbar splanchnic nerves.

a. Postganglionic cells lie in aorticorenal ganglia and course within the renal plexus to the kidneys.

b. Sympathetic nerves principally supply the renal vasculature.

(1) With excessive sympathetic stimulation, urine production ceases.

(2) Sympathectomy, interrupting input to the aorticorenal ganglion, produces renal vasodilation and transient renal diuresis.

2. Pain derived from the kidney and upper ureter is referred to the T12 to L2 dermatomal distribution (lumbar and inguinal regions as well as the anterosuperior thigh).

I. RENAL FUNCTION

1. One of the functions of the kidney is the conservation of minerals and the maintenance of ionic balance in the body fluids.
   a. Each kidney contains 10^6 or more nephrons, which comprise the basic morphologic and physiologic units of the kidney.
      (1) An ultrafiltrate of plasma (provisional urine) exudes through the glomerular capillaries into Bowman's capsule, the beginning of the nephron.
      (2) Along the proximal tubule, select substances are absorbed from the provisional urine, resulting in osmotic uptake of water. In addition, some substances are secreted into the provisional urine by the tubular epithelium.
      (3) In the distal tubule, sodium and water may or may not be absorbed, depending on the circulating concentrations of antidiuretic hormone and aldosterone.
      (4) The fluid that remains passes into the collecting ducts as urine.
      (5) About 500 collecting ducts, distributed within 9 to 14 pyramids in the normal adult kidney, terminate in 7 to 9 minor calyces.
   b. Although approximately 150 to 200 L of provisional urine are produced each day, reabsorption results in only 1 to 2 L of urine to be voided by micturition.
2. The kidneys produce vasoactive substances, which control blood pressure.

J. CLINICAL CONSIDERATIONS

1. A kidney with a stenotic or occluded renal artery, or one that is nonfunctional with respect to production of urine, will produce an overabundance of angiotensin, with resultant systemic hypertension.
2. The renal outflow tract can be visualized radiographically by intravenous pyelography (IVP). Contrast material, injected intravenously, is excreted by the kidney and concentrated in the outflow tract where it appears radiopaque. X-rays taken at intervals demonstrate this process.
3. Urinary components (calcium compounds and urea) may salt out in the calyces, and concretions (**nephroliths**) may develop. Usually, nephroliths are small enough to pass through the ureter and be eliminated. However, they may become large enough to lodge within and obstruct a ureter. Large staghorn concretions may completely occlude a calyx.

II. THE URETERS

A. The ureters are the excretory ducts between the kidneys and the urinary bladder (Fig. 18-3).

1. The **renal pelvis**, lying in the renal sinus, narrows at the ureteropelvic junction to form the ureter.
2. Each ureter descends retroperitoneally within the **periureteral sheath**, which is an extension of Gerota's fascia, anterior to the psoas muscle.
3. After crossing the common iliac vessels, each ureter dives over the brim of the deep pelvis.
   a. The ureters receive vascular twigs from the various iliac arteries.
   b. At this location, the ureter may be affected by aneurysms of the iliac artery.
   c. Because this is the narrowest part of the ureter, calculi may lodge here, with excruciating pain referred to the inguinal and pubic regions, the thigh, and the anterior scrotum or labia (L1–L2).

B. RELATIONS

1. The **right ureter**:
   a. Is posterior to the descending portion of the duodenum.
   b. Is posterior to the root of the mesentery proper, containing the ileocolic and right colic vessels.
   c. Is posterior to the right gonadal vessels, which contribute important vascular twigs to the right ureter.
2. The **left ureter**:
   a. Lies posterior to the left colic vessels.
   b. Lies adjacent to the left gonadal vessels, which contribute important vascular twigs to the left ureter.
   c. Lies posterior to the sigmoid mesocolon, containing the sigmoidal arteries and superior rectal artery.
3. Within the deep pelvis, each ureter courses retroperitoneally in the endopelvic fascia anterior to the sacrum.

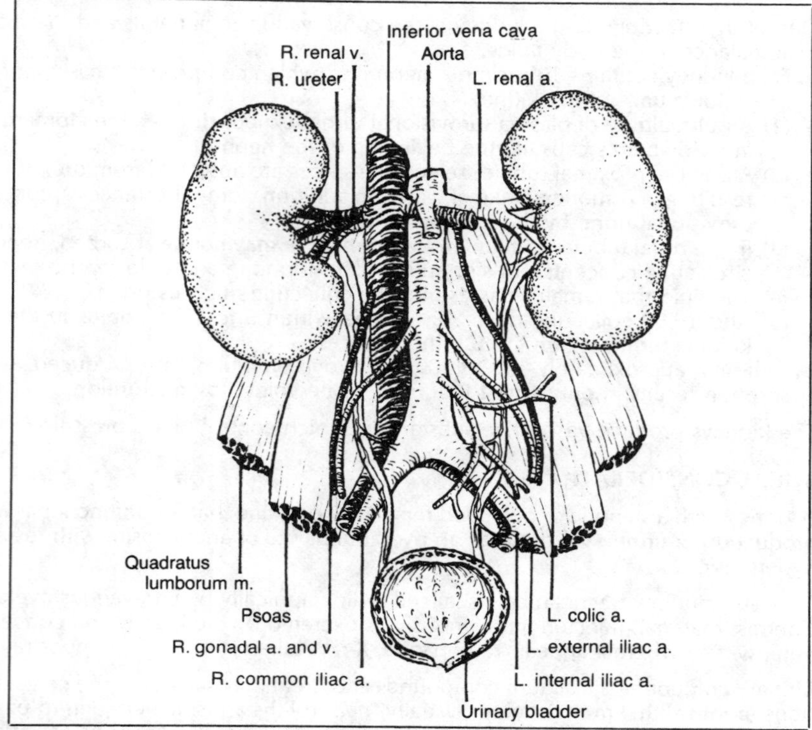

Figure 18-3. *The ureters and urinary bladder.* The blood supply to the ureters is from numerous sources.

a. In males, the ureters pass inferior to the vas deferens.
b. In females, the ureters pass inferior to the cardinal ligaments and uterine vessels, where they may be inadvertently clamped or sectioned along with uterine vessels.
c. The ureters converge to enter the bladder posteroinferiorly.
d. The diameter of the ureter decreases as the ureter passes through the bladder wall, providing another point at which a nephrolith may become lodged.

C. BLOOD SUPPLY TO THE URETERS

1. The blood supply to the ureters is very diffuse and variable.

2. Each ureter is serviced by small arterial twigs from the renal arteries, aorta, small arteries of the posterior abdominal wall, gonadal arteries, and common and internal iliac arteries, as well as the inferior vesicle arteries (see Fig. 18-3).

3. Mobilization of a ureter, or even traction upon a ureter during surgery, should be avoided, because interruption of the delicate blood supply results in ischemic necrosis of the ureter.

D. INNERVATION OF THE URETERS

1. The **least splanchnic nerve** from T12, with synapses in the aorticorenal ganglion, supplies the sympathetic innervation to the renal pelvis.

2. The **lumbar splanchnic nerves** from L1 to L2 supply sympathetic innervation to the abdominal and pelvic portions of the ureter.

3. The **pelvic splanchnic nerves** from S2 to S4 supply parasympathetic innervation to the entire ureter.

4. Visceral afferent nerve fibers from the various regions of each ureter travel to the spinal cord via the closest splanchnic nerve. Pain originating from specific regions of the ureter is referred accordingly.

E. CLINICAL CONSIDERATIONS

1. The ureters display intrinsic peristaltic activity.

a. The ureters contain a thick muscular wall, composed of circular and longitudinal layers, which moves the urine toward the bladder by waves of contraction (**urination**).

(1) A bolus of urine, distending ureteral walls, results in muscular contraction above the bolus, and relaxation below.

(2) The peristaltic waves occur one to six times per minute.

b. Sectioning a ureter interrupts the physiologic continuity of peristalsis, but after re-anastomosis, a new peristaltic wave is initiated below the site.

2. Nephroliths.

a. The ureter narrows at the junction between the renal pelvis and the ureter proper, at the point where the ureter crosses the pelvic brim, and again where the ureter passes through the wall of the bladder.

(1) A stone may lodge or pass slowly through the ureter in these three regions.

(2) There is ureteral distension (hydronephrosis) proximal to the stone.

b. The distension causes excruciating pain (renal colic), which is purported to be among the most intense pains experienced by humans (see Table 17-2).

(1) Upper ureteral obstruction and distension causes pain to be referred to the lumbar region (T12, L1).

(2) Middle ureteral obstruction and distension causes pain to be referred toward inguinal and pubic regions, as well as to the anterior scrotum or mons pubis and to the superoanterior thigh (L1 and L2).

(3) Lower ureteral obstruction and distension cause pain to be referred to the perineum and sometimes to the leg (S2–S4).

c. When a kidney stone obstructs a ureter, hydronephrosis results. If the stone fails to move despite copious imbibition, hydronephrosis may produce kidney damage and thus require treatment.

(1) Instrumentation can be passed through the urethra and urinary bladder into the obstructed ureter to fragment the stone.

(2) Alternatively, the ureter may be approached by surgical incision.

3. Renal and ureteral anomalies are rather frequent (4%).

a. Anomalies in renal development include **polycystic kidneys**, in which the collecting tubules fail to join a calyx; **horseshoe kidney**, in which the two kidneys are joined across the midline; **lobated kidneys**, which maintain their fetal pattern of segmentation; and **ectopic kidneys**, which may be located in the pelvis or both on the same side of the body.

b. There are numerous variations possible with respect to the ureter, with numerous branching patterns.

c. Dual ureters have a slower flow of urine than normal ureters and are therefore more susceptible to infection, which can spread to the kidneys from the urinary bladder.

4. Voiding the urinary bladder is micturition, not urination.

III. ADRENAL (SUPRARENAL) GLANDS

A. The adrenal glands are located anteromedial to the kidneys, between the kidneys and the diaphragm.

B. THE ADRENAL PARENCHYMA is divided into two zones.

1. The **adrenal cortex**:

a. Is derived from the mesoderm of the embryonic urogenital ridge.

b. Produces three classes of steroid hormones:

(1) Mineralocorticoids (aldosterone) necessary for salt metabolism.

(a) Deficiency of mineralocorticoids results in Addison's disease.

(b) Adrenal hyperplasia with excessive mineralocorticoid secretion results in Conn's syndrome.

(2) Glucocorticoids (such as cortisone).

(a) These affect metabolism, especially that of the connective tissues.

(b) Adrenal hyperplasia with excessive glucocorticoid production leads to Cushing's syndrome.

(3) Male and female sex hormones.

(a) Hyperplasia with excessive secretion of sex hormones produces the adrenogenital syndrome. Depending upon the hormone produced, the result is either masculinization of females or feminization of males.

(b) Sex hormones produced by the adrenals are metabolized in the liver; therefore, liver dysfunction in males (often associated with alcohol-related liver disease) results in gynecomastia and the development of a feminine escutcheon.

2. The **adrenal medulla**:
   a. Is derived from embryonic neural crest tissue, which also gives rise to sympathetic ganglia.
   b. Consists primarily of chromaffin (pheochrome) cells.
   c. Is innervated largely by the greater splanchnic nerves.
   d. Secretes epinephrine into the bloodstream in response to cholinergic stimulation.

C. BLOOD SUPPLY TO THE ADRENALS

1. Adrenal arteries (see Fig. 18-2).
   a. Each inferior phrenic artery gives off one or more **superior suprarenal arteries**.
   b. The aorta gives off the **middle suprarenal arteries** on each side.
   c. Each renal artery gives off one or more **inferior suprarenal arteries**.

2. Suprarenal veins (see Fig. 18-2).
   a. One adrenal vein exits at the hilus of each adrenal gland.
   b. The **right adrenal vein** usually drains into the inferior vena cava.
   c. The **left adrenal vein** usually drains into the left renal vein.

D. CLINICAL CONSIDERATIONS

1. **Pheochromocytoma** is a tumor of the adrenal medulla. Excessive bursts of epinephrine and norepinephrine, which result in paroxysms of hypertension, may be released from these tumors upon sympathetic activation or even abdominal palpation.

2. Great care must be taken during adrenalectomy. The suprarenal vein must be ligated as soon as practical before manipulation of the gland so that catecholamines do not escape into the circulation. The right adrenal gland is more difficult to approach surgically than the left because it is in part posterior to the inferior vena cava.

3. There may be ectopic adrenal tissue.
   a. Ectopic adrenal tissue usually consists of both cortex and medulla.
   b. Pheochromocytomas may occur anywhere along the embryonic urogenital ridge.

IV. THE POSTERIOR ABDOMINAL WALL

A. The posterior abdominal wall is formed by the diaphragm, the bilateral quadratus lumborum muscle, and the thoracolumbar fascia.

B. THE DIAPHRAGM (Fig. 18-4)

1. The diaphragm is a dome-shaped muscular sheet separating the thoracic from the abdominal cavity. The upper surface is covered by parietal pleura and pericardium; the lower surface is covered by parietal peritoneum.

2. Origins of the diaphragm.
   a. The sternal origin is the posterior surface of the xiphoid process.
   b. The costal origin is comprised of the inner surfaces of the costal margin (costal cartilages 7–9 as well as the tips of ribs 10–12).
   c. The lumbar origins are two crura arising from the lateral aspects of the upper three or four lumbar vertebrae.

3. The muscle fascicles insert into the **central tendon** of the diaphragm.

4. Diaphragmatic contraction pulls the central tendon caudally, thereby increasing thoracic volume and decreasing thoracic pressure.

5. Weak areas.
   a. The **sternocostal triangle** (foramen of Morgagni) is located between the sternal and costal portions.
      (1) This area transmits superior epigastric vessels.
      (2) It is a potential site for herniation (usually acquired).
   b. The **lumbocostal triangle** (foramen of Bochdalek) is located between costal and lumbar portions.
      (1) This area may be nonmuscular, with only two layers of serous membrane separating the thoracic from the abdominal cavity.
      (2) It is a potential site for herniation of a peritoneal sac of the abdominal viscera.
      (3) Although the stomach is the most commonly herniated viscus, nearly every abdominal organ (except the sigmoid colon) has been reported herniated into the thorax.
      (4) Diaphragmatic hernia is 8 to 10 times more common on the left than on the right.

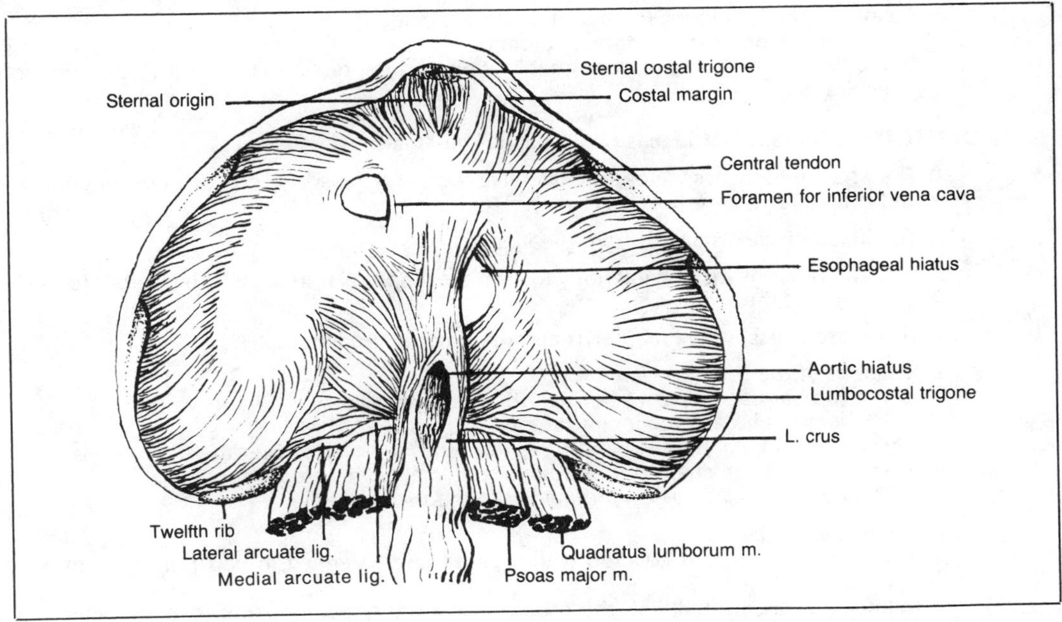

Figure 18-4. *The diaphragm.* The inferior surface of the diaphragm with the three hiatuses is depicted.

c. The **lateral arcuate ligament** is formed by the free edge of the diaphragm between its attachments at the tip of the 12th rib and the transverse process of L1. Beneath it runs the quadratus lumborum muscle.

d. The **medial arcuate ligament** is formed by the free edge of the diaphragm between the transverse process of L1 and vertebral bodies of L1 and L2. Beneath it runs the most cephalad portion of the psoas muscle.

6. Hiatuses.

a. The **aortic hiatus** is located in the midline, between the diaphragmatic crura, and transmits the aorta, azygos veins, sympathetic trunk, thoracic duct, and occasionally, splanchnic nerves.

b. The **caval hiatus** (foramen of the inferior vena cava) is located to the right of the midline and transmits the inferior vena cava.

c. The **esophageal hiatus** transmits the esophagus and vagal trunks.

(1) The esophageal hiatus is located slightly left of the midline in the muscular portion of the diaphragm.

(2) It is the site of hiatal hernia.

(a) Hiatal hernia is relatively common (1 percent), accounting for 98 percent of all diaphragmatic hernias.

(b) The cardiac end of the stomach may herniate through this hiatus into the thorax (sliding hiatal hernia).

7. Innervation of the diaphragm.

a. The diaphragm develops in the cervical region along with the heart.

b. Motor innervation (except for the crural regions) is provided by the phrenic nerves.

c. Sensory innervation is dual.

(1) Sensation from the larger central region is carried by phrenic nerves to segments C3 to C5 of the spinal cord. Irritation of the diaphragmatic pleura or diaphragmatic peritoneum is referred to the shoulder region.

(2) Sensation from the peripheral region is supplied by intercostal nerves. Pain from the periphery of the diaphragm is referred to the costal margin of the abdominal wall.

C. THE QUADRATUS LUMBORUM MUSCLE is the most lateral muscle of the posterior abdominal wall (see Fig. 18-4; Fig. 18-5).

1. This muscle originates from the iliac crest and transverse processes of vertebrae L3 to L5.

2. It passes beneath the lateral arcuate ligament of the diaphragm to insert into the inferior border of the 12th rib and the transverse processes of L1.

3. Actions of the quadratus lumborum muscle include:
   a. Lateral flexion of the vertebral column.
   b. Fixation of the 12th rib during inspiration and depression of the 12th rib during forced expiration.

D. THE ILIOPSOAS MUSCLE has two separate muscular heads.

1. The **psoas muscle** originates from lateral aspects and transverse processes of vertebral bodies and discs T12 to L4 (see Figs. 18-4 and 18-5).
2. The **iliac muscle** originates from the iliac fossa.
3. Iliac and psoas heads fuse inferiorly to form the **iliopsoas muscle**, which inserts into the lesser trochanter of the femur.
4. This muscle acts to flex the pelvis and the thigh at the hip joint.
5. Relations of the psoas muscle.
   a. The lumbar plexus passes through the psoas muscle.
   b. The kidneys lie anterolateral to the psoas muscle.
   c. The ureters lie anterior to and cross the psoas muscle to enter the deep pelvis.
   d. The sympathetic chain is located anteromedial to the psoas muscle.
   e. The common iliac vessels are medial to the psoas muscle.
6. The **psoas fascia**.
   a. The psoas muscle is covered with a dense fascia derived in part from the lumbodorsal fascia (see Fig. 18-5).
   b. Infections associated with the psoas muscle or vertebral column may be limited by the psoas fascia and track into the thigh.

E. THE LUMBODORSAL (THORACODORSAL) FASCIA (see Fig. 18-5)

1. The posterior abdominal wall lateral to the quadratus lumborum muscle consists largely of **lumbodorsal fascia**.
2. The lumbodorsal fascia is formed by fusion of the posterior aponeuroses of the internal oblique and transverse muscles of the abdomen.

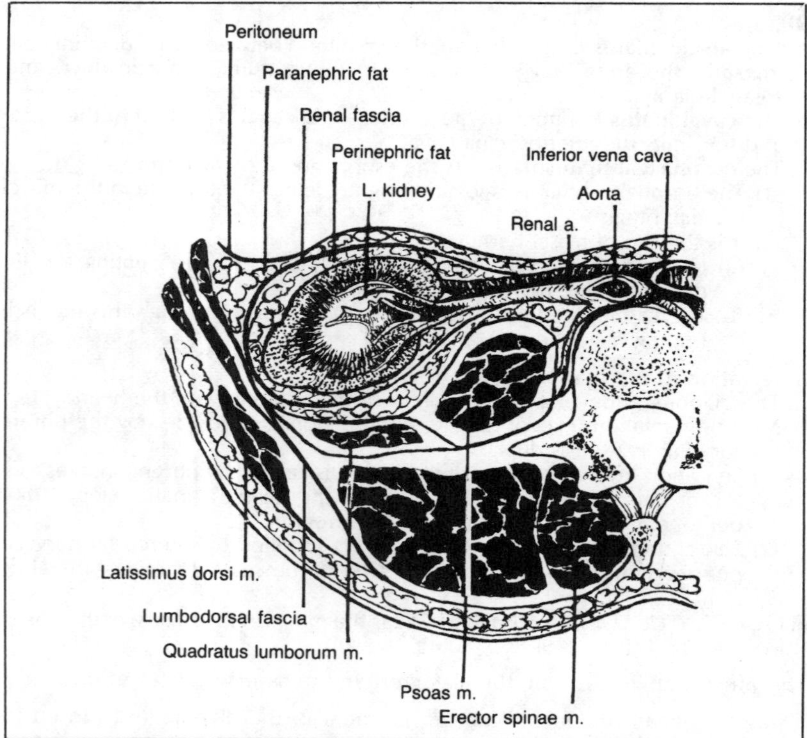

Figure 18-5. *The posterior abdominal wall and renal fascia.*

3. This fascia splits into three leaves that form two muscular compartments.
   a. The posterior leaf inserts into the vertebral spinous processes and provides the site of origin for the latissimus dorsi muscle.
   b. The middle leaf inserts into the tip of the vertebral transverse processes. With the posterior leaf, it encloses the erector muscle of the spine.
   c. The anterior leaf inserts midway along the transverse processes.
      (1) With the middle leaf, it encloses the quadratus lumborum muscle.
      (2) The psoas fascia arises about midway along the middle leaf.

4. The **lumbar triangle** (Petit's).
   a. The lumbar triangle is bounded by the posterolateral free edge of the external oblique muscle, the anterolateral border of the latissimus dorsi muscle, and the superior aspect of the iliac crest.
   b. The floor of the trigone (triangle) consists of the lumbodorsal fascia and, sometimes, fibers of the internal oblique muscle.
   c. This region may be the site of a rare lumbar hernia.
   d. Since the abdominal wall is thinner here, it is used in surgical approaches to the kidneys and ureters.

F. THE POSTERIOR ABDOMINAL VESSELS

1. The **aorta** (see Fig. 18-3).
   a. The aorta enters the abdomen via the aortic hiatus between the diaphragmatic crura.
   b. Relations.
      (1) The aorta is located on the anterior aspect of the vertebral column, deviating slightly to the left during passage toward the bifurcation.
      (2) It lies to the left of the inferior vena cava.
      (3) The aorta is associated with the celiac and superior mesenteric plexuses and the included ganglia, as well as with the aortic, inferior mesenteric, and hypogastric plexuses.
   c. Branches of the abdominal aorta, before bifurcating into the common iliac arteries, include inferior phrenic arteries, the celiac artery, the superior mesenteric artery, renal arteries, gonadal arteries, numerous twigs to the ureters, the inferior mesenteric artery, lumbar arteries, and the middle sacral artery.
      (1) Aneurysms of the aorta commonly involve the origins of the aortic branches. A large aneurysm may erode into an adjacent vertebra and produce back pain as the principal symptom.
      (2) Occlusion of the aorta superior to the renal arteries is usually fatal, because the kidneys require far more blood than can flow through collateral vessels. Slow occlusion below the renal arteries results in progressive development of collateral pathways.
         (a) A major shunt between the internal thoracic artery and the inferior epigastric artery will develop, with enlargement of the intercostal and abdominal arteries.
         (b) The marginal artery of Drummond may hypertrophy if occlusion occurs between the origins of the superior and inferior mesenteric arteries.

2. The **inferior vena cava** (see Fig. 18-3).
   a. The inferior vena cava lies to the right of the aorta along the anterior aspect of the vertebral column.
   b. This vein receives the common iliac veins, lumbar veins, right gonadal vein, renal veins, right adrenal vein, right inferior phrenic vein, and hepatic veins.
   c. It enters the thorax via the foramen of the inferior vena cava.
   d. The inferior vena cava develops from three distinct embryologic venous pathways that fuse and atrophy in various places.
      (1) There are numerous collateral pathways and variations.
      (2) The vein may be absent, its drainage functions served by an enlarged azygos system entering the thorax via the aortic hiatus in the diaphragm.

G. LYMPHATIC CHANNELS

1. The **common iliac nodes**:
   a. Are adjacent to the common iliac vessels.
   b. Receive drainage from the lower limbs, the perineum, and the gluteal region via the **inguinal nodes**.
   c. Drain into the aortic nodes.

2. The **aortic (lumbar) nodes**:
   a. Lie adjacent to the aorta and receive drainage from the posterior abdominal wall and iliac nodes.

b. Form the lumbar lymphatic trunk, often one on each side of the aorta.
c. Drain into the cisterna chyli.

3. The **cisterna chyli**:
a. Receives drainage from the intestinal lymphatic trunk and from the celiac and superior mesenteric nodes, as well as from the lumbar lymphatic trunks.
b. Lies posterior and a little to the right of the aorta, usually between the crura of the diaphragm at the aortic hiatus.
c. Represents the lower expanded portion of the thoracic duct.
(1) The thoracic duct courses through the posterior mediastinum between the aorta and the vertebral column.
(2) It empties into the left subclavian vein near the junction of this vein with the internal jugular vein.

H. THE NERVES OF THE POSTERIOR ABDOMINAL WALL

1. **Sympathetic nerves** (Fig. 18-6).
a. The paired **sympathetic trunks** run in a groove between the psoas muscle and the vertebral column.
(1) Each sympathetic chain consists of ganglia (approximately one for each vertebral level) and interconnecting nerve fibers.
(2) The five lumbar ganglia may be fused to varying extents.
b. **Rami communicantes** provide pathways between the spinal cord and the lumbar ganglia.
(1) Only lumbar ganglia L1 and L2 (occasionally L3) receive white rami from the spinal cord.
(a) White rami contain myelinated presynaptic (preganglionic) nerve fibers, as well as visceral afferent fibers.
(b) The sympathetic nerves of the white rami synapse either in the paravertebral or prevertebral ganglia.
(2) Every lumbar ganglion gives off a gray ramus.
(a) Gray rami contain unmyelinated postsynaptic (postganglionic) nerve fibers.
(b) These postganglionic sympathetic fibers return to their respective spinal nerves to subserve peripheral autonomic functions, such as vasoconstriction, piloerection, and perspiration.
c. **Lumbar splanchnic nerves** innervate portions of the gastrointestinal tract, kidneys, and ureters.
(1) Each lumbar splanchnic nerve is composed of presynaptic and postsynaptic nerve fibers and also carries visceral afferent nerves.
(2) The splanchnic nerves join the **aortic plexus**, which:
(a) Communicates superiorly with the superior mesenteric and celiac plexuses.
(b) Communicates inferiorly with the hypogastric plexus, which conveys sympathetic pathways into the pelvis for urinary and sexual function.
(c) Contains small sympathetic ganglia, which provide postsynaptic fibers to the descending colon, sigmoid colon, and ureters.
(3) Sympathetic nerves to the viscera are primarily vasoconstrictors.
(4) Lumbar splanchnic pathways and the associated white rami communicantes convey to the lumbar region of the spinal cord afferent pain fibers from the descending colon and sigmoid colon, as well as from the upper and middle portions of the ureters. Pain originating from these structures is referred to lumbar, inguinal, and pubic regions, as well as to the anterosuperior portion of the thigh (L1–L2).

2. **Somatic nerves** of the posterior abdominal wall innervate the abdominal musculature and convey sensation from the skin and parietal peritoneum (see Fig. 18-6).
a. The **lumbar plexus** is the upper portion of the **lumbosacral plexus**.
b. Many of the roots of the lumbar plexus pass through the psoas muscle.
c. This plexus of somatic nerves is formed from spinal nerves as the extremities develop from the segmental dermatomes and myotomes (see Chapter 4 III C).
(1) The posterior division of a somatic plexus is equivalent to the lateral branch of a spinal nerve.
(2) The anterior division of a somatic plexus is equivalent to the continuation of a spinal nerve distal to the lateral branch.
d. Major nerves of the lumbar plexus.
(1) The **iliohypogastric nerve** (T12 + L1, anterior).
(a) This nerve emerges from the lateral side of the psoas muscle to run within the musculature of the abdominal wall.
(b) Its iliac branch is sensory to the upper gluteal region.

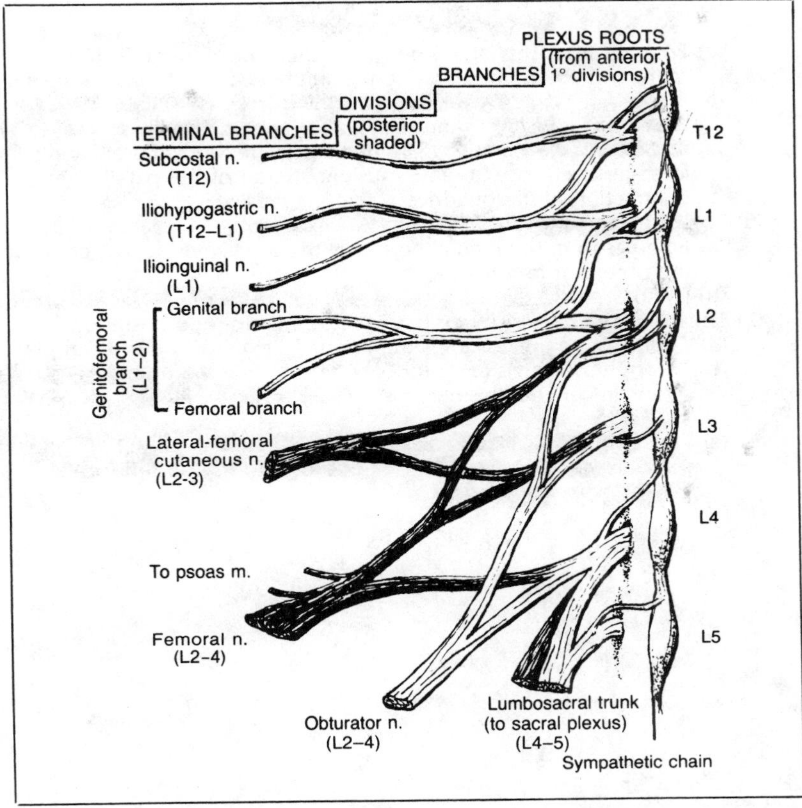

Figure 18-6. *The lumbar plexus.* The lumbar plexus is formed by the anterior primary rami of the lumbar spinal nerves. The anterior divisions of the lumbar plexus are *light*, and the posterior divisions are *dark*. The lumbosacral trunk from L4 to L5 joins the sacral plexus. Spinal nerves T12, L1, and L2 have both white and gray rami communicantes. Spinal nerves L3, L4, and L5 have only gray rami communicantes.

- **(c)** Its hypogastric branch is motor to the internal oblique and transverse muscles of the abdomen and sensory to the pubic (hypogastric) region.
- **(d)** The iliohypogastric nerve supplies afferent and efferent limbs for the **abdominal reflex**, whereby stroking the lower abdominal wall produces a rippling of the underlying abdominal muscles.

(2) The **ilioinguinal nerve** (L1, anterior):

- **(a)** Emerges from the lateral border of the psoas muscle and runs parallel to, and sometimes anastomoses with, the iliohypogastric nerve. Its terminal branch accompanies the spermatic cord through the superficial inguinal ring.
- **(b)** Is motor to the internal oblique and transverse muscles of the abdomen and sensory to the upper medial part of the thigh, the root of the penis or mons pubis, and the anterior scrotum or superior portion of the labia majora.

(3) The **genitofemoral nerve** (L1 + L2, anterior).

- **(a)** This nerve emerges anterior to the psoas muscle and runs along its anterior surface.
- **(b)** Its genital branch (the external spermatic nerve) passes through the inguinal canal. This branch is motor to the cremaster muscle (the efferent limb of the **cremaster reflex** in men) and sensory to the anterior scrotum or superior portion of the labia majora.
- **(c)** Its femoral branch (the lumboinguinal nerve) passes into the thigh beneath the inguinal ligament, medial to the psoas muscle. This branch is sensory to the anterior superior aspect of the thigh and provides the afferent limb of the **cremaster reflex** in men, whereby stroking the anteromedial thigh produces elevation of the testes within the scrotum.

(4) The **lateral femoral cutaneous nerve** (L2 + L3, posterior):

- **(a)** Emerges lateral to the psoas muscle, runs across the iliac fossa, and passes beneath the inguinal ligament close to the anterior superior iliac spine.

(b) Is sensory to the lateral aspect of the thigh.

(5) The **femoral nerve** (L2 + L3 + L4, posterior):

(a) Emerges from the lateral border of the psoas muscle just before the inguinal ligament and enters the thigh by passing beneath the inguinal ligament.

(b) Is motor to the anterior aspect of the thigh, sensory to the anterior and medial aspects of the thigh, and, by its saphenous branch, sensory to the anterior and medial aspects of the leg. This nerve provides both afferent and efferent limbs of the **knee-jerk reflex**, whereby transient stretch of the patellar tendon produces brief contraction of the quadriceps femoris muscle.

(6) The **obturator nerve** (L2 + L3 + L4, anterior):

(a) Emerges from the medial border of the psoas muscle and enters the thigh through the obturator foramen.

(b) Is motor, and occasionally sensory, to the medial aspect of the thigh.

(7) The **lumbosacral trunk nerve** (L4 + L5, anterior and posterior).

(a) This large connector joins the lumbar plexus with the sacral plexus.

(b) The posterior division contributes to the common peroneal portion of the sciatic nerve. In general, this nerve is sensory and motor to the anterior leg and dorsum of the foot.

(c) The anterior division contributes to the tibial portion of the sciatic nerve. In general, this nerve is sensory and motor to the posterior thigh, posterior leg, and plantar aspect of the foot.

Part IV Abdominal Region

STUDY QUESTIONS

Directions: Each question below contains five suggested answers. Choose the **one best** response to each question.

1. Which of the following statements concerning the superficial inguinal ring is true?

(A) It is formed in part by the falx inguinalis
(B) It is formed in part by the rectus sheath
(C) It is a perforation in the internal oblique muscle
(D) It is a perforation in the transversalis fascia
(E) None of the above statements is true

2. Stimulation of the cremaster muscle draws the testis up from the scrotum toward the superficial inguinal ring. These afferent impulses are carried centrally by

(A) an anterior primary ramus
(B) the iliohypogastric nerve
(C) the subcostal nerve
(D) the vagus nerve
(E) a white ramus communicans

Questions 3–10

A 43-year-old advertising executive is brought to the emergency room with searing pain to the midback and "coffee grounds" hematemesis. The present illness began shortly after eating a heavy lunch, which included two drinks. He relates a history of periodic attacks of heartburn, nausea, and pain (pointing to his epigastric region). These attacks are often relieved by antacids or upon eating. He smokes two packs of cigarettes a day and admits to moderate social use of alcohol. An EKG is normal. Upon examination the patient exhibits abdominal rigidity, but no region elicits particularly marked tenderness upon palpation.

3. All of the following structures refer pain via the greater splanchnic nerve to the epigastric region EXCEPT the

(A) abdominal esophagus
(B) descending duodenum
(C) gallbladder
(D) ileum
(E) stomach

4. The sharp searing midback pain is explained by stimulation of neurons traveling along the

(A) greater splanchnic nerve
(B) lesser splanchnic nerve
(C) least splanchnic nerve
(D) lumbar splanchnic nerves
(E) spinal nerves (T12–L2)

5. Perforation of a peptic ulcer is suspected, and the patient is prepared for immediate surgery. Under general anesthesia the skin and anterior wall of the rectus sheath are incised in a paramidline incision from the xiphoid process to the umbilicus. The rectus abdominis muscle is retracted laterally, and the posterior wall of the rectus sheath is incised. As a result of the incision and manipulation of the rectus abdominis muscle, which of the following results might be expected?

(A) Ischemic necrosis of the rectus abdominis muscle *superior* to the umbilicus
(B) Ischemic necrosis of the rectus abdominis muscle *inferior* to the umbilicus
(C) Paralysis of the rectus abdominis muscle in the region *medial* to the incision of the sheath
(D) Paralysis of the rectus abdominis muscle in the region *lateral* to the incision of the sheath
(E) None of the above results can be expected

6. Upon exploration of the abdominal cavity, a slight amount of blood and other fluid is observed in the pouch of Morison and was aspirated. The pouch of Morison (subhepatic and hepaticorenal recesses) is directly continuous with

(A) the inferior mesenteric space
(B) the infracolic compartment
(C) the left paracolic gutter
(D) the lesser sac
(E) none of the above

7. No sign of perforation was seen on the anterior aspects of the stomach or duodenum. An incision was made in the membrane between the stomach and liver, avoiding the blood vessels of the lesser curvature as well as those of the portal triad. This incision passed through

(A) the falciform ligament
(B) the gastrohepatic ligament
(C) the greater omentum
(D) the hepatoduodenal ligament
(E) none of the above

8. The lesser sac was aspirated and a ¼ cm perforation was seen on the superior-posterior aspect of the pyloric antrum. The ulcer had eroded a small arterial branch, which was bleeding. The artery involved in this ulceration is most likely a branch of

(A) the esophageal artery
(B) the left gastric artery
(C) the left gastroepiploic artery
(D) a short gastric artery
(E) none of the above arteries

9. The surgeon required more exposure in order to resect the ulcer. He extended the previously made opening in the lesser omentum toward the esophagus whereupon the field suddenly filled with bright red blood. The surgeon quickly gained control of the bleeding. Both ends of the artery were clamped and ligated. Within a few moments he noticed that a portion of the left lobe of the liver blanched, indicating ischemia. The ligated artery most likely was the

(A) left gastric artery
(B) common hepatic artery
(C) right gastric artery
(D) right hepatic artery
(E) left gastroepiploic artery

10. The gastric ulcer was successfully resected. The surgical pathology examination was negative for malignancy. The postoperative course was stormy with spiking temperatures, but the patient gradually improved. The reason that the patient did not succumb to liver necrosis in this instance is abundant anastomoses exist between the distal portion of the left gastric artery and all of the following EXCEPT the

(A) esophageal arteries
(B) right gastric artery
(C) short gastric arteries
(D) right hepatic artery

(end of group question)

Questions 11–18

A 34-year-old man came into the emergency room with excruciating abdominal pain in the left flank and inguinal region. A urine sample was positive for blood. While the patient was waiting for x-rays, he complained that the pain was worse and had moved into the anterior and medial aspects of the left thigh as well as to the pubic region. The "KUB" (kidneys, ureters, and bladder) flat plate indicated a possible calculus in the lower left quadrant.

11. The initial pain in the flank and inguinal region resultant from a nephrolith is due to afferent pain fibers, which travel via the

(A) greater splanchnic nerve
(B) iliohypogastric nerve
(C) least splanchnic nerve
(D) lumbar splanchnics
(E) vagus nerve

12. The subsequent renal colic referred to the medial and anterior aspects of the left thigh as well as to the pubic region is the result of afferent pain fibers, which travel via the

(A) genitofemoral nerve
(B) ilioinguinal nerve
(C) lesser splanchnic nerve
(D) lumbar splanchnic nerves
(E) vagus nerve

13. The patient was admitted for observation and possible surgery. The pain remained in the lateral thigh and pubic regions and was unremitting. From this particular pattern of renal colic and the anatomy of the renal system, it is suspected that a stone has lodged

(A) at the junction of the renal pelvis and ureter
(B) at midureter as it passes beneath the gonadal vessels
(C) at the pelvic brim
(D) in the intramural portion of the ureter where it penetates the bladder
(E) in the urethra

14. The patient is scheduled for surgery. Under general anesthesia the peritoneal cavity is opened by incision of the skin, fascia, and musculature of the abdominal wall from the iliac crest to the pubis above and parallel to the left inguinal ligament. All of the following layers are severed in the anterior abdominal incision EXCEPT

(A) Camper's fascia
(B) deep fascia
(C) Gerota's fascia
(D) Scarpa's fascia
(E) transversalis fascia

15. Two large nerves are encountered and preserved in the process of making the abdominal incision. The location of these nerves is

(A) between the internal oblique and external oblique muscles
(B) between the transverse abdominis and internal oblique muscles
(C) in the superficial fascia
(D) in the transversalis fascia
(E) within the rectus sheath

16. The surgeon mobilized the descending bowel from the posterior abdominal wall. He gently lifted the bowel and separated its mesentery from the peritoneum covering the dorsal body wall. This reestablished mesentery contained the

(A) left lumbar arteries
(B) left renal artery
(C) left testicular artery
(D) middle colic artery
(E) none of the above arteries

17. With the descending colon and its mesentery drawn to the midline, the ureter was visible beneath the peritoneum. The stone was palpated and removed through a small longitudinal incision. Care was taken to preserve the blood supply to the ureter. The blood supply to the ureter is obtained from all of the following vessels EXCEPT the

(A) common iliac artery
(B) gonadal artery
(C) inferior mesenteric artery
(D) inferior vesicle arteries
(E) renal artery

18. The abdominal incision was closed in layers. The patient was allowed up on the first postoperative day and was discharged on the tenth day. As a result of the location and direction of the incision, healing might be expected to be accompanied by

(A) paralysis of a portion of the rectus abdominis muscle
(B) minimal scarring
(C) ischemia to the rectus abdominis muscle
(D) significant weakness of a portion of the lateral abdominal wall
(E) none of the above results

(end of group question)

Directions: Each question below contains four suggested answers of which **one or more** is correct. Choose the answer

A if **1, 2, and 3** are correct
B if **1 and 3** are correct
C if **2 and 4** are correct
D if **4** is correct
E if **1, 2, 3, and 4** are correct

19. Structures that might be encountered within the layers of the abdominal wall while performing an appendectomy at McBurney's point (midway between the anterior-superior iliac spine and the umbilicus), using the McBurney muscle splitting incision, include the

(1) inferior epigastric artery
(2) lateral femoral cutaneous nerve
(3) aberrant obturator artery
(4) iliohypogastric nerve

20. Which of the following statements characterize Meckel's diverticulum?

(1) It is located within 3 feet of the ileocecal valve
(2) It is usually lined by gastric mucosa
(3) It is present in about 3 percent of the population
(4) It is a remnant of the urachus

SUMMARY OF DIRECTIONS

A	B	C	D	E
1, 2, 3 only	1, 3 only	2, 4 only	4 only	All are correct

21. Considering the incidence of lymphatic metastatic carcinoma from the rectum, knowledge of its lymphatic drainage is important. Near the middle rectal valve, the lymphatic drainage is to the

(1) inferior mesenteric nodes
(2) internal iliac nodes
(3) superficial inguinal nodes
(4) superior mesenteric nodes

22. After radical esophagectomy (total removal of esophagus), a functional reanastomosis can be made between the pharynx and stomach, using a living segment of transverse colon brought up into the thorax and neck on a mesenteric pedicle containing the original blood vessels and nerves. Important considerations in performing this operation would be to

(1) anastomose the hepatic flexure end of the transverse colon to the stomach and the splenic flexure end to the pharynx
(2) take care not to sever the pelvic splanchnic nerves
(3) preserve the esophageal venous plexus
(4) preserve the middle colic vessels to the transverse colon

ANSWERS AND EXPLANATIONS

1. The answer is E. (*Chapter 15 X B 1*) The superficial inguinal ring is an opening in the external oblique aponeurosis. The superficial ring transmits the spermatic cord in males and the round ligament of the uterus in females. In males the internal oblique muscle contributes the cremaster muscle. The deep ring is a perforation in the transversalis fascia.

2. The answer is A. [*Chapter 4 III C 4 b; Chapter 15 X D 2 c (3)–(4); Chapter 18 IV H 2 d(3)*] The afferent limb of the cremaster reflex is carried by the femoral branch of the genitofemoral nerve. This is the continuation of an anterior primary ramus of a spinal nerve.

3. The answer is D. (*Chapter 16 VI E 2 a (2); Chapter 17 III H 2 d; XV B, Table 17-2*) The visceral afferent neurons from the lower esophagus, stomach, duodenum, gallbladder, and bile duct all travel through the celiac plexus. These neurons gain access to the greater splanchnic nerve, pass through the ganglia of the sympathetic chain (levels 5–9), reach the associated spinal nerves via white rami communicantes, and enter the spinal cord via the dorsal root. The visceral afferents from the terminal duodenum, jejunum, ileum, ascending colon and transverse colon travel through the superior mesenteric plexus then along the lesser splanchnic nerve.

4. The answer is E. [*Chapter 16 VI E 2 a (1)*] The parietal peritoneum is innervated by twigs from the anterior primary rami of the spinal nerves. Thus, sharp localized pain is an indication of peritonitis.

5. The answer is E. (*Chapter 15 IX A 1 a–c*) Because the blood supply enters the rectus abdominis muscle from the superior and inferior ends and because the innervation enters this muscle from the lateral aspect, the incision described will compromise neither the vascular supply nor the innervation.

6. The answer is D. [*Chapter 16 VI D 3 c (2) (a)*] The pouch of Morison communicates with the lesser sac via the epiploic foramen, with the supracolic compartment, the right paracolic gutter, and the subphrenic space. It does not communicate directly with the infracolic compartment or the left paracolic gutter.

7. The answer is B. (*Chapter 16 III C; V E 3 b*) The lesser omentum runs between the foregut and the liver. The portion between the stomach and liver, the gastrohepatic ligament, is relatively avascular. The portion between the duodenum and liver, hepatoduodenal ligament, contains the portal triad, which includes the major blood supply to the liver.

8. The answer is E. [*Chapter 17 III F 1 c; XII A 3 c (1)*] The most probable arteries involved in pyloric peptic ulceration are the right gastric, common hepatic, gastroduodenal, and right gastroepiploic arteries. In addition, ulceration of the posterior wall of the body of the stomach may involve the splenic artery with profuse bleeding.

9. The answer is A. (*Chapter 17 V E 2 a (5), b*) An aberrant left hepatic artery frequently arises from the left gastric artery. Since the arteries supplying the liver are end-arteries, ligation of an aberrant artery usually will produce ischemia of a portion of the liver.

10. The answer is D. (*Chapter 17 III F 2 a; V E 1–3*) There are few anastomoses between the left and right lobes of the liver, thus precluding anastomotic flow from the right hepatic artery. Abundant anastomoses among the left gastric artery, the esophageal arteries, right gastric artery, and short gastric branches of the splenic artery hypertrophy so that the aberrant left hepatic artery receives sufficient blood to supply the left lobe of the liver.

11. The answer is C. (*Chapter 17 Table 17-1; Chapter 18 II D 1*) The renal calyces, renal pelvis, and most of the proximal portion of the ureters generally receive afferent innervation from spinal segment T12 via the least splanchnic nerve. Since the somatic portions of spinal segment T12 supply the lumbar and inguinal regions of the abdominal wall, pain is referred accordingly.

12. The answer is D. (*Chapter 17 Table 17-1; Chapter 18 II D 2*) The middle section of the abdominal ureters generally receive innervation from spinal segments L1–L2. As such the afferent pathway is via the lumbar splanchnic nerves. Since the somatic portions of spinal segments L1 and L2 supply the pubic region and anterior aspects of the scrotum or labia majora, as well as the anterior and medial portions of the thigh, pain is referred accordingly.

13. The answer is C. (*Chapter 18 II A 3 c*) The ureters are narrowest at the pelvic brim where they cross the iliac vessels and enter the deep pelvis. Thus, the probability of a nephrolith lodging in a ureter at this point is high.

14. The answer is C. (*Chapter 15 III; VIII; Chapter 16 VI E 1 c; Chapter 18 II A 2*) An incision through the anterior abdominal wall passes through Camper's fascia (the superficial layer of the superficial fascia), Scarpa's fascia (the deep layer of the superficial fascia), the deep fascia investing the musculature, and the transversalis fascia before reaching the peritoneum. Gerota's fascia or the renal fascia comprises the false capsule of the kidney. After gaining access to the peritoneal cavity, the periureteral sheath, an extension of Gerota's fascia, is incised to gain direct access to the ureter.

15. The answer is B. [*Chapter 4 III C 4 b; Chapter 15 VII A 2, B 2; Chapter 18 IV H 2 d (1)*] The nerves of the anterior abdominal wall are located between the internal oblique and transverse abdominis layers and their aponeurotic extensions. The two large nerves encountered in the incision would be the iliohypogastric and the ilioinguinal nerves.

16. The answer is E. (*Chapter 16 VI B; Chapter 18 II B 2*) During development, the mesentery of the descending colon becomes adherent to the dorsal body wall of the peritoneal cavity anterior to the left ureter and the left gonadal neurovascular bundle. As such, the left colic artery with its sigmoidal and superior rectal branches lie in the former descending mesocolon anterior to the left ureter.

17. The answer is C. (*Chapter 18 II C 2*) The ureter obtains its blood supply from twigs of numerous vessels, including the renal, gonadal, common iliac, internal iliac, and inferior vesicle arteries. While the left colic artery comes into close proximity of the ureter, there are normally no anastomoses between this vessel and those of the ureter.

18. The answer is B. (*Chapter 2 I D*) In the lower abdominal wall, the cleavage lines run in a direction nearly parallel to the inguinal ligament. An incision approximating this direction would gape less and produce minimal scarring.

19. The answer is D (4). [*Chapter 15 IX A 3 b; Chapter 17 X B 2 h (2)*] The ilioinguinal nerve lies between the internal oblique and transverse abdominis muscles just superior to the inguinal ligament. At the level of McBurney's point, the inferior epigastric artery has entered the rectus sheath. The lateral femoral cutaneous nerve, lying on the iliacus muscle in the iliac fossa, escapes from the pelvis by passing beneath the inguinal ligament close to the anterior-superior iliac spine. An aberrant obturator artery crosses the femoral ring where it may be encountered during repair of a femoral hernia.

20. The answer is A (1, 2, 3). (*Chapter 16 IV D 5; Chapter 17 IX F 3*) An ileal diverticulum (Meckel's) is a remnant of the vitelline duct, which connects the midgut to the yolk sac during early development. The urachus is a remnant of the duct, which connects the bladder to the alantois. A "rule of threes" applies to Meckel's diverticulum: it is usually 3 inches long, within 3 feet of the ileocecal valve, and present in about 3 percent of the population. It is usually lined with gastric mucosa, the secretions of which often produce ileal ulceration.

21. The answer is A (1, 2, 3). (*Chapter 17 X E 8 a–c; XI C 3; XIV D 1 b*) The lymphatic drainage from the midrectum is along the inferior mesenteric, internal iliac, and external iliac nodes. In general, the lymphatic drainage follows the arteries. Thus, carcinoma of the rectum may disseminate widely within the abdomen and pelvis as well as to the inguinal nodes.

22. The answer is D (4). [*Chapter 17 X C 3–4; XII B 2 e, 3–4, D 3 b (1)*] The transverse colon receives its blood supply by the middle colic branch of the superior mesenteric artery and its parasympathetic innervation by the vagus nerve which sends axons along the perivascular plexus. Failure to preserve the vascular pedicle would result in denervation and necrosis.

Part V
Back

19
The Back

I. THE VERTEBRAL COLUMN

A. GENERAL STRUCTURE

1. The vertebral column is composed of 33 bones (**vertebrae**), divided into cervical, thoracic, lumbar, sacral, and coccygeal regions. Vertebrae differ in appearance according to region.

2. There are 23 to 24 **intervertebral disks.**
   a. The disks form **amphiarthroses (symphyses)** between the vertebral bodies.
      (1) In the cervical region there are no disks between the occiput and the atlas, nor between the atlas and the axis.
      (2) Remnants of disks occur within the sacrum and coccyx.
   b. The disks vary in shape, closely corresponding to the associated vertebral bodies, and permit a limited amount of movement between adjacent vertebra.

3. The vertebral column is stabilized by ligaments, which run between the vertebral bodies, the vertebral arches, the transverse processes, and the spinous processes.

B. SPINAL CURVATURES (Fig. 19-1)

1. The **sacral curvature** (S1 to coccyx):
   a. Is concave anteriorly.
   b. Is characteristic of the fetus and lower animal forms and, as such, is considered a primary curvature.

2. The **lumbar curvature** (T12 to L5):
   a. Is concave posteriorly.
   b. Appears between the ages of 12 and 18 months.
   c. Tends to be more pronounced in females than in males.
   d. Presents a greater potential for instability than other regions.
   e. Represents the major adaptation to upright posture.

3. The **thoracic curvature** (T2 to T12):
   a. Is concave anteriorly.
   b. Is characteristic of the fetus and lower animal forms and, as such, is considered a primary curvature.

4. The **cervical curvature** (C2 to T2):
   a. Is concave posteriorly.
   b. Appears between the late intrauterine period and 3 to 4 months after birth.

5. The **normal curvatures** are due almost entirely to the shape of the intervertebral disks, rather than the vertebral bodies.

6. **Abnormal curvatures.**
   a. **Kyphosis** ("hunchback") is an exaggerated thoracic curvature that usually results from congenital deformities, pathologic erosion, or traumatic collapse of one or more vertebral bodies.
   b. **Lordosis** ("swayback") is a greater than normal lumbar curvature. Most commonly it is due to compression of the posterior portion of the lumbar intervertebral disks as compensation for a forward shift in the center of gravity, such as that accompanying the late stages of pregnancy or occurring in obesity.
   c. **Scoliosis** is a lateral curvature. The causes of scoliosis include congenital malformation, pathologic erosion, and unilateral weakness or paralysis of vertebral musculature. Most

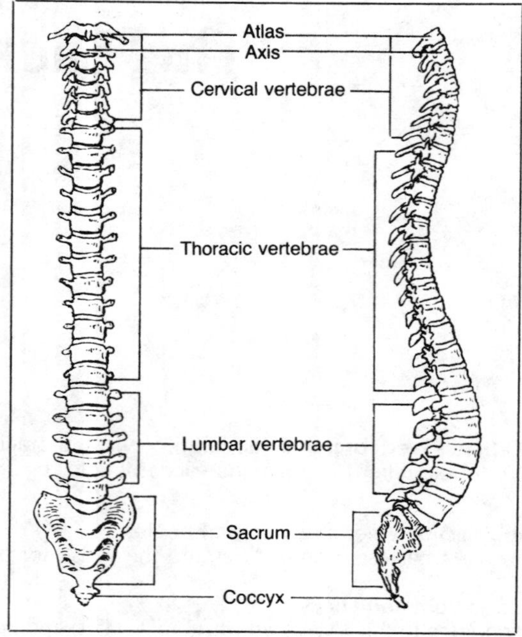

Figure 19-1. *The vertebral column.* The anterior and lateral aspects of the spine illustrate the major differences between the vertebra of the six regions. The primary sacral and thoracic curvatures as well as the secondary lumbar and cervical curvatures are evident in the lateral view.

commonly it is idiopathic (of unknown causes). A compensatory curve in the opposite direction in another region of the spine usually accompanies scoliosis.

(1) As scoliosis progresses, there is a concomitant rotation toward the convex side. This deformity progressively interferes with mechanical respiration.

(2) Scoliosis is amenable to surgical correction, preferably in the late teens and definitely prior to the age of 30.

C. STRUCTURE OF THE TYPICAL VERTEBRA (see Fig. 19-3)

1. The **body** forms the major supportive portion of the vertebra.

a. The body is approximately cylindrical.

b. As an adaptation to upright posture each subsequent body must support a greater proportion of the mass of an individual; therefore, bodies increase in bulk toward the sacrum.

c. The bodies of most vertebrae articulate with the superjacent and subjacent bodies through intervertebral disks that form symphyses (amphiarthroses).

d. The superior and inferior articular processes of most vertebrae articulate with the superjacent and subjacent vertebrae by synovial joints (diarthroses).

2. The **vertebral arch** forms the posterior portion of the vertebra.

a. The **pedicles** arise posterolaterally from the vertebral body and form the foot processes of the vertebral arch.

b. The **laminae** are supported by the pedicles and fuse in the midline to form the roof of the vertebral arch.

3. Vertebral projections.

a. The **spinous process** arises from the vertebral arch where the laminae fuse. It serves as a lever for the attachment of vertebral musculature and ligaments.

b. The **transverse processes** arise on each side of the vertebral arch at the point at which the pedicles and laminae fuse. They serve as levers for the attachment of vertebral musculature and ligaments, as well as points of articulation for the necks of ribs 2 through 10 in the thoracic region.

c. The **costal processes** arise on each side of the vertebral body anterior to the pedicle. They form the ribs in the thoracic region and contribute to the transverse processes of the cervical and lumbar vertebrae, as well as to the alae (wings) of the sacrum.

d. The **superior articular processes** (zygapophyses) are located on the superior surface of the pedicles.

(1) These form diarthrodial articulations with inferior articular processes of the vertebra immediately above.

(2) These processes generally face posteriorly in the cervical region, posterolaterally in the thoracic region, medially in the lumbar region, and posteriorly in the sacrum.

e. The **inferior articular processes** (zygapophyses) arise from the inferior surfaces of the pedicles.
  (1) These form diarthrodial articulations with the superior articular processes of the vertebra immediately below.
  (2) These processes generally face anteriorly in the cervical region, anterolaterally in the thoracic region, and laterally in the lumbar region, but become anterior by the fifth lumbar vertebra.

4. The foramina.
  a. The **vertebral foramen** is formed by the posterior surface of the vertebral body and the vertebral arch.
    (1) Successive foramina form the **vertebral canal.**
    (2) The vertebral foramina contain the spinal cord and its meningeal coverings as well as the associated vessels embedded in adipose tissue.
  b. ***Inter*****vertebral foramina** are located between adjacent vertebrae.
    (1) The intervertebral foramen on each side is formed by a deep notch in the inferior surface of each pedicle of one vertebra and a shallow notch in the superior surface of each pedicle of the vertebra immediately below.
    (2) These foramina transmit the spinal nerves as they exit the vertebral canal.

D. REGIONAL VARIATION IN VERTEBRAL CHARACTERISTICS

1. **Cervical vertebrae (7).**
  a. The **atlas** is the first vertebra (Fig. 19-2).
    (1) This vertebra, lacking a body and a spinous process, is composed of an anterior arch, the vertebral arch, and paired transverse processes.
    (2) It articulates superiorly with the occiput of the skull (the atlantooccipital joint).
      (a) Two large concave facets receive the occipital condyles.
      (b) The atlantooccipital joint functions in flexion/extension (± 15°).
      (c) There is no intervertebral disk at the atlantooccipital joint, which accounts for the extensive range of movement here.
    (3) The atlas articulates inferiorly with the axis.
  b. The **axis** is the second vertebra (see Fig. 19-2).
    (1) The second vertebra is similar to the typical cervical vertebrae except that it has a superior projection, the **dens**, and large flat superior articular surfaces which are nearly horizontal.

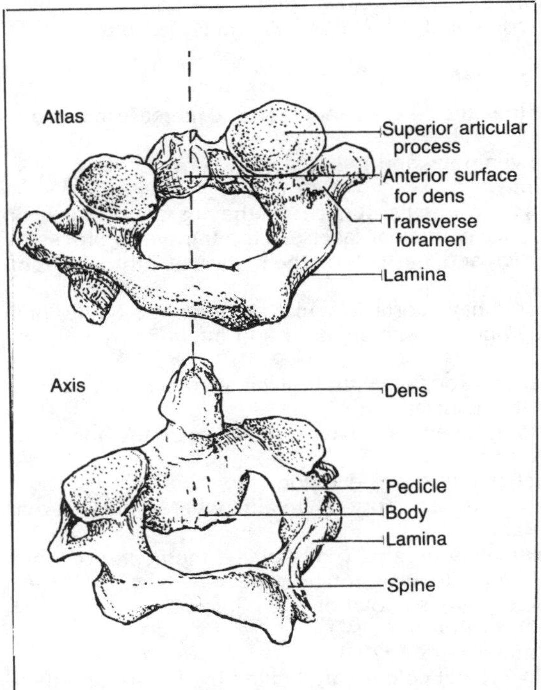

Figure 19-2. *The atlas and axis.* The atlas articulates with the occipital condyles of the cranium and has no body or spinous process. The odontoid process (dens) of the axis, about which rotation occurs, represents the misplaced body of the atlas.

(2) It articulates superiorly with the atlas both at the dens, which projects into the atlas, and at the zygapophyses.

(a) The dens represents the body of the atlas, which has become separated from the first vertebra and fused with the second.

(i) The dens, stabilized within the atlas by several ligaments, serves as a pivot about which rotation occurs.

(ii) Fracture and posterior dislocation of the dens may crush the spinal cord at the level of the first cervical vertebra, with subsequent death from paralysis of all respiratory musculature.

(b) The atlantoaxial joint functions primarily in rotation (± 45°).

(c) There is no intervertebral disk at the atlantoaxial joint, which accounts for the extensive range of rotational movement here.

(3) The axis articulates inferiorly with the third cervical vertebra through an intervertebral disk.

c. **Typical cervical vertebrae** (see Fig. 19-3).

(1) Cervical vertebrae are relatively small but increase in size toward the thoracic region.

(2) They have a characteristic **transverse foramen** in each transverse process.

(a) The transverse foramen represents a hiatus that results from incomplete fusion of the costal and transverse processes.

(b) Except for the seventh cervical vertebra, the transverse foramen transmits the **vertebral artery**, which provides collateral circulation to the brain.

(3) The cervical spinous processes angle sharply downward and are frequently bifid. The spinous process of the seventh cervical vertebra, the **vertebra prominens**, is a most useful surface landmark.

(4) The cervical articular processes have a characteristic orientation.

(a) These articular facets are cup-shaped and lie in a more-or-less transverse plane.

(b) The superior articular surfaces generally face posteriorly, whereas the inferior articular surfaces generally face anteriorly.

(c) These joints have three degrees of freedom, allowing rotation, flexion/extension, and lateral flexion (bending).

(d) The orientation of and the great mobility at the cervical zygapophyses predispose to dislocation.

d. Summary of movements over the cervical region.

(1) Flexion: 40°/extension: 90°, primarily at the atlantooccipital joint.

(2) Lateral flexion: 35° to 40°, primarily between the second through seventh cervical vertebrae.

(3) Rotation: ± 45° to 50°, primarily at the atlantoaxial joint.

(4) The combination of flexion/extension and lateral flexion permits circumduction of the head.

2. **Thoracic vertebrae** (Fig. 19-3).

a. The 12 thoracic vertebrae are larger than the cervical vertebrae and increase in size toward the lumbar region.

b. The thoracic spinous processes are very long and oriented caudally.

c. The costal processes develop into ribs.

d. The bodies and transverse processes have articular surfaces for the ribs (diarthrodial joints).

(1) The upper 10 thoracic vertebrae have articular facets on the transverse processes for the necks of the ribs as well as articular demifacets on the bodies near the origins of the pedicles for the heads of the associated ribs.

(2) The eleventh and twelfth vertebrae have costal articular facets on the bodies only.

(3) The second through ninth ribs articulate with superior and inferior vertebrae at the level of the intervertebral disk.

(a) The bodies of the first and tenth through twelfth thoracic vertebrae have complete facets for the articulation of the associated ribs.

(b) The second through ninth thoracic vertebrae have demifacets on the bodies for articulations with the associated ribs.

e. The thoracic zygapophyses have a characteristic orientation.

(1) The paired superior articular processes face posterolaterally, whereas the inferior articular processes face anterolaterally.

(2) The articular surfaces lie approximately on an arc of a circle, the center of which is located in the vertebral body, thereby allowing rotation.

(a) Over the 12 thoracic vertebrae there is a total of ± 35° of rotation.

(b) There is some flexion (40°) and extension (20°).

(c) There is a small amount of lateral flexion (20°).

(3) The movements of the thoracic vertebral column are limited by the ribs and their attachment to the sternum.

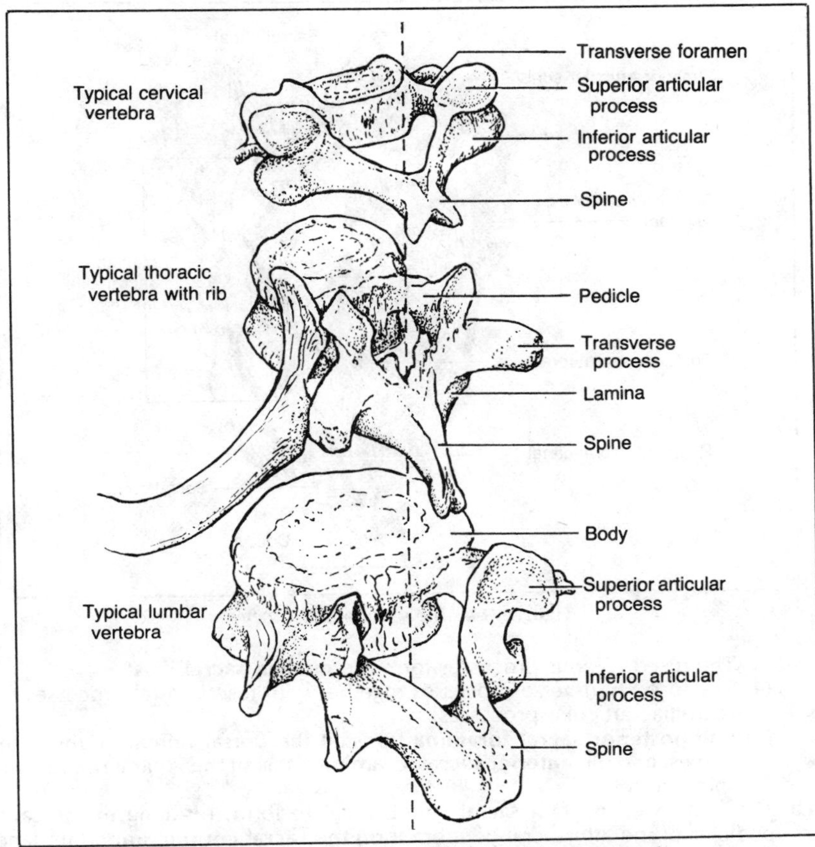

Figure 19-3. *Typical cervical, thoracic, and lumbar vertebrae.*

3. **Lumbar vertebrae** (see Fig. 19-3).
   a. The five lumbar vertebrae are very large and heavy, becoming progressively larger toward the sacral region.
   b. The spinous processes are stubby and project directly posterior. This allows considerable extension as well as access to the intervertebral canal with a needle (**lumbar puncture**).
   c. The lumbar articular processes have a characteristic orientation.
      (1) They lie on nearly parasagittal planes. The articular surfaces lie on an arc of a circle, the center of which is located near the tips of the spinous processes.
      (2) The superior articular surfaces face medially, whereas the inferior articular surfaces face laterally. By the level of the fifth lumbar vertebra, however, the inferior facets become directed anteriorly.
         (a) This orientation limits rotation over the lumbar region to $\pm 5°$.
         (b) It, however, allows considerable flexion (60°)/extension (35°).
         (c) There is a moderate amount of lateral flexion (20°).
         (d) The combination of flexion/extension and lateral flexion allows a considerable amount of circumduction of the trunk.
         (e) The articular facets for the fifth lumbar vertebra are directed anteriorly and engage the posteriorly directed superior processes of the first sacral vertebra.

4. **Sacral vertebrae** (Fig. 19-4).
   a. The five sacral vertebrae fuse to form the sacrum.
      (1) The articular facets for the first sacral vertebra are directed posteriorly, thereby preventing the lumbar vertebrae from slipping forward on the anteriorly slanted body of the sacrum.
      (2) Four lines of fusion are evident on the anterior surface.
         (a) There may be a rudimentary disk between the first and second sacral vertebrae.
         (b) Rarely, there may be sacralization of the fifth lumbar vertebra whereby that vertebra becomes partially or completely fused with the sacrum.

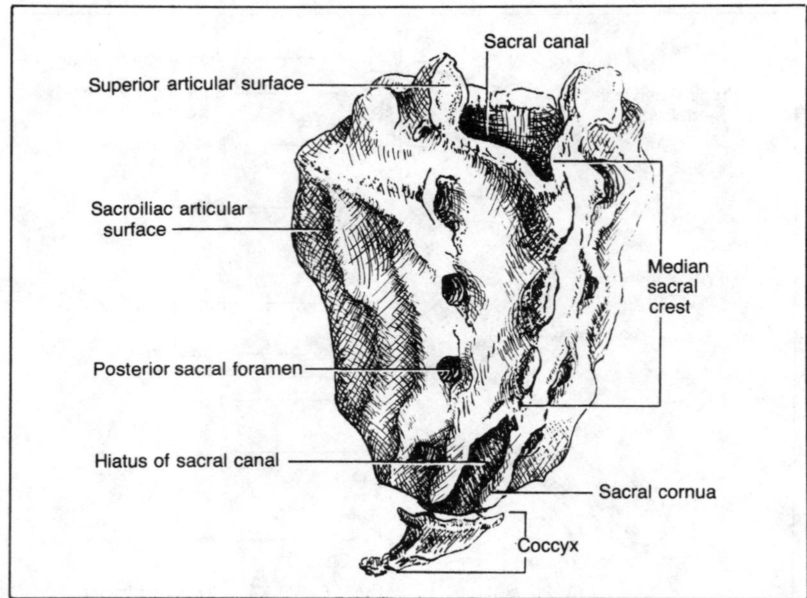

Figure 19-4. *The sacrum and coccyx.*

(3) The fused spinous processes form the **median sacral crest.**

(4) The fused transverse processes together with fused costal processes form the **alae** or sacroiliac articular processes.

(5) The **posterior sacral foramina** transmit the dorsal primary rami of the sacral spinal nerves, and the **anterior sacral foramina** transmit the ventral primary rami of the sacral spinal nerves.

b. The lamina of the fifth sacral vertebra fail to form, resulting in the **sacral hiatus.** The pedicles of the fifth sacral vertebra form the **sacral cornua**, important landmarks used in locating the sacral hiatus for the administration of *caudal anesthesia.*

5. Coccygeal vertebrae (see Fig. 19-4).

a. The four or five coccygeal vertebrae fuse to form the coccyx, but there may be a disk between the first and second coccygeal vertebrae.

b. The first coccygeal vertebra articulates with the sacrum. The sacrococcygeal joint usually has a small intervertebral disk but may fuse in some individuals.

c. The first (and sometimes the second) coccygeal vertebra has rudimentary pedicles that form the coccygeal cornua.

d. The second through fourth coccygeal vertebrae represent rudimentary vertebral bodies.

e. Fracture of the fused coccygeal joints or of a fused sacrococcygeal joint, or arthritis in these joints, causes a great deal of pain (**coccygodynia**).

6. Clinical considerations.

a. Compression fractures of the vertebral bodies may result from pathology or trauma. The internal collapse of the vertebral body may not only result in kyphosis or scoliosis but also may reduce the size of the intervertebral foramen and increase the possibility of spinal nerve compression.

b. Fracture of the pedicles dissociates a vertebral body from the stabilizing influences of the associated zygapophyses. Such fractures predispose to misalignment of the vertebral column, with risk of compression of the spinal cord as well as damage to the exiting spinal nerves.

(1) Because the vertebral canal is relatively wide in the cervical region, a small displacement may not compress the spinal cord.

(2) Because the vertebral canal is relatively narrow in the thoracolumbar region, paraplegia is a common sequel to a thoracic fracture dislocation.

c. Spondylolisthesis.

(1) The large angle (140°) between the axes of the fifth lumbar vertebra and the sacrum at the lumbosacral joint, as well as the considerable angle (41°) of the body of the first sacral vertebra from the horizontal, imposes enormous forces upon the superior articular processes of the first sacral vertebra.

(2) Spondylolisthesis is an anterior displacement of the vertebral column, usually at the lumbosacral joint. This may be due to:
(a) Congenital defect of the lamina of L5 (**spondylolysis**), found in 5 percent of the population.
(b) Fracture of the pedicles or inferior articular processes of the fifth lumbar vertebra.
(c) Fracture of the lamina or superior articular processes of the sacrum.
(3) Since the center of gravity normally passes through the lumbosacral joint, gravity causes the vertebral column to slide forward on the intervertebral disk at the lumbosacral joint, with the anterior longitudinal ligament supporting the entire vertebral structure.
(4) Spondylolisthesis usually involves compression of spinal nerves S1 and S2, producing symptoms of sciatica.

E. THE INTERVERTEBRAL DISKS

1. The functions of the intervertebral disks include:
   a. The binding of the vertebrae together (**amphiarthroses**).
   b. Permission of movement between adjacent vertebrae.
   c. The uniform distribution of forces over the entire surface of the vertebra regardless of the degree of flexion/extension, rotation, or lateral flexion.

2. The disks conform to surfaces of apposed vertebral bodies.
   a. The intervertebral disks comprise about 25 percent of the total length of the vertebral column.
   b. In the cervical region the disks are small and thin, but by the lumbar region the disks have become progressively larger and thicker.
   c. Differences in thickness anteriorly and posteriorly define the secondary vertebral curvatures (i.e., they are thicker anteriorly in the lumbar and cervical regions).

3. Each disk is composed of an outer fibrocartilaginous portion and an inner gelatinous central portion (see Fig. 19-5).
   a. The **anulus fibrosus** forms the outer portion of the intervertebral disk.
      (1) This structure is composed of lamelliform connective tissue bands. Each band is oriented in a different direction, providing considerable strength.
      (2) The anulus fibrosus is firmly attached to and joins adjacent vertebrae, forming a symphysis type of amphiarthrosis.
      (3) It supports the central nucleus pulposus.
   b. The **nucleus pulposus** forms the central portion of the intervertebral disk.
      (1) This structure is a remnant of the embryologic **notochord**.
      (2) It is surrounded and supported by the anulus fibrosus.
      (3) The nucleus pulposus is composed of mucinous tissue with fibrous and mucopolysaccharide complexes, which contribute to its high osmotic pressure.
      (4) The extremely high water content of the disk (70 to 80 percent) accounts for the fact that a person may be as much as 1 cm taller upon rising in the morning than at the end of a strenuous day.
      (5) The nucleus pulposus functions as a noncompressible but deformable pad, which distributes weight evenly over a larger surface area and allows considerable movement.

4. Clinical considerations.
   a. **Disk degeneration.**
      (1) Degeneration is associated with chronic dehydration of the nucleus pulposus.
      (2) Loss of water and mucopolysaccharide leads to a narrowing of intervertebral space and reduces the capacity of the disk to act as a cushion between vertebrae.
      (3) Narrowing of the intervertebral disks results in diminished stature. It also decreases the size of the intervertebral foramina, thereby increasing the possibility of spinal nerve compression.
      (4) Occasionally, degeneration arthroses (spondylosis) produces progressive calcification at the superior and inferior margins of the vertebral bodies to form **osteophytes**. These may compress the spinal nerves, especially in the cervical region with resultant arm pain, or cause soft tissue irritation, especially in the lumbar region with resultant back pain.
   b. **Disk herniation** ("slipped disk").
      (1) Herniation is the prolapse (extrusion) of the nucleus pulposus through the anulus fibrosus.
      (2) Because of the lumbar curvature and the considerable mass of the body superior to this region, herniation usually occurs between the fourth and fifth lumbar vertebrae or be-

tween the fifth lumbar and first sacral vertebrae. Disk herniation also occurs frequently in the cervical region.

(3) Rarely, prolapse may occur into the superjacent or subjacent vertebral bodies (Schmorl's node).

(4) Most commonly, herniation is directed posterolaterally. The presence of the anterior longitudinal ligament and the posterior longitudinal ligament reinforces the underlying anulus fibrosus.

(5) Posterolateral prolapse impinges upon the spinal nerve of the next lower vertebral level, causing symptoms associated with the dermatomic and myotomic distributions of that nerve. This is true in *both* the cervical and the lumbar regions, but for different reasons.

(a) In the cervical region, there are eight cervical nerves, but seven vertebrae; thus, herniation of the disk between the fourth and fifth cervical vertebrae impinges upon spinal nerve C5, which exits through the intervertebral foramen formed by the fourth and fifth cervical vertebrae (see III C 1).

(b) In the lumbar region, the deeply notched pedicles of the vertebrae are located high on the body and the intervertebral disks are thicker; thus, herniation of the disk between the fourth and fifth lumbar vertebrae does not harm spinal nerve L4, which exits just superior to the disk. However, the herniation compresses spinal nerve L5 as it descends across the disk between the fourth and fifth lumbar vertebrae to its intervertebral foramen, producing the symptoms of **sciatica**. If the herniation is more substantial and posteriorly directed, it may involve one or more spinal nerves below the level of herniation; if very large, it may involve some spinal nerves on the contralateral side as well.

(c) In addition to the pain referred to the distribution of the compressed spinal nerve, local pain produced from the stretched anulus fibrosus initiates painful spasm of the back muscles.

F. LIGAMENTS OF THE VERTEBRAL COLUMN (Fig. 19-5)

1. The **supraspinous ligament** connects spinous processes at the tips and forms the **ligamentum nuchae** in the cervical region.

2. The **interspinous ligaments** run between spinous processes.
   - **a.** These ligaments limit the range of motion of the vertebrae.
   - **b.** Small tears in these ligaments, the result of hyperextension trauma, may be the basis for "whiplash" injury.

3. The **ligamentum flavum** (*flavum, L.* yellow) stretches between adjacent laminae. Except for a midline gap, the paired ligaments, together with the vertebral arch, complete the posterior aspect of the vertebral canal.
   - **a.** Unlike the other ligaments, the ligamentum flavum is composed of elastic tissue.
   - **b.** Upon traumatic hyperextension of the neck, the capacity of the ligamentum flavum to take up slack may be exceeded, and the resultant buckling may injure the spinal cord.

4. The **anterior longitudinal ligament** connects the vertebral bodies and intervertebral disks anteriorly. This ligament:
   - **a.** Is a very broad band and runs from the sacrum to the occipital bone.
   - **b.** Resists the gravitational tendency toward increased lordosis.
   - **c.** Reinforces the anulus fibrosus anteriorly and directs herniation posteriorly.
   - **d.** Can be utilized to splint fractured vertebrae when the trunk is cast in extension.

5. The **posterior longitudinal ligament** connects the vertebral bodies and intervertebral disks posteriorly. This ligament:
   - **a.** Is denticulate in shape, wider posterior to the vertebral body and narrower posterior to the intervertebral disk.
   - **b.** Resists the gravitational tendency toward increased kyphosis.
   - **c.** Supports the anulus fibrosus posteriorly and directs herniation of the disk laterally.

II. SOFT TISSUES OF THE BACK

A. THE SUPERFICIAL MUSCULATURE OF THE BACK is associated with the pectoral and pelvic girdles.

B. THE POSTERIOR MUSCLES OF THE VERTEBRAL COLUMN are innervated by dorsal primary rami of spinal nerves and are arranged in three layers. The most superficial layer runs upward

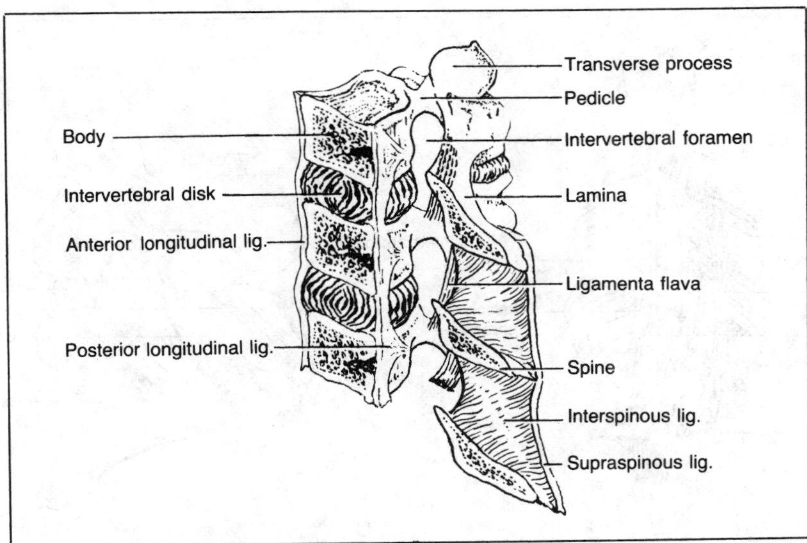

Figure 19-5. *The ligaments of the vertebral column.*

and obliquely outward, the middle layer runs parallel to the vertebral column, and the innermost layer runs upward and obliquely inward. The actions of these muscles include extension, rotation, and lateral flexion.

1. The **spinotransverse group** originates from spinous processes and the nuchal ligament, runs superiorly and obliquely laterally, and inserts onto either transverse processes or the base of the skull (Fig. 19-6).
   - **a.** The **splenius capitis** runs from spines of the seventh cervical through the fourth thoracic vertebrae and the nuchal ligament, to insert into the superior nuchal line and mastoid process of the skull.
   - **b.** The **splenius cervicis** runs from spines of the third through sixth thoracic vertebrae to the transverse processes of the second through fourth cervical vertebrae.
   - **c.** These muscles, acting unilaterally, rotate the head and neck toward the same side and, acting bilaterally, elevate the head and extend the neck.
2. The **sacrospinalis group (erector spinae)** originates from the sacrum, iliac crests, and spinous processes of the lumbar and lower thoracic vertebrae; runs parallel to the vertebral column; and inserts into ribs and transverse processes (see Fig. 19-6).
   - **a.** The **iliocostalis muscles** comprise the most lateral portion of the erector spinae muscle.
     - **(1)** The iliocostalis lumborum runs from sacrum and iliac crests to the inferior borders of the lower six ribs.
     - **(2)** The iliocostalis thoracis runs from the superior borders of the lower six ribs to the inferior borders of the upper six ribs.
     - **(3)** The iliocostalis cervicis runs from the superior borders of the upper six ribs to the inferior borders of the transverse processes of the fourth through sixth cervical vertebrae.
   - **b.** The **longissimus** comprises the middle portion of the erector spinae muscle (see Fig. 19-6).
     - **(1)** The longissimus thoracis runs from the sacrum to the inferior borders of the transverse processes and associated ribs of the lower four thoracic vertebrae.
     - **(2)** The longissimus cervicis runs from the superior borders of the transverse processes of the upper six thoracic vertebrae to the inferior borders of the transverse processes of the second through sixth cervical vertebrae.
     - **(3)** The longissimus capitis runs from the superior borders of the second through sixth cervical vertebrae to the mastoid process of the skull.
   - **c.** The **spinalis** comprises the medial portion of the erector spinae muscle and is poorly developed.
     - **(1)** The spinalis thoracis, which is the most highly developed portion, runs between spinous processes.
     - **(2)** The spinalis cervicis is inconsistently present.
     - **(3)** The spinalis capitis is usually considered the medial portion of the semispinalis capitis muscle.
   - **d.** The sacrospinalis group of muscles acting unilaterally, flex the vertebral column to the same side and, acting bilaterally, extend the vertebral column.

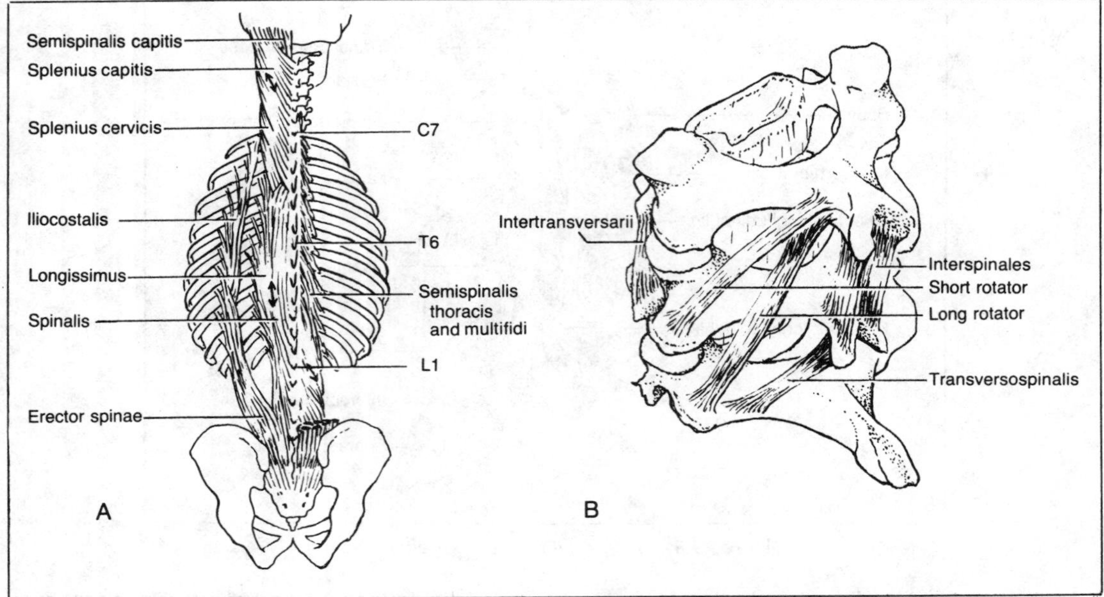

Figure 19-6. *The muscles of the back. A, The long muscles. B, The short muscles.*

3. The **transversospinal group (transversospinalis)** originates from transverse processes, runs superiorly and obliquely medially, and inserts into spinous processes (see Fig. 19-6). The deeper muscles traverse shorter distances.
   a. The **semispinalis muscle** lies beneath the erector spinae muscle and generally passes over five or more vertebrae (see Fig. 19-6).
      (1) The semispinalis thoracis runs from the transverse processes of the lower seven thoracic vertebrae to the spines of the upper four thoracic vertebrae.
      (2) The semispinalis cervicis runs from the transverse processes of the upper six thoracic vertebrae to the spines of the second through fifth cervical vertebrae.
      (3) The semispinalis capitis overlies the semispinalis cervicis and runs from the transverse processes of the fourth cervical through the fourth thoracic vertebrae to the inferior nuchal line of the skull.
      (4) The medial portion of this muscle, originating from spinous processes, is termed the spinalis capitis.
   b. The **multifidus muscles (multifidi)** lie deep to the semispinalis muscle and generally pass over two to three vertebrae (see Fig. 19-6).
      (1) Except for their shorter bundles, they are nearly indistinguishable from the semispinalis muscle.
      (2) The multifidus muscles are best developed in the lumbar and cervical regions.
   c. The **long rotators** run from the transverse processes to the spinous processes two vertebrae above.
   d. The **segmental muscles** run between adjacent vertebrae (see Fig. 19-6).
      (1) The **short rotators** run between the transverse and spinous processes of adjacent vertebrae.
      (2) The **interspinales muscles** run between spinous processes and are best developed in the cervical region.
      (3) The **intertransversarii muscles** run between transverse processes and are especially well developed in the lumbar and cervical regions.
   e. The **transversospinalis muscles**, acting unilaterally, rotate the neck and trunk to the *opposite* side and, acting bilaterally, extend the vertebral column.

4. The **suboccipital muscles** act to extend the head.
   a. The inferior oblique capitis runs from the spine of the axis to the transverse process of the atlas.
   b. The superior oblique capitis runs from the transverse process of the atlas to the occiput.
   c. The posterior rectus capitis major runs from the spine of the axis to the occiput.
   d. The posterior rectus capitis minor runs from the spine of the atlas to the occiput.

5. The **posterior muscles**, innervated by dorsal primary rami of the spinal nerves, variously function in extension, lateral flexion, and rotation. In addition, tonic activity of this musculature adjusts the posture so that the center of gravity of the individual is maintained directly over the sacrum.

C. THE ANTERIOR VERTEBRAL MUSCLES

1. The **scalene muscles** run from transverse processes of cervical vertebrae to the first and second ribs.
2. The **longus colli muscle** runs from the fourth through sixth cervical vertebrae to the occiput.
3. The **lateral rectus capitis muscle** runs between the transverse process of the atlas and the jugular process of the occiput.
4. The **quadratus lumborum muscle** runs from the iliac crests to the inferior borders of the twelfth ribs.

D. FUNCTIONAL CONSIDERATIONS

1. In addition to major movement of the vertebral column, the normal function of the vertebral musculature is the making of fine adjustments to keep the center of gravity of the body over the first sacral vertebra.
2. When a person stands erect and balanced with a 10-kg (22-pound) weight in each hand, 20 kg is distributed evenly over each vertebral disk.
3. If, instead, the 20 kg is held 20 cm anterior to the vertebral column, the forward shift in the center of gravity requires contraction of the back muscles in order to keep the individual erect.
   a. The muscles that insert into the tips of the spinous processes (i.e., those having the maximum mechanical advantage because of a 2-cm lever arm) exert 200 kg of counterforce (20 kg × 20 cm/2 cm = 200 kg).
   b. This counterforce, pulling downward on one side of the vertebra, is additive to the weight of the object lifted. Thus the total force distributed over each intervertebral disk is 220 kg.
4. If, in addition, the individual supports the 20-kg weight with flexed arms and with the trunk flexed forward so that the weight extends 50 cm anterior to the fifth lumbar vertebra, the back muscles must generate 500 kg of counterforce (20 kg × 50 cm/2 cm = 500 kg).
   a. This muscular pull downward on the vertebral spine is additive to the weight of the object lifted, so that 520 kg must be distributed over each lumbar intervertebral disk.
   b. Since a disk may rupture when forces exceed 500 to 800 kg, the individual is clearly placed in physical jeopardy.

III. THE SPINAL CORD

A. GENERAL STRUCTURE

1. The **spinal cord** is an extension of the brain and a component of the **central nervous system.**
   a. The **spinal nerves** arise from the spinal cord (Fig. 19-7).
   b. The region of rootlets that comprises a spinal nerve is termed a **spinal segment**.
2. The spinal cord extends from the **medulla oblongata** at the level of the first cervical vertebra and terminates as the **conus medullaris** at the level of the intervertebral disk between the first and second lumbar vertebrae.
   a. Until the third fetal month, the spinal cord is as long as the vertebral canal, and both extend to the level of the fourth sacral vertebra.
   b. After the fourth month the vertebral column outgrows the spinal cord, so that the cord ultimately ends at the upper lumbar levels. However, the spinal roots proceed through the dural sac to exit at the appropriate intervertebral foramina. The resultant distribution of spinal roots forms the **cauda equina** (*L.* horse's tail).
   c. In most individuals, the spinal cord terminates between the first and second lumbar vertebrae. Because the spinal cord extends below the second lumbar vertebra in 1 percent of the population, neither lumbar puncture nor spinal anesthesia should be attempted above the level of the fourth lumbar vertebra.
   d. Because the spinal cord is considerably shorter than the vertebral column, serious injury at a particular vertebral level produces spinal cord injury at a different (lower) segmental level.
3. The spinal cord and its meningeal coverings lie in an osseofibrous canal (the neural or **vertebral canal**).

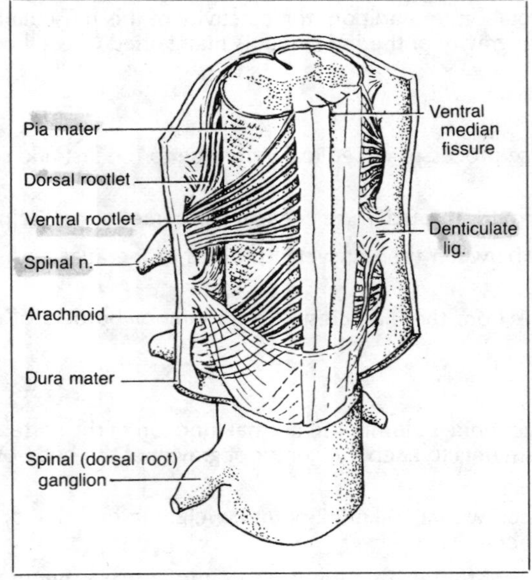

Figure 19-7. *The spinal cord and meninges.*

a. The vertebral canal is formed by the vertebral body, pedicles and laminae of the vertebrae, the intervertebral disks, and the ligamentum flavum.
b. The **epidural space**, between the boundaries of the vertebral canal and the dura mater, is filled with loose fatty connective tissue and contains the extensive **vertebral venous plexus.**

4. The spinal cord is suspended in **cerebrospinal fluid** (CSF).
   a. CSF is secreted into the ventricles of the brain by the choroid plexuses. It passes into the subarachnoid space, bathing the brain and spinal cord. It is absorbed from the subarachnoid space into the superior sagittal venous sinuses of the cranium. There is thus a natural circulation of CSF.
   b. CSF functions as a shock absorber, reducing the possibility of cord injury from contact with the boundaries of the vertebral canal during sudden traumatic acceleration or deceleration.
   c. CSF is sampled and examined to assist in the diagnosis of disease or pathologic conditions relating to the brain and spinal cord. The presence of red blood cells is indicative of a bleed, such as a ruptured aneurysm, whereas a large number of polymorphonuclear leukocytes is indicative of bacterial infection, such as meningococcal meningitis.

5. The spinal cord is composed of gray matter and white matter.
   a. Nerve cell bodies comprise the **gray matter.**
      (1) The neurons located dorsally in the gray matter tend to be small internuncials interposed in reflex arcs.
      (2) The neurons located laterally in the gray matter tend to be autonomic neurons. These leave by the ventral root and innervate smooth muscle and secretory cells.
      (3) The neurons located ventrally in the gray matter tend to be the **lower motor neurons.** These leave the spinal cord in the ventral roots and innervate striated muscle.
      (4) The neurons located medially and dorsally in the gray matter tend to be second-order neurons in sensory pathways. The axons of these cells enter the white matter and ascend to the brain stem.
   b. Bundles of nerve axons, organized into tracts, form the **white matter.**
      (1) **Motor tracts** are composed of **upper motor neurons**, which originate in the brain and descend in the spinal cord to make synaptic contact with the **lower motor neurons.**
      (2) **Sensory tracts** consist of bundles of axons.
         (a) Some sensory tracts are continuations of the peripheral sensory nerves, which have their cell bodies in the **dorsal root ganglion**, coursing toward the brain stem.
         (b) Other sensory tracts arise from the neurons of the gray matter of the spinal cord, upon which peripheral sensory nerves make synaptic contact.
         (c) These tracts convey different sensory modalities to the brain stem. Some are crossed, whereas others ascend on the ipsilateral side to cross in the brain stem.

6. Functional considerations.
   a. Other peripheral nerves either make synaptic contact directly or do so through internuncial neurons with lower motor neurons to form the basic **reflex arc** (see Fig. 4-3).
   b. Spinal nerve rootlets enter and leave the cord at every segmental level. Thus, the amount of white matter in the spinal cord diminishes from the cervical region to the conus medullaris. Because of the large number of reflexes associated with the extremities, the amount of gray matter increases in the cervical and lumbosacral regions.
   c. The sensory and motor tracts are somatotopically organized. Each sensory tract carries specific modalities, and each motor tract correlates with specific motor functions. Thus, a lesion that produces a specific combination of functional deficits at any level can be localized with precision. However, since the neurons of the spinal cord do not regenerate, often there is little that can be done with this knowledge other than the prevention of further loss.

B. THE MENINGES (see Fig. 19-7)

1. The **dura mater** (*L.* hard mother) is the outermost meningeal layer. This layer:
   a. Is a fibrous membrane in the form of a tubular sheath.
   b. Extends from the foramen magnum of the cranium (where it fuses with the periosteum of the cranial vault) to the level of the second sacral vertebra.
   c. Terminates as the **coccygeal ligament**, a dural covering of the **filum terminale.**
   d. Extends into the intervertebral foramina, surrounds spinal nerves, and fuses with the periosteum of the vertebrae.
   e. Defines the **epidural space** and the **subdural space.**

2. The **arachnoid** (*L.* spiderlike) is the intermediate meningeal layer.
   a. The arachnoid layer is loosely adherent to the dura.
      (1) It is separated from the dura by a potential space (the **subdural space**).
      (2) The arachnoid layer does *not* contain CSF.
   b. The arachnoid is attached to the underlying pia mater by numerous **arachnoid trabeculae.**
      (1) The considerable cavity between the arachnoid and pial layers defines the **subarachnoid space.**
      (2) The subarachnoid space contains cerebrospinal fluid.
      (3) This space from the conus medullaris (approximately L1) to the level of the second sacral vertebra is termed the **dural sac** (a misnomer).
      (4) It is the dural sac into which a needle is inserted in the lumbar puncture (spinal tap) and for the induction of spinal anesthesia.

3. The **pia mater** (*L.* tender mother) is the innermost meningeal layer.
   a. The pial layer closely invests the spinal cord. It contains a plexus of small blood cells that supply the neural tissue.
   b. The **denticulate ligament** supports the entire spinal cord in the center of the subarachnoid space. It consists of 21 pairs of lateral reflections of the pia, which attach to the dura. Thus the spinal cord is tethered in CSF.
   c. The **filum terminale** is an extension of the pia beyond the conus medullaris to the end of the dural sac at the level of the second sacral vertebra, where it becomes covered with dura and is termed the **coccygeal ligament.**

C. THE NERVE ROOTS (Fig. 19-8)

1. There are 31 pairs of **spinal nerves:** eight cervical (C1 through C7 above each respective cervical vertebra and C8 below the seventh cervical vertebra), twelve thoracic, five lumbar, five sacral, and one coccygeal (each below the respective vertebra).
   a. Each dorsal and ventral root is formed by six to eight rootlets. The sensory rootlets arise in linear fashion from the posterolateral sulci and the motor rootlets arise from the ventrolateral sulci.
   b. Spinal nerves are formed by a dorsal (sensory) and a ventral (motor) root.

2. As a result of unequal growth of the neural canal and spinal cord, there is a progressive descending obliquity of the nerve roots, forming the **cauda equina.**
   a. Only in the cervical region do the segments of the spinal cord correspond in position to the level of the corresponding vertebrae.
   b. Below the cervical region, each spinal nerve from a given cord segment travels inferiorly (sometimes several inches as do the sacral rootlets) before exiting at the appropriate intervertebral foramen.
   c. The nerve roots of the cauda equina float in CSF. Therefore, a needle introduced into the dural sac will displace the roots without damage.

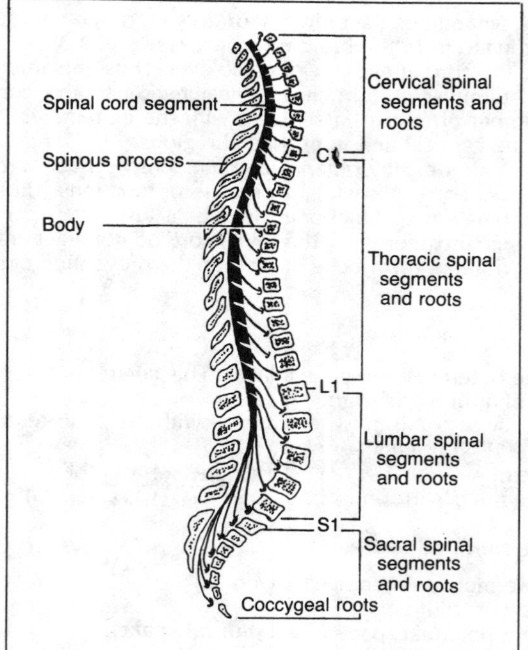

Figure 19-8. *Schematic diagram of the spinal cord and roots.* The spinal cord usually terminates just caudal to the L1 vertebral body below which the spinal roots form the cauda equina as each root courses to its respective intervertebral foramen.

3. The spinal nerve leaves the vertebral canal through the intervertebral foramen. Each spinal nerve bifurcates into a dorsal primary ramus and a ventral primary ramus (see Fig. 4-3).
   a. The dorsal primary rami supply the dermatomes and myotomes of the median portion of the back.
   b. The ventral primary rami supply the dermatomes and myotomes of the lateral and anterior portions of the body as well as the extremities in their entirety.

D. VASCULATURE OF THE SPINAL CORD

1. The **arterial supply** to the spinal cord is mainly by branches of the vertebral, deep cervical, intercostal and lumbar arteries.
   a. These vessels give off spinal branches.
   b. The anterior and posterior spinal branches of the vertebral artery give off ascending and descending branches, which run in the pia mater and fuse to form the midline **anterior spinal artery** and two **posterior spinal arteries**
   c. The spinal branches of the other vessels bifurcate into **anterior** and **posterior radicular arteries.**
      (1) Most of the radicular arteries supply the dorsal and ventral roots and terminate in the pia associated with these roots.
      (2) In the lower cervical and upper lumbar regions, six to eight larger radicular arteries reach the spinal cord where they course in the pia mater.
         (a) Some anterior radicular arteries reach the anterior median fissure of the spinal cord and divide into ascending and descending branches. These anastomose with each other to continue the **anterior spinal artery**, which supplies the ventromedial aspects of the spinal cord.
         (b) Some posterior radicular arteries reach the posterolateral sulcus. Their ascending and descending branches anastomose to continue the **posterior spinal arteries**, which supply the dorsal and lateral aspects of the spinal cord.
2. The veins of the spinal cord form a plexus in the pia mater.
   a. Two median longitudinal veins, two anterolateral longitudinal veins, and two posterolateral longitudinal veins are usually distinguishable.
   b. These veins drain into the **vertebral venous plexus**, which lies in the fatty connective tissue of the epidural space.
   c. The **vertebral venous plexus** drains into **vertebral veins**, which generally accompany the arteries.

d. The vertebral venous plexus communicates superiorly with the venous sinuses of the cranium and inferiorly with the venous sinuses of the deep pelvis.

E. CLINICAL CONSIDERATIONS

1. **Lumbar puncture (spinal tap)** is utilized to obtain samples of CSF for laboratory analysis.
   - **a.** Lumbar puncture is facilitated by the fact that the spinal cord rarely extends below the third lumbar vertebra and that the spinous processes of the lower lumbar vertebrae project directly posterior, thereby providing access to the vertebral canal.
   - **b.** When a patient is placed on the side in decubitus position, thereby maximally flexing the vertebral column, the space between the spinous processes of adjacent lumbar vertebrae is widened further.
   - **c.** A line tangential to the highest points of the iliac crests passes through the level of the fourth lumbar vertebra or the interspace between the fourth and fifth lumbar vertebrae.
   - **d.** Local anesthetic is applied over the interspace between the fourth and fifth vertebrae, and a spinal needle is inserted through skin, superficial fascia, supraspinous ligament, interspinous ligament, between the paired flaval ligaments, and through the epidural space, the dura, and the arachnoid to enter the subarachnoid space, which contains CSF.

2. **Spinal anesthesia** is utilized to block the roots of the spinal nerves *within* the dural sac.
   - **a.** The procedure is the same as that for lumbar puncture.
   - **b.** The number of segmental levels anesthetized is controlled by the amount of anesthetic injected and the position of the patient. During spinal anesthesia, the operating table must not be tilted so that the patient's head is below the horizontal.

3. **Epidural anesthesia** is utilized to block the spinal nerves in the epidural space in the lumbar region (i.e., external to the dural sac).
   - **a.** The procedure is similar to that for lumbar puncture except that the dura is not penetrated.
   - **b.** The number of nerves anesthetized above or below the injection site is controlled by the amount of anesthetic injected into the connective tissue of the epidural space.

4. **Caudal (epidural) anesthesia** is utilized to block the spinal nerves in the epidural space (i.e., external to the dural sac).
   - **a.** A needle inserted between sacral cornua penetrates the sacrococcygeal ligament to gain access to the sacral hiatus. The anesthetic is infiltrated into the epidural space around the dural sac.
   - **b.** The degree of ascent of anesthesia can be controlled (e.g., for parturition), and afferent pain fibers from the perineum and cervix running in the sacral nerves can be blocked (saddle block), leaving unaffected the lumbar fibers that control the abdominal musculature, as well as those involved in reflex uterine contractions.

Part V Back

STUDY QUESTIONS

Directions: Each question below contains five suggested answers. Choose the **one best** response to each question.

1. Which of the following structures pass through the posterior sacral foramina?

(A) Anterior primary rami
(B) Filum terminale
(C) Sensory rootlets
(D) Spinal nerves
(E) None of the above structures

2. In the adult vertebral column, the spinal cord usually terminates between

(A) T12 and L1
(B) L1 and L2
(C) L2 and L3
(D) L3 and L4
(E) L4 and L5

3. The vertebral venous plexus is found in

(A) the dural sac
(B) the epidural space
(C) the subarachnoid space
(D) the subdural space
(E) none of the above

4. The dura mater terminates caudally as the

(A) coccygeal ligament
(B) denticulate ligament
(C) dural sac
(D) filum terminale
(E) conus medullaris

5. The most important landmark for the preferred location for a spinal tap or induction of spinal anesthesia is the

(A) iliac crests
(B) posterior-superior iliac spines
(C) prominent spinous process of T12
(D) sacral cornua
(E) sacral hiatus

Directions: Each question below contains four suggested answers of which **one or more** is correct. Choose the answer

A if **1, 2, and 3** are correct
B if **1 and 3** are correct
C if **2 and 4** are correct
D if **4** is correct
E if **1, 2, 3, and 4** are correct

6. Characteristics of the atlas include

(1) an odontoid process
(2) flat inferior articular facets
(3) a prominent spinous process
(4) bowl-shaped superior articular facets

7. Which of the following structures are formed by or receive a contribution from the costal processes of the developing vertebrae?

(1) The transverse processes of the cervical vertebrae
(2) The ribs of the thoracic region
(3) The alar plates of the sacrum
(4) The sacral cornua

SUMMARY OF DIRECTIONS

A	B	C	D	E
1, 2, 3 only	1, 3 only	2, 4 only	4 only	All are correct

8. The transversospinalis group includes which of the following intrinsic back muscles?

(1) Long and short rotators
(2) Semispinalis
(3) Multifidi
(4) Splenius cervicis

9. Support for the nucleus pulposus is provided by which of the following structures?

(1) The anterior longitudinal ligament
(2) The annulus fibrosus
(3) The posterior longitudinal ligament
(4) The ligamentum flavum

10. Large volumes of cerebrospinal fluid are found in which of the following spaces?

(1) Epidural space
(2) Dural sac
(3) Subdural space
(4) Subarachnoid space

ANSWERS AND EXPLANATIONS

1. The answer is E. [*Chapter 19 I D 4 a (5)*] The sacral spinal nerves leave the vertebral canal via intervertebral foramina, which bifurcate almost immediately into anterior and posterior foramina. The dorsal primary rami pass through the posterior sacral foramina to innervate the back muscles and provide sensation to the dorsal portion of the dermatomes.

2. The answer is B. (*Chapter 19 III A 2 c*) In most individuals, the spinal cord extends from the medulla oblongata at the level of the first cervical vertebrae and terminates as the conus medullaris at the level of the first lumbar intervertebral disk (between vertebrae L1 and L2). Only in 1 percent of the population does the conus medullaris extend beyond the L2 level.

3. The answer is B. (*Chapter 19 III A 3 b, B 1 e, 2 a–b*) The vertebral venous plexus lies in the loose fatty connective tissue of the epidural space. It receives drainage from the spinal vertebral plexus and drains through the intervertebral veins. It provides a collateral route for venous drainage of the pelvis and as such is a pathway for the spread of pelvic malignancies.

4. The answer is A. (*Chapter 19 III B 1 c*) From the dural sac, the dura mater extends to the second sacral vertebra as the coccygeal ligament, covering the pial filum terminale. The denticulate ligament supports the entire spinal cord in the center of the subarachnoid space.

5. The answer is A. (*Chapter 19 III E 1 c*) The preferred site for a spinal tap is between vertebrae L4 and L5. The intertubercular plane through the iliac crests passes through the level of the fourth lumbar vertebra or the fourth intervertebral disk. The sacral cornua define the sacral hiatus, which is used for caudal anesthesia. The spinous process of T12 is most difficult to locate and is therefore, not a landmark. The posterior-superior iliac spines lie at the L5 to S1 level.

6. The answer is C (2, 4). (*Chapter 19 I D 1 a–b*) The atlas, the first cervical vertebra, has large concave superior facets, which articulate with the occipital condyles. Considerable flexion/extension occurs at the atlantooccipital joint. The inferior facets, which articulate with the axis, are rather flat, permitting considerable rotation at the atlantoaxial joint. The atlas has neither a body nor a prominent spinous process, the former being represented by the odontoid process of the axis.

7. The answer is A (1, 2, 3). [*Chapter 19 I D 1 c (2) (a), 2 c, 4 a (4)*] The costal processes of the developing vertebrae participate in the formation of the transverse processes of the cervical vertebrae, the ribs, and the alar plates of the sacrum. The sacral cornua are the pedicles of the fifth sacral vertebra where the lamina have failed to form.

8. The answer is A (1, 2, 3). (*Chapter 19 II A 1, 3*) While all of the muscles listed are primarily rotators of the vertebral column, the splenius cervicis and splenius capitis muscles are members of the more superficial spinotransverse group, which originate on the spinous processes and diverge superolaterally to insert into transverse processes. The semispinalis, multifidi, and rotators comprise the deeper transversospinalis group, which originate on transverse processes and converge superomedially onto spinous processes.

9. The answer is A (1, 2, 3). (*Chapter 19 I E 3 a, F 4–5*) The annulus fibrosus forms the outer portion of the intervertebral disk and provides the principal support for the more gelatinous nucleus pulposus. Additional support is provided by the anterior and posterior longitudinal ligaments. Disk herniation occurs most frequently in the posterolateral aspect of the disk where these ligaments do not meet.

10. The answer is C (2, 4). [*Chapter 19 III A 3 b, 4, B 2 a (1), b (2)–(3)*] Cerebrospinal fluid is contained in the subarachnoid space. Below the termination of the conus medullaris, the subarachnoid space is termed the dural sac. The subdural space is only a potential space, and the epidural space contains the vertebral venous plexus.

Part VI
Pelvic Region

20
The Pelvis

I. OSSEOUS PELVIS

A. THE PELVIS is formed by the bilateral **innominate bones**, the medial **sacrum**, and the **coccyx**.

1. The **coxal** (innominate or hip) **bone** on each side of the pelvis is formed by the fusion of three separate bones—the **ilium**, **ischium**, and **pubis**; these join at the **acetabulum** (*L.* vinegar cup), which is the fossa of the hip joint.
   - **a.** The **ilium** is the most superior portion of the coxal bone (Fig. 20-1).
     - **(1)** A flared **alar plate** provides attachment for back and thigh musculature, whereas the concave inner surface, the iliac fossa, provides attachment for the iliacus muscle.
     - **(2)** The posterior portion of the ala, the auricular surface, articulates with the sacrum at the **sacroiliac joint**.
     - **(3)** At the free superior edge of the ala the **iliac crest**, which terminates at the antero- and posterosuperior iliac spines, provides attachment for back and abdominal musculature as well as several named ligaments.
     - **(4)** The **greater sciatic notch** of the ilium provides an aperture through which neurovascular as well as muscular structures pass from the deep pelvis into the thigh.
       - **(a)** The **piriformis muscle** originates from the anterior aspect of the sacrum, passes through the greater sciatic notch, and inserts at the base of the greater trochanter of the femur.
       - **(b)** That portion of the **greater sciatic foramen** superior to the piriformis muscle, the **suprapiriformis recess**, contains the **superior gluteal neurovascular bundle**.
       - **(c)** That portion of the greater sciatic foramen inferior to the piriformis muscle, the **infrapiriformis recess**, contains:
         - **(i)** The **inferior gluteal neurovascular bundle**.
         - **(ii)** A portion of the **sciatic nerve**, which may occasionally pass through the piriformis muscle. In such cases, compression of the nerve by the muscle may produce sciatica.
         - **(iii)** The **pudendal neurovascular bundle**.
     - **(5)** The **iliopectineal line** describes the pelvic inlet and provides the boundary between the greater pelvis and the deep pelvis.
   - **b.** The **ischium** is the most inferior portion of the coxal bone (see Fig. 20-1).
     - **(1)** The **ischial ramus** fuses with the inferior pubic ramus.
     - **(2)** The **ischial tuberosity**, a palpable landmark, provides an attachment for muscles of the posterior thigh.
       - **(a)** This tuberosity also provides the distal attachment for the **sacrotuberous ligament**.
       - **(b)** When standing, the gluteus maximus muscle on each side covers the ischial tuberosity; when sitting, the gluteus maximus muscles move laterally so that one sits on the ischial tuberosities.
     - **(3)** The **ischial spine** provides attachment for the **sacrospinous ligament**, which bridges the greater sciatic notch just superior to the sacrotuberous ligament to form the greater sciatic foramen.
     - **(4)** The **lesser sciatic notch** of the ischium lies between the ischial tuberosity and the ischial spine; it provides an aperture through which neurovascular as well as muscular structures pass from the deep pelvis to the thigh and perineum.
       - **(a)** The lesser sciatic notch with the sacrospinous and sacrotuberous ligaments forms the **lesser sciatic foramen**.

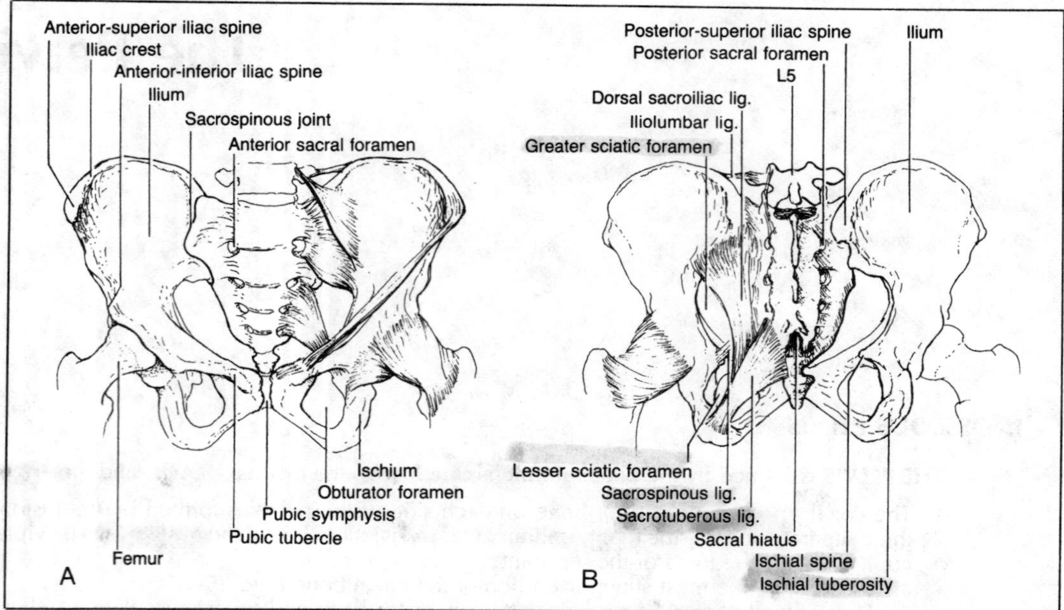

Figure 20-1. *The pelvis. A,* Anterior aspect of the female pelvis. The anterior ligaments supporting the left lumbosacral joint are included. *B,* Posterior aspect of the female pelvis. The posterior ligaments supporting the left lumbosacral joint are included.

(b) The **obturator internus muscle** originates on the internal aspect of the ilium and pubis, passes through the lesser sciatic foramen, and inserts at the base of the greater trochanter of the femur.

(c) The **pudendal neurovascular bundle** arises in the deep pelvis and passes through the lesser sciatic foramen to reach the perineum.

c. The **pubis** is the most anterior portion of the coxal bone (see Fig. 20-1A).

(1) The **superior pubic ramus** fuses with the ilium.

(2) The **inferior pubic ramus** fuses with the ischial ramus.

(3) Anteriorly, left and right pubic bones articulate at the **pubic symphysis**.

(a) The pubic symphysis, a true amphiarthrosis, is composed of hyaline cartilage articular surfaces separated by a fibrocartilage disk.

(b) This joint relaxes under the influence of progesterone in the weeks prior to parturition, allowing the birth canal to widen.

2. The **sacrum** forms the posterior portion of the pelvis (see Fig. 20-1).

a. The five sacral vertebrae fuse to form the sacrum.

(1) Four lines of fusion are evident on the anterior surface.

(2) The fused spinous processes form the **medial sacral crest**.

(3) The fused transverse processes together with fused costal processes form the **alae**, or sacroiliac articular processes.

(4) The **posterior sacral foramina** transmit the dorsal primary rami of the sacral spinal nerves; the **anterior sacral foramina** transmit the ventral primary rami of the sacral spinal nerves.

b. The laminae of the fifth sacral vertebra fail to form, resulting in the **sacral hiatus**. The pedicles of the fifth sacral vertebra form the **sacral cornua**, which are important landmarks used in locating the sacral hiatus for the administration of *caudal anesthesia*.

3. The **sacroiliac joint** is formed by the articulation of the alar plates of the sacrum with the ilium (see Fig. 20-1B).

a. Because there is a synovial cavity within it, the sacroiliac joint is a diarthrosis.

(1) Although there is always slight mobility at this joint, movement is greatest in the pregnant female in the weeks prior to parturition.

(2) Movement at this joint decreases with age; ankylosis usually is complete by age 50.

b. As a result of ligamentous reinforcement, this joint is probably among the strongest diarthrodial joints in the body. These ligaments oppose the rotational effect of gravity at the sacroiliac joint.

(1) The **sacrotuberous ligament** runs from the sacrum to the ischial tuberosity.
  (a) This ligament strongly reinforces the sacroiliac joint.
  (b) It transmits directly to the sacrum the forces generated by the hamstring muscles of the posterior thigh.

(2) The **sacrospinous ligament** runs from the sacrum to the ischial spine.
  (a) This ligament closes the greater sciatic notch and bounds the lesser sciatic foramen.
  (b) The boundaries of the lesser sciatic foramen include the sacrotuberous ligament, lesser sciatic notch, and sacrospinous ligament.

(3) The **iliolumbar ligament** runs between the iliac crest and the transverse processes of the fifth lumbar vertebra.

(4) The **dorsal sacroiliac ligaments** run between the posterosuperior iliac spine and the dorsum of the sacrum.

(5) The **ventral sacroiliac ligament** runs across the sacroiliac joint on the anterior surface of the pelvis.

(6) The **interosseous ligaments** run between the articular surfaces. The tendency of the sacrum to rotate forward tightens these ligaments, locking the joint.

c. "Low back" pain is often misdiagnosed as "sacroiliac sprain" or "lumbago." There have been essentially no documented instances of sacroiliac sprain. Low back pain is usually either of muscular origin or due to a herniated disk at the L4 or L5 level.

4. The **coccyx**, or **tail bone**, forms the inferoposterior portion of the pelvis.
  a. The four or five coccygeal vertebrae fuse to form the coccyx, but there may be a disk between the first and second coccygeal vertebrae.
  b. The first coccygeal vertebra articulates with the sacrum. The sacrococcygeal joint usually has a small intervertebral disk, but may fuse in some individuals.
  c. The first (and sometimes second) coccygeal vertebra has rudimentary pedicles, which form the coccygeal cornua.
  d. The second through fourth coccygeal vertebrae represent rudimentary vertebral bodies.
  e. Fracture of the fused coccygeal joints or of a fused sacrococcygeal joint, or arthritis in these joints, causes the painful condition, *coccygodynia*.
  f. In the male the coccyx tends to be directed more anteriorly toward the pubis, whereas in the female the coccyx tends to be slightly more vertical, thereby providing a larger pelvic outlet.

B. **THE BONY PELVIS** is made up of the sacrum and pelvic girdle; the enclosed pelvic cavity is an extension of the abdominal cavity.

1. The **major (greater** or **false) pelvis** is that portion of the bony pelvis between the iliac crests and the pelvic brim.
  a. Laterally, the **alar plates** (wings) of the ilia form the walls of the major pelvis.
  b. Posteriorly, the major pelvis is bounded by the lumbar vertebrae, L3, L4, and L5.
  c. Anteriorly, the major pelvis is bounded by the anterior abdominal wall.
  d. This region contains portions of the gastrointestinal tract; with the exceptions of a gravid uterus and the fundus of the full urinary bladder, no pelvic organs lie in the major pelvis.

2. The **minor (lesser** or **true) pelvis** lies inferior to the pelvic brim.
  a. The **pelvic inlet** demarcates the minor from the major pelvis.
  b. The **pelvic cavity** forms the birth canal and contains portions of the gastrointestinal tract (loops of ileum, sigmoid colon, rectum, and anal canal) as well as the lower portion of the urinary tract and certain reproductive organs.
  c. The **pelvic outlet** is closed by the pelvic diaphragm and covered by the perineum.

3. The pelvic inlet.
  a. The pelvic inlet is defined by the sacral promontory and the linea terminalis of the innominate bone. The **linea terminalis** includes the:
    (1) Pubic crest.
    (2) Iliopectineal line.
    (3) Arcuate line of the ilium.
  b. Measurements of the female pelvic inlet.
    (1) **Conjugate diameters.**
      (a) The **true conjugate** (about 10 cm) is the anteroposterior diameter from the sacral promontory to the superior margin of the pubic symphysis. This can be measured only radiographically in a lateral projection.
      (b) The **diagonal conjugate** (about 10.5 cm) is measured from the sacral promontory to the inferior margin of the pubic symphysis.

(c) The **obstetric conjugate** is the least anteroposterior diameter from the sacral promontory to a point a few millimeters below the superior margin of the pubic symphysis.

(2) The **oblique diameter** (about 12.5 cm) is measured from the sacroiliac joint to the contralateral iliopectineal eminence.

(3) The **transverse diameter** (13.5 cm) is the widest distance across the pelvic brim.

4. The pelvic outlet.

a. The pelvic outlet is defined by the coccyx, ischial tuberosities, inferior pubic ramus, and pubic symphysis. It is closed by the pelvic diaphragm and reinforced by the urogenital diaphragm.

b. Measurements of the female pelvic outlet:

(1) The **transverse diameter** is between the ischial tuberosities. It is approximately as long as a clenched fist is wide.

(2) The **transverse midplane diameter** (about 10.5 cm) is measured between the ischial spines. If this diameter is less than 9.5 cm, there is nearly a 50 percent chance that surgical intervention will be necessary during labor.

(3) The **anteroposterior (sagittal) diameter** (about 13.5 cm) is measured from the lower margin of the pubic symphysis to the sacrococcygeal joint.

c. Pelvimetry, although controversial, can be a valuable diagnostic tool in cases of cephalopelvic disproportion.

5. Characterization of the pelvis.

a. The pelvis may be classified according to the shape of the inlet.

(1) The normal male pelvis is *android* with a heart-shape due to the prominent sacrum.

(2) The normal female shape is *gynecoid*.

(3) An *anthropoid pelvis* has a lengthened anteroposterior diameter and a shortened transverse diameter.

(4) A *platypoid pelvis* is oval with a shortened anteroposterior diameter and a lengthened transverse diameter.

(5) A pelvis may be *asymmetric* due to scoliosis, osteomalacia, or congenital malformation.

b. Sex differences.

(1) The bones of the female pelvis are thinner than those of the male, the pelvic cavity is less funnel-shaped, the surface for articulation of the sacrum with L5 is smaller, and the subpubic angle is almost 90°.

(2) A summary of sexual differences is found in Table 20-1.

Table 20-1. Differences between the Male and Female Pelvis

	Male	**Female**
Size	Smaller	Larger
Greater sciatic notch	Narrower	Wider
Inferior pubic angle	Narrower	Wider
False pelvis	Deep	Shallow
Pelvic inlet	Android	Gynecoid
Pelvic outlet	Smaller	Larger

II. PELVIC MUSCULATURE

A. THE PELVIC FLOOR (pelvic diaphragm) [Fig. 20-2]

1. The pelvic floor, composed of the **coccygeus** and **levator ani muscles**, supports the pelvic viscera and constitutes the principal mechanism for fecal continence.

2. The **coccygeus (ischiococcygeus)** muscle forms a small portion of the pelvic floor and is a rudimentary "tail-wagger."

a. This simple muscle originates from the ischial spine.

b. It inserts onto the lateral borders of the fourth sacral to the second coccygeal vertebrae, as well as along the **sacrospinous ligament**, which may be regarded as a degenerated portion of the coccygeus muscle.

3. The levator ani muscle forms most of the pelvic floor.

a. The complex levator ani muscle is divided into three parts.

(1) The **iliococcygeus muscle**, a rudimentary "tail-wagger":

(a) Arises laterally from the **ischial spine** and the **arcus tendineus** (white line) of the **obturator fascia**.

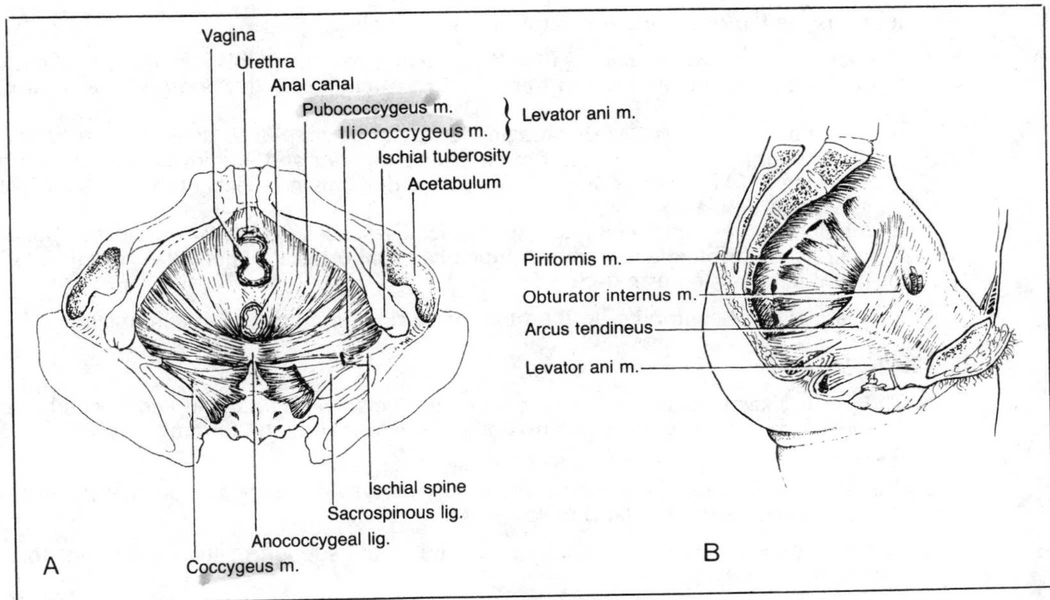

Figure 20-2. *The pelvic floor. A,* Inferior view of the female pelvis showing the muscles of the pelvic diaphragm. *B,* Sagittal section of the female pelvis showing the muscles of the pelvic diaphragm.

(b) Inserts onto the coccyx.

(2) The **pubococcygeus muscle**, also a rudimentary "tail-wagger":

(a) Arises from the arcus tendineus, laterally, and the **pubic arch**, anteriorly.

(b) Inserts primarily onto the coccyx and into the anococcygeal raphe.

(3) The **puborectalis muscle** is the most medial portion of the levator ani muscle.

(a) Arising from the pubic arch medial to the pubococcygeus muscle, the fibers loop around the rectum to form the **rectal sling**.

(b) Some fibers of the puborectalis muscle loop posterior to the vagina in the female and posterior to the prostate in the male to insert into the central tendon of the perineum. These fibers are considered by some to be separate muscles, the pubovaginalis and puboprostaticus.

b. The **urogenital hiatus** is a midline defect in the levator ani muscle through which pass the rectum, the urethra, and in the female, the vagina. The urogenital hiatus is reinforced inferiorly by the **urogenital diaphragm**.

4. Functional considerations.

a. The pelvic floor supports the urinary bladder and participates in urinary continence.

b. The rectal sling, by drawing the rectum anteriorly, is the principal component for the maintenance of fecal continence.

c. In the female, the pelvic floor stabilizes the vagina and indirectly participates in the support of the uterus.

d. The muscles of the pelvic floor are innervated by twigs from the sacral plexus.

B. **Other muscles**, such as the iliacus, obturator internus, and piriformis, originate within the pelvis but are associated with the lower extremity and act across the hip joint (see Fig. 20-2B).

III. PELVIC PERITONEUM AND FASCIAE

A. THE VISCERAL (ENDOPELVIC) FASCIA

1. The endopelvic fascia, a continuation of the extraperitoneal connective tissue layer, ensheathes viscera and forms septa between the organs.

2. Condensations of endopelvic fascia provide ligamentous support for the pelvic adnexa and the neurovascular structures that supply the pelvic organs.

B. THE PARIETAL FASCIA

1. The parietal fascia is a continuation of the transversalis fascia into the pelvis.

2. It covers the piriformis and obturator internus muscles.
3. This fascial layer, which attaches from the arcuate line of the pubis and ilium, thickens over the obturator internus muscle to form the **arcus tendineus**, the origin of the levator ani muscle.
   a. At the arcus tendineus, a **diaphragmatic fascial layer** splits to cover both superior and inferior surfaces of levator ani muscle as the superior and inferior fasciae of the pelvic diaphragm. The superior fascia of the pelvic diaphragm reflects onto the pelvic viscera as **visceral fascia**.
   b. Inferior to the arcus tendineus, the parietal fascial layer splits around the pudendal neurovascular bundle to form the **pudendal canal** (Alcock's) along the medial surface of the obturator internus muscle.
4. This fascial layer contributes to the **superior fascia** of the **urogenital diaphragm**.

C. PELVIC LIGAMENTS

1. The pelvic ligaments are thickened continuities between the parietal and visceral fasciae; in some locations they are condensations of connective tissue forming the neurovascular sheaths.
2. These condensations of connective tissue support pelvic viscera and provide pathways for the associated neurovascular bundles.
3. The pelvic ligaments are especially important in the support of the uterus and, to some extent, the urinary bladder.

D. Peritoneum overlies portions of the parietal and visceral endopelvic fasciae.

IV. PELVIC VASCULATURE

A. THE COMMON ILIAC ARTERIES AND VEINS (Fig. 20-3)

1. The left and right common iliac arteries arise with the bifurcation of the **aorta** at the level of the fourth lumbar vertebra.
2. The left and right common iliac veins join to form the **inferior vena cava**.
3. The common iliac vessels bifurcate at the level of the fifth lumbar intervertebral disc to form **internal** and **external iliac** vessels.
4. With the exception of small branches to the psoas muscle, twigs to the ureters, and an occasional iliolumbar artery or accessory inferior renal artery, the common iliac arteries give rise to no major branches.

B. THE EXTERNAL ILIAC ARTERY AND VEIN (see Fig. 20-3)

1. The external iliac artery and vein course in the greater pelvis just above the pelvic brim.
2. The external iliac artery gives off the **inferior epigastric artery** and **deep iliac circumflex artery** in the vicinity of the inguinal ligament.
3. The external iliac vessels exit the pelvis beneath the inguinal ligament and become the **femoral artery** and **vein**.

C. THE INTERNAL ILIAC ARTERY AND VEIN (see Fig. 20-3)

1. The internal iliac arteries supply the walls of the pelvis, the pelvic viscera, and the perineum.
2. The branches of the internal iliac vessels are extremely variable in site or origin, arrangement, number, and position.
3. The **posterior trunk** of the **internal iliac artery** provides only somatic branches.
   a. The **iliolumbar artery** courses upward behind the external iliac vessels as far as the medial border of the psoas muscle, where it divides into an iliac branch and a lumbar branch.
      (1) The lumbar branch supplies the psoas and quadratus lumborum muscles as well as the lower lumbar vertebra and dural sac.
      (2) The iliac branch supplies the iliacus muscle. Its anastomoses with the distributions of the superior gluteal artery, iliac circumflex artery, and lateral femoral circumflex artery provide collateral circulation about the hip.
      (3) The iliolumbar artery and vein are especially vulnerable to tearing in posterior pelvic fractures with resultant severe hemorrhage.

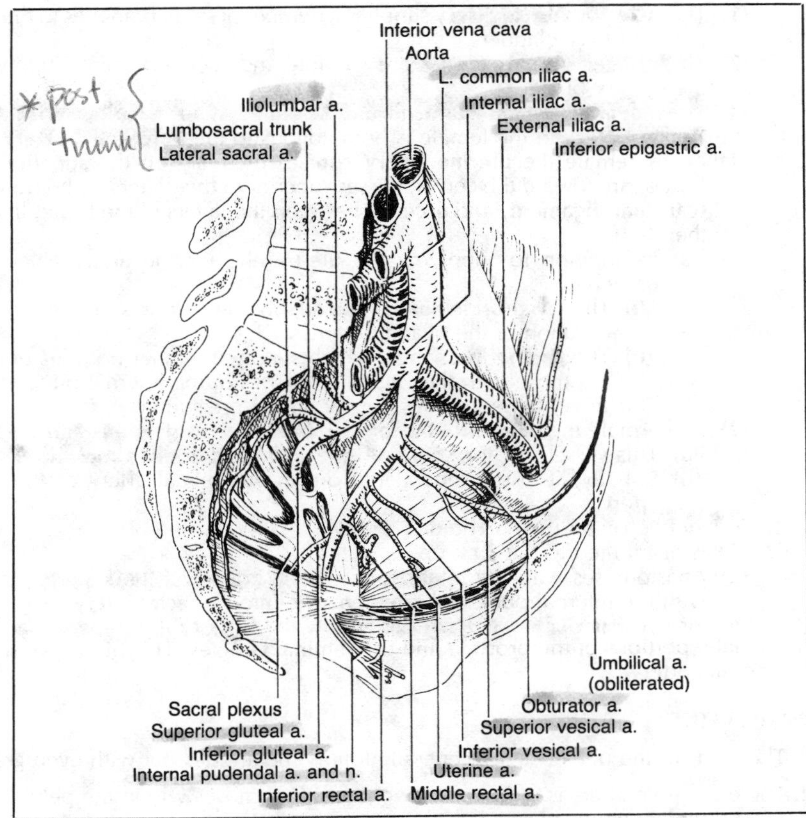

Figure 20-3. *The vasculature of the pelvis.*

b. The **lateral sacral artery** supplies the region of the sacrum.

c. The **superior gluteal artery** is the largest branch of the internal iliac artery in the adult.

(1) This artery leaves the pelvic cavity via the suprapiriform portion of the **greater sciatic foramen**.

(2) It supplies portions of the gluteus maximus, gluteus medius, and gluteus minimus muscles and sends twigs to the hip joint.

4. The **anterior trunk** of the **internal iliac artery** supplies both somatic and visceral structures.

a. The **inferior gluteal artery** supplies somatic structures.

(1) Inside the pelvic cavity this vessel supplies portions of the coccygeus, piriformis, and levator ani muscles.

(2) This artery leaves the pelvic cavity via the infrapiriform portion of the greater sciatic foramen.

(3) Outside the pelvis this vessel supplies portions of the gluteus maximus muscle, the lateral rotator muscles, and the hip joint.

b. The **internal pudendal artery** supplies the somatic structures of the perineum.

(1) This vessel exits the pelvis via the infrapiriform portion of the greater sciatic foramen.

(2) It enters the ischiorectal fossa of the perineum via the **lesser sciatic foramen**.

(3) The branches of the internal pudendal artery are discussed in Chapter 21 IV E 1.

c. The **obturator artery** supplies somatic structures in the anteromedial thigh.

(1) This vessel runs ventrally along the pelvic wall, medial to the obturator fascia, and leaves the pelvic cavity via the obturator canal to supply the adductor muscles and the artery of the ligamentum teres.

(2) Within the pelvic cavity, this vessel provides twigs to the iliacus muscle.

(3) The obturator artery may arise as an aberrant obturator artery from the inferior epigastric artery (30 percent). Such an anomalous artery may complicate surgery in the region of the femoral ring.

d. The **umbilical artery** retains a lumen for a short distance beyond the internal iliac artery and gives off one or two branches to the urinary bladder.

(1) The **superior vesical artery** supplies a number of small branches to the cranial portion of the urinary bladder.

(2) The **middle vesical arteries** are variable and, when present, supply the fundus of the urinary bladder.

(3) The **medial umbilical ligaments** are the obliterated remnants of the **umbilical artery**.

e. The **uterine artery** in the female is homologous to the **deferential artery** in the male.

(1) In the female the **uterine artery** courses medially on the superior surface of the levator ani toward the cervix, passes superior to the ureters in the **transverse cervical (cardinal) ligament**, and ascends between the layers of the broad ligament to reach the uterus.

(a) In addition to supplying the uterus, the uterine artery gives off two major branches.

(i) The **tubal branches**, which supply the oviducts, anastomose with the distribution of the ovarian arteries.

(ii) The **vaginal branches**, which supply the inner portions of the vagina, anastomose with the vaginal branches of the internal pudendal artery.

(b) This artery hypertrophies greatly during pregnancy.

(2) In the male the **deferential artery** supplies the vas deferens and epididymis.

(a) This vessel anastomoses with the spermatic artery about the testes.

(b) Usually, it arises separately from the umbilical artery or the inferior vesical artery.

f. The **middle rectal (hemorrhoidal) artery:**

(1) Is distributed to the rectum.

(2) Anastomoses with the inferior mesenteric artery via the superior rectal artery and with the internal pudendal artery via the inferior rectal artery.

g. One or two **inferior vesical arteries** supply the neck of the urinary bladder, and in the male, portions of the prostate and the seminal vesicles. They may give rise to the deferential artery.

D. PELVIC VEINS

1. The veins of the pelvic cavity generally follow the arteries but with even greater variation.

2. There are several areas of important venous anastomoses within the pelvic cavity.

a. The **rectal venous plexus** lies about the rectum.

(1) Abundant anastomoses exist between superior rectal, middle rectal, and inferior rectal (hemorrhoidal) veins.

(2) These anastomoses connect the hepatic portal system with systemic drainage.

b. The **vesical venous plexus** lies about the base of the bladder and, in the male, the prostate.

(1) This venous plexus receives drainage from the penis or clitoris via the deep dorsal vein and from the urinary bladder, as well as the vas deferens, seminal vesicle, and prostate in the male.

(2) Because drainage from this plexus is diffuse and without valves, cancer of the prostate may spread via several hematogenous routes, with the potential for metastatic rests along the courses.

(a) The principal drainage is into the internal iliac vein.

(b) Collateral drainage occurs into the **vertebral venous plexus** within the neural canal, the obturator veins, which frequently drain into the external iliac veins, and the rectal plexus.

E. PELVIC LYMPHATICS

1. The lymphatic drainage of the pelvis is extremely variable.

2. The drainage generally follows the tributaries of the internal iliac veins to the pelvic wall.

3. Anastomoses occur widely with the abdominal lymphatics around the rectum and with the superficial lymphatic drainage of the perineum.

4. The lymphatic drainage is of major surgical concern in procedures involving carcinoma of the pelvic viscera.

V. THE LUMBOSACRAL PLEXUS

A. THE LUMBOSACRAL PLEXUS provides the somatic innervation to the pelvis and lower extremities.

1. The **lumbar plexus** (T12–L4) lies in the posterior abdominal wall and iliac fossa (see Chapter

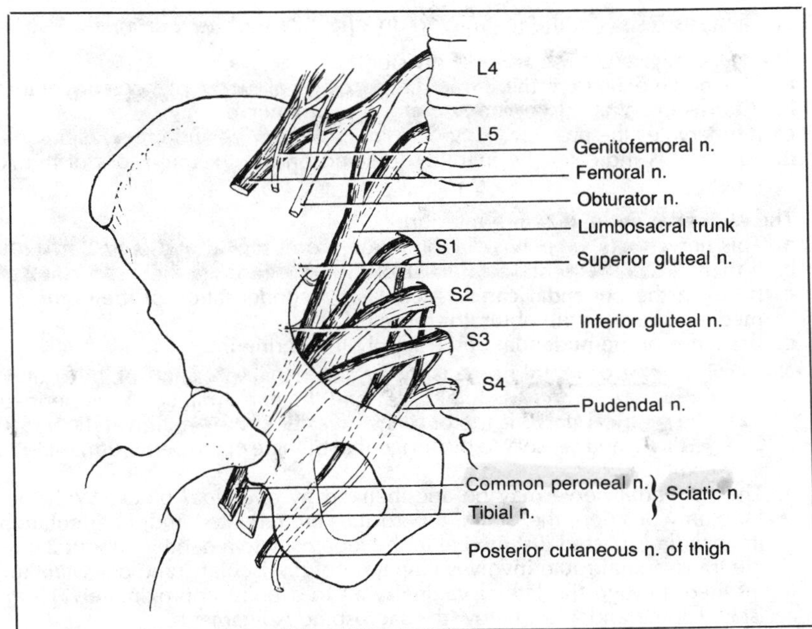

Figure 20-4. *The sacral plexus.* The anterior divisions (*light*) and posterior divisions (*shaded*) of the sacral plexus are indicated.

18 IV H). The **lumbosacral trunk** (L4–L5, anterior and posterior divisions) contributes to the sacral plexus.

a. These roots join the lumbar plexus with the sacral plexus.
b. The anterior division of the lumbosacral trunk contributes the **tibial** portion of the **sciatic nerve.** In general, this nerve is sensory and motor to the posterior aspects of the thigh and leg as well as to the plantar surface of the foot.
c. The posterior division of the lumbosacral trunk contributes the **superior** and **inferior gluteal nerves** as well as the **common peroneal** portion of the sciatic nerve. In general, these nerves are sensory and motor to the posterior hip as well as to the anterior leg and to the dorsum of the foot.

2. The **sacral plexus** (L4–S3) lies in the minor pelvis and supplies the gluteal region, posterior thigh, leg, and foot as well as the perineum.

3. The pelvic splanchnic nerves (nervi erigentes) arise from spinal nerves S2–S3 or S3–S4.
   a. These consist of parasympathetic preganglionic fibers.
   b. They also contain visceral afferent fibers from specific portions of the pelvic viscera.

4. The **coccygeal plexus** (S4–Cxl) supplies the terminal portion of the vertebral column.

B. MAJOR BRANCHES OF THE SACRAL PLEXUS (Fig. 20-4)

1. The **superior gluteal nerve** (L4–S1, posterior).
   a. This nerve exits the pelvis via the suprapiriform portion of the greater sciatic foramen.
   b. It supplies the gluteus medius and gluteus minimus muscles.
   c. Severance of this nerve results in "abductor lurch," a rolling (Trendelenburg) gait due to loss of abductive power in the hip.

2. The **inferior gluteal nerve** (L5–S2, posterior).
   a. This nerve exits via the infrapiriform region of the greater sciatic foramen.
   b. It supplies the gluteus maximus muscle.
   c. Severance of this nerve results in "abductor lurch," a rolling (Trendelenburg) gait due to loss of abductive power in the hip.

3. The **common peroneal nerve** (L4–S2, posterior).
   a. This nerve exits via the infrapiriform region of the greater sciatic foramen.
   b. It forms the posterior component of the sciatic nerve.
   c. It innervates the anterior aspect of the leg and the dorsum of the foot.

d. Paralysis results in the inability to dorsiflex the foot (foot drop) and to evert the foot.

4. The **tibial nerve** (L4–S3, anterior division).

a. This nerve exits from the infrapiriform region of the greater sciatic foramen.
b. It forms the anterior component of the sciatic nerve.
c. It innervates the posterior region of the thigh and leg and the plantar surface of the foot.
d. Paralysis is indicated by inability to stand on the toes and loss of the Achilles tendon reflex.

5. The **pudendal nerve** (S2–S4, anterior).

a. This nerve exits the pelvis via the infrapiriform region of the greater sciatic foramen.
b. It then enters the ischiorectal fossa of the perineum via the lesser sciatic foramen.
c. It lies in the **pudendal canal (Alcock's)**, a condensation of the obturator fascia on the medial surface of the obturator muscle.
d. Branches of the pudendal nerve supply the perineum.

(1) The **inferior rectal nerve** is motor to the external anal sphincter and sensory to the inferior portion of the anal canal and the anal triangle of the perineum.
(2) The **perineal nerve** is motor to the muscles of the urogenital diaphragm and external genitalia and sensory to the urogenital triangle of the perineum, including the external genitalia.

e. The pudendal nerve may be anesthetized by *pudendal block*. With the patient in the lithotomy position, the ischial tuberosities are palpated and, after cutaneous anesthesia, the needle is passed just medial to the tuberosity to a depth of about 2 cm. Alternatively, the transvaginal route involves palpation of the ischial spine per vaginam; the needle is inserted through the lateral vaginal wall to a point approximately 1 cm medial to the ischial spine and 1 cm below the sacrospinous ligament.

21
The Perineum

I. THE PERINEUM

A. PERINEAL BOUNDARIES

1. The perineum is defined as the area between the **ischial tuberosities**, extending from the **pubis** to the **coccyx**.
2. The perineum covers the **pelvic outlet**.
3. It is pierced by anal and urogenital canals.

B. The perineum is divisible by the **transverse diameter** (a line connecting the ischial tuberosities) into the **anal triangle** posteriorly and the **urogenital triangle** anteriorly (Fig. 21-1).

1. Each perineal triangle has special musculature.
2. Both perineal triangles share the levator ani muscle, the same innervation, and the same blood supply.
3. The principal structures of the anal triangle are the anal canal and the anus; the anus is the terminal portion of the gastrointestinal tract.
4. The principal structures of the urogenital triangle are the external genitalia. Although the differences between the superficial structures of the urogenital triangle of the male and female are obvious, the structures have homologous counterparts in both sexes.

II. DEVELOPMENT OF THE EXTERNAL GENITALIA

A. THE INDIFFERENT STAGE (weeks 3 to 6)

1. In the perineum **cloacal folds** form on either side of the **cloacal membrane**, which separates the lumen of the hindgut from the perineal region.
2. Anteriorly, the cloacal folds fuse to form the **genital tubercle**.
3. By the sixth week, the **urorectal septum** divides the cloaca into the **anal canal** and the **urogenital sinus**.
   a. This septum divides the cloacal membrane into the **urogenital membrane** and the **anal membrane**. The cloacal sphincter is similarly divided, becoming the urogenital diaphragm and the external anal sphincter.
   b. The urogenital septum also divides the cloacal folds into **urethral folds** and **anal folds**.
   c. The fusion of the urogenital septum with the cloacal membrane forms the **central tendon** of the perineum (**perineal body**).
4. **Genital swellings (labial-scrotal folds)** develop lateral to the urethral folds.
5. Shortly afterward, the urogenital and anal membranes degenerate, opening the alimentary and urogenital canals to the perineum.

B. THE DEFINITIVE EXTERNAL GENITALIA OF THE MALE

1. Rapid elongation of the genital tubercle produces the **phallus**.
   a. During this growth process, the urethral folds are drawn anteriorly, producing the **urogenital (urethral) groove**.

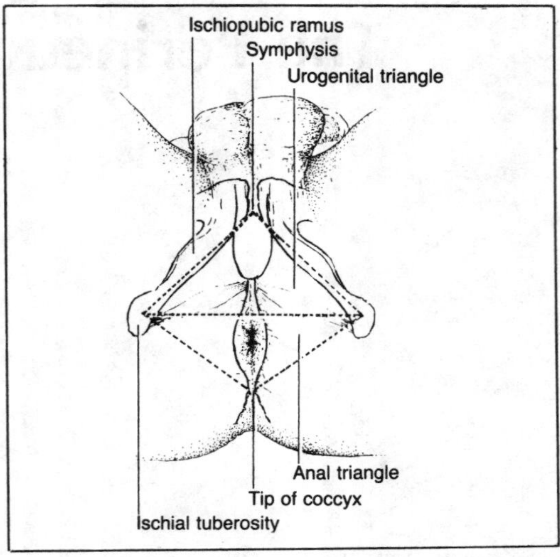

Figure 21-1. *The perineum.* The boundaries and landmarks of the perineum are indicated. The perineum is divided into anal and urogenital triangles by the transverse diameter of the pelvic outlet.

- **b.** By the twelfth week, the urethral folds fuse over the urethral groove to form the **penile urethra**.
- **c.** During the sixteenth week, the **glans penis** differentiates and the **external urethral meatus** connects with the penile urethra at the **fossa navicularis**.
- **d.** The skin of the body of the penis grows over the glans penis as the **prepuce**. Initially the prepuce fuses to the glans, but it becomes separated at birth or shortly thereafter.

2. The **scrotal folds** migrate somewhat posteriorly and fuse, forming the scrotal septum; each fold forms one-half of the definitive scrotum.

- **a.** The male gonads, or **testes**, develop in the cranial region of the posterior abdominal wall. Toward the end of the ninth week the testes begin a retroperitoneal migration toward the iliac fossa (see Fig. 22-4).
  - **(1)** A fascial condensation, the **gubernaculum testis**, runs from the inferior pole of each testis, through the inguinal canals, to the respective scrotal fold.
  - **(2)** The gubernaculum normally fails to lengthen during the subsequent period of rapid body growth, which results in a shift of the gonads into the major pelvis.
- **b.** By 13 weeks, the testes lie in the inguinal region but retain the original neurovascular connections.
- **c.** At about this time an evagination of the coelom, the **processus vaginalis (vaginal process)**, penetrates the abdominal wall of the inguinal region and enters the scrotal fold, drawing with it some of the layers of the abdominal wall.
- **d.** During the seventh month the testes commence their descent through the inguinal canal, posterior to the vaginal process.
- **e.** In the scrotum the vaginal process in the vicinity of the testes forms the visceral and parietal layers of the **tunica vaginalis**.
- **f.** The neck of the vaginal process normally fuses by birth or shortly thereafter.

C. THE DEFINITIVE EXTERNAL GENITALIA OF THE FEMALE

1. The genital tubercle elongates slightly to form the clitoris.

- **a.** The urethral folds do not fuse, but develop into the **labia minora**.
- **b.** The urogenital groove remains as the **vestibule**.

2. The genital swellings form the **labia majora**.

3. The female gonads, the **ovaries**, develop in the cranial portion of the posterior abdominal cavity. Subsequent to the ninth week, the ovaries migrate into the minor pelvis (see Fig. 22-4).

- **a.** A fascial condensation runs from the inferior pole of each ovary, through the inguinal canals, to the respective labial fold.
- **b.** This ligament normally fails to lengthen during the subsequent period of rapid body growth, which results in a shift of the gonads into the major pelvis.
- **c.** A portion of this ligament is incorporated into the developing uterine wall, which thus divides the ligament into a cranial **ovarian ligament** and a caudal **round ligament of the ovary**.

d. By 13 weeks, the ovaries lie in the minor pelvis but retain the original neurovascular connections that form the **suspensory ligament of the ovary**.
e. At about this time an evagination of the coelom, the **processus vaginalis (canal of Nuck)**, penetrates the abdominal wall of the inguinal region and enters the labial fold.
f. The neck of the vaginal process normally fuses by birth or shortly thereafter.

III. THE ANAL TRIANGLE

A. **THE ISCHIORECTAL FOSSA** is the region about the anal canal (Fig. 21-2).

1. The boundaries of the ischiorectal fossa on each side include:
   a. The **obturator internus muscle**, anterolaterally.
   b. The **levator ani muscle**, superiorly.
   c. The sacrotuberous ligament and the overlying **gluteus maximus muscle**, posterolaterally.

2. An **anterior horn** extends forward into the urogenital triangle on each side superior to the **urogenital diaphragm**.

3. The ischiorectal fossae communicate contralaterally posterior to the anal canal.

4. These fossae contain considerable fat separated by stringy connective tissue fibers and septa. Tension in the gluteus maximus muscles compresses the fat of the ischiorectal fossa around the anal canal, contributing to fecal continence; however, the fat allows dilation of the canal during defecation when the muscles are relaxed.

5. The ischiorectal fossae contain **inferior hemorrhoidal (rectal) nerves** and **vessels**.
   a. These are branches of the internal pudendal vessels and the pudendal nerve.
   b. They supply the **external anal sphincter**.
   c. Inadvertent severing of the nerves in the ischiorectal fossa results in *sphincteric incontinence* during brief peristaltic waves.

6. Abscesses in the ischiorectal fossa may become large or may even extend to the contralateral side around the anus (*''horseshoe'' abscess*).

B. **THE ANAL CANAL** (see Fig. 17-18)

1. The anal canal is that part of the alimentary canal inferior to the pelvic diaphragm.
   a. The anal canal is directed posteriorly and inferiorly from the **rectum** to the **anal verge**, is 3 to 4 cm long, and is open to the exterior via the **anus**.
   b. The most medial fibers of the **pubococcygeus muscle** (the puborectalis or rectal sling) form the **perineal flexure** (''carrying angle'') of the rectum. Thus the anal canal is directed downward and backward from the **rectum** to the **anal verge**.

2. Internal structure (see Fig. 17-18).
   a. The anal canal can be divided into an upper two-thirds and a lower third, representing a division between the visceral and somatic portions.
      (1) The upper two-thirds of the anal canal belongs to the intestine with respect to the mucosa, blood supply, and autonomic innervation.
      (2) Only the lower third belongs to the perineum with respect to mucosa, blood supply, and autonomic innervation.
   b. The mucosa of the upper two-thirds is thrown into longitudinal folds about 1 cm long, the **anal columns** (Morgagni's). The bases of the anal columns are joined together by small semilunar folds of tissue, the **anal valves**.
      (1) The bases of the anal columns and intervening anal valves define the **pectinate line**, which approximates the division between visceral and somatic portions of the anal canal.
         (a) Superior to the pectinate line, the epithelium is insensitive to touch, whereas below it the epithelium is extremely sensitive.
         (b) Venous anastomoses between the portal system draining the rectum and the systemic system draining the anal canal are potential sites for varicosities (hemorrhoids).
      (2) Between the bases of the anal columns, behind the anal valves, are small blind sacs—the **anal sinuses (crypts)**—into which the **anal glands** open.
      (3) The anal glands are rudimentary circumanal glands.
   c. The mucosal lining of the lower portion of the anal canal is squamous epithelium.

3. The **anal sphincters**.
   a. The anal canal is always closed, except during the passage of feces or flatus. The sphincteric mechanism has a remarkable ability to distinguish between and separate flatus from feces.

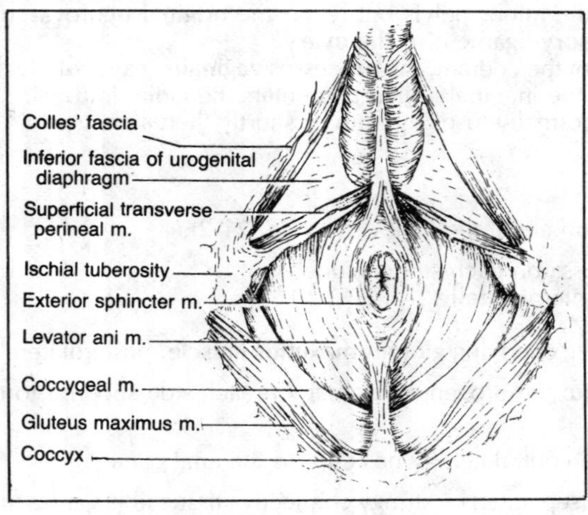

Figure 21-2. *The anal triangle.* The fat of the ischiorectal fossa has been removed to show the underlying pelvic floor and external anal sphincter.

(1) The entire anal canal is enclosed in a sphincteric tube of muscle, the **internal anal sphincter**, which is a continuation of the circular muscular layer of the intestine.

(a) The intrinsic tone of the involuntary internal anal sphincter maintains this closure.

(b) This sphincter is composed of involuntary smooth muscle, which is inhibited by the approach of a peristaltic wave.

(2) The **external anal sphincter** is striated muscle that is under voluntary control via the rectal branches of the pudendal nerve.

(a) The external sphincter is a complex structure composed of three distinct layers (see Fig. 21-2):

(i) The **subcutaneous** portion, which is also known as the **corrugator ani muscle**, is a remnant of the panniculus carnosa, a primitive superficial muscle that also gives rise to the platysma muscle about the lower face and neck.

(ii) The **superficial portion**.

(iii) The **deep portion**.

(b) The external anal sphincter can maintain a voluntary tonic contracture for about 20 to 30 seconds, a duration that provides the voluntary contraction necessary to counter the passage of a peristaltic wave when the internal anal sphincter relaxes.

b. Continued anal continence is a function of the **rectal sling**, a portion of the levator ani muscle.

C. THE VASCULAR SUPPLY TO THE ANAL TRIANGLE

1. The perineum is supplied in large part by the **internal pudendal artery** (see Figs. 21-8 and 21-13).

a. Its **inferior hemorrhoidal (rectal) branches** supply the anal triangle.

b. These vessels form anastomotic connections with the middle hemorrhoidal (rectal) artery.

2. The perineum is drained in large part by the **internal pudendal vein**.

a. The **inferior hemorrhoidal (rectal) veins** parallel the inferior hemorrhoidal arteries.

b. These veins form extensive anastomotic connections with the **middle hemorrhoidal (rectal) veins**.

D. THE INNERVATION OF THE ANAL TRIANGLE is by the **inferior rectal branch** of the **pudendal nerve**.

1. This nerve is motor to the external anal sphincter and sensory to the inferior portion of the anal canal and the integument of the anal triangle (see Figs. 21-8 and 21-13).

2. Severance of the inferior rectal branch results in paralysis of the external anal sphincter and may result in momentary embarrassments as peristaltic waves reach the terminal portion of the gastrointestinal tract.

E. FUNCTIONAL CONSIDERATIONS

1. Defecation is prevented primarily by the **puborectalis** muscle, which cants the lower part of the rectum forward (the **rectal sling**), effectively kinking the lumen (the **carrying angle**). Additionally, when standing erect the mass of fat within the ischiorectal fossa is compressed against the anal canal anteriorly and laterally by the location and tonic activity of the buttocks.

2. When the feces stored within the sigmoid colon are suddenly moved into the rectum, dilatation of the rectal ampulla occurs and the urge to evacuate the rectum is perceived.
3. Voluntary effort (constant tonus in the puborectalis and, occasionally, contraction of the external anal sphincters and gluteus maximus muscles) is required at this point to prevent defecation.
4. When the time and place are propitious, compression of the anal canal by the ischiorectal fat is released by suitable anatomic positioning; the puborectalis muscle is relaxed, thereby allowing the rectum to straighten and descend slightly; and the anal sphincters are relaxed, the internal according to the Bayliss-Starling law and the external voluntarily.
5. The muscular movements of the terminal portion of the alimentary canal, assisted by gravity, evacuate the rectum. This may be accompanied by the Valsalva maneuver, which increases the intraabdominal pressure and facilitates the expulsion of feces.
6. After passage of each fecal mass, the puborectalis muscle reestablishes the carrying angle, and the anal sphincters contract, thereby restoring anal continence.

F. CLINICAL CONSIDERATIONS

1. The anal glands empty into the base of the anal crypts behind the anal valves. These glands are prone to infection and may give rise to fistulous tracts that require surgical correction.
2. Because there are rich anastomoses between the hepatic portal system and the systemic venous drainage, portal hypertension often results in *hemorrhoidal varices*.
   a. Hemorrhoids above the **pectinate line** are classified as internal.
      (1) The innervation above this line is similar to that of the rest of the gut; there are pressure receptors but no definitive pain receptors (i.e., there is an urge to defecate when feces enter the rectum).
      (2) Internal hemorrhoids are painless because the epithelium is innervated by visceral afferents.
      (3) Hemorrhoids in this region are likely to be large, silent, and dangerous and may be noticed only when they become large enough to prolapse through the anus or bleed profusely. Such bleeding may be sufficient to produce anemia.
   b. Hemorrhoids below the pectinate line are termed external.
      (1) Conversely, the anal canal below the pectinate line is innervated somatically via the rectal (hemorrhoidal) branches of the pudendal nerve.
      (2) These hemorrhoids are painful because the epithelium is innervated by rectal branches of the pudendal nerve.
      (3) External hemorrhoids are so exquisitely sensitive that even small hemorrhoids in this region demand attention.

IV. THE MALE UROGENITAL TRIANGLE

A. THE EXTERNAL GENITALIA, composed of the **penis** and **scrotum**, are the most obvious features of the urogenital triangle in the male (Fig. 21-3).

1. The **penis**.
   a. Because the penis is suspended along the abdominal wall in lower mammals, the anterior surface in humans is termed the dorsal surface.
   b. The penis is composed of three bodies of vascular erectile tissue (Fig. 21-4).
      (1) The central erectile **corpus spongiosum (corpus cavernosum urethrae)** is ventrally situated, terminates as the glans penis, and contains the penile urethra.
      (2) There is an erectile corpus cavernosum on either side, dorsally.
   c. The **radix**, or root of the penis, composed of the crura of the corpora cavernosa and the **bulb** of the corpus spongiosum, is located in the superficial perineal pouch.
   d. The **body** of the penis is formed by the fusion of the vascular erectile structures as they leave the perineum.
   e. Each erectile structure or corpus is surrounded by a distinctive connective tissue layer, the **tunica albuginea**.
      (1) The tunica albuginea, which surrounds the corpora cavernosa, is very dense and not particularly elastic. This greatly impedes venous return and results in the extreme turgidity of these structures when the erectile tissue becomes engorged with blood.
      (2) Because the central corpus spongiosum is surrounded by a less dense layer of

connective tissue, the corpus spongiosum, with its terminal portion, the glans penis, does not become excessively turgid on erection, thereby permitting the passage of the ejaculate.

f. The **suspensory ligament** arises from the **linea alba** and inserts into the deep fascia of the penis (see Fig. 21-5).

g. The **penile urethra** extends along the penis within the corpus spongiosum (see Fig. 21-5).

(1) This portion of the urethra begins in the superficial pouch just as the membranous urethra penetrates the urogenital diaphragm.

(2) After about 2.5 cm, it receives the ducts from the **bulbourethral glands** (Cowper's).

(3) The **urethral glands** (Littre's), which are mucous glands, line this segment of the urethra.

(4) Terminally, the penile urethra widens within the glans penis, forming the **fossa navicularis**.

(5) The penile urethra is homologous to the vestibule in the female.

(6) Introduction of instruments into the male urethra must be accomplished with care, with the penis straightened into an approximately erect position.

h. The penis is covered by skin and fascia.

(1) The superficial fascia of the abdominal wall extends over the body of the penis, but it is not attached to the glans penis.

(a) The superficial layer of the superficial fascia loses its fat and fuses with the deep layer of the superficial fascia to form the superficial fascia of the penis.

(b) The variable extension of the integument over the glans penis is the **prepuce** or **foreskin**.

(c) Superficial fascia does not extend over the glans penis, which is covered by mucous epithelium tightly bound to the underlying tunica albuginea.

(2) An extension of the deep fascia that invests the muscles at the base of the penis forms the **deep fascia** (Buck's) of the penis.

i. The penis is innervated by the terminal branches of the pudendal nerve.

2. The **scrotum** (see Fig. 21-3).

a. Skin of the scrotum is thin with very little fat, a condition that is important in maintaining a lower testicular temperature.

b. The **dartos layer**, or **tunic**, is formed by the fusion of the superficial and deep layers of superficial fascia.

(1) It is innervated by the **ilioinguinal, genitofemoral**, and **dorsal nerves** of the penis.

(2) It contains smooth muscle and functions in temperature regulation.

(a) Although core body temperature is 37°C., the scrotal temperature is 33.9°C.

(b) When cold, the scrotum is contracted and wrinkled, bringing the testes into close contact with the body to conserve heat.

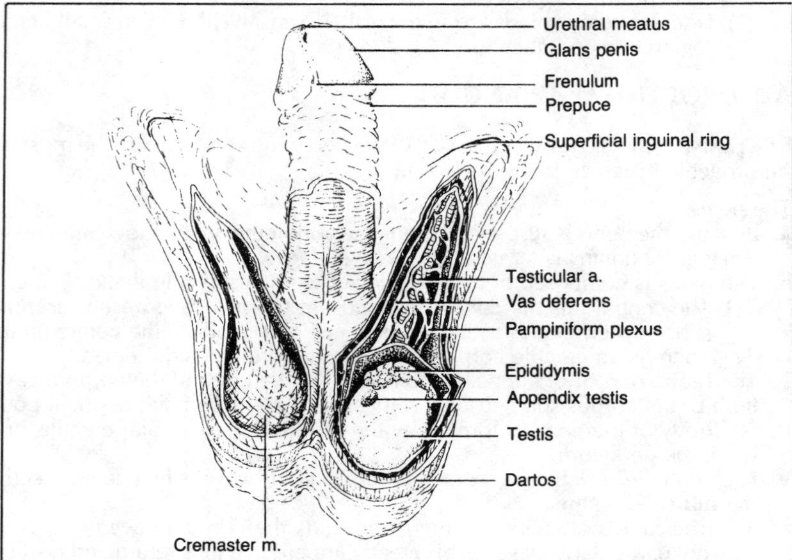

Figure 21-3. *The male external genitalia.* The contents of the left scrotum and spermatic cord are indicated.

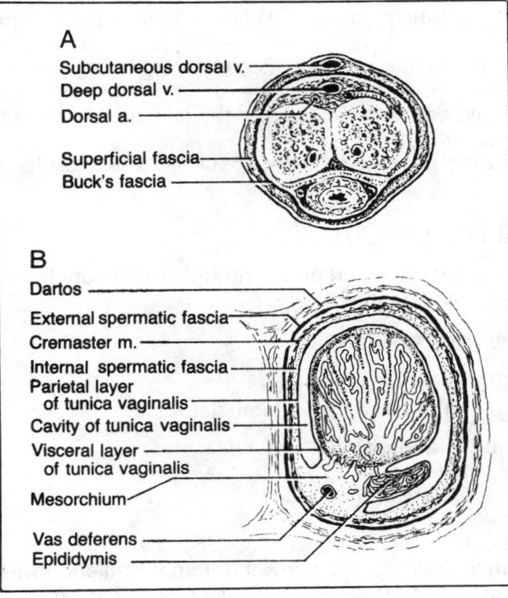

Figure 21-4. *A, The penis.* A cross section of the penis showing the fascial layers and erectile bodies. *B, The testis.* A cross section through the scrotum and testis.

(c) When warm, the scrotum is flaccid and distended to dissipate heat.

3. The **spermatic cord** (see Fig. 21-3).

a. The external spermatic fascia.

(1) The external spermatic fascia is derived from the deep (investing) fascia of the **external oblique muscle**.

(2) This forms the outermost layer of the spermatic cord.

b. The **cremaster muscle** and fascia.

(1) The **cremaster muscle** is derived from the **internal oblique muscle** and its investing fascia.

(2) It is innervated by the genitofemoral nerve, which provides the motor limb of the *cremaster reflex*.

c. The **internal spermatic fascia**.

(1) The internal spermatic fascia is derived from the **transversalis fascia**.

(2) This layer contains the testicular neurovascular bundle and the vas deferens.

(a) Each **testicular artery** arises from the abdominal aorta.

(b) The **pampiniform plexus** of veins coalesces into each **testicular vein**.

(i) The right testicular vein drains into the inferior vena cava, whereas the left testicular vein drains into the left renal vein.

(ii) Varices of the pampiniform plexus are common (80 percent). Fully 90 percent of all varices are on the left side. This is because local venous hypertension resultant from the sigmoid colon, which contains stored feces, compresses the testicular vein.

(c) The artery of the vas deferens provides some collateral circulation to the testes and scrotum.

d. The **tunica vaginalis**.

(1) The tunica vaginalis is a remnant of the **processus vaginalis**.

(a) The processus vaginalis, an evagination of the peritoneal cavity into the scrotum, is usually occluded in the adult spermatic cord. A patent processus vaginalis predisposes to indirect (congenital) hernias.

(b) Partial occlusion of a processus vaginalis can result in fluid accumulation—*hydrocele processus vaginalis*—which can not be distinguished from a hernia or incompletely descended testis until surgery is performed.

(2) Anterior and lateral to the testes, the processus vaginalis remains patent, forming the tunica vaginalis, which represents a detached portion of the peritoneal cavity within the scrotum.

(a) Although it is known that the testes develop retroperitoneally and descend into the scrotal sacs retroperitoneally, there is debate as to whether the adult testis is peritoneal or retroperitoneal, with resultant confusion of terminology.

(i) The **parietal layer** of the tunica vaginalis is equivalent to parietal peritoneum.

(ii) The **visceral layer** of the tunica vaginalis is equivalent to visceral peritoneum.

(iii) The **mesorchium**, the mesentery of the testis, is the reflection of the parietal layer to become the visceral layer.

(b) Fluid accumulation within the peritoneal cavity about the testes results in *hydrocele*.

B. THE FASCIA OF THE UROGENITAL TRIANGLE (Fig. 21-5)

1. The underlying muscular and fascial structures support and contribute to the function of the external genitalia as well as reinforce the abdominal cavity.

2. The **superficial layer of the superficial perineal fascia**:
   - **a.** Is a fatty layer just beneath the dermis.
   - **b.** Is continuous with Camper's fascia of the abdominal wall.
   - **c.** Passes across the perineum to become the fat of the ischiorectal fossa.

3. The **deep layer of the superficial perineal fascia** (Colles' fascia):
   - **a.** Is a membranous layer that retains sutures.
   - **b.** Is continuous with Scarpa's fascia of the abdominal wall.
   - **c.** Is attached as follows:
     - **(1)** Laterally, it is attached to the ischiopubic rami (see Fig. 21-2).
     - **(2)** Anteriorly, it is continuous with Scarpa's fascia of the abdominal wall and fuses with the superficial layer of the superficial fascia, which passes over the penis as the **superficial fascia of the penis** (see Fig. 21-5A).
     - **(3)** Medially, it inserts into the adventitia of the urethra.
     - **(4)** Inferiorly, it fuses with the superficial layer of the superficial fascia and passes over the scrotum, where it contributes to the **dartos layer**.
     - **(5)** Posteriorly, it passes inferior (superficial) to the **superficial transverse perineal muscle** to the posterior edge of the urogenital triangle; then it runs superiorly to the posterior edge of the **deep transverse perineal muscle**.
       - **(a)** The midline fibers join the **central tendon (perineal body)**.
       - **(b)** The more lateral fibers merge with the inferior and superior leaflets of the fascia of the urogenital diaphragm, which run anteriorly toward the pubis, enclosing the deep transverse perineal muscle. Thus these two leaflets define the **deep perineal pouch** (see Fig. 21-7).
         - **(i)** The superficial (inferior) layer is termed the **inferior fascia of the urogenital diaphragm**, or the **perineal membrane**.

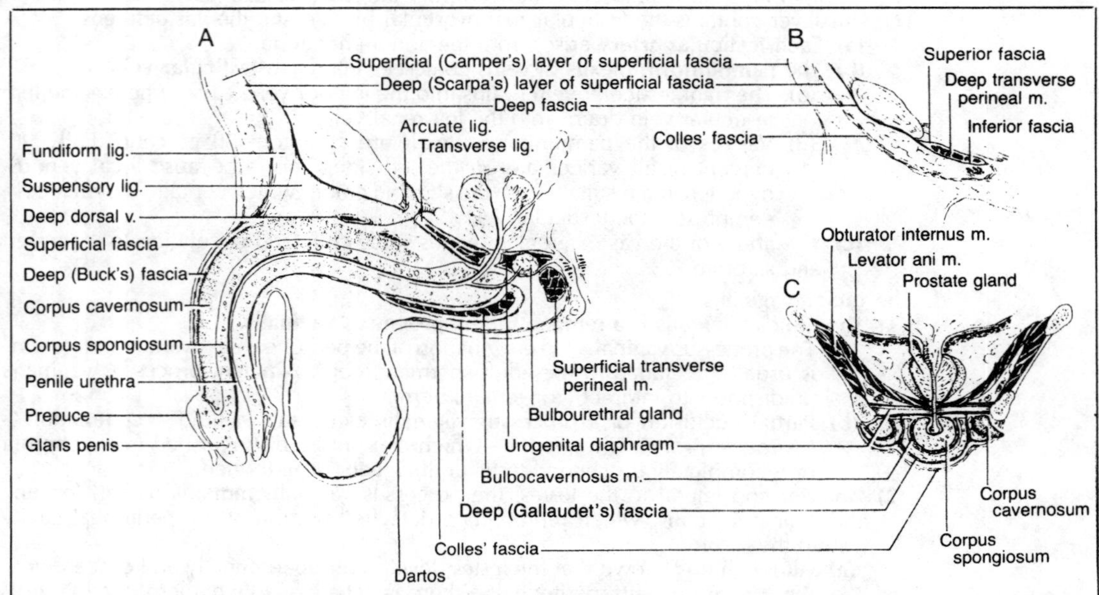

Figure 21-5. *The perineal fascia in the male. A,* Midsagittal section. *B,* Parasagittal section. *C,* Coronal section.

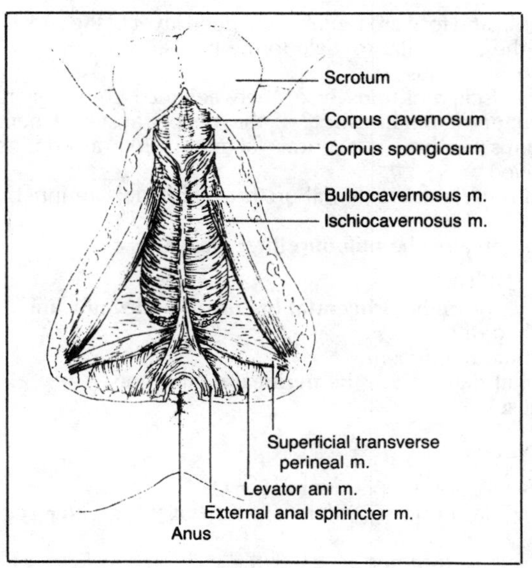

Figure 21-6. *The superficial perineal pouch in the male.*

(ii) The deep (superior) layer is termed the **superior fascia of the urogenital diaphragm**.

4. The **deep layer of investing fascia** (Gallaudet's) covers the muscles of the superficial pouch (Fig. 21-5C).

C. THE SUPERFICIAL PERINEAL POUCH (SPACE) [Fig. 21-6]

1. The **superficial perineal pouch** is defined as that region between Colles' fascia and the inferior fascia of the urogenital diaphragm (perineal membrane).
2. The contents of the superficial perineal pouch include:
   a. The **superficial transverse perineal muscle**.
      (1) This muscle may be absent unilaterally or bilaterally.
      (2) It arises on each side from the ischial tuberosities.
      (3) Each inserts into the perineal body (central tendon).
      (4) Although it is a very small and weak muscle, it probably acts to stabilize the perineum.
      (5) It is a palpable surgical landmark that delineates the posterior extent of the superficial pouch.
   b. The **ischiocavernosus muscles** and **crura** of the penis.
      (1) These structures arise from the ischial tuberosities and adjacent rami of the ischia.
      (2) On each side a crus, composed of erectile tissue, forms a radix of the penis.
      (3) Each crus is covered by an ischiocavernosus muscle.
         (a) These muscles insert onto the penis just distal to the point at which the crura join to form the body of the penis.
         (b) These muscles are innervated by the perineal branches of the pudendal nerve.
   c. The **bulbocavernosus (bulbospongiosus) muscles** and **bulb** of the penis.
      (1) The bulb of the **corpus spongiosum** in the male is composed of erectile tissue.
      (2) In the male the bulbospongiosus muscles overlie the bulb of the penis.
         (a) The muscles arise from the central tendon, posteriorly, and from the **median raphe**.
         (b) Each inserts into the deep fascia of the penis just after the crura fuse.
         (c) They are covered by the deep investing fascia (Gallaudet's) in the superficial pouch, which becomes the deep fascia (Buck's) on the penile shaft.
         (d) Once termed the compressor urethrae, the bulbospongiosus muscle functions to expel the final drops of urine from the penile urethra. Because the muscle does not extend beyond the very base of the penile shaft, it is rather ineffectual in this function.

D. THE DEEP PERINEAL POUCH (SPACE) [Fig. 21-7]

1. The deep perineal pouch is between the inferior and superior fascial planes of the urogenital diaphragm.

a. The deep pouch is formed by the split and forward return of the deep layer of the superficial fascia from the posterior edge of the urogenital triangle to the pubis.

b. It contains the **deep transverse perineal** muscle.

(1) The deep transverse perineal muscle runs transversely between the ischiopubic rami.

(2) The **external urethral sphincter**, a modification of the deep transverse perineal muscle about the **membranous urethra**, is a prime factor in urethral continence once desire to void is perceived.

(3) The deep transverse perineal muscle and external urethral sphincter are innervated by the pudendal nerve.

c. The deep pouch in the male also contains the **bulbourethral glands**.

2. The **urogenital diaphragm** (see Fig. 21-5C).

a. The **urogenital diaphragm** lies inferior to the **urogenital hiatus** of the **levator ani**.

b. The urogenital diaphragm consists of the:

(1) **Superior fascia** of the urogenital diaphragm.

(2) **Inferior fascia** of the urogenital diaphragm, the **perineal membrane**.

(3) **Deep transverse perineal muscle.**

E. THE VASCULAR SUPPLY OF THE PERINEUM

1. The **internal pudendal artery** supplies most of the perineum (Fig. 21-8).

a. This artery leaves the pelvic cavity via the infrapyriform portion of the **greater sciatic foramen**.

b. It then passes through the **lesser sciatic foramen** to enter the ischiorectal fossa of the perineum, where it courses in the **pudendal canal** (Alcock's) on the medial side of the obturator internus muscle.

c. The internal pudendal arteries give rise to the:

(1) **Inferior rectal (hemorrhoidal) arteries**.

(2) **Superficial perineal arteries**, including:

(a) The **transverse perineal branch**, which supplies the superficial and deep perineal pouches.

(b) The **posterior scrotal** or **labial branch**.

(3) **Deep perineal arteries**, which supply blood to the erectile tissues of the penis.

(a) The **dorsal arteries** of the penis lie beneath the deep fascia (Buck's) and external to the tunica albuginea.

(b) The **deep (central) arteries**, located within the corpora cavernosa, supply those erectile tissues.

(c) The **bulbar arteries** of the bulb of the penis supply erectile tissue of the corpus spongiosum and glans penis.

2. The **external pudendal arteries**, which arise from the femoral arteries, supply the anterior superficial portion of the perineum and the subcutaneous tissue of the penis.

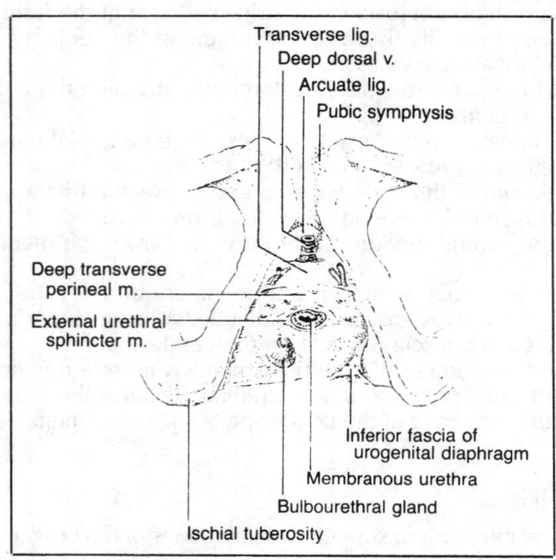

Figure 21-7. *The deep perineal pouch in the male.*

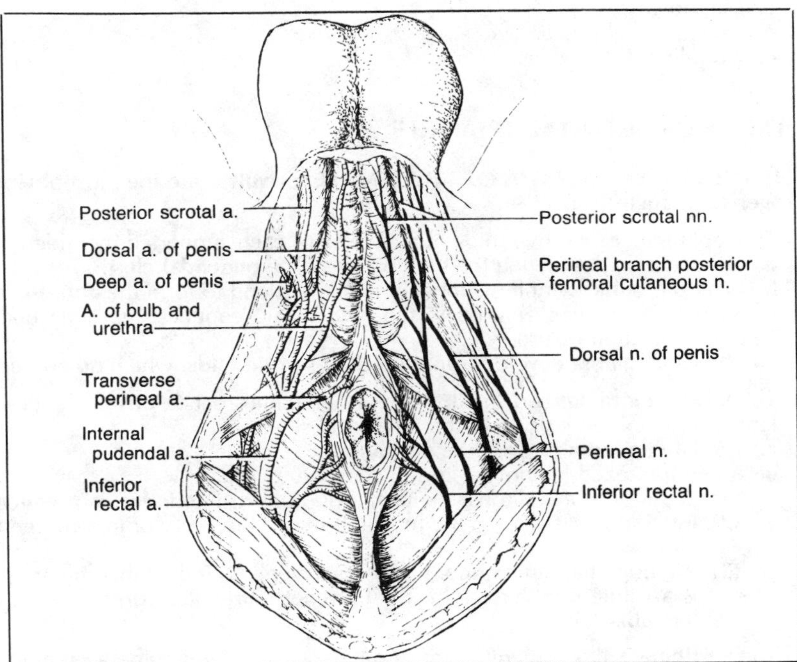

Figure 21-8. *The vasculature and innervation of the male perineum.*

F. THE VENOUS DRAINAGE OF THE PERINEUM

1. The veins of the perineum generally parallel the internal pudendal artery.
2. The **deep dorsal vein** of the penis penetrates the urogenital diaphragm between the **arcuate** and **transverse ligaments** and drains into the **prostatic venous plexus** (see Fig. 21-7).
3. The **superficial dorsal veins** of the penis and the veins of the anterior scrotum drain into the **external pudendal veins**, which enter the femoral vein.

G. INNERVATION OF THE UROGENITAL TRIANGLE is by the **perineal branch** of the **pudendal nerve** (see Fig. 21-8).

1. The pudendal is a mixed nerve, containing the general somatic afferent (GSA), general somatic efferent (GSE), and sympathetic (GVE) nerves as components.
2. The branches of the pudendal nerve are named according to the arteries that they accompany.
3. On the dorsum of the penis the terminal branch of the pudendal nerve lies between the deep penile fascia (Buck's) and the tunica albuginea and supplies the dorsum and glans of the penis.
4. The perineum may be blocked by anesthesia of the pudendal nerve as it courses along the pudendal canal. A needle is introduced just medial to the ischial tuberosity and directed toward the ischial spine.

H. CLINICAL CONSIDERATIONS

1. Urinary extravasations.
   - **a.** Perforation of the urethra superior to the urogenital diaphragm, resulting from trauma or disease, may be intrapelvic, intraabdominal, intraperitoneal, or extraperitoneal.
   - **b.** Perforations of the penile urethra produce extravasation of urine into the superficial pouch. From this space the urine may extravasate along the abdominal wall (beneath Scarpa's fascia), into the scrotum (under the dartos tunic), and onto the penis (either superficial or beneath Buck's fascia).
2. Congenital anomalies of the penis.
   - **a.** Hypospadias (incomplete fusion of the penile urethra with the opening on the ventral surface).
   - **b.** Epispadias (with an opening on the dorsal surface).

c. Micropenis.
d. Congenital absence.
e. Doubling of the glans.

V. THE FEMALE UROGENITAL TRIANGLE

A. THE EXTERNAL GENITALIA OF THE FEMALE, or **vulva**, are the most obvious feature of the urogenital triangle (Fig. 21-9).

1. The **labia majora** are two folds of hirsute skin, each supported by underlying fat pads.
   a. The labia majora define the boundaries of the **pudendal cleft**.
   b. Over the pubic symphysis the labia majora blend to form the **anterior commissure**.
   c. Across the midline anterior to the anus a fold connects the labia majora to form the **posterior commissure**.
   d. The labia majora develop from the **labial-scrotal folds**, which do not fuse in the female.
2. The **labia minora** are two small, hairless, sometimes pendulous, folds located between the labia majora on either side of the urogenital sinus.
   a. The labia minora enclose the **vestibule**.
   b. Anteriorly, each labium minora divides into two parts.
      (1) Superior to the clitoris, the lateral portions fuse to form the **prepuce of the clitoris**.
      (2) Inferior to the clitoris, the medial portions fuse inferior to form the **frenulum of the clitoris**.
      (3) Although the labia minora appear to blend with the labia majora posteriorly, there is a subtle transverse fold, the **fourchette** (frenulum of the labia), which connects the labia minora.
3. The **vestibule** is the urogenital sinus bordered by the labia minora, frenulum of the clitoris, and fourchette.
   a. The **external urethral ostium** is located anterosuperior to the vaginal opening approximately 2 cm inferior to the glans clitoris.
   b. The **introitus** (vaginal opening), the predominant feature of the posterior region of the vestibule, is covered incompletely by the **hymen**.
      (1) An imperforate hymen, which is rare, must be surgically opened prior to the first menses.
      (2) The hymen is usually stretched, and sometimes torn, during the first coitus.
      (3) If partially intact, it is always torn during the first vaginal childbirth.
   c. The **fossa navicularis** (vestibular fossa of the vagina) is the region between the vaginal opening and the fourchette.
   d. The **greater vestibular glands** (Bartholin's):
      (1) Are located as a pair in the superficial perineal pouch on either side of the vaginal opening.
      (2) Secrete mucus.
      (3) Are homologous to the bulbourethral glands (Cowper's) in the male.
   e. The **paraurethral glands** (Skene's):
      (1) Open on either side of the external urethral ostium.
      (2) Are prone to infection by pathogenic organisms.
      (3) Are homologous to the prostate in the male.
   f. The lesser vestibular glands also empty into the vestibule.

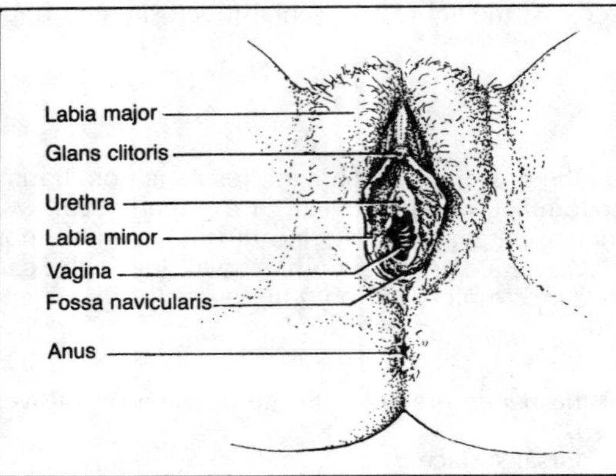

Figure 21-9. *The female external genitalia.*

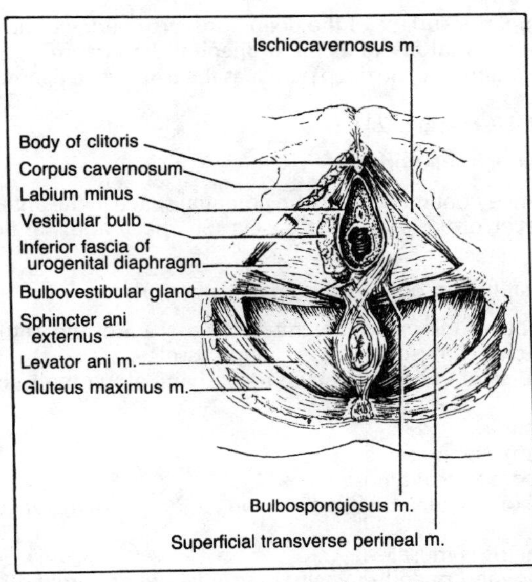

Figure 21-10. *The superficial perineal pouch in the female.*

g. In contrast to the female's, the male vestibule fuses to form the penile urethra.

4. The **clitoris**, about 2.5 cm long, is not usually freely projecting.
   a. The **body** of the clitoris is comprised of three bodies of erectile tissue.
      (1) Fusion of crura of the corpora cavernosa—two bodies of erectile tissue—forms the body of the clitoris.
      (2) The vestibular bulbs—two other bodies of erectile tissue—unite to form the thread-like commissure of the clitoris, which terminates in an expansion, the glans clitoris.
      (3) The clitoris is attached to the pubic symphysis by a suspensory ligament.
      (4) An **ischiocavernosus muscle** overlies each crus.
   b. The **glans clitoris** is composed of erectile tissue that caps the body of the clitoris.
      (1) It is formed by the merging of the **vestibular bulbs**.
      (2) It is homologous to the glans penis in the male.
      (3) There are numerous sensory nerve endings converging on the glans.
   c. The **prepuce** or **foreskin** of the clitoris is formed by the fusion of the lateral parts of the labia minora.
   d. The **frenulum** of the clitoris is formed by the fusion of the medial parts of the labia minora.

B. FASCIA OF THE FEMALE UROGENITAL TRIANGLE

1. The underlying muscular and fascial structures support and contribute to the function of the external genitalia as well as reinforce the abdominal cavity.

2. The **superficial layer of the superficial fascia.**
   a. This fatty layer is continuous with Camper's layer of the abdominal wall and gives form to the mons pubis and labia majora.
   b. The fat is lost as this layer extends into the labia minora.

3. The **deep layer of the superficial fascia (Colles' fascia).**
   a. This membranous layer is continuous with Scarpa's fascia of the anterior abdominal wall.
   b. Relations of Colles' fascia in the female.
      (1) Laterally, it is attached to the ischiopubic rami.
      (2) Anteriorly, it is continuous with Scarpa's fascia.
      (3) Medially, it inserts into the adventitia of the urethra and vagina.
      (4) Inferiorly, it passes deep to the labia majora.
      (5) Posteriorly, it delimits the urogenital triangle by passing superficial to the **superficial transverse perineal muscle**.
         (a) The midline fibers insert into the adventitia of the vagina and the **central tendon of the perineum**.
         (b) The more lateral fibers reach the posterior edge of the **deep transverse perineal muscle** before merging with the fascia of the urogenital diaphragm.
            (i) This fascial reflection over the inferior surface of the deep transverse perineal muscle contributes to the **inferior** or **external fascia of the urogenital diaphragm** (see Fig. 21-10).

(ii) The fascia over the superior surface of the deep transverse perineal muscle forms the **superior** or **internal fascia of the urogenital diaphragm**.
(iii) These two layers fuse anterior to the deep pouch as the **transverse ligament**.

C. THE SUPERFICIAL PERINEAL SPACE (POUCH) (Fig. 21-10)

1. Colles' fascia bounds the superficial pouch inferiorly.
2. Anteriorly, the superficial perineal space is continuous with the potential space in the anterior abdominal wall between the deep layer of the superficial fascia (Scarpa's) and the deep investing fascia.
3. The inferior fascia of the urogenital diaphragm bounds this space superiorly.
4. The superficial pouch contains the roots of the external genitalia, the crura of the clitoris, and the vestibular bulbs with the associated muscles, nerves, and vessels.
   a. The **corpora cavernosa**, paired bodies of vascular erectile tissue lying along the ischiopubic rami:
      (1) Fuse to form the body of the clitoris.
      (2) Are covered by the ischiocavernosus muscles.
      (3) Are homologous to the male corpora cavernosa.
   b. The **vestibular bulbs**, paired masses of vascular erectile tissue on either side of the vaginal introitus under the labia minora:
      (1) Are covered by the bulbospongiosus muscles.
      (2) Tend to spread the labia minora and open the vaginal introitus, when turgid.
      (3) Are homologous to the fused halves of the bulb of the penis in the male.
   c. The **ischiocavernosus muscles** overlie the clitoral crura.
      (1) Their origin is the medial side of the ischial tuberosities and ischial rami.
      (2) They insert into the margin of the pubic arch and into each crus of the clitoris.
   d. The **bulbospongiosus (bulbocavernosus) muscles** overlie the vestibular bulbs.
      (1) Their origin is the anterior portion of the central tendon.
      (2) The medial fascicles attach to the deep fascia of the dorsum of the clitoris; the lateral fascicles attach to the inferior fascia of the urogenital diaphragm.
   e. The **superficial transverse perineal muscle.**
      (1) Its origin is along the anterior portions of the ischial tuberosities.
      (2) It inserts into the central tendon of the perineum (perineal body).
      (3) Its action is to stabilize the central tendon of the perineum.
   f. The **greater vestibular (vulvovaginal) glands** (Bartholin's):
      (1) The ducts of these mucous glands open into the vestibule.
      (2) These glands are homologous to the bulbourethral glands in the male.
   g. The **central tendon (perineal body)** [see Fig. 21-12] is a common point of attachment of connective tissue and muscles of the anal and urogenital triangles.

D. THE DEEP PERINEAL SPACE (POUCH) [Fig. 21-11]

1. The deep perineal pouch is delineated by the superior and inferior fascial layers of the urogenital diaphragm.

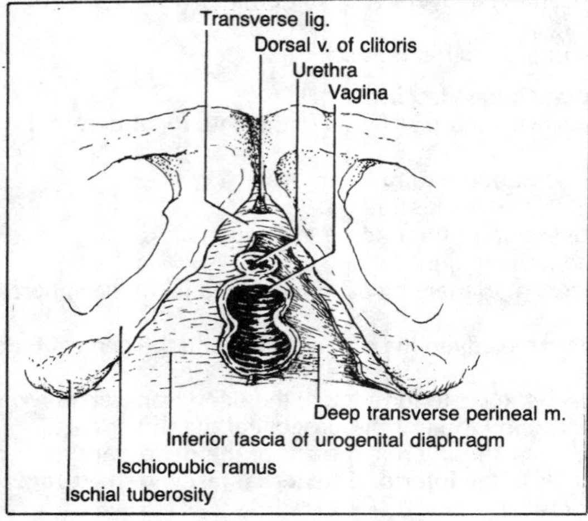

Figure 21-11. *The deep perineal pouch in the female.*

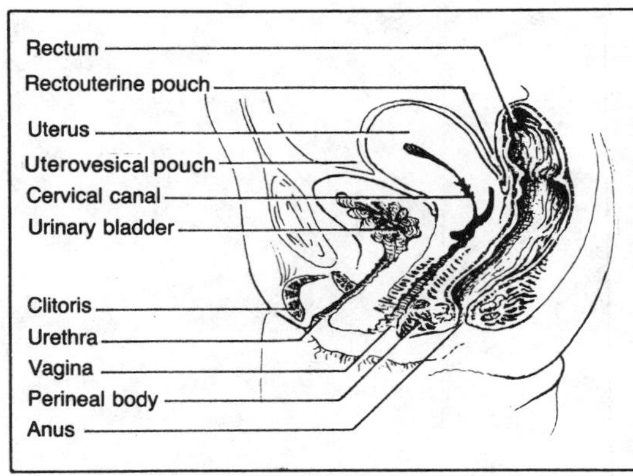

Figure 21-12. *Midsagittal section through the female genitalia.*

2. In the female the deep space contains the deep transverse perineal muscle and the external urethral sphincter.
3. The superior and inferior fascial layers and the deep transverse perineal muscle form the **urogenital diaphragm**, which lies across the perineum and supports the urogenital hiatus in the levator ani muscle.
   a. The **inferior fascia of the urogenital diaphragm (perineal membrane**, triangular ligament) bounds the deep pouch inferiorly.
   b. The **superior fascia of the urogenital diaphragm** bounds the deep pouch superiorly.
   c. The attachment of the superior and inferior fascias of the urogenital diaphragm to the ischiopubic rami limits the deep pouch laterally.
   d. The merging of the inferior and superior fascias of the urogenital diaphragm anterior to the deep transverse perineal muscle forms the **transverse perineal ligament**.
      (1) The **arcuate ligament** behind the pubic symphysis is the anterior portion of this layer.
      (2) The **deep dorsal vein** of the clitoris runs through a hiatus between the transverse and arcuate ligaments.
   e. The superior and inferior fascial layers of the urogenital diaphragm merge posterior to the deep transverse perineal muscle and join with Colles' fascia (deep layer of the superficial perineal fascia).
4. Contents of the deep perineal pouch (see Fig. 21-11).
   a. The **deep transverse perineal muscle**:
      (1) Originates in the ischiopubic ramus.
      (2) Inserts into the central raphe of the urogenital diaphragm tendon, the adventitia of the vaginal wall, and the central tendon of the perineum; some fibers are continuous with the muscles surrounding the anal canal.
   b. The **sphincter urethrae (external urethral sphincter)** is a modified portion of the deep transverse perineal muscle.
      (1) These fascicles arch anterior to the urethra, but do not pass posterior to the urethra and encircle that structure, as in the male, because at this level the urethra is embedded in the adventitia of the anterior vaginal wall (Fig. 21-12). This anatomic arrangement predisposes toward urinary stress incontinence, especially after trauma associated with difficult parturition.
      (2) Some fibers of the sphincter insert into the vaginal wall.
      (3) The external urethral sphincter is innervated by the pudendal nerve.
   c. The dorsal artery and nerve of the clitoris, terminal branches of the pudendal neurovascular bundle, course through the deep perineal pouch.

E. THE VASCULATURE OF THE FEMALE UROGENITAL TRIANGLE (Fig. 21-13)

1. The arterial supply.
   a. The **internal pudendal artery** supplies most of the external genitalia.
      (1) The **transverse perineal branch** runs parallel to the superficial transverse perineal muscle and gives off the **bulbar branches**, which supply the vestibule and erectile tissues of the vestibular bulbs.
      (2) The **posterior labial branches** supply the posterior portion of the vulva.
      (3) The **deep (central) artery** of the clitoris, located in the corpora cavernosa, supplies the erectile tissues of those bodies.

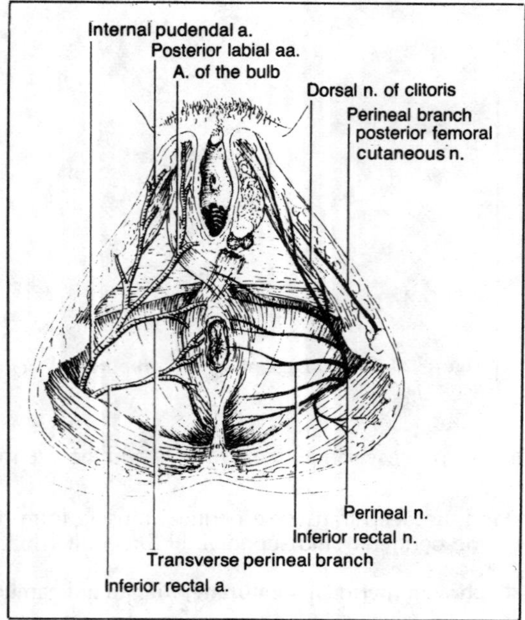

Figure 21-13. *The vasculature and innervation of the female perineum.*

(4) The **dorsal artery** of the clitoris, which lies between the deep fascia of the clitoris and the tunica albuginea, supplies the superficial aspects of the clitoris.

b. The **external pudendal branches** of each femoral artery supply the anterior aspects of the mons, anterior portions of the labia majora, and subcutaneous tissue of the clitoris.

2. The venous return:
   a. With one exception, the venous drainage of the female urogenital triangle generally parallels the arterial supply.
   b. The **deep dorsal vein** of the clitoris passes between the arcuate and transverse ligaments of the urogenital diaphragm to drain into the vaginal venous plexus (see Fig. 21-11).

3. The pathways of lymphatic drainage of the perineum are of considerable clinical importance because of the high prevalence of cervical carcinoma.
   a. The lymphatics of the vulva and inferior vagina drain largely into the superficial inguinal nodes.
   b. The superficial lymphatics of the clitoris drain toward the superficial inguinal nodes, whereas the deep vessels drain into the internal iliac nodes.
   c. The middle vagina generally drains toward the internal iliac nodes.
   d. The upper vagina generally drains toward the internal iliac and common iliac nodes.

F. INNERVATION OF THE FEMALE UROGENITAL TRIANGLE (see Fig. 21-12)

1. Branches of the **pudendal nerve** supply most of the perineum.
   a. The pudendal nerve arises from spinal segments S2–S4.
      (1) The **posterior labial branches** supply the skin of the labia majora and minora.
      (2) The **dorsal nerve** of the clitoris supplies that structure.
      (3) The **inferior rectal branches** supply most of the anal triangle.
   b. *Pudendal block:* Pain fibers from the perineum may be blocked by injecting an anesthetic in the vicinity of the pudendal nerve in the perineum. The needle is inserted through the posterolateral vaginal wall, just beneath the pelvic diaphragm and angled toward the ischial tuberosity. Within a centimeter of the tuberosity, the needle will be in the vicinity of the **pudendal canal**.

2. The **ilioinguinal** and **genitofemoral nerves**, which arise from L1 and L2, supply the anterior part of each labium as well as the corresponding half of the mons pubis.

3. **Perineal branches** of the posterior femoral cutaneous nerve arise from spinal segments S2–S4 and supply the posterolateral region of the urogenital triangle.

G. CLINICAL CONSIDERATIONS

1. The bulbovestibular glands are prone to infection (*Bartholin cysts*).

2. **Episiotomy** may be performed, in which the perineum is sectioned prior to parturition to prevent uncontrolled tearing. Suturing an incision is preferable to repairing a ragged tear.
   a. **Midline episiotomy.**
      (1) The incision is into the posterior vaginal wall and carried posteriorly in the midline through the fossa navicularis to divide the central tendon.
      (2) The incision does not extend into the deep fibers of the external anal sphincter.
      (3) The advantage of the midline episiotomy is that it is relatively bloodless and painless because no major vessels or nerves are transected. However, this incision provides a limited expansion of the birth canal with a slight possibility of tearing the anal sphincters.
   b. **Mediolateral episiotomy.**
      (1) The incision into the posterior vaginal wall is carried posterolaterally into the ischiorectal fossa.
      (2) This incision incises skin over the ischiorectal fossa, bulbospongiosus muscle, superficial transverse perineal muscle, fascia and muscle of the urogenital diaphragm, and transverse perineal branches of the internal pudendal artery and pudendal nerve.
      (3) The advantage of the mediolateral episiotomy is that it allows greater expansion of the birth canal into the ischiorectal fossa. However, there is an increased risk of infection due to contamination of the ischiorectal fossa, and the incision is more difficult to close layer by layer.

22
The Pelvic Viscera

I. THE PELVIC PORTION OF THE GASTROINTESTINAL TRACT

A. THE RECTUM is that part of the large intestine, about 13 cm (5 in) long, between the sigmoid colon and the anal canal (Fig. 17-18).

1. The rectum is a retroperitoneal structure located immediately superior to the pelvic diaphragm.
   a. The upper limit of the rectum is where the sigmoid mesocolon disappears.
   b. The lower limit is the pelvic floor.

2. The **ampulla**, which lies just above the pelvic floor, is the widest part of the rectum and is capable of considerable distension.
   a. Because the feces are stored in the sigmoid colon, the ampulla is usually empty.
   b. Movement of feces into the rectal ampulla from the sigmoid colon generates the sensation of rectal fullness and the urge to defecate.

3. The rectum pierces the pelvic diaphragm (the levator ani muscle) to become the anal canal.
   a. The **rectal sling** is formed by the **puborectalis muscle**, which is the most medial portion of the **pubococcygeus muscle** of the **levator ani** group.
      (1) The puborectalis muscle fibers loop posterior to the rectum, and some fibers attach to the rectum.
      (2) Tension in this muscle cants the rectum forward.
   b. The rectal sling is the single most important factor in fecal continence.

B. Within the rectum are three transverse **rectal folds** (valves of Houston), formed by the inner three layers of the intestinal wall.

1. The high incidence of rectal, colonic, and prostatic carcinoma makes digital and sigmoidoscopic examination important in the post-middle-aged male.
   a. A majority of rectal neoplasms can be discovered by digital palpation.
   b. Carcinomatous spread posteriorly from the rectum may involve the sacral plexus, with pain distribution down the leg. Anterior spread may involve pelvic viscera, such as the prostate, bladder, uterus, or vagina.

2. Prolapse of the rectum through the anus is often related to damage of the levator ani muscle, usually the result of obstetric trauma.

C. The arterial supply, venous return, and lymphatic drainage of the rectum are of considerable importance.

1. The rectum receives branches from the inferior mesenteric artery via the **superior rectal (hemorrhoidal) artery**, from the internal iliac artery via the **middle rectal (hemorrhoidal) arteries**, and from the internal pudendal arteries via the **inferior rectal (hemorrhoidal) arteries**. Anastomotic connections are abundant in the submucosa between tributaries of these arteries (Fig. 17-20).

2. The venous return from the rectum is along similarly named vessels, between which anastomoses are especially rich (see Fig. 17-20).
   a. Portal hypertension (accompanying cirrhosis) or local compression of the inferior mesenteric artery (because of chronic constipation with dilation of the sigmoid colon or because of the presence of the fetus during the later stages of pregnancy) results in the shunting of blood from the rectum and anal canal into the systemic venous system.

b. This shunting results in *hemorrhoids,* which are variceal dilatations of the submucosal anal and perianal venous plexuses.
c. Hemorrhoidal varices may rupture during the evacuation of feces and produce considerable blood loss. Chronically bleeding hemorrhoids may lead to anemia.
d. Hemorrhoids are amenable to treatment by ligature or cautery.

3. Because the lymphatic drainage of the rectum parallels the various arterial pathways, metastatic carcinoma of the rectum may be widely disseminated within the abdomen and pelvis as well as to the inguinal nodes. In addition, hematogenous spread to the liver frequently occurs.

II. THE PELVIC PORTION OF THE URINARY SYSTEM

A. THE URETERS

1. The ureters convey urine from the kidneys to the urinary bladder.
   a. To reach the urinary bladder the ureters pass anterior to the psoas muscles and common iliac vessels to enter the minor pelvis.
   b. They run along the posterolateral wall of the deep pelvis in which they lie extraperitoneally in the endopelvic fascia.
   c. In the male they pass posterior to the **vas deferens** (see Fig. 22-2).
   d. In the female they pass beneath the **cardinal ligament** and **uterine vessels** (see Fig. 22-1).
   e. They converge to enter the urinary bladder posteroinferiorly, where they define the upper lateral limits of the **trigone**.
      (1) Because the bladder is muscular, the intramural portion of each ureter is narrow. This is a point at which nephroliths may lodge.
      (2) The oblique course of each ureter through the bladder wall functions as a check valve to prevent reflux of urine from the urinary bladder into a ureter.

2. The pelvic ureters (aside from the common bile duct) are the structures most often abused during abdominal and pelvic surgical procedures.
   a. Injuries must be recognized and repaired at once to prevent urine extravasation and possible subsequent peritonitis.
   b. The female ureters are frequently injured due to inadvertent clamping, ligation, or sectioning along with the uterine vessels.
   c. Careful identification and subsequent avoidance of the ureters is a major objective of abdominopelvic surgery.

3. Innervation of the ureters.
   a. The proximal portion of each ureter is innervated by autonomic nerves that arise from T12 and run in the least splanchnic nerve. As such, pain from this region refers pain to the lumbar region.
   b. In the vicinity of the pelvic brim, the ureter is innervated by autonomic nerves that arise from L1 and L2 and run in the lumbar splanchnic nerves. As such, pain from this region refers to the inguinal and pubic regions as well as to the lateral and anterior aspects of the thigh.
   c. The terminal portion of the ureter is innervated by autonomic nerves that arise from S2 to S4 and run in the pelvic splanchnic nerves. As such, pain from this portion refers to the perineum as well as the posterior thigh and leg.

B. THE URINARY BLADDER

1. The urinary bladder is located immediately behind the pubic symphysis and the superior pubic rami (see Figs. 22-2 and 22-8).
   a. The **fundus** of the urinary bladder is extraperitoneal.
      (1) This region expands freely, rising above the pubic crest.
      (2) It normally accommodates 250 to 300 ml, but may accommodate up to 500 ml.
   b. The **base** of the urinary bladder rests on the pelvic floor.
      (1) Internally, the base of the urinary bladder has a relatively unexpandable portion, the **trigone**, which the ureters enter and the urethra leaves.
      (2) The base is associated with the prostate gland in the male.
   c. The anterior surface of the urinary bladder defines the **retropubic space**, a potential space filled with endopelvic fascia.
      (1) The retropubic surgical approach to the prostate is through this space.
      (2) Urethral trauma superior to the urogenital diaphragm or rupture of the urinary bladder results in extravasation of urine into the retropubic space.

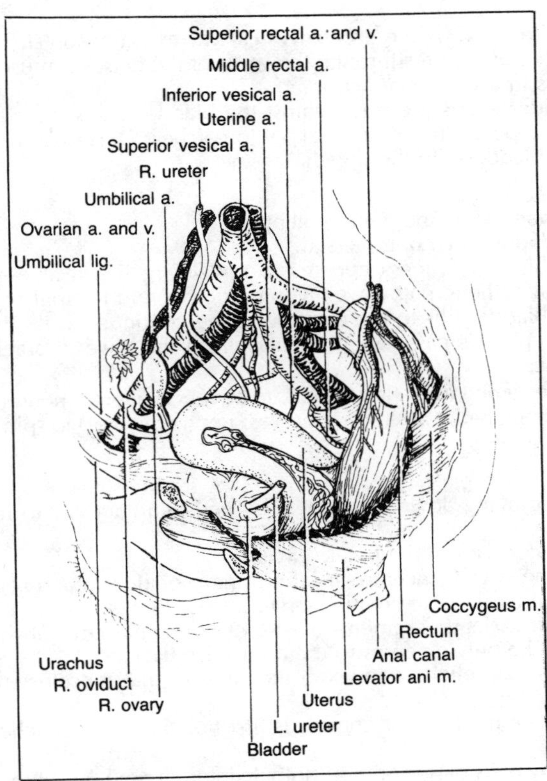

Figure 22-1. *The female pelvic viscera.* A parasagittal section indicates the relationships between the urinary bladder, uterus, and rectum as well as the major pelvic vasculature.

d. The posterior surface.
   (1) In the female the posterior surface of the urinary bladder is related to the **vesicouterine pouch** and the **vesicovaginal septum** (Fig. 22-1).
   (2) In the male the posterior surface is related to the **rectovesical pouch, rectovesical septum** (Denonvilliers' fascia), and **seminal vesicles**. The rectovesical septum is in part formed by embryonic fusion of the inferior portion of the rectovesical pouch.

e. The urinary bladder is supported by several pelvic structures and condensations of endopelvic fascia.
   (1) It is supported by the urogenital diaphragm and is stabilized by the penetration of the urogenital diaphragm by the urethra.
   (2) There is a supportive condensation of endopelvic fascia at the base of the urinary bladder.
   (3) The ***medial* umbilical ligaments**, remnants of the **umbilical arteries**, support the urinary bladder anterolaterally.
   (4) The ***median* umbilical ligament** supports the bladder superiorly.
      (a) The **urachus** forms as a tubular duct between the cloaca and the placental allantois.
      (b) The urinary bladder forms in the base of the urachus.
      (c) The distal portion of the urachus solidifies to become the median umbilical ligament.

2. The structure of the urinary bladder.

a. The walls of the urinary bladder are composed of a substantial layer of smooth muscle, the **detrusor muscle**.

b. The mucosal lining is composed of transitional epithelium.
   (1) The transitional epithelium, unique to the urinary system, becomes attenuated as the bladder fills and allows considerable stretching.
   (2) This layer is thicker and thrown into folds in the empty bladder.

c. The **trigone** is a triangular area at the base of the bladder.
   (1) Because it is relatively indistensible, it has few mucosal folds when the bladder is empty.
   (2) It is defined posterosuperiorly by the ostia of the ureters (about 5 cm apart) and inferiorly by the urethra.

(3) The ostia of the ureters are protected by the **ureterovesical valves** (Sampson's).
(a) The angular penetration of the ureters through the walls of the detrusor muscle within the trigone results in a valve-like action.
(b) Urine in the filling bladder exerts pressure against the bladder walls, thereby compressing the anterior aspect of the intramural portion of the ureters to prevent reflux of urine from the bladder into the ureters.

3. Innervation of the bladder (see Fig. 22-3).
a. The motor innervation to the bladder is via the nervi erigentes.
b. The afferent pathways from the bladder are controversial.
(1) The sensory perception for bladder fullness appears to travel largely via afferent neurons that lie along the sympathetic pathways (hypogastric nerve) to spinal segments T12–L2. Also, as the bladder fills to the extreme, urine continues to fill the ureters; this transient hydronephrosis causes additional pain in the lower thoracic and upper lumbar dermatomes.
(2) The sensory limb of the bladder-emptying reflex appears to travel via afferent neurons that lie along the parasympathetic pathways (pelvic nerves, nervi erigentes) to spinal segments S2–S4.

4. Functional considerations.
a. Urination is the process of urine moving down the ureters by peristaltic activity to fill the urinary bladder.
(1) Upon filling, the urinary bladder rises in the pelvic cavity.
(a) Afferent impulses from the stretch receptors of the urinary bladder reach segments S2–S4 of the spinal cord via the pelvic nerve.
(b) Autonomic efferent impulses leave segments S2–S4 of the spinal cord at the same level and travel by the pelvic nerves to the detrusor muscle.
(c) Stretch in the detrusor muscle elicits a reflex contraction of that muscle, enhancing the urge to void.
(d) In the normal, average male the bladder content is first noted when it reaches 100 to 150 ml.
(2) There does not appear to be any physiologic or morphologic internal urethral sphincter to close the opening to the prostatic urethra. The initial urinary sphincteric function has been debated intensely. It seems that physiologic sphincteric action in the empty to partially full bladder is due to the longitudinal folding of the mucosa in the region of the trigone. During ejaculation, contraction of the smooth muscle in the prostate gland and the peristaltic contraction of the prostate urethra function as an internal urethral sphincter.
(3) Upon filling, the mucosa of the trigone area around the ureteral outlet is stretched, thereby opening the prostatic urethra so that urinary continence of the partially full to full urinary bladder is maintained by action of the **external urethral sphincter**.
(a) The external urethral sphincter is formed about the membranous urethra from the adjacent circularly arranged fibers of the deep transverse perineal muscle and is under voluntary control via the deep perineal branches of the pudendal nerve.
(b) In addition, the most medial fibers of the **puborectalis muscle** of the pelvic floor contribute to the support of the bladder and may contribute some to urinary continence by raising the bladder, thereby attenuating the urethral lumen.
(c) Distension becomes uncomfortable in the male at a volume of 350 to 400 ml, beyond which painful sensations are experienced.
(d) If unrelieved, involuntary micturition occurs in the male at about 500 ml.
(e) In the female these values are considerably lower because the bladder is smaller and the external urinary sphincter is incomplete as a result of the penetration of the urogenital diaphragm by the vagina and the intimate association between the urethra and vagina.
b. Micturition is the process of emptying the urinary bladder through the urethra. It is a reflex action, initiated by the stretching of the detrusor muscle.
(1) When the puborectalis muscle is relaxed in anticipation of micturition, the bladder descends slightly; this process relieves the stretching that attenuates the lumen of the initial segment of the urethra.
(2) When the external urethral sphincter is allowed to relax, the bladder commences to void, often assisted by a Valsalva maneuver, which raises the intra-abdominal pressure.
(3) At the termination of micturition, strong contractions of the external urethral sphincter, pelvic floor, and in the male, bulbocavernosus muscles expel the residual urine from the urethra, and continence is restored.

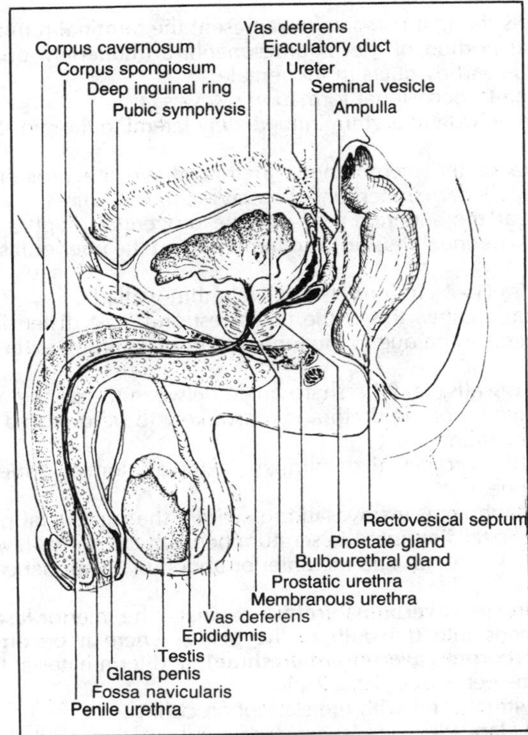

Figure 22-2. *The male pelvic viscera.* A midsagittal section depicts the lower portion of the male genitourinary tract.

(4) The detrusor muscle relaxes with the absence of stretch.

(5) Thus the bladder is under the influence of the autonomic nervous system (probably the parasympathetic division) via reflex arcs along the pelvic nerve with overriding voluntary control of the external urethral sphincter via the pudendal nerve.

5. Clinical considerations.

a. Transurethral cystoscopy is performed to inspect the mucosa of the urinary bladder and prostate. It may be combined with procedures to remove ureteral calculi and bladder stones. It is also an integral part of the transurethral approach to prostatectomy.

b. Patent urachus.

(1) A completely patent urachus (very rare) allows reflux of urine through the umbilicus.

(2) Approximately 33 percent of all males have to some extent a patent lumen of the urachus. Severance of such a patent urachus by a transverse midline incision can result in urine extravasation into the peritoneal cavity.

c. Pelvic fractures may damage the urinary bladder, urethra, rectum, uterus, vagina, nerves, and, especially blood vessels, with a potential for severe internal hemorrhage.

(1) The urinary bladder is most vulnerable in pelvic fracture, particularly when distended, with resulting intrapelvic and subperitoneal extravasation of urine.

(2) Following severe pelvic fractures, immediate surgical repairs to soft tissues may be essential.

C. THE URETHRA

1. The **male urethra** is divided into prostatic, membranous, and penile segments (Fig. 22-2).

a. The **prostatic urethra** is the portion that begins at the vesical neck at the apex of trigone, extends through the prostate gland, and terminates at the superior fascia of the urogenital diaphragm.

(1) This portion of urethra has the thickest walls.

(2) The **urethral crest (crista urethralis)** is located on the posterior wall of this segment.

(a) The numerous **prostatic ducts** open into the prostatic urethra on either side of the urethral crest.

(b) The **colliculus seminalis (verumontanum)** is the widest portion of the urethral crest.

(c) In the midline of the verumontanum lies the **prostatic utricle (uterus masculinus)**, an invagination about 5 mm deep.

(i) The prostatic utricle is thought by some to represent the terminal remnant of the fused terminal portion of the paramesonephric (müllerian) ducts, which form the uterus and oviducts in the female.

(ii) This structure is prone to occasional formation of cysts.

(d) Ejaculatory ducts enter the prostatic urethra immediately lateral to the prostatic utricle (see Fig. 22-2).

(3) Injury to the prostatic urethra or the membranous urethra above the urogenital diaphragm may result in intrapelvic, extraperitoneal extravasation of urine.

b. The **membranous urethra** begins at the superior fascia of the urogenital diaphragm, extends through the deep transverse perineal muscle, and terminates at the inferior fascia of the urogenital diaphragm.

(1) This short portion of the urethra (1–2 cm long) is relatively thin-walled.

(2) Except for the external urethral meatus, this is the narrowest and least distensible portion of the urethra. This seems to be due to the tone of the surrounding **external urethral sphincter**.

(3) Due to its narrowness, delicate walls, and the sharp angle between the penile and membranous portions, this segment of the urethra is most likely to be ruptured by trauma or passage of instrumentation.

(a) If damage occurs above the urogenital diaphragm, extravasation of urine is intrapelvic and extraperitoneal.

(b) If damage is below this diaphragm, extravasation is within the superficial perineal space, with possible extension into the scrotum beneath the dartos layer, the penis superficial to Buck's fascia, and the anterior abdominal wall beneath Scarpa's fascia.

c. The extrapelvic terminal **penile (spongy, cavernous) urethra** begins at the inferior fascia of the urogenital diaphragm; extends into the bulb of the penis, where it becomes surrounded by **corpus spongiosum (corpus cavernosum urethrae)**; and terminates at the external urethral meatus of the glans penis (see Fig. 22-2).

(1) It varies in length among individuals and with the state of erection.

(2) The ducts of the **bulbourethral glands** (Cowper's) enter the penile urethra just below the urogenital diaphragm.

(3) The penile urethra enlarges into the **fossa navicularis** just prior to termination at the **external urethral meatus**.

(4) Damage to the penile urethra may result in extravasation of urine.

(a) If the deep penile fascia (Buck's) remains intact, extravasation occurs deep to this fascial layer.

(b) If the deep penile fascia is damaged, extravasation of urine occurs between the deep and superficial penile fascial layers, with possible extension beneath Colles' fascia into the superficial perineal space, beneath the dartos fascia into the scrotum, and beneath Scarpa's fascia into the anterior abdominal wall.

2. The **female urethra** (see Fig. 22-8).

a. The female urethra is approximately 3.75 cm in length.

b. It corresponds to the prostatic and membranous portions of the male urethra.

c. It is fused to the adventitia of the anterior vaginal wall.

(1) Because of this physical relationship, the fibers of the deep transverse perineal muscle do not pass posterior to the urethra; thus the external urethral sphincter in the female is incomplete. This explains in part the much higher incidence of stress incontinence among women, especially if the urogenital diaphragm is damaged during parturition.

(2) The intimate relationship of the urethra to the vagina predisposes the urethra to injuries and subsequent cystocele associated with difficult parturition.

d. The **urethral glands** open into the urethra; the **paraurethral** glands (Skene's) open into the vestibule adjacent to the urethral orifice.

(1) These glands are homologous to the prostate.

(2) They tend to be especially prone to infection.

e. The urethra opens into the **vestibule** between the **labia minora**.

III. INNERVATION OF THE PELVIC VISCERA

A. THE SYMPATHETIC INNERVATION TO THE PELVIC VISCERA arises from spinal segments T12–L2 (Fig. 22-3).

1. White rami communicantes carry the preganglionic sympathetic motor neurons from the lowest thoracic and upper lumbar spinal nerves to the sympathetic chain of paravertebral ganglia.

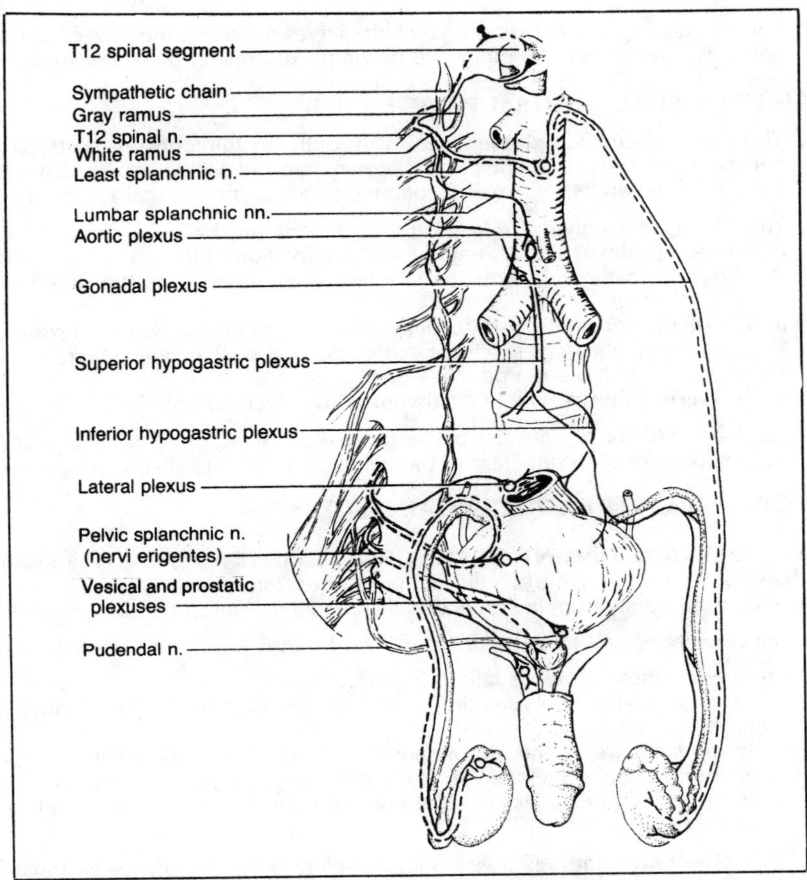

Figure 22-3. *The innervation of the male bladder, reproductive tract, and genitalia.* The sympathetic pathways arise from the lower thoracic and upper lumbar spinal levels (there are no white rami below L2) and reach the pelvic viscera via thoracic and lumbar splanchnic nerves and then the hypogastric plexuses. The parasympathetic pathways arise from the midsacral spinal levels and reach the pelvic viscera via the pelvic splanchnic nerves. Visceral afferent fibers (*dashed*) from the pelvic viscera travel specifically along either one or the other autonomic pathways, producing specific patterns of referred pain. The pudendal nerve provides somatic innervation to and from the perineum.

2. Some of these neurons may synapse within the paravertebral ganglia before leaving via a lumbar splanchnic nerve to join the **aortic plexus**. Those neurons that do not synapse in the paravertebral ganglia seem to synapse in small ganglia located along the aortic plexus.
3. The **superior hypogastric plexus** arises from the **aortic plexus** at the bifurcation of the aorta. This pathway brings sympathetic nerves into the pelvis.
4. The **inferior hypogastric plexuses (left and right hypogastric nerves)** are formed by the bifurcation of the superior hypogastric plexus at the pelvic brim.
   a. These sympathetic pathways run along the anterior surface of the sacrum toward the rectum.
   b. They provide the sympathetic contribution to the **pelvic plexus (lateral pelvic plexus)** on the walls of the rectum.

B. THE PARASYMPATHETIC INNERVATION TO THE PELVIC VISCERA arises from spinal segments S2–S4 (see Fig. 22-3).

1. Presynaptic parasympathetic neurons arise from spinal segments S2 and S3 *or* S3 and S4 (seldom S2, S3, *and* S4) and leave the spinal nerves in the vicinity of the anterior sacral foramina to form the **pelvic nerves (nervi erigentes, pelvic splanchnic nerves)**.

2. These parasympathetic pelvic splanchnic nerves run along the lower sacrum to provide the parasympathetic contribution to the pelvic plexus (lateral pelvic plexus).

C. THE PELVIC PLEXUS (LATERAL PELVIC PLEXUS)

1. This autonomic plexus on the lateral walls of the rectum contains postsynaptic sympathetic neurons from the hypogastric plexus, presynaptic and postsynaptic parasympathetic neurons from the nervi erigentes, small autonomic ganglia, and visceral afferent neurons.
2. The lateral plexus gives rise to several autonomic plexi.
   a. The **rectal plexus** provides autonomic innervation to the rectum and anal canal as well as parasympathetic innervation in a retrograde direction to the large colon as far as the splenic flexure.
   b. The **uterovaginal** or **prostatic plexus** gives rise to the **cavernous plexuses**, which convey parasympathetic nerves to the erectile tissues as well as sympathetic nerves to the glands associated with the reproductive tract.
   c. The **vesical plexus** is associated with urinary bladder function.
3. Surgical interference with the pelvic plexus, as for example in rectal resection, results in varying degrees of dysfunction of the pelvic and abdominal structures innervated.

IV. DEVELOPMENT OF THE UROGENITAL DUCTS

A. THE EARLY DEVELOPMENT OF THE GENITOURINARY SYSTEM. The development of the kidneys proceeds through three distinct pairs of excretory organs that develop sequentially in the abdominal cavity from the cranial to the caudal regions of the urogenital ridge (Fig. 22-4).

1. The **pronephros** develops by the fourth week but is never completely functional.
2. The **mesonephros** develops late in the fourth week.
   a. Rows of parallel nephrons drain into the **mesonephric (wolffian) duct**, which empties into the cloaca.
   b. By the sixth week, the mesonephros consists of elongated organs projecting into the coelomic cavity on either side of the midline of the posterior abdominal wall.
   c. By the eighth week, most of the mesonephros has degenerated, but the mesonephric duct remains.
3. The **metanephros**, the definitive kidney, develops during the fifth week and is fully functional by the eighth week.
   a. It develops from two primordia:
      (1) A **metanephric diverticulum** arises from the posterior wall of the caudal region of the mesonephric duct.
      (2) The **metanephric mass** arises from the posterior region of the urogenital ridge.
   b. The growth of the metanephric duct into the metanephric mass results in the formation of several kidney lobes.

B. DURING THE INDIFFERENT STAGE (weeks 6 to 8) there are no apparent differences between the developing male and female (see Fig. 22-4A).

1. By the sixth week, two pairs of genital ducts have formed.
   a. Mesonephric (wolffian) ducts are urinary tracts exiting from the primitive mesonephros and draining into the unpartitioned cloaca.
   b. Paramesonephric (müllerian) ducts arise parallel to the mesonephric ducts.
      (1) Cranially, the paramesonephric ducts open directly into the coelomic cavity.
      (2) Caudally, these ducts fuse just prior to entering the cloaca to form the **uterovaginal primordium**.
2. In the male, the **mesonephric** ducts form the genital tract and the paramesonephric ducts largely degenerate.
3. In the female, the **paramesonephric** ducts form the oviducts and uterus and the mesonephric ducts largely degenerate.
4. The **urinary bladder** is formed by partition of the cloaca by the urorectal septum into an anterior **urogenital sinus** and a posterior **anorectal canal**.
   a. The most cranial portion of the urogenital sinus forms the urinary bladder.
      (1) The cranial portion of the urogenital sinus is continuous with the **allantois**, a diverticulum of the hindgut into the placenta.
      (2) When the lumen of the allantoic duct is obliterated, it becomes the cord-like **urachus**,

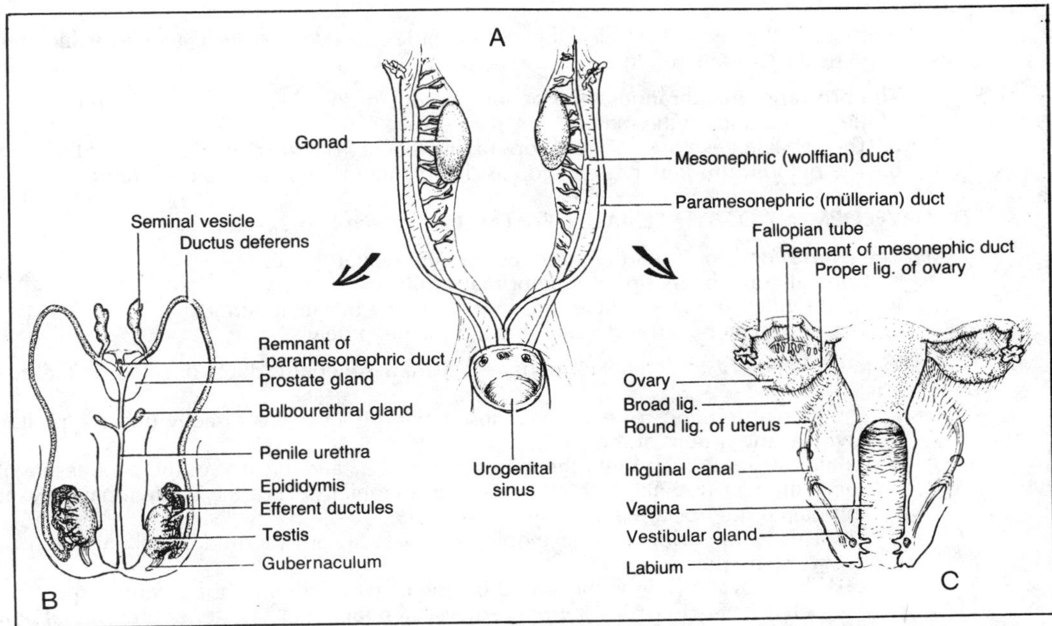

Figure 22-4. *The development of the gonads and genital ducts. A,* In the indifferent stage both mesonephric and paramesonephric ducts lead to the urogenital sinus. The gonads are located high in the abdomen. *B,* In the developing male the mesonephric ducts develop into the vas deferens and epididymis. The testes descend into the scrotum. *C,* In the female the paramesonephric ducts partially fuse to form the uterus, while the unfused portions form the oviducts. The ovaries descend into the deep pelvis.

connecting the superior pole of the urinary bladder to the umbilicus and forming the **median umbilical ligament**.

b. The middle portion of the urogenital sinus forms the **urethra** in the female, which is equivalent to the membranous and prostatic portions of the urethra in the male.

c. The inferior portion of the urogenital sinus remains as the definitive urogenital sinus, which persists as the **vestibule** in the female and forms the **penile urethra** in the male on fusion of the urethral folds.

d. With the partition of the cloaca by the urorectal septum, the mesonephric ducts enter the urogenital sinus.

e. As the walls of the terminal region of the mesonephric ducts become incorporated into the developing urinary bladder, the metanephric diverticula separate from the mesonephric ducts so that the former enter the urinary bladder whereas the latter enter the urethra.

C. DEVELOPMENT OF THE MALE GENITAL TRACT (see Fig. 22-4B)

1. The **testes** develop from the mesonephric (urogenital) ridge.

a. Seminiferous cords develop into **seminiferous tubules, straight tubules,** and the **rete testis**.

b. Between the seminiferous cords, interstitial mesenchyme differentiates into hormone-producing **interstitial cells** (Leydig's).

2. Under the influence of the fetal testicular hormones, the mesonephric ducts develop.

a. As the mesonephros degenerates, some mesonephric tubules are incorporated into the developing genital duct system.

(1) Some of the mesonephric tubules become the **efferent tubules**, which open into the portion of the mesonephric duct that develops into the epididymis.

(2) Some of these mesonephric tubules fail to contact the epididymis, yet fail to degenerate. These become the vestigial **paradidymis** and the appendix epididymis.

b. Most of the more cranial portions of the paramesonephric duct degenerate, except for a small cranial portion that remains as the **appendix testis**.

c. The greater part of the mesonephric duct forms the **vas deferens**.

d. A diverticulum toward the caudal end of each mesonephric duct forms each **seminal vesicle**.

e. Distal to the seminal vesicles, the mesonephric duct becomes the **ejaculatory duct**, which joins the prostatic urethra.

3. The **prostatic**, **membranous**, and **penile** portions of the urethra form the terminal portion of the genital tract in the male.
a. The prostate develops as five groups of diverticula from the cranial portion of the urethra.
b. The **bulbourethral glands** develop as diverticula of the membranous urethra.

D. DEVELOPMENT OF THE FEMALE GENITAL TRACT (see Fig. 22-4C)

1. The **ovaries** develop from the mesonephric (urogenital) ridge.
a. Cortical cords break up into the **primary follicles**.
b. All 2 million primary follicles are formed during this fetal period.
c. The developing ovaries descend toward the deep pelvis.

2. In the absence of testicular hormone, the paramesonephric ducts develop as the genital tract.
a. The cranial terminus of each paramesonephric duct opens directly into the peritoneal cavity as the ostium of the oviduct.
b. As the ovaries descend into the minor pelvis, the caudal fusion of the paramesonephric ducts (the uterovaginal primordium) extends cranially. Thus the original paired vaginas and paired uteri become a single midline tube.
(1) This fusion brings the developing oviducts toward the midline, lifting them away from the pelvic walls.
(2) This elevation forms the **broad ligament**, which divides the inferior aspect of the pelvic cavity into an anterior **uterovesical pouch** and a posterior **uterorectal pouch**.
c. The fused paramesonephric ducts form the corpus of the uterus and cervix.
d. Two groups of vestigial mesonephric tubules remain in the vicinity of the ovary, the epoophoron and the paroophoron.
e. The mesonephric ducts degenerate except for the most cranial and most caudal portions, which persist as the appendix vesiculosa and the Gartner's ducts, respectively.

3. Development of the vagina.
a. The vagina develops along the posterior wall of the urogenital sinus from two solid sinovaginal bulbs that fuse in the midline.
b. The termination of the fused paramesonephric ducts (uterovaginal primordium) joins the sinovaginal bulbs.
c. The vaginal lumen is formed by cavitation of the sinovaginal bulbs, except for the most inferior portion, which persists as the **hymen**; the hymen separates the vaginal cavity from the vestibule until the perinatal period.

V. THE MALE REPRODUCTIVE TRACT

A. THE TESTES (Fig. 22-5)

1. Each testis is approximately 4.5 cm × 3 cm × 2.7 cm.

2. Within the scrotum, each testis is surrounded by the **tunica albuginea**, a tough connective tissue layer.

3. Each testis is divided into approximately 300 lobules, each of which contains approximately 1 to 4 **seminiferous tubules**.
a. The seminiferous tubules are approximately 75 cm long. Thus there are approximately 750 m of tubules per testis.
b. Waves of spermatogenesis (development of spermatids) and subsequent spermiogenesis (maturation of spermatids into spermatozoa) occur along the seminiferous tubules.
c. The seminiferous tubules lead into about 25 **straight tubules**, which coalesce in the **rete testes** to form 15 to 20 **efferent ductules**; these in turn pierce the tunica albuginea to enter the head of the **epididymis**.

4. Blood supply to the testes (see Fig. 21-3).
a. Each **testicular artery** arises from the aorta below the respective renal arteries, courses through the spermatic cord, and enters the testis via the mesorchium.
b. There is a rich collateral blood supply to the testes.
(1) The **artery** of the **ductus deferens** arises from the internal iliac artery and is homologous to the uterine artery.
(2) The **cremaster artery** arises as a branch of the inferior epigastric artery.
(3) The **external pudendal arteries** arise as branches of the femoral arteries.

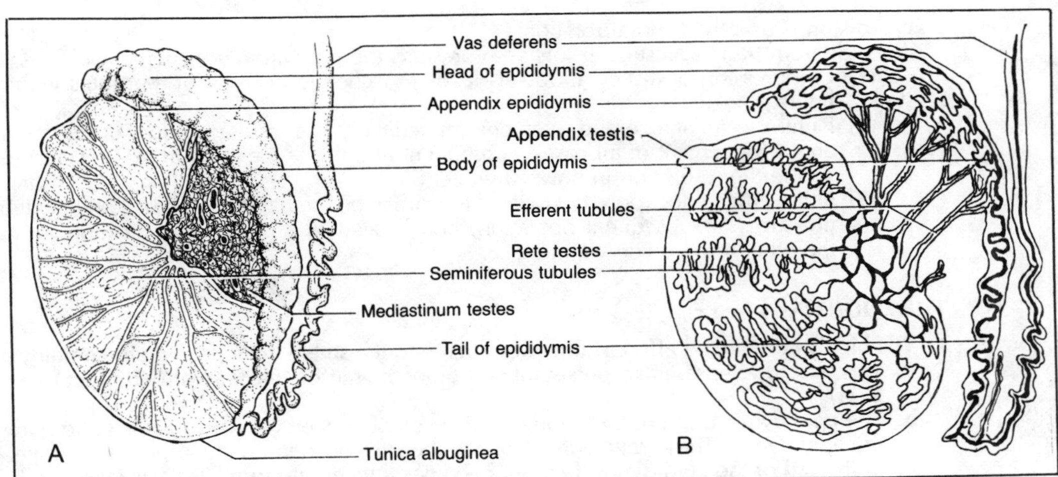

Figure 22-5. *The testis. A,* This parasagittal section depicts the testicular parenchyma and the undissected epididymis. *B,* The duct system is schematized.

5. Venous return from the testes (see Fig. 21-3).
   a. The **testicular (internal spermatic) veins** provide the principal drainage of the testes.
      (1) The 8 to 10 veins of the **pampiniform plexus** in the lower end of the spermatic cord unite to form each testicular vein.
         (a) The pampiniform plexus functions as a **countercurrent heat exchanger**.
         (b) This plexus is frequently affected by **varicocele**.
            (i) Fully 90 percent of all varices are on the left side.
            (ii) Apparently this is due to local venous hypertension resultant from compression of the sigmoid colon, which contains stored feces, on the left testicular vein.
            (iii) The condition is often painful, affects temperature regulation, and may result in decreased sperm viability.
            (iv) The best resolution is through ligation of the testicular vein superior to the deep inguinal ring so that the blood returns via collaterals. This procedure alleviates the back pressure resultant from the hydrostatic columns of venous blood within the abdomen.
      (2) The right testicular vein drains into the inferior vena cava, whereas the left testicular vein drains into the left renal vein.
      (3) The lymphatic drainage of the testis parallels the spermatic vein.
   b. Collateral venous drainage is via the **external pudendal veins**, **posterior scrotal veins**, **cremaster veins**, and veins of the vas deferens.
   c. The lymphatics from the testes parallel the testicular veins to drain directly into the para-aortic lymph nodes. This explains in part why testicular seminoma disseminates widely and rapidly.

6. Innervation of the testes (see Fig. 22-3).
   a. The motor role of the autonomic nerves to the testes is uncertain.
      (1) Sympathetic nerves from the aortic and renal plexuses innervate the testes.
      (2) Parasympathetic nerves from the prostatic plexus supply the vas deferens and epididymis.
   b. Afferent nerves from the testis (see Fig. 22-3).
      (1) Afferent fibers travel to the spinal cord along the sympathetic pathways, parallel to the testicular vessels.
      (2) These pass through the aorticorenal plexus, thence along the lesser and least splanchnic nerves to spinal segments T10–T12, as well as along the lumbar splanchnic nerves to spinal segments L1–L2.
      (3) Pain originating in the testes is referred to the middle and lower abdominal wall.

7. **Testicular torsion.**
   a. Rotation (torsion) of the testes may occur from 90° to 360° about the spermatic cord.
   b. Precipitating factors seem to be strong contractions of the cremaster muscle and the presence of a large connection between the parietal and visceral layers of the tunica vaginalis, the "mesentery of the testes."

c. Torsion is usually external rotation.
d. Aside from trauma, testicular torsion represents the only urologic emergency.
  (1) Compression of testicular vessels leads to ischemic necrosis of the testes within 6 hours.
  (2) Failure to recognize and intervene immediately has resulted in loss of the testis in about 80 percent of all cases, subsequent atrophy of the testis in 10 percent more, and fertile resolution in only 10 percent.
  (3) In the emergency room the twisted testis may be medially rotated by gentle external pressure. At subsequent operation, both sides will be sewn into position because the other testis is very valuable.

B. THE EPIDIDYMIS (see Fig. 22-5)

1. The highly convoluted **efferent ductules** open into the single duct, forming the epididymis.
  a. This structure represents a persistent portion of the **mesonephric (wolffian) duct** (see Fig. 22-4).
  b. The epididymis is about 6 cm long and compactly coiled to form the **head** and **body** of the epididymis. These segments comprise the sites of sperm maturation and storage.
  c. In the **tail** of the epididymis the duct becomes thicker and forms the **vas deferens.**

2. Maturation and storage of the spermatozoa occurs in the epididymis.

3. The blood supply to the epididymis is predominantly by the **deferential artery**.

4. Innervation of the epididymis.
  a. The parasympathetic and visceral innervation of the epididymis is by the **nervi erigentes**.
  b. Afferent nerves from the epididymis and vas deferens travel along the vas deferens to the prostatic plexus, thence to the lateral pelvic plexuses, nervi erigentes, and sacral nerves to reach spinal segments S2–S4. Pain originating in the epididymis, such as that accompanying epididymitis, is referred to the distribution of sacral nerves 2–4.

C. THE VAS DEFERENS

1. The vas deferens represents the persistent portion of the **mesonephric (wolffian) duct** (see Fig. 22-4).

2. Relations of the vas deferens:
  a. The vas deferens is a direct continuation of the duct of the epididymis.
  b. It has a very muscular wall, which accounts for the cord-like composition.
  c. The external course of the vas deferens.
    (1) From the inferior pole of the testis the vas deferens passes into the center of the spermatic cord, which is the usual site for vasectomy.
    (2) It transits the inguinal canal and enters the abdominal cavity by passing through the deep inguinal ring located lateral to the inferior epigastric artery.
  d. The internal course of the vas deferens.
    (1) In the abdominal cavity the vas deferens lies within the transversalis fascia.
    (2) At the deep inguinal ring it immediately diverges from the **testicular neurovascular bundle** and passes over the **external iliac vessels** and **obturator neurovascular bundle** to enter the deep pelvis.
    (3) In the pelvic cavity the vas lies within endopelvic fascia.
    (4) It passes anterosuperior to the **ureters** to converge toward the prostatic urethra at the base of the urinary bladder.
  e. The vas deferens may be double on occasion, an important factor if vasectomy is to be successful.

3. The **ampulla** (Fig. 22-6).
  a. The ampulla of the vas deferens is an enlarged, sacculated portion near the base of the **prostate gland**.
  b. The **seminal vesicles** join the vas deferens at the point at which the ampulla becomes the ejaculatory duct.

4. The **ejaculatory duct** (see Fig. 22-6; Fig. 22-7).
  a. The ejaculatory duct is the narrow (0.5 mm), thin-walled terminal portion of the vas deferens.
  b. It passes through the parenchyma of the prostate gland.
  c. It joins the **prostatic urethra** by an opening on the **colliculus seminalis (verumontanum)** on either side of the **prostatic utricle**.

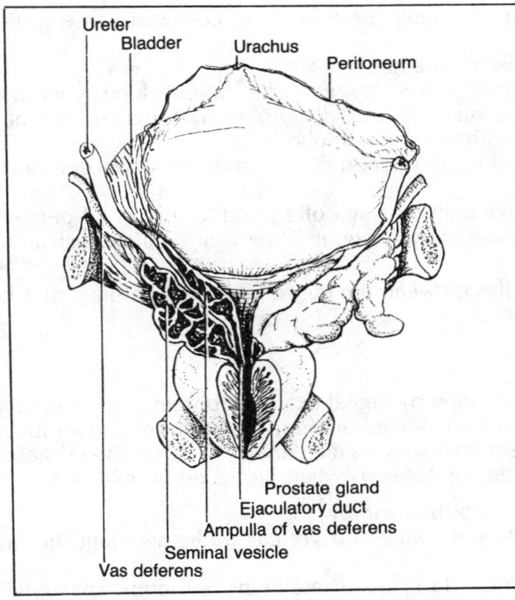

Figure 22-6. *The accessory glands of the male.* The posterior aspect of the urinary bladder, seminal vesicles, and prostate gland is indicated with the terminal portion of the left vas deferens, the left seminal vesicle, and the prostate gland all dissected open.

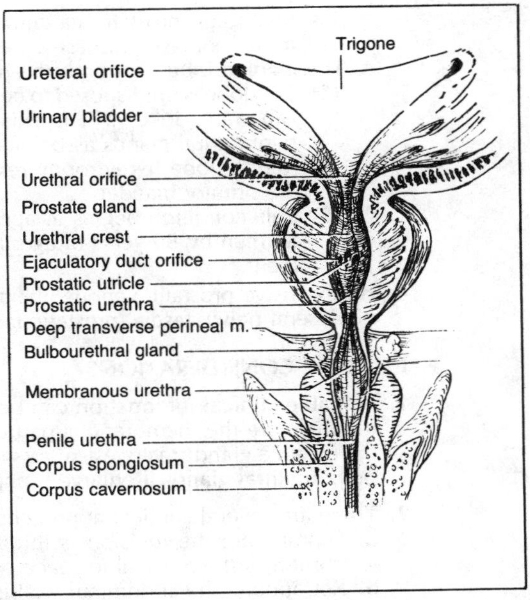

Figure 22-7. *The male urethra.*

D. THE SEMINAL VESICLES (see Fig. 22-6)

1. The seminal vesicles are secretory glands.
   a. These glands do *not* function for sperm storage.
   b. They contribute seminal fluid, which contains fructose and choline. Fructose, which is not produced anywhere else in the body, provides a forensic determination for the occurrence of rape. However, choline crystals form the preferred basis for the determination of the presence of semen (the Florence test).

2. Each seminal vesicle is a dilated convoluted tube about 4 cm long (9 cm if straightened).

3. Located at the base of the urinary bladder, posteriorly, they may be palpated per rectum if swollen and inflamed.

E. THE PROSTATE GLAND

1. The prostate is a compound tubuloalveolar gland.
   a. It secretes 0.5 to 2 ml of fluid per day, which contributes 10 to 20 percent of the seminal fluid of the ejaculate.
   b. This fluid contains citric acid, acid phosphatase, prostaglandins, and fibrinogen.

2. The relations of the prostate gland.
   a. The prostate gland lies between the base of the urinary bladder and upper surface of the levator ani and deep transverse perineal muscles (see Fig. 22-7).
   b. The anterior surface is adjacent to the **retropubic space** (Retzius').
   c. The posterior surface is adjacent to the seminal vesicles and the **rectovesical septum (prostatoperitoneal membrane, Denonvilliers' fascia)**. It is palpable per rectum.

3. The prostate gland forms as five diverticula of the prostatic urethra; the prostatic parenchyma is subdivided into five lobes.
   a. The **left** and **right lateral lobes** are extensive and include what was formerly termed an anterior lobe.
   b. The **left** and **right posterior lobes** also include the apex of the gland.
      (1) These lobes are most predisposed to *malignant transformation*.
      (2) Malignant cells are found in nearly 50 percent of the prostate glands of all males over the age of 50 years. This is the most common malignancy in the adult male.

(3) Metastatic prostatic carcinoma represents the third most common cause of death from neoplastic diseases.

c. The **median lobe** surrounds the prostatic urethra.

(1) This lobe is predisposed to *benign prostatic "hypertrophy"* (middle lobe hypertrophy, Albarrán's lobe, BPH). This condition is actually *proliferative hyperplasia* of the periurethral glands associated with the median lobe.

(2) Middle lobe hypertrophy results in obstruction of the urethra and visceral neck of the urinary bladder.

(3) This condition begins at approximately the age of 45 and occurs in 80 percent of all men by 80 years of age. However, only about 10 percent require surgical treatment.

4. An extensive prostatic venous plexus lies between the dense prostatic capsule and a layer of visceral pelvic fascia (prostatic fascia).

F. CLINICAL CONSIDERATIONS

1. Valuable clinical information can be obtained by digital examination per anus. Palpable in the male are the membranous part of a catheterized urethra, posterior and lateral lobes of the prostate gland, rectovesical fossa, seminal vesicles if enlarged, bladder when distended, bulbourethral glands if enlarged, and ductus deferens when displaced or enlarged.

2. There are several surgical approaches for *prostatectomy.*

a. Suprapubic: the incision is through abdominal and vesical walls, avoiding the major portion of the retropubic space.

b. Retropubic: the abdominal wall is incised to gain access to the retropubic space, which is then enlarged by pushing the bladder posteriorly to expose the prostate.

c. Perineal: the incision is through the central tendon and the medial portion of the levator ani muscle. The dissection is anterior to the rectovesical fascia (Denonvilliers'), which "separates the air and the water." The rectum is pushed posteriorly and the urethra pulled forward to visualize the prostate.

d. Transurethral: removal of hypertrophic tissue in conjunction with cystoscopy.

3. Because the afferent nerve pathways from the testis accompany the testicular arteries and veins, testicular pain is referred to the lower thoracic and upper lumbar dermatomes.

4. Because the afferent nerve pathways from the epididymis accompany the nervi erigentes, the pain associated with epididymitis is referred to the middle sacral dermatomes.

VI. THE FEMALE REPRODUCTIVE TRACT

A. THE PERITONEUM AND PELVIC MESENTERIES

1. Serous **parietal peritoneum** lines the pelvic cavity. **Visceral peritoneum** covers those pelvic viscera that protrude into the pelvic cavity. The reflections of parietal peritoneum to visceral peritoneum are **mesenteries.**

2. The **mesometrium (broad ligament)** is the mesentery of the uterus.

a. It is formed as the parietal peritoneum is reflected off the walls of the pelvic cavity to course over the uterus and uterine tubes; this mesentery connects lateral margins of the uterus with the side wall of the pelvis.

b. It partitions the peritoneal cavity of the pelvis into two so-called pouches (Fig. 22-8).

(1) The **uterovesical pouch** is an extension of the pelvic peritoneal cavity between the anterior surface of the uterus and the posterior surface of the urinary bladder.

(2) The **rectouterine** or **rectovaginal pouch** (Douglas') is an extension of the pelvic peritoneal cavity between the posterior surfaces of the uterine cervix and the anterior surface of the rectum. The floor of this pouch is apposed to the posterior fornix of the vagina.

c. Three peritoneal folds are continuations of the broad ligament (see Fig. 22-11).

(1) The **mesosalpinx** supports the uterine tube.

(a) Beginning at the juncture of the mesovarium, this mesentery represents the superior limit of the broad ligament.

(b) It contains the tubal branches of the uterine vessels as well as the vestigial epoophoron and paroophoron.

(2) The **mesovarium** supports the ovary.

(a) This mesentery is a posterior extension of the broad ligament.

(b) It becomes continuous with the serosa or "germinal epithelium" of the ovary.

(3) The **infundibulopelvic ligament (suspensory ligament of the ovary)** contains the **ovarian artery** and **ovarian vein** as well as lymphatics and nerves.

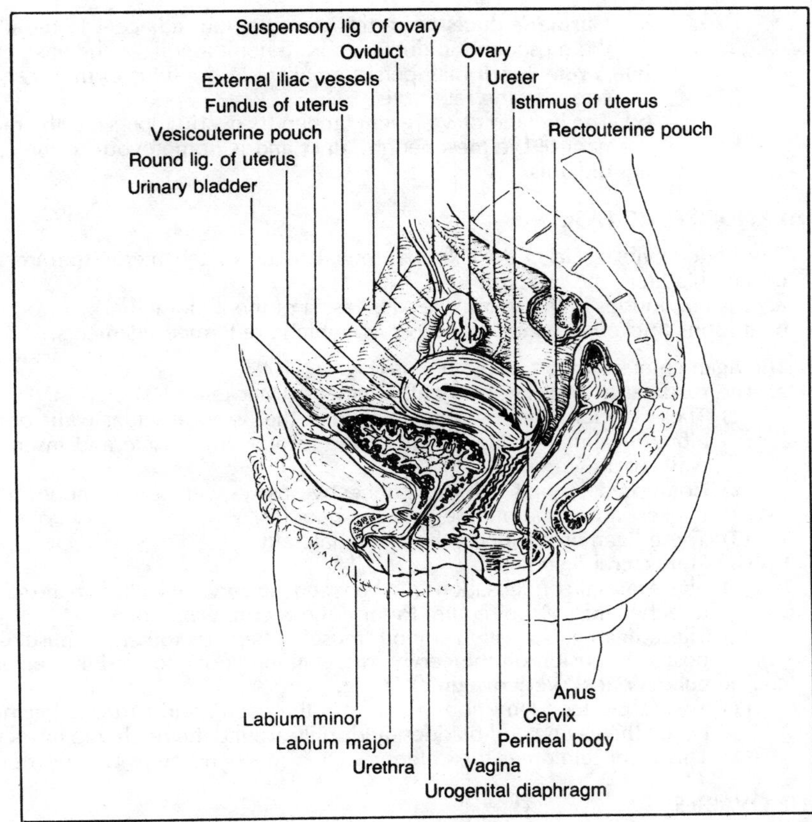

Figure 22-8. *The female pelvic viscera.* This midsagittal section depicts the lower portion of the female urinary tract and the genital tract.

(a) This mesentery is an elevation of peritoneum between the pelvic brim and the mesovarium.
(b) It is formed by the descent of the ovary into the deep pelvis.

d. Contents of the broad ligament (Fig. 22-9).
(1) A **uterine tube** lies in the superior free edge of the broad ligament on each side of the uterus.
(2) The ovary is attached by the mesovarium to the posterior aspect of the broad ligament.
(3) The **ovarian ligament (proper ligament of the ovary)** runs between the medial pole of the ovary and the uterine cornu, just below the origin of the uterine tube.
(4) The **round ligament of the uterus** runs from the lateral uterine border near the point of attachment of the ovarian ligament, between the anterior and posterior leaflets of the broad ligament, and through the inguinal canal (Nuck's) to fuse with the dermis of the labium major.
(5) The uterine neurovascular bundle lies at the base of the broad ligament within the transverse cervical (cardinal) ligament and in the mesometrium adjacent to the uterus.
(6) The ovarian vessels lie within the suspensory ligament of the ovary and the mesovarium, with branches in the mesosalpinx.
(7) Lymphatics and nerve fibers course through the broad ligament.
(8) Several vestigial structures lie within the broad ligament.
(a) The **ovarian** and **round ligaments** are homologous to the gubernaculum testis in the male.
(b) Although the mesonephric (wolffian) ducts degenerate in the female, remnants of them may be sites of cyst formation.
(i) The **epoophoron** constantly lies in the lateral portion of the mesosalpinx and mesovarium, represents residual mesonephric tubules, and is homologous to the epididymis.
(ii) The **paroophoron** variably lies more medially in the mesosalpinx, also represents residual mesonephric tubules, and is homologous to the paradidymis.

(iii) **Gartner's ducts** lie in the mesometrium adjacent to the uterus, represent the mesonephric ducts, and are homologous to the vas deferens.
(iv) A **rete ovarii** may persist adjacent to the hilus of the ovary and is homologous to the rete testis.
(v) The hydatid of Morgagni (appendix ovarii) represents the most cranial remnant of the mesonephric duct and is homologous to the appendix of epididymis.

B. THE ENDOPELVIC FASCIA

1. The endopelvic fascia is the extraperitoneal tissue of the uterus **(parametrium)**, vagina, urinary bladder, and rectum.
   a. It is continuous with the transversalis fascia of the abdomen.
   b. It forms thickened bands in the pelvis, which are termed ligaments.
2. The ligamentous support of the uterus (Fig. 22-10).
   a. The **cardinal (transverse cervical) ligaments** (Mackenrodt's).
      (1) These fascial condensations arise from the osseomuscular walls of the deep pelvis along the lines of attachment of the levator ani muscle and insert into the lateral walls of the cervix.
      (2) Composed of white fibrous connective tissue with some smooth muscle, they are thickest about the uterine vascular bundles.
      (3) These ligaments support the cervix laterally.
   b. The **uterosacral ligaments.**
      (1) These fascial condensations run between the cervical walls and the sacrum, diverging to either side of the rectum to form the **rectouterine folds**.
      (2) These ligaments contain smooth muscle fibers (rectouterine muscles) and maintain posterior tension on the cervix, drawing the cervix toward the rectum.
   c. The **pubocervical (vesicovaginal) fascia.**
      (1) This fascial condensation arises from the cervix and cardinal ligament, passes between the vagina and bladder, and splits around the urethra to insert onto the pubis.
      (2) This layer reinforces the anterior vaginal wall and helps to prevent cystocele.

C. THE OVARIES

1. The ovaries are the female gonads.
2. Each almond-shaped ovary is approximately 3 cm × 1.5 cm × 1 cm.
3. Position and attachments (see Figs. 22-9 and 22-11).

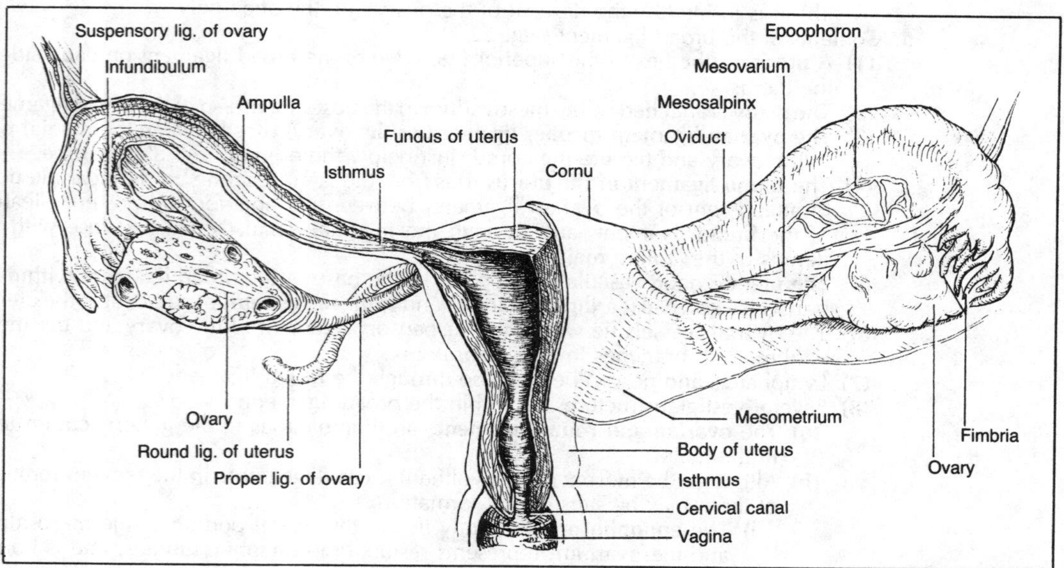

Figure 22-9. *The female reproductive tract.* The posterior aspect of the female reproductive tract is shown. The ligamentous supports of the uterus, right oviduct, and right ovary are depicted intact. The left side of the uterus, left oviduct, and left ovary are shown dissected.

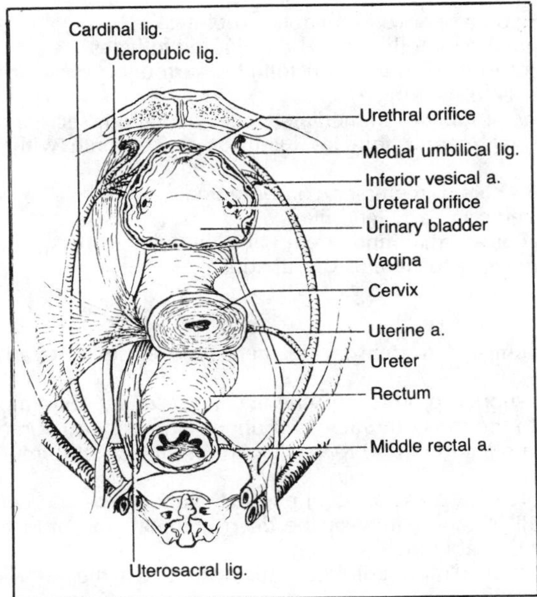

Figure 22-10. *The ligamentous support of the cervix.* Anterior, lateral, and posterior condensations of pelvic fascia support the cervix. The close relationship between the ureters and uterine arteries in the cardinal ligament is depicted.

a. The ovaries are located on the posterior side of the broad ligament and project into the **ovarian fossa**, a triangular depression on each side of the minor pelvic cavity bounded by the external iliac vessels and the internal iliac vessels.
  (1) The **mesovarium** attaches each ovary to the broad ligament.
  (2) The **infundibulopelvic (suspensory) ligament** attaches each ovary to the lateral pelvic wall and contains the ovarian neurovascular bundle.
  (3) The **ovarian ligament (proper)** runs within the broad ligament from the inferior medial pole of each ovary to the uterus in the vicinity of the cornu and is homologous to the cranial portion of the **gubernaculum** in the male.

b. Generally, one fimbria of the oviduct infundibulum—the **ovarian fimbria**—attaches to the ovary.

4. The vasculature of the ovary.
  a. The ovarian arterial supply (see Figs. 22-1 and 22-12).
    (1) The **ovarian arteries** arise from the abdominal aorta in the vicinity of the renal arteries and run inferiorly in the suspensory (infundibulopelvic) ligament and mesovarium to reach the ovary.
    (2) The ovarian and tubal branches of the **uterine arteries** anastomose with the ovarian artery in the mesosalpinx and can supply the ovary if the suspensory ligament is transected or the ovarian artery occluded.
  b. Venous drainage from the ovary.
    (1) The ovarian veins drain asymmetrically.
      (a) The **right ovarian vein** drains directly into the inferior vena cava near the renal vein.
      (b) The **left ovarian vein** drains into the left renal vein.
    (2) Venous twigs from the ovaries also drain along the tubal branches of the uterine veins.
  c. The principal lymphatic drainage of the ovaries is toward the aortic nodes in the vicinity of the renal arteries.

5. Ovarian innervation (see Fig. 22-13).
  a. Although it is known that the autonomic innervation of the gonads is from the aortic plexus, the role of the autonomic nerves with regard to the ovary is unclear.
  b. Visceral afferent fibers from the ovary run along the sympathetic pathways to spinal segments T12–L2. Intractable ovarian pain may be alleviated by transecting suspensory ligaments, which contain the afferent (GVA) fibers.

6. The internal structure of the ovary (see Fig. 22-9).
  a. The ovarian **cortex** is the functional portion of the ovary.
    (1) The ovarian surface is covered by modified peritoneum ("germinal epithelium") or serosa, which continues from the ovarian hilus as the mesovarium.

(2) In the mature adult the cortex contains several notable structures.
(a) There are a multitude of immature follicles and developing follicles.
(b) There may be one or more mature (graafian) follicles surrounding maturing oocytes. These structures secrete estrogen.
(c) Corpora lutea, the remains of follicles immediately after ovulation, secrete estrogen and progesterone. Numerous follicles luteinize each month without reaching ovulation.
(d) Corpora albicantia are the scars of regressed corpora lutea.
(3) All the primordial follicles develop in early fetal life.
(4) After menopause the ovary becomes small and atretic, and the follicles disappear.
b. The ovarian **medulla** contains connective tissue and vasculature.

7. Clinical considerations.
a. Ovarian torsion.
(1) An abnormally long mesovarium and suspensory ligament predisposes to ovarian torsion (rare).
(2) Torsion compresses the blood supply, producing ischemia and resultant abdominal symptoms with pain referred to the lower thoracic and upper lumbar dermatomes.
(3) This condition occurs as a complication in 10 to 20 percent of all ovarian tumors.
b. Ectopic ovaries.
(1) The normal caudal migration of the ovaries may continue so that the ovaries descend to the anterior abdominal wall in the vicinity of the deep inguinal ring, into the inguinal canal or, very rarely, the labia majora.
(2) Ovaries that fail to descend are termed retroperitoneal and are located in the posterior abdominal wall superior to the kidneys.
(3) Ovaries may occasionally prolapse into the pouch of Douglas.
c. Congenital absence of one or both gonads is usually associated with a unilateral defect of the genital system (very rare).
d. Numerous types of ovarian tumors are common.

D. THE UTERUS

1. The **oviducts** or **uterine tubes** (Fallopius') represent the unfused portions of the paramesonephric (müllerian) ducts.
a. Approximately 10 cm long, each uterine tube begins at an ostium, which opens freely to the peritoneal cavity.
b. They are located at the upper free edge of the broad ligament and are supported by **mesosalpinx** (Fig. 22-11).
c. The oviduct is divided into three regions (see Fig. 22-9).
(1) The **infundibulum**, 2 mm in diameter, opens into the peritoneal cavity via the **internal ostium**.
(a) The ostium is fimbriated.
(b) One fimbria, the **ovarian fimbria**, usually adheres to the ovary.
(2) The **ampulla**, 2.5 mm in diameter, is thin-walled, tortuous, and constitutes the major portion of the oviduct.
(3) The **isthmus**, 1 mm in diameter, is thick-walled and straight.
(4) The **intramural** portion, less than 1 mm in diameter, lies within the uterine wall and communicates with the uterine cavity.
d. The structure of the uterine tube.
(1) The serosa is visceral peritoneum.
(2) The muscularis is comprised of outer longitudinal and inner circular layers of smooth muscle.
(3) The mucosa is ciliated columnar epithelium interspersed with mucous secretory cells.
e. Functional considerations.
(1) The fimbriae generally sweep over the ovarian surface so that an expelled ovum is directed into the infundibular ostium.
(2) Fertilization of the ova usually occurs in the ampullary portion of the uterine tube.
(3) The ova are propelled along the oviduct by peristaltic movement and ciliary action.
f. Clinical considerations.
(1) The female genital canal is a direct communicating pathway from the urogenital sinus of the perineum into the peritoneal cavity and is a pathway for infection.
(2) *Salpingitis*—inflammation of the oviduct frequently associated with *pelvic inflammatory disease (PID)*—is perhaps the most common cause of female sterility. If fertilization does occur, the zygote may fail to traverse a subsequently scarred or narrowed uterine tube, with resultant ectopic tubal implantation.

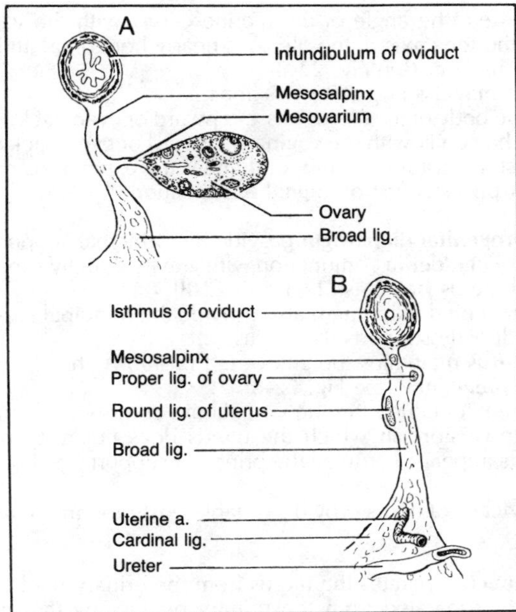

Figure 22-11. *The ligamentous support of the ovary and oviduct. A,* This lateral parasagittal section through the infundibulum of the oviduct shows the mesosalpinx and mesovarium as extensions of the broad ligament. *B,* This more medial parasagittal section through the isthmus of the oviduct depicts the proper ligament of the ovary within the broad ligament. The relationship of the uterine artery to the ureter in the cardinal ligament at the base of the broad ligament also is indicated.

(3) Ectopic implantation.
- **(a)** Because the oviducts open directly into the peritoneal cavity, an ovum may be fertilized and the zygote implanted in the peritoneal cavity or even in the ovary, but ectopic implantations are most common in the ampullary portion of the uterine tube (tubal pregnancy).
- **(b)** In the case of tubal implantation, rupture of the tube with hemorrhage and expulsion of the embryo into the peritoneal cavity usually occurs between the fourth week (if in the isthmus) and the tenth week (if in the ampulla). This is a gynecologic emergency because the patient may die of internal hemorrhage.

(4) Uterosalpingography, in which radiopaque dye is injected into the uterus and oviducts, is used to determine the patency of the uterine tube.

2. The **uterus** represents the fused portions of the paramesonephric (müllerian) ducts.

a. The uterus, although variable in size, is normally 8 cm × 2.5 cm × 5 cm in the nulliparous woman and somewhat larger in the multiparous woman.

b. The uterus is divided into four regions (see Fig. 22-9).

(1) The **fundus** is the region of the uterus superior to the uterine cornua. This region contributes most of the upper uterine segment during pregnancy.

(2) The **cornu** on each side defines the point of entrance of a uterine tube.

(3) The **body (corpus)** of the uterus is the region inferior to the cornu and superior to the cervix.
- **(a)** The uterine cavity is shaped like an inverted flattened triangle.
- **(b)** The **isthmus** is the dividing line between the cervix and the uterus. It corresponds with the internal os.
- **(c)** The isthmus becomes incorporated into the uterus during pregnancy as the lower uterine segment. This is the preferred site for surgical delivery by caesarian section.

(4) The **cervix** is the narrow inferior portion of the uterus that projects into the vagina.
- **(a)** It is divided into three regions.
  - **(i)** The **internal os** marks the junction of the cervical canal with the uterine body.
  - **(ii)** The **cervical canal**, about 2.5 cm long, lies between the internal and external ostia.
  - **(iii)** The **external os** is the opening of the cervical canal into the vagina.
- **(b)** Changes in the cervix occur during pregnancy.
  - **(i)** The cervix becomes softer, changing in consistency from feeling like a "nose" to feeling like "lips."
  - **(ii)** The glands of the cervical canal secrete a protective mucous plug.

c. Uterine positions:

(1) The normal uterine position is *anteflexed* (uterus bent forward on itself at the level of the internal os) and *anteverted* (the angle of the uterine cervix with the vagina normally is about 90°). Thus the long axis of the uterus is nearly horizontal and the body of the uterus lies on the bladder (see Fig. 22-8).

(2) In some individuals the uterus may assume other positions.

(a) *Retroflexion:* the axis of the body of the uterus passes upward or even backward.

(b) *Retroversion:* the axis of the cervix with the vagina is upward or even backward so that the uterus may rest against the rectum, and either the external os or the posterior lip of the cervix presents first on digital examination.

d. Uterine support:

(1) The **levator ani muscle** and **urogenital diaphragm** provide the principal support for the urinary bladder. The **urinary bladder** in conjunction with gravity directly supports the anteverted and anteflexed uterus (see Figs. 22-1 and 22-8).

(2) The levator ani muscle and urogenital diaphragm also provide the principal support for the vagina, which then indirectly supports the uterus.

(3) The **round ligament of the uterus** maintains the anteverted position. This ligament stretches considerably during pregnancy (see Fig. 22-8).

(4) Each **cardinal ligament** with the contained uterine vasculature contributes variable support in this region. In retroversion, in which the uterus does not rest on the bladder, the cardinal ligaments appear to provide the principal support (see Fig. 22-10).

(5) The uterosacral and pubocervical ligaments provide variable posterior and anterior support, respectively.

e. Uterine relations (see Fig. 22-8).

(1) Anteriorly, the **uterovesical pouch** separates the uterus from the urinary bladder.

(a) The supravaginal cervix is separated from the urinary bladder by the pubocervical (vesicovaginal) fascia.

(b) The infravaginal cervix lies posterior to the **anterior fornix** of the vagina.

(2) Posteriorly, the uterine body is separated from the rectum by the **rectouterine** or **rectovaginal pouch** (Douglas'). The most caudal portion of the rectouterine pouch fuses to form the **rectouterine septum** between the vagina and rectum; this layer corresponds to the rectovesical fascia in the male.

(3) Lateral to the uterus lies the broad ligament and its contents. The ureter is approximately 1 cm lateral to the supravaginal cervix.

f. Vasculature of the uterus (Fig. 22-12).

(1) The blood supply to the uterus is principally by the uterine arteries.

(a) These vessels arise from each internal iliac artery.

(b) They run in the transverse cervical ligament at the base of the broad ligament.

(c) They cross immediately anterior to the ureters in the transverse cervical ligament, which is a most important surgical consideration.

(d) The branches of the uterine artery include:

(i) The **ascending branch**, which gives off the **tubal branch** and anastomoses with the ovarian artery.

(ii) The **descending (vaginal) branch**, which anastomoses with other vaginal branches and with the perineal branches of the internal pudendal artery.

(e) During pregnancy the uterine artery becomes hypertrophic, with an enormous volume of blood flowing to the uterus and placenta.

(2) The blood return from the uterus is along similarly named veins, which generally parallel the arterial supply.

(3) Lymphatic drainage of the uterus also parallels the blood supply.

(a) The lymph from the uterine fundus drains mainly along the ovarian vessels to the aortic nodes.

(b) The uterine body generally drains along the internal iliac vessels, but some lymphatics follow the round ligament to the inguinal nodes.

(c) The cervix drains in three directions.

(i) The principal drainage is along the uterine vessels toward the internal iliac nodes.

(ii) There are also lymph pathways along the internal pudendal vessels to the internal iliac nodes.

(iii) A small but significant pathway drains laterally in the parametrium toward the obturator and external iliac nodes.

(iv) Other pathways course through the deep pelvis toward the sacral nodes.

(d) Due to widespread lymphatic drainage, invasive carcinoma of the cervix may require extensive lymphatic dissection.

g. Uterine innervation (Fig. 22-13).

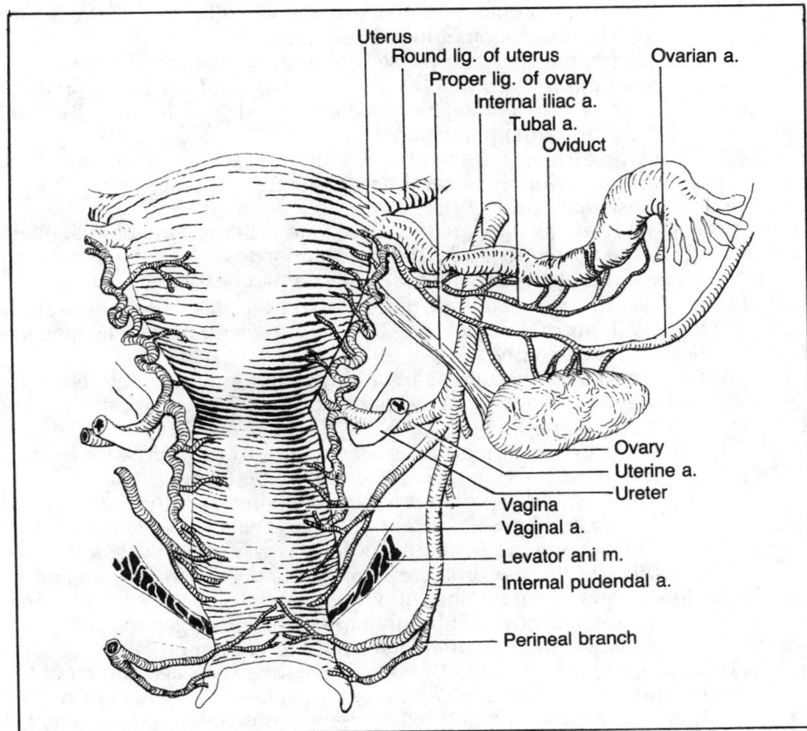

Figure 22-12. *The vasculature of the female reproductive tract.* The ovarian, uterine, and pudendal arteries supply the female tract and genitalia with anastomoses occurring between the ovarian and uterine arteries as well as between the uterine arteries and deep perineal branches of the pudendal arteries.

(1) The autonomic innervation to the uterus is derived as the uterovaginal plexus from the lateral pelvic plexus.

(2) Sympathetic neurons originate from the lateral horns of the T12–L2 or L3 segments of the spinal cord.

- **(a)** These neurons leave the spinal nerve via the white rami communicantes to gain access to the sympathetic chain, where they may synapse.
- **(b)** The second-order neurons leave the sympathetic chain via **lumbar splanchnic nerves** and course in sequence through the aortic plexus, superior hypogastric plexus, inferior hypogastric plexus, lateral pelvic plexus, and uterovaginal plexus to reach the uterus.

(3) Parasympathetic neurons originate from the S2, S3, and S4 segments of the spinal cord.

- **(a)** These neurons leave the spinal nerves in the **pelvic splanchnic nerves (nervi erigentes)**.
- **(b)** These nerves pass through the lateral pelvic plexus and uterovaginal plexus to reach the uterus, where they synapse with a second-order neuron.

(4) Visceral afferents from the uterus generally take two pathways to the spinal cord.

- **(a)** From the body of the uterus and uterine tubes the course of the visceral afferents is parallel to the sympathetic pathways. Hence pain associated with uterine spasm and salpingitis is *referred* to the dermatomes from T12–L2 (i.e., the midback, the inguinal and pubic regions, and the anterior thigh).
- **(b)** From the cervical region the course of the visceral afferents is along the nervi erigentes. Thus pain associated with cervical dilation is *referred* to sacral dermatomes S2–S4 (i.e., the perineum, gluteal region, posterior thigh, and leg).

h. Structure of the uterus.

(1) The **endometrium** (columnar mucosa) undergoes cyclic changes with the menstrual cycle.

- **(a)** During the menstrual phase, when the corpus luteum is regressing and the progesterone titers are decreasing, there is a sloughing of endometrium.

(b) During the proliferative phase, under the influence of estrogen from the maturing follicle, the endometrium is reformed.
(c) During the secretive phase, under the influence of progesterone and estrogen from the new corpus luteum, there is glandular proliferation and secretion as well as proliferation of the endometrial blood vessels so that the mucosa is ready for an implantation.

(2) The **myometrium**, forming the bulk of the uterine wall, is smooth muscle.
(3) The **mesometrium** is **visceral peritoneum**.

i. Clinical considerations.

(1) Palpable per vaginam are the cervix and ostium of the uterus, the vagina, the body of the uterus if retroverted, the rectouterine fossa, and under certain pathologic conditions, the ovary, uterine tubes, and broad ligament.
(2) Because of the location of the ureters in the transverse cervical ligament, there is the possibility of division of the ureter when clamping and ligating the uterine arteries during hysterectomy.
(3) Compression of the ureters by a growth in the pelvis may result in hydronephrosis and uremia, a frequent cause of death.
(4) Uterine prolapse.

(a) Retroversion of the uterus predisposes toward prolapse because the weight of the uterus is no longer supported by the bladder.

(i) First-order prolapse is defined as retroversion, with the cervix protruding to an abnormal extent into the vagina (procidentia).
(ii) Second-order prolapse is defined as protrusion of the cervix from the vagina.
(iii) Third-order prolapse is complete eversion of the vagina.

(b) Trauma to the uterine support (particularly to the round ligament of the uterus) as a result of difficult childbirth, as well as general relaxation of connective tissues in old age, may predispose to uterine prolapse.

(5) During parturition pain from dilation of the cervix and stretch of the perineum may be appropriately blocked, although the thoracic and lumbar nerves controlling abdominal musculature and reflex uterine contraction are not affected.

(a) Caudal anesthesia: the needle is inserted between the sacral cornua, penetrating the sacrococcygeal ligament across the sacral hiatus to gain access to the vertebral canal. The anesthetic is infiltrated into the epidural space external to the dural sac, and the spinal nerves are anesthetized as they leave the dura. This is most effective for the sacral nerves.
(b) Epidural anesthesia: similar to caudal anesthesia, except the needle is inserted into the L4–L5 interspace and the anesthetic is infiltrated into the epidural space about the dural sac. This is most effective for the lower lumbar nerves.
(c) Spinal anesthesia: similar to epidural anesthesia, except the needle is inserted through the dural sac and the anesthetic infiltrated directly into the subarachnoid space in which the nerve roots are anesthetized. This is very effective for the lumbar and sacral nerves but is more difficult to control and subject to more complications and side effects.

E. THE VAGINA (see Fig. 22-8)

1. From its opening at the vestibule of the vulva, the vagina passes superiorly and posteriorly through the pelvic diaphragm and surrounds the cervix of the uterus.

2. The size of the vagina is somewhat variable.

a. The anterior wall, which is approximately 7.5 cm long, terminates anterior to the projecting cervix as the **anterior fornix**.
b. The posterior wall, approximately 10.5 cm long, terminates in the **posterior fornix** and is related to the **rectouterine pouch** (Douglas').
c. The lateral walls terminate in **lateral fornices** on either side of the projecting cervix.
d. Normally, the vaginal lumen is *H*-shaped.
e. Relations of other structures to the vagina.

(1) Anterior to the upper portion of the vagina lies the uterus. The midvagina apposes the base of the bladder with the lower half of the urethra embedded in the adventitia of the lower vaginal wall.
(2) Posteriorly, the posterior fornix of the vagina is separated from the rectum by the uterorectal pouch (Douglas'), the upper quarter of the posterior vaginal wall is separated from the rectum by the uterorectal septum, and the lower vagina is separated from the anal canal by the **perineal body**.
(3) Laterally, the vagina is flanked by the levator ani muscle and endopelvic fascia. The ureters and uterine arteries lie close to the lateral fornices.

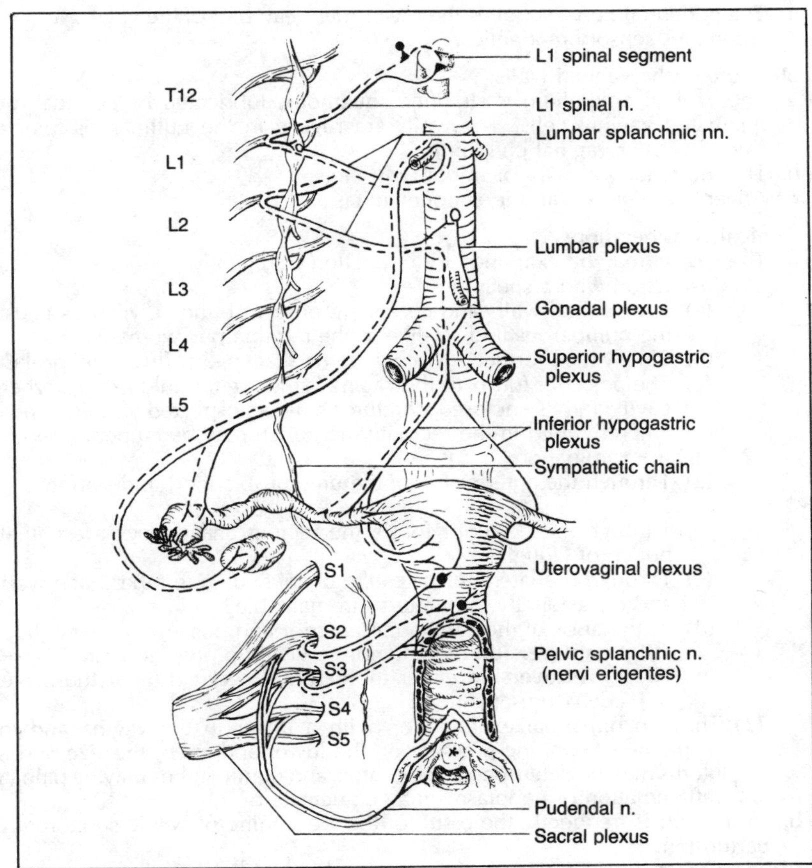

Figure 22-13. *The innervation of the female reproductive tract and genitalia.* The sympathetic pathways arise from the lower thoracic and upper lumbar spinal levels (there are no white rami below L2) and reach the pelvic viscera via thoracic and lumbar splanchnic nerves and then the hypogastric plexuses. The parasympathetic pathways arise from the midsacral spinal levels and reach the pelvic viscera via the pelvic splanchnic nerves. Visceral afferent fibers (*dashed*) from the pelvic viscera travel specifically along either one or the other autonomic pathways, producing specific patterns of referred pain. The pudendal nerve provides somatic innervation to and from the perineum.

- **(4)** The upper two-thirds of the vagina is within the pelvic cavity (i.e., superior to the pelvic floor).
- **(5)** The lower third of the vagina is within the perineum (i.e., inferior to the pelvic floor). It communicates with the exterior via the **introitus**.

3. The vaginal vasculature (see Fig. 22-12).

- **a.** The arterial supply is via the **vaginal branch** of the **uterine artery**, occasional **vaginal branches** of the internal iliac arteries, possible twigs from the **middle rectal arteries**, and branches from the **internal pudendal arteries**.
- **b.** The vaginal plexus of veins drains into the internal iliac vein.
- **c.** The lymphatic drainage of the vagina is varied.
  - **(1)** The upper third of the vagina drains into external and internal iliac nodes.
  - **(2)** The middle third drains into the internal iliac nodes.
  - **(3)** The lower third drains into the internal iliac nodes as well as into the superficial inguinal nodes.

4. Innervation to the vagina (see Fig. 22-13).

- **a.** The uterovaginal plexus of nerves, a spray from the lateral pelvic plexus, supplies the upper two-thirds of the vagina with sympathetic, parasympathetic, and afferent fibers. The visceral afferents from this region travel along the nervi erigentes so that pain may be referred to the dermatomal distribution of spinal segments S2–S4.

b. The pudendal nerve supplies the lower (perineal) part of the vagina and mediates all the cutaneous sensory modalities.

5. Structure of the vaginal walls.

a. The vaginal epithelium is stratified squamous, lubricated by cervical mucus and desquamated vaginal cells. The walls are rugose in the nulliparous female but become smoother after vaginal childbirth.

b. The muscular walls are of smooth muscle.

c. Adventitia blends with the endopelvic fascia.

6. Clinical considerations.

a. The vagina may be examined by inspection or digitally.

(1) Inspection with a speculum.

(a) The vaginal walls and cervix are observed (the cervical os is usually round in the nulliparous and slit-like in the multiparous woman).

(b) Biopsy is performed, cytologic smears taken, and first-order prolapse is detected.

(c) The posterior fornix of the vagina is the site for culdoscopy, whereby the pelvic cavity and its enclosed structures may be inspected visually, and culdocentesis, whereby fluid in the rectouterine pouch may be tapped.

(2) Digital examination:

(a) Through the anterior fornix the urethra, bladder, and symphysis pubis are palpable.

(b) Through the posterior fornix the rectum and coccyx are palpable across the pouch of Douglas.

(c) Through the lateral fornices the ovaries, uterine tubes, side wall of the pelvis, and occasionally the ureters, are palpable.

(d) At the apex of the vagina the anterior and posterior cervical lips are palpable.

(i) In **anteversion** the anterior cervical lip presents first.

(ii) In **retroversion** either the **external cervical os** or the posterior lip of the cervix presents first.

(3) Through bimanual examination with a finger in the vagina and the other hand palpating or exerting pressure on the lower abdomen, the size and position of the uterus may be determined, the ovaries and uterine tubes may be palpated, and pelvic inflammation or neoplasms may be detected.

b. Herniation is frequently the result of the weakening of pelvic support structures during parturition.

(1) Cystocele is herniation of the urinary bladder through the anterior vaginal wall into the vaginal lumen.

(2) Rectocele is herniation of the rectum through the posterior vaginal wall into the vaginal lumen.

c. Anomalies and variations.

(1) Uterine and vaginal duplication occurs to variable extents.

(a) **Didelphis**, a complete double uterus and double vagina, is very rare.

(b) **Uterus duplex**, or double uterus with a single vagina, is rather uncommon.

(c) **Bicornuate** or **septate uterus**, a double or partitioned fundus, is not uncommon.

(d) **Uterus arcuatus** has no internal septum, but there is a depression on the external surface of the fundus.

(2) Unicornuate uterus with complete absence of the uterine tube on one side of the uterus is rare.

(3) Vaginal anomalies are mainly associated with failure of the vaginal plate to canalize.

(a) Rectovaginal fistulas are commonly coincident with an imperforate anus or perineal trauma.

(b) Vesicovaginal fistulas (rare) are associated with continuous urinary discharge from the vagina.

(c) Urethrovaginal fistulas are associated with urinary discharge from the vagina only upon micturition.

VII. FUNCTIONAL CONSIDERATIONS

A. PRECOITAL ACTIVITY of variable duration and intensity is usually required to establish a receptive psychological attitude before sexual congress occurs.

1. In the female this stimulation results in secretion from the cervical glands, erection of the clitoris with accompanying enlargement of the vestibular bulb, swelling of the vaginal walls, and secretion from the vestibular glands.

2. In the male a similar stimulation results in the erection of the penis, emission (movement

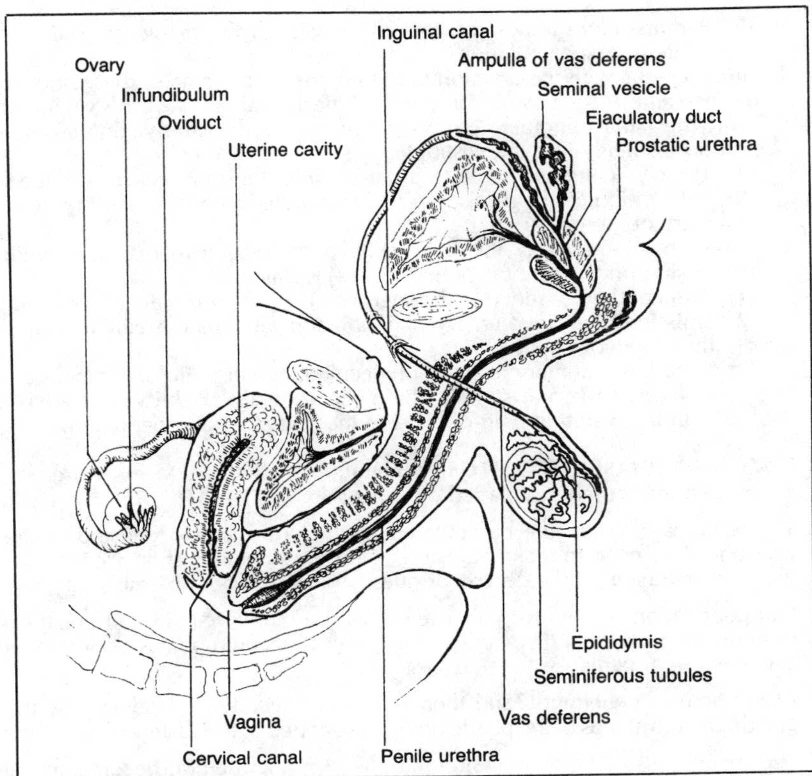

Figure 22-14. *Coitus.* The second (penetration) phase of coitus (after Da Vinci).

of the sperm from the testes and epididymis to the terminal portion of the vas deferens), and secretion from the glands associated with the urethra.

a. When an individual is sexually aroused by the appropriate stimulus, parasympathetic volleys traveling via the pelvic nerves to the cavernous plexuses cause a relaxation of the smooth muscle associated with the arteries supplying the erectile tissue.

- **(1)** These arteries represent the terminal branches of the internal pudendal artery, which supplies the penis or clitoris.
- **(2)** The consequent widening of these arteries results in an increased blood flow into the cavernous spaces.
- **(3)** The increase in blood flow is especially pronounced because the arteries contain longitudinal ridges that partially occlude the lumina when the circular muscularis is slightly tensed.
- **(4)** The corpora have a spongy structure with irregular endothelial-lined vascular spaces that are essentially collapsed in the flaccid penis or clitoris and vestibular bulbs. With increased blood flow, the corpora fill, expand, and become turgid.

b. However, erection is a vascular event that involves not only increased inflow but also reduced efflux, of blood.

- **(1)** As the erectile tissue expands, the peripheral veins are compressed against the enveloping tunica albuginea, which effectively impedes drainage of blood from the cavernous sinuses.
- **(2)** Concomitantly, the bulbocavernosus and ischiocavernosus muscles, which envelop the roots of the penis or clitoris, contract periodically under voluntary control, enhancing the establishment of turgor.
- **(3)** Consequently, the penis or clitoris enlarges, elongates, and becomes hard.

c. Although the turgor of the corpora cavernosa in the male is extreme, the corpus spongiosum (corpus cavernosum urethrae) and its extension, the glans penis, do not become as firm.

- **(1)** This is due to the thinner and more elastic tunica albuginea surrounding this erectile body, which not only allows for a more flexible vaginal penetration but offers some resiliency when the penis strikes the cervix or wall of the posterior fornix during coitus.

(2) Perhaps more important, the less turgid corpus spongiosum allows the urethra to dilate during ejaculation.

d. Simultaneous with the erection of the clitoris in the female, the vestibular bulbs expand by the same mechanism, compressing the lateral walls of the vagina and causing the overlying labia minora to part. The anterior wall of the vagina swells somewhat and protrudes slightly into the vestibule.

(1) The cervical glands secrete profusely into the upper reaches of the vagina.

(2) The vestibular glands also secrete mucinous material at the introitus, facilitating the entry of the penis.

e. Under sexual stimulation the male genital ducts become increasingly motile, which results in emission (movement of sperm to the ejaculatory duct).

(1) Seminal fluid is added to the sperm cells by the seminal vesicles and prostate gland.

(2) Emission is an involuntary phenomenon, apparently regulated by the parasympathetic nervous system.

(3) If there is voluntary restraint of ejaculation over a prolonged period of time, in spite of repeated sexual stimuli, the muscular walls of the vas deferens go into spasm, with the resulting pain referred to the abdomen and perineum.

B. THE SECOND PHASE OF COITUS, penetration of the vagina by the erect penis, can be accomplished in a number of varied postures (Fig. 22–14).

1. The postures of coitus are particularly significant if there is repeated failure to effect conception or the need to make adjustments for an especially short vagina, an unusually long penis, retroversion of the cervix, or obesity of one or both partners.
2. Full penetration of the erect penis elevates the mobile cervix and creates a pocket in the posterior fornix so that the average 9-cm vagina accommodates a substantial portion of the average 15-cm penis.
3. At the time of penetration and thereafter, male secretions produced by the bulbourethral glands and the glands of the penile urethra are added to the vestibular and cervical secretions.
4. The sperm cells move by emission into the prostatic urethra, become activated by seminal fluid, and are motile.
5. Coital movements accentuate the stimulation of the penis, clitoris, and vaginal walls while emotional responses may be intensified and glandular secretions may be augmented.
6. When a particular degree of stimulation is reached, and when the partners so desire, the physical and emotional process of orgasm may occur.

C. THE THIRD, OR ORGASMIC, PHASE OF COITUS

1. In the male, orgasm is concomitant with ejaculation, which is brought about by sympathetic activity transmitted along the hypogastric nerve and lateral pelvic plexus and then through the prostatic and cavernous plexuses.

a. This sympathetic outflow results in ejaculation with the spasmotic and rhythmic contraction of the vas deferens, seminal vesicles, prostate gland, and urethra.

b. This is accompanied by apparently involuntary contractions of the bulbocavernosus and ischiocavernosus muscles as well as the muscles of the urogenital diaphragm, pelvic floor, and external anal sphincter, all innervated by the pudendal nerve.

c. Although the onset of ejaculation may be controlled, once started, ejaculation cannot be interrupted.

d. The ejaculate (2 to 5 cc) is discharged into the upper part of the vagina where, depending on the coital posture, a seminal pool may form. Regardless of position, the semen have access to the external cervical os.

2. In the female, orgasm is concomitant with contractile spasms of the uterus, apparently initiated by sympathetic nerves of the uterovaginal plexus.

a. The uterine contractions tend to draw sperm cells into the cervical canal.

b. Additionally, orgasm may involve contraction of the bulbocavernosus and ischiocavernosus muscles as well as the muscles of the urogenital diaphragm, pelvic floor, and external anal sphincters.

D. THE FOURTH PHASE OF COITUS (POSTCOITAL RESOLUTION)

1. Owing to the sympathetic outflow to the genital areas, which is now dominant in the male and female, the periarterial muscle increases its tone, thereby reducing the blood flow to the erectile tissues of the penis, clitoris, and vestibular bulbs, which quickly become flaccid.

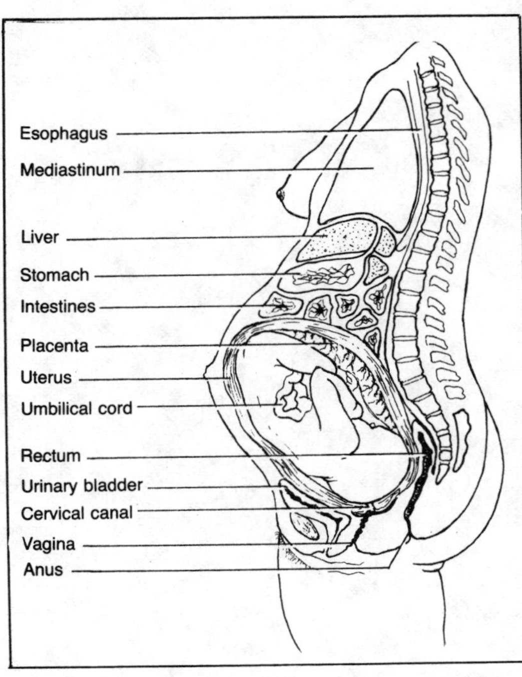

Figure 22-15. *The pregnant uterus.* Fetal development at approximately 35 weeks showing the relationships within the abdominal cavity.

2. Usually, but not always, the male (and perhaps the female) enters into a refractory period of variable duration during which a second erection is not possible, probably due to the results of sympathetic innervation during orgasm.
3. In the absence of physical contraception, once the sperm cells enter the cervical canal it takes approximately 30 to 60 minutes at a swim-rate of 2 mm per minute to traverse the uterine cavity and enter the ostia of the oviducts.
   a. Once in the oviduct, the sperm are capacitated (attain the ability to penetrate an ovum).
   b. Somewhere in the distal third of the oviduct an ovum, if present, may be fertilized by the penetration of one sperm cell, and procreation of the species continues.

E. IMPLANTATION

1. Three to four days are required after fertilization for the zygote to reach the uterus.
2. At this time the endometrium is in the secretory phase and implantation may occur.
3. With the development of the chorionic membranes, human chorionic gonadotropin maintains the corpus luteum so that the endometrium remains in the secretory stage.
4. Development and growth of the fetus requires 38–40 weeks (Fig. 22-15).

Part VI Pelvic Region

STUDY QUESTIONS

Directions: Each question below contains five suggested answers. Choose the **one best** response to each question.

1. In addition to the vas deferens, structures normally found within the internal spermatic fascia of the spermatic cord include all of the following EXCEPT the

(A) testicular artery
(B) pampiniform venous plexus
(C) deferential artery
(D) testicular nerves
(E) genitofemoral nerve

2. The remnant of the abdominal cavity present in the adult scrotum is the

(A) gubernaculum
(B) ductus deferens
(C) spermatic cord
(D) tunica vaginalis
(E) none of the above

3. In the female, the middle rectal (hemorrhoidal) artery has an anastomotic connection with which of the following arteries?

(A) Internal pudendal
(B) Obturator
(C) Superior gluteal
(D) Umbilical
(E) Uterine

4. The female structure occupying the same position in the inguinal region that the vas deferens occupies in the male is the

(A) broad ligament of the uterus
(B) transverse cervical ligament
(C) proper ligament of the ovary
(D) round ligament of the uterus
(E) suspensory ligament of the ovary

Questions 5–8

A 36-year old male complained to his family physician of occasional dull throbbing pain associated with the right testis and scrotum. Examination indicated varicocele of the pampiniform plexus. The examining physician remarked to the patient that he had, in all probability, had the condition for most of his adult life and that it should not be a bother to him. The patient emphatically stated that it had arisen within the last few months. The physician was inclined to pass off the complaint, but since the patient was bothered by the pain, surgery was considered.

5. Factors that might be considered by the examining physician include the fact that varicocele of the pampiniform plexus on the right side is associated with

(A) a long, redundant mesorchium
(B) situs inversus abdominis
(C) testicular torsion
(D) hydrocele processus vaginalis
(E) none of the above

6. It was decided that the testicular vein would be ligated just cranial to the deep inguinal ring. It is expected that this procedure would

(A) eliminate pressure by the column of venous blood
(B) reduce countercurrrent heat exchange
(C) reduce blood flow through the testes
(D) reduce sperm viability
(E) abolish the cremasteric reflex

7. During the early stages of the surgical operation, in which an incision was made parallel to the inguinal ligament, a retroperitoneal mass was observed in the vicinity of the lower pole of the right kidney. This growth encroached on the right testicular vein and impeded the flow so that it did not drain normally into the

(A) inferior vena cava
(B) internal iliac vein
(C) hepatic portal vein
(D) right renal vein
(E) right suprarenal vein

8. Since the tumor appeared to be of renal origin, it was decided that a computed tomography scan, an arteriogram, and an intravenous pyelogram should be performed before removal was attempted. Thus, the testicular vein was ligated as planned. After ligation, venous blood from the testes will return via all of the following collateral pathways EXCEPT the

(A) external pudendal veins
(B) cremaster veins
(C) veins of the vas deferens
(D) deep dorsal veins of the penis
(E) posterior scrotal veins

(end of group question)

Directions: Each question below contains four suggested answers of which **one or more** is correct. Choose the answer

A if **1, 2, and 3** are correct
B if **1 and 3** are correct
C if **2 and 4** are correct
D if **4** is correct
E if **1, 2, 3, and 4** are correct

9. Distinguishing features of the female pelvis include

(1) an oval pelvic inlet as compared to heart-shaped in the male
(2) a shallow false pelvis with flaring of the ilia as compared to the deep false pelvis with more vertical ilia in the male
(3) a larger pelvic outlet than the male
(4) a narrow subpubic angle between inferior pubic rami as compared to the wide angle in the male

10. Correct statements about the ischiorectal fossa include which of the following?

(1) It is filled with adipose tissue traversed by irregular connective tissue septa
(2) It is continuous with the superficial (fatty) layer of superficial fascia of the perineum and abdominal wall
(3) It contains inferior hemorrhoidal vessels and nerves
(4) It extends between the pelvic and urogenital diaphragms

11. External hemorrhoids develop in branches of the hemorrhoidal vein and are characterized by

(1) profuse bleeding
(2) being larger than internal hemorrhoids
(3) a location superior to the pectinate line
(4) pain

12. Structures found in the male superficial perineal space include the

(1) ischiocavernosus muscle
(2) corpus spongiosum
(3) bulbospongiosus muscle
(4) bulbourethral glands (Cowper's)

13. Correct statements about the ischiocavernosus muscle include which of the following?

(1) It contracts to impede venous drainage from corpus cavernosus
(2) It inserts into the central tendon of the perineum
(3) It lies in the superficial perineal space
(4) It receives its motor innervation from nervi erigentes

14. Which of the following statements correctly characterizes the external urethral sphincter?

(1) It receives innervation by the pudendal nerve
(2) It has identical anatomical structure in both the male and female
(3) It is under voluntary control
(4) It is a portion of the pelvic diaphragm

15. Characteristics of the bulbourethral glands (Cowper's) include which of the following?

(1) They are homologous with the paraurethral glands (of Skene)
(2) They secrete into the penile urethra prior to and during ejaculation
(3) They contribute fructose to seminal fluid
(4) They lie in the deep perineal space

16. In the female perineum, the vestibule has which of the following characteristics?

(1) It receives both the urethra and the vagina
(2) It is bordered by the labia majora
(3) It receives drainage from the greater vestibular and paraurethral glands
(4) It receives sensory innervation from the nervi erigentes

17. True statements about the ovary include which of the following?

(1) Its sensory fibers follow the lateral pelvic plexus back to nervi erigentes
(2) Its arteries are found in the suspensory ligament of the ovaries
(3) It is supported by a fold of serosa called the mesosalpinx
(4) Its arteries are direct branches of the aorta

18. Sensory fibers for pain from the body of the uterus travel to the spinal cord via the

(1) plexus around the uterine arteries
(2) white rami communicantes of L3 and L4
(3) lateral pelvic plexus and inferior hypogastric plexus
(4) nervi erigentes

19. Benign prostatic hypertrophy (BPH) results in obstruction of the prostatic urethra by enlargement of

(1) the posterior lobe
(2) the anterior lobe
(3) either lateral lobe
(4) the median lobe

ANSWERS AND EXPLANATIONS

1. The answer is E. *(Chapter 21 IV A 3)* The testicular neurovascular bundle, the ductus deferens, and a remnant of the processus vaginalis lie within the internal spermatic fascia of the spermatic cord. The genital branch of the genitofemoral nerve lies in the cremaster layer beneath the external spermatic fascia.

2. The answer is D. *[Chapter 21 IV A 3 d (1)]* The tunica vaginalis of the testis is a remnant of the processus vaginalis, which extended from the abdominal cavity into the scrotum. The gubernaculum, a condensation of connective tissue between the inferior pole of the testis and the scrotum, guides the descent of the gonad. As the testis descends through into the scrotum, it drags with it components of the abdominal wall which form the spermatic cord.

3. The answer is A. *[Chapter 20 IV C 3 f (2)]* The middle rectal artery anastomoses with the superior rectal branch of the inferior mesenteric artery and with the inferior rectal branch of the internal pudendal artery. The middle rectal vein makes similar anastomotic connections.

4. The answer is D. *[(Chapter 22 VI A 2 d (4)]* The gubernaculum fails to lengthen during development in the male, drawing the testis into the scrotum so that the vas deferens comes to lie in the inguinal canal. In the female the counterpart of the gubernaculum is the combined proper ligament of the ovary and the round ligament of the uterus. The latter passes through the inguinal canal to the deep fascia of the labia majora.

5. The answer is B. *[Chapter 21 IV A 3 c (2) (b)]* Varicocele of the pampiniform plexus on the right side is very uncommon because this condition usually results from compression of the testicular vein by the full sigmoid colon. In cases of situs inversus, where the sigmoid colon is on the right side, the varicocele also would tend to be on the right.

6. The answer is A. *[Chapter 22 V A 5 a (1) (b)]* Testicular vein ligation at the inguinal ligament eliminates the hydrostatic pressure produced by the column of venous blood in the abdomen. Blood return from the testes is through anastomotic pathways. While a varicocele interferes with the countercurrent heat exchange mechanism, thereby raising scrotal temperature and decreasing sperm viability, the ligation reverses this effect.

7. The answer is A. *[Chapter 22 V A 5 a (2)]* The right testicular vein normally drains into the inferior vena cava. The left spermatic vein joins the left renal vein.

8. The answer is D. *(Chapter 22 V A 5 b)* The collateral venous drainage of the testes is primarily along the veins of the vas deferens, but there are anastomoses with the external pudendal veins, the posterior scrotal veins, and the cremaster veins (branches of the veins of the anterior abdominal wall). The deep dorsal vein of the penis, draining the deep structure of the penis, does not offer the possibility for anastomotic connections.

9. The answer is A (1, 2, 3). *(Chapter 20 I B 5 b; Table 20-1)* In comparison with the male pelvis, that of the female is larger, has a shallower false pelvis, a more oval inlet, a larger outlet, and a wider inferior pubic angle.

10. The answer is E (all). *(Chapter 21 III A 2–5; IV B 2 c)* The fat that fills the ischiorectal fossa is a continuation of the superficial layer of the superficial fascia of the perineum and abdominal wall. Within this layer are the inferior rectal (hemorrhoidal) neurovascular bundles. The anterior horn of the ischiorectal fossa extends between the pelvic and urogenital diaphragms.

11. The answer is D (4). *(Chapter 21 III F 2 b)* The portion of the anal canal inferior to the pectinate line is innervated by the inferior rectal branches of the pudendal nerve. Hemorrhoids in this region are painful and demand early attention. Internal hemorrhoids, above the pectinate line, are painless and tend to bleed profusely, frequently causing anemia.

12. The answer is A (1, 2, 3). *(Chapter 21 IV C 2, D 1 c)* The superficial perineal space contains the root of the penis and the associated bulbospongiosus and ischiocavernosus muscles. The bulbourethral glands are located in the deep perineal space.

13. The answer is B (1, 3). *[Chapter 21 IV C 2 b; Chapter 22 VII A 2 b (2)]* The ischiocavernosus muscle, lying in the superficial pouch, contracts periodically under voluntary control, thereby enhancing the establishment of turgor in the erect penis. It is innervated by the perineal branch of the pudendal nerve. It inserts just distal to the point at which the crura join to form the body of the penis or into the margin of the pubic arch and into each crus of the clitoris.

14. The answer is B (1, 3). *[Chapter 21 IV D 1 b (2), V D 4 b)]* The external urethral sphincter, that portion of the deep transverse perineal muscle adjacent to the urethra, is innervated by the internal pudendal nerve and is therefore under voluntary control. In the male, the external urethral sphincter completely surrounds the membranous urethra; in the female, because the urethra is embedded in the anterior vaginal wall, this sphincter is incomplete and less effective. It is part of the urogenital diaphragm.

15. The answer is C (2, 4). *(Chapter 21 IV D 1 c, V A 3 e; Chapter 22 V D)* The bulbourethral glands, lying in the deep perineal space, discharge into the penile urethra prior to and during ejaculation. The prostate gland is homologous to the paraurethral glands in the female. The fructose component of seminal fluid is a product of the seminal vesicles.

16. The answer is B (1, 3). *(Chapter 21 V A 3, F 1 a)* The vestibule is bounded by the labia minora. It receives the urethra and the vagina, as well as the ducts of the paraurethral and vestibular glands. The innervation of the vestibule is by the perineal branches of the pudendal nerve.

17. The answer is C (2, 4). *(Chapter 22 VI C 3 a, 4 a (1), 5 b)* The blood supply to the ovaries is primarily by the ovarian arteries (direct branches of the aorta), which run in the suspensory ligament. In addition to the suspensory ligament, the mesovarium and the proper ligament of the ovary provide support. Afferent innervation is to the aortic plexus and thence along the least splanchnic and lumbar splanchnic nerves to the spinal levels; thus, pain is referred to the T12, LI, and L2 dermatomes.

18. The answer is B (1, 3). *[Chapter 22 VI D 2 g (2) (b), (4)]* The afferent fibers innervating the uterine body travel along the uterine plexus and about the uterine vessels, the lateral pelvic plexus, the inferior and superior hypogastric plexuses, and the aortic plexus to reach the lumbar splanchnic nerves which arise from spinal segments L1 and L2. (There are no white rami and thus no lumbar splanchnic nerves below L2.) Pain from the uterus is referred to the inguinal and pubic regions, as well as to the lateral and anterior aspects of the thigh.

19. The answer is D (4). *[Chapter 22 V E 3 c (1)]* The median (middle) lobe of the prostate gland is most commonly involved in proliferative hyperplasia (benign prostatic hypertrophy), which obstructs the prostatic urethra with retention of urine in the bladder. The posterior lobe is most commonly involved in malignant transformation.

Part VII
Lower Extremity

23
The Gluteal Region

I. INTRODUCTION

A. BASIC PRINCIPLES

1. The basic pattern of the human lower extremity may be approximated by imagining the primitive position: the appendage is abducted until it is horizontal with the soles facing forward so that the lower extremity approximates the pelvic fin of a fish (see Fig. 4-3).
   a. The anterior side is the **primitive ventral surface**; the posterior side is the **primitive dorsal surface**.
   b. An imaginary line drawn through the long axis of the extremity will differentiate between a cephalad **preaxial border** and a caudal **postaxial border**.
2. From this original fin-like position of the lower extremity, flexing the knee 90° and extending the ankle 90° approximates the amphibian/reptilian lower extremity.
3. In mammals, humans in particular, the amphibian/reptilian knee has rotated 90° cephalad, bringing the lower extremity alongside the body wall, so that the primitive ventral surface faces cephalad. There is a concomitant pronation of the distal portion.
4. With assumption of an erect posture by extension of the hip, there is an effective additional 90° rotation so that the primitive dorsal surface of the lower extremity faces anteriorly and the primitive ventral surface faces posteriorly.
   a. The arrangement of nerves in the limbs reflects the developmental origin as a horizontal bud (see Fig. 4-2).
   b. Each spinal nerve innervates a sensory discrete area, the dermatome.
      (1) In the region of the lower extremity, the dermatomes are drawn distally and appear to march down the preaxial side and back along the postaxial side (see Fig. 4-3).
      (2) While there is generally overlap between sequential dermatomes, there is *no* overlap across the axial lines.
   c. Each spinal nerve innervates a discrete muscle group, and in the lower extremity, there is an orderly preaxial to postaxial progression of muscles of myotomal origin and, therefore, sequential innervation (see Fig. 4-3).
5. While the human upper and lower extremities are homologous, structural differences between the upper and lower extremities reflect different functions. The leg portion of the lower extremity becomes fixed in the pronated position and loses the ability to supinate.
   a. The lower extremity of the human retains its original role of support and propulsion at the expense of dexterity and intrinsic mobility.
   b. The extensive fusion within the pelvic girdle and the greater stability of the joints are adaptations to weight bearing.

B. THE LOWER EXTREMITY can be divided into five regions.

1. The **pelvic girdle.**
   a. The **ilium.**
   b. The **ischium.**
   c. The **pubis.**
2. The **thigh**.
   a. The **femur.**

3. The **leg.**
   a. The **tibia.**
   b. The **fibula.**

4. The **ankle.**
   a. The proximal row of **tarsal bones.**
   b. The distal row of **tarsal bones.**

5. The **foot.**
   a. The **metatarsal bones.**
   b. Three rows of **phalanges.**

II. THE HIP REGION

A. BONY LANDMARKS

1. The **pelvis.**
   a. Iliac crest.
   b. Anterior-superior iliac spine.
   c. Inguinal ligament.
   d. Pubic tubercle.
   e. Pubic symphysis.
   f. Ischiopubic ramus.
   g. Ischial tuberosity.
   h. Posterior-superior iliac spine.
   i. Ischial spine.

2. The **femur.**
   a. Femoral head
   b. Greater trochanter
   c. Femoral shaft.

B. THE OSSEOUS PELVIS (review Chapter 20 I A).

1. The pelvic (hip) girdle consists of a coxal (innominate or hip) bone on either side of the sacrum (see Fig. 20-1).
   a. The coxal (innominate or hip) bone on each side of the pelvis is formed by the fusion of three separate bones, the ilium, pubis, and ischium.
   b. The thigh, leg, ankle, and foot (the free portion of the lower extremity) are suspended from the axial skeleton by the pelvic girdle (the embedded portion of the lower extremity).

2. The pelvic girdle has three articulations.
   a. Proximally, the pelvic girdle articulates with the axial skeleton through the **sacroiliac joints.**
      (1) The **sacroiliac joint** is formed by the articulation of the alar plates of the sacrum with the ilium.
      (2) This attachment of the lower extremity to the axial skeleton is primarily ligamentous.
      (3) Unlike the pectoral girdle, there normally is almost no movement between the pelvic girdle and the axial skeleton so that the pelvic girdle is extremely stable.
   b. Anteriorly, left and right coxal bones articulate at the **pubic symphysis.**
   c. Distally, each coxal bone articulates with the femur at the **iliofemoral (hip) joint.**
      (1) The **ilium, pubis**, and **ischium**, join at the **acetabulum** (*L.* vinegar cup), which is the fossa of the hip joint (Fig. 23-1).
         (a) The hemispherical acetabulum faces laterally but slightly posteriorly and slightly inferiorly.
            (i) It is just short of a true hemisphere, which contributes to the great stability of the joint.
            (ii) There is a discontinuity in the inferior rim, the **acetabular notch**.
            (iii) The depth of the acetabulum is increased by a lip of fibrocartilage, the **acetabular labrum**, which bridges the acetabular notch as the **transverse acetabular ligament.**
         (b) The **ligamentum teres** originates from the transverse ligament over the acetabular notch and inserts into the **fovea** of the femoral head.
      (2) The acetabulum is lined by horseshoe-shaped hyaline cartilage.
         (a) The bone behind the articular surface is thick with the bony trabeculae arranged along the lines of force.
         (b) The rough nonarticular medial wall of the acetabulum enclosed by the limbus of the articular cartilage is termed the **acetabular fossa**. This wall is a thin sheet of bone.

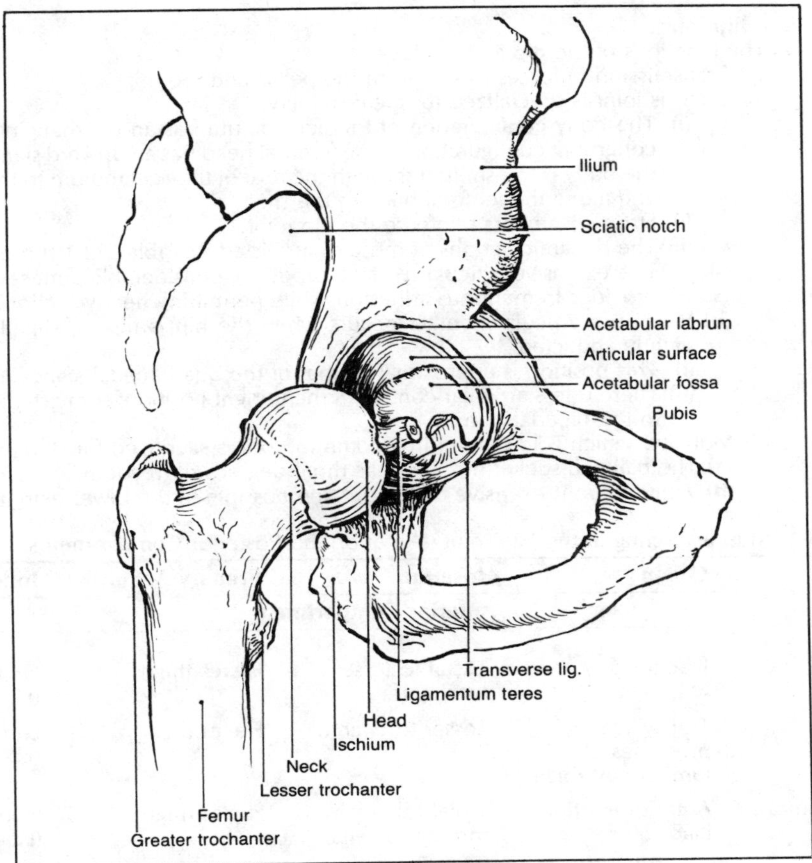

Figure 23-1. *The hip joint.* The femur articulates with the pelvic bone at the acetabulum. The ligamentum teres has been cut and the femur disarticulated.

3. The **femur** is the long bone of the thigh (see Fig. 23-1).

a. The proximal end of the femur consists of head, neck, greater and lesser trochanters, and shaft.

(1) The **head** is approximately two-thirds of a sphere.

(a) It fits deeply into the acetabulum.

(b) It is covered with hyaline articular cartilage, except at the **foveal notch,** which receives the **ligamentum teres**.

(c) The epiphyseal plate demarcates the head from the neck.

(2) The **neck** joins the head to the shaft.

(a) It is normally angled upward at about 125° —slightly more in males, less in females.

(b) It is normally angled forward (anteverted) by about 15°.

(3) The **shaft** bows slightly anteriorly.

(a) The **greater trochanter** on the lateral surface projects above the junction with the neck and provides attachment for the gluteus medius and minimus muscles as well as for the piriformis muscle.

(b) The **lesser trochanter** is situated on the medial surface distal to the junction of the neck and receives the insertion of the iliopsoas muscle.

(c) The **intertrochanteric crest** runs obliquely along the posterior surface of the shaft between the trochanters.

b. The femur transmits huge forces between the pelvis and tibia.

(1) The bony trabeculae of the femoral head, neck, and trochanters are arranged along the lines of prevailing force, effectively transmitting these forces to the compact bone of the shaft.

(2) This trabecular arrangement attains greatest development and strength during the active period of life and is reduced by bone resorption during senesence, thereby predisposing to fractures.

4. The **hip joint.**

a. The functions of the hip joint include:

(1) Transmission of forces to and from the pelvis and the thigh.

(a) This joint is specialized for great stability.

(i) The bony configuration of the acetabulum has an overhanging slant, and the congruent configuration of the femoral head has an upward slant. The weight of the body is transmitted through the roof of the acetabulum to the femoral head and along the femoral neck to the shaft.

(ii) Strong ligaments reinforce the hip joint.

(iii) The dynamic actions of muscles are listed in Tables 23-1 through 23-4.

(iv) The extensive articular surface produces considerable surface tension, which provides tremendous adhesion while permitting nearly frictionless gliding.

(b) Maximum articular contact occurs when the hip is flexed, slightly everted, and slightly abducted.

(i) This position is approximately that of the quadrapedal stance.

(ii) There is less articular contact in the upright position; some stability is sacrificed to the bipedal stance.

(2) Mobility, which is important in locomotion but is sacrificed for stability.

(a) The ball-and-socket hip joint has three degrees of freedom.

(b) While there is exensive circumduction possible in the lower extremity, it is not as

Table 23-1. Muscles Acting at the Hip Joint—Anterior and Posterior Compartments

Muscle	Origin	Insertion	Primary Action	Innervation
		Anterior Compartment		
Iliopsoas:				
Iliacus	Iliac fossa	Distal to lesser trochanter	Flexes thigh	Femoral n. (L2–L4, posterior)
Psoas	Transverse processes of lumbar vertebrae	Lesser trochanter	Flexes thigh	Lumbar nn. (L2–L5, posterior)
Rectus femoris	Anterior-inferior iliac spine	Tibial tuberosity through patella tendon	Flexes thigh	Femoral n. (L2–L4, posterior)
Sartorius	Anterior-superior iliac spine	Medial tibial head	Flexes, abducts, and laterally rotates thigh	Femoral n. (L2–L3, posterior)
Pectineus	Iliopectineal line and pubis	Base of lesser trochanter	Flexes and adducts thigh	Femoral n. (L2–L3, posterior) Obturator n. (L2–L3, anterior)
		Posterior Compartment		
Gluteus maximus	Posterosuperior ilium and sacrum, associated ligaments and fascia	Gluteal tuberosity of femur and iliotibial tract	Extends, abducts, and laterally rotates thigh	Inferior gluteal n. (L5–S2, posterior)
Adductor magnus:				
Posterior belly	Ischial tuberosity	Adductor tubercle of femur	Adducts, extends, and aids medial rotation of thigh	Tibial n. (L4–L5, anterior)
Semitendinosus	Ischial tuberosity	Medial surface of proximal tibia	Extends and aids medial rotation of thigh	Tibial n. (L5–S1, anterior)
Semimembranosus	Ischial tuberosity	Posterior side of of tibial condyle	Extends and aids medial rotation of thigh	Tibial n. (L5–S1, anterior)
Biceps femoris:				
Long head	Ischial tuberosity	Lateral side of fibular head	Extends and aids lateral rotation of thigh	Tibial n. (L5–S2, anterior)

great as in the shoulder joint, where muscles (not ligaments and bones) stabilize the joint.

b. The **joint capsule** (Fig. 23-2).

(1) The joint capsule attaches to the margins of the acetabular lip and transverse acetabular ligament then extends like a sleeve to the base of the femoral neck.

(2) The inner surface of the capsule is **synovial membrane**.

(3) The outer capsular layer is fibrous.

(a) Some fibers run circumferentially about the neck, the **zonula orbicularis**.

(b) Other groups of fibers are reflected along the neck as **retinacula**, carrying blood vessels to the synovium and femoral head.

(4) A small pouch of synovium, the **psoas bursa**, protrudes through the capsule anteriorly.

(5) External to the fibrous capsule are three fibrous condensations that form the capsular ligaments.

(a) The **iliofemoral ligament** (Y-ligament of Bigelow):

(i) Runs from the anterior-inferior iliac spine to the length of the intertrochanteric line, anteriorly.

(ii) Is especially thick at the edges, giving the appearance of an inverted 'Y'. The lateral limb is termed the **iliotrochanteric band**.

(iii) Resists hyperextension and excess lateral rotation.

(b) The **pubofemoral ligament:**

(i) Runs from the iliopubic ramus to the inferior surface of the intertrochanteric crest.

Table 23-2. Muscles Acting at the Hip Joint—Abductor and Adductor Groups

Muscle	Origin	Insertion	Primary Action	Innervation
		Abductor Group		
Gluteus medius	Superolateral surface of ilium	Greater trochanter	Abducts and aids medial rotation of thigh	Superior gluteal n. (L4–S2, posterior)
Gluteus minimus	Lateral surface of ilium	Greater trochanter	Abducts and aids medial rotation of thigh	Superior gluteal n. (L4–S1, posterior)
Tensor fasciae latae	Anterior-superior iliac spine and iliac crest	Iliotibial tract and lateral fibular head	Abducts and flexes the thigh; rotates thigh medially	Superior gluteal n. (L4–S1, posterior)
Piriformis	Anterior sacrum	Greater trochanter	Laterally rotates thigh; aids abduction of thigh	Nn. to piriformis (S1–S2, posterior)
		Adductor Group		
Adductor magnus:				
Anterior belly	Ischial ramus and pubis	Distal portion of linea aspera	Adducts, flexes, and laterally rotates thigh	Obturator n. (L2–L4, anterior)
Posterior belly	Ischial tuberosity	Adductor tubercle	Adducts, extends, and medially rotates thigh	Tibial n. (L4–L5, anterior)
Adductor longus	Pubis between crest and symphysis	Linea aspera	Adducts thigh, aids flexion and medial rotation of thigh	Obturator n. (L2–L3, anterior)
Adductor brevis	Body and inferior pubic ramus	Proximal portion of linea aspera	Adducts and flexes thigh	Obturator n. (L2–L4, anterior)
Pectineus	Iliopectineal line and pubis	Base of lesser trochanter	Adducts and flexes thigh	Femoral n. (L2–L3, anterior) Obturator n. (L2–L3, anterior)
Gracilis	Ischiopubic ramus	Medial side of tibial head	Adducts and flexes thigh	Obturator n. (L3–L4, anterior)

(ii) Is rather thin, and a tear may result in continuity between the joint capsule and the iliopectineal bursa.
(iii) Resists excess abduction.
(c) The **ischiofemoral ligament:**
(i) Runs from the posterior surface of the acetabulum to the inner aspect of the greater trochanter and the lateral inner surface of the intertrochanteric crest.
(ii) Is the thinnest of the capsular ligaments.
(6) Posterior tilting of the pelvis with assumption of the bipedal posture twists and tightens the capsular ligaments.
(a) Flexion tends to unwind the ligaments so that the hip is least stable in the flexed position.
(b) Posterior dislocation is thus more frequent, as when the flexed thigh comes violently into contact with the dashboard during a car accident.
(7) The **ligamentum teres:**
(a) Runs from the acetabular notch to the femoral fovea.
(b) Plays little, if any, role in the stability of the hip.
(c) Carries the **artery of the ligamentum teres**, which is inconstant but may be an important blood supply to the head.

C. MUSCULATURE OF THE HIP

1. Movement of the thigh at the hip joint is accomplished by muscles that run from the axial skeleton or pelvic girdle across the hip joint to the femur, tibia, or fibula (Fig. 23-3; see Fig. 23-4).

2. There are four groups of muscles acting at the hip joint (see Table 23-1).
a. The primary flexors are mostly primitive dorsal muscles, which lie in the anterior thigh.
b. The primary abductors are mostly primitive dorsal muscles which include the gluteal muscles of the buttock.
c. The primary extensors are mostly primitive ventral muscles (the hamstring group), which lie in the posterior thigh.
d. The primary adductors are mostly primitive ventral muscles and include the adductor group of the medial thigh.
e. In addition, most hip muscles have secondary actions that produce either medial or lateral rotation.

3. All muscles that cross the hip joint act on the thigh.
a. Flexion/extension occurs about a transverse axis through the femoral head.
(1) Flexion (90° with knee extended, 120° with knee flexed) is accomplished by a muscle

Table 23-3. Muscles Acting at the Hip Joint—Lateral Rotators

Muscle	Origin	Insertion	Primary Action	Innervation
Gluteus maximus	Posterosuperior ilium and sacrum	Gluteal tuberosity of femur and iliotibial tract	Extends, abducts, and laterally rotates thigh	Inferior gluteal n. (L5–S2, posterior)
Piriformis	Anterior sacrum	Greater trochanter	Laterally rotates thigh; aids abduction of thigh	Nn. to piriformis (S1–S2, posterior)
Obturator externus	Pubic and ischial rami	Greater trochanter	Laterally rotates thigh; aids abduction of thigh	Obturator n. (L3–L4, anterior)
Gemellus superior	Ischial spine	Greater trochanter through obturator internus tendon	Laterally rotates thigh	Nn. to obturator internus (L5–S2, anterior)
Obturator internus	Pubic and ischial rami	Greater trochanter	Laterally rotates thigh	Nn. to obturator internus (L5–S2, anterior)
Gemellus inferior	Ischial tuberosity	Greater trochanter through obturator internus tendon	Laterally rotates thigh	Nn. to quadratus femoris (L4–S1, anterior)
Quadratus femoris	Ischial tuberosity	Distal to greater trochanter	Laterally rotates thigh	Nn. to quadratus femoris (L4–S1, anterior)

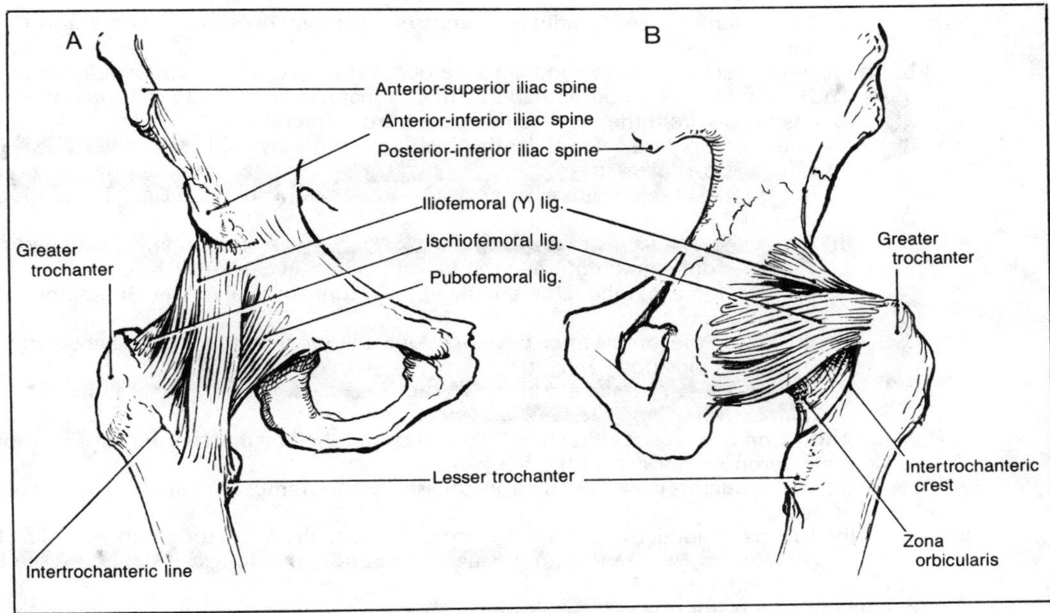

Figure 23-2. *The fibrous joint capsule of the hip joint. A,* Anterior. *B,* Posterior.

group with a line of action passing anterior to the transverse axis of the hip. The major flexor muscle of the hip is the iliopsoas, assisted by the sartorius, rectus femoris, pectineus, and tensor fasciae latae, and to a lesser extent by the adductor longus (Fig. 23-4; see Table 23-1).

(a) With the knee extended, flexion is limited primarily by tension in the hamstring muscles.

(b) With the knee flexed, the hip may be flexed another 30° until it is stopped by apposition of the thigh to the abdominal wall.

(2) Extension (15°) is accomplished by muscles with lines of action that pass posterior to the transverse axis of the hip. These major extensors of the hip include two groups (see Fig. 23-4 and Table 23-1):

(a) The gluteus maximus muscle is the strongest extensor and the one used principally in climbing stairs and rising from the seated position.

(b) The posterior belly of the adductor magnus muscle and the hamstring group (the

Table 23-4. Muscles Acting at the Hip Joint—Medial Rotators

Muscle	Origin	Insertion	Primary Action	Innervation
Gluteus medius	Superolateral surface of ilium	Greater trochanter	Abducts and aids medial rotation of thigh	Superior gluteal n. (L4–S2, posterior)
Gluteus minimus	Lateral surface of ilium	Greater trochanter	Abducts and aids medial rotation of thigh	Superior gluteal n. (L4–S1, posterior)
Semitendinosus	Ischial tuberosity	Medial surface of proximal tibia	Extends and aids medial rotation of thigh	Tibial n. (L5–S1, anterior)
Semimembranosus	Ischial tuberosity	Posterior side of tibial condyle	Extends and aids medial rotation of thigh	Tibial n. (L5–S1, anterior)
Adductor magnus:				
Posterior belly	Ischial tuberosity	Adductor tubercle of femur	Adducts, extends, and aids medial rotation of thigh	Tibial n. (L4–L5, anterior)

biceps femoris, semitendinosus, and semimembranosus muscles) are also extensors.

b. Abduction/adduction occurs about an anteroposterior axis through the femoral head.

(1) Abduction (30°) is accomplished by muscles that pass superior to the anteroposterior axis, generally from the ilium to the greater trochanter.

(a) The abductors are flat muscles that overlap in three layers. These include the tensor fasciae latae, gluteus maximus (superficially), and the gluteus medius, which overlies the gluteus minimus (the most anterior and deep) [see Fig. 23-5 and Tables 23-1 to 23-4].

(b) The abductors are extremely important in pelvic stability during walking because, with the foot planted on the ground, they maintain the pelvis level.

(i) This force of abductor contraction is approximately three times the body weight.

(ii) Within the joint, a force develops equal to four times the body weight (abductor action plus body weight).

(c) The tensor fasciae latae and the gluteus maximus, the superficial muscles of the buttock, insert into the **iliotibial tract**.

(2) Adduction of the abducted limb (20°) is accomplished by muscles that pass inferior to the anteroposterior axis of the hip joint.

(a) The adductors generally run from the ischiopubic ramus to the medial aspect of the femur.

(b) The major adductors of the thigh are included in the adductor group, including the adductor magnus, adductor longus, adductor brevis, and gracilis muscles (see Fig. 23-5 and Table 23-2).

c. Rotation occurs about a vertical axis through the femoral head.

(1) Lateral rotation (eversion) [30°] is accomplished by muscles that act posterior to the vertical axis.

(a) The principal lateral rotators comprise the pelvotrochanteric group (the piriformis, obturator internus, superior and inferior gemelli, and the obturator externus) assisted by the quadratus femoris. The posterior portion of the gluteus maximus and the sartorius muscles also have external rotatory function.

(b) These muscles pass nearly transversely, generally from the ischium or through its greater sciatic notch to insertions in the vicinity of the greater trochanter and intertrochanteric crest.

(2) Medial rotation (inversion) [60°] is accomplished by muscles that act anterior to the vertical axis of the hip joint. There are no specific internal rotators, but several muscles have secondary medial rotatory actions such as the anterior portion of the gluteus minimus and tensor fasciae latae.

d. Circumduction is produced by simultaneous movement about two or more axes.

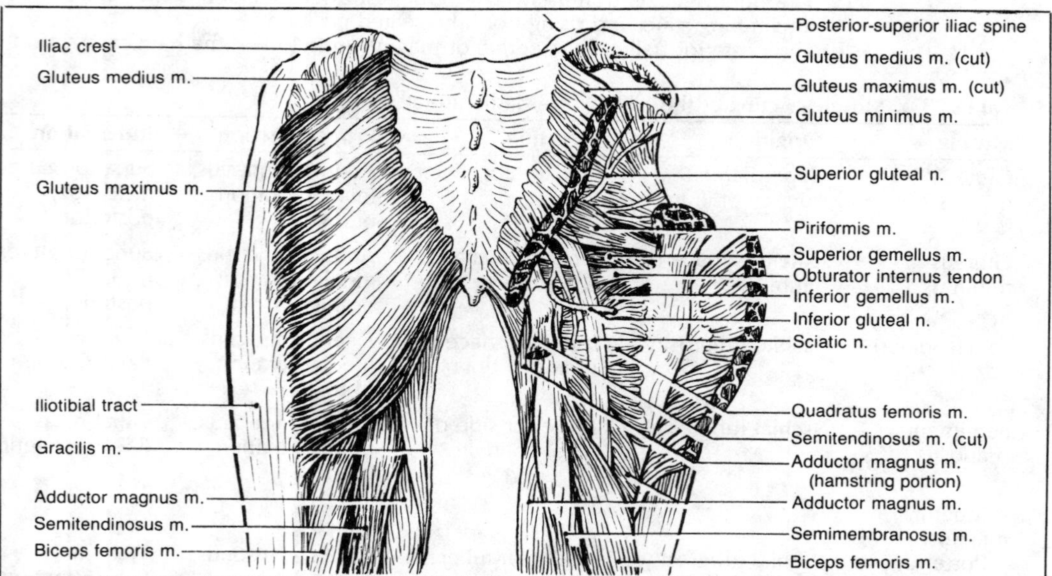

Figure 23-3. *The principal tensor muscles of the thigh.* The right femoral neurovascular bundles lie in the femoral triangle.

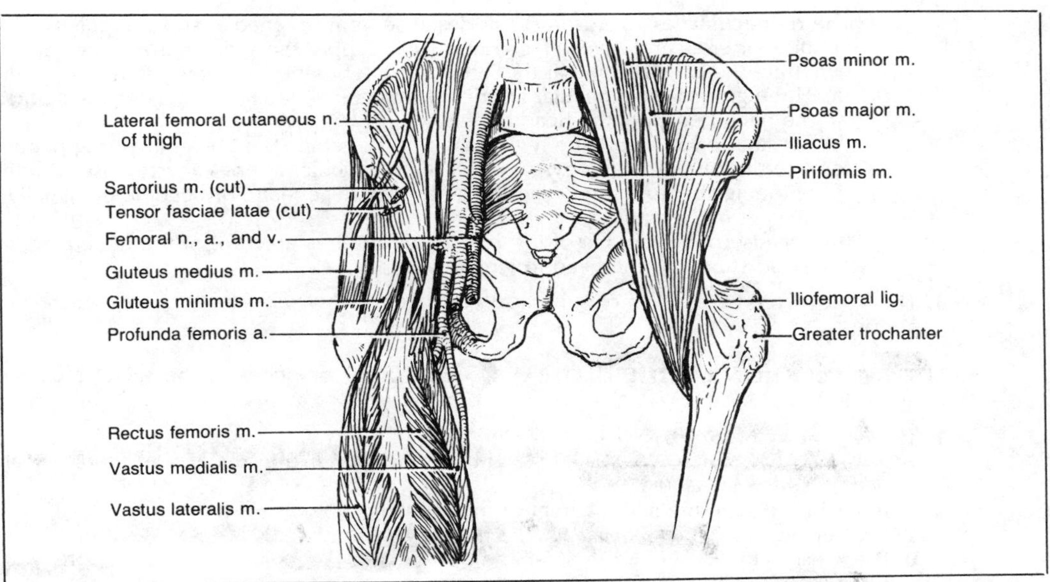

Figure 23-4. *The gluteal musculature.* The right sciatic nerve passes through the medial inferior quadrant of the buttock.

D. THE VASCULATURE OF THE GLUTEAL REGION AND HIP (see Chapter 20 IV).

1. The hip region receives blood supply from branches of the internal and external iliac arteries (Fig. 23-5).

a. The **superior gluteal artery** is a branch of the posterior trunk of the internal iliac artery.

(1) It leaves the pelvic cavity via the suprapiriform portion of the greater sciatic foramen along with the superior gluteal nerve, just deep to the posterior edge of the gluteus medius muscle.

(2) The main trunk passes anteriorly deep to the gluteus medius muscle.

b. The **inferior gluteal artery** is a branch of the anterior trunk of the internal iliac artery.

(1) It leaves the pelvic cavity via the infrapiriform region of the greater sciatic foramen along with the inferior gluteal nerve, pudendal nerve, and sciatic nerve.

(2) It passes anteriorly on the deep surface of the gluteus maximus muscle.

(3) It supplies portions of the gluteus maximus muscle, the lateral rotators of the hip, and and the most proximal portions of the hamstring group and sends twigs to the hip joint.

(4) It anastomoses with branches of the profundus femoris artery, forming part of the **cruciate anastomosis.**

c. The **obturator artery** is most often a branch of the anterior trunk of the internal iliac artery.

(1) Frequently (30 percent) this artery arises from the inferior epigastric artery and descends to the obturator foramen close to the femoral ring where it may complicate surgical repair of a femoral hernia.

(2) It leaves the pelvic cavity via the obturator foramen.

(3) It supplies the adductor group and sends twigs to the joint, including the artery of the ligamentum teres.

d. The **femoral artery**, the continuation of the external iliac artery, gives off the **profunda femoris artery** from which the **medial** and **lateral circumflex femoral arteries** arise.

(1) These supply the proximal portions of the more lateral and anterior muscles of the thigh and are the major supply to the trochanters, neck, and head of the femur.

(2) The two circumflex arteries pass transversely to anastomose with each other and with branches of the first perforating artery (inferiorly) and with the inferior gluteal artery (superiorly), forming the **cruciate anastomosis** on the posterolateral surface of the femur.

2. The blood supply to the proximal femur is an important consideration (see Fig. 23-5).

a. The nutrient artery, a branch of the profunda femoris artery, supplies the shaft.

b. The obturator, medial circumflex femoral, and lateral circumflex femoral arteries supply the trochanters and the neck and head of the femur.

(1) The **artery of the ligamentum teres**, a branch of the obturator artery, is of principal importance in children, in whom it supplies the region of the head proximal to the epiphyseal plate.

(2) The **retinacular (extracapsular) arteries** arise from the medial and lateral femoral circumflex arteries and give off branches that supply the greater trochanter and that penetrate the joint capsule at the intertrochanteric line to supply the neck.
(a) Three groups of retinacular arteries, with great variability, run along the surface of the neck to supply the head.
(b) These arteries supply the region distal to the epiphyseal plate in young people.
(c) After fusion of the epiphyseal plate, the retinacular arteries anastomose with those of the ligamentum teres, and the latter may degenerate or become marginally significant in the adult.
(d) Because of the course of the retinacular arteries, fracture of the neck places the blood supply to the head in jeopardy.

3. The **venous return** consists of superfical and deep pathways. The deep veins parallel the arteries.

E. THE INNERVATION OF THE GLUTEAL REGION is by branches of the sacral plexus (see Chapter 20 V).

1. The **superior gluteal nerve** (L4–S1, posterior).
   a. This nerve exits the pelvis via the suprapiriform portion of the greater sciatic foramen along with the superior gluteal vessels.
   b. It lies in the supermedial quadrant of the buttock.
   c. It innervates the gluteus medius, gluteus minimus, and tensor fasciae latae muscles.
   d. Palsy results in "abductor lurch," a rolling (Trendelenburg) gait because of loss of stability of the pelvis.

2. The **inferior gluteal nerve** (L5–S2, posterior).
   a. This nerve exits the pelvis via the infrapiriform portion of the greater sciatic foramen, along with the inferior gluteal vessels and the sciatic nerve.
   b. It lies in the inferolateral quadrant of the buttock.
   c. It innervates the gluteus maximus muscle.
   d. Palsy of this nerve results in difficulty in rising from a seated position and in climbing stairs, due to weakness of hip extension.

3. The **sciatic nerve.**
   a. It passes through the gluteal region but does not contribute to the gluteal innervation.
   b. It exits from the pelvis via the infrapiriform recess of the greater sciatic foramen, along with the inferior gluteal neurovascular bundle.
   c. It courses in an arc, starting halfway between the ischial tuberosity and the iliac spine. It passes over the neck of the femur halfway between the ischial tuberosity and the greater trochanter and lies in the inferomedial quadrant of the buttock.

F. STRUCTURES LEAVING THE DEEP PELVIS THROUGH THE GREATER SCIATIC NOTCH include (from cranial to caudal) the superior gluteal neurovascular bundle, the piriformis muscle, the inferior gluteal neurovascular bundle, the sciatic nerve, the internal pudendal vessels, and the pudendal nerve (see Figs. 23-3 and 23-4).

G. CLINICAL CONSIDERATIONS

1. A stable hip is essential as a fulcrum for the action of abduction. In **congenital dislocation** of the hip, the femoral head lies outside of the acetabulum, and a failure of abduction results in the waddling (Trendelenburg) gait.

2. In children, the femoral head gets its arterial supply from the artery of the ligamentum teres. Elevated intracapsular pressure, as a result of **septic arthritis**, may compress the artery and produce ischemic necrosis of the head.

3. **Fractures of the femoral neck** occur frequently, particularly in the elderly.
   a. Intracapsular fractures may be subcapital or midcervical.
      (1) In the elderly with resorption and weakening of the bony trabeculae, these fractures may be the result of a fall or the result of muscle spasm with the fall subsequent to the fracture.
      (2) Intracapsular fractures endanger the blood supply to the proximal fragment, so that nonunion and avascular necrosis may result, depending upon the degree to which the retinacular arteries are torn and upon the effectiveness of the artery of the ligamentum teres.
      (3) So common is this complication that in an elderly patient the initial treatment of choice is joint replacement.

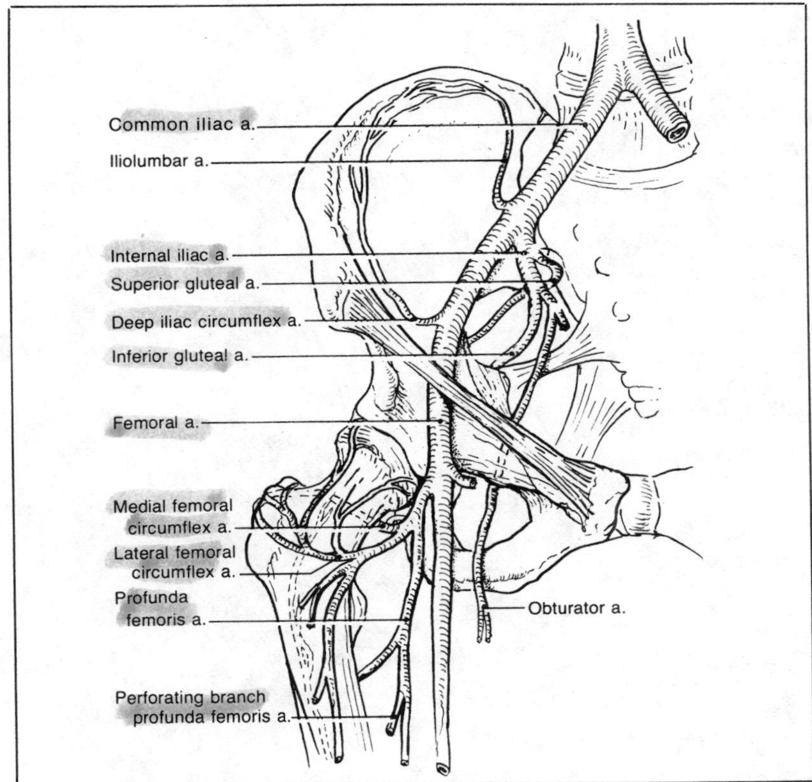

Figure 23-5. *The vasculature of the thigh and gluteal region.*

b. Extracapsular fractures, intertrochanteric and subtrochanteric, occur in a region with abundant anastomotic blood supply and present few problems with healing once reduction and internal fixation have been achieved.

c. The greater trochanter and lesser trochanter have separate ossification centers and may be avulsed along the epiphyseal plates prior to age 18 or 19, especially in heavy teenage males.

d. The femoral shaft is especially strong, and fractures of the proximal femoral shaft are the result of violent trauma.

4. Dislocations of the hip.

a. Because the hip joint is so stable, dislocation is usually produced by trauma severe enough to fracture the acetabulum.

b. The hip joint is least stable in the flexed position, which slackens the ligaments of the fibrous capsule.

(1) In the more common posterior dislocation, the femoral head comes to lie posterior to the iliofemoral ligament, either superior or inferior to the obturator internus tendon. The sciatic nerve is usually damaged.

(2) In anterior dislocation, the femoral head comes to lie anterior to the iliofemoral ligament, either superiorly against the superior public ramus or inferiorly in the obturator foramen.

c. If the articular capsule is torn as a result of dislocation, the blood supply to the head could be jeopardized.

5. Osteoarthritis.

a. Progressive degeneration of the articular cartilage results in pain and limited range of movement.

b. A cane used as support will be held in the hand opposite the affected hip. When the unaffected leg is raised to step, the arm and cane balance the trunk. Abductor action to stabilize the pelvis is therefore unnecessary and the weight-bearing joint is relieved of as much as 75 percent of the compressive loading and the resultant pain.

24 The Thigh and Knee Joint

I. THE THIGH

A. BONY LANDMARKS (Fig. 24-1).

1. The distal **femur.**
   a. Shaft.
   b. Medial epicondyle.
   c. Lateral epicondyle.
2. The **patella**
3. The proximal **tibia.**
   a. Medial condyle
   b. Anterior lip of the tibial plateau
   c. Tibial tubercle
   d. The medial surface of the shaft
   e. Medial malleolus
4. The proximal **fibula.**
   a. Head
   b. Lateral collateral ligament
   c. Lateral malleolus

B. THE DISTAL FEMUR

1. A ridge, the **linea aspera**, runs along the posterior aspect of the femur.
2. Distally, the femoral shaft broadens transversely to provide a stable weight-bearing articulation with the tibia.
   a. The articular surface is divided inferiorly and posteriorly by a deep **intercondylar notch (fossa)** between the lateral and medial condyles.
   b. The **trochlear groove**, the patellar articular surface, bridges between the condyles and gives the articular surfaces of the anterior aspect of the distal femur a U-shaped appearance. However, the patellar articular surface is separated from the tibial articular surface by a faint groove in each condyle.
   c. The **lateral condyle** is wider than the medial condyle. It lies along the principal line of force and is probably more important in weight bearing.
   d. The **medial condyle** has a longer tibial articular surface than the lateral condyle.
      (1) This results in a lateral rotation of the tibia with reference to the femur during the terminal phase of extension.
      (2) The medial condyle usually projects slightly further distally than the lateral condyle, producing the physiologic inward angle at the knee, valgus, which is more pronounced in the female (170°) than in the male (175°).
         (a) Exaggerated inward angling is termed *genu valgum* or "knock-knees."
         (b) Exaggerated outward angling is termed *genu varum* or "bow-legs."
   e. The condyles form part of a spiral so that the radii of the curvature forms a spiral-shaped evolute of rotation rather than a simple axis.
      (1) As the knee is extended, the radii of curvature of both condyles increase. The tibia is effectively pushed distally. This tends to tighten the ligaments that connect the tibia and femur.
         (a) The radii are shorter posteriorly (17 mm medial, 12 mm lateral).

(b) The radii are longer distally (38 mm medial, 60 mm lateral).

(2) In flexion the transverse axis of rotation passes more posteriorly through the femoral condyles. Ligaments that attach to the femur anteriorly are therefore taut in flexion.

(3) In extension the transverse axis of rotation passes more anteriorly through the femoral condyles. Ligaments that attach to the femur posteriorly are therefore taut in extension.

f. The **lateral epicondyle**, a bony ridge lateral to the lateral condyle, serves as the attachment for the **lateral (fibular) collateral ligament** and the lateral head of the gastrocnemius muscle.

g. The **medial epicondyle**, a bony ridge medial to the medial condyle, serves as a point of attachment for the **medial (tibial) collateral ligament** and the medial head of the gastrocnemius muscle. Superior to the medial epicondyle is the **adductor tubercle**, where the posterior belly (hamstring portion) of the adductor magnus attaches.

C. THE PROXIMAL TIBIA

1. Medial and lateral tibial condyles form the tibial plateau, which articulates with the femur, the **femorotibial joint**.
   a. The **medial tibial condyle** is concave in both the frontal and sagittal planes. Like the medial femoral condyle, it is longer in the anteroposterior diameter.
   b. The **lateral tibial condyle** is concave in the frontal plane but convex in the sagittal plane and more nearly round.
   c. Between the tibial condyles, the **intercondylar eminence** lodges between the femoral condyles and acts as a pivot for rotation about a vertical axis.
   d. The major lack of congruency between the tibial and femoral condyles is compensated for by the fibrocartilage menisci, which become slightly distorted as flexion/extension occurs.

2. The **proximal (superior) tibiofibular joint.**
   a. The tibial head articulates with the fibula laterally, just inferior to the lateral condyle.
      (1) It is a gliding joint with a small synovial cavity.
      (2) There is extremely limited movement.
   b. An **interosseous membrane** connects the fibula and tibia along their shafts.
      (1) The connective tissue fibers run inferiorly and laterally between the interosseous borders of the tibia to the interosseous border of the fibula.
      (2) The interosseous membrane functions as a shock absorber and helps to stabilize the ankle joint.

3. On the anterior tibial surface of the lateral condyle, an elevation serves as insertion for the iliotibial tract.

4. The **tibial tuberosity**, the insertion for the patellar tendon, lies inferior to the head on the anterior surface of the shaft.

5. The tibial shaft is triangular in shape and tapers distally towards the medial malleolus.

D. THE PATELLA is a sesamoid bone in the tendon of the quadriceps femoris muscle group.

1. It is triangular in shape with the apex directed inferiorly.

2. It articulates with the femur at the **femoropatellar joint**.

3. The **patellar tendon** (often termed ligament) runs between the apex and the tibial tuberosity.
   a. Developing within the quadriceps femoris tendon, the patella is encapsulated in dense connective tissue, the **patellar retinaculum**.
      (1) The patellar retinaculum extends inferolaterally and inferomedially to insert into the tibial condyles, representing a broad tendinous insertion for the quadriceps femoris.
      (2) It reinforces the capsule of the knee joint.
   b. The subcutaneous bursae lie anterior to the knee.
      (1) The **prepatellar bursa** is superficial to the patella and deep to the epithelium.
      (2) The **infrapatellar bursa** lies between the tibial tuberosity and the epithelium.
      (3) These are subject to inflammation with repeated abuse—housemaid's knee (prepatellar bursitis) and vicar's knee (infrapatellar bursitis).

4. The **femoropatella joint** is a gliding joint.
   a. A vertical ridge divides the posterior surface of the patella into medial and lateral articular surfaces.
   b. The lateral and medial articular surfaces of the patella glide on the **trochlear groove**, the patellar articular surface of the femur.
   c. In flexion the tension from the quadriceps femoris tendon and the patellar retinaculum tends to keep the patella deep in the trochlear groove between the femoral condyles.

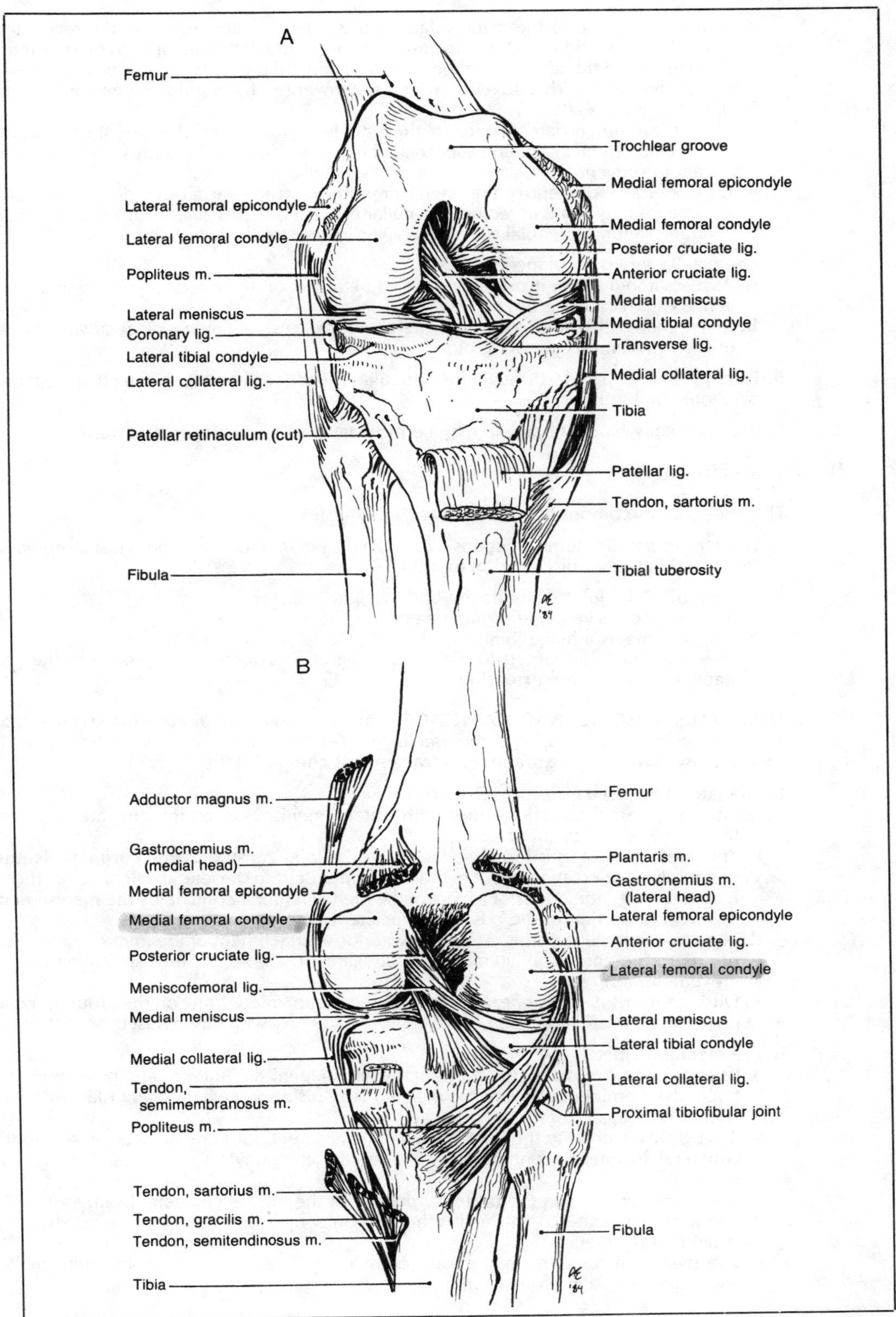

Figure 24-1. *The knee joint. A,* Anterior. *B,* Posterior.

d. In extension, due to the usual valgus at the knee, the line action of the rectus femoris, vastus lateralis, and vastus intermedius muscles (through the quadriceps femoris tendon to the patella and thence through the patellar tendon to the tibial tuberosity) is somewhat lateral to the patella. This, together with a shallower trochlear groove in extension, tends to draw the patella laterally.

(1) The prominent lateral ridge of the trochlear groove and the lateral epicondyle, the patellar retinaculum, and especially the line of action of the vastus medialis muscle all act to prevent lateral dislocation of the patella.

(2) Underdevelopment of the lateral ridge of the trochlear groove or exaggerated genu valgum may result in recurrent patellar dislocation. This can be corrected surgically by transplanting the tibial tuberosity slightly more medially and inferiorly.

5. The patella serves two functions.

a. As a sesamoid bone, it obviates wear and attrition on the quadriceps tendon as it passes across the trochlear groove.

b. It lengthens the lever arm and thereby increases the mechanical advantage of the quadriceps femoris muscle group.

6. Fracture of the patella through the articular surface will result in eventual degenerative patellofemoral arthritis.

7. A congenitally bipartate patella may be misdiagnosed as a patellar fracture.

II. THE KNEE JOINT

A. The knee joint is composed of two joints (Fig. 24-2).

1. The **femoropatellar joint** functions as a pulley (trochlea) for the quadriceps femoris muscle group and has been discussed before (I D 4).

2. The **femorotibial joint** supports the body weight.

a. It has an extensive range of movement.

b. It approximates a hinge joint.

c. The knee joint is adapted to weight bearing at any degree of flexion, but the greatest stability occurs in full extension.

B. THE MENISCI (SEMILUNAR CARTILAGES) are composed of fibrocartilage rings that are triangular in cross section. The menisci incompletely divide the femorotibial joint into a **suprameniscal compartment** and an **inframeniscal compartment**.

1. The **lateral meniscus** is nearly O-shaped.

a. The anterior and posterior horns on the lateral meniscus insert into the lateral tubercle of the intercondylar eminence.

b. The lateral meniscus is loosely attached to the joint capsule by the **coronary ligament**.

c. There is little or no attachment of the lateral meniscus to the lateral collateral ligament, but it is attached to the posterior aspect of the medial femoral condyle by the **meniscofemoral ligament**, a portion of the posterior cruciate ligament.

d. The close insertions of this meniscus, with loose attachment of the joint capsule and little or no attachment to the lateral collateral ligament, allows this meniscus a fair amount of movement.

e. During the last degrees of extension with the foot planted firmly on the ground, the femur pivots and the lateral femoral condyle slides anteriorly with this meniscus to lock the knee.

2. The **medial meniscus** is C-shaped.

a. Its ends insert external to the horns of the lateral meniscus anterior and posterior to the intercondylar eminence, giving the medial meniscus a somewhat greater diameter than the lateral meniscus.

b. It is tightly attached to the joint capsule by a ligament (formerly named the **short internal collateral ligament**), running from the medial epicondyle to the lateral margin of the meniscus.

c. It is tightly attached to the posterior portion of the medial collateral ligament.

d. The widely separated points of attachment confer upon this meniscus a severely restricted range of movement.

e. The medial attachment to the joint capsule and medial collateral ligament make this meniscus especially prone to injury.

C. THE JOINT CAPSULE AND LIGAMENTS

1. The **synovial capsule** of the knee joint is extensive.

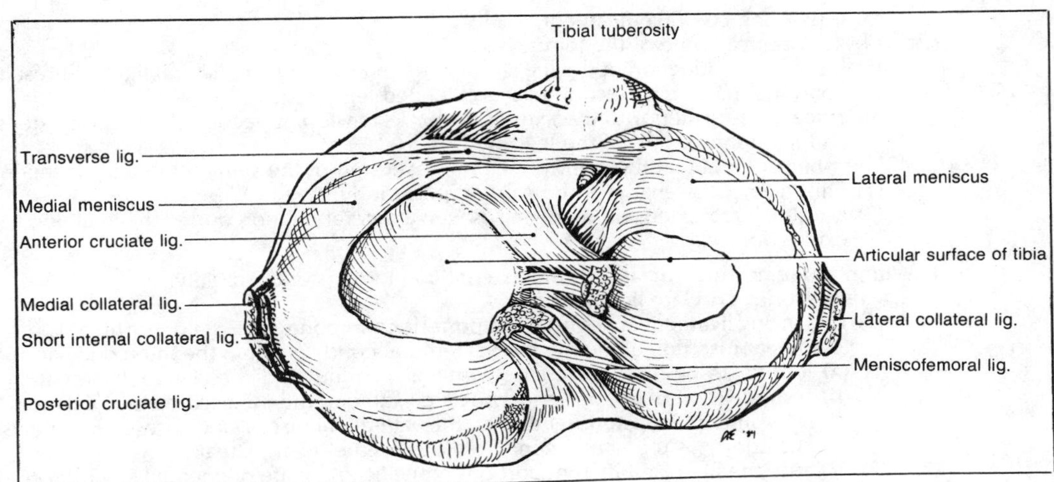

Figure 24-2. *The articular cartilages and ligaments of the knee joint.*

a. The synovial cavity proper is more or less inserted about the margins of the articular surfaces.
b. Medially, the capsule attaches to the medial meniscus.
c. Posteriorly, the capsule invaginates into the intercondylar notch so that the cruciate ligaments are technically extracapsular.
d. The **suprapatellar bursa** extends proximally from the capsule between the quadriceps femoris tendon and the femur.
   (1) Effusions of the knee joint will distend this bursa and will be evident upon examination (water on the knee).
   (2) When postinflammatory adhesions develop in the suprapatellar bursa, the mobility of the patella with tibial extension is decreased (stiff knee).
e. Numerous other synovial bursae about the knee may or may not communicate with the joint capsule.
f. When effusion distends the joint capsule, the knee assumes approximately 20° of flexion.

2. The **fibrous capsule** of the knee is rather weak.
   a. The **short internal collateral ligament**, a thickening of the fibrous capsule deep to the medial collateral ligament, extends from the medial epicondyle to the medial meniscus.
   b. The **coronary ligament** is the thickened periphery of the capsule that loosely attaches the two menisci to the tibia. Frequently (60 percent) it is thickened anteriorly to form the **transverse ligament**, which connects the anterior margins of the menisci.

3. Four strong external ligaments reinforce the fibrous capsule of the knee joint.
   a. The **patellar retinaculum**, extending into the tibial condyles, reinforces the joint capsule medially and laterally while the patellar tendon reinforces the capsule anteriorly.
   b. The **medial (tibial) collateral ligament.**
      (1) Phylogenetically, this ligament is the degenerated tendon of insertion of the posterior portion of the adductor magnus muscle.
      (2) It extends from the medial femoral epicondyle and widens to insert onto the shaft of the tibia below the level of the tibial tuberosity.
      (3) It is broad and fan-shaped, so that a portion of this ligament can be effective at any degree of flexion.
      (4) The posterior portion is fused with the fibrous joint capsule and is attached to the medial meniscus.
      (5) It prevents abduction of the leg at the knee. A torn medial collateral ligament results in abnormal passive abduction of the extended leg.
      (6) Since it lies posterior to the axes of flexion/extension, it becomes taut upon extension and thus limits extension of the leg.
   c. The **lateral (fibular) collateral ligament.**
      (1) Phylogenetically, this ligament is the degenerated tendon of origin of the peroneus longus muscle.
      (2) It extends between the lateral femoral epicondyle to the head of the fibula. It is not quite parallel to the medial collateral ligament.

(3) It is strong cord-like ligament.
(4) It is relatively free of the joint capsule.
(5) It prevents adduction of the leg at the knee. A torn lateral collateral ligament results in abnormal passive adduction of the extended leg.
(6) Since it lies posterior to the axes of flexion/extension, it becomes taut upon extension and lax upon flexion of the leg.

d. The **oblique popliteal ligament** is a lateral extension of the semimembranosus tendon.
(1) This ligament strengthens the capsule posteriorly.
(2) It resists hyperextension of the leg as well as lateral rotation during the final degrees of extension.

4. Within the joint, the **cruciate ligaments** confer anteroposterior stability.

a. The **posterior cruciate ligament.**
(1) This strong ligament arises from the posterior intercondylar fossa of the tibia. It fans out to a linear insertion along the medial femoral condyle within the intercondylar notch.
(a) It tends to be more or less perpendicular to the medial collateral ligament.
(b) A subdivision of the posterior cruciate ligament, the **meniscofemoral ligament**, inserts into the posterior horn of the lateral meniscus and draws this meniscus posteriorly as the lateral condyle slides on the tibial plateau.
(2) Because the fibers in different portions of the ligament are of unequal length, some are always in a stage of tension. Fibers inserting anteriorly on the femoral condyle tend to be taut in flexion whereas those inserting posteriorly tend to be taut in extension.
(3) The posterior cruciate ligament is a key stabilizer of the knee joint; it checks anterior movement of the femur on the tibial plateau (posterior movement of the tibia with respect to the femur).
(a) In the flexed weight-bearing knee, such as upon descending stairs or walking downhill, this ligament prevents the femur from sliding forward on the tibial plateau.
(b) A torn posterior cruciate results in posterior displacement of the tibia (the posterior drawer sign).
(c) A torn posterior cruciate will produce a perceived instability such that the individual upon descending stairs will lead with the opposite leg for each step.

b. The anterior cruciate ligament.
(1) This strong ligament arises from the anterior intercondylar fossa of the tibia and runs posterolaterally to insert onto the lateral femoral condyle posteriorly within the intercondylar notch.
(a) It tends to be perpendicular to the lateral collateral ligament.
(b) It is shorter than the posterior cruciate ligament.
(2) It checks posterior movement of the femur on the tibial plateau (anterior movement of the tibia with reference to the femur).
(a) Because the fibers in different portions of the ligament are of unequal length, some are always taut regardless of the degree of flexion.
(b) Since this ligament inserts posteriorly on the femur, it maximally tightens upon extension.
(c) If torn, the leg may not lock upon full extension.
(d) A torn anterior cruciate ligament produces the anterior displacement on the tibia (the anterior drawer sign). However, this instability may be noted only occasionally when the knee suddenly gives way.

5. Rotational stability of the knee joint is conferred by the collateral and cruciate ligaments.
a. Due to the direction of the cruciate ligaments (slightly coiled inward), lateral rotation of the femur tightens the coils, preventing further lateral femoral rotation of the subterminally extended knee.
b. Due to the direction of the collateral ligaments (slightly coiled outward), medial rotation of the femur tightens the coils and prevents further medial rotation.

6. Because extension of the knee is checked by the collateral and cruciate ligaments, in the erect posture with the knee extended and locked, the weight of the body at the knee is stabilized primarily by ligaments.

III. THE MUSCULATURE OF THE THIGH

A. THE FASCIA LATA is the deep fascia of the thigh.

1. Laterally, the fascia lata is thickened from the iliac crest to the lateral tibial condyle, forming the **iliotibial tract (band)**.

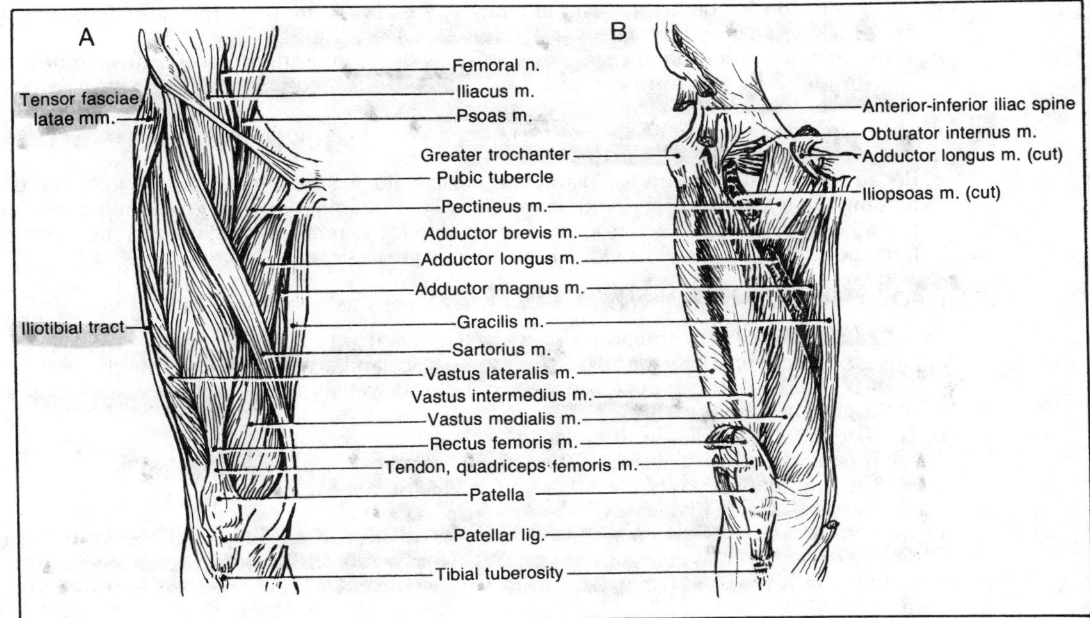

Figure 24-3. *The musculature of the anterior compartment of the thigh. A,* The superficial muscles. *B,* The deep muscles.

a. The iliotibial tract is the tendon of insertion for the tensor fasciae latae muscle and the gluteus maximus muscle.

b. Because the tensor fasciae latae and gluteus maximus muscles originate on the pelvis and pass over the hip joint, they function in extension of the thigh.

c. Because the iliotibial tract passes over the knee joint, the tensor fasciae latae and gluteus maximus muscles function in both extension and flexion of the leg at the knee joint.

(1) When the knee is extended, the line of action of the iliotibial tract passes anterior to the axis of rotation of the knee joint and the tensor fasciae latae and gluteus maximus muscles therefore act to extend the knee.

(2) When the knee is slightly flexed, the line of action of the iliotibial tract moves posterior to the axis of rotation for flexion/extension of the knee joint and the tensor fasciae latae and gluteus maximus muscles become flexors of the knee joint.

(3) This is evident when a person, standing erect and somewhat relaxed, has the knees knocked from behind. The line of action of the iliotibial tract is shifted across to the axis of rotation for flexion/extension so that the tensor fasciae latae and gluteus maximus muscles, which act to keep the knee extended in the normal erect posture, now act to flex the knee and produce the disconcerting sinking.

2. Fascial septa from the fasciae latae divide the thigh musculature into anterior and posterior compartments.

a. The **lateral intermuscular septum** arises from the posterior border of the iliotibial tract and inserts onto the lateral lip of the linea aspera. This provides an important insertion onto the femur for the tensor fasciae latae and the gluteus maximus muscles.

b. The **medial intermuscular septum** arises from the anteromedial border of the iliotibial tract and attaches to the medial lip of the linea aspera.

c. Because the leg rotates anteriorly during development, the primitively dorsal musculature comes to lie in the anterior compartment. Likewise, the primitively ventral musculature comes to lie in the posterior compartment.

B. THE ANTERIOR COMPARTMENT MUSCULATURE is divided into two groups.

1. The **iliopsoas group** acts at the hip (Fig. 24-3).

a. This group consists of the iliacus and psoas muscles.

b. They arise from the iliac fossa and lumbar spine and insert into the lesser trochanter on the the shaft of the femur, anteromedially.

c. They function in flexion of the hip and vertebral column.

d. They are innervated by twigs from the second through fourth lumbar nerves.

2. The **anterior group** acts primarily at the knee (see Fig. 24-3).

a. This group consists of the sartorius and quadriceps femoris muscles, the latter consisting of the rectus femoris and the three vasti (lateralis, intermedius, and medialis).
b. They generally arise from the anterior portion of the ilium (sartorius and rectus femoris) or the femoral shaft (the vasti) and insert into the proximal tibia through the patella, except that the sartorius inserts into the medial tibial condyle.
c. They function primarily to extend the knee, except the sartorius, which flexes the knee.
d. They are innervated by the femoral nerve.
e. Because the rectus femoris is a biarticular muscle (flexing the thigh at the hip joint and extending the leg at the knee joint), its efficiency at one joint depends upon the position of the other joint—that is, when the hip is extended, the rectus femoris is stretched and its force-generating capacity to extend the knee is greatest (see Chapter 3 III C).

C. THE POSTERIOR COMPARTMENT MUSCULATURE consists of two groups.

1. The **adductor group** acts primarily at the hip (Fig. 24-4).
   a. This group consists of the gracilis, adductor longus, and adductor brevis muscles and the anterior portion of the adductor magnus muscle, as well as the obturator externus muscle and a portion of the pectineus muscle.
   b. The adductor group lies in three layers.
      (1) The anterior layer consists of the gracilis, pectineus, and adductor longus.
      (2) The intermediate layer consists of the adductor brevis muscle.
      (3) The adductor magnus muscle lies most posteriorly.
   c. The adductor group generally arises from the inferior pubic rami and inserts into the femoral shaft medially, except the gracilis, which inserts into the medial tibial condyle.
   d. Its function is primarily adduction of the hip; only the gracilis has an action at the knee joint.
   e. It is innervated by the obturator nerve.
2. The **hamstring group** acts both at the hip and the knee (see Fig. 24-4).
   a. This group consists of the semitendinosus, the semimembranosus, the long head of the biceps femoris, and the posterior portion of the adductor magnus.
   b. It arises from the ischial tuberosity. Phylogenetically, the sacrotuberous ligament is considered to be a remnant of the hamstring group. The group inserts into the tibial condyles, except the deep portion of the adductor magnus, which inserts onto the adductor tubercle of the medial femoral epicondyle.
      (1) The semitendinosus and semimembranosus muscle tendons join the **pes anserinus** (*L.* goose foot) and the gracilis tendon to insert just distal to the medial tibial condyle.
      (2) The long and short heads of the biceps femoris muscle insert laterally onto the head of the fibula.

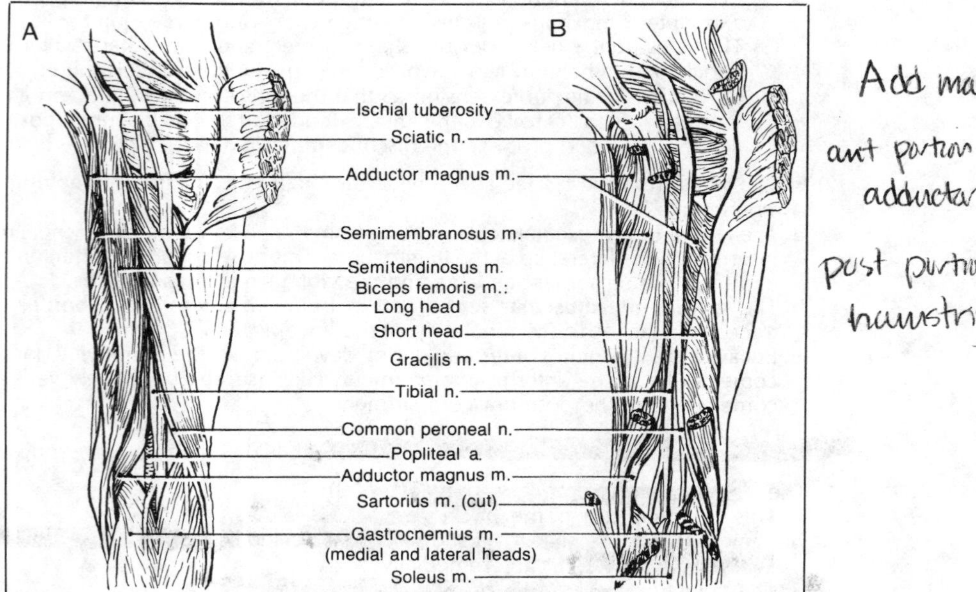

Figure 24-4. *The muscles of the posterior compartment of the thigh. A,* The superficial muscles. *B,* The deep muscles.

Add magn
ant portion - adductor
post portion hamstring

c. Because the hamstrings cross the hip joint posteriorly, they extend the thigh; because the semitendinosus, semimembranosus, and biceps femoris cross the knee joint posteriorly, they flex the leg.

d. The hamstring group is innervated by the **tibial nerve**, the anterior portion of the **sciatic nerve**. The short head of the biceps femoris is not technically a part of the hamstring group and is innervated by the **common peroneal nerve**, the posterior portion of the sciatic nerve.

e. Because the hamstring muscles are biarticular (extending the thigh at the hip and flexing the leg at the knee), their efficiency at the knee is greatest when the hip is flexed (Chapter 3 III C).

IV. MOVEMENTS AT THE KNEE JOINT

A. THE AXES OF THE KNEE JOINT

1. There are three groups of muscles acting at the knee joint.
   a. The primary extensors are primitive dorsal muscles that come to lie in the anterior compartment of the thigh (Table 24-1).
   b. The primary flexors are primitive ventral muscles that come to lie in the posterior compartment of the thigh (see Table 24-1).

Table 24-1. Thigh Muscles Acting at the Knee Joint—Anterior Compartment

Muscle	Origin	Insertion	Primary Action	Innervation
		Anterior Compartment		
Rectus femoris	Anterior-inferior iliac spine	Tibial tuberosity through patellar tendon	Extends knee	Femoral n. (L2–L4, posterior)
Vastus lateralis	Proximal femur lateral to linea aspera	Tibial tuberosity through patellar tendon	Extends knee	Femoral n. (L2–L4, posterior)
Vastus medialis	Proximal femur medial to linea aspera	Tibial tuberosity through patellar tendon	Extends knee	Femoral n. (L2–L4, posterior)
Vastus intermedius	Anterior aspect of proximal femur	Tibial tuberosity through patellar tendon	Extends knee	Femoral n. (L2–L4, posterior)
Sartorius	Anterior-superior iliac spine	Medial tibial side of tibial head	Flexes knee	Femoral n. (L2–L3, posterior)
		Posterior Compartment		
Semitendinosus	Ischial tuberosity	Medial surface of proximal tibia	Flexes knee	Tibial n. (L5–S1, anterior)
Semi-membranosus	Ischial tuberosity	Posterior side of tibial condyle	Flexes knee	Tibial n. (L5–S1, anterior)
Biceps femoris: Long head	Ischial tuberosity	Lateral side of fibular head	Flexes knee	Tibial n. (L5–S2, anterior)
Short head	Femoral shaft	Lateral side of fibular head	Flexes knee	Common peroneal n. (L5–S2, posterior)
Gracilis	Ischiopubic ramus	Medial side of tibial head	Flexes knee	Obturator n. (L3–L4, anterior)
Gluteus maximus	Posterosuperior ilium and sacrum	Gluteal tuberosity and linea aspera of femur and iliotibial band to fibular head	Extends extended knee and flexes flexed knee	Inferior gluteal n. (L5–S2, posterior)
Tensor fasciae latae	Anterior-superior iliac spine and iliac crest	Iliotibial band to fibular head	Extends extended knee and flexes flexed knee	Superior gluteal n. (L4–S1, posterior)

c. Secondary flexors, primitive ventral muscles of the posterior compartment of the leg, include the popliteus and plantaris muscles, as well as the two heads of the gastrocnemius muscle.

2. All muscles that cross the knee act on the leg.

a. Flexion/extension (140°) occurs about a transverse axis through the femoral condyles.

(1) Flexion is accomplished by the hamstring group, which has a line of action that passes posterior to the axis of the knee joint. The major flexors of the hip include the hamstring muscles of the posterior compartment of the thigh and the sartorius, gracilis, tensor fasciae latae, and gluteus maximus of the anterior, anteromedial, and abductor compartments, respectively.

(2) Extension is accomplished by the quadriceps femoris (rectus femoris and the three vasti), which has a line of action anterior to the axis of the knee joint. In addition, upon full extension, the line of action of the tensor fasciae latae and gluteus maximus muscles through the iliotibial tract passes anterior to the axis of rotation of the joint, maintaining the knee in the locked position.

b. Rotation (20°) occurs about a vertical axis that passes through the intercondylar tubercle. Rotation occurs concomitant with flexion/extension. While the knee cannot be actively rotated, the flexed knee may be passively rotated through 70°.

(1) Lateral tibial rotation occurs during the terminal phase of extension, mainly as a result of the differences in the lengths and shapes of the medial and lateral condyles and the limits on movement imposed by the cruciate and collateral ligaments.

(2) Medial tibial rotation, unlocking the knee, requires contraction of the **popliteus muscle**. Once the knee is unlocked, the flexors can initiate flexion.

B. DYNAMIC ACTION OF THE KNEE (Fig. 24-5)

1. The knee joint does not function as a simple hinge joint.

a. The axis of rotation changes through an evolute.

b. Separate and distinct movements occur in the suprameniscal and inframeniscal compartments.

(1) In the suprameniscal compartment, the femoral condyles slide within the meniscal surfaces as a result of rotation about the transverse axis.

(2) In the inframeniscal compartment, the femoral condyles glide and pivot on the tibial condyles.

(3) The combined sliding and gliding movements produce rolling.

c. These motions are not independent but occur in conjunction with flexion/extension, the proportion of each dictated by the ligaments, the shapes of the condyles, and the lengths of the articular surfaces.

d. The ligaments are arranged so that the femur and tibia are always held in close apposition with stability throughout the range of movement.

2. The phases of extension.

a. The so-called **"rolling" phase**:

(1) With the knee flexed, the anterior and posterior cruciate ligaments tether the femoral condyle on the tibial plateau. As the knee begins to extend, the points of attachment of the cruciate ligaments change relative to each other, and the condyles roll anteriorly on the tibial plateau.

(2) The movements involve rotation about the transverse axis of the knee with the femoral condyles sliding in the suprameniscal compartment and an axis-less forward gliding on the tibial plateau in the inframeniscal compartment (see Fig. 24-5A).

b. The so-called **"sliding" phase**:

(1) As extension continues, the axis of rotation moves anteriorly, and the increasing radii of curvature ensures that the cruciate ligaments never slacken.

(2) Toward full extension, the anterior cruciate ligament and the posterior fibers of the posterior cruciate ligament prevent further forward gliding of the femur on the tibial plateau, resulting in rotation in the suprameniscal compartment only (see Fig. 24-5A).

c. The so-called **"terminal"** or **"spin" phase**:

(1) Just before terminal extension, the lateral condyle reaches the limit of its articular surface and stops sliding; the medial condyle continues to rotate on its longer articular surface. The lateral condyle thus glides forward around the axis formed by the taut anterior cruciate ligament (see Fig. 24-5B).

(2) The result is medial rotation of the femur in the inframeniscal compartment about a vertical axis through the intercondylar eminence. (Some sources have the vertical axis passing through the lateral femoral condyle, which would be correct if the lateral femoral condyle only pivoted and did not slide anteriorly during the terminal phase of extension.)

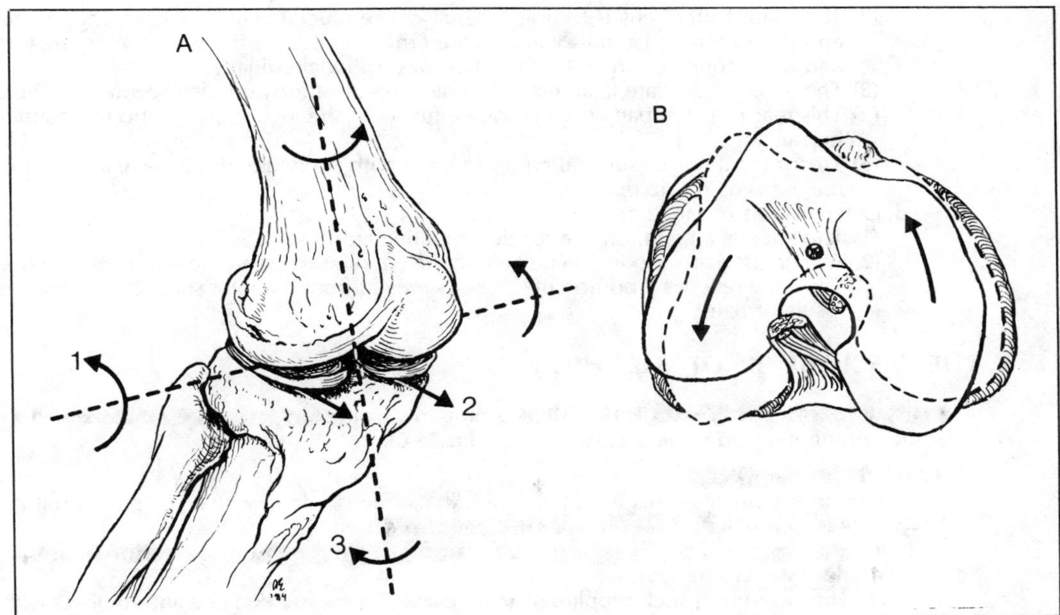

Figure 24-5. *The dynamic action of the knee joint. A,* During extension, rotation occurs in the suprameniscal compartment about the transverse axis *(1)* and forward gliding occurs in the inframeniscal compartment (2). The medial rotation of the terminal phase of extension occurs about a vertical axis (3). *B,* Looking down on the right tibial plateau, in terminal extension the relative movement of the femur relative to the tibia is indicated by the *arrow,* with an axis of rotation about the anterior cruciate ligament.

- **(3)** This medial rotation of the femur tightens the oblique popliteal ligament, as well as the medial and lateral collateral ligaments, so that the knee becomes locked or screwed home.
- **(4)** At terminal extension, the angle between the femur and tibia slightly exceeds 180°.
  - **(a)** The weight of the body passes anterior to the axis of the knee and tightens the knee ligaments.
  - **(b)** While muscular activity is unnecessary to maintain the knee's extended locked position while standing, the tensor fasciae latae and gluteus maximus stabilize the knee and ensure the locked position.

3. Clinical considerations.

a. Meniscal tears.

- **(1)** During the rolling movements of flexion and extension the menisci are drawn anteriorly or posteriorly. The menisci can be torn if they fail to follow the movements of the femoral condyles.
  - **(a)** Violent hyperextension leads to transverse meniscal tears or detachment of the anterior horns.
  - **(b)** Violent twisting movements, which combine lateral displacement with lateral rotation, pull the medial meniscus toward the center of the joint, where it may be trapped and crushed by the medial femoral condyle. This event may lead to a longitudinal (bucket-handle) tear or partial detachment of the medial meniscus.
- **(2)** A torn or detached meniscus that fails to follow the movements of the condyles can be wedged between the condylar surfaces so that the knee becomes locked in partial flexion with full extension impossible.

b. Ligamentous sprains and tears.

- **(1)** Because the knee joint represents a compromise between stability and mobility, there is great potential for sprains (partial tears) and tears. Violent hyperextension, abduction, and adduction may sprain or completely tear one or more collateral or cruciate ligaments.
- **(2)** Torn ligaments produce knee instability.

c. Combined knee injury, such as the "unhappy triad," is the result of fixation of a semiflexed leg with violent abduction and lateral rotation such as occurs as the result of "clipping" in football or the "caught-edge" fall in skiing.

- **(1)** The medial collateral ligament is torn by excessive abduction and, since this ligament is attached to the joint capsule, the medial capsular ligaments are torn as well.

(2) The medial meniscus is torn as a result of the medial collateral ligament attachment and the excessive lateral rotation. Often this meniscus is drawn between the femoral and tibial condyles and either crushed or torn longitudinally.
(3) The anterior cruciate ligament is torn as a result of forward displacement of the tibia.
(4) This tear is accompanied by bloody effusion in the joint capsule and the connecting bursae.
(5) Treatment involves surgical reconstruction with every attempt to save as much of the meniscus as possible.

d. Osteoarthritis of the knee.
(1) The knee is a common site for degenerative arthritis.
(2) A cane used as support will be held in the hand on the same side as the affected knee. The cane relieves a portion of the body weight from the joint surface and reduces the resultant pain.

V. THE VASCULATURE OF THE THIGH

A. THE ARTERIAL SUPPLY TO THE THIGH is primarily by the femoral and obturator arteries with some contribution from the gluteal vessels (Fig. 24-6).

1. The **obturator artery.**

a. This artery usually arises from the **internal iliac artery** deep within the pelvis, but it may also arise aberrantly from the inferior epigastric artery.

b. It exits the pelvic cavity with the obturator nerve through the **obturator foramen** and divides into two branches.
(1) The **anterior branch** supplies muscles of the adductor group and anastomoses with the medial circumflex branch of the profunda femoris artery.
(2) The **posterior branch** mainly supplies the hamstring group and gives off the **acetabular branch** from which the **artery of the ligamentum teres** arises.

2. The **femoral artery.**

a. This is the continuation of the **external iliac artery**.

b. It enters the **femoral triangle** of the thigh by passing through the vascular compartment between the inguinal ligament and iliopubic ramus (see Chapter 15 X F 1).
(1) The boundaries of the femoral triangle are the inguinal ligament, sartorius muscle, and adductor longus muscle.
(2) The femoral pulse is palpable within the femoral triangle.

c. Within the femoral triangle, the femoral artery gives rise to several subcutaneous branches.
(1) The **superficial epigastric artery** supplies the hypogastric region of the abdomen.
(2) The **external pudendal arteries** supply the anterior pudendal region.

d. The **deep femoral (profunda femoris) artery**, the largest branch of the femoral artery, also arises within the femoral triangle. It is initially lateral to the femoral artery then it passes posterior to the femoral artery and comes to lie on the psoas muscle before passing into the posterior compartment.
(1) The **lateral circumflex femoral artery**.
(a) This arises from the lateral side of the deep femoral artery.
(b) It divides into three branches:
(i) The **ascending branch**, which passes superiorly and anastomoses with the inferior gluteal artery.
(ii) The **transverse branch**, which passes around the femur to anastomose with the medial femoral circumflex artery at the **cruciate anastomosis**.
(iii) The **descending branch**, which anastomoses both with the descending genicular branch of the femoral artery and with the superior lateral genicular branch of the popliteal artery.
(2) The **medial femoral circumflex artery**.
(a) This arises from the posterior aspect of the deep femoral artery.
(b) It divides into three branches:
(i) The **ascending branch**, which passes to the gluteal muscles.
(ii) The **transverse branch**, which passes posterior to the femur to anastomose with the transverse branch of the lateral femoral circumflex artery at the **cruciate anastomosis**.
(iii) The **descending branch**.
(3) Four **perforating arteries** supply the major portion of the hamstring muscles.

e. The **descending genicular artery**.
(1) This artery arises medially from the femoral artery in the distal portion of the thigh.
(2) It divides into two branches:

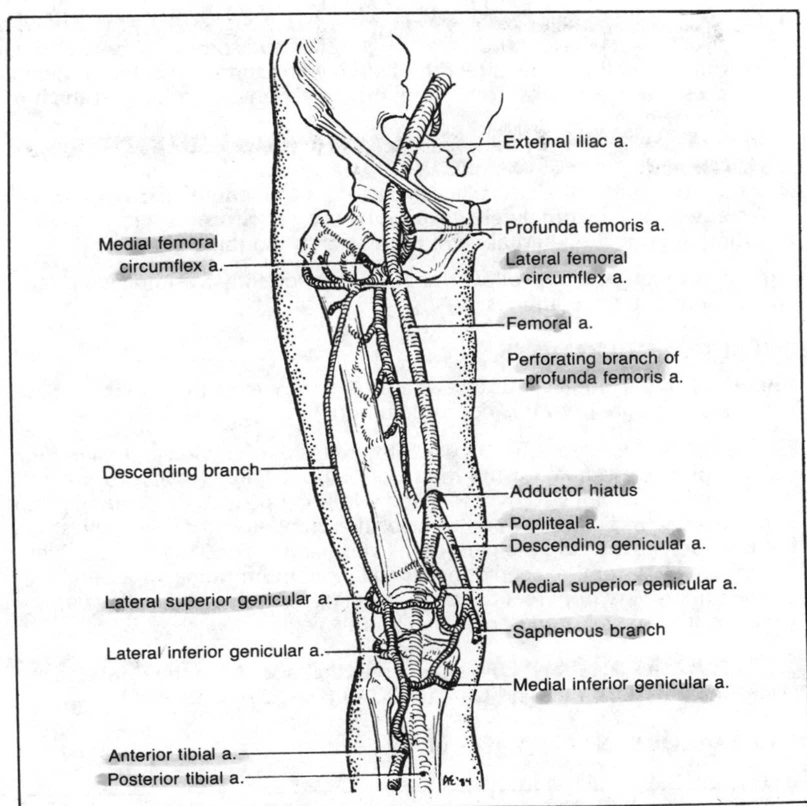

Figure 24-6. *The vasculature of the thigh.*

(a) The **saphenous branch**, which accompanies the saphenous nerve to supply the superficial tissues of the medial aspect of the leg.

(b) The **articular branch**, which descends medial to the knee and anastomoses with the medial superior genicular artery.

f. The femoral artery becomes the **popliteal artery** upon entering the **adductor canal** (Hunter's) which passes between the adductor and hamstring portions of the adductor magnus muscle.

3. The **popliteal artery.**
   a. It descends through the popliteal fossa, where a pulse may be palpated by compressing the artery against the popliteus muscle when the knee is flexed.
   b. The popliteal artery gives off several branches that participate in the **geniculate anastomoses**, which provide an abundant collateral circulation around the knee.
      (1) The **lateral superior genicular branch** anastomoses with both the descending branch of the lateral femoral circumflex artery and the lateral inferior genicular artery.
      (2) The **medial superior genicular branch** anastomoses with both the descending genicular artery and the medial inferior genicular branch.
      (3) The **lateral inferior genicular branch** anastomoses with both the lateral superior genicular branch and the anterior tibial recurrent branch of the anterior tibial artery.
      (4) The **medial inferior genicular branch** anastomoses with the medial superior genicular branch.
      (5) The **medial genicular branch** supplies the capsule of the knee joint.
      (6) The **sural arteries** descend superficially in the posterior compartment of the leg.

B. THE VENOUS RETURN of the thigh is by way of superficial and deep veins.

1. The **great saphenous vein** is the principal superficial vein of the thigh.
   a. This vein lies in the superficial fascia on the medial aspect of the thigh and leg, passing on the flexor sides of the ankle, knee, and hip joints.
      (1) Perforating veins communicate between the saphenous vein and the deep veins of the leg and thigh.

(2) Valves in the saphenous vein direct the flow upward and protect that vein from the hydrostatic pressure produced by the standing column of blood. In addition, valves in the perforating veins direct the flow inward and protect the saphenous vein from the pressure that results from the pumping action of the leg and thigh muscles upon the deep veins.

b. In the femoral triangle, the saphenous vein receives superficial epigastric, superficial circumflex iliac, and external pudendal veins.

c. It then penetrates the fascia lata through the **saphenous hiatus**, an opening in the fasciae latae, which is incompletely covered by the **cribriform fascia**.

d. Where it joins the femoral vein, there is an important valve.

2. The deep veins generally follow the arteries, frequently as subdivided venae comitantes accompanying the large arteries.

C. CLINICAL CONSIDERATIONS

1. Aneurysm of the popliteal artery where the artery exits the adductor canal in the popliteal fossa (Hunter's aneurysm) is not uncommon.

2. The course of the deep veins between the muscles of the leg and thigh forms a deep venous pump which is essential for the return of blood, against gravity, to the heart.

a. Old age and venous stasis associated with long periods of standing predispose the superficial veins to dilation, which reduces the competency of the valves.

b. Incompetent valves in the perforating veins allow blood to escape at high pressure into the saphenous system, exacerbating the dilation and forming varicosities.

c. Treatments vary from identification and ligation of the perforating veins with incompetent valves to removal of the entire saphenous vein.

VI. INNERVATION OF THE THIGH is by nerves that arise from the **lumbar plexus** and the **sacral plexus** (see Chapter 18 IV H 2 and Chapter 20 V) [Figs. 24-7 and 24-8].

A. THE CUTANEOUS INNERVATION of the thigh.

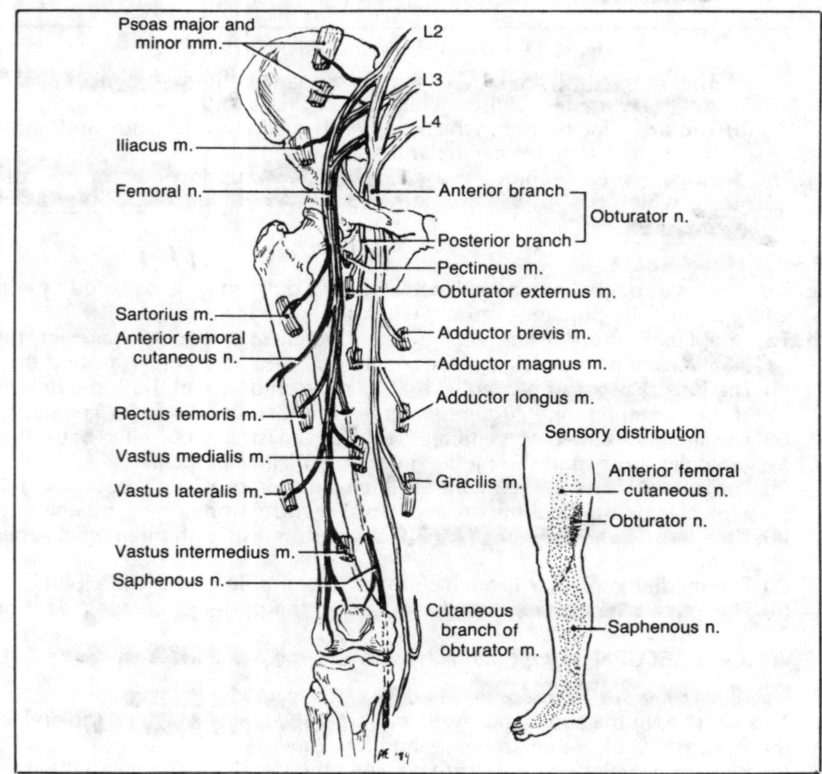

Figure 24-7. *The innervation of the anterior compartment of the thigh.* The distribution of the femoral and obturator nerves.

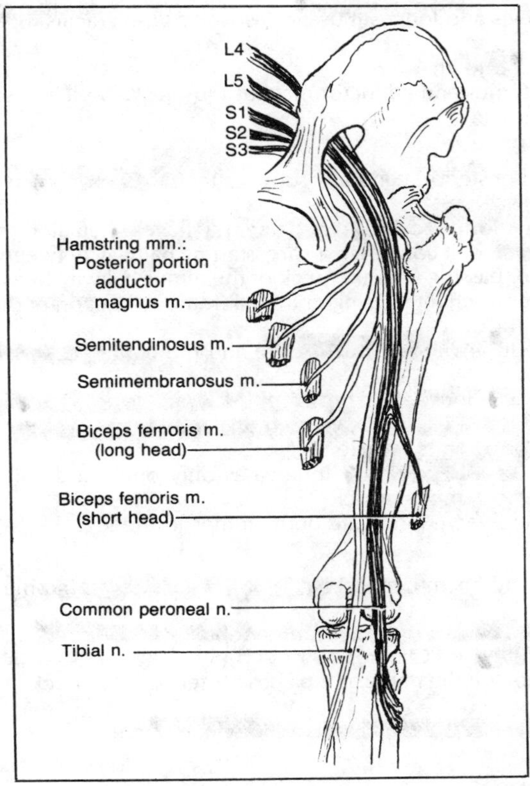

Figure 24-8. *The innervation of the posterior compartment of the thigh.* The distribution of the common peroneal and tibial divisions of the sciatic nerve.

1. The **femoral branch** of the **genitofemoral nerve** (L1–L2, anterior) is sensory to the medial aspect of the proximal thigh.
2. The **lateral femoral cutaneous nerve** (L2–L3, posterior) is sensory to the lateral aspect of the thigh.
3. The **posterior femoral cutaneous nerve** (S1–S3, anterior and posterior) is sensory to most of the posterior aspect of the thigh.
4. Cutaneous branches of the **femoral nerve** (L2–L4, posterior) innervate the anterior aspect of the thigh.

B. INNERVATION OF THE ANTERIOR COMPARTMENT of the thigh is primarily by the **femoral nerve** (L2–L4, posterior) [see Fig. 24-7].

1. It exits the greater pelvis through the muscular compartment inferior to the inguinal ligament.
   - **a.** It gives off the **anterior femoral cutaneous nerve**, which innervates the dermatomes of the anterior aspect of the thigh.
   - **b.** Motor branches innervate the sartorius, rectus femoris, vastus lateralis, vastus intermedius, and vastus medialis muscles.
   - **c.** It forms the afferent and efferent limbs of the knee-jerk reflex, whereby sudden stretch of the patellar tendon causes reflex contraction of the quadriceps femoris muscles.
2. Injury to the femoral nerve is usually the result of trauma to the femoral triangle. Such an injury will result in weakness of hip flexion and inability to extend the knee.

C. INNERVATION OF THE POSTERIOR COMPARTMENT is by the **obturator nerve**, a derivative of the lumbar plexus, and by the **sciatic nerve**, a derivative of the sacral plexus. (see Fig. 24-8).

1. The **obturator nerve** (L2–L4, anterior) generally supplies the adductor muscles.
   - **a.** It exits the deep pelvis via the obturator foramen, along with the obturator artery, and immediately divides.
     - **(1)** The **anterior (superficial) branch:**
       - **(a)** Is sensory to a small area of the distal medial aspect of the thigh.

(b) Innervates the adductor brevis and longus muscles, as well as the gracilis muscle.

(2) The **posterior (deep) branch:**

(a) Sends sensory twigs to the hip joint.

(b) Innervates the obturator externus and adductor magnus muscles as well as a small portion of the pectineus muscle.

b. Injury to the obturator nerve is rare.

2. The **sciatic nerve** (L4–S3, anterior and posterior) generally supplies the hamstring group as it passes through the thigh.

a. It exits from the pelvis via the infrapiriform recess of the greater sciatic foramen along with the inferior gluteal neurovascular bundle. It courses in an arc, starting halfway between the ischial tuberosity and the iliac spine, passing over the neck of the femur halfway between the ischial tuberosity and the greater trochanter, lying in the inferomedial quadrant of the buttock.

b. The **sciatic nerve** is formed by the fusion of the **tibial nerve** and the **common peroneal nerve** (See Fig. 24-8).

(1) The **tibial nerve** (L4–S3, anterior) innervates the semimembranosus and semitendinosus muscles and the long head of the biceps femoris muscle, as well as the posterior portion of the adductor magnus muscle.

(2) The **common peroneal nerve** (L4–S2, posterior) innervates only one muscle in the thigh, the short head of the biceps femoris muscle.

(3) Cutaneous branches of the sciatic nerve innervate dermatomes in the leg.

3. Clinical considerations.

a. The sciatic nerve may be injured by an intramuscular injection into the inferomedial quadrant of the buttock.

b. Injury is most commonly the result of herniation of an intervertebral disk compressing the spinal roots which comprise this nerve [see Chapter 19 I E 4 b (5)].

c. Injury can also be the result of traction on the nerve as it passes posterior to the neck of the femur.

d. Sciatica is pain referred to the dermatomal distribution of the sciatic nerve.

25
The Anterior Leg and Dorsal Foot

I. THE LEG

A. DEFINITION. The leg is that part of the lower extremity that lies between the knee and ankle.

1. Because the muscles of the leg act across the ankle joint as well as the joints of the foot, the leg, ankle, and the foot are best studied as a functional unit.
2. The primitive dorsal musculature of the leg comes to lie anteriorly.
3. The extensor muscles of the anterior and lateral compartments comprise the dorsiflexors (extensors) of the ankle and the extensors of the pedal digits.
4. The foot is composed of seven tarsal bones in two rows, five metatarsal bones in one row, and (except for the great toe) three rows of phalanges.

B. BONY LANDMARKS

1. The **tibia.**
   a. Medial condyle.
   b. Anterior lip of the tibial plateau.
   c. Tibial tubercle.
   d. The medial surface of the shaft.
   e. Medial malleolus.
2. The proximal **fibula.**
   a. Head.
   b. Lateral collateral ligament.
   c. Lateral malleolus.
3. The **tarsal bones.**
   a. Calcaneus (heel).
   b. Navicular bone.
   c. Metatarsal bones.
   d. Phalanges.

C. BONES OF THE LEG (see Figs. 24-1 and 25-1)

1. The distal **tibia.**
   a. This bone corresponds to the radius of the forearm (recall that the upper and lower extremities rotated in opposite directions, phylogenetically).
   b. The tibial tubercle receives the attachment of the patellar tendon (ligament).
   c. The medial surface of the tibia is subcutaneous.
   d. The shaft narrows then expands distally as the **medial malleolus**.
   e. The distal end articulates with the talus at the **talocrural (ankle) joint**.
      (1) The articular surface is trapezoid in shape (wider anteriorly) and concave.
      (2) The articular surface continues medially as the articular surface of the medial malleolus.
   f. Laterally, the **fibular notch** accommodates the **distal tibiofibular joint**.
   g. The medial malleolus is notched posteriorly by the tendon of the tibialis posterior muscle.
2. The distal **fibula.**

a. This bone corresponds to the ulna of the forearm. In the process of developmental pronation and fixation in the pronated position, it has lost contact with the femur and thus takes no part in the knee joint.
b. The shaft is narrow. A slightly bulbous head articulates with the lateral tibial condyle at the **proximal tibiofibular joint**.
c. The **styloid process** of the head attaches the lateral collateral ligament.
d. Distally, the shaft expands slightly as the **lateral malleolus**.
(1) The lateral malleolus extends further distally than the medial malleolus.
(2) It is subcutaneous.
(3) The medial surface is tightly bound to the tibia at the **distal tibiofibular joint**.
(4) On the medial surface of the malleolus is a triangular articular facet for the talus.
(5) A deep malleolar fossa, in which the posterior tibiofibular ligament attaches, lies posterior to the articular facet.
(6) The posterior surface of the malleolus is grooved by the tendons of the peroneus longus and brevis muscles.

3. The **interosseous membrane.**
a. This tough fibrous membrane connects the adjacent borders of the tibia and fibula for their entire lengths.
b. The most distal region of the membrane is thickened, forming the **interosseous ligament**.
c. This membrane and ligament stabilize the tibiofibular joints.

4. The **tibiofibular joints** are syndesmoses where little movement takes place (see Chapter 3 II D 1 a).
a. There is no contact between the fibula and tibia at the distal tibiofibular joint.
b. The tibia and fibula are forced apart by the trochlea of the talus intervening between the medial and lateral malleoli.
c. The distal tibiofibular joint is reinforced by the tibiofibular ligaments so that the malleoli grip the trochlea tali like pincers.
(1) The **anterior tibiofibular ligament** runs in an inferolateral direction.
(2) The **posterior tibiofibular ligament** also runs in an inferolateral direction but is considerably stronger. It is especially thick inferiorly, where it forms the **inferior transverse ligament**, which contains a large proportion of elastic fibers.
d. Sprain or rupture of the transverse ligaments produces diastasis of the ankle, whereby the talus is no longer held tightly by the pincer grip of the medial and lateral malleoli, resulting in ankle instability.

D. BONES OF THE FOOT include the seven **tarsal bones** (tarsos, G. flat) and the five **metatarsal bones** (Fig. 25-1).

1. The proximal tarsal group consists of three bones, not in a row.
a. The **talus** (astragalus) has a rounded head, a neck, and a cuboid body.
(1) The head articulates anteriorly with the navicular bone and inferiorly with the calcaneus.
(2) The neck has a deep groove inferiorly, the **sulcus tali**, in which the interosseous ligament of the talocalcanean joint attaches.
(3) The body has a number of articular surfaces.
(a) There is a flat inferior articular surface for the calcaneus.
(b) Superiorly, the **trochlea tali** articulates with the tibia. The trochlea is wider anteriorly.
(c) Laterally and medially, the body of the talus articulates with the malleoli.
(4) Medial and lateral tubercles on the small posterior surface are separated by a groove for the flexor hallucis longus tendon.
(5) Numerous ligaments, but no muscles, attach to the talus.
b. The **calcaneus** is the largest tarsal bone and forms the heel.
(1) It has three superior articular surfaces for the talus; the sulcus calcanei separates the middle and posterior articular surfaces.
(2) The anterior surface articulates with the cuboid bone.
(3) The notch in the anteromedial edge is bridged by the stout **calcaneonavicular (spring) ligament**, which runs between the calcaneus and navicular bones.
(4) The calcanean attachment of the spring ligament forms a prominent shelf, the **sustentaculum tali**.
(a) With the spring ligament, the sustentaculum supports the head of the talus and helps maintain the longitudinal plantar arch.
(b) The sustentaculum tali is grooved inferiorly by the tendon of the flexor hallucis longus muscle.
(5) An anterior tubercle provides the attachment of the **long plantar ligament**.
(6) The posteroinferior edge of the calcaneus has a lip, more prominent medially and laterally, for the attachment of the short plantar-flexor muscles of the foot.

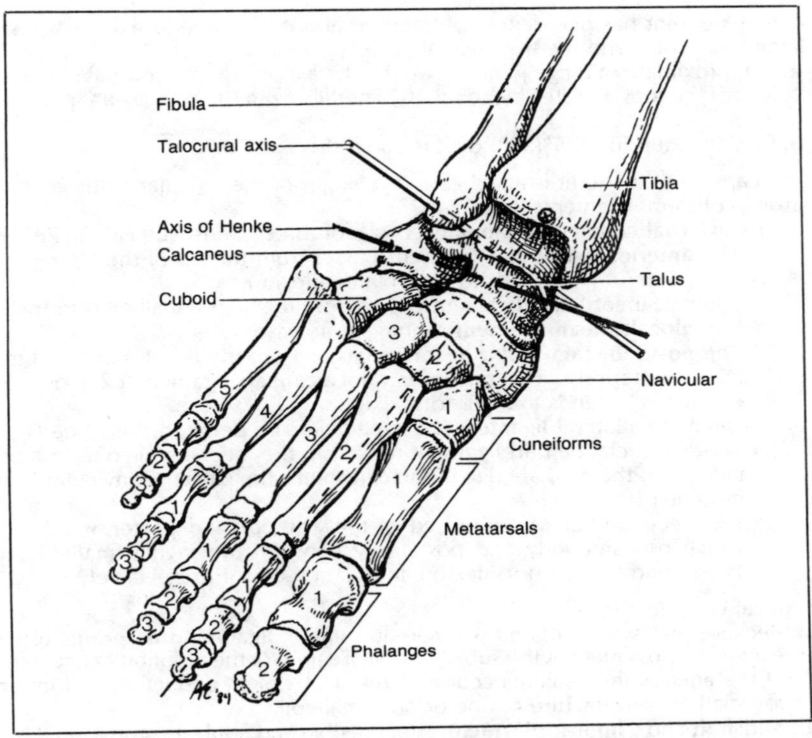

Figure 25-1. *The bones of the distal leg and foot.*

(7) The posterior surface has a horizontal ridge for the insertion of the **calcanean (Achilles) tendon**.
(8) A prominent trochlea on the lateral surface separates the tendons of the peroneus longus and brevis muscles.

c. The **navicular (scaphoid) bone** articulates with all of the tarsal bones, except the calcaneus, to which it is strongly connected by the spring (plantar calcaneonavicular) ligament.
(1) It is curved posteriorly to receive the head of the talus at the apex of the arch of the foot.
(2) It has a prominent tuberosity, which is the principal attachment of the tibialis posterior muscle. Occasionally, it is a separate sesamoid bone.

2. The distal tarsal row consists of four bones.
a. The three **cuneiform bones** (**medial, intermediate,** and **lateral**) articulate with the navicular bone, proximally, and the first three metatarsal bones, distally.
b. The **cuboid bone** is lateral to the navicular bone and the cuneiform bones. It articulates posteriorly with the calcaneus at the **calcaneocuboid joint**, and with the fourth and fifth metatarsals distally.

3. The **metatarsals** and **phalanges** (*phalanx, G.* row of soldiers) are rather similar to the metacarpals and phalanges of the hand.

II. THE ANKLE AND TARSAL JOINTS

A. THE TALOCRURAL (ANKLE) JOINT

1. The socket for the talocrural joint is formed by the tibia, the two malleoli, and the inferior transverse ligament. These structures form a mortise for the tendon of the trochlea tali.
a. The weight of the body is transmitted from the tibia to the talus, which distributes the weight anteriorly and posteriorly within the foot.
b. The center of gravity in the erect posture passes somewhat anterior to the ankle joint.
(1) The wedge-shaped trochlea tali is forced between the malleoli upon dorsiflexion, thus increasing the inherent stability of the joint.
(2) The strength of the tibiofibular ligaments prevents the malleoli from separating.
(3) The fan-shaped collateral ligaments keep the trochlea tali in the joint socket.

2. The ankle joint has one degree of freedom about a transverse axis that passes through the lateral malleolus and the trochlea tali (see Fig. 25-1).
   a. It approximates a hinge joint. However, the axis is not in a coronal plane, so that the long axis of the foot deviates by about 10° laterally from the parasagittal plane in full dorsiflexion.
   b. Flexion/extension (25°/35°) occurs about this axis.
3. The capsule of the joint is attached to the edges of the articular cartilage and supported by strong collateral ligaments (Fig. 25-2).
   a. The **external collateral ligaments** consist of three bands (see Fig. 25-2A).
      (1) The **anterior talofibular ligament** passes from the tip of the lateral malleolus to the talus, anteriorly. It tends to limit plantar-flexion.
      (2) The **calcaneofibular ligament** passes from the lateral malleolus to the calcaneus with the **talocalcanean ligament** running at its base.
      (3) The **posterior talofibular ligament** passes from the tip of the lateral malleolus to the talus, posteriorly. The **posterior talocalcanean ligament** extends this band to the calcaneus. It tends to limit dorsiflexion.
   b. The **medial collateral ligament** has superficial and deep portions (see Fig. 25-2B).
      (1) The superficial **deltoid ligament** runs from the medial malleolus to a broad line of attachment, the medial edge of the calcaneus, the calcaneonavicular ligament, and the navicular bone.
      (2) The deep portion is comprised of the **anterior** and **posterior talotibial ligaments**, which run anteriorly and posteriorly between the medial malleolus and the talus. These tend to limit dorsiflexion and plantar-flexion, respectively.
4. Clinical considerations.
   a. Because the talocrural joint is inherently stable, **ligamentous sprains** usually result from excessive movement at the subtalar joint. Tearing of the tibiofibular ligaments (true sprains of the ankle joint) usually occur as a result of violent adduction and internal rotation in association with fracture of one or both malleoli.
   b. **Midshaft** and **"boot-top" fractures** of the tibia and fibula are common. Usually the bones override somewhat due to muscular pull.
   c. **Distal fractures of the tibia and fibula** are the result of severe external rotation, and abduction may fracture the tibia just proximal to the malleoli. This is a typical skiing fracture, particularly with the new plastic boots, because the bones rotate within the boot.

B. SUBTALAR JOINT is formed where the talus rests on the calcaneus.

1. There are three planar joints at which rotational gliding motions occur.
2. The talus and calcaneus are connected by three major ligaments:
   a. The **lateral talocalcanean ligament**.
   b. The **medial talocalcanean ligament**.
   c. The **interosseous talocalcanean ligament**, which runs from deep within the sulcus tali to the sulcus calcanei and provides an axis of rotation about which movement in the subtalar joint occurs.
3. The calcaneofibular and the deltoid ligaments also traverse this joint.
4. The movement of this joint can be described as abduction/adduction. However, because the anterior talocalcanean articulation is part of the transverse tarsal joint, the movements are conjoined so that abduction/adduction does not occur in pure form.
5. Excessive movements in this joint will produce ligamentous sprains. Severe trauma causes more widespread injury and may produce fractures of the malleoli and ligamentous tears involving the talocrural joint.
   a. Abduction stresses the medial aspect of the joint.
   b. Adduction stresses the lateral aspect of the joint.

C. THE TRANSVERSE TARSAL JOINT consists of two conjoined joints.

1. The **talocalcaneonavicular joint** is formed by the articulation of the ovoid head of the talus with a socket formed by the navicular bone, the spring (plantar calcaneonavicular) ligament, and the anterior articular surface of the calcaneus.
   a. This is a ball-and-socket joint.
   b. It is supported by strong ligaments.
      (1) The talonavicular ligament and dorsal calcaneonavicular ligament support the joint dorsally.

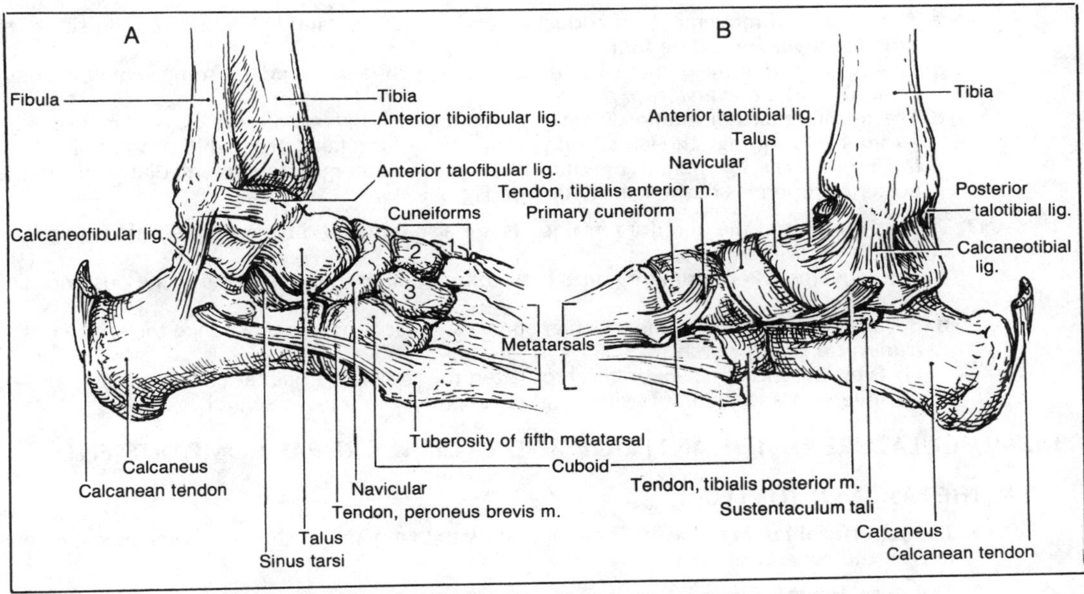

Figure 25-2. *The ligaments of the ankle joint. A,* Lateral. *B,* Medial.

(2) The **plantar calcaneonavicular (spring) ligament** supports the head of the talus. Laxity of this ligament results in "fallen arches" or "flat feet."

(3) Numerous other ligaments provide additional support.

c. Movement in this joint can be described as a rotational supination and pronation.

(1) Because the talus and calcaneus also participate in the subtalar joint, the movements are conjoined so that pronation/supination does not occur in pure form.

(2) A slight amount of conjoined dorsiflexion/ plantar-flexion also occurs in this joint.

2. The **calcaneocuboid joint** is part of the midtarsal joint.

a. This is a gliding joint at which movement is an accommodation to the movements that occur conjointly at the subtalar and talocalcaneonavicular joints.

b. It is supported by the **long plantar ligament**.

D. THE CUNEONAVICULAR, CUNEOCUBOID, INTERCUNEIFORM, AND TARSOMETATARSAL JOINTS

1. They are all supported by a series of strong dorsal and plantar ligaments.

2. They permit small gliding movements of accommodation that change the shape of the **transverse plantar arch**.

E. THE METATARSOPHALANGEAL JOINTS

1. These joints are supported by **deep transverse metatarsal ligaments** and **collateral ligaments**.

2. They have two degrees of freedom, permitting flexion/extension and abduction/adduction.

a. Unlike the hand and as an accommodation to walking, extension of the pedal digits (55°) is greater than flexion (35°).

b. Also unlike the hand, abduction/adduction take place about a long axis through the second toe.

c. Unlike the thumb, the great toe is not capable of opposition.

F. THE INTERPHALANGEAL ARTICULATIONS

1. These are basic hinge joints with one degree of freedom, permitting flexion and extension.

2. They are supported by collateral ligaments.

G. CONJOINED MOVEMENTS OF THE FOOT

1. The subtalar and transverse tarsal joints (the posterior tarsal joints) are mechanically linked by common joints so that they function together with but one degree of freedom.

a. The conjoined movements of adduction and supination (and to some extent dorsiflexion) produce **inversion** of the foot.
b. The conjoined movements of abduction and pronation (and to some extent plantar-flexion) produce **eversion** of the foot.
c. The resultant axis of rotation (Henke's) about which inversion/eversion occurs passes inferomedially through the lateral side of the neck of the talus, through the sinus tali, along the interosseous ligament to the sinus calcanei, and emerges from the medial aspect of the posterior tubercles of the calcaneus (see Fig. 25-1).

2. The movements at the anterior tarsal joints are also conjoined because the articulations are mechanically linked.
   a. The movements are primarily ones of accommodation to movements in the posterior tarsal joints.
   b. Because of the shape of the articular surfaces, these movements change the shape of the **transverse plantar arch**.
      (1) Eversion and dorsiflexion tend to flatten the transverse plantar arch.
      (2) Inversion and plantar-flexion tend to heighten the transverse arch.

III. MUSCULATURE OF THE ANTERIOR AND LATERAL CRURAL COMPARTMENTS

A. THE FASCIA OF THE LEG

1. The **superficial fascia** of the leg is continuous with that of the thigh. It supports the cutaneous nerves and superficial veins.

2. The **deep (crural) fascia** is continuous with the fasciae latae of the thigh posteriorly. It also would be continuous anteriorly if both did not attach to the patella.
   a. Anteromedially, the crural fascia is attached to the medial surface of the tibia.
   b. Laterally, two septa run to the anterior and posterior borders of the fibula, the anterior and posterior intermuscular septa.
      (1) The **posterior intermuscular septum**, the fibula, the interosseous membrane, and the tibia divide the leg into **anterior/lateral** and **posterior compartments.**
      (2) The **anterior intermuscular septa** subdivide the extensor compartment into an **anterior compartment** and a **lateral compartment**.
   c. At the ankle, the crural fascia condenses to form retinacula for the tendons of the extensor, flexor, and peroneal muscles (Fig. 25-3).
      (1) The **superior extensor retinaculum** (transverse crural ligament).
         (a) This retinaculum passes between the distal shafts of the tibia and fibula.

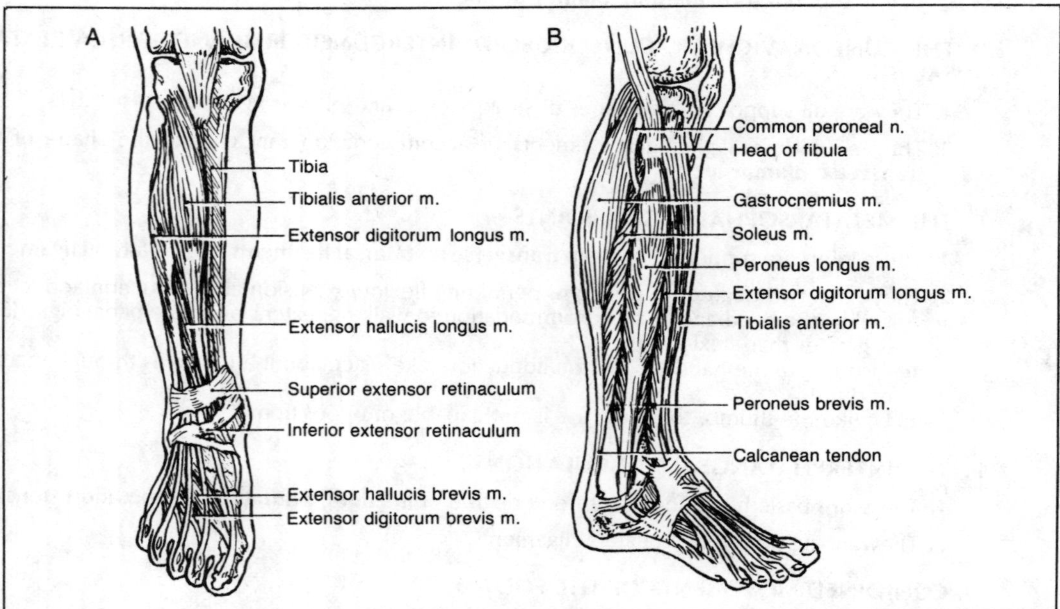

Figure 25-3. *The musculature of the anterior and lateral crural compartments. A,* The anterior compartment muscles. *B,* The lateral compartment muscles.

(b) It contains the tibialis anterior, the extensor hallucis longus, the extensor digitorum longus tendon, and the peroneus tertius muscle.

(2) The **inferior extensor retinaculum** is Y-shaped.

(a) It diverges from a common lateral attachment on the calcaneus to both the medial malleolus and the deep fascia of the medial aspect of the foot.

(b) Unlike the superior extensor retinaculum, the inferior extensor retinaculum is subdivided into compartments for the individual tendons.

(3) The **peroneal retinacula.**

(a) The **superior peroneal retinaculum** runs from the lateral malleolus to the calcaneus.

(b) The **inferior peroneal retinaculum** is a continuation of the inferior extensor retinaculum over the lateral surface of the calcaneus. It is attached to the peroneal trochlea between the peroneus longus and the peroneus brevis.

B. THE ANTERIOR COMPARTMENT (Table 25-1)

1. The anterior compartment of the leg contains the dorsiflexors of the ankle as well both invertors and evertors of the foot.
   a. The muscles of this group arise from the lateral tibial condyle, as well as from the proximal two-thirds of the tibial and fibular shafts and the interosseous membrane.
   b. This group includes four muscles (see Figs. 25-3 and 25-5).
      (1) The **tibialis anterior** inserts on the medial aspect of the medial cuneiform bone.
      (2) The **extensor hallucis longus** inserts into the distal phalanx of the great toe.
      (3) The **extensor digitorum longus** divides into four slips, which insert into the **extensor expansion (hood)** of the second through fifth digits. The extensor expansion then attaches over the dorsal surfaces of the middle and distal phalanges.
      (4) The **peroneus tertius** inserts into the bases of the fourth and fifth metatarsals.
2. This group of muscles is innervated by the deep branch of the common peroneal (anterior tibial) nerve.

Table 25-1. Anterior and Lateral Compartment Muscles Acting at the Ankle and Foot Joints

Muscle	Origin	Insertion	Primary Action	Innervation
		Anterior Compartment		
Tibialis anterior	Proximal half of anterior tibia	Medial cuneiform bone and base of first metatarsal	Dorsiflexion and inversion	Deep peroneal n. (L4–L5, posterior)
Extensor hallucis longus	Front of fibula	Base of second phalanx of great toe	Extends terminal phalanx of great toe	Deep peroneal n. (L4–S1, posterior)
Extensor digitorum longus	Lateral condyle of tibia, interosseous membrane, and fibula	Second and third phalanges of second to fifth toes	Extends metatarso-phalangeal joints	Deep peroneal n. (L4–S1, posterior)
Peroneus tertius	Anterior fibula	Base of fifth metatarsal	Dorsiflexes and everts foot	Deep peroneal n. (L5–S1, posterior)
		Lateral Compartment		
Peroneus longus	Proximal half of fibula	Medial cuneiform bone and base of first metatarsal	Plantar-flexion and eversion of foot	Superficial peroneal n. (L5–S1, posterior)
Peroneus brevis	Distal half of fibula	Base of fifth metatarsal	Plantar-flexion and eversion of foot	Superficial peroneal n. (L5–S1, posterior)
		Dorsum of Foot		
Extensor hallucis brevis	Calcaneus	Base of proximal phalanx of great toe	Extends proximal phalanx of great toe	Deep peroneal n. (L4–S1, posterior)
Extensor digitorum brevis	Calcaneus	Extensor hood of second to fourth toes	Extends metatarso-phalangeal joints of second, third, and fourth toes	Deep peroneal n. (L4–S1, posterior)

C. THE LATERAL COMPARTMENT (see Table 25-1)

1. The lateral compartment of the leg contains peroneal muscle.
   a. The muscles of this group arise from the fibula.
   b. This group includes two muscles (Fig. 25-4).
      (1) The **peroneus longus** tendon grooves the posterior surface of the lateral malleolus. It passes over the calcaneus below the peroneal tubercle and turns medially in a groove on the inferior surface of the calcaneus to insert into the medial cuneiform bone and the first metatarsal bone.
      (2) The **peroneus brevis** tendon passes superior to the peroneal tubercle to insert into the base of the fifth metatarsal (see Fig. 25-2A).
2. The peroneal muscles evert and plantar-flex the foot.
3. They are innervated by the **superficial branch** of the common peroneal nerve.

D. MUSCLE ACTIONS are determined by the location of the tendons relative to the axis of rotation of the ankle and that of the foot (see Table 25-1).

1. **Flexion/extension** occurs about a transverse axis through the ankle joint (Fig. 25-4).
   a. Muscles that pass anterior to this axis dorsiflex the ankle; those that pass posterior to this axis plantar-flex the ankle.
   b. All of the muscles of the anterior compartment pass anterior to this axis and are dorsiflexors.
   c. Both muscles of the lateral compartment pass behind the lateral malleolus and are thus posterior to this axis so that they plantar-flex the ankle.
2. **Inversion/eversion** occurs about the complex axis of Henke (see Fig. 25-4).

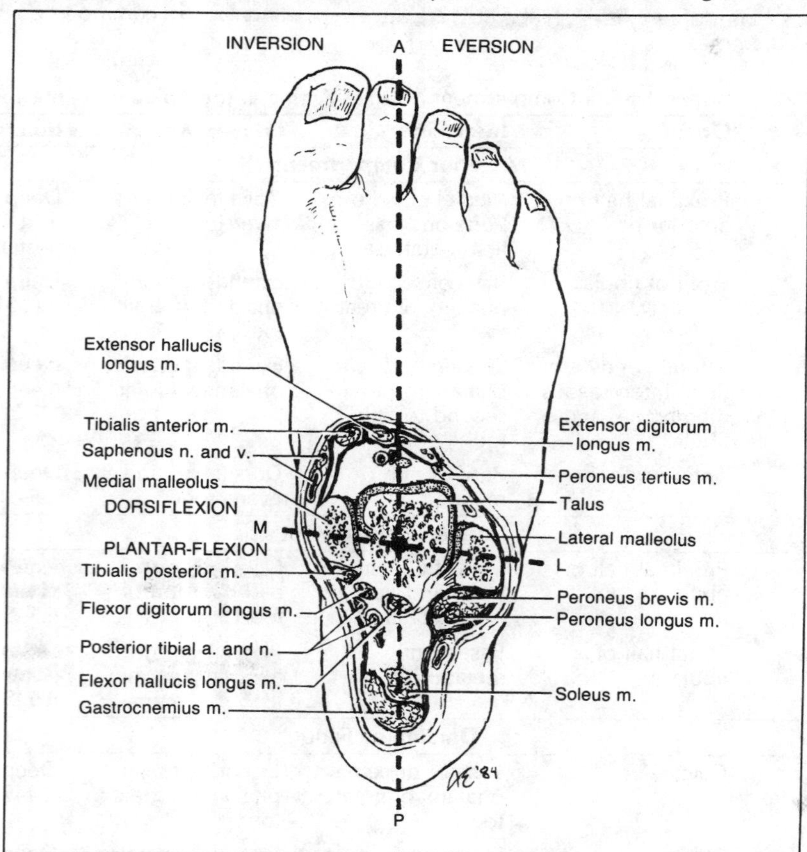

Figure 25-4. *The action of the crural muscles at the ankle joint.* Those muscles passing anterior to the transverse axis of the ankle joint dorsiflex the foot; those passing posterior to this axis plantar-flex the foot. Those muscles passing medial to the resultant axis (A-P) of the subtalar and midtarsal joints invert the foot; those passing lateral to this axis evert the foot.

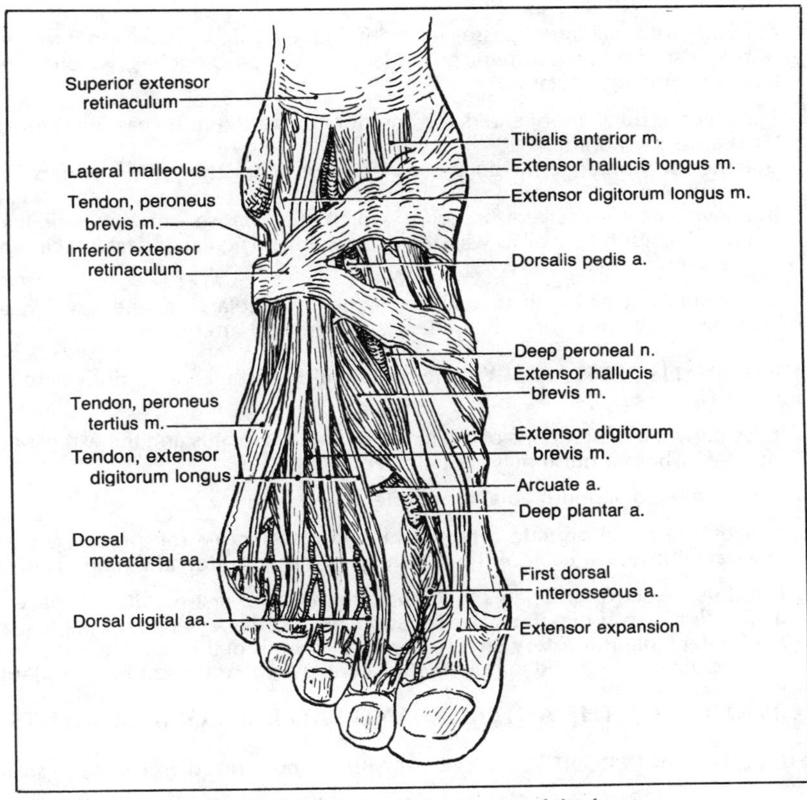

Figure 25-5. *The muscles of the dorsum of the foot.*

a. Muscles that pass medial to this axis are inverters; muscles that pass lateral to it are everters.
b. The muscles of the anterior compartment are both inverters (tibialis anterior and extensor hallucis longus) and everters (extensor digitorum longus and peroneus tertius).
c. Both muscles of the lateral compartment are strong everters.

IV. THE MUSCULATURE OF THE DORSUM OF THE FOOT

A. THE CRURAL (DEEP) FASCIA of the foot is divided into two layers by the extensor tendons of the foot. The two layers fuse at the margins of the foot.

B. THE DORSUM of the foot has two short dorsiflexor muscles, unlike the back of the hand (see Fig. 25-2 and Table 25-1).

1. The **extensor digitorum brevis muscle:**
a. Arises from the calcaneus and is divided into three slips, which insert into the lateral margins of the extensor hoods of the second, third, and fourth toes.
b. Assists the action of the extensor digitorum longus in dorsiflexion of the metatarsophalangeal joints.
c. Is usually (78 percent) innervated by the deep peroneal nerve but may be innervated by the superficial peroneal nerve.

2. The **extensor hallucis brevis muscle:**
a. Arises, along with the extensor digitorum brevis, from the calcaneus and inserts into the base of the proximal phalanx of the great toe.
b. Is an extensor of the proximal phalanx of the great toe with an innervation similar to the extensor digitorum brevis.

V. THE VASCULATURE OF THE ANTERIOR AND LATERAL COMPARTMENTS AND DORSUM OF THE FOOT

A. The popliteal artery gives rise to the anterior tibial artery in the popliteal fossa (Fig. 25-6).

1. This anterior tibial artery passes inferiorly in the popliteal fossa and into the superior portion of the posterior compartment of the leg. It gives off branches, which anastomose with the lateral inferior genicular artery.
2. The anterior tibial artery enters the anterior compartment by passing through a gap in the interosseous membrane.
   a. It then accompanies the deep peroneal nerve along the anterior surface of the interosseous membrane.
   b. It gives rise to recurrent branches, which anastomose with the genicular branches of the popliteal artery as well as with the descending genicular and lateral circumflex branches of the deep femoral artery.
3. At the ankle it gives off medial and lateral malleolar branches, which anastomose with branches of the posterior tibial artery and peroneal artery.

B. THE DORSALIS PEDIS ARTERY is the continuation of the anterior tibial onto the dorsum of the foot (see Fig. 25-6).

1. It lies between the tendons of the extensor hallucis longus and the extensor digitorum longus muscles, where a pulse may be palpated.
2. It gives off lateral and medial tarsal arteries.
3. It gives rise to an **arcuate artery**, which courses across the metatarsals to give off **dorsal metatarsal arteries**, each of the which bifurcates into **dorsal digital arteries**.
4. The dorsalis pedis terminates by bifurcating over the first metatarsal space.
   a. The **first dorsal metatarsal artery** supplies the adjacent sides of the first and second toes.
   b. The **deep plantar artery** dives between the heads of the first dorsal interosseous muscle to the plantar aspect and joins with the lateral plantar artery to form the **plantar arterial arch**.

VI. INNERVATION OF THE ANTERIOR AND LATERAL COMPARTMENTS

A. THE COMMON PERONEAL NERVE (L4–S3), the posterior division of the **sciatic nerve**, inner-

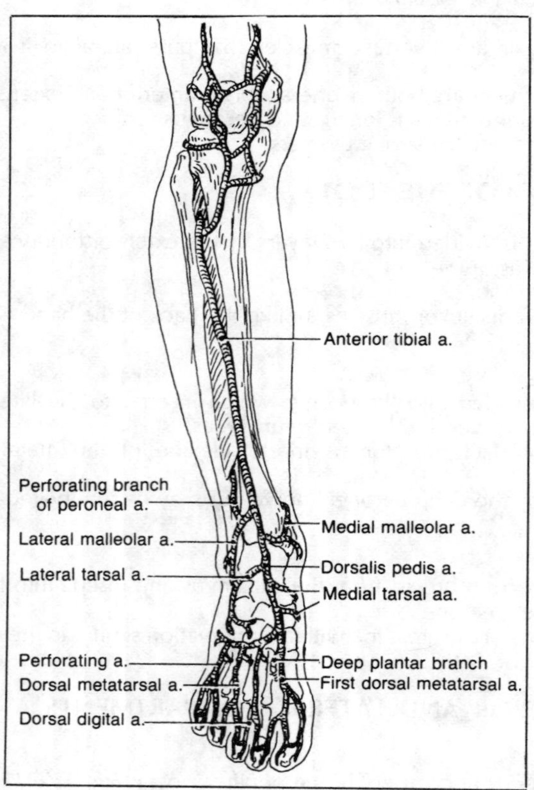

Figure 25-6. *The vasculature of the anterior crural compartment and dorsum of the foot.*

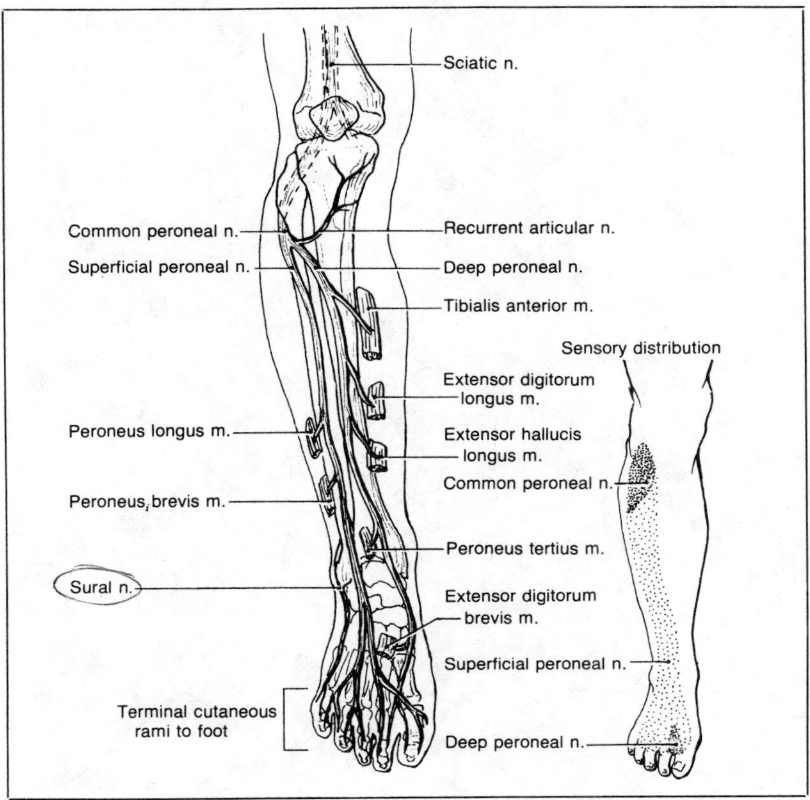

Figure 25-7. *The distribution of the common peroneal nerve in the leg and foot.*

vates the anterior and lateral compartments of the leg. High in the popliteal fossa, the common peroneal nerve diverges from the tibial nerve (Fig. 25-7).

1. As it courses laterally in the popliteal fossa, the common peroneal nerve gives rise to one or more articular branches, and the sural communicating nerve contributes the lateral half of the sural nerve.
2. It passes posterior to the head of the fibula (from which it gets its name: peroneus, *L.* pin) and turns anteriorly across the neck of that bone, where it divides into three branches.
   a. The lateral sural cutaneous branch innervates the lateral aspect of the leg.
   b. The **superficial peroneal nerve** innervates the muscles of the lateral compartment, and a terminal cutaneous branch innervates the dorsum of the foot.
   c. The **deep peroneal nerve** innervates the muscles of the anterior compartment, and a terminal cutaneous branch innervates a small area between the first and second toes (see Fig. 25-7).

B. CLINICAL CONSIDERATIONS

1. Lesion of the common peroneal nerve results in foot drop—that is, inability to dorsiflex the foot or to stand back on the affected heel, as well as loss of sensation along the lateral aspect of the leg and dorsum of the foot. Peroneal nerve palsy produces a characteristic gait with "foot-slap."
2. Proximal injury to this nerve may result from herniation of an intervertebral disk or intramuscular injection into the inferomedial quadrant of the buttock.
3. Distal injury to this nerve may result from pressure where it passes over the fibula. This pressure can be as frank as blunt trauma or fracture of the proximal fibula, as subtle as constant pressure, such as that exerted by the edge of a plaster cast which terminates just below the knee, or even as innocuous as sitting with the knees crossed for a prolonged period.

26
The Posterior Leg and Plantar Foot

I. THE PRIMITIVE VENTRAL ASPECTS OF THE LEG, ANKLE, AND FOOT

A. The primitive ventral aspects of the leg and foot can be considered as a functional unit.

B. The bony landmarks (see Chapter 25 I B).

C. The bones of the leg (see Chapter 25 I C).

D. The bones of the foot (see Chapter 25 I D).

II. THE MUSCULATURE OF THE POSTERIOR CRURAL COMPARTMENT

A. THE FASCIA

1. The **superficial fascia** of the leg is continuous with that of the thigh. It supports the cutaneous nerves and superficial veins.

2. The **deep (crural) fascia** is continuous with the fascia lata of the thigh posteriorly and envelops the muscles of the leg.

a. The tibia, the interosseous membrane, the fibula, and the posterior intermuscular septum (a septum of the crural fascia) together form a barrier that divides the leg into **anterolateral** and **posterior compartments**.

b. Another septum of the crural fascia, the **transverse crural septum**, subdivides the **posterior compartment** into superficial and deep compartments.

c. On the medial side of the ankle, the crural fascia condenses to form a retinaculum for the tendons of the flexor muscles. The **flexor retinaculum (tarsal tunnel)** passes between the medial malleolus and the calcaneus.

(1) It is subdivided into four compartments.

(2) It tethers the tibialis posterior muscle, the flexor digitorum longus muscle, the posterior tibial artery and nerve, and the flexor hallucis longus. The mnemonic here is **T**om, **D**ick, **AN**d **H**arry.

3. The deep fascia continues onto the foot.

B. THE FLEXOR MUSCLES OF THE CALF are divided into superficial and deep groups by the transverse crural septum (Fig. 26-1 and Table 26-1).

1. The **triceps surae** comprise the superficial group.

a. This group consists of three muscles.

(1) The two heads of the **gastrocnemius muscle** arise from the posterior aspects of the femoral epicondyles.

(2) The small **plantaris muscle** arises from the distal femur.

(3) The **soleus muscle** arises from the proximal posterior surfaces of the fibula and tibia.

b. The tendons unite to form the **calcanean (Achilles) tendon**, which inserts into the calcanean tuberosity. Achilles tendinitis, caused by repeated stress upon heel strike, may progress to degeneration and rupture of the tendon.

c. These muscles are innervated by the tibial nerve (L4–L5) and plantar-flex the foot (see Fig. 25-4).

2. The deep group consists of four muscles (Fig. 26-2; see Table 26-1).

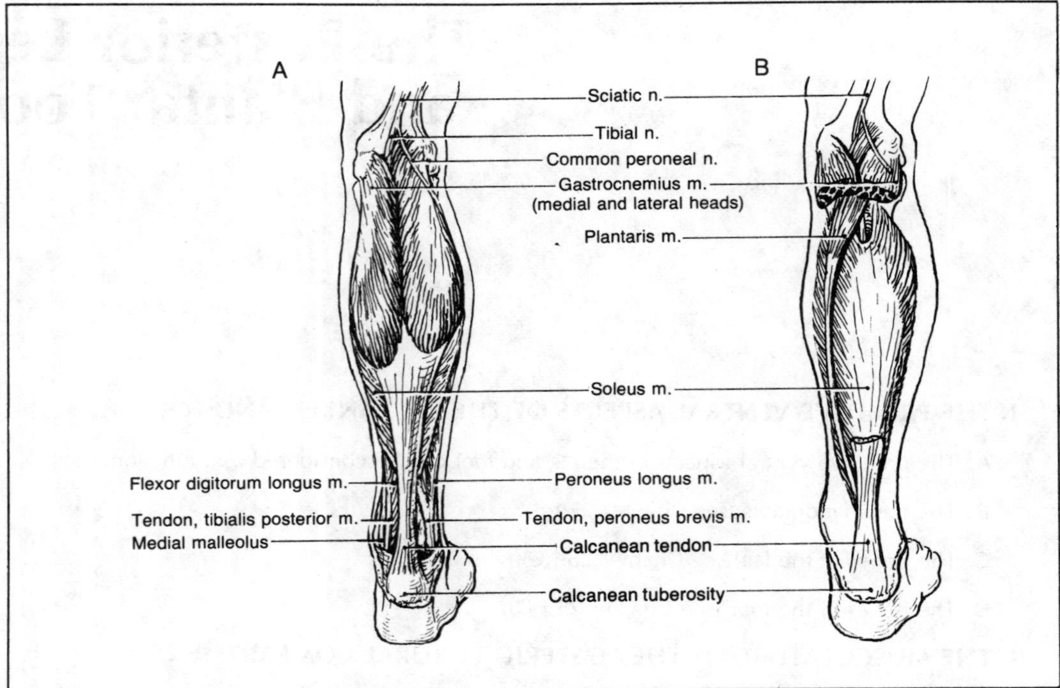

Figure 26-1. *The superficial musculature of the posterior crural compartment.* A, The gastrocnemius muscle. B, The soleus and plantaris muscles.

Table 26-1. Posterior Compartment Muscles Acting at the Ankle and Foot Joints

Muscle	Origin	Insertion	Primary Action	Innervation
		Superficial Group		
Gastrocnemius	Posterior surfaces of medial and lateral femoral condyles	Calcaneus	Plantar-flexion	Tibial n. (L5–S2, anterior)
Soleus	Proximal tibia, interosseous membrane, and fibula	Calcaneus	Plantar-flexion	Tibial n. (L5–S2, anterior)
Plantaris	Lateral supra-condylar ridge	Calcaneus	Plantar-flexion	Tibial n. (L4–S1, anterior)
		Deep Group		
Popliteus	Lateral femoral condyle	Medially on proximal posterior tibia	Medially rotates and flexes knee	Tibial n. (L4–S1, anterior)
Flexor hallucis longus	Distal third of fibula	Phalanges of great toe	Flexes great toe	Tibial n. (L5–S2, anterior)
Tibialis posterior	Posterior shafts of tibia, fibula, and interosseous membrane	Navicular and medial cuneiform bones	Plantar-flexion and inversion of foot	Tibial n. (L5–S1, anterior)
Flexor digitorum longus	Posterior surface of tibial shaft	Distal phalanges of second through fifth toes	Flexes distal phalanges of second through fifth toes	Tibial n. (L5–S2, anterior)

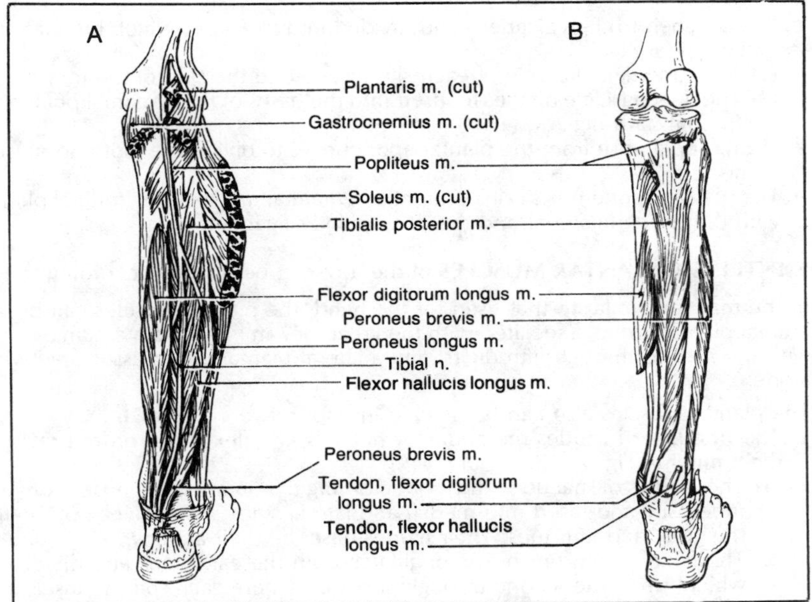

Figure 26-2. *The deep musculature of the posterior crural compartment.* A, The flexor digitorum longus and flexor hallucis longus muscles. B, The tibialis posterior muscle.

a. The **popliteus muscle** forms the floor of the popliteal fossa.

(1) It arises from the lateral femoral condyle and inserts into the posterior medial aspect of the proximal tibia (see Fig. 24-1B).

(2) It acts either to medially rotate the tibia or to laterally rotate the femur, depending upon which bone is fixed, thereby unlocking and initiating flexion of the knee.

b. The **flexor hallucis longus** originates along the posterior aspect of the fibula and inserts into the distal phalanx of the great toe.

(1) The tendon of this muscle is associated with two sesamoid bones where it passes under the distal head of the first metatarsal.

(2) It is powerful and plays an important role in the propulsive (toe-off) phase of gait.

c. The **flexor digitorum longus** arises from the interosseous membrane and posterior surface of the tibia then divides into four slips, which insert into the distal phalanges of the second through fifth toes.

d. The **tibialis posterior muscle**, the deepest of the posterior group, arises from the tibia, interosseous membrane, and fibula to insert into the cuneiform bones with slips to the second through fourth metatarsals. **Shin splints** are most likely tendinitis of the tibialis posterior origin at the tibia and interosseous membrane. This condition can progress to a periostitis. The cause is likely an imbalance that may be corrected by using different running shoes or orthotics.

3. Actions of the muscles of the posterior compartment (see Fig. 25-4).

a. Muscles that pass posterior to the transverse axis of the talocrural joint plantar-flex the ankle. These include all of the muscles of the posterior and lateral compartments.

b. Muscles that pass medial to the resultant axis of the combined subtalor and midtarsal joints (the axis of Henke) invert the foot. All of the muscles of the posterior compartment are invertors.

III. THE MUSCULATURE OF THE PLANTAR FOOT

A. THE FASCIA

1. The **superficial fascia** of the leg is continuous with that of the foot. It supports the cutaneous nerves and superficial veins.

2. The **plantar (deep) fascia** of the foot is continuous with the crural fascia of the leg.

a. It is thickened centrally as the **plantar aponeurosis**.

b. It contains numerous septa, which run between the skin and deep fascia, limiting the mobility of the skin of the sole of the foot.

c. It is attached to the calcaneus and divided into five slips, which bifurcate into superficial and deep fascicles.
  (1) The superficial fascicle of each slip inserts into the skin of the toes.
  (2) The deep fascicle divides to insert into the bases of the proximal phalanx of the second through fifth digits.

d. Strong septa arise from the plantar aponeurosis to divide the foot into muscular compartments.

e. The plantar aponeurosis is instrumental in maintaining the longitudinal plantar arch, along with the spring ligament and the deep plantar ligament.

B. THE INTRINSIC PLANTAR MUSCLES of the foot can be divided according to two methods.

1. In the manner similar to that used for the hand, the plantar muscles can be grouped into a medial plantar group associated with the great toe, an intermediate plantar group associated with the second through fifth digits, and a lateral plantar group associated with the fifth toe alone.

2. The plantar muscles also can be divided into four layers (Table 26-2).
  a. The **first layer** includes the abductor hallucis, the flexor digitorum brevis, and abductor digiti minimi (Fig. 26-3A).
    (1) The abductor hallucis and abductor digiti minimi both arise from the calcanean tuberosity and insert into medial and lateral sides, respectively, of the first phalanx of the great and fifth toes. Their names describe their actions.
    (2) The flexor digitorum brevis originates from the calcaneus and divides into four slips which run to the second through fifth toes, where each splits to insert into the sides of the middle phalanx. In a manner similar to that of the flexor digitorum superficialis of the upper extremity, the split insertions of the flexor digitorum brevis provide passageways for the tendons of the flexor digitorum longus.
  b. The **second layer** includes the quadratus plantae and four lumbrical muscles, as well as the tendon sheaths of two extrinsic muscles (see Fig. 26-3B).
    (1) The quadratus plantae (accessory flexor digitorum) runs from the calcaneus to the distal phalanges of the second through fifth toes.
    (2) The lumbricals are numbered (1–4) from the medial side of the foot.
      (a) They arise from the tendons of the flexor digitorum longus and insert into the dorsal digital expansions over the proximal phalanges.

Table 26-2. Intrinsic Muscles of the Plantar Foot

Muscle	Origin	Insertion	Primary Action	Innervation
		First Layer		
Abductor hallucis	Medial tubercle of calcaneus	Base of first phalanx of great toe	Abducts great toe	Tibial n. (L5–S1, anterior)
Flexor digitorum brevis	Medial tubercle of calcaneus	Splits to sides of middle phalanx of second through fifth toes	Flexes middle phalanx of second through fifth toes	Tibial n. (L5–S1, anterior)
Abductor digiti minimi	Calcaneus	Lateral side of first phalanx of fifth toe	Abducts fifth toe	Tibial n. (S1–S2, anterior)
		Second Layer		
Quadratus plantae	Calcaneus	Tendon of flexor digitorum longus	Flexes distal phalanx of second through fifth toes	Tibial n. (S1–S2, anterior)
Lumbricales	Tendons of flexor digitorum longus	Medial sides of phalanx of second through fifth toes	Flex proximal phalanx and extend middle and distal phalanges of second through fifth toes	Tibial n. (L5–S2, anterior)

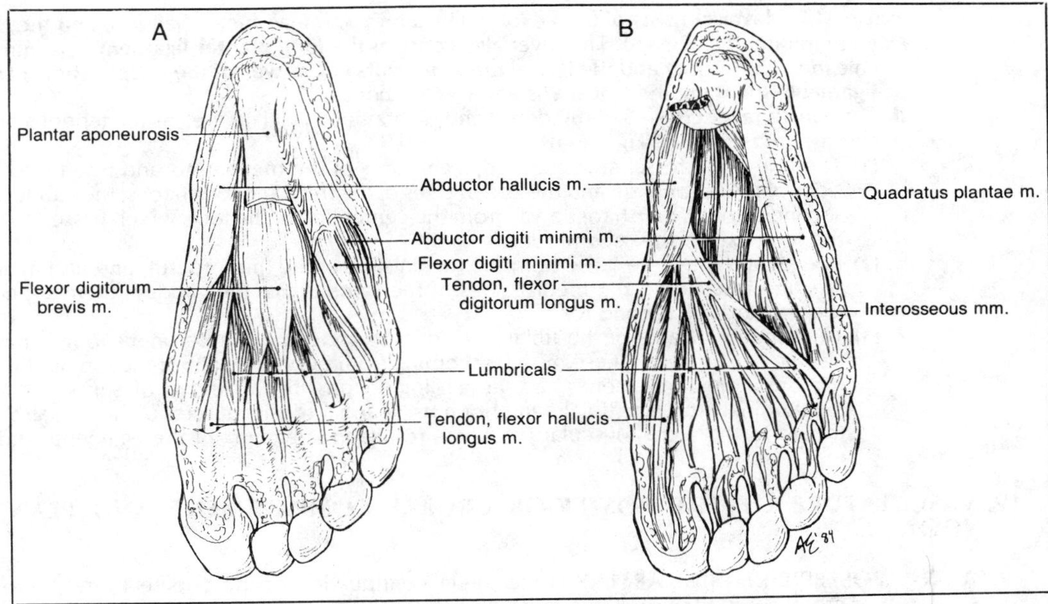

Figure 26-3. *The superficial muscles of the plantar foot.* A, The first layer. B, The second layer.

(b) They flex the metatarsophalangeal joints and extend the proximal and distal interphalangeal joints.

(3) The tendon sheaths of the flexor hallucis longus and flexor digitorum longus muscles lie in this layer. The powerful flexor hallucis longus, passing under the sustentaculum tali and the first metatarsal, bowstrings beneath and thereby supports the longitudinal plantar arch.

Table 26-2. Continued

Muscle	Origin	Insertion	Primary Action	Innervation
		Third Layer		
Flexor hallucis brevis	Cuneiforms and cuboid	Base of first phalanx of great toe	Flexes proximal phalanx of great toe	Tibial n. (L5–S1, anterior)
Adductor hallucis:				
Oblique head	Calcaneus	Base of first phalanx of great toe	Adducts great toe	Tibial n. (S1–S2, anterior)
Transverse head	Metatarsophalangeal ligaments			
Flexor digiti minimi brevis	Base of fifth metatarsal	Base of first phalanx of fifth toe	Flexes proximal phalanx of fifth toe	Tibial n. (S1–S2, anterior)
		Fourth Layer		
Interossei:				
Dorsal (4)	Between adjacent sides of metatarsals	First and second to sides of second toe, third and fourth to sides of third and fourth toes	Abduct second through fourth toes from the second toe	Tibial n. (S1–S2, anterior)
Plantar (3)	Medial sides of third through fifth metatarsals	Medial side of corresponding toe	Adduct third through fifth toes toward the second toe	Tibial n. (S1–S2, anterior)

c. The **third layer** consists of the flexor hallucis brevis, the adductor hallucis, and the flexor digiti minimi (Fig. 26-4A). This layer also contains the **long plantar ligament** (spanning the calcanean tuberosity and the lateral three metatarsals), as well as the deeper **short plantar ligament** (spanning the calcaneus and cuboid bones).

d. The **fourth layer** consists of the dorsal and plantar interossei, as well as the tendons of two extrinsic muscles (see Fig. 26-4B).

(1) The dorsal interossei arise from adjacent sides of the metatarsals and insert into both sides of the second toe and the lateral sides of the third and fourth toes; they abduct the second through fourth toes away from the center line of the foot, which passes through the long axis of the second toe.

(2) The plantar interossei arise from the medial side of the third, fourth, and fifth metatarsals and insert onto the medial sides of the same digits; they adduct their respective toes toward the second toe.

(3) The tendon sheaths of the tibialis posterior muscle and the peroneus longus muscle also lie in this layer. These muscles, coming from opposite sides of the foot and inserting onto the same bones, act as a sling to provide a degree of support for the longitudinal arch. The tibialis anterior muscle and intrinsic muscles of the foot also contribute to the arch. Muscular weakness results in stretching of the ligaments and flat feet.

IV. VASCULATURE OF THE POSTERIOR CRURAL COMPARTMENT AND PLANTAR FOOT

A. **THE POSTERIOR TIBIAL ARTERY** is the direct continuation of the popliteal artery into the posterior crural compartment. It descends medially in the posterior compartment of the leg, giving off several branches (Fig. 26-5).

1. In the upper third of the leg, the posterior tibial artery gives rise to the **peroneal artery**, which descends laterally in the posterior compartment between the tibialis posterior and flexor hallucis longus muscles.

   a. It gives off several branches to the muscles of the posterior and lateral crural compartments.

   b. It anastomoses with the posterior tibial artery.

   c. It gives rise to the **posterior lateral malleolar branch**, which passes lateral to the ankle to anastomose with the anterior lateral malleolar branch of the anterior tibial artery.

   d. A lateral calcanean branch supplies that aspect of the heel.

   e. Occasionally, the peroneal artery is especially large and gives rise to the dorsalis pedis artery.

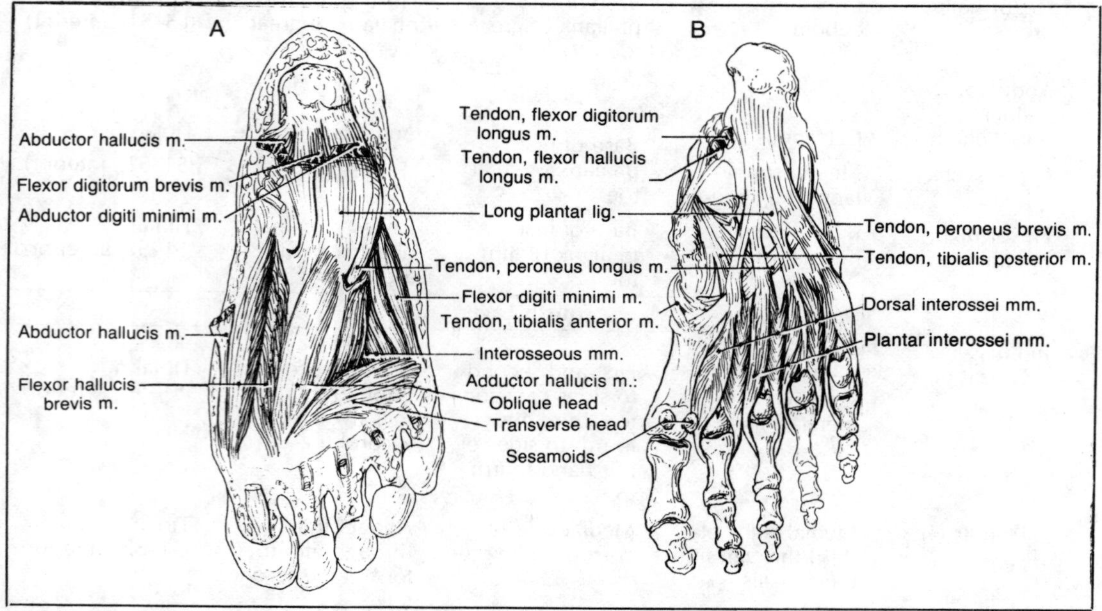

Figure 26-4. *The deep muscles of the plantar foot.* A, The third layer. B, The fourth layer with the deep ligaments.

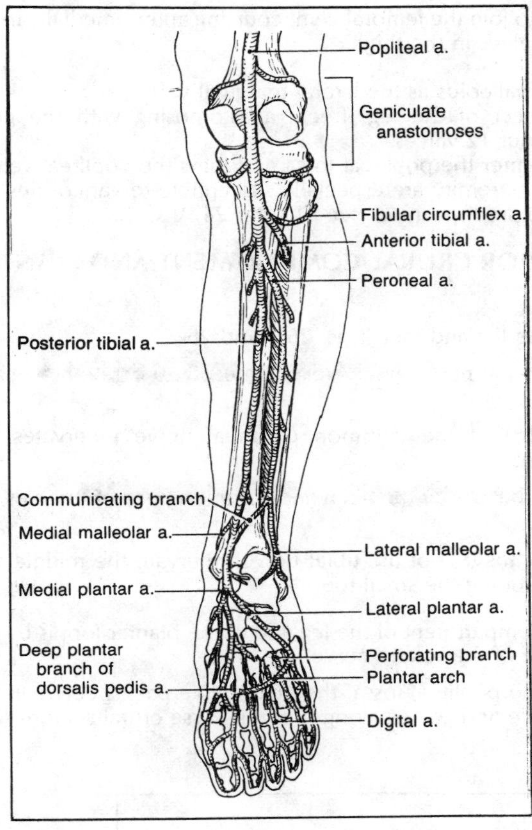

Figure 26-5. *The vascular supply to the posterior and lateral crural compartments and the plantar foot.*

2. The **posterior medial malleolar branch** passes medially about the ankle to anastomose with the anterior medial malleolar branch of the anterior tibial artery.

3. A medial calcanean branch supplies that aspect of the heel.

4. The posterior tibial artery continues posterior to the medial malleolus, where a pulse is normally palpable, and bifurcates into the plantar arteries.

B. THE PLANTAR ARTERIES (see Fig. 26-5)

1. The larger **lateral plantar artery** deviates laterally, across the long plantar arch and then turns medially across the bases of the metatarsals.
   a. It joins with the deep plantar branch of the dorsalis pedis artery in the first interosseous space to form the **plantar arterial arch**.
   b. It gives off numerous branches.
      (1) The proper digital artery to the lateral side of the fifth toe.
      (2) Four **plantar metatarsal arteries**.
         (a) Each bifurcates into proper digital arteries to supply the sides of adjacent toes.
         (b) Each communicates with the corresponding branch of the dorsal metatarsal artery.

2. The smaller **medial plantar artery** supplies the medial side of the plantar portion of the foot, terminating along the lateral side of the great toe.

C. THE VENOUS RETURN of the foot and leg is by way of superficial and deep veins.

1. The **superficial veins** of the leg are the great saphenous vein and the small saphenous vein.
   a. The **great saphenous vein:**
      (1) Arises by the coalescence of a venous network into the **medial marginal vein** of the dorsal foot.
      (2) Passes, with the saphenous nerve, anterior to the medial malleolus, where it is readily accessible for venipuncture or insertion of venous lines (see Fig. 25-4).
      (3) Ascends along the medial aspect of the leg and passes posterior to the medial condyles of the knee. It contains about 12 valves in the leg.

(4) Passes through the fossa ovalis to join the femoral vein, coursing anteriomedially in the thigh. It contains about eight valves in the thigh.

b. The **small saphenous vein:**

(1) Begins posterior to the lateral malleolus as the lateral marginal vein.

(2) Ascends along the lateral aspect of the leg, freely anastomosing with the great saphenous vein. It contains about 12 valves.

(3) Penetrates the crural fascia to enter the popliteal fossa and joins the popliteal vein.

c. The superficial veins of the lower extremity are especially susceptible to varicosities, for which there seems to be a genetic predisposition (see Chapter 24, V C 2).

V. THE INNERVATION OF THE POSTERIOR CRURAL COMPARTMENT AND PLANTAR FOOT

A. The cutaneous sensory innervation of the leg and foot (Figs. 26-6 and 26-7).

1. The **saphenous branch** (L4) of the femoral nerve innervates the medial aspect of the leg and foot, including the great toe.

2. The **superficial peroneal branch** (L5) of the common peroneal nerve innervates the anterolateral aspect of the leg and foot.

3. The **sural nerve** (S2), formed by contributions of the tibial and common peroneal nerves, innervates the posterior aspect of the leg.

4. The **lateral** and **medial plantar branches** (S1) of the tibial nerve innervate the middle and lateral aspects of the plantar foot, including the small toe.

B. The motor innervation to the posterior compartment of the leg and of the plantar foot is by the **tibial nerve** (L4–S3), the anterior division of the **sciatic nerve**.

1. Separating from the sciatic nerve in the popliteal fossa, the tibial nerve passes between the two heads of the gastrocnemius muscle and passes along the transverse crural septum (Fig. 26-6).

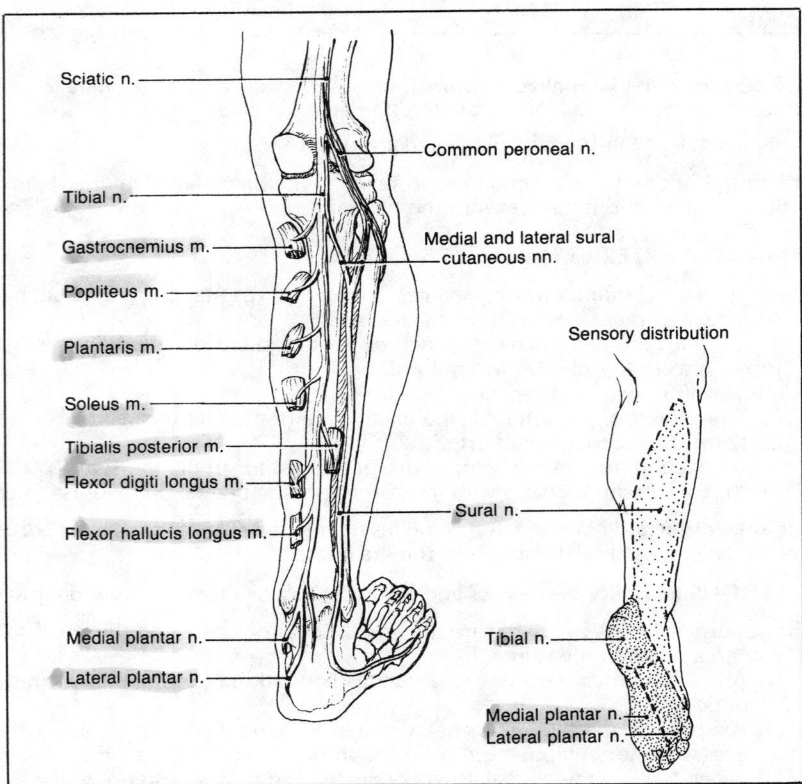

Figure 26-6. *The distribution of the tibial nerve in the leg.*

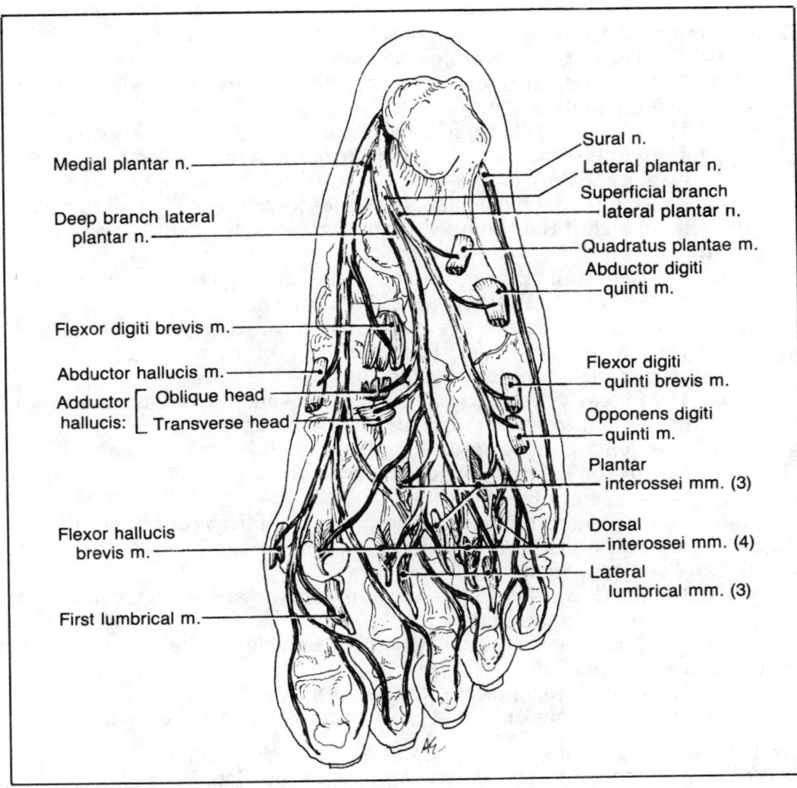

Figure 26-7. *The distribution of the tibial nerve in the foot.*

2. It gives off branches to the musculature of the superficial and deep posterior compartments and accompanies the posterior tibial vessels posterior to the medial malleolus as it enters the foot.
   a. The **medial plantar nerve** (L4–L5) supplies the skin and the intrinsic muscles of the medial side of the plantar foot (Fig. 26-7).
   b. The **lateral plantar nerve** (S1–S2) supplies the skin and the intrinsic muscles of the lateral side of the plantar foot (see Fig. 26-7).

C. CLINICAL CONSIDERATIONS

1. Injuries to the tibial nerve are less frequent than injuries to the common peroneal nerve because the course is deeper and generally more protected.
2. Herniated vertebral disks in the lumbar region may compress the sacral roots, and trauma at the popliteal fossa of medial malleolus may injure the tibial nerve.
   a. A high lesion involves wasting of calf musculature and inability to stand on tiptoe. There is loss of the ankle jerk reflex (L4–L5) when the Achilles tendon is tapped.
   b. A low lesion, such as medial malleolar fracture or tarsal tunnel syndrome (produced by trapment of the nerve under the flexor retinaculum) results in inability to abduct or adduct the toes (which some people cannot normally accomplish) and loss of plantar sensitivity.

VI. AMBULATION

A. **IN GAIT** the center of gravity is repeatedly moved forward, resulting in a momentary imbalance. The individual moves a lower limb in an attempt to regain stability. This results in an alternate shifting of dynamic equilibrium from one foot to the other.

B. **NORMAL GAIT** has been described as the translation of the body's center of gravity through space with a minimum expenditure of energy.

1. In normal gait, each lower limb passes through cycles consisting of alternating stance and swing phases.

a. The **stance phase** (66 percent) consists of three parts:

(1) Heel-strike or the restraining portion.

(a) The hip abductors and adductors stabilize the pelvic girdle over the newly weight-bearing limb as the heel hits the ground.

(b) The quadriceps femoris stabilizes the slightly flexed knee.

(c) The gluteus maximus and hamstrings extend the femur, propelling the center of gravity forward.

(d) Relaxation of the dorsiflexors permits lowering of the foot to midstance, transferring the weight from the heel to the plantar arch.

(2) Midstance.

(a) The center of gravity is momentarily directly over the limb and distributed evenly by the plantar arch to the heel and metatarsals.

(b) The dorsiflexors pull the tibia forward, moving the center of gravity anteriorly.

(c) The triceps surae begin to lift the heel off the ground.

(3) Toe-off or the propulsive portion.

(a) The triceps surae maximally lift the heel and move the body forward by transferring the weight to the metatarsals and phalanges.

(b) The quadriceps femoris extends the slightly flexed knee.

(c) The flexor hallucis longus gives a final strong plantar-flexion, which pushes the body off from the great toe.

(d) The momentum raises the limb free from the ground into the swing phase.

b. The **swing phase** (33 percent) consists of two parts:

(1) Early swing portion:

(a) The hip flexors accelerate the limb forward and the knee flexes passively, lifting the foot from the ground.

(b) The tibialis anterior and synergistic dorsiflexors keep the foot clear of the ground.

(2) Late swing portion:

(a) The dorsiflexors position the foot for heel-strike.

(b) The hamstrings decelerate the limb just prior to heel-strike.

2. Ambulation consists of reciprocally alternating cycles of gait.

a. Walking is defined as gait in which there are overlapping stance phases.

b. Running has no double-stance phase (i.e., left heel-strike with right toe-off).

C. PATHOLOGICAL GAIT is an attempt either to minimize pain or to reduce expenditure of energy by compensation.

1. Heel-strike transmits force to the hip joint, so patients with hip disease avoid heel-strike when walking.

2. Disease in the first metatarsophalangeal joint inhibits toe-off, thereby altering normal gait.

Part VII Lower Extremity

STUDY QUESTIONS

Directions: Each question below contains five suggested answers. Choose the **one best** response to each question.

1. Which of the following actions takes place during the final phase of knee joint extension?

(A) The femur glides forward on the medial tibial condyle
(B) The medial femoral condyle pivots on the tibial plateau
(C) The leg undergoes medial rotation
(D) The popliteus muscle is stretched
(E) None of the above actions

2. Meniscal tears in the knee joint usually result from which of the following circumstances?

(A) Compression
(B) Hyperextension
(C) Hyperflexion
(D) Rotation in partial flexion
(E) Rotation in full extension

3. Tingling, painful, or itching sensations in the lateral region of the thigh may occur in the older overweight individual as a result of a bulging abdomen compressing a nerve beneath the inguinal ligament. Which of the following nerves is involved?

(A) Anterior femoral cutaneous
(B) Femoral branch of the genitofemoral
(C) Genital branch of the genitofemoral
(D) Ilioinguinal
(E) Lateral femoral cutaneous

4. The dorsalis pedis artery is most commonly a continuation of which of the following arteries?

(A) Anterior tibial
(B) Lateral plantar
(C) Medial plantar
(D) Peroneal
(E) Posterior tibial

5. Which of the following muscles inserts onto the tuberosity of the fifth metatarsal bone?

(A) Abductor digiti minimi
(B) Peroneus brevis
(C) Peroneus longus
(D) Tibialis anterior
(E) Tibialis posterior

6. The contraction of the plantar interossei muscles between the fourth and fifth toes results primarily in

(A) abduction of the fourth toe
(B) adduction of the fourth toe
(C) adduction of the fifth toe
(D) flexion of the interphalangeal joints of the fourth toe
(E) flexion of the interphalangeal joints of the fifth toe

Questions 7–9

The great saphenous vein needs to be cannulated for an intravenous line in a presurgical patient.

7. A convenient site for insertion of an intravenous line in the great saphenous vein is

(A) anterior to the medial malleolus
(B) inferior to the inguinal ligament
(C) posterior to the lateral malleolus
(D) medial to the popliteal fossa
(E) none of the above

8. During the cutdown and preparation of the vein for insertion of the cannula, the patient experiences pain radiating along the medial border of the dorsum of the foot. Which of the following nerves was accidently included in a ligature during the cannulation procedure?

(A) Lateral sural
(B) Medial femoral cutaneous
(C) Medial sural
(D) Saphenous
(E) Superficial peroneal

9. Considering the pain distribution along the medial border of the dorsum of the foot, which spinal level is represented in the saphenous nerve?

(A) L2
(B) L3
(C) L4
(D) L5
(E) S1

(end of group question)

Directions: Each question below contains four suggested answers of which **one or more** is correct. Choose the answer

A if **1, 2, and 3** are correct
B if **1 and 3** are correct
C if **2 and 4** are correct
D if **4** is correct
E if **1, 2, 3, and 4** are correct

Questions 10–12

A person experiences weakness when climbing stairs. As part of the neurological examination, the muscle action of the thigh is tested by asking the patient to extend the thigh against resistance.

10. Muscles that are active in producing the indicated movement include the

(1) semitendinosus
(2) gluteus maximus
(3) semimembranosus
(4) gluteus medius

11. A patient is placed in the supine position with the hip and knee joints extended. The patient is asked to abduct the limb against resistance supplied by the examiner. This tests which of the following muscles?

(1) Gluteus medius
(2) Semitendinosus
(3) Gluteus minimus
(4) Semimembranosus

12. The neurological examination reveals that the knee-jerk reflex is normal and that the patient can stand on his heels and toes. From this result, it can be concluded that the innervation involved in this patient's problem includes which of the following nerves?

(1) Common peroneal
(2) Femoral
(3) Tibial
(4) Inferior gluteal

(end of group question)

13. A tumor of the piriformis muscle that compresses the superior gluteal nerve as it passes through the suprapiriform recess of the greater sciatic foramen would produce paralysis of which of the following muscles?

(1) Gluteus minimus
(2) Tensor fasciae latae
(3) Gluteus medius
(4) Gluteus maximus

14. During full active extension of the knee joint, the ligaments that are tightened include the

(1) tibial collateral
(2) fibular collateral
(3) patellar
(4) anterior cruciate

15. After lesion of the tibial portion of the sciatic nerve some flexion still may be possible at the knee joint. The muscles responsible for this remaining flexion include the

(1) long head of the biceps femoris
(2) gracilis
(3) gastrocnemius
(4) short head of the biceps femoris

16. Factors that contribute to the stability of the ankle joint include the

(1) deltoid ligament
(2) trapezoidal shape of the trochlea tali
(3) calcaneotibular ligament
(4) calcaneonavicular ligament

17. Muscles attaching to the plantar surface of the distal (third) phalanx of the second through fifth toes include the

(1) group of dorsal and plantar interossei
(2) flexor digitorum brevis
(3) group of lumbricals
(4) flexor digitorum longus

Directions: The group of questions below consists of lettered choices followed by several numbered items. For each numbered item select the **one** lettered choice with which it is **most** closely associated. Each lettered choice may be used once, more than once, or not at all.

Questions 18-20

For each of the muscle compartments listed below, select the action most likely to be associated with it.

(A) Dorsiflexors of the foot
(B) Evertors of the foot
(C) Both
(D) Neither

18. The muscles of the anterior crural compartment

19. The muscles of the lateral crural compartment

20. The muscles of the posterior crural compartment

ANSWERS AND EXPLANATIONS

1. The answer is D *(Chapter 24 IV A 2 b)* During the terminal phase of extension, the femur rotates medially about the anterior cruciate ligament, so that the medial femoral condyle glides backward and the popliteus muscle is stretched. The medial rotation of the thigh is equivalent to a lateral rotation of the leg.

2. The answer is D. *(Chapter 24 IV B 3 a)* Most meniscal tears are the result of lateral rotation of the partially flexed leg. Because the knee joint is most stable in full extension, meniscal tears are unlikely in this position. Since the full flexion is not compatible with adequate mobility or support, it is a rarely used posture and unlikely to cause problems.

3. The answer is E. *(Chapter 24 VI A 2)* The lateral femoral cutaneous nerve passes between the inguinal ligament and the iliopubic ramus near the anterior-superior iliac spine. Here the nerve may be entrapped, producing sensory symptoms along the lateral aspect of the thigh.

4. The answer is A. *(Chapter 25 V B; Chapter 26 IV A 1 e)* The dorsalis pedis artery usually is a continuation of the anterior tibial artery. Occasionally, the dorsalis pedis artery arises from the peroneal artery, a branch of the posterior tibial artery.

5. The answer is B. *(Chapter 25 Table 25-1; Figure 25-2A)* The peroneus brevis inserts into the tubercle of the fifth metatarsal. Because it is a powerful evertor, forceful inversion avulses the tendinous insertion.

6. The answer is C. *[Chapter 26 III B 2 d (2)]* The plantar interossei adduct the toes toward the axis of the second toe. The dorsal interossei abduct the toes away from the second toe. The plantar and dorsal interossei also extend the proximal and distal interphalangeal joints and flex the metatarsophalangeal joints.

7. The answer is A. *[Chapter 26 IV C 1 a (2)].* The preferred site for saphenous vein cannulation is where the vein passes immediately anterior to the medial malleolus. The lesser saphenous vein, which is not usually used for venipuncture, passes posterior to the lateral malleolus. The saphenous vein passes medial to the popliteal fossa and enters the femoral vein at the level of the inguinal ligament.

8. The answer is D. *(Chapter 26 V A 1)* The saphenous nerve comes into the proximity of the great saphenous vein posterior to the medial tibial condyle. It accompanies the vein along the anteromedial aspect of the leg and into the foot anterior to the medial malleolus.

9. The answer is C. *(Chapter 24 Figure 24-7; Chapter 4 Figure 4-3)* The saphenous nerve, the terminal cutaneous portion of the femoral nerve, contains the contribution from the L4 spinal root. The medial aspect of the foot is generally innervated by L4, and the lateral aspect is generally innervated by S1.

10. The answer is A (1, 2, 3). *(Chapter 23 II C 3 a (2); Table 23-1)* The hamstring group and the gluteus maximus are all extensors of the thigh. The gluteus medius is an abductor.

11. The answer is B (1, 3). *(Chapter 22 II C 3 b (1); Table 23-2)* Abduction of the thigh is primarily by the gluteus medius and minimus muscles, as well as the tensor fasciae latae muscle. The semimembranosus and semitendinosus muscles, the principal extensors of the thigh, are weak adductors.

12. The answer is D (4). *(Chapter 23, II E 2 d; Table 23-1)* The inferior gluteal nerve innervates the gluteus maximus muscle, and paralysis of this produces weakness in climbing stairs and in rising from a seated position. The normal knee jerk indicates that the femoral nerve is not involved, and the ability to stand on the heels and toes indicates that the tibial and common peroneal nerves, respectively, are not involved.

13. The answer is A (1, 2, 3). *(Chapter 23 II E 1)* The superior gluteal nerve innervates the gluteus medius and minimus muscles, as well as the tensor fasciae latae muscle. The gluteus maximus muscle is innervated by the inferior gluteal nerve.

14. The answer is E (all). *[Chapter 24 III C 3; IV B 2 c (1) (3)]* During active extension, the quadriceps femoris pulls on the tibia through the patellar ligament. As the leg comes into full extension, the collateral ligaments become taut, as do the cruciate ligaments, with the anterior cruciate ligament providing the axis about which the femur rotates medially.

15. The answer is C (2, 4). *(Chapter 24 Table 24-1)* The gracilis muscle, which acts to flex the knee, is innervated by the obturator nerve, and the short head of the biceps muscle, which is also a knee flexor, is innervated by the common peroneal nerve. As such, paralysis of the "true" hamstring group by tibial nerve palsy does not abolish the ability to flex the knee.

16. The answer is A (1, 2, 3). *[Chapter 25 II A 1 b (1), (3)]* The stability of the ankle joint is provided primarily by ligaments, including the deltoid ligament (between the medial malleolus of the tibia and the naviculum and calcaneus) and the lateral ligament (between the lateral malleolus of the fibula and the talus and calcaneus). The trapezoidal shape of the trochlea tali increases stability in dorsiflexion. The calcaneonavicular (spring) ligament supports the talus and helps maintain the longitudinal plantar arch.

17. The answer is D (4). *(Chapter 26 II B 2 c; Table 26-1; Table 26-2)* Only the flexor digitorum longus reaches the plantar surface of the distal phalanx of toes two through five. The flexor digitorum brevis splits at its insertion to the plantar surface of the middle phalanges, so that deeper flexor digitorum longus may reach the distal phalanx. The lumbricals and interossei insert into the dorsal expansions of the pedal digits.

18–20. The answers are 18-C, 19-B, 20-D. *(Chapter 25 III D; Figure 25-4; Chapter 26 II B 3)* The muscles of the anterior crural compartment, passing anterior to the transverse axis of the ankle joint and medial to the resultant axis of the subtalar and transverse tarsal joints, dorsiflex and invert the foot. The muscles of the lateral crural compartment, passing posterior to the transverse axis of the ankle joint and lateral to the resultant axis of the subtalar and transverse tarsal joints, plantar-flex and evert the foot. The muscles of the posterior crural compartment, passing posterior to the transverse axis of the ankle joint and medial to the resultant axis of the subtalar and transverse tarsal joints, plantar-flex and invert the foot.

Part VIII
Somatic Neck and Neurocranium

27
The Posterior Cervical Triangle

I. INTRODUCTION

A. The head and neck may be divided, developmentally, functionally, and medically, into an anterior visceral portion and a posterior somatic portion.

1. The **somatic portion** comprises the neurocranium and the cervical vertebral column with the associated musculature.

2. The **visceral portion** comprises derivatives of the upper end of the primitive gut and its associated branchial (gill) structures.

3. The course of the sternomastoid muscle approximates the separation of the somatic and visceral portions of the neck and, as such, divides the neck into an **anterior triangle** and a **posterior triangle**.

B. The **fascia** of the head and neck.

1. The **superficial fascia** in the head and neck regions contains loose connective tissue.

a. The more superficial portions contain variable amounts of adipose tissue.

b. The deeper portions enclose the voluntary muscles of facial expression. This group is represented by the platysma muscle in the neck.

c. In the head and face, the superficial fascia is frequently continuous with and inseparable from the deep fascia overlying bone.

2. A plane of **deep cervical fascia** surrounds the neck and gives off septa, which separate the visceral and somatic portions. This fascial layer can be subdivided into four parts.

a. The **superficial (investing) layer of the deep cervical fascia** ensheaths the neck beneath the superficial fascia (Fig. 27-1).

(1) It encircles the neck, attaching to the ligamentum nuchae posteriorly.

(2) Superiorly, it is attached along the mandible, mastoid process, and ligamentum nuchae. Inferiorly, it finds attachment along the acromion, the clavicle, and the manubrium sterni.

(3) This layer splits several times to individually ensheath the trapezius, omohyoid, and sternomastoid muscles.

(a) Anteriorly between the sternomastoid muscles, the split layers fail to reunite, producing the small **suprasternal space** (Burns'), which contains the anterior jugular veins and an occasional lymph node.

(b) Superiorly, it splits to ensheath the parotid gland and the submandibular glands; the deep layer of the parotid fascia is thickened to form the **stylomandibular ligament**.

b. The **pretracheal fascia** represents a middle fascial layer, which originates as septa from the superficial layer of the deep fascia (see Fig. 27-1).

(1) This layer passes deep to the infrahyoid (strap or ribbon) muscles and anterior to the trachea.

(2) Superiorly, it attaches to the cricoid cartilage; inferiorly, this layer continues into the middle mediastinum, where it fuses with the fibrous pericardium.

(3) Anteriorly, the thyroid gland is surrounded by and held in place by the pretracheal fascia.

c. The **carotid sheath** is a fascial condensation about the carotid arteries, the internal jugular vein, and the vagus nerve (CN X) [see Fig. 27-1].

(1) The pretracheal fascia and the superficial layer of the deep cervical fascia both send septa to the carotid sheath.

(2) The sympathetic chain and superior cervical ganglion lie posterior to the carotid sheath, anterior to the prevertebral fascia.

d. The **prevertebral fascia** encircles the vertebral column and its associated muscles, defining the somatic portion of the neck (see Fig. 27-1).

(1) After originating from the cervical spinous processes, it passes between the trapezius and the intrinsic muscles of the cervical spine.

(2) While in the posterior triangle, it is apposed to the superficial layer of the deep cervical fascia.

(a) The spinal accessory nerve (CN XI) and lymphatics lie between these two layers.

(b) Low in the posterior triangle, the prevertebral layer invests the scalene muscles, leaving a space between them and the superficial layer of the deep cervical fascia.

(i) This space is occupied by the subclavian and external jugular veins, the transverse cervical and suprascapular arteries, and the omohyoid muscle.

(ii) As the nerves to the upper limb pass between the anterior and middle scalene muscles, they carry a continuation of the prevertebral fascia, which forms the **axillary sheath**.

(3) Anteriorly, the prevertebral fascia is attached to the transverse processes of the cervical vertebrae. It then runs to the contralateral transverse process, enclosing the longus colli muscles, the anterior intrinsic muscles of the vertebral column.

(4) Inferiorly, the prevertebral fascia extends into the posterior mediastinum.

e. The **fascial planes** form potential spaces in the neck. Superficial or deep extension of infection tends to be limited by the fascial planes but may track up and down the neck along the resultant spaces (see Fig. 27-1).

(1) The **visceral compartment** is that region bounded by the the pretracheal and prevertebral fascial planes. It is subdivided into two spaces.

(a) The **pretracheal space** is deep to the pretracheal fascia, anterior to the esophagus, and descends into the superior mediastinum. (Infection within the pretracheal space usually comes to the surface in the suprasternal space or may track into the superior mediastinum.)

(b) The **retrovisceral space** is posterior to the esophagus, anterior to the prevertebral fascia, and descends into the posterior mediastinum. (Infection in the retrovisceral space may track deep into the mediastinum.)

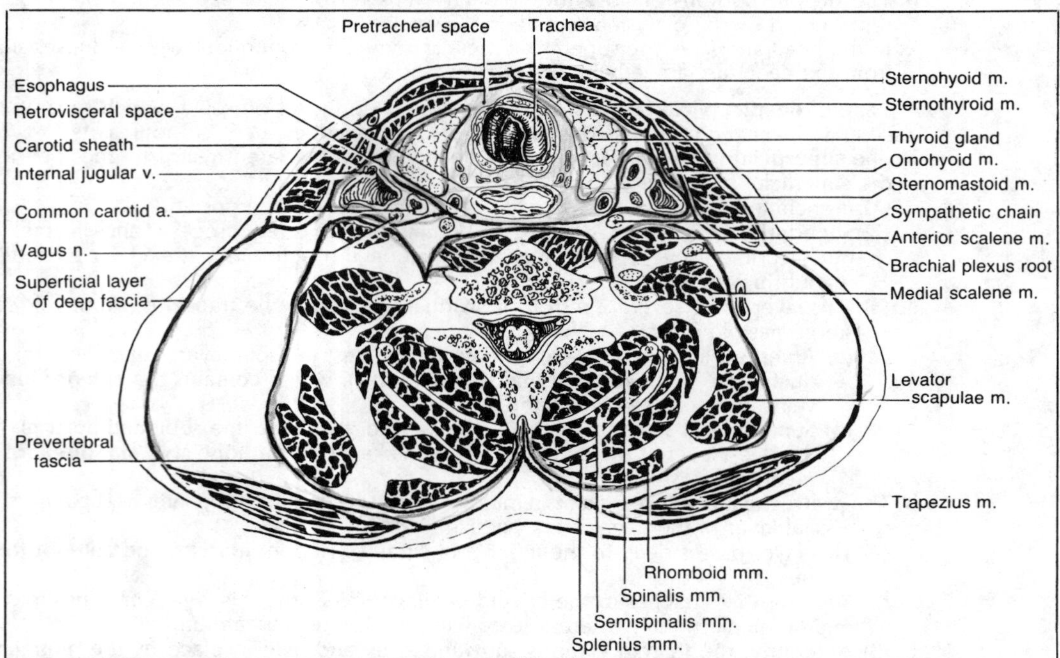

Figure 27-1. *The deep cervical fascia.* The superficial layer of the deep cervical fascia invests the trapezius, sternomastoid, and infrahyoid muscles. The pretracheal fascia invests the visceral structures. The carotid sheath contains the carotid arteries, internal jugular vein, and vagus nerve. The prevertebral fascia invests the musculature of the somatic neck.

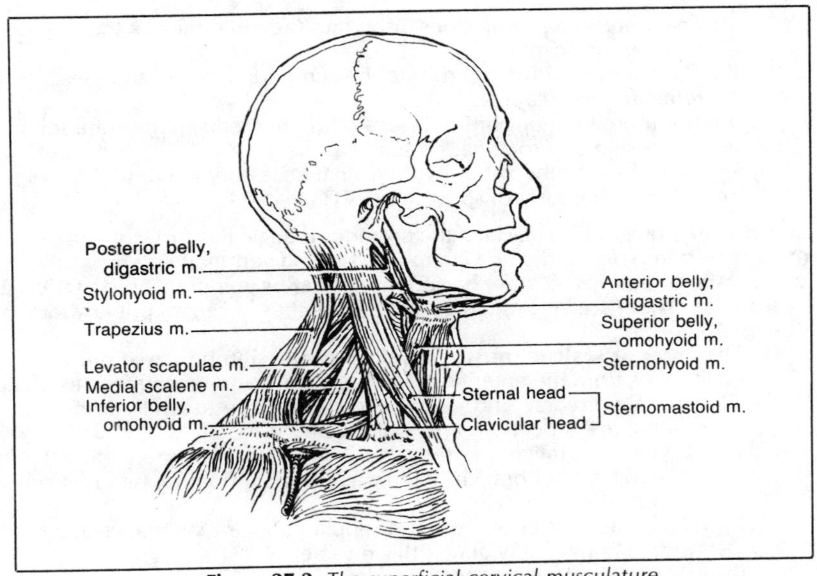

Figure 27-2. *The superficial cervical musculature.*

(2) Infection deep to the prevertebral layer will find its way up and down within the somatic compartment of the neck or come to the surface in the posterior triangle.

II. THE SOMATIC NECK

A. THE BOUNDARIES OF THE POSTERIOR TRIANGLE of the somatic neck are the trapezius muscle, posteriorly, the sternomastoid muscle, anteriorly, and the clavicle, inferiorly (Fig. 27-2).

1. The sternomastoid muscle lies approximately superficial to the division between the somatic neck and the visceral neck.

2. Structures in the posterior cervical triangle include muscles and nerves associated with the cervical vertebral column.

3. The posterior cervical triangle is essentially the superior aspect of the pectoral girdle, so many of the structures of the posterior triangle are en route to the upper limb.

B. THE CERVICAL VERTEBRAE (see Chapter 19 I D 1)

C. THE CERVICAL MUSCULATURE

1. The **superficial muscles** of the posterior cervical triangle are enclosed in the superficial layer of the deep cervical fascia, are derived from one continuous sheet, and are associated with the pectoral girdle.
   a. The **sternomastoid (sternocleidomastoid) muscle** divides the neck into anterior and posterior triangles (see Fig. 27-2).
      (1) It has a narrow tendinous head from the manubrium sterni and a flat muscular head from the medial third of the clavicle.
      (2) It inserts onto the lateral aspect of the mastoid process of the skull.
      (3) It rotates and laterally flexes the head (i.e., it apposes the side of the head to the ipsilateral shoulder).
         (a) This action is characterized in patients with spasm of the sternomastoid—**spasmotic torticollis**.
         (b) **Congenital torticollis** is produced by fibrosis of the sternomastoid, usually the result of hyperextension injury of the muscle upon difficult parturition.
      (4) The sternomastoid muscle is innervated by the spinal accessory nerve (CN XI) as well as by twigs from the ventral rami of spinal nerves C2 and C3.
   b. The **trapezius muscle** derives its name from its bilateral appearance (see Fig. 27-2).
      (1) It arises from the superior nuchal line of the occiput and the external occipital protuberance as well as the from ligamentum nuchae and the spinous processes of vertebrae C7 to T12.
      (2) Its insertion into the pectoral girdle is extensive.

(a) The more superior portions insert into the distal third of the clavicle, elevating and rotating the scapula.
(b) The middle portion inserts into the acromion process and the spine of the scapula, retracting the scapula.
(c) The more inferior portion inserts onto the medial part of the spine of the scapula, depressing and rotating the scapula.

(3) The trapezius is innervated by the spinal accessory nerve (CN XI) as well as by twigs from the ventral rami of spinal nerves C3 and C4.

2. Beneath the **prevertebral fascia** of the posterior triangle lies a row of muscles which comprise the **intrinsic muscles** of the cervical portion of the vertebral column.
   a. The **lateral muscles** include the scalenes (anterior, middle, and posterior) and the levator scapulae. These are innervated by lateral branches of ventral primary rami of spinal nerves (Fig. 27-3).
      (1) The **anterior scalene muscle** lies deep to the sternomastoid muscle.
         (a) It arises from the anterior tubercles of the transverse processes of the third to sixth cervical vertebrae and inserts into the scalene tubercle of the first rib, thus intercalating the subclavian artery and vein.
            (i) The first branches of the subclavian artery (i.e., the vertebral artery and the inferior thyroid branch of the thyrocervical trunk) pass superiorly on the surface of this muscle.
            (ii) The superficial cervical and superascapular branches of the thyrocervical trunk pass horizontally across this muscle.
         (b) Its action is to elevate the first rib.
         (c) Roots of the cervical plexus and of the brachial plexus emerge at the posterolateral edge of this muscle.
            (i) It is innervated by twigs from the ventral primary rami of spinal nerves C4 to C6.
            (ii) The phrenic nerve courses anterior to this muscle to enter the thorax.
      (2) The **middle scalene muscle:**
         (a) Lies posterior to the roots of the brachial plexus.
         (b) Arises from the posterior tubercles of the transverse process of the cervical vertebrae and inserts along the first rib posterior to the subclavian groove.
         (c) Acts to raise the first rib.
         (d) Is innervated by twigs from the ventral primary rami of spinal nerves C3 to C8.
         (e) Is traversed by the dorsal scapula nerve.
      (3) The **posterior scalene muscle** is actually that part of the middle scalene that inserts along the outer surface of the second rib.
      (4) The **levator scapulae:**
         (a) Acts on the pectoral girdle.
         (b) Crosses the middle of the posterior triangle.
         (c) Arises from the posterior tubercles of the transverse processes of the upper four cervical vertebrae and inserts into the superior angle of the scapula.

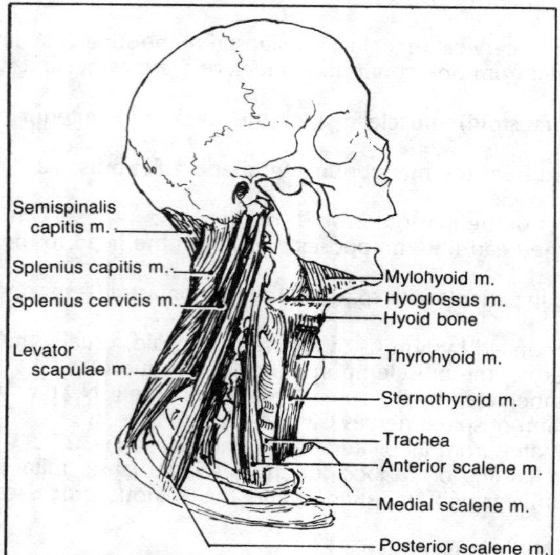

Figure 27-3. *The lateral cervical musculature.*

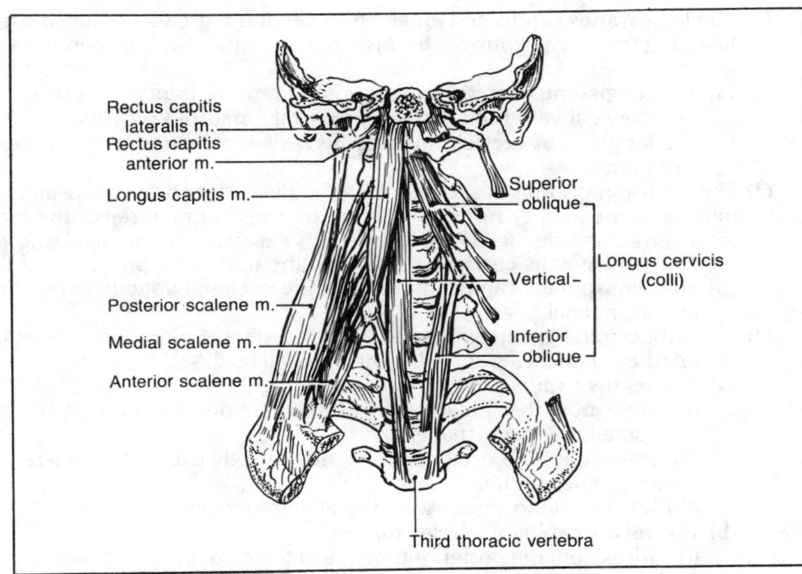

Figure 27-4. *The anterior cervical musculature.*

- **(d)** Acts to raise the shoulder.
- **(e)** Is innervated by twigs from the ventral primary rami of spinal nerves C3 and C4 and occasionally by a twig from the dorsal scapula nerve (C5), which runs along its deep surface.

b. The **anterior muscles** include the longus colli, longus capitis, rectus capitis anterior, and rectus capitis lateralis (Fig. 27-4). These are innervated by anterior branches of the ventral primary rami.

- **(1)** The **longus cervicis (colli) muscle**, which primarily flexes the vertebral column, arises in three parts.
  - **(a)** The superior oblique portion runs from the anterior tubercles of the transverse processes of the third to sixth cervical vertebrae to the spine of the axis.
  - **(b)** The vertical portion runs from the bodies of the upper thoracic vertebrae to insert onto the bodies of the upper cervical vertebrae.
  - **(c)** The inferior oblique portion runs from the bodies of the upper thoracic vertebrae to the anterior tubercles of the transverse processes of the fourth and fifth cervical vertebrae.
- **(2)** The **longus capitis muscle** arises from the anterior tubercles of the transverse processes of the third through sixth cervical vertebrae; it inserts onto the base of the occipital bone. It primarily flexes the head.
- **(3)** The **rectus capitis anterior** arises from the lateral side of the atlas and inserts into the basilar part of the occipital bone. It flexes the head at the atlanto-occipital joint.
- **(4)** The **rectus capitis lateralis** arises from the transverse processes of the atlas; it inserts into the jugular process of the occipital bone. It stabilizes the atlanto-occipital joint.

c. The **posterior muscles** are composed of the splenius group, the semispinalis, and the suboccipital group. These are innervated by dorsal primary rami of spinal nerves.

- **(1)** The **splenius muscles** comprise the external layer (spinotransverse group) of the intrinsic muscles of the back (Fig. 27-5). Acting bilaterally, they extend the head and neck; acting singly, they turn the head toward the ipsilateral shoulder. They are therefore synergists of the contralateral sternomastoid muscles.
  - **(a)** The **splenius capitis muscle:**
    - **(i)** Arises from the spines of the upper thoracic vertebrae and the lower portion of the ligamentum nuchae.
    - **(ii)** Emerges from beneath the trapezius muscle in the posterior triangle and inserts into the mastoid process of the skull.
  - **(b)** The **splenius cervicis muscle:**
    - **(i)** Arises from the spines of the third through sixth thoracic vertebrae.
    - **(ii)** Lies deep to the splenius capitis and inserts onto the transverse processes of the upper four cervical vertebrae.

(2) The **longissimus capitis** and **longissimus cervicis** muscles, comprising the intermediate layer (sacrospinalis group) of intrinsic musculature, run between transverse processes (see Fig. 27-5).
   (a) The **longissimus capitis muscle** arises from the transverse processes of the lower four cervical vertebrae and inserts into the mastoid process.
   (b) The **longissimus cervicis muscle** runs between the transverse processes of the neck and thorax.

(3) The **semispinalis muscles**, comprising the deep layer (transversospinalis group) of the intrinsic musculature, run between the transverse processes of the thoracic and cervical vertebrae (see Fig. 27-5). They act as extensors of the head and neck.
   (a) The **semispinalis cervicis** inserts onto the transverse processes of the atlas.
   (b) The **semispinalis capitis** inserts onto the occipital bone between the superior and inferior nuchal lines.

(4) The **suboccipital group** comprises the deepest layer of intrinsic musculature and is innervated by branches from spinal nerve C1 (Fig. 27-6).
   (a) The **rectus capitis posterior major:**
      (i) Arises from the spine of the axis and then diverges from its counterpart, leaving a small triangular space.
      (ii) Inserts onto the occipital bone immediately caudal to the lateral part of the inferior nuchal line.
      (iii) Acts to extend the head at the atlanto-occipital joint.
   (b) The **rectus capitis posterior minor:**
      (i) Arises from the posterior tubercle of the atlas and inserts onto the occipital bone immediately caudal to the medial part of the inferior nuchal line.
      (ii) Acts to extend the head at the atlanto-occipital joint.
   (c) The **obliquus capitis inferior:**
      (i) Arises from the spine of the axis and inserts into the transverse process of the atlas.
      (ii) Rotates the head to the ipsilateral side at the atlantoaxial joint.
   (d) The **obliquus capitis superior:**
      (i) Arises from the transverse process of the atlas and inserts between the superior and inferior nuchal lines of the occiput.
      (ii) Rotates the head to the ipsilateral side at the atlantoaxial joint and extends the head at the atlanto-occipital joint.
   (e) The **suboccipital triangle** on each side is bounded by the rectus capitis posterior major and the two obliquus muscles.
      (i) The floor of this triangle is formed by the atlanto-occipital membrane, which represents the missing posterior articulation between the axis and atlas.
      (ii) It contains the posterior arch of the atlas, the primary dorsal rami of the first cervical nerve, and the vertebral artery.

3. Clinical considerations.

a. The scalene muscles and the sternomastoid muscle assist extreme inspiratory effort by

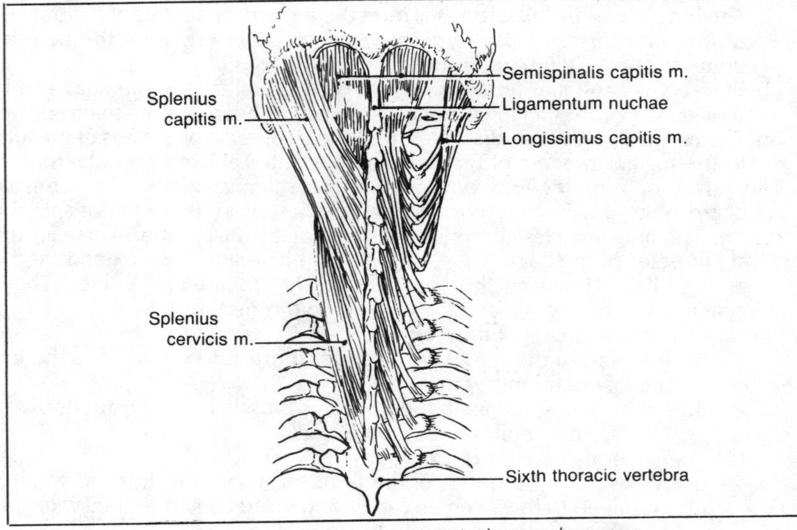

Figure 27-5. *The posterior cervical musculature.*

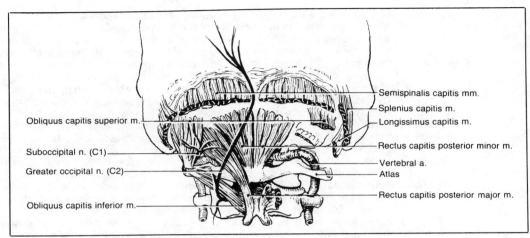

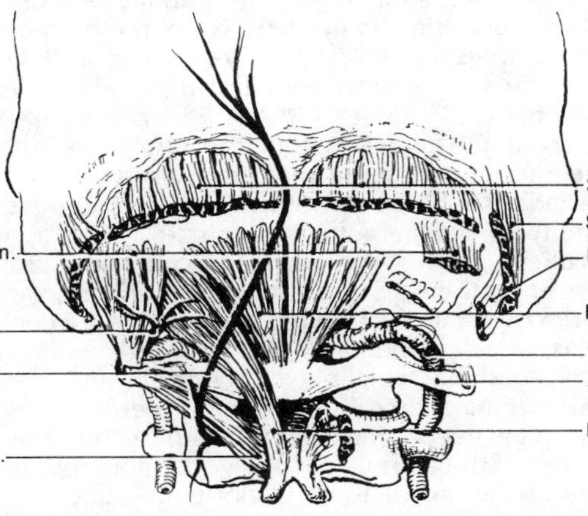

Figure 27-6. *The suboccipital muscles and triangle.*

elevating the first and second ribs. Use of these muscles accompanied by labored breathing is a sign that a patient is in extreme respiratory distress, as in an asthma attack.

b. The subclavian artery and brachial plexus may be subject to compression where they pass over the first rib. The cause may be a cervical rib or spasm of the scalene muscles—**scalene syndrome**. The results are ischemia of the limb and pain along the distribution of the affected nerves.

c. The relations of the scalenus anterior are important when inserting a central venous line into the subclavian vein.

(1) The needle must be guided medially, approximating the long axis of the clavicle, to reach the posterior surface where the vein runs over the first rib.

(2) If the large-bore needle is passed directly perpendicular to the clavicle, then the subclavian artery and the cervical pleura are at risk of being punctured.

D. THE NERVES OF THE POSTERIOR TRIANGLE

1. The **dorsal primary rami** of the cervical nerves innervate the instrinsic musculature of the back and the dorsal portions of the cervical dermatomes.

a. The dorsal branch of C1 (**suboccipital nerve**) supplies the suboccipital muscles and, unlike the other spinal nerves, has no sensory root (see Fig. 27-6).

b. The medial branch of C2 forms the **greater occipital nerve**, which pierces the semispinalis and trapezius to supply sensory innervation to the posterior regions of the scalp (see Fig. 27-6).

c. The medial branch of C3 forms the small **third (least) occipital nerve**, which also innervates the posteroinferior portion of the scalp.

2. The **ventral primary rami** of the cervical nerves pass along a shallow groove on the superior surface of the cervical transverse processes and come to lie in the space between the anterior and middle scalene muscles (Fig. 27-7).

a. The **ventral primary rami of nerves C1 to C4** intermingle in a predictable pattern to form the **cervical plexus**, which gives rise to cutaneous nerves, muscular branches, and communicating branches.

(1) The **cutaneous nerves** emerge into the posterior triangle at the posterior border of the sternomastoid muscle.

(a) The **lesser occipital nerve** (C2) ascends to the scalp posterior to the auricle.

(b) The **great auricular nerve** (C2–C3) passes vertically by the angle of the jaw before bifurcating about the auricle.

(c) The **transverse cervical (colli) nerve** (C2–C3) crosses the sternomastoid horizontally to innervate the anterior triangle of the neck.

(d) The **supraclavicular nerves** (C3–C4) supply the dermatomes of the shoulder and upper chest.

(2) Muscular branches.

(a) Small branches pass from the cervical plexus to all muscles arising from the ventral and lateral aspects of the cervical vertebral column, as well as to the trapezius, sternomastoid, and diaphragm.

(b) The **ansa cervicalis** innervates the strap muscles of the visceral neck.

(i) A communicating branch from C1 joins the hypoglossal nerve (CN XII). Some of the C1 fibers leave the hypoglossal nerve as the **superior root** of the **ansa cervicalis** (whence the former name **descendens hypoglossi**) to innervate most of the infrahyoid muscles. Other C1 fibers continue with the hypoglossal nerve to innervate the geniohyoid, thyrohyoid, and sternohyoid muscles.

(ii) The **inferior root** of the **ansa cervicalis** is formed by the union of branches from C2 and C3 (whence the former name **descendens cervicalis**) and innervates the omohyoid and sternothyroid muscles.

(3) The sympathetic chain sends **gray rami communicantes** to each cervical root (with perhaps the exception of C1), carrying secretomotor, vasomotor, and pilomotor fibers.

b. The **ventral primary rami of the spinal nerves C4 to C8** contribute to the formation of the **brachial plexus**, which is primarily involved with innervation of the upper extremity. Several branches of the brachial plexus course through the posterior triangle.

(1) The **dorsal scapular nerve** (C5, posterior) pierces the scalenus medius muscle, runs posteriorly on the deep surface of the levator scapulae muscle, and descends along the medial border of the scapula to supply the rhomboid muscles.

(2) The **suprascapular nerve** (C5–C6, posterior) runs along the base of the posterior triangle just superior to the clavicle. It innervates the supraspinatus muscle before passing through the great scapular notch to innervate the infraspinatus muscle.

3. The **spinal accessory nerve** (CN XI) [see Fig. 27-7].

a. It has an atypical course.

(1) Rootlets from C1 to C5 unite within the vertebral canal and pass superiorly through the foramen magnum into the cranial cavity and exit through the jugular foramen.

(2) As the spinal accessory nerve transits the jugular foramen, it is joined for a short distance by the **cranial accessory nerve** before the latter splits off to join the vagus nerve to be distributed to the recurrent laryngeal nerve.

(3) From the jugular foramen, the spinal accessory nerve courses laterally and passes posterior (70 to 80 percent) to the jugular vein.

(4) It descends across the medial surface of the stylohyoid and digastric muscles to enter the deep surface of the sternomastoid muscle, to which it sends branches.

(5) It emerges at the posterior border of the sternomastoid muscle in the posterior triangle and passes superficial to the levator scapulae. Its position approximates the perpendicular bisector of a line joining the mastoid process to the angle of the mandible.

(6) It passes deep to the trapezius muscle, to which it sends branches, about 5 cm superior to the clavicle.

(7) At the superior angle of the scapula, it accompanies the dorsal scapular artery, and they run together along the medial border of the scapula superficial to the rhomboid muscles.

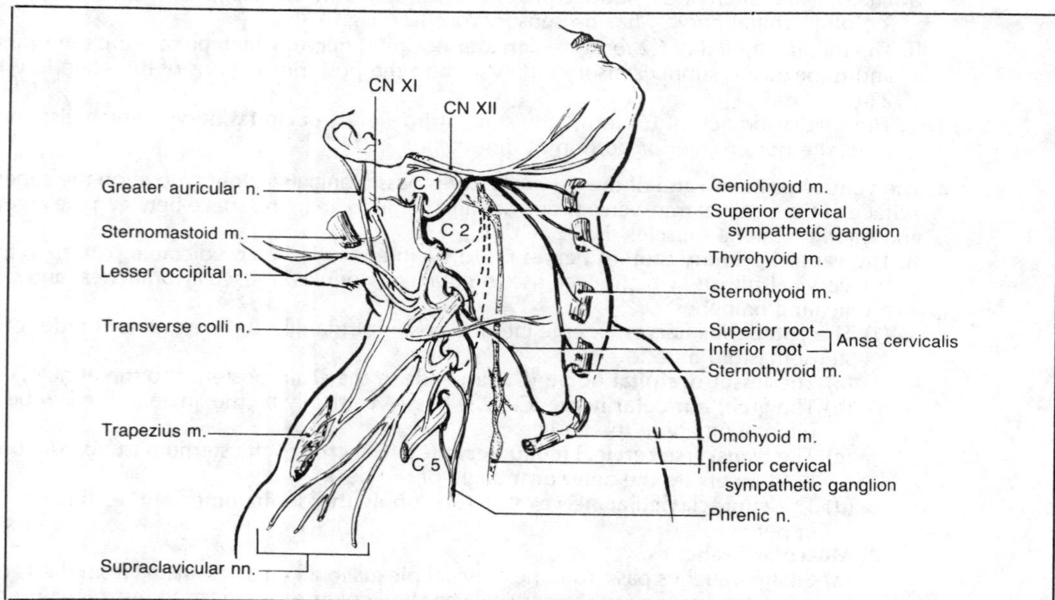

Figure 27-7. *The cervical plexus.*

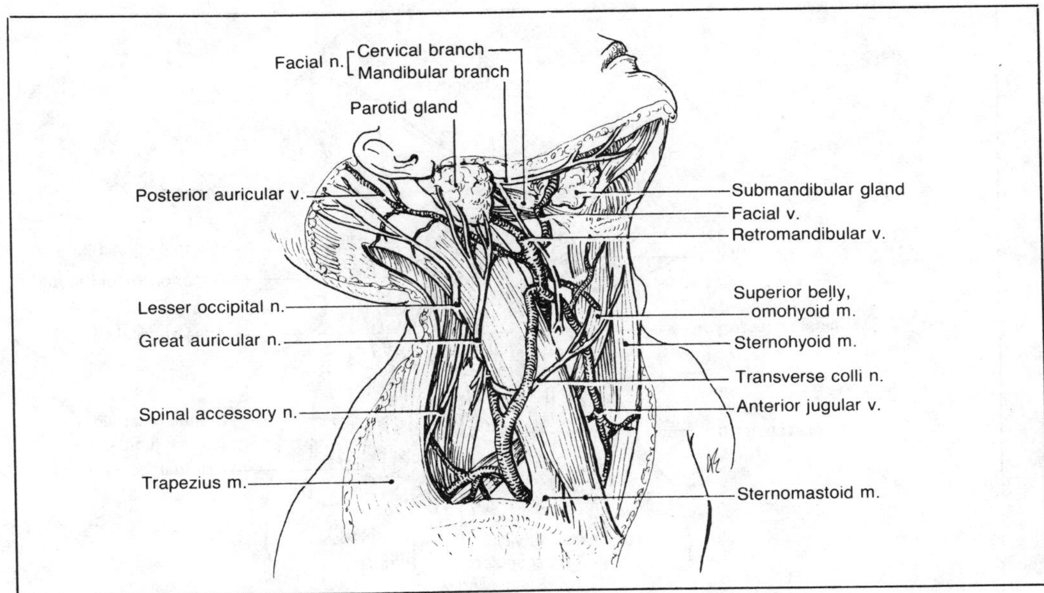

Figure 27-8. *The superficial structures of the posterior cervical triangle.*

b. The spinal accessory must be avoided when exploring the posterior triangle. Paralysis of the spinal accessory nerve results in an inability to shrug the shoulder. Swollen lymph nodes in the posterior triangle may irritate this nerve, producing **spasmodic torticollis**.

E. THE VASCULATURE OF THE POSTERIOR TRIANGLE

1. The **occipital artery** contributes a profuse blood supply to the scalp (Fig. 27-9).
  - **a.** This large vessel is a branch of the external carotid artery.
  - **b.** It is crossed near its origin by the hypoglossal nerve.
  - **c.** It then passes along the inferior border of the posterior belly of the digastric muscle, deep to the sternomastoid muscle.
  - **d.** It grooves the medial surface of the mastoid bone, where it may be torn in a fracture of the base of the skull, producing a pathognomonic hematoma—Battle's sign.
  - **e.** It emerges from behind the sternomastoid muscle at the apex of the posterior triangle.

2. The **superficial cervical artery** is a branch of the thyrocervical trunk (Fig. 27-9).
  - **a.** It passes anterior and lateral to the anterior scalene muscle and across the base of the posterior triangle to reach the edge of the trapezius muscle, where it turns superiorly after giving rise to the dorsal scapular artery.
  - **b.** It anastomoses with a branch of the greater occipital artery.
  - **c.** Occasionally (30 percent), the **superficial cervical artery** and the **dorsal scapular artery** arise from a common trunk, which is called the **transverse cervical artery**.

3. The **suprascapular artery** is also a branch of the thyrocervical trunk (see Fig. 27-9).
  - **a.** It runs laterally in company with the suprascapular nerve.
  - **b.** The suprascapular artery passes superior to the superior transverse ligament of the scapular notch; the suprascapular nerve passes inferior to it.

4. The **external jugular vein** is formed at the angle of the jaw by the union of the posterior division of the **retromandibular vein** and the **posterior auricular vein** (see Fig. 27-8).
  - **a.** It tends to be superficial and variable.
  - **b.** It receives four tributaries.
    - **(1)** The suprascapular vein accompanies the like-named artery.
    - **(2)** The superficial cervical vein accompanies the like-named artery.
    - **(3)** The anterior jugular vein from the anterior triangle.
    - **(4)** The posterior jugular vein from the apex of the posterior triangle.
  - **c.** The external jugular vein passes in the superficial fascia and posterior to the middle of the clavicle, where it joins the subclavian vein.
  - **d.** The external jugular vein may be demonstrated by obstructing its drainage with light finger pressure or by having the patient perform Valsalva's maneuver.
  - **e.** Careful observation of the external jugular will usually reveal venous pressure waves. The

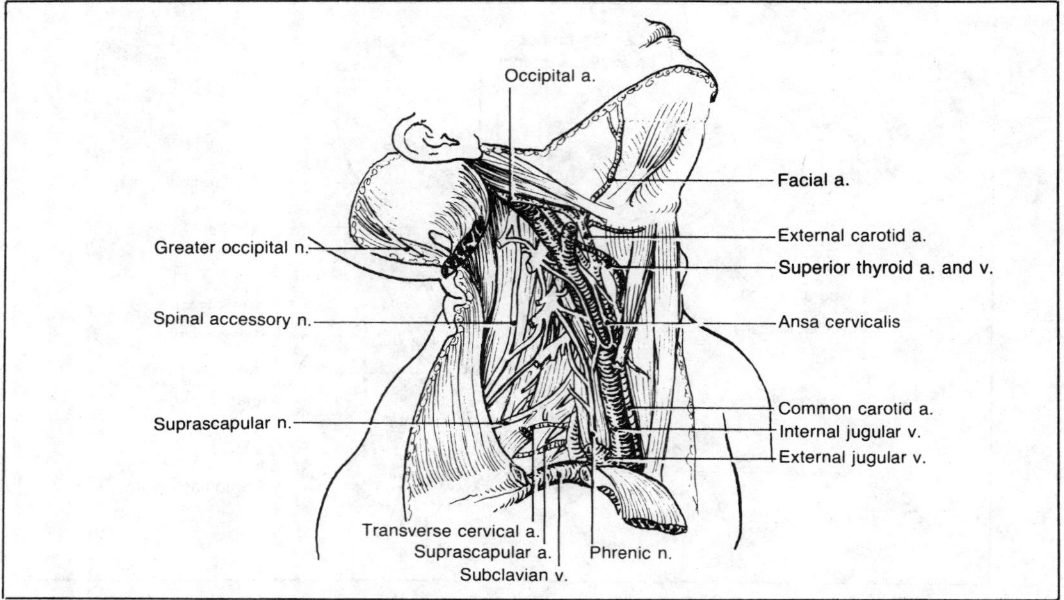

Figure 27-9. *The deeper structures of the posterior cervical triangle.*

level of the column of blood in the external jugular vein is a useful indication of right atrial pressure.

F. THE LYMPHATICS OF THE POSTERIOR TRIANGLE

1. The **superficial cervical lymph nodes** lie along the external jugular vein.
2. The **occipital nodes**, at the apex of the posterior triangle, and the **retroauricular nodes**, over the mastoid process, drain the posterior scalp either through the superficial cervical nodes or through the deep cervical nodes (related to the internal jugular vein).
3. The **nodes of the neck** are numerous and accessible. A lump in the posterior triangle is a common presentation of malignancy within the lymphatic drainage. No physical examination is complete without palpation of the lymph nodes of the neck.
4. The **thoracic duct** passes anterior to the insertion of the scalenus anterior to drain into one of the great veins at the root of the neck on the left side.
5. The bronchomediastinal trunks also pass into the base of the neck and enter the great veins on their respective sides. Tumors of the abdomen and thorax may therefore involve the nodes about the subclavian vein.
6. The breast drains via the axillary nodes into the subclavian trunk in the base of the neck.

28
The Neurocranium

I. INTRODUCTION

A. THE SKULL is the skeleton of the head.

B. THE CRANIUM is the portion of the skull without the mandible.

1. The cranium may be divided, for the purpose of description and study, into an anterior visceral portion and a posterior somatic portion.
   a. The **neurocranium** is the **somatic portion** of the cranium. It consists of eight bones.
   b. The **facial cranium** is made up of derivatives of the upper end of the primitive gut and its associated branchial (gill) structures. It consists of 14 bones.

2. The **calvaria** (*L.* skullcap) is the vault of the neurocranium that serves to cover and protect the cerebral hemispheres of the brain.
   a. The brain is protected by meninges, which attach to the calvaria and to the brain.
   b. The brain is suspended in a pool of cerebrospinal fluid (CSF) confined to the subarachnoid space between arachnoid and pial meningeal layers.

3. Applied to the inferior surface of the cranium, anteriorly, are various sensory organs and structures derived from the oral end of the primitive gut.

4. Many nerves arising from the brain pierce the base of the neurocranium to reach the face and the visceral neck. Conversely, many blood vessels in the face and the visceral neck traverse the bones of the neurocranium to reach the inner aspect of the cranial vault.

5. The neurocranium is covered by the scalp.

II. THE SCALP

A. THE SCALP comprises five layers, which correspond to the letters of the word itself: **S**kin, **C**onnective tissue, **A**poneurosis, **L**oose connective tissue, and **P**eriosteum of the calvaria (Fig. 28-1).

1. The **skin** over the scalp has the greatest concentration of hair, hair follicles, and sebaceous glands of the body.
   a. The distribution of the long, thick, visible terminal hairs is sexually determined. Androgens produce a variable loss of terminal hairs (pattern baldness) in a wide sagittal band from the forehead to the lambdoid suture.
   b. The scalp is a particularly common site for sebaceous cysts, formed by obstructed sebaceous glands.
   c. The scalp receives both somatic and branchiomeric innervation.
      (1) The posterior half of the scalp is innervated by somatic nerves: the **greater occipital nerve** (C2, posterior), the **lesser occipital nerve** (C2–3, anterior) and the **least (third) occipital nerve** (C3, posterior).
      (2) The anterior half of the scalp (the **forehead**) is innervated by branches of the **trigeminal nerve** (CN V): the supraorbital and supratrochlear nerves (from CN V_1), as well as the zygomaticotemporal nerves (from CN V_2) and the auriculotemporal nerve (from CN V_3).

2. The **subcutaneous tissue** of the scalp is dense and binds the skin strongly to the underlying epicranial aponeurosis.
   a. It contains the rich and widely anastomotic arterial supply to the scalp.

(1) The major blood supply to the scalp is provided by branches of the **external carotid artery**, including the **occipital**, **retroauricular**, and **superficial temporal arteries**.
(2) Some of the scalp is supplied by the **ophthalmic branch** of the **internal carotid artery**, including the **supraorbital artery** and **supratrochlear artery**, which accompany nerves of the same name around the orbital margin into the forehead.

b. Wounds of the scalp bleed profusely but heal well.

c. Swelling as a result of bleeding or edema in this layer is limited by the denseness of the connective tissue, and therefore appears as a firm tender lump.

3. The **occipitofrontalis muscle** and the **epicranial aponeurosis** (galea aponeurotica) comprise the subcutaneous muscle layer of the scalp.

a. Several flat muscles arise from the periphery of the calvaria to insert into the epicranial aponeurosis.
(1) The paired **frontalis muscles** have no direct bony attachment. They arise above the eyebrows within the dense superficial fascia, which is in turn bound by fascial septa to the supraorbital ridges of the frontal bones. Contraction of the frontalis muscles raises the eyebrows, as in the expression of surprise.
(2) The paired **occipitalis muscles** arise from the highest nuchal line of the occipital bone and the mastoid process of the temporal bone. Contraction of the occipitalis muscles draws the scalp posteriorly.
(3) The frontalis and occipitalis muscles, together with the common tendon, form the **occipitofrontalis muscle**.
(4) The **anterior** and **superior auricularis muscles** (but not the **posterior auricularis muscle**) converge from a broad origin on the epicranial aponeurosis to an insertion at the base of the auricle. Some individuals have developed rather remarkable control over these muscles.
(5) All of these muscles are derivatives of primitive branchiomeric musculature and are innervated by the facial nerve (CN VII).

b. The epicranial aponeurosis is under tension from the various muscles, and therefore deep lacerations of the scalp gape widely.

4. The **loose connective tissue** between the epicranial aponeurosis and the periosteum permits considerable movement of the scalp.

a. This forms the **subaponeurotic space**.

b. Bleeding in the subaponeurotic space (**extracranial hematoma**) can extend over the cranium. It can extend posteriorly to the highest nuchal line; anteriorly into the eyelids to produce the "black-eye," and laterally to the temporal line.
(1) Extracranial hematoma frequently occurs in conjunction with normal childbirth, the lumpy clot being resorbed within a few weeks.
(2) In conjunction with a depressed cranial fracture, the pressure from arterial bleeding into the closed subaponeurotic space can exacerbate the depressed fracture, compressing the brain or even driving bone fragments into the brain.

c. The loose connective tissue is the plane of separation in any injury that tears the scalp from the calvaria (e.g., that produced by the scalping knife of the American Indian).

5. At the sutures the **periosteum** is fused firmly with the bone and with the periosteum of the adjacent bone, thus limiting the subperiosteal space.

B. THE VEINS OF THE SCALP

1. The superficial veins generally parallel the arteries.

2. The veins of the deeper layers communicate with the diploic veins of the cranium which, in turn, communicate, via **emissary veins**, with the dural sinuses within the cranial vault.

a. Because none of the deep veins in the subaponeurotic space have valves, infection may spread along these veins into the cranial cavity.

b. Hence the term "danger space" is applied to the subaponeurotic layer.

III. THE NEUROCRANIUM

A. THE NEUROCRANIUM CONSISTS OF EIGHT BONES

1. The neurocranium forms from three midline endochondral bones and three pairs of lateral dermal (intramembranous) bones. However, the most anterior of the lateral pairs (the frontal bones) fuse during development into a single bone.

a. The unpaired midline bones include:
(1) The **ethmoid bone**, anteriorly.
(2) The **sphenoid bone**, centrally.

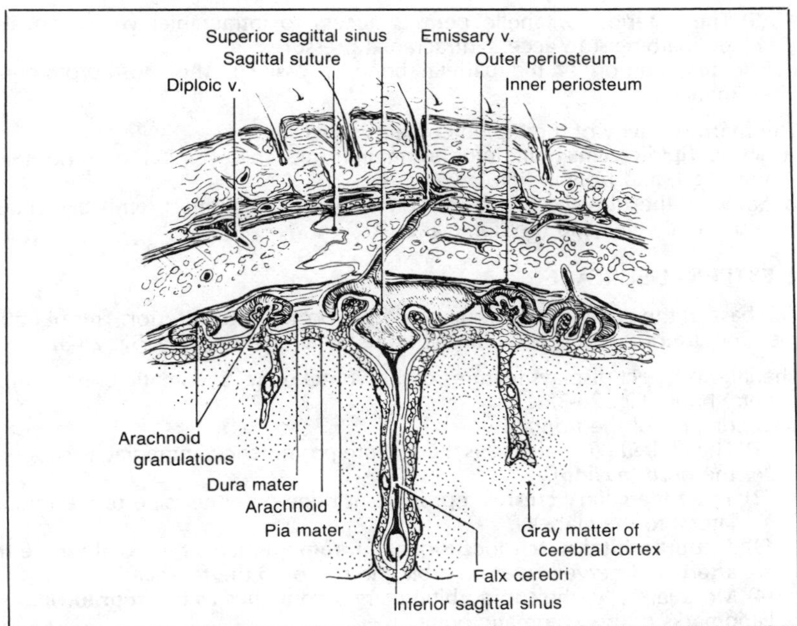

Figure 28-1. *The scalp, cranium, and meninges in coronal section.*

(3) The **occipital bone**, posteriorly.

b. The paired lateral bones include:
(1) The fused **frontal bone**, anterolaterally.
(2) The **temporal bones**, laterally.
(3) The **parietal bones**, superiorly and lateroposteriorly.

2. While some of the bones of the cranium (such as the frontal) fuse early in life (synchondroses), most interdigitate at unyielding sutures.
   a. In children the sutures permit centrifugal growth of flat bones.
   b. In adults the boundaries of the various bones have little significance but provide useful landmarks for the identification of related structures.

3. The vault of the neurocranium is for the most part provided by the articulation of the flat (squamous) bones.
   a. The sagittal axis of the base of the neurocranium is formed by three bones: the ethmoid bone, the body of the sphenoid bone, and the basal portion of the occipital bone.
   b. The calvaria is formed by five bones. The frontal bone and the petrous portions of the temporal bones all have squamous projections which form the walls of the calvaria. The roof is provided by the flat parietal bones on either side of the midline.
   c. Posteriorly, the squamous portion of the occipital bone bridges the space between the petrous portions of the temporal bones.

4. The development and growth of the skull.
   a. The squamous bones are formed by intramembranous ossification early in fetal life.
   b. These plates "float" within the periosteal membranes on the surface of the developing brain and do not articulate with each other.
   c. During passage through the birth canal, the bony plates ride over one another to some extent, facilitating parturition.
   d. In infancy, the bony plates resolve into compact inner and outer **tables** separated by cancellous **diploë**, and the bones either fuse with each other at synchondroses or form sutures.
   e. In an infant there are gaps (fontanelles) in the calvaria at the rounded corners of the parietal bones (Fig. 28-2).
      (1) The anterior gap is the **anterior fontanelle**.
      (2) The posterior gap is the **posterior fontanelle**.
      (3) There are also anterolateral and posterolateral fontanelles.
      (4) The anterior fontanelle is diamond-shaped and the posterior is triangular. This enables the obstetrician to determine the orientation of the fetal head by vaginal examination during labor.

(5) The anterior fontanelle permits access to intracranial veins and also enables the perinatologist to access intracranial pressure.

f. The first portion of the parietal bone to ossify is the most prominent—the **parietal eminence**.

5. The marrow cavity of the diploë contains veins.
   a. These **diploic veins** drain into four main trunks: the occipital, posterior temporal, anterior temporal, and frontal diploic veins.
   b. Some of the diploic veins form communicating channels (**emissary veins**) between the superficial veins and the dural sinuses.

B. THE EXTERNAL SURFACE

1. The base of the cranium slopes upward from posterior to anterior. This oblique surface forms the juncture of the neurocranium and the facial cranium (see Fig. 28-3).
2. The lateral aspect displays features of the frontal bone, zygomatic bone, temporal bone, and parietal bone (Fig. 28-3).
   a. Landmarks of the frontal bone.
      (1) The **glabella** (*L.* hairless) is the smooth prominence immediately superior to the root of the nose (**nasion**).
      (2) The **superciliary crest** is a prominent ridge on either side of the midline just superolateral to the glabella.
      (3) A **supraorbital notch** (occasionally **foramen**) marks the point where the supraorbital artery and nerve leave the orbit and run onto the forehead.
      (4) More laterally, the supraorbital margin continues as the **supraorbital ridge**.
   b. Landmarks of the zygomatic bone.
      (1) The orbital margin turns sharply inferiorly along the frontal process of the zygomatic bone.
      (2) The temporal process of the zygomatic bone projects posteriorly to fuse with the zygomatic frontal process of the temporal bone, forming the **zygomatic arch**.
      (3) The zygomatic arch bridges the **temporal fossa** and conceals the base of the cranium from lateral view.
   c. Landmarks of the temporal bone.
      (1) At the temporal root of the zygomatic arch, the **articular tubercle** lies just anterior to the **condylar notch**.
         (a) These form a sigmoidal articular surface for the condylar process of the mandible.
         (b) An articular disk is interposed between the bones of the **temporomandibular joint**.
      (2) A flat, nonarticular **tympanic plate** intervenes between the condylar notch and the **external auditory meatus**.

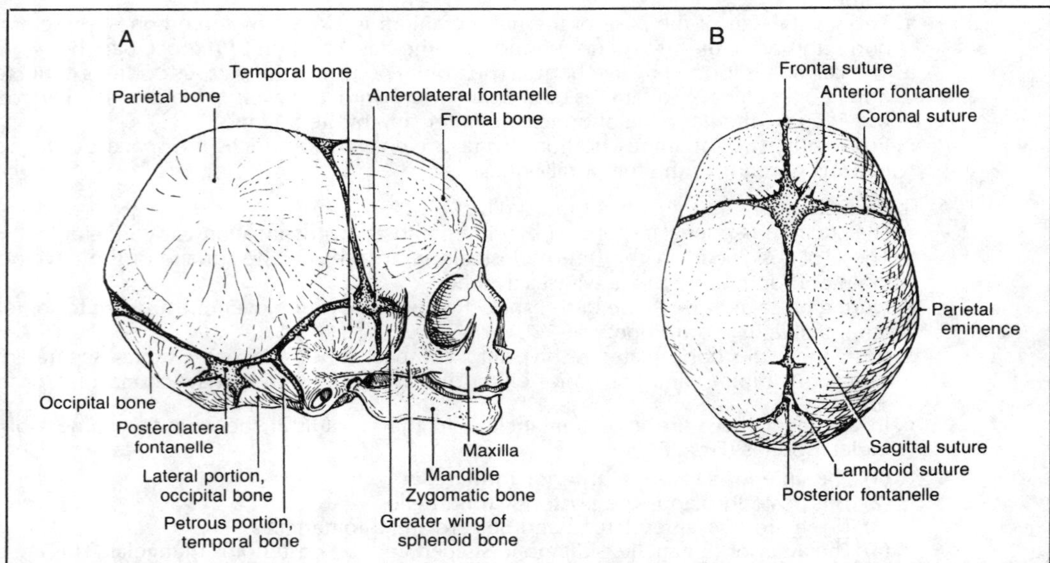

Figure 28-2. *The skull of the newborn. A,* Lateral aspect. Note the relative sizes of the neurocranium and facial skeleton, the fontanelles, and the absence of a mastoid process. *B,* Superior aspect.

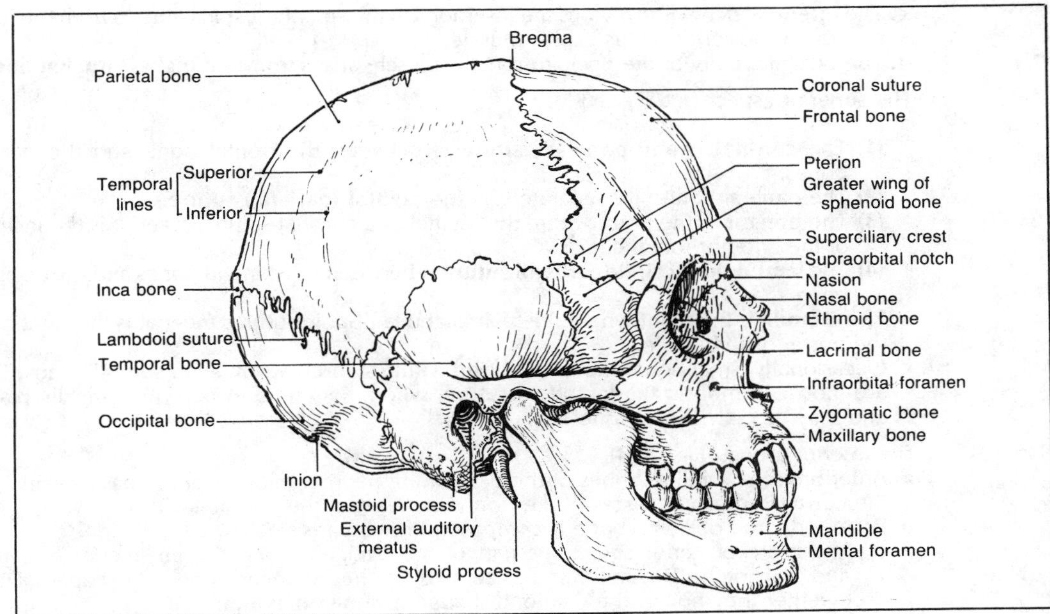

Figure 28-3. *The lateral aspect of the skull.*

(3) Superior to the external auditory meatus and mastoid process, the zygomatic arch continues as a bony ridge, the **supramastoid crest**.
   (a) A small **suprameatal spine** lies at the posterosuperior edge of the external auditory meatus. This spine and the supramastoid crest bind the small **suprameatal triangle** which is a surgical landmark for the **mastoid antrum**.
   (b) The supramastoid crest passes along the lateral aspect of the neurocranium as the **temporal line** from which the temporalis muscle originates.

(4) Inferior to the external auditory meatus the long **styloid process** projects at a right angle to the base of the cranium.
   (a) This process gives rise to several muscles of the visceral neck, as well as the stylohyoid and stylomandibular ligaments.
   (b) It is a remnant of the second branchial arch.

(5) The stout **mastoid process** lies posterolateral to the styloid process.
   (a) The mastoid process is not present at birth but develops as a result of traction from the sternomastoid muscles.
   (b) The splenius capitis and longissimus capitis muscles insert onto its posterolateral aspect.
   (c) The mastoid process is filled with air cells which communicate, via the **mastoid antrum**, with the cavity of the middle ear. The spread of a middle ear infection into the mastoid air cells is difficult to treat because of poor drainage. Radical surgery involves scraping and extirpation of the infected air cells (mastoidectomy).
   (d) On the internal aspect, the mastoid air cells are related to the sigmoid sinus.
   (e) There are foramina for a few emissary veins posterior to the mastoid process.

d. The squamous temporal and frontal bones do not meet; the small gap is filled by the greater wing of the sphenoid bone.
   (1) The **pterion** is the point at which the sphenoid bone joins the parietal bone.
   (2) In infants it is the location of the anterolateral fontanelle.
   (3) It is located two fingerbreadths above the midpoint of the zygomatic arch.
   (4) Its intracranial surface is grooved and occasionally tunneled by the middle meningeal vessels, so that a blow to the side of the head, which either fractures the skull or tears the periosteum at this point, produces rapid intracranial hemorrhage (epidural hematoma).

3. The posterior aspect (see Figs. 28-3 and 28-4).
   a. The attachment of the ligamentum nuchae to the occipital bone is marked by the **external occipital protuberance (inion)** and a thin **external occipital crest**, extending in the sagittal plane to the foramen magnum.
   b. The **superior nuchal line** connects the inion with the mastoid process and gives origin to the trapezius muscle.

c. The **inferior nuchal line** divides the insertions of the splenius capitis muscle from the insertion of the posterior rectus capitis muscle.
d. The occipital belly of the occipitofrontalis muscle arises from the **highest nuchal line**.

4. The superior aspect (see Fig. 28-3).
   a. Sutures.
      (1) The **coronal (frontoparietal) suture** is between the frontal bone and the parietal bones.
      (2) The parietal bones are separated by the **sagittal (parietal) suture.**
      (3) The **bregma** is the point where the sagittal and coronal sutures meet; it is the location of the anterior fontanelle.
      (4) The **lambdoid (occipitoparietal) suture** is between the parietal bones and the occipital bone.
   b. The **lambda** is the point where the sagittal and lambdoid sutures meet; it is the location of the posterior fontanelle.
   c. Occasionally, small squamous bones form within sutures, wormian or sutural bones. They are most common in the lambdoid suture, where they tend to be symmetrically placed and may be called Inca bones.
5. The inferior aspect (Fig. 28-4).
   a. Anteriorly, the midline bones of the base of the neurocranium roof the nasal cavity, and the corrugated orbital plates of the frontal bones roof the orbital cavities.
   b. Posteriorly, the occipital bone is composed of two parts.
      (1) The more posterior squamous part of the occipital bone has an inferior triangular midline notch. The anterior basilar part of the occipital bone is Y-shaped. Fused together they bound the diamond-shaped **foramen magnum**.
      (2) Along the anterior edges of the foramen magnum (the basilar part of the occipital bone) are paired **occipital condyles** for articulation with the atlas.
      (3) The **hypoglossal (anterior condylar) canal**, emerging beneath the anterolateral edge of each condyle, transmits the hypoglossal nerve (CN XII).
      (4) A **posterior condylar canal**, at the posterolateral end of each condyle, transmits an emissary vein.
   c. The petrous portions of the temporal bones consist of two columns of dense bone that abut the anterolateral edges of the limbs of the basioccipital bone.
      (1) Each petrous portion of the temporal bone and the adjacent edges of the squamous portions of the temporal and occipital bones contain numerous foramina.
      (2) The region immediately inferior to the petrous portion of the temporal bone lies between the neck of the mandible and the pharynx.
      (3) The important bony landmarks lie in parallel rows that run obliquely along the petrous portion of the temporal bone.

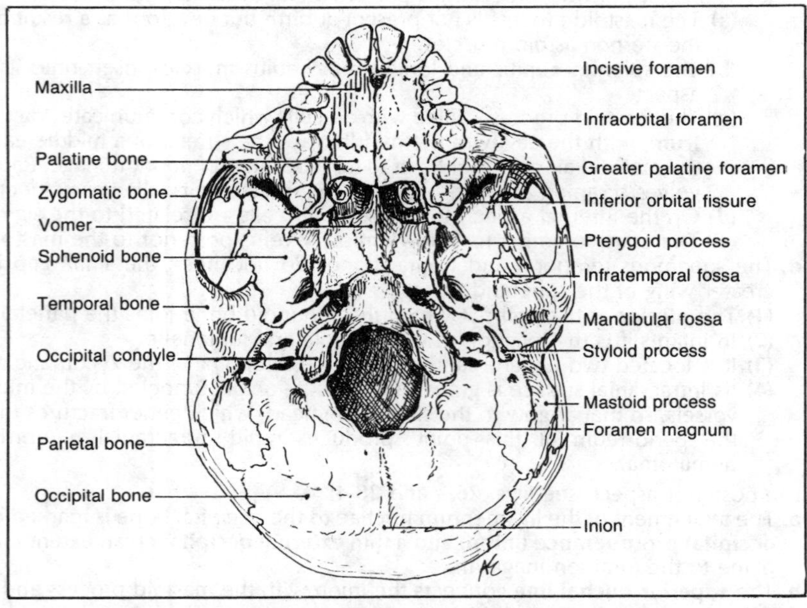

Figure 28-4. *The inferior aspect of the cranium.*

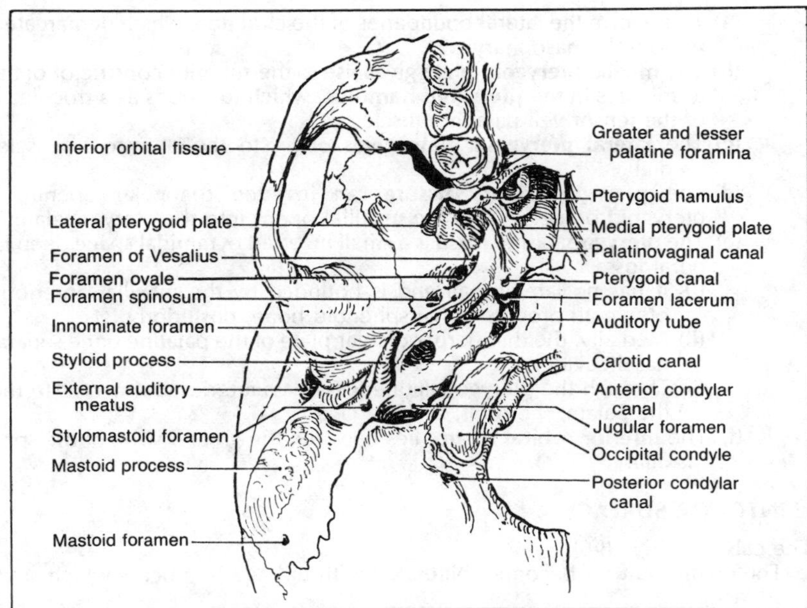

Figure 28-5. *The foramina and landmarks of the infratemporal fossa.*

(a) The middle row lies in the anterior edge of the petrous bone (Fig. 28-5).

- **(i)** The **mastoid process** is grooved on its medial side for the digastric muscle and for the occipital artery.
- **(ii)** The **tympanic plate** forms the sheer vertical anterior wall of the external auditory meatus of the middle ear.
- **(iii)** The middle ear cleft continues anteriorly as the **tympanic (eustachian) tube**. The bony portion emerges from the anteromedial end of the tympanic plate. It is continued by a cartilaginous portion which runs in a groove along the anterolateral margin of the petrous portion of the temporal bone.

(b) The posteromedial row (see Fig. 28-5).

- **(i)** The **jugular foramen** runs posteriorly into the cranial cavity. It transmits the internal jugular vein, glossopharyngeus nerve (CN IX), vagus nerve (CN X), and spinal accessory nerve (CN XI).
- **(ii)** The **carotid canal** curves sharply anteriorly along the petrous temporal bone. At the ragged apex of the temporal bone, the canal wall becomes discontinuous as the **foramen lacerum**. The artery continues undeviated to enter the cranial cavity alongside the body of the sphenoid bone.
- **(iii)** The anterior margin of the foramen lacerum is bounded by the root of the greater wing of the sphenoid bone.

(c) The anterolateral row.

- **(i)** The **petrosquamous fissure** lies between the petrous and squamous portions of the temporal bone. Here the superior and inferior walls of the middle ear cleft overlap anteriorly. A tiny plate of **tegmen tympani** projects into the petrosquamous fissure from above, dividing it into a **petrotympanic fissure** (that transmits the chorda tympani nerve) and a **squamotympanic fissure**.
- **(ii)** At the posterolateral angle of the greater wing of the sphenoid is the **spinous process**.
- **(iii)** The **foramen spinosum** transmits the middle meningeal artery.
- **(iv)** The **foramen ovale** transmits the motor and sensory roots of the mandibular division of the trigeminal nerve (CN V), together with the accessory meningeal artery, the lesser superficial petrosal nerve, a recurrent meningeal branch of the trigeminal nerve, and an emissary vein.
- **(v)** The **scaphoid fossa** lies at the base of the **medial pterygoid plate** of the sphenoid bone and marks the site of origin of the tensor veli palatini muscle.
- **(vi)** The **pterygoid (vidian) canal** lies at the base of the scaphoid fossa. It transmits the vidian nerve (formed by the greater superficial petrosal nerve and the deep petrosal nerve) to the pterygopalatine fossa.

(4) The **pterygoid processes** arise from the inferior surface of the greater wing of the sphenoid bone.

(a) These form the lateral boundaries of the **choanae**, which demarcate the nasal cavity from the nasopharynx.
(b) The **medial pterygoid plate** gives rise to the superior constrictor of the pharynx and terminates in the **pterygoid hamulus**, which functions as a trochlea for the tendon of the tensor veli palatini muscle.
(c) The **lateral pterygoid plate** gives origin to the pterygoid muscles (lateral and medial).
(d) The **pterygomaxillary fissure**, an inverted triangular opening between the pterygoid processes and the maxilla, opens into the pterygopalatine fossa.
(e) The **pterygopalatine fossa** is a small inverted pyramidal space deep to the infratemporal fossa.
  (i) It has no lateral wall and is bounded by the maxilla, anteriorly, and by the pterygoid process of the sphenoid bone, posteriorly.
  (ii) Medially, the thin perpendicular plate of the palatine bone separates it from the nasal cavity.
  (iii) Through the pterygopalatine fossa pass arteries and nerves to the nasal cavity, the palate, the orbit, and the face.
(f) The **inferior orbital fissure** lies between the greater wing of the sphenoid and the maxilla.

C. THE INTERIOR SURFACE

1. The calvaria (Fig. 28-6).
  a. The cranial sutures become obliterated with old age, a process which starts at the inner surface.
  b. The inner surface of the calvaria is deeply grooved and occasionally tunneled by the meningeal vessels.
  c. A **frontal crest** provides attachment for the falx cerebri, a midsagittal fold of dura mater.
    (1) It diminishes and becomes the **sagittal groove**, which deepens posteriorly to accommodate the **sagittal sinus**.
    (2) On either side of the midline are **granular lacunae**.
      (a) These accommodate the venous lacunae, into which CSF drains from arachnoid granulations.
      (b) They become larger and deeper with age.
    (3) The sagittal groove becomes a raised gutter over the occipital bone.
      (a) At the **internal occipital protuberance** it meets similar gutters, converging from both sides in the transverse plane.
      (b) This marks the **confluence of sinuses** (torculus of Herophilus).

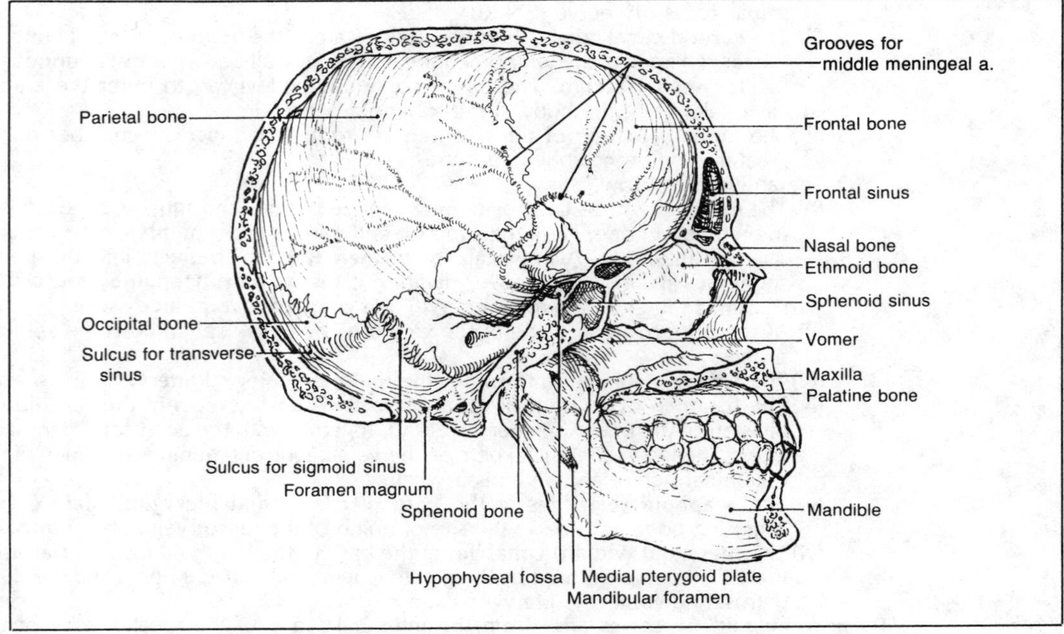

Figure 28-6. *The lateral aspect of the interior skull.*

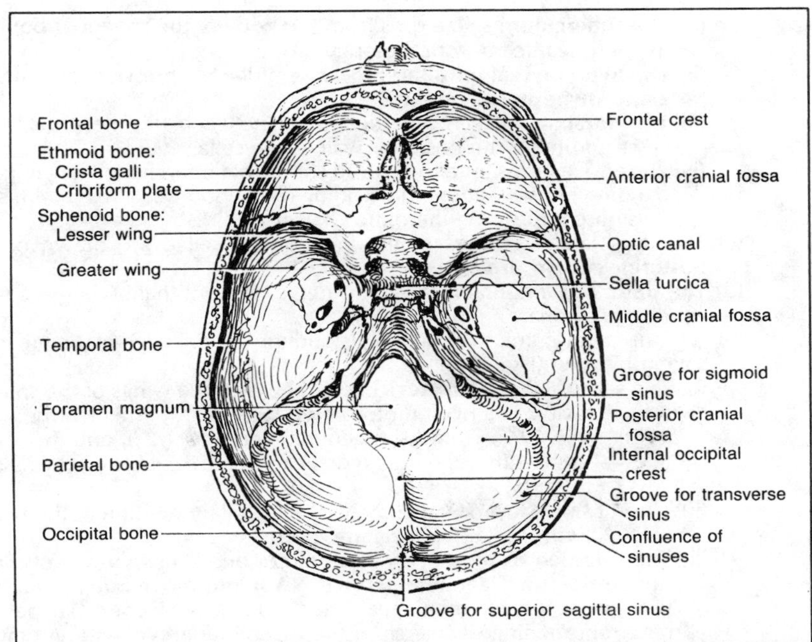

Figure 28-7. *The floor of the cranium.*

d. The **internal occipital crest**, below the confluence of sinuses, provides attachment for the falx cerebelli.

2. The floor of the cranial cavity (see Figs. 28-6; Fig. 28-7).

a. In the midline, the floor of the cranial cavity declines in a continuous convex curve, interrupted only by the **tuberculum sellae** and the **dorsum sellae** of the sphenoid bone.

b. Laterally the **cranial fossae** (anterior, middle, and posterior) decline in steps from anterior to posterior.

(1) The **anterior cranial fossa**, formed by the ethmoid, frontal, and sphenoid bones, contains the frontal lobes of the brain.

(a) The anterior fossa is bounded anteriorly and laterally by the frontal bone.

(b) The **crista galli** of the ethmoid bone provides attachment for the falx cerebri.

(i) Between the crista galli and the crest of the frontal bone is a small **foramen cecum**, transmitting an emissary vein that passes from the veins of the frontal sinus and nasal cavity to the superior sagittal sinus.

(ii) On either side of the crista galli lie the olfactory bulbs of the brain.

(iii) The thin **cribriform plate** of the ethmoid bone is perforated on each side by 15 to 20 foramina, through which the olfactory nerves pass from the nasal mucosa to each olfactory bulb.

(c) The orbital plates of the frontal bone form the floor of the anterior fossa and the roof of the orbit.

(d) The smooth arc of the lesser wing of the sphenoid bone demarcates the sharp posterior margin of the anterior cranial fossa.

(i) Anteriorly, the lesser wing meets the orbital plate.

(ii) Laterally, the greater wing of the sphenoid bone fuses with frontal, parietal, and temporal bones to close the pterion.

(iii) Medially, the lesser wing forms the **anterior clinoid process**.

(iv) The space between the greater and lesser wings is the **superior orbital fissure**.

(2) The **middle cranial fossa**, formed by the sphenoid and temporal bones, is deeper than the anterior fossa and is related to the pituitary gland and the temporal lobes of the brain (see Fig. 28-7).

(a) The **optic groove (sulcus chiasmatis)** grooves the sphenoid bone between the **optic foramina** and the tuberculum sellae. It is named because it contains the optic chiasm, rather than the optic nerve.

(b) The **sella turcica** is named for its fanciful resemblance to a premedieval Turkish saddle with its anterior tuberculum sellae, central fossa, and posterior dorsum sellae.

(i) The **tuberculum sellae** rises from the body of the sphenoid bone, dividing the hypophyseal fossa from the optic groove.
(ii) The **hypophyseal (pituitary) fossa**, a midline depression, contains the **pituitary gland (hypophysis cerebri)**.
(iii) The **dorsum sellae** is a vertical plate between the the posterior clinoid processes, forming the posterior wall of the sella.
(iv) The above terms tend to be used somewhat interchangeably (e.g., ". . .an expanding pituitary adenoma eroding the hypophyseal fossa with supracellar extension compressing the optic chiasm. . . .").

(c) The dorsolateral corners of the flat dorsum sellae are drawn out to form the **posterior clinoid processes**.
(d) The **optic canal** contains the optic nerve, the ophthalmic artery, and the central vein of the retina.
(e) A group of foramina form an arc around the medial edge of the greater wing of the sphenoid bone (Fig. 28-8).
(i) The **superior orbital fissure** passes between the wings of the sphenoid into the orbital cavity. The oculomotor nerve (CN III), the trochlear nerve (CN IV), the ophthalmic division of the trigeminal nerve (CN V_1), and the abducens nerve (CN VI) all pass through the medial end of the superior orbital fissure to gain access to the orbit.
(ii) The **foramen rotundum** conveys the maxillary division of the trigeminal nerve (CN V_2) into the pterygopalatine fossa.
(iii) The **foramen ovale** transmits the motor and sensory roots of the **mandibular division** of the **trigeminal nerve** (CN V_3) into the infratemporal fossa, together with an accessory meningeal artery, the lesser superficial petrosal nerve, a recurrent meningeal branch of the trigeminal nerve, and an emissary vein.
(iv) The **foramen spinosum** transmits the **middle meningeal artery**.
(v) There are other small, variable foramina that lie along this arc (e.g., the emissary foramen, the lacrimal foramen, the foramen of Vesalius, and the innominate canal).

(f) The **carotid canal** passes through the body of the petrous portion of the temporal bone to emerge lateral to the dorsum sellae.
(i) The **foramen lacerum** is formed by superior and inferior defects in the carotid canal (i.e., the petrous temporal bone fails to meet the sphenoid bone). In the floor of the middle cranial fossa, the foramen lacerum is covered by a layer of fibrocartilage and periosteum.
(ii) No structure passes all the way through the foramen lacerum.

(g) Most of the anterior wall of the middle cranial fossa is formed by the greater wing of the sphenoid bone.
(h) The posterior margin of the middle cranial fossa is formed by the superior margin of the petrous temporal bone lateral to the posterior clinoid process.
(i) The roof of the middle ear cavity is the thin **tegmen tympani**. It is easy to understand how a severe middle ear infection may erode into the cranial cavity.
(ii) The anterior semicircular canal is accommodated by the **arcuate eminence**.
(iii) The superior edge of the petrous portion of the temporal bone is sharp and grooved for the superior petrosal sinus.
(iv) At the apex of the petrous portion of the temporal bone is a depression for the trigeminal (semilunar) ganglion and, medial to that, a groove for the abducens nerve (CN VI).

(3) The **posterior cranial fossa**, the deepest and largest of the fossae, is formed by the temporal and occipital bones and contains the cerebellum and brain stem (see Figs. 28-6 and 28-7).
(a) The **clivus** of the basal portion of the occipital bone climbs from the foramen magnum to the dorsum sellae.
(b) The **transverse sinus** and **sigmoid sinus** deeply groove the walls of the posterior cranial fossa. The sigmoid sinus joins the **inferior petrosal sinus** to become the **superior bulb** of the **internal jugular vein** as it exits the cranial cavity through the **jugular foramen**.
(c) On the posterior surface of the petrous portion of the temporal bone are numerous minor fossae: a fossa for the endolymphatic sac, one for the canaliculus (aqueduct) of the cochlea, and the subarcuate fossa.
(d) Several foramina pass through the basal walls of the posterior cranial fossa.
(i) The **internal auditory meatus** emerges through the posteromedial surface of the petrous portion of the temporal bone, transmitting the facial nerve (CN VII) and the vestibulocochlear nerve (CN VIII).

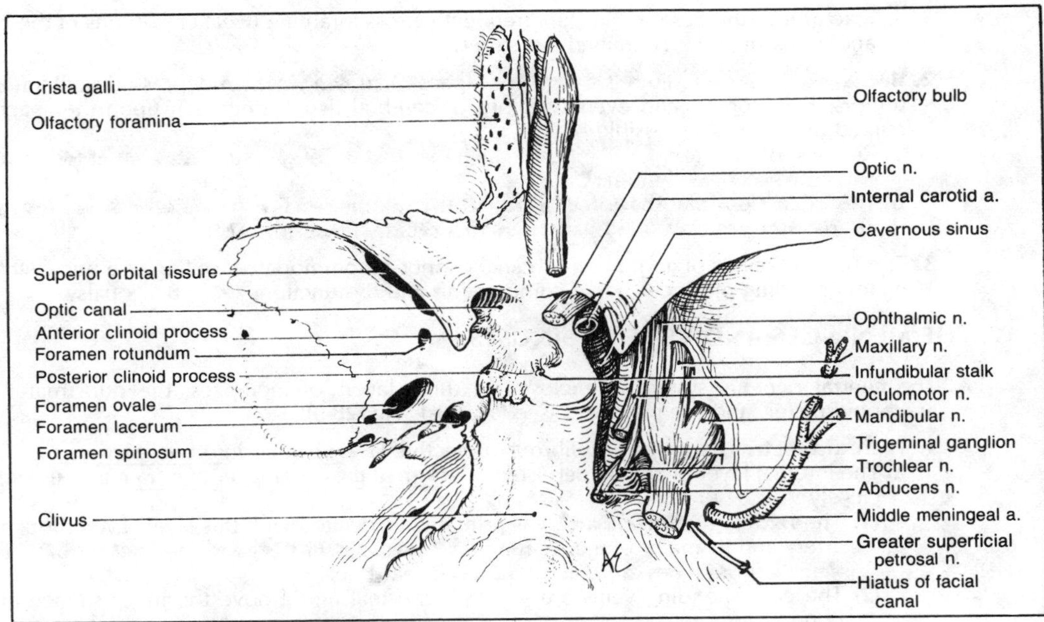

Figure 28-8. *The foramina of the middle cranial fossa.*

(ii) The **jugular foramen** transmits the internal jugular vein through its posterior part. The glossopharyngeus nerve (CN IX), the vagus nerve (CN X), and the spinal accessory nerve (CN XI) all pass through the jugular foramen anterior to the internal jugular vein.

(iii) The **anterior condylar (hypoglossal) canal** transmits the hypoglossal nerve (CN XII).

(iv) At the **foramen magnum**, the brain stem is continuous with the spinal cord. The spinal roots of the spinal accessory nerve (CN XI) and the vertebral arteries pass into the cranial cavity through this foramen.

(v) The **posterior condylar canal** transmits an emissary vein from the sigmoid sinus to the veins of the pericranium.

D. SUMMARY OF THE FORAMINA OF THE NEUROCRANIUM

1. The anterior cranial fossa is pierced by numerous **olfactory foramina** in the cribriform plate of the ethmoid bone.
2. The major foramina of the middle cranial fossa all pass through the sphenoid bone:
   a. The **optic canal**: CN II, the ophthalmic artery, and the central vein of the retina.
   b. The **superior orbital fissure**: CN III, CN IV, CN V_1, CN VI, and the ophthalmic vein.
   c. The **foramen rotundum**: CN V_2.
   d. The **foramen ovale**: CN V_3.
   e. The **foramen spinosum**: the middle meningeal artery.
   f. The **foramen lacerum**: nothing traverses it completely.
3. The major foramina of the posterior cranial fossa pass through the temporal or occipital bones:
   a. The **internal auditory meatus**: CN VII and CN VIII.
   b. The **jugular foramen**: CN IX, CN X, CN XI, and the jugular vein.
   c. The **hypoglossal (anterior condylar) canal**: CN XII.
   d. The **foramen magnum**: roots of CN XI (retrograde), the brain stem, and vertebral arteries.

E. CLINICAL CONSIDERATIONS

1. Fracture of the cranium may involve the scalp, dura, cranial nerves, and cerebral tissue.
   a. The prominence of the parietal bones makes them most susceptible to fracture. The occipital bones are less often fractured.
   b. Bone fragments may tear the venous sinuses, dura, or a meningeal vessel or even lacerate the brain.

c. Fractures of the base of the skull frequently cross foramina (weaker portions of the bone) and jeopardize the contained structures.

2. Because the neurocranium is a rigid box, space-occupying lesions (depressed skull fractures, tumors, hematomas, and even swellings of cerebral tissue) encroach upon the space occupied by the incompressible brain.
a. Regions of the temporal lobes may be forced through the tentorial notch, causing oculomotor nerve dysfunction.
b. The brain stem may be forced through the foramen magnum, compressing nerves and arteries; the resultant medullary ischemia rapidly results in death.

3. The bony confines of a foramen or canal cannot accommodate swelling of a nerve. Inflammatory swelling produces nerve compression and dysfunctions (e.g., Bell's palsy).

IV. THE MENINGES AND VENOUS SINUSES

A. The central nervous system is enclosed in three layers of meninges: the dura mater, the arachnoid mater, and the pia mater (see Fig. 28-1; Fig. 28-9).

1. The **dura mater** is a tough and fibrous protective layer divided into two portions.
a. The internal layer of cranial periosteum is termed the outer layer of dura mater. It adheres closely to the bone.
(1) The **middle meningeal artery** is primarily associated with this layer. Even though this artery and its major branches run between the periosteal and meningeal dural layers, the small branches are within the periosteal layer.
(2) The corresponding veins are wholly periosteal and groove the inner surface of the bone.
b. The inner layer of dura mater is the true meningeal dura mater. It adheres in most places to the outer layer, explaining the confusion when this structure was named about 2300 years ago.
(1) Folds of the inner dural layer (thus two layers thick) extend deeply between the two major divisions of the brain.
(a) The **falx cerebri** (*falx, L.* sickle) lies in the midsagittal plane between the cerebral hemispheres.
(i) It attaches to the crista galli and the frontal crest, anteriorly.
(ii) Superiorly, the two layers (one from either side of the cranial vault) split and attach to the lateral edges of the sagittal groove to enclose the **superior sagittal sinus**.
(iii) The falx cerebri joins the tentorium cerebelli posteriorly.
(b) The small **falx cerebelli** lies in the midsagittal plane, attaching to the internal occipital crest.
(c) The **tentorium cerebelli** (*tendo, L.* to stretch) lies in a more or less transverse plane between the cerebellum and the cerebrum.
(i) It attaches around the occipital bone and along the ridge of the petrous portion of the temporal bone.
(ii) The midline **tentorial notch (incisura tentorii)** fits snugly about the midbrain and divides the posterior cranial fossa from the rest of the cranial cavity.
(iii) Laterally, the two layers split and attach to the calvarial walls to enclose the **transverse sinus**.
(iv) Along the ridge of the petrous portion of the temporal bone, the two layers split to enclose the **superior petrosal sinus**.
(v) At the apex of the petrous portion of the temporal bone, the inferior layer evaginates into the middle cranial fossa as the **trigeminal cave** (Meckel's) for the trigeminal ganglion.
(vi) Medial to this, the corner of the tentorium cerebelli twists so that the free edge continues anteriorly toward its attachment on the anterior clinoid process, and the attached border continues medially to the posterior clinoid process.
(d) The **diaphragma sellae** covers the hypophyseal fossa except for a small hiatus through which the infundibular stalk and the hypophyseal portal veins pass.
(2) Along the attached borders of the dural folds, the two layers diverge somewhat before joining with the periosteal dura.
(a) These spaces, bounded by two leaves of meningeal dura and with a base of periosteal dura, are lined by endothelium to form **dural venous sinuses** (see Figs. 28-1 and 28-9).
(b) The structure of the sinus prevents collapse, so that if a sinus is torn by a bone fragment from a compound depressed cranial fracture or accidently incised at surgery,

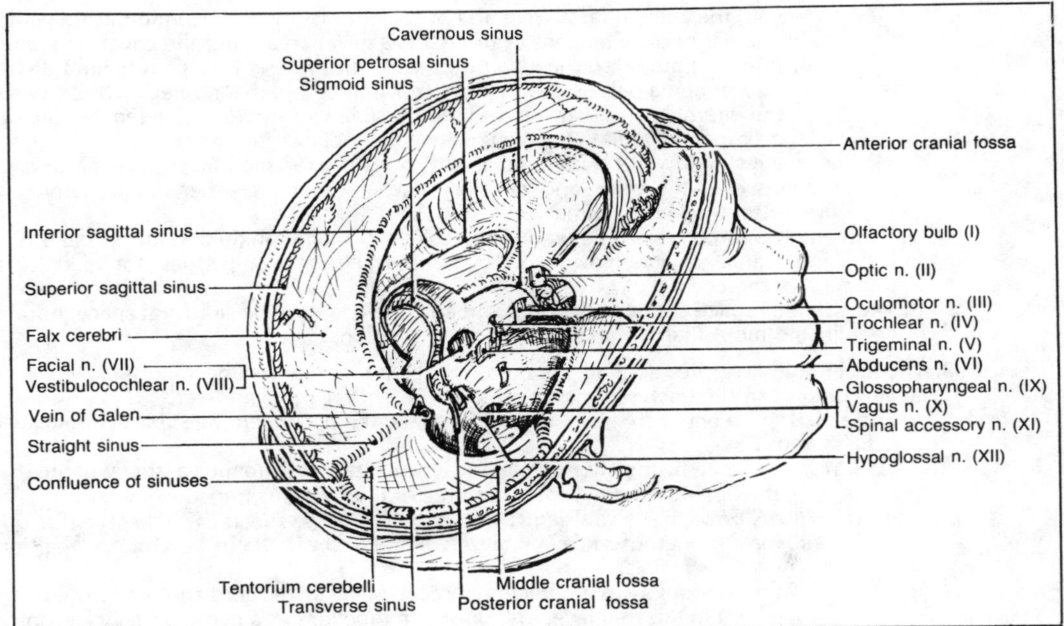

Figure 28-9. *The dural folds, venous sinuses, and cranial nerves.* Note that the cranial nerves pass through the meninges in numerical order.

air can be sucked into the venous system sufficient to fill the right ventricle with froth and cause death.

(c) The major sinuses at dural folds.

(i) The **superior sagittal sinus** lies in the root of the falx cerebri.

(ii) The **inferior sagittal sinus** lies in the free edge of the falx cerebri.

(iii) A small vein (misnamed the **great vein of Galen**), which drains the posterior aspect of the midbrain and unites with the inferior sagittal sinus to form the **straight sinus**, which runs along the intersection of the falx cerebri and the tentorium cerebelli.

(iv) The small **occipital sinus** lies in the root of the falx cerebelli.

(v) The superior sagittal, straight, and occipital sinuses, which meet at the **internal occipital protuberance** to form the **confluence of sinuses (torculus of Herophilus)**.

(vi) The **left** and **right transverse sinuses** drain from the confluence. The superior sagittal sinus drains predominantly into the right transverse sinus; the straight and occipital sinuses drain predominantly into the left transverse sinus.

(vii) The **transverse sinuses** lie along the edge of the tentorium cerebelli. On reaching the petrous portion of the temporal bone, they turn sharply inferiorly as the **sigmoid sinus**, which becomes the internal jugular vein.

(viii) The **superior petrosal sinus** enters the transverse sinus at the point where it becomes the sigmoid sinus.

(3) Numerous small sinuses (e.g., the inferior petrosal, sphenoparietal, and cavernous sinuses) are unrelated to dural folds (Fig. 28-10). Of these, only the **cavernous sinus** is of major importance.

(a) The cavernous sinus is like a plexus of sinuses lying lateral to the body of the sphenoid bone, extending from the superior orbital fissure to the petrous portion of the temporal bone.

(b) Venous flow through the cavernous sinus is a slow percolation.

(i) It receives the ophthalmic vein, the central vein of the retina and the middle and inferior cerebral veins, as well as emissary veins from the deep face and visceral neck.

(ii) It drains into the superior and inferior petrosal sinuses.

(c) The carotid artery and abducens nerve (CN VI) run through the cavernous sinus medially within sheaths of endothelium. Cranial nerves III, IV, V_1, and V_2 lie more laterally along the wall of the cavernous sinus (Fig. 28-11).

(i) Because the ophthalmic veins drain a small area of the face via connections with the angular facial vein and because emissary veins communicate with the visceral neck, infections in these areas may spread into the cavernous sinus.

(ii) Infection may result in cavernous sinus thrombosis with pain behind the eye, cranial nerve palsies affecting eye movement (ophthalmoplegia), thrombosis of the internal carotid artery with the possibility of stroke, and edema of the optic disk (papilledema) from venous engorgement of the retina.

(4) The meningeal layer of dura continues through the foramen magnum. The dura also continues over the optic and olfactory tracts. About the other cranial nerves it is represented by perineurium.

c. From anterior to posterior, the cranial nerves pass through the dura in numerical order to leave the cranial cavity. However, this numerical relationship does not hold for the foramina of the bony skull.

d. The potential space between the dura and arachnoid layers, the **subdural space**, normally contains no more than a thin film of serous secretion.

2. The **arachnoid layer** lies deep to the dura, to which it adheres somewhat.

a. Fine strands of subarachnoid tissue cross to the underlying pia mater.

b. The actual space between the arachnoid layer and the pia mater, the **subarachnoid space**, is filled with CSF.

(1) CSF is produced by the choroid plexuses, which project into the ventricles of the brain.

(a) CSF pressure is usually somewhat greater than venous pressure.

(b) It flows through the ventricular system, entering the subarachnoid space through a midline **foramen of Magendie** and two lateral **foramina of Luschka** in the roof of the fourth ventricle.

(c) CSF flows through the subarachnoid space of the spinal cord and brain. The brain is suspended in this thin layer of CSF, which functions as a hydraulic shock absorber.

(d) It reenters the circulatory system through the **arachnoid villi** into the venous sinuses, primarily the superior sagittal sinus.

(i) These small nipple-like tufts of arachnoid project through the inner layer of dura into the venous sinuses.

(ii) When CSF pressure is higher than venous pressure, there is drainage of CSF into the venous system.

(iii) When venous pressure is greater than CSF pressure, the minute tubules are compressed, preventing seepage of blood into the CSF, where it is an extreme irritant.

(iv) The flow of CSF may be interrupted by congenital malformations or by the scarring associated with the healing of meningitis. The fluid accumulates as hydrocephalus. In children with hydrocephalus, the neurocranium enlarges and the fontanelles fail to close. This condition can be treated by early placement of a shunt to drain the CSF into the venous system or into the peritoneal cavity.

(2) The major distributing arteries to the brain course in the subarachnoid space.

(a) A vascular bleed from a major artery—for example, a ruptured cerebral aneurysm—will disseminate widely in the subarachnoid space.

(b) The presence of blood in the CSF from a lumbar tap is evidence of arterial bleeding.

3. The **pia mater** is a delicate layer that intimately follows the surface of the brain through all of its convolutions. It descends into sulci and ensheaths the blood vessels as they enter the brain parenchyma.

B. CLINICAL CONSIDERATIONS

1. Intracranial bleeding creates a space-occupying hematoma.

a. The raised intracranial pressure causes headache, nausea, depressed consciousness, and edema of the optic disc.

b. Tentorial herniation compresses the oculomotor nerve to extraocular and intraocular muscles. One of the first signs of temporal lobe herniation is dilation of the pupil on the side of the lesion then bilateral dilation as continuing herniation involves the contralateral nerve. This is the reason for monitoring pupillary responses in cases of head injury.

c. Compression of the medulla causes anoxia to the vital centers of the brain stem and subsequent death.

2. Epidural hematoma.

a. Epidural hematoma is most commonly the result of fracture of the temporal or parietal bones.

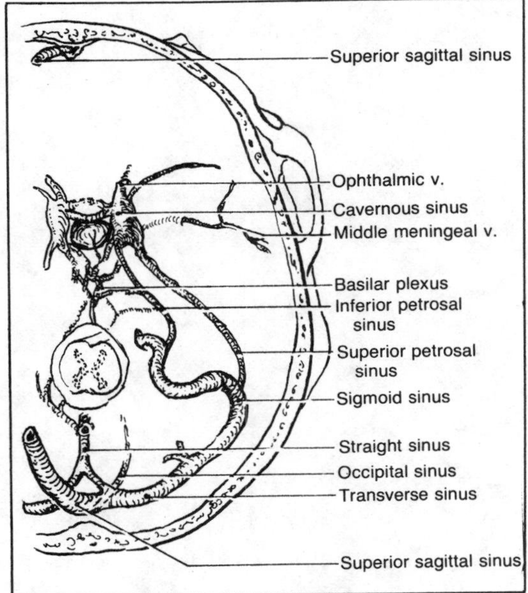

Figure 28-10. *The venous sinuses of the cranium.*

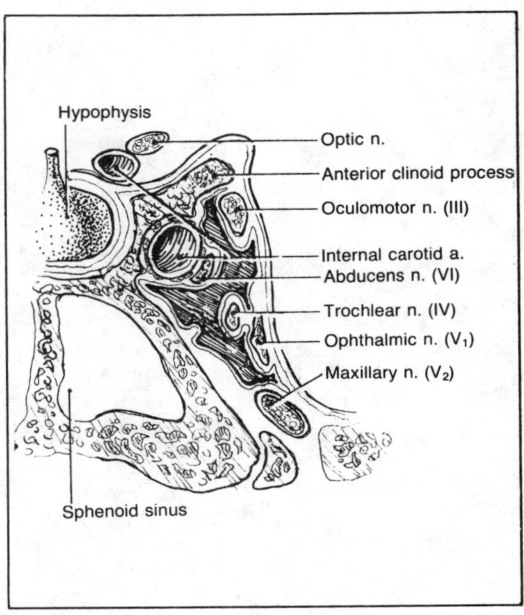

Figure 28-11. *The contents of the cavernous sinus.*

(1) The middle meningeal vessels, lying in the meningeal groove, may be torn (see Fig. 28-6).
  (a) Bleeding may be either arterial or venous.
  (b) Arterial bleeding is rapid and at arterial pressure.
  (c) Symptoms of brain compression generally occur within three hours.

(2) Usually, head trauma is followed by a transient loss of consciousness (concussion), which is followed by a lucid interval that lasts from a few minutes to several hours. Individuals often recover from the initial concussion only to die hours later from an undiagnosed epidural hematoma—the "talk-and-die syndrome."

b. Emergency craniotomy is indicated to drain the hematoma, thereby relieving the pressure on the brain, and to effect hemostasis.

3. Subdural hematoma.

a. As the larger cerebral veins pierce the arachnoid layer and cross the subdural space to gain access to the dural sinuses, they are relatively unsupported and vulnerable.
  (1) As a result of a blow to the head or even a sudden jarring, the veins may be torn in the subdural space.
  (2) Bleeding extends beneath the dura and superficial to the arachnoid; it seldom appears in the CSF if the arachnoid is intact.

b. The venous ooze is often very slow, and, because CSF production is a function of intracranial pressure, the effects of the subdural hematoma are in part compensated for by a reduction in CSF production. Hence the characteristic chronic fluctuation of symptoms.

4. Subarachnoid hemorrhage occurs by the rupture of a cerebral artery, which has been weakened by atheroma or idiopathic (berry) aneurysm.

a. There is usually a history of sudden severe headache with onset correlated to strenuous activity.

b. Blood is present in a lumbar tap.

5. Pial hemorrhage.

a. Cerebral trauma that causes the brain to strike the cranial surfaces may bruise the small vessels of the pia and the brain tissue. This not only occurs directly under the blow (coup), but often at the opposite side of the brain (contracoup) as a result of rebound.

b. The aftermath of hemorrhage is atrophy, degeneration, and corresponding loss of function, illustrated by the "punch-drunk" boxer.

29 The Brain

I. INTRODUCTION

A. THE BRAIN consists of a central stem partially surrounded by two expansions, the cerebrum and the cerebellum.

1. The **brain stem** is continuous with the **spinal cord** within the foramen magnum.
   a. In the posterior cranial fossa, the brain stem lies on the basioccipital bone.
   b. It passes through the **tentorial notch (incisura tentorii)** of the tentorium cerebelli to reach the middle cranial fossa.
   c. In the middle cranial fossa, it lies on the body of the sphenoid bone.

2. The **cerebellum** extends dorsally from the brain stem and fills the posterior cranial fossa.

3. The **cerebrum** occupies the greater portion of the middle and anterior cranial fossa.
   a. The temporal lobes protrude laterally, resting on the sphenoid and temporal bones in the middle cranial fossa.
   b. The frontal lobes rest on the orbital plates of the frontal bone in the anterior cranial fossa.

B. THE CORTEX consists of at least 10^{15} individual neurons that are interconnected in specific patterns.

1. Some neurons make synaptic contact with several thousand other neurons; others are more specific and synapse with very few neurons.

2. In the adult, neurons may die if subjected to damage or ischemia.
   a. Neuronal death is accompanied by alterations in function.
   b. In many instances, function may be lost or altered permanently because brain tissue does not regenerate.

C. THE CRANIAL NERVES arise from the ventral and lateral surfaces of the brain stem and exit the cranium through foramina in the floor. They pass between branches of the arterial supply to the brain.

II. THE CEREBRUM (FOREBRAIN) consists of the cerebral hemispheres (telencephalon) and the diencephalon.

A. THE LEFT AND RIGHT CEREBRAL HEMISPHERES consist of cerebral cortex, white matter, and the basal ganglia (Fig. 29-1).

1. Originally related to the olfactory bulbs, the cerebral hemispheres expanded superiorly and posteriorly during development.

2. Their numerous connecting axon pathways (the internal capsule) flank the thalamus and hypothalamus (diencephalic structures). The interconnections between the cerebral hemispheres and the thalamus provide a very close functional association.

3. The **cerebral hemispheres** are separated by a deep cleft, the **longitudinal cerebral fissure**, into which the falx cerebri projects.
   a. Each cerebral hemisphere may be divided, like any other part of the central nervous system (CNS), into white and gray matter.
      (1) The **gray matter**, covering the surface, consists of nerve cells (neurons) arranged in vertical functional columns with a complex circuitry.

(a) Input into the columns is principally (99 percent) from cortical association areas and the thalamus, with only a small (1 percent) input from subcortical areas.
(b) The output of each functional column is primarily to other cortical areas, to the thalamus, and to the basal ganglia, as well as to the cerebellum. The neurons of the column influence the lower motor neurons of the cranial nerve nuclei and the spinal cord.

(2) The deeper **white matter** consists of tracts or fascicles of axons from the input and output neurons.

b. The surface is folded into **gyri** (ridges) separated by **sulci** (grooves). These cortical folds increase the surface area, thereby increasing the number of functional columns of neurons that can be accommodated.

(1) While the pattern of gyri and sulci is quite variable, some are remarkably consistent in location and provide convenient landmarks.

(2) Several of the sulci are deeper than the others and may be used to demarcate various lobes, some of which are named for the bones under which they lie.
(a) The **frontal lobe.**
(b) The **temporal lobe.**
(c) The **parietal lobe.**
(d) The **occipital lobe.**
(e) The **limbic lobe.**
(f) The **central (insular) lobe.**

c. The subdivision of the cerebral cortex into lobes is useful because it closely approximates the division of function (see Fig. 29-1).

(1) The developing cerebral hemisphere folds itself around the ridge of the sphenoid, forming a deep groove between the frontal and temporal lobes, the **lateral fissure (Sylvius's)**.
(a) The **parietal** and **frontal lobes** lie superior to this fissure.
(b) The **temporal lobe** lies inferior to this fissure.

(2) A **central sulcus** (fissure of Rolando) extends in a coronal plane from the posterior end of the sylvian fissure over the lateral surface and for a short distance on the medial surface in the longitudinal fissure.
(a) The **frontal lobe** lies anterior to this sulcus.
(b) The **parietal lobe** lies posterior to this sulcus.

(3) The **parieto-occipital fissure** more or less defines the boundary between the parietal and occipital lobes, especially on the medial surface.

(4) The **calcarine fissure**, only discernible in the medial aspect, bisects the occipital lobe in a transverse plane.

d. The functions of the hemispheres.

(1) In general each cerebral hemisphere deals with the contralateral side of the body. Usually one cerebral hemisphere (the left in 98 percent of all individuals) is "dominant." However, some higher functions are lateralized to one or the other hemisphere in a predictable way.
(a) The left hemisphere in most individuals is concerned with verbal, calculating, and analytical thinking as well as interpretation of speech, stereognosis, and motor function to the right hand.
(b) The right hemisphere in most individuals is the seat of nonverbal, spatial, temporal, and synthetic function, appreciation of art and music, and motor function to the left hand. Lesions affecting the right hemisphere may produce loss of visuospatial awareness such that paralysis of the left side is compounded by a tendency to ignore the paralyzed part.

(2) The **primary cortical areas** are regions most directly related to specific functions.
(a) These usually lie adjacent to major sulci.
(b) In some cortical areas (especially the precentral and postcentral gyri), the neurons are **somatotopically** arranged. Areas of the body may be mapped onto the corresponding points on the surface of the cortex.

(3) The **secondary cortical (association) areas** lie adjacent to the primary areas and are concerned with a higher level of organization and integration. The functional areas are really much more diffuse than the regions defined in the classical description and overlap one another.

e. The functions of the lobes.

(1) The **frontal lobe**, lying anterior to the central sulcus and lateral fissure, serves motor function, speech, cognition, and the highest levels of affective behavior (see Fig. 29-1).
(a) The **primary motor area (precentral gyrus)** of the frontal lobe contains the cell bodies of primary motor pathways.
(i) These project down the pyramidal pathways to the contralateral lower motor

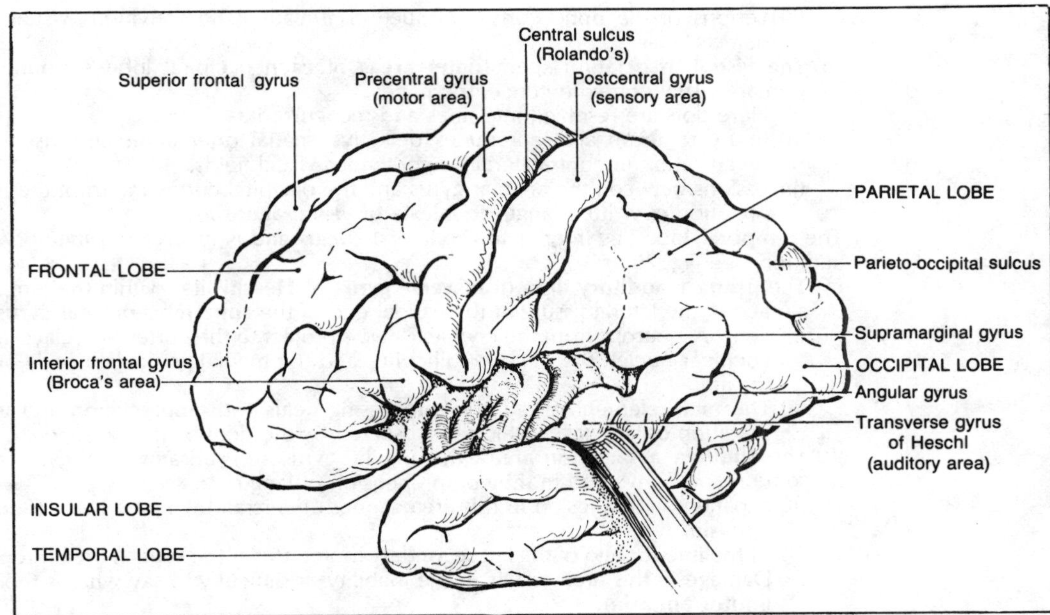

Figure 29-1. *The cerebrum.* A lateral view of the left (dominant) cerebral hemisphere. The lateral fissure has been spread open by traction on the temporal lobe to show the underlying insular lobe.

neurons in the brain stem and spinal cord. They also influence the lower motor neurons less directly through relays in the brain stem.

(ii) Connections from the ventral lateral and ventral anterior nuclei of the thalamus convey modulating influences from the basal ganglia and from the cerebellum to this motor area.

(iii) This area has a high degree of somatotopic organization (the motor homunculus) [Fig. 29-2]. Damage to specific locations correlates with loss of specific function (**spastic upper motor neuron paralysis**) on the opposite side of the body.

(b) The **supplementary motor area** of the medial surface of the frontal lobe (**prefrontal gyrus**) is somehow involved in the integration of voluntary movements.

(2) The **parietal lobe**, between the central sulcus and the parieto-occipital fissure, is involved with somatosensory processing (see Fig. 29-1).

(a) The **primary sensory area (postcentral gyrus)** of each parietal lobe receives general sensation from the contralateral side of the body via relays in the ventral posterolateral and posteromedial nuclei of the thalamus.

(i) This area has a high degree of somatotopic organization (the sensory homunculus) [see Fig. 29-2].

(ii) Damage to specific locations correlates with **paresthesias** on the opposite side of the body.

(b) The **sensory association area** posterior to the postcentral gyrus of the parietal lobe integrates tactile stimuli to create an understanding of shape and form, thus enabling the recognition of an object.

(c) The **supramarginal gyrus** of the parietal lobe is involved in higher perceptual mechanisms. Lesions to this area result in **tactile agnosia** (the inability to identify objects by feel).

(3) The **occipital lobe** lies posterior to the parieto-occipital sulcus and contains the visual cortex (see Fig. 29-1).

(a) The **primary visual areas (striate cortex)** lie on either side of the calcarine sulcus in the occipital lobe.

(i) The retinal pathways that convey the visual fields of each eye project to the lateral geniculate bodies of the thalamus in such a manner that the left visual field for each eye is projected to the right lateral geniculate body (see V A 2 c below).

(ii) From the lateral geniculate body of the thalamus, the pathways pass posteriorly to the occipital cortex in the optic radiation.

(iii) Lesions of the optic cortex produce contralateral homonymous visual field defects.

(b) The **visual association (peristriate) areas** of each occipital lobe surround the primary visual cortex in concentric bands.

(i) Here dots are resolved into lines and recognized as shapes.

(ii) Injury to these areas produces defective spatial orientation and visual disorganization in contralateral homonymous visual fields.

(iii) Lesions between the angular gyrus and the occipital cortex interrupt the interpretation of written language (**alexia** or **visual agnosia**).

(4) The **temporal lobe** lies inferior to the lateral sulcus and is involved in memory and audition (see Fig. 29-1).

(a) The **primary auditory area (transverse gyrus of Heschl)** lies within the temporal surface of the lateral fissure at the caudal end of the **superior temporal gyrus**.

(i) Sensory neurons from the cochlear nerve project to the cortex via relays in the cochlear nuclei, the inferior colliculus, and the medial geniculate body of the thalamus.

(ii) The main element of auditory processing deals with appreciation and interpretation of language, which is subserved by the dominant hemisphere.

(b) The **auditory association areas** (the angular gyrus and supramarginal gyri) are in some manner involved in the comprehension of language.

(i) A patient with a lesion in this area cannot understand what he hears (**sensory aphasia**).

(ii) At the inferior end of the left secondary motor area is the **speech area** (Broca's). Damage in this area results in the inability or difficulty to say what is thought (**motor aphasia**).

(5) The **insular lobe** lies at the base of the lateral fissure and seems to be involved in the autonomic nervous system (see Fig. 29-1).

(6) The **limbic lobe** lies deep in the sagittal fissure (cingulate gyrus and angular gyrus) and is concerned with olfaction, regulation of the viscera, primitive emotions, and behavioral activity (see Fig. 29-4).

f. Projections of the cerebral cortex.

(1) Commissural fibers pass between hemispheres above the thalamus as the **corpus callosum** (Fig. 29-3; see Fig. 29-4).

(a) The corpus callosum contains over 10^6 crossing axons.

(b) These connections are important because the hemispheres have different capabilities.

(c) Cutting the corpus callosum results in a **split-brain** individual, in whom information

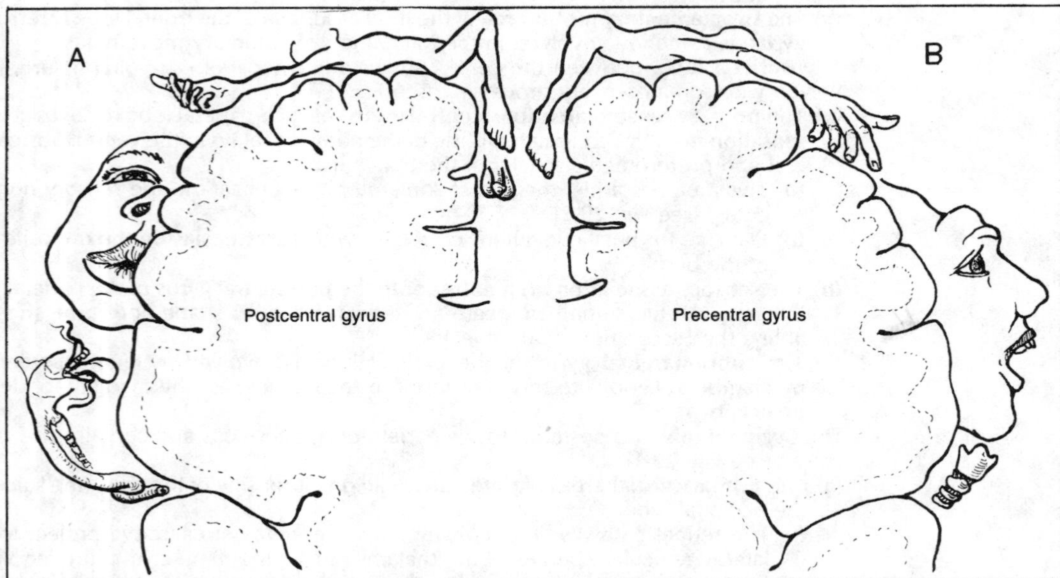

Figure 29-2. *The somatotopic organization of the cerebral cortex. A,* Coronal section through the postcentral gyrus. The relative sizes of the cortical areas associated with the various regions of the body are depicted by the sensory homunculus. *B,* Coronal section through the precentral gyrus. The relative sizes of the cortical areas that innervate the muscles of the various regions of the body are depicted by the motor homunculus.

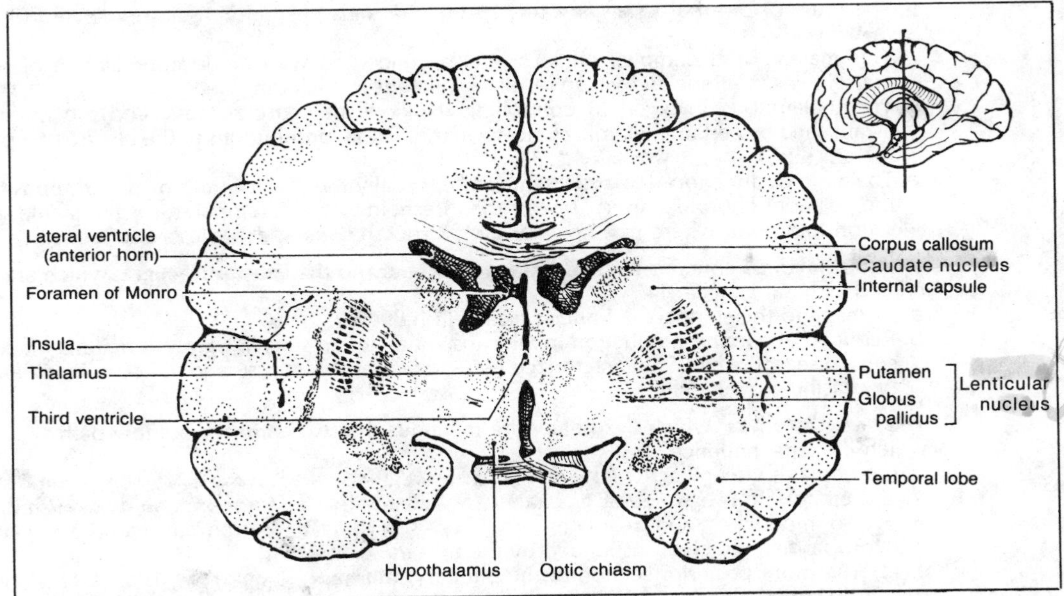

Figure 29-3. *Coronal section through the cerebrum.* The section is through the basal ganglia, just anterior to the thalamus.

cannot be relayed to the contralateral side for specialized association (e.g., a common object placed in one hand cannot be recognized or matched with a similar object in the other hand).

(2) The primary motor areas project via the **corticobulbar pathways** to the lower motor neurons in the brain stem nuclei and via the **pyramidal pathways** to lower motor neurons in the gray matter of the spinal cord. They also influence the lower motor neurons less directly through relays in the brain stem, the **extrapyramidal pathways.**

B. THE BASAL GANGLIA consist of several large nuclei deep within the base of the cerebral hemispheres (see Fig. 29-4).

1. These nuclei include:
   - **a.** The **caudate nucleus.**
   - **b.** The **lenticular nucleus.**
     - **(1)** The **putamen.**
     - **(2)** The **globus pallidus.**
   - **c.** The **subthalamic nucleus.**
   - **d.** The **substantia nigra.**
2. Many areas of the cerebral cortex project to the basal ganglia. The outflow from the basal ganglia is primarily to the thalamus, whence they project via a feedback circuit to the motor areas of the cerebral cortex.
3. The functions of the basal ganglia seem to involve the modulation of the motor outflow of the cortex, smoothing voluntary actions.
4. Dysfunction of the basal ganglia produces **release dyskinesias**, such as tremors, choreas, ballism, and athetosis.
   - **a.** Cerebral primary motor areas project through the **internal capsule** (between the caudate nucleus and thalamus, medially, and the lenticular nucleus, laterally) before passing into the **cerebral peduncle.**
   - **b.** Small lesions here, caused by thrombosis or hemorrhage of the striate arteries, have devastating effects (**stroke**).

C. THE DIENCEPHALON, the most rostral portion of the brain stem, consists predominantly of the thalamus, with small epithalamic, subthalamic, and hypothalamic regions (see Fig. 29-3).

1. The **thalamus**, with over 25 separate nuclei, serves as a major synaptic relay station.
   - **a.** All afferent modalities (except olfaction) synapse in the thalamus on the way to the cerebral cortex.

b. The output from the basal ganglia is also to the thalamus before being projected to the motor cortex.
c. This massive and complex nucleus is also concerned with crude appreciation of subconscious sensations and vague levels of awareness, including sleep.
d. The thalamus is involved in controlling the level of consciousness, and the rhythms established by certain thalamic nuclei form the major contributions to the electroencephalogram (EEG).
e. Thalamic dysfunction (**thalamic syndrome**), usually the result of tumor or thrombosis of the posterior choroidal artery, lowers the threshold for pain, temperature, and tactile sensation to a point where non-noxious stimuli produce marked unpleasant effects.

2. The **subthalamus** contains the **subthalamic nucleus** and the **substantia nigra**, which are included as basal ganglia.
   a. Damage to the subthalamic nuclei results in **ballism**.
   b. Reduced dopamine production in the substantia nigra results in **paralysis agitans** (Parkinson's disease), which is characterized by rigidity, tremors at rest, and difficulty in initiating or ceasing movements.

3. The **hypothalamus**, which lies below the thalamus, has two primary outflow pathways by which it exerts influence.
   a. It modulates visceral activities, such as thermoregulation, appetite, thirst, sexual impulses, and emotion through neural mechanisms mediated by the autonomic nervous system.
   b. It also modulates and regulates the release of hormones from the **pituitary gland (hypophysis)**, which is suspended by the thin **infundibulum**.
      (1) The more posterior section of the pituitary, the **neurohypophysis**, is an extension of the hypothalamus and releases antidiuretic hormone (vasopressin) and milk let-down-factor (oxytocin) into the bloodstream.
      (2) The more anterior section, the **adenohypophysis**, is a pharyngeal derivative (**Rathke's pouch**).
         (a) It releases various trophic hormones, under hypothalamic control, which regulate the major endocrine axes.
         (b) It is prone to the formation of benign tumors, which erode the walls of the pituitary fossa and may impinge on the optic chiasm, causing visual field loss.

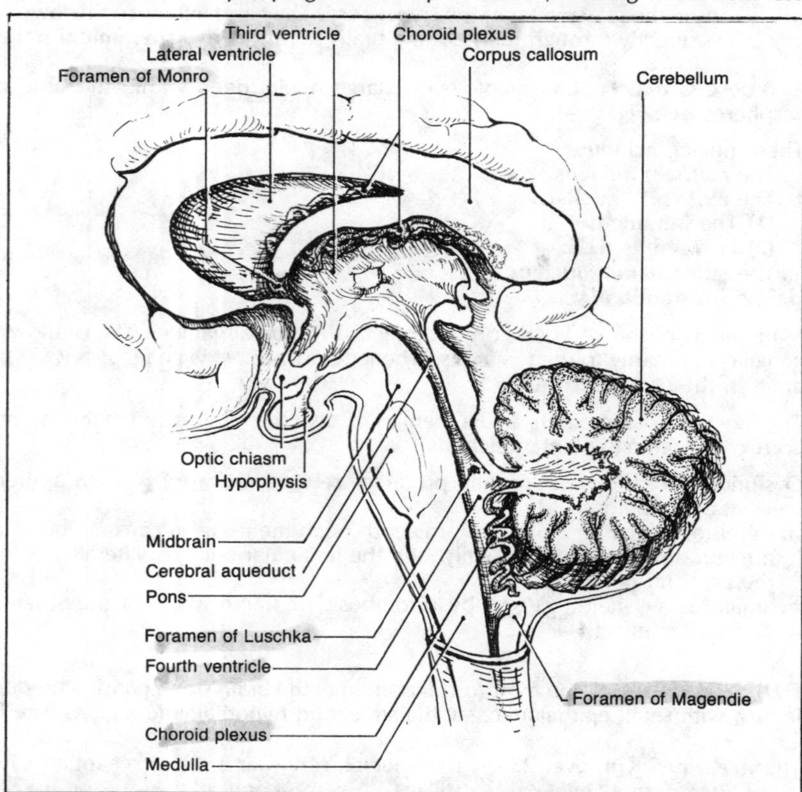

Figure 29-4. *Sagittal section through the brain, brain stem, and cerebellum.*

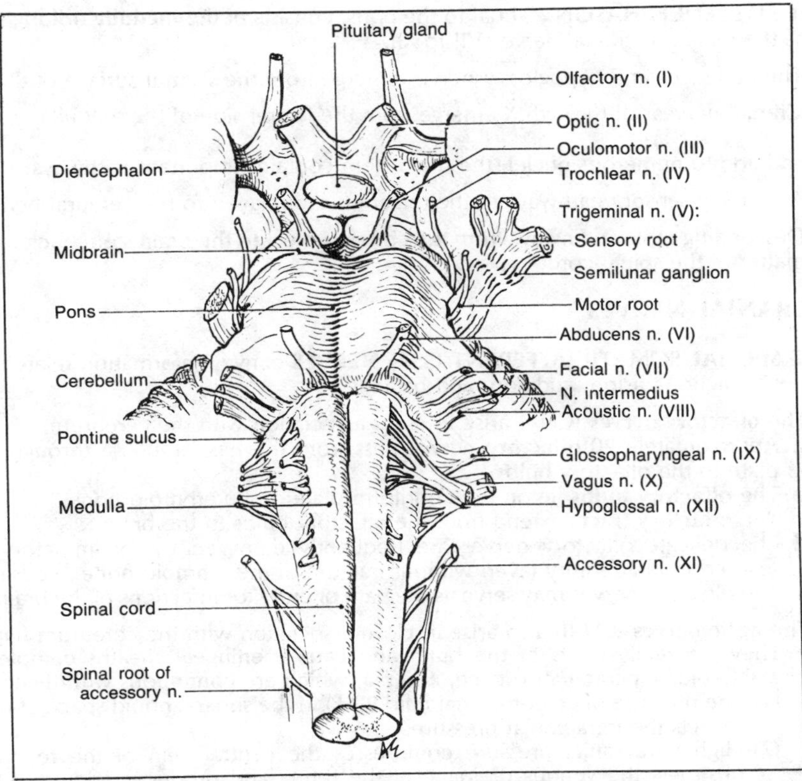

Figure 29-5. *The brain stem.* The anterior aspect of the brain stem and the cranial nerves.

III. THE MIDBRAIN COMPRISES THE MESENCEPHALON

A. THE MESENCEPHALON lies inferior to the tentorium cerebelli (see Fig. 29-3; Fig. 29-4).

B. THE MIDBRAIN is associated, through the substantia nigra and the basal ganglia, with modulation of movement. Its **tectum** (roof) also contains centers for auditory and visual reflexes.

C. It contains the nuclei of the oculomotor nerve and the trochlear nerve (see Fig. 29-7).

1. The oculomotor nerve (CN III) emerges from the anterior aspect of the midbrain (Fig. 29-5).
2. The trochlear nerve (CN IV) emerges from the posterior aspect of the midbrain (see Fig. 29-5).

IV. THE HINDBRAIN

A. THE METENCEPHALON consists of the cerebellum and pons (see Fig. 29-4).

1. The brain stem has its own dorsal expansion, the **cerebellum**.
   a. The expression of cerebellar function is ipsilateral.
   b. The cerebellum is concerned with the coordinated contraction and relaxation of agonist and antagonistic muscle groups to produce smooth precise movements.
   c. Cerebellar dysfunction produces jerky uncoordinated movements (**ataxia**).
2. The **pons** has three pairs of **cerebellar peduncles**.
   a. The middle peduncles project fibers to the cerebellum from the pontine nuclei, which receive input from the cerebral cortex.
   b. The superior peduncles primarily relay to the thalamus.
   c. The inferior peduncles project relay fibers from the peripheral proprioceptors and brain stem to the cerebellum and from the cerebellum to the brain stem.
3. The trigeminal nerve emerges from the pons laterally, and the abducens nerve emerges from the anterior aspect of the pontomedullary junction (see Fig. 29-5).

B. THE MYELENCEPHALON, caudal to the pons, consists of the **medulla oblongata**, which contains the nuclei of cranial nerves VII to XII.

1. The abducens and hypoglossal nerves emerge from the ventral surface of the medulla.

2. Cranial nerves VII through X emerge from the lateral side of the medulla.

C. In addition to numerous nuclei, the brain stem contains important pathways.

1. Ascending sensory pathways to the thalamus are relayed to the cerebral hemispheres.

2. Descending motor pathways from the cortex project to the brain stem nuclei and to the gray matter of the spinal cord.

V. THE CRANIAL NERVES

A. THE SPECIAL SOMATIC AFFERENT (SSA) NERVES convey information related to the special senses, olfaction, vision, audition, and balance.

1. The **olfactory nerves** (CN I) arise in close association with the cerebrum.

a. Approximately 20 **olfactory nerves** pass from the nasal mucosa through the cribriform plate to the olfactory bulbs.

b. The **olfactory bulbs** lie on the cribriform plate of the ethmoid bone.

c. The **olfactory tracts** extend from the olfactory bulbs to the brain.

d. The delicate olfactory nerves are frequently damaged by inflammatory processes or severed in head injury (even without fracture of the ethmoid bone), causing **anosmia**.

e. The olfactory nerves may serve as the path of entry for infections of the brain or meninges.

2. The **optic nerves** (CN II) also arise in close association with the cerebrum (Fig. 29-6).

a. They are really tracts of the brain and carry meningeal sheaths complete with dura, arachnoid, subarachnoid space, and pia, which are continuous with those of the brain.

(1) The pressure of cerebrospinal fluid (CSF) in the subarachnoid space of the optic nerve reflects the intracranial pressure.

(2) High intracranial pressure compresses the central vein of the retina and thereby obstructs the venous drainage of the retina and produces edema at the optic disc (**papilledema**).

b. By convention, the portion of the optic tract anterior to the optic chiasm is often termed the optic nerve.

(1) The true equivalent of optic nerves are represented in the retina by the **bipolar cells**, which contact the rod and cone receptor cells.

(2) The ganglion cells of the retina, the optic "nerve," the optic tracts, and the lateral geniculate bodies are diencephalic structures.

c. The visual pathways (see Fig. 29-6):

(1) The **lens** of each eye projects an inverted and reversed visual field onto each retina. Thus, lateral visual fields project onto the nasal halves of each retina; medial visual fields project onto the temporal halves of each retina.

(2) The optic nerves leave the retina at the **blind spot** and course to the optic chiasm. Lesions of the optic nerve anterior to the chiasm will cause **total blindness** in one eye.

(3) At the **optic chiasm**, pathways from both retinas are sorted such that similar visual fields from each eye are collected on the opposite side of the brain.

(a) This is accomplished by crossing of the pathways from the nasal half of each retina at the optic chiasm; the pathways from the temporal half of each retina remain ipsilateral.

(b) The optic chiasm is in close relation to the pituitary fossa, so that suprasellar extension of a pituitary tumor may press on the optic chiasm and interrupt input from the nasal retina of both eyes to produce loss of both temporal visual fields (**bitemporal hemianopsia** or **tunnel vision**).

(4) From the optic chiasm, the optic tracts containing common visual fields pass posteriorly to enter the **lateral geniculate body** of the thalamus.

(5) From each lateral geniculate body of the thalamus, the pathways project through the temporal lobe along the **optic radiations** to the primary visual area of each occipital lobe. Lesions of the optic tracts, of the optic radiations, or of the occipital cortex result in **cortical blindness** with **contralateral homonymous hemianopsia** (i.e., loss of the contralateral half of the visual fields of both eyes).

(6) There are also connections with midbrain nuclei that are involved with reflex eye movements.

d. The optic nerve is one of the regions affected by the widespread demyelination of the central nervous system, which characterizes disseminated sclerosis.

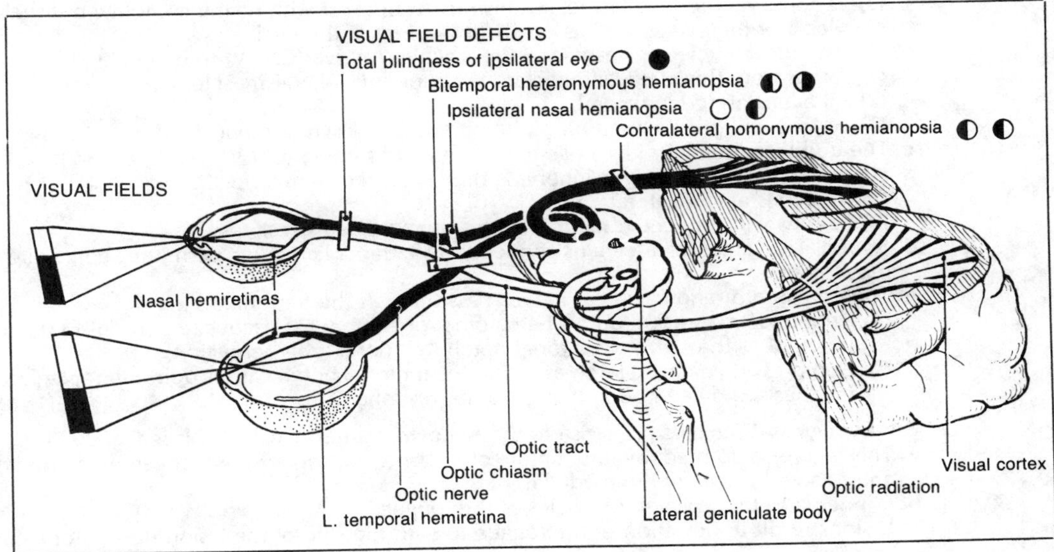

Figure 29-6. *The optic pathways.* The locations of typical lesions and the resultant blindness are shown.

3. The **vestibular** and **cochlear nerves** (CN VIII) arise in association with the hindbrain.
   a. The **vestibular nuclei** are involved with balance and equilibrium.
      (1) Through connections to the nuclei of the external ocular muscles, they enable the eye to fix on an object when the head is moving.
      (2) These nuclei also project to the cervical spinal cord and enable the head to remain stable when the body moves.
      (3) Abnormal ocular movements (**nystagmus**) are associated with vestibular dysfunction.
   b. The **cochlear nuclei** relay acoustic information to the medial geniculate body of the thalamus for subsequent relay to the primary auditory cortex.

B. **CRANIAL NERVES III THROUGH XII** arise from the brain stem. They may contain motor, sensory, and parasympathetic fibers, as well as special motor and sensory fibers.

1. The motor and sensory innervation of the branchial arch structures have characteristics of visceral and somatic systems and are referred to as special visceral. The nuclei of these fibers also occupy an intermediate position, anatomically.

2. Within the spinal cord, the motor and sensory areas have specific locations.
   a. The somatic sensory neurons relay to the posterior (dorsal) horn.
   b. The motor nuclei are located ventrally and laterally.
      (1) The somatic motor neurons are in the anterior (ventral) horn.
      (2) The visceral motor cells are intermediate in the lateral horn.

3. In the brain stem the basic arrangement is modified by the size of the fourth ventricle, which separates the dorsal horns. The dorsal horns are connected by the thin roof (tegmentum) as though the walls never came together to fuse in the dorsal midline.
   a. The somatic sensory nuclei are dorsolateral.
   b. The motor nuclei are arranged in three longitudinal columns that extend through the brain stem (see Figs. 29-7 and 29-8).

C. **THE GENERAL SOMATIC EFFERENT (GSE) NERVES**, arising from the column of somatic motor nuclei, are equivalent to motor roots and innervate muscle derived from embryologic somites.

1. The **oculomotor**, **trochlear**, and **abducens nerves** (CN III, IV, and VI) innervate the extraocular muscles.
   a. The **oculomotor nucleus** gives rise to the somatic component of the **oculomotor nerve** (CN III) [Fig. 29-7].
      (1) Axons from this nucleus innervate five of the extraocular muscles (levator palpebrae superioris, superior rectus, medial rectus, inferior rectus, and inferior oblique).
      (2) The axons of each nerve are both ipsilateral (medial rectus) and contralateral.

(3) Lesions will produce **lateral strabismus** (**wall eyes**) with nearly complete **ophthalmoplegia** of the eye as well as **ptosis** (**drooping**) of the eyelid.

b. The **trochlear nucleus** gives rise to the **trochlear nerve** (CN IV) [see Fig. 29-7].
(1) Axons from this nucleus innervate the superior oblique muscle.
(2) All axons are contralateral.
(3) Lesions will produce a **diplopia** upon looking down and out.

c. The **abducens nucleus** gives rise to the **abducens nerve** (CN VI) [see Fig. 29-7].
(1) Axons from this nucleus innervate the lateral rectus muscle.
(2) All axons are ipsilateral.
(3) Lesions will produce a **medial strabismus** (**crossed eyes**).

d. A medial longitudinal fasciculus connects the extraocular nuclei, permitting conjugate eye movements.

e. The oculomotor and trochlear nerves pass through the tentorial notch.
(1) An intracranial mass (tumor, hematoma, or edema) may squeeze portions of the temporal lobes through the tentorial notch (temporal lobe herniation).
(2) Compression of cranial nerves within the tentorial notch causes **ophthalmoplegia** with mydriasis (a dilated pupil that does not respond to light).

2. The **hypoglossal nucleus** gives rise to the **hypoglossal nerve** (CN XII) [see Fig. 29-7].
a. This nerve is formed by the union of a series of rootlets, which emerge from the anterolateral sulcus of the medulla.
b. It innervates the ipsilateral muscles of the tongue.
c. Lesions result in deviation of the tongue toward the side of the lesion upon protrusion.

D. THE SPECIAL VISCERAL EFFERENT (SVE) NERVES, arising from the most lateral column of motor nuclei, innervate muscles derived from branchial arch derivatives (i.e., the musculature associated with the embryonic gills) [see Fig. 29-7].

1. The **trigeminal motor nucleus**, located in the mid pons, gives rise to the branchiomotor component of the **trigeminal nerve** (CN V) [see Fig. 29-7].
a. It is the most rostral special visceral nucleus.
b. The axons pass through the trigeminal ganglion and exit the cranium via the foramen ovale to join the mandibular nerve (CN V_3).

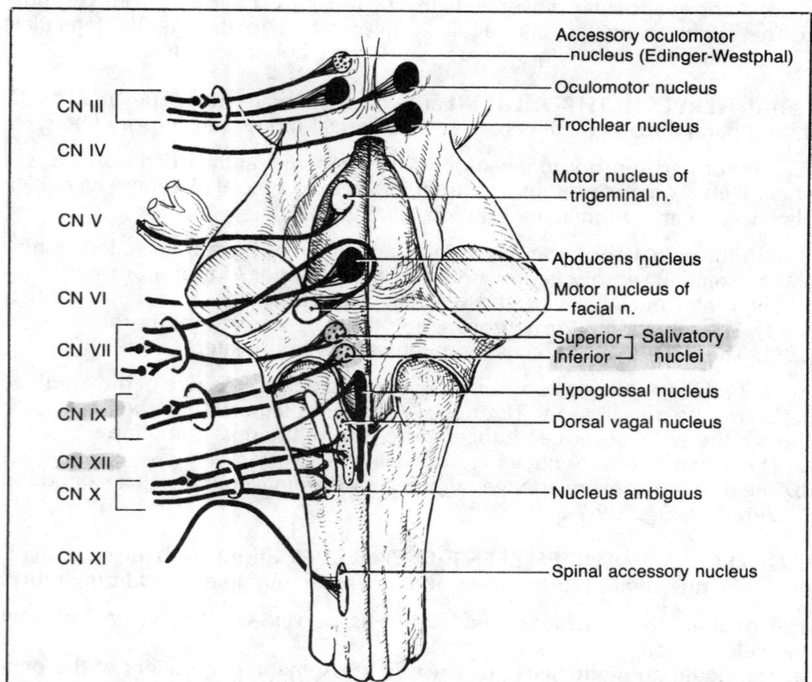

Figure 29-7. *The motor nuclei of the brain stem.* The somatic motor nuclei (*unshaded*), the branchiomeric motor nuclei (*shaded*), and the parasympathetic nuclei (*stippled*) contribute axons to the motor components of the cranial nerves.

c. It innervates the ipsilateral muscles of mastication.

2. The **facial motor nucleus** gives rise to the branchiomotor component of the **facial nerve** (CN VII) [see Fig. 29-7].
   a. This nucleus lies just caudal to the trigeminal motor nucleus.
   b. The **facial nerve** innervates the ipsilateral muscles of facial expression.

3. The **nucleus ambiguus** gives rise to the branchiomotor component of the **glossopharyngeal, vagus,** and **cranial accessory nerves** (see Fig. 29-7).
   a. The **glossopharyngeal nerve** (CN IX) innervates the ipsilateral stylopharyngeus muscle.
   b. The **vagus nerve** (CN X) innervates the ipsilateral pharyngeal muscles via the pharyngeal plexus, as well as the ipsilateral cricothyroid muscle via the superficial branch of the superior laryngeal nerve.
   c. The **cranial accessory nerve** (CN XI), which briefly joins the spinal accessory nerve, innervates the ipsilateral muscles of the larynx (except the cricothyroid muscle) via the recurrent laryngeal nerve.

4. The lateral cell column of the upper five or six cervical segments contributes rootlets to the **spinal accessory nerve** (CN XI) by passing into the cranial cavity through the foramen magnum.
   a. It joins the cranial accessory nerve within the jugular canal.
   b. After passing through the jugular foramen, the two divisions separate.
   c. The spinal accessory nerve innervates the ipsilateral sternomastoid and trapezius muscles.

E. **THE GENERAL VISCERAL EFFERENT (GVE) NERVES,** arising from the intermediate column of motor nuclei, are the **parasympathetic nuclei** (see Fig. 29-7).

1. The **accessory oculomotor nucleus** (of Edinger-Westphal), the most rostral parasympathetic nucleus, gives rise to the parasympathetic component of the **oculomotor nerve** (CN III) [see Fig. 29-7].
   a. The preganglionic parasympathetic neurons from this nucleus travel via the oculomotor nerve (CN III) to the ciliary ganglion.
   b. The postganglionic axons of the ciliary ganglion course to the sphincter and ciliary muscles of the pupil.
   c. This pathway controls pupillary constriction and accommodation to near vision.

2. The **superior salivatory nucleus** gives rise to the parasympathetic secretomotor fibers of the **facial nerve** (CN VII) [see Fig. 29-7].
   a. The preganglionic axons travel with the **nervus intermedius** of the facial nerve to the pterygopalatine ganglion (via the greater superficial petrosal and vidian nerves) and the submandibular ganglion (via the chorda tympani and lingual nerve).
   b. The postganglionic axons of the pterygopalatine ganglion supply the lacrimal, nasal, and palatine glands.
   c. The postganglionic axons of the submandibular ganglion supply the submandibular gland, the sublingual gland, and the glands of the anterior tongue.

3. The **inferior salivatory nucleus** gives rise to the parasympathetic secretomotor fibers of the **glossopharyngeal nerve** (CN IX) [see Fig. 29-7].
   a. The preganglionic axons initially travel with the glossopharyngeal nerve, then with the tympanic nerve (Jacobson's) and the lesser superficial petrosal nerve to reach the otic ganglion.
   b. The postganglionic axons of the otic ganglion then supply the parotid gland.

4. The **dorsal vagal nucleus**, the most caudal of the cranial parasympathetic nuclei, gives rise to the parasympathetic component of the **vagus nerve** (CN X) [see Fig. 29-7].
   a. It provides the long preganglionic parasympathetic fibers, which synapse in small ganglia associated with the viscera of the thorax and much of the abdomen.
   b. The postganglionic neurons from the small parasympathetic ganglia are secretomotor and stimulate visceral musclar activity.

F. **THE GENERAL VISCERAL AFFERENT (GVA) NERVES** convey sensation from the thoracic and abdominal viscera to the brain stem.

1. The **solitary nucleus** receives input from the **glossopharyngeal nerve** with cell bodies in the **inferior glossopharyngeal (petrosal) ganglion** to mediate the gag and carotid reflexes (see Fig. 29-8).

2. The **dorsal nucleus of the vagus** receives input from the **vagus nerve** with cell bodies in the **inferior vagal (nodosal) ganglion** to mediate the visceral reflexes involving the vagus nerve (e.g., cough, cardiac, and respiratory reflexes) [see Fig. 29-7].

G. THE SPECIAL VISCERAL AFFERENT (SVA) NERVES convey taste sensation to the brain stem.

1. The **nucleus solitarius** receives taste input from the lingual, glossopharyngeal, and vagus nerves.
   a. Taste from the anterior two-thirds of the tongue finds its way via the **lingual nerve**, to the **chorda tympani** to the the **facial nerve** (CN VII). The cell bodies of these afferent neurons are located in the **geniculate ganglion** and synapse in the solitary nucleus (Fig. 29-8).
   b. Taste from the posterior surface of the tongue is carried via the **glossopharyngeal nerve** (CN IX). The cell bodies of these afferent neurons are located in the **inferior glossopharyngeal (petrosal) ganglion** and synapse in the nucleus solitarius (see Fig. 29-8).
   c. Taste from the epiglottis is carried via the **vagus nerve** (CN X). The cell bodies of these afferent neurons are located in the **inferior vagal (nodosal) ganglion** and synapse in the nucleus solitarius (see Fig. 29-8).
2. From the solitary nucleus the second neurons convey taste sensation to the thalamus.

H. THE GENERAL SOMATIC AFFERENT (GSA) NERVES are equivalent to dorsal roots.

1. The **trigeminal nuclei** receive somatic afferent sensation from the trigeminal, glossopharyngeal, and vagus nerves.
   a. The **trigeminal nerve** (CN V) is responsible for the general sensation from the face, the anterior scalp, the mouth, the pharynx, the nose, and most of the ear.
      (1) The cell bodies of most of these afferent neurons are located in the **trigeminal (gasserian, semilunar) ganglion**.
      (2) Each of the modalities of pain, touch, and temperature is associated with a different brain stem nucleus where the second-order neurons are located.
         (a) The **principal nucleus of V** in the pons (see Fig. 29-8) receives fibers carrying touch and two-point discrimination.
         (b) The **mesencephalic nucleus of V** (see Fig. 29-8) mediates proprioception (e.g., the **jaw-jerk reflex**). This nucleus contains first-order sensory neurons, the only exception to the rule that the first-order neurons are located in dorsal root ganglia or cranial equivalents.
         (c) The **spinal nucleus of V**, which extends into the cervical spinal cord (see Fig. 29-8), receives fibers conveying pain, light touch, and temperature sensation. This nucleus is somatotopically organized such that neurons from the ophthalmic nerve enter most caudal while those from the mandibular nerve enter most rostral.

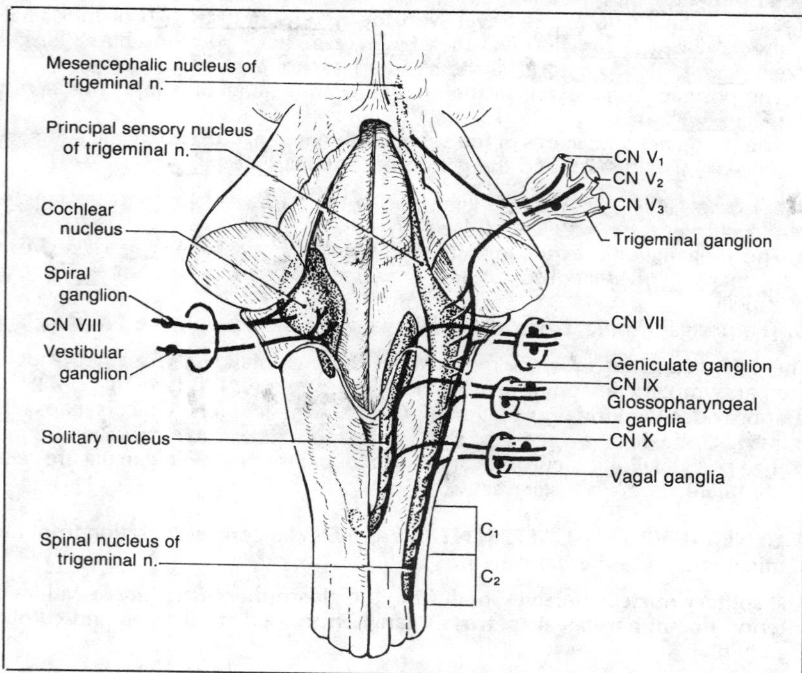

Figure 29-8. *The sensory nuclei of the brain stem.* The sensory components of the cranial nerves terminate in sensory nuclei.

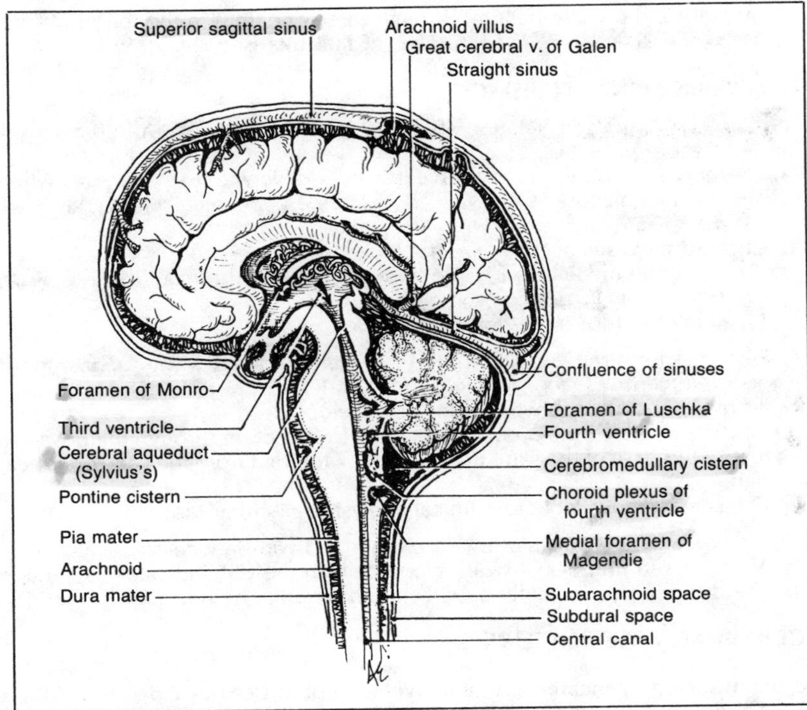

Figure 29-9. *The ventricular system of the brain.* Cerebrospinal fluid (CSF), secreted by the choroid plexuses of the ventricles, enters the subarachnoid space through three foramina in the roof of the fourth ventricle. The brain is suspended in this layer of CSF, which is resorbed into the venous system at the arachnoid villi.

b. The **glossopharyngeal nerve** is responsible for a small area of somatic sensation within the external auditory meatus.

(1) The cell bodies of these afferent neurons are located in the **superior glossopharyngeal (jugular) ganglion**.

(2) The fibers synapse in the spinal nucleus of V (see Fig. 29-8).

c. The **vagus nerve** (CN X) is responsible for a small area of somatic sensation on the posterior surface of the ear and a small area within the external auditory meatus.

(1) The cell bodies are located in the **superior vagal (jugular) ganglion**.

(2) The fibers synapse in the spinal nucleus of V (see Fig. 29-8).

VI. THE VENTRICULAR SYSTEM

A. The ventricles represent the rostral end of the central canal of the neural tube. The terminal expansion of the neural tube is divided into two lateral ventricles, the midline third ventricle, the cerebral aqueduct, and the midline fourth ventricle.

1. The **lateral ventricles**.

a. These are cavities within each cerebral hemisphere (see Fig. 29-3).

b. Each has three horns that extend into the frontal, temporal, and occipital lobes.

c. Each opens into the third ventricle through the **interventricular foramen** (Monro's).

2. The **third ventricle**.

a. This cleft-like cavity lies between the thalami (see Figs. 29-3 and 29-9).

b. Posteriorly, it is continuous with the **cerebral aqueduct**.

3. The **iter** or **cerebral aqueduct (Sylvius's)** lies in the midbrain and connects the third and fourth ventricles (Fig. 29-9).

4. The **fourth ventricle**.

a. This cavity lies in the pons beneath the cerebellum.

b. Posteriorly, it extends into the **central canal** of the spinal cord.

c. It communicates with the **cisterna magna** of the subarachnoid space (located between the medulla and the cerebellum) through three openings in the roof: the median **foramen of Magendie** and two lateral **foramina of Luschka** (see Fig. 29-9).

B. THE CEREBROSPINAL FLUID (CSF)

1. CSF is formed by a combination of ultrafiltration and active transport across the vascular tufts of the choroid plexuses.
   a. The nervous tissue around the ventricles is deficient at several sites where the pia mater and ependyma fuse. Overlying vascular mesenchyme invades this membrane to form choroid plexuses.
   b. **Choroid plexuses** occur in several locations:
      (1) Along the medial wall of the body and inferior horns of the lateral ventricles.
      (2) In the roof of the third ventricle.
      (3) Along the roof of the fourth ventricle.
2. CSF escapes from the ventricles into the **cisterna magna** of the subarachnoid space through three foramina in the roof of the fourth ventricle.
   a. If flow is obstructed, hydrocephalus develops.
   b. Fracture of the ethmoid bone may allow CSF to drain through the nose—**rhinorrhea**; fracture of the base of the skull may allow CSF to drain from the ear—**otorrhea**, a most serious sign.
   c. CSF is readily available for clinical analysis by lumbar tap.
3. CSF is normally a clear fluid low in protein and with few cells.
   a. An increased number of white blood cells in the CSF indicates bacterial meningitis.
   b. Blood in the CSF indicates subarachnoid (arterial) hemorrhage.

VII. THE CEREBRAL VASCULATURE

A. The brain is a highly metabolic organ. While it comprises but 2 percent of the body weight, it utilizes about 20 percent of the blood supply.
   1. It requires a sufficient and constant circulation of blood.
   2. Ischemia rapidly leads to loss of consciousness, followed by convulsions and (since the neurons begin to die after 4 minutes) brain death.

B. Blood to the brain is supplied by paired internal carotid arteries and paired vertebral arteries (Fig. 29-10).
   1. The **carotid arteries**.
      a. The right **common carotid artery** originates from the brachiocephalic artery, while the left common carotid artery arises directly from the arch of the aorta.
      b. In the neck each common carotid bifurcates into an **external carotid artery**, which supplies the face and neurocranium, and an **internal carotid artery**, which generally supplies the brain.
         (1) At the carotid bifurcation, the **carotid sinus** monitors the blood pressure, and the **carotid body** monitors the oxygen partial pressure (PO_2).
         (2) These arteries are instrumental in reflex control of cardiac rate and output.
            (a) The carotid branch of CN IX is the afferent limb.
            (b) The vagus nerve is the efferent limb.
   2. The **internal carotid artery** ascends in a stepwise manner from the mouth of the carotid canal just medial to the styloid to the anterior clinoid process of the sphenoid.
      a. Within the carotid canal, it is surrounded by a plexiform sheath of **postganglionic sympathetic fibers.**
      b. At the apex of the petrous bone, it turns up and medially along the body of the sphenoid, where it turns anteriorly again within the cavernous sinus.
      c. On leaving the cavernous sinus, it arches backwards beneath the anterior clinoid process and the optic nerve.
      d. An **ophthalmic branch** accompanies the optic nerve and gives off the central retinal artery, which enters the optic nerve.
      e. The **posterior communicating branch** passes posteriorly above the oculomotor nerve.
      f. The internal carotid artery divides into anterior and middle cerebral arteries.
         (1) The **anterior cerebral artery** passes dorsally around the corpus callosum at the base of the sagittal sulcus in company with its fellow of the opposite side. The two are connected by a small **anterior communicating branch**. They supply the medial surface of the frontal and parietal cerebral cortex.

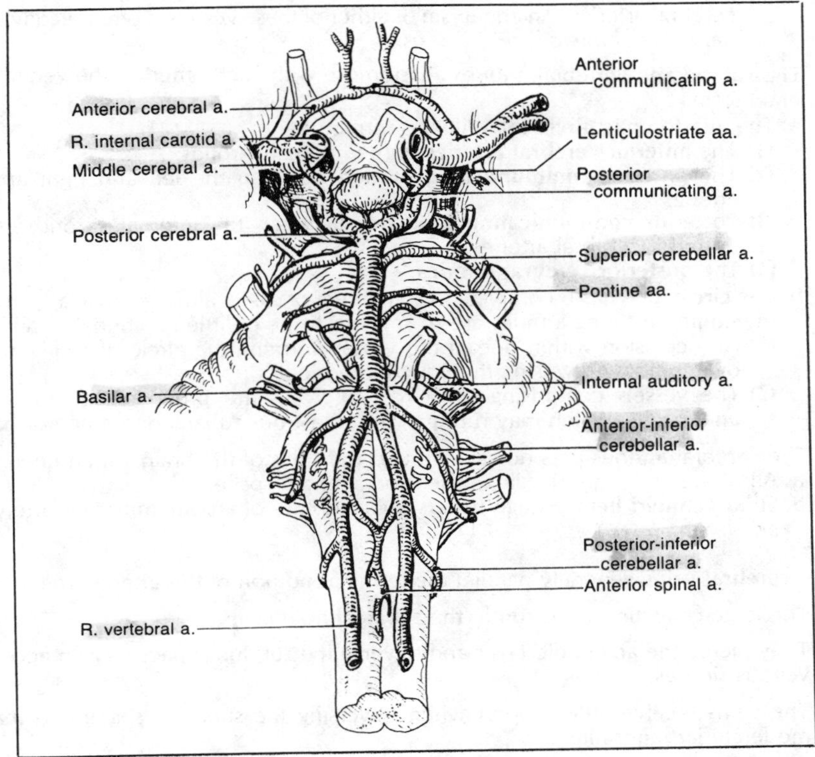

Figure 29-10. *The arterial supply to the brain.*

(2) The **middle cerebral artery** passes superiorly in the lateral fissure.
 (a) It supplies most of the lateral surface of the cerebral cortex.
 (b) **Lenticulostriate arterial branches** supply the basal ganglia, the thalamus, and the internal capsule. They are commonly the site of hemorrhage or thrombosis that causes the familiar pattern of contralateral paralysis, termed a **stroke** after its sudden and devastating nature.

3. The **vertebral arteries** arise from the subclavian arteries.
 a. They pass through the transverse foramina of the sixth through first cervical vertebrae.
 b. They pierce the dura mater and arachnoid at the atlanto-occipital membrane.
 c. They pass through the foramen magnum and ascend on the clivus of the basioccipital bone, giving off several branches.
  (1) The **posterior-inferior cerebellar arteries** pass laterally over the inferior surface of the cerebellum.
   (a) The posterior branches pass inferior to the cerebellum.
   (b) These, in turn, give off **posterior spinal arteries**.
  (2) The **anterior spinal artery** is formed by the union of a small branch from each vertebral artery. It runs down the midline in the anterior median fissure of the medulla.
 d. At the pontomedullary junction, the vertebral arteries unite to form the **basilar artery**, which gives off numerous branches.
  (1) The **anterior-inferior cerebellar artery** hooks around the abducens nerve and passes between the vestibulocochlear nerve and the glossopharyngeal nerve.
   (a) Each anterior branch passes laterally superior to the cerebellum.
   (b) Each has a small **labyrinthine branch**.
  (2) There are numerous small **pontine branches**.
  (3) At the anterior border of the pons, the basilar artery gives off the **superior cerebellar arteries**.
  (4) At this point the basilar artery bifurcates into the **posterior cerebral arteries**. Each of these arteries runs laterally along the pons to supply the tentorial surface of the occipital lobe.
  (5) The oculomotor nerve (CN III) emerges between the superior cerebellar and posterior

cerebral arteries. An aneurysm of either of these vessels is often readily apparent as the nerve is compressed.

4. The carotid and vertebral arteries anastomose with each other on the ventral surface of the brain stem.
   a. The anastomotic **circle (of Willis)** is formed by:
      (1) The **anterior cerebral arteries** of the internal carotids.
      (2) The **anterior communicating branch** between the left and right anterior cerebral arteries.
      (3) **Posterior communicating branches** between the internal carotid arteries and the posterior cerebral arteries.
      (4) The **posterior cerebral arteries**.
   b. The **circle of Willis** is complete in 90 percent of cases, although in a large number of these the communicating arteries are so small as to be of little functional significance.
      (1) An occlusion within one of the vessels forming the circle of Willis may be partially overcome by reversing the flow of the blood.
      (2) The vessels comprising the circle of Willis are prone to the formation of **berry aneurysms**, which may rupture with disastrous subarachnoid hemorrhage.

5. The cerebral arteries pass dorsally around the sides of the brain within deep fissures.
   a. All of the major arteries lie in the subarachnoid space.
   b. **Subarachnoid hemorrhage** occurs upon rupture of an atheromatous artery or of a berry aneurysm.

C. The **cerebral veins** generally parallel the terminal portion of the arterial tree in the pial layer.

1. The large collecting veins run in the subarachnoid layer.

2. They pierce the arachnoid layer and traverse the subdural space to gain access to the dural venous sinuses.

3. The veins receive little support when traversing the subdural space and may be torn by moderate jarring trauma.

4. Hemorrhage into the subdural space, **subdural hematoma**, has no pathway of escape and may not pass across the subarachnoid layer to present in the CSF.

D. Clinical considerations.

1. In many victims of **stroke**, there is a stepwise progression of disease prior to a major crippling attack.
   a. This may involve transient episodes such as dizziness, weakness, and loss of memory.
   b. Angiograms can be used to localize the site of blockage.
   c. Blockages or aneurysms of the major cerebral vessels may be treated surgically.

2. **Cerebral hemorrhage** may be caused by a rupture of cerebral vessels within the brain parenchyma (stroke) or by penetrating trauma. There is no surgical cure.

30
The Orbit, Eye, and Ear

I. THE ORBIT (ORBITA, ORBITAL CAVITY)

A. The bony orbit nearly surrounds and protects the eyeball (Fig. 30-1).

1. It is a pyramidal cavity with the axis directed about 23° from the sagittal plane.

a. The superior wall, the floor of the anterior cranial fossa, is formed largely by the frontal bone, with a minor contribution from the sphenoid bone.

(1) Laterally, the roof contains a depression for the lacrimal gland, the **lacrimal fossa**.

(2) The **supraorbital notch** or **foramen** at the supraorbital margin transmits the supraorbital branch of the frontal nerve.

(3) Anteriorly, the frontal bone contains the **frontal sinuses**.

b. The medial wall (in the sagittal plane) is largely composed of the ethmoid bone and the small lacrimal bone.

(1) That portion of the medial wall that separates the orbit from the underlying ethmoid air cells—the **lamina papyracea**—is very thin.

(2) The anterior and posterior ethmoid foramina transmit neurovascular structures from the orbit to the nasal cavity and paranasal sinuses.

(3) Anteriorly in the lacrimal bone, the fossa of the lacrimal sac (**lacrimal groove**) continues as the **nasolacrimal canal**, which contains the nasolacrimal duct and terminates in the inferior nasal meatus.

c. The lateral wall (about 45° from the sagittal plane) is formed largely by the zygomatic bone and the greater wing of the sphenoid bone.

d. The floor of the orbit is formed by the maxilla, with a minute contribution from the palatine bone.

(1) The floor is traversed by the **infraorbital groove**, which transmits the infraorbital branch of the maxillary nerve. This groove becomes the **infraorbital canal** just before reaching the infraorbital margin and exits via the **infraorbital foramen**.

(2) The floor of the orbit is also the roof of the maxillary sinus.

2. At the apex of the bony orbit are three apertures (see Fig. 30-1).

a. The **optic canal (foramen):**

(1) Connects the orbit with the middle cranial fossa.

(2) Contains the optic nerve and the ophthalmic artery.

b. The **superior orbital fissure:**

(1) Connects the orbit with the middle cranial fossa.

(2) Transmits several neurovascular structures (see Fig. 29-8).

(a) The oculomotor nerve (CN III).

(i) Superior division.

(ii) Inferior division.

(b) The trochlear nerve (CN IV).

(c) Three branches of the ophthalmic division of the trigeminal nerve (CN V_1).

(i) The lacrimal nerve.

(ii) The frontal nerve.

(iii) The nasociliary nerve.

(d) The abducens nerve (CN VI).

(e) The ophthalmic vein.

c. The **inferior orbital fissure:**

(1) Connects the orbit with the pterygopalatine fossa.

(2) Transmits the infraorbital branch of the maxillary division of the trigeminal nerve.

(3) Is spanned in large part by the involuntary **orbitalis muscle** (Müller's).

d. The **superior** and **inferior orbital fissures** are continuous at their medial end around the orbital surface of the greater wing of the sphenoid bone.

B. THE EYELIDS (PALPEBRAE)

1. The eyelids protect the anterior aspect of the eye (see Fig. 30-2).
a. They enclose a potential space, the **conjunctival sac**, which opens into the skin of the face at the palpebral fissure.
b. The **palpebral fissure** is bounded by the upper and lower palpebral margins.
c. The **palpebral margins** meet at the medial and **lateral canthi**.

2. The upper eyelid is the larger and more mobile.

3. Each lid contains a **tarsal plate** of dense connective tissue, which maintains the shape of the lid and to some extent protects the underlying eyeball (see Fig. 30-2).
a. The superior and inferior tarsal plates merge to form the **medial palpebral ligament** and the **lateral palpebral ligament**, which insert into their respective orbital margins.
b. The **orbital septum** is a fascial sheet that attaches the tarsal plates to the superior and inferior orbital margins.
c. The levator palpebrae superioris muscle attaches to the tarsal plate of the superior eyelid.

4. The subcutaneous tissue of the eyelids is very loose and may swell significantly with collection of fluid (puffy-eyelids) or extravasated blood (black eye).

5. The eyelid contains three muscles.
a. The palpebral portion of the **orbicularis oculi muscle** runs beneath the skin of the eyelid.
(1) The palpebral portion is a muscle of facial expression.
(2) Contraction of this portion produces the blink.
(3) The fibers of the palpebral portion are so arranged that when they shorten they move towards geodesic lines—the shortest distance between two points over the surface of a sphere—and thus close the palpebral fissure.
(4) Along with the other muscles of facial expression, the palpebral portion of the orbicularis oculi is innervated by the facial nerve (CN VII).
(a) This nerve forms the efferent limb of the blink reflex.
(b) Paralysis of the palpebral portion of the orbicularis oculi results in inability to close the eyelids. This is a major problem in Bell's (facial nerve) palsy because when the eyes cannot blink the conjunctiva dries and the cornea may ulcerate.
b. The **levator palpebrae superioris** is an extraocular muscle (see Figs. 30-2 and 30-6).
(1) This muscle arises from the apex of the orbit.
(2) It inserts into the upper eyelid and acts to draw the lid upwards when the eyeball is elevated.
(3) It widens into an aponeurosis, which inserts into the superior tarsal plate.
(4) It is innervated by the superior division of the oculomotor nerve. Paralysis produces **ptosis**—inability to lift the lid when the gaze is directed upward.
c. The **superior tarsal muscle** (Müller's) is composed of smooth (involuntary) muscle.
(1) It runs from the superior border of the tarsal plate of the upper lid to the tendon of the levator palpebrae superioris muscle.
(2) The superior tarsal muscle is innervated by sympathetic nerves that have their preganglionic cell bodies in the upper thoracic levels and their postsynaptic cell bodies in the superior cervical ganglion.
(3) It accentuates the opening of the palpebral fissure under sympathetic stimulation (''wide-eyed fear'').
(4) A lesion involving the cranial sympathetics (Horner's syndrome) with paralysis of this muscle produces a slight drooping of the lid, **pseudoptosis**.

6. The margins of the eyelids contain a double or triple row of hairs, as well as numerous sebaceous tarsal (meibomian) glands that lubricate the lids, sebaceous glands (Zeis') associated with the hair follicles, and sweat glands (Moll's).
a. The tarsal glands occasionally become infected (**acute meibomianitis**).
b. Infection of the glands of Zeis produces the common sty.

7. The inner surface of the eyelid and the eyeball is lined with the **conjunctivae**.
a. The **palpebral conjunctiva** of the inner surface of the eyelids is reflected off the bases of the lids onto the eyeball (see Fig. 30-6).
(1) Superior and inferior **conjunctival fornices** are formed by these reflections.
(2) The lacrimal ducts drain into the superior fornix.
(3) The palpebral conjunctiva is richly supplied with blood. It is normally pink but becomes palid in anemia.

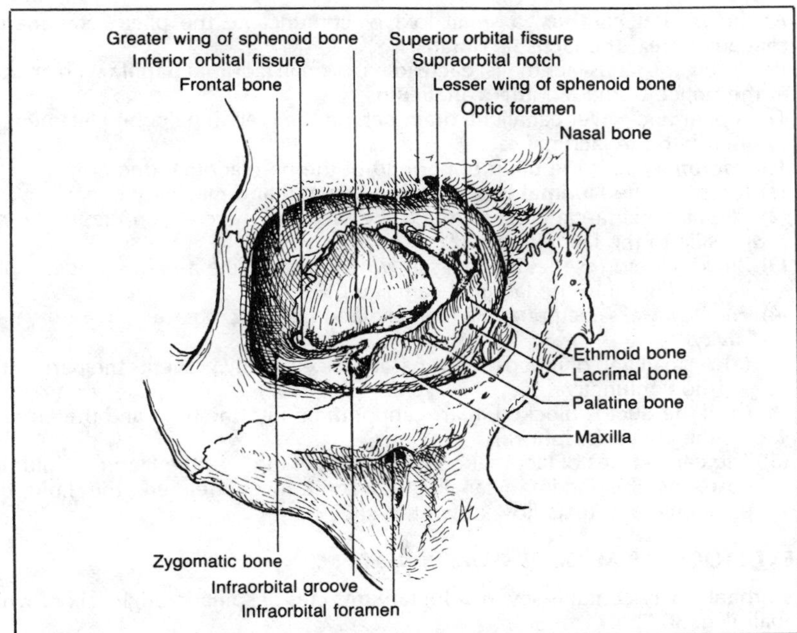

Figure 30-1. *The bones of the right orbit.*

b. The **bulbar conjunctiva** protects the cornea.

(1) It is transparent and contains a great number of small blood vessels, which produce the "blood-shot eye" or "pink-eye" when dilated due to allergy, irritation, or inflammation.

(2) The conjunctiva is innervated by the ophthalmic and maxillary divisions of the trigeminal nerve (CN V).

(a) This is the afferent limb of the blink reflex.

(b) The conjunctiva is very sensitive and any foreign object in contact with the conjunctiva causes blinking.

(c) Because of the interconnections in the brain stem, reflex blinking normally occurs in both eyes simultaneously—the consensual blink reflex.

(d) The afferent and efferent portions of the facial and trigeminal nerves are routinely tested in this manner.

(i) If both eyes blink when the right eye is touched but not when the left eye is touched, the problem must lie with the left trigeminal nerve.

(ii) Conversely, if only the right eye blinks when either eye is touched, the problem must lie with the left facial nerve.

8. The **lacrimal glands** are located in the superior lateral region of the orbit (Fig. 30-2).

a. The lacrimal gland has an intimate relationship with the tendon of the levator palpebrae superioris muscle. The major portion of the lacrimal gland lies above it, while a smaller portion lies beneath it. Thus, movement of the eyelids tends to milk the gland, so that continuous lubrication is provided and the conjunctiva is kept moist.

b. About a dozen ductules drain the lacrimal gland into the lateral region of the superior conjunctival fornix.

c. The lacrimal gland is controlled by parasympathetic innervation.

(1) Presynaptic parasympathetic neurons arise from the **superior salivatory nucleus** (see Fig. 29-7).

(2) The preganglionic axons travel with the **nervus intermedius** of the facial nerve, the greater superficial petrosal nerve, and the nerve of the vidian canal to the pterygopalatine ganglion in the pterygopalatine fossa.

(3) The postganglionic axons of the **pterygopalatine ganglion** travel along the maxillary nerve, its zygomatic branch, and a communicating branch to the lacrimal branch of the ophthalmic nerve, finally reaching the lacrimal gland.

d. With cholinergic stimulation, tears flush across the eyeball and is drawn by capillary action along the palpebral fissure.

9. The medial end of the palpebral margin is hairless (see Fig. 30-2).

a. At the **medial canthus**, a small fold of conjunctiva, the plica semilunaris, encloses a triangular area, the **lacus lacrimalis**.

b. Bordering the lacus lacrimalis, each lid has a small **lacrimal papilla** with an apical punctum at the orifice of the **lacrimal canaliculus**.

c. The upper and lower canaliculi drain behind the medial palpebral ligament and then join to drain into the lacrimal sac.

d. The **lacrimal sac** is the blind upper end of the nasolacrimal duct.

(1) It rests in the **lacrimal groove** of the lacrimal and maxillary bones.

(2) The lacrimal fascia passes around the sac from the crest on the frontal process of the maxilla to the lacrimal crest.

(3) Blinking deforms the lacrimal sac, which, acting like a bulb syringe, aspirates the collecting fluid into the lacrimal canaliculi.

(4) The lacrimal sac empties via the **nasolacrimal duct** into the inferior meatus of the nasal cavity.

(a) A small flap of mucosa at its nasal orifice usually prevents the percolation of air into the conjunctival sac.

(b) If the duct is blocked, tears cannot drain into the nose and therefore run out onto the cheek (**epiphora**).

(5) The early stages of lacrimation involve "sniffling," as the lacrimal fluid drains into the nose. As the lacrimal canaliculi become overwhelmed, the fluid overflows the palpebrae and tears flow down the cheeks.

C. THE EXTRAOCULAR MUSCULATURE

1. The orbital cavity contains seven voluntary muscles of somatic origin, six of which move the eyeball (Fig. 30-3).

2. The eyeballs, suspended within the orbit by a fibrous capsule, are very mobile, with movement occurring about three axes.

a. Elevation and depression occur about the transverse axis through the equator of the eyeball.

b. Abduction and adduction occur about a vertical axis through the poles.

c. Intorsion and extorsion occur about an anteroposterior axis through the pupil. Intorsion is thus a medial rotation of the upper portion of the eye toward the midline.

3. The **anulus tendineus** (Zinn's) lies at the apex of the periorbita (see Fig. 30-3).

a. It gives rise to the four rectus muscles.

b. The optic nerve, ophthalmic artery, superior and inferior divisions of the oculomotor nerve, abducens nerve, and nasociliary branch of the ophthalmic nerve enter the orbital cavity through the anulus tendineus as they leave the optic canal and the superior orbital fissure.

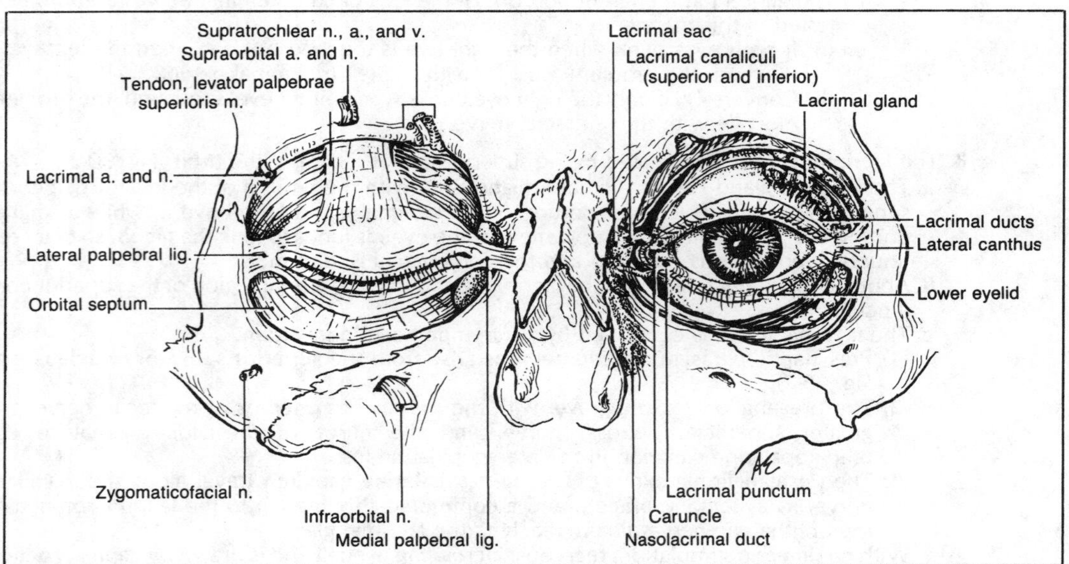

Figure 30-2. *The palpebrae and lacrimal apparatus.* The structure of the eyelids and the insertion of the levator palpebrae superioris muscle of the right eye and the lacrimal apparatus of the left eye.

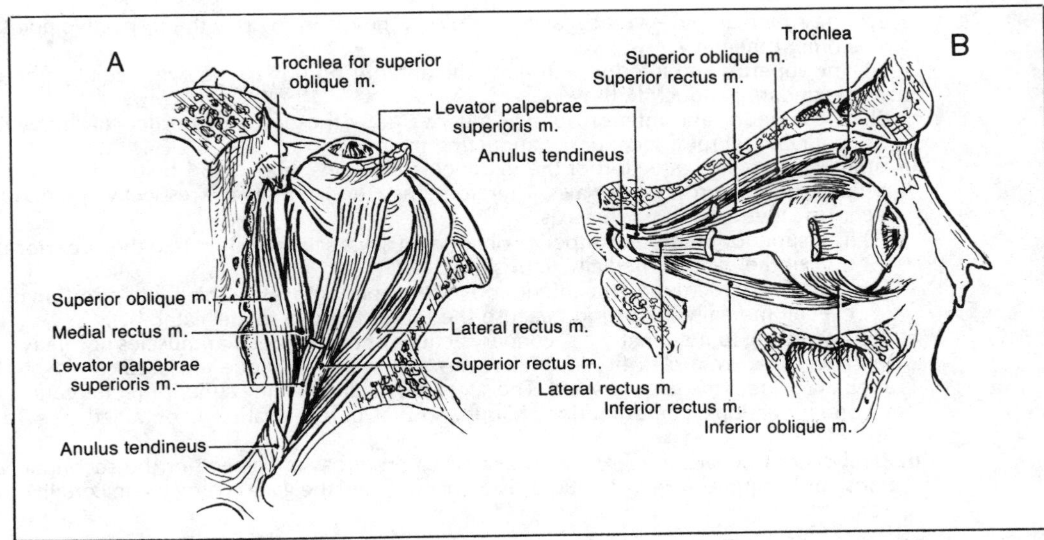

Figure 30-3. *The extrinsic muscles of the right eye. A,* Viewed from above. *B,* Viewed from the right side.

c. The ophthalmic vein and trochlear nerve, as well as the frontal and lacrimal branches of the ophthalmic nerve, pass superior to the anulus tendineus as they pass through the superior orbital fissure.

4. The four rectus muscles diverge from the anulus tendineus to insert into the sclera of the eyeball (see Fig. 30-3).

a. The **superior rectus muscle.**

(1) This muscle inserts onto the anterosuperior aspect of the eyeball.

(2) It primarily elevates the eyeball.

b. The **inferior rectus muscle.**

(1) This muscle inserts onto the anteroinferior aspect of the eyeball.

(2) It primarily depresses the eyeball.

c. The **lateral rectus muscle** inserts onto the anterolateral aspect of the eyeball and abducts the eye.

d. The **medial rectus muscle** inserts onto the anterolateral aspect of the eyeball and adducts the eye.

5. Three other muscles have different origins (see Fig. 30-3).

a. The **superior oblique muscle.**

(1) This muscle originates from the sphenoid bone just superomedial to the anulus tendineus.

(2) It has a unique course: Its tendon passes through a fibrous trochlea (*L.* pulley) attached to the frontal bone at the anterosuperior aspect of the medial orbital margin. The tendon then passes backward to insert onto the posterolateral quadrant of the superior surface of the eyeball.

(3) Thus, this muscle draws the posterior portion of the eyeball upward, depressing the eye.

b. The **inferior oblique muscle.**

(1) This muscle originates from the anteroinferior aspect of the medial orbital margin.

(2) It seems to represent the distal portion of a primitive muscle that paralleled the superior oblique muscle. It passes backward to insert onto the posterolateral quadrant of the inferior surface of the eyeball.

(3) Thus, this muscle draws the posterior portion of the eyeball downward, elevating the eye.

c. The **levator palpebrae superioris muscle.**

(1) This muscle originates just superior to the anulus tendineus.

(2) It inserts into the tarsal plate of the superior eyelid.

(3) It acts to draw the eyelid upwards when the eyeball is elevated.

6. The actions of the extraocular muscles (Table 30-1).

a. Because the bony orbit is directed somewhat laterally and because the pupil normally is directed anteriorly, the lines of action of the extraocular muscles, above and below the eye, pass medial to the vertical axis. Therefore, they tend to draw the respective points of attachment medially.

(1) The superior and inferior recti draw the anterior portion of the eye medially, thereby secondarily adducting the eye.

(2) The superior and inferior oblique muscles draw the posterior portion of the eyeball medially, thereby secondarily abducting the eye.

b. Similarly, the lines of action of the extraocular muscles, above and below the eye, pass medial to the anteroposterior axis. Therefore, they tend to draw the respective points of attachment medially about this axis.

(1) The superior rectus and superior oblique muscles secondarily rotate the **superior** portion of the eyeball medially (intorsion).

(2) The inferior rectus and the inferior oblique secondarily rotate the inferior portion of the eyeball medially (extorsion because the top of the eye rotates laterally).

c. It is interesting to note that the secondary actions of the extraocular muscles normally tend to cancel. For example, the superior rectus and inferior oblique are used to elevate the eyeball to direct the gaze upward. The secondary actions of the superior rectus cancel the secondary actions of the inferior oblique, so that pure elevation is obtained (see Table 30-1).

d. Paralysis of one or more extraocular muscles produces imbalance of the secondary actions, and complex forms of double vision occur when the gaze is directed in certain directions.

7. The **fascia bulbi** (Tenon's capsule) forms a connective tissue socket in which the eyeball is suspended.

a. The extraocular muscles pass through Tenon's capsule to insert into the sclera of the eyeball.

b. The sheaths of the medial and lateral rectus tendons are thickened, forming the medial and lateral **check ligaments**, which are so named because they check the action of opposite muscles.

c. These fascial condensations continue around the inferior surface of the eyeball to form a fascial hammock that suspends the eyeball between the medial and lateral margins of the bony orbit.

D. INNERVATION OF THE ORBIT

1. Nerves of the extraocular musculature

a. The **oculomotor nerve** (CN III) [Fig. 30-4; see Fig. 29-7].

(1) Axons from the oculomotor nucleus innervate five of the extraocular muscles.

(a) The **superior division** innervates the levator palpebrae superioris and the superior rectus muscles.

(b) The **inferior division** innervates the medial rectus, the inferior rectus, and the inferior oblique muscles.

(2) Axons from the accessory oculomotor nucleus (Edinger-Westphal) contribute the

Table 30-1. Extraocular Musculature

Muscle	Origin	Insertion	Primary Action	Secondary Actions	Innervation
Superior rectus	Anulus tendineus	Anterosuperior aspect	Elevation	Adduction and intorsion	Oculomotor (CN III)
Inferior oblique	Lacrimal bone	Posterolateral inferior quadrant	Elevation	Abduction and extorsion	Oculomotor (CN III)
Inferior rectus	Anulus tendineus	Anteroinferior aspect	Depression	Adduction and extorsion	Oculomotor (CN III)
Superior oblique	Sphenoid bone	Posterolateral superior quadrant	Depression	Abduction and intorsion	Trochlear (CN IV)
Medial rectus	Anulus tendineus	Anteromedial aspect	Adduction	. . .	Oculomotor (CN III)
Lateral rectus	Anulus tendineus	Anterolateral aspect	Abduction	. . .	Abducens (CN VI)
Levator palpebrae superioris	Sphenoid bone	Tarsal plate of upper eyelid	Raises upper eyelid	. . .	Oculomotor (CN III)

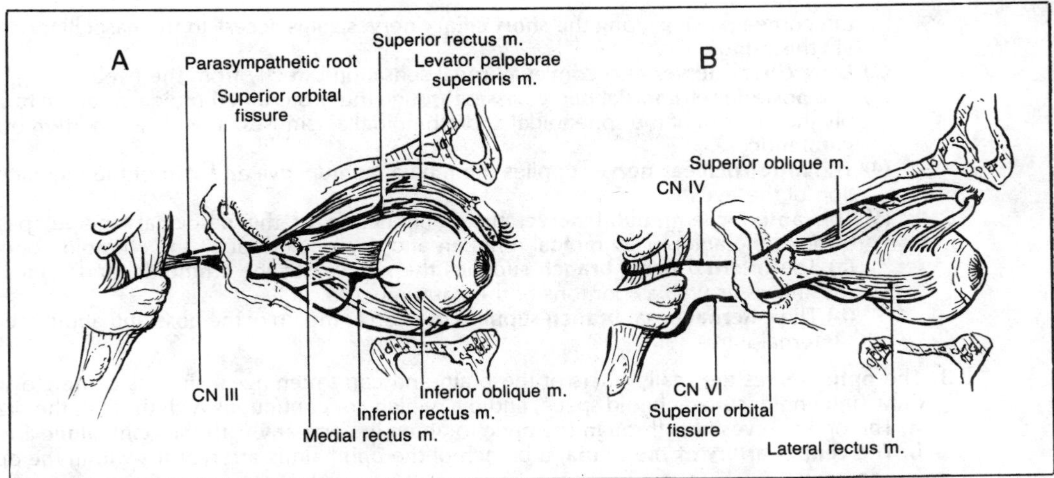

Figure 30-4. *The innervation of the extrinsic muscles of the eye. A,* The oculomotor nerve. *B,* The trochlear and abducens nerves.

parasympathetic (general visceral efferent) component of the oculomotor nerve (see Fig. 29-7).

(a) The preganglionic parasympathetic neurons leave the oculomotor nerve via the **parasympathetic (motor) root** to reach the ciliary ganglion.

(b) The postganglionic axons of the **ciliary ganglion** course through the **short ciliary nerves** to the iridial sphincter for pupillary constriction and to the ciliary muscle for accommodation to near vision.

(3) Lesion of the oculomotor nerve produces lateral strabismus (**cocked eye** or **wall eye**) and nearly complete ophthalmoplegia of the eye as well as ptosis (drooping) of the eyelid. In addition, the pupil is dilated, and there is inability to accommodate for near vision.

b. The **trochlear nerve** (CN IV) [see Fig. 29-7].

(1) Axons arise from the trochlear nucleus and innervate the superior oblique muscle.

(2) Lesions produce diplopia upon looking down. Individuals with diplopia usually experience difficulty and apprehension upon descending stairs.

c. The **abducens nerve** (CN VI) [see Fig. 29-7].

(1) Axons arise from the abducens nucleus and innervate the lateral rectus muscle.

(2) Lesions produce medial strabismus (**crossed eyes**). The diplopia is minimal when looking toward the side opposite the lesion.

d. A medial longitudinal fasciculus in the midbrain connects the extraocular nuclei, permitting conjugate eye movements and convergence.

e. The oculomotor and trochlear nerves pass through the incisura tentoria. An intracranial mass (tumor, hematoma, or edema) may squeeze portions of the temporal lobes through the tentorial notch (temporal lobe herniation), compressing the oculomotor nerve within the tentorial notch and causing **ophthalmoplegia** and **mydriasis** (dilated pupil).

2. The **ophthalmic division** of the **trigeminal nerve** (CN V) provides the sensory branches within and about the orbit (Fig. 30-5).

a. The **lacrimal nerve** runs superolaterally within the orbit.

(1) It supplies the lateral portion of the eyelids and conjunctiva.

(2) Proximally, it contains no secretomotor fibers to the lacrimal gland.

(3) Distally, it communicates with the zygomatic branch of the maxillary nerve and receives postsynaptic parasympathetic secretomotor fibers from the pterygopalatine ganglion.

b. The **frontal nerve**, the largest branch, bifurcates.

(1) The **supraorbital nerve** runs just superior to the axis of the orbit.

(a) It passes through the **supraorbital notch** or **foramen**.

(b) It supplies the skin of the eyelid, forehead, and scalp.

(2) The **supratrochlear nerve** runs more medially and supplies the skin of the medial portions of the eyelids and the central portion of the forehead.

c. The **nasociliary nerve** runs superomedially within the orbit and becomes the anterior ethmoidal nerve.

(1) A **sensory root** (**ramus communicans**) connects to the ciliary ganglion. Sensation from the cornea passing along the **short ciliary nerves** gains access to the nasociliary nerve via this ramus.

(2) **Long ciliary nerves** also convey corneal sensation directly from the eyeball.

(3) The **posterior ethmoidal nerve** passes through the posterior ethmoidal foramen to supply the mucosa of the sphenoidal and ethmoidal air sinuses, as well as a portion of the nasal mucosa.

(4) The **infratrochlear nerve** supplies the medial conjunctiva and skin of the medial portion of the eyelids.

(5) The **anterior ethmoidal nerve**, the continuation of the nasociliary nerve, passes through the anterior ethmoidal foramen and descends through the ethmoid bone.

(a) The **internal nasal branch** supplies the mucosa of the ethmoidal and frontal air sinuses as well as portions of the nasal mucosa.

(b) The **external nasal branch** supplies the skin at the tip of the nose and about the external nares.

3. The **optic nerves** are really tracts of the brain and carry meningeal sheaths complete with dura, arachnoid, subarachnoid space, and pia, which are continuous with those of the brain.

a. The optic nerves pass through the optic foramen in company with the ophthalmic artery.

b. The **central artery of the retina**, a branch of the ophthalmic artery, runs within the optic nerve.

4. Autonomic innervation of the orbit.

a. Sympathetic pathways.

(1) The preganglionic neurons lie in the upper thoracic segments of the spinal cord.

(a) These axons leave the spinal nerves and gain access to the sympathetic chain via the **white rami communicantes**.

(b) The axons then course superiorly along the **sympathetic chain** to reach the superior cervical ganglion.

(2) From the **superior cervical ganglion**, the postsynaptic neurons course in the carotid plexus within the adventitia of the internal carotid artery. From the **carotid plexus** the sympathetic fibers may reach the eye by a number of routes.

(a) The sympathetic fibers generally continue along the internal carotid artery and the ophthalmic artery and its posterior ciliary branches to gain access to the orbit and eyeball.

(b) In the cavernous sinus, sympathetic fibers may pass to the ophthalmic nerve and its nasociliary branch and thence along the long ciliary nerves to the eyeball or, alternatively, along the sensory root (ramus communicans) of the ciliary ganglion and then along the short ciliary nerves to the eyeball.

(3) Injury to the cervical sympathetic chain results in **Horner's syndrome**. The signs include:

(a) Miosis (pupillary constriction) due to unopposed action of the parasympathetic nerves.

(b) Pseudoptosis (a slight drooping of the eyelid) due to paralysis of the superior tarsal muscle (Müller's).

(c) Slight enophthalmos (retraction of the eyeball into the orbit) due to paralysis of the orbitalis muscle (Müller's), which spans the inferior orbital fissure.

(d) Anhidrosis (lack of sweating).

(e) Flushing due to release from the vasoconstrictive effects of the sympathetic nerves.

b. Parasympathetic pathways.

(1) Presynaptic neurons from the accessory oculomotor nucleus (Edinger-Westphal's) course in the oculomotor nerve (CN III).

(a) The preganglionic parasympathetic neurons leave the oculomotor nerve via the **parasympathetic (motor) root**.

(b) They then reach the **ciliary ganglion**, which is located just lateral to the optic nerve.

(2) The postganglionic axons of the ciliary ganglion course through the **short ciliary nerves**.

(a) They course to the iridial sphincter for pupillary constriction and to the ciliary muscle of the pupil for accommodation to near vision.

(b) Some nonparasympathetic pathways may pass though the ciliary ganglion to gain access to or from the short ciliary nerves.

(i) Sympathetic neurons from the ophthalmic artery may pass through a **sympathetic root** and continue through the ciliary ganglion to reach the eyeball via the short ciliary nerves.

(ii) Sensory fibers from the cornea that pass along the short ciliary nerves continue through the ciliary ganglion to reach the nasociliary nerve via the **sensory root** (ramus communicans).

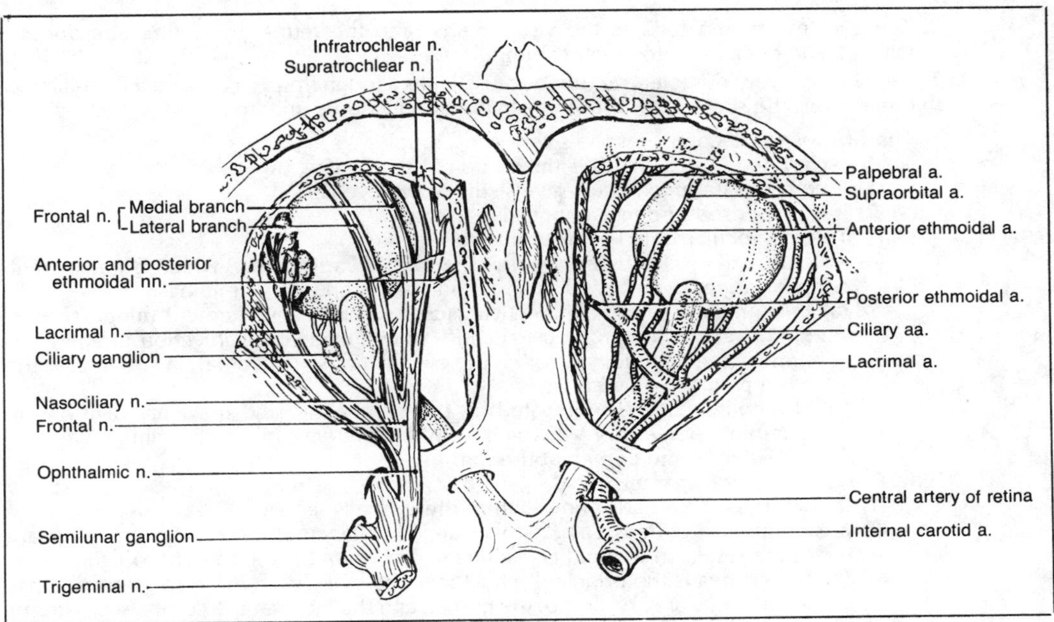

Figure 30-5. *The nerves and vessels of the orbit.* The ophthalmic division of the left trigeminal nerve, the right ophthalmic artery.

E. THE ORBITAL VASCULATURE

1. The **ophthalmic artery** is a branch of the internal carotid artery (see Fig. 30-5).
   - a. It enters the orbit through the **optic canal** beneath the optic nerve.
   - b. It supplies the eyeball through the **central artery of the retina**, as well as by anterior ciliary and posterior ciliary branches.
   - c. Within the orbit it winds around the medial surface of the optic nerve in company with the nasociliary nerve.
   - d. The branches correspond to the branches of the ophthalmic division of the trigeminal nerve: ethmoidal, supratrochlear, supraorbital, and lacrimal, as well as anterior and posterior ciliary.
2. The **ophthalmic vein**.
   - a. The **superior ophthalmic vein** drains the superior regions of the orbit, eyelids, and forehead. It anastomoses with the terminal twigs of the angular vein, a branch of the facial vein.
   - b. The **inferior ophthalmic vein** drains the inferior regions of the orbit and eyelids, as well as the **vorticose veins** from the eyeball. It also anastomoses with the terminal twigs of the angular vein and the pterygoid plexus.
   - c. The superior and inferior ophthalmic veins leave the orbit via the superior orbital fissure to drain, either separately or by one trunk, into the cavernous sinus.
   - d. The anastomoses between the angular and ophthalmic veins may result in spread of infection from the periorbital and perinasal regions to the cavernous sinus.
     - (1) Inflammatory thrombosis of the cavernous sinus interferes with venous drainage of the retina and thereby results in engorgement of the retinal arteries, followed by retinal ischemia and ultimate blindness.
     - (2) Because the ophthalmic, oculomotor, trochlear, and abducens nerves also pass through the cavernous sinus, infection here may result in ophthalmoplegia, mydriasis, and sensory deficits in the periorbital region.

II. THE EYEBALL (BULBUS OCULI)

A. The eyeball is formed as an outgrowth of the brain.

1. The two layers of the retina of the adult eye represent an extension of the brain.
2. Other parts of the eyeball, such as the lens, are derived from ectoderm.

3. The eye inverts and focuses the visual fields onto the retina; the retina transduces electromagnetic radiation into electrical impulses.

B. The eyeball is a very durable structure because it has a tough fibrous coat and a fluid-filled cavity that maintains the shape and distributes hydraulic pressures uniformly.

1. The **fibrous tunic** comprises the sclera and cornea.

a. The **sclera** (*skleros, G.* hard) is the white of the eye (Fig. 30-6).

(1) It comprises approximately five-sixths of the eyeball.

(2) It is composed of dense connective tissue.

(3) The extraocular muscles insert into the sclera.

(4) Anteriorly, the sclera is covered by the **bulbar conjunctiva**, which is transparent and contains a great number of small blood vessels and nerve endings.

(a) Inflammation of the conjunctiva causes vascular engorgement (**blood-shot eye** or **pink eye**). Although the usual cause of conjunctivitis is infection or allergy, occasionally the pathology is within the eyeball (iritis, glaucoma), in which case there is often pain or impaired vision.

(b) The conjunctiva is innervated by the ophthalmic and maxillary divisions of the trigeminal nerve (CN V), which form the afferent limb of the blink reflex.

(5) The sclera joins the cornea at the **limbus**.

b. The **cornea** is transparent.

(1) It comprises the anterior one-sixth of the eyeball (see Fig. 30-6).

(2) It is composed of dense, regularly arranged collagen fibers; the extreme regularity of the tissue results in a liquid-crystalline structure that is transparent to light.

(a) The cornea is the principal refractor of the eye, much more so than the lens. The small radius of curvature of the cornea and the fact that the cornea separates media of two different refractive indices (air and aqueous humor) result in strong bending of light waves.

(b) The shape of the cornea and the anteroposterior diameter of the eye determines the focal point.

(i) Insufficient corneal refraction that cannot be corrected by the lens results in a focal point behind the retina—**hyperopia** or **hypermetropia**. The result is farsightedness because distant objects can be focused on the retina by accommodation of the lens.

(ii) Too much corneal refraction results in a focal point in front of the retina—**myopia**. The result is nearsightedness because close objects focus on the retina without accommodation by the lens.

(iii) Irregularities in the shape of the cornea produce variation in the focal points—**astigmatism**.

(3) The cornea is covered by **corneal epithelium**, a continuation of the conjunctiva.

(4) Injury or inflammation of the cornea (**keratitis**) resolves with healing but with a loss of

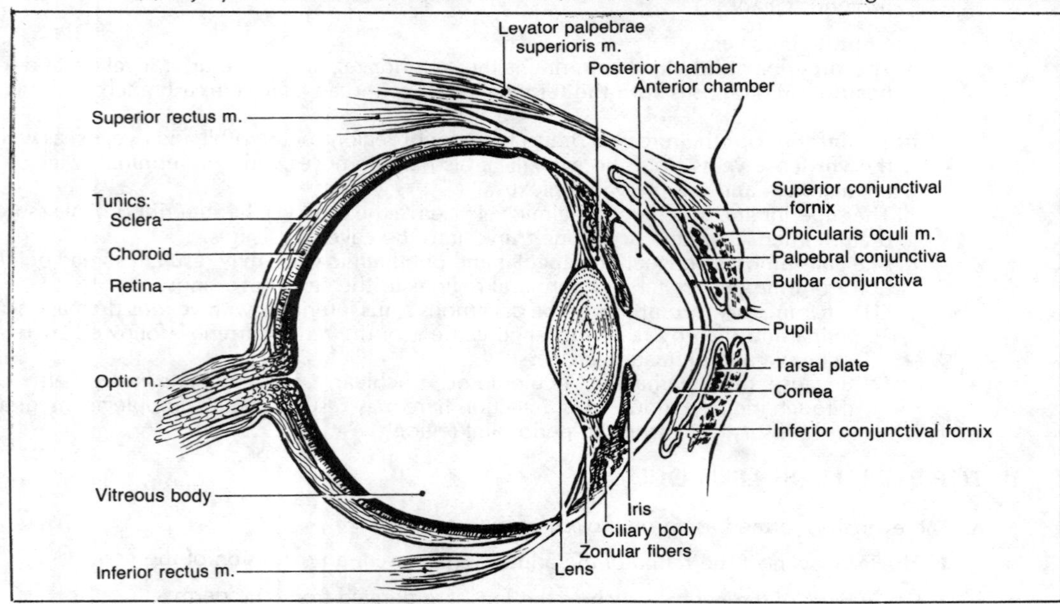

Figure 30-6. *The eyeball and conjunctival sac.*

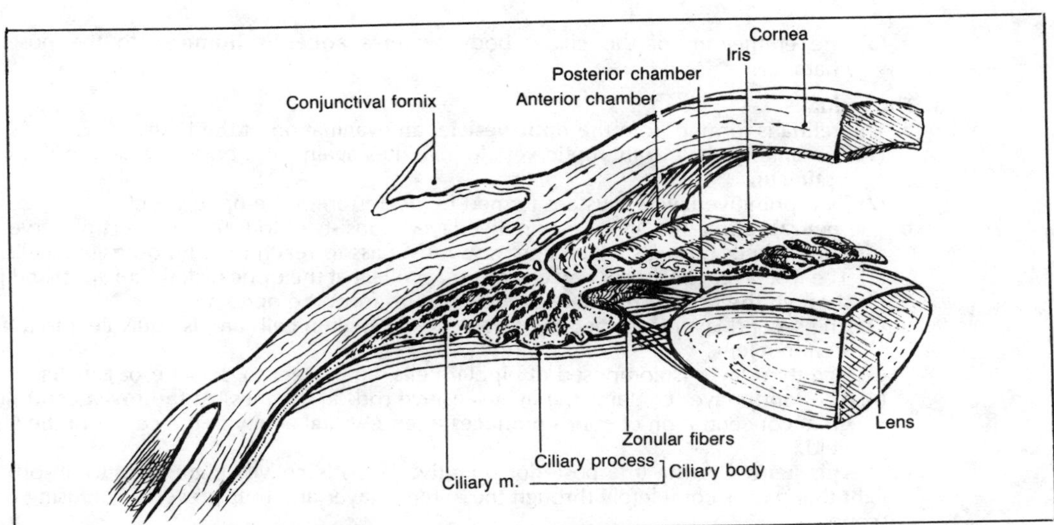

Figure 30-7. *The limbus of the eyeball and the ciliary body.*

the high degree of organization and, consequently, loss of transparency. Corneal scars can be treated by corneal transplant (keratoplasty).

2. The **vascular tunic.**
   a. The **choroid layer** consists primarily of blood vessels supplied by the short ciliary arteries and drained by the vorticose veins (see Fig. 30-6).
   b. The **ciliary body** is the anterior continuation of the choroid layer (Fig. 30-7).
      (1) The ciliary body suspends the lens by a multitude of **zonular fibers** (Zinn's), which insert into the capsule of the lens.
         (a) The normal elastic tension exerted on the capsule of the lens through the zonular fibers tends to flatten the lens, so that there is minimal refraction of light rays—far accommodation.
         (b) The **ciliary muscle** runs from the base of the ciliary body to the scleral spur at the limbus. Contraction of this muscle draws the ciliary body anteriorly, thereby releasing tension on the zonular fibers. This movement allows the elastic lens to assume a more spherical shape, so that there is increased refraction and convergence of light rays—near accommodation.
         (c) Accommodation is controlled by complex central pathways and effected by the parasympathetic nerves of the accessory oculomotor nucleus (Edinger-Westphal's).
            (i) These nerves travel to the orbit along the oculomotor nerve (CN III), leave via the parasympathetic root, and synapse in the ciliary ganglion.
            (ii) The postsynaptic fibers travel along the short ciliary nerves to reach the eyeball and ciliary body.
      (2) The **iris** arises from the ciliary body, anteriorly.
         (a) It divides the space between the cornea and lens into **anterior** and **posterior chambers**.
         (b) It contains variable amounts of pigment.
            (i) If it is heavily pigmented in the anterior layer, the result is brown eyes.
            (ii) If the pigment is only in the posterior layer, the iris appears blue or gray because light refracts within the unpigmented anterior layer.
         (c) The iris contains primitive myoepithelial cells surrounding a central aperture, the **pupil**.
            (i) The **sphincter pupillae** consist of myoepithelial cells arranged in a circle under parasympathetic control.
            (ii) The **dilator pupillae** consist of radially arranged myoepithelial cells under sympathetic control.
         (d) The size of the pupil is normally a reflex response to the intensity of light reaching the retina.
            (i) Injury to the oculomotor nerve releases parasympathetic influence and produces a dilated pupil (mydriasis).
            (ii) Injury to the upper thoracic spinal cord or to the cervical sympathetic chain (Horner's syndrome) releases sympathetic influence and produces a constricted pupil (**miosis**).

(3) The epithelium of the ciliary body secretes **aqueous humor** into the posterior chamber.

3. The **retina**.
   a. The retina is formed from the **optic vesicle**, an evagination of the brain.
      (1) The anterior half of the optic vesicle involutes against the posterior half, forming the **optic cup**.
      (2) Two primitive layers thus are formed by occlusion of the optic vesicle.
   b. The **neural retina** (the anterior primitive layer) consists of four major layers of nerve and supporting cells through which light rays must pass to reach the photosensitive cells.
      (1) The first layer consists of nerve axons that collect at the **optic disk** (**blind spot**) and pass through the **cribriform plate** of the sclera to form the optic nerve.
      (2) The second layer is formed by the so-called ganglion cells and is equivalent to a brain stem nucleus.
      (3) The third layer is composed of bipolar cells, equivalent to dorsal root ganglia.
      (4) The fourth layer contains the light-sensitive rods and cones. At the **fovea centralis**, a great concentration of cones produces intense visual acuity in the center of the visual field.
   c. The **pigmented retina** (the posterior primitive layer) is heavily pigmented to absorb any light that passes completely through the anterior layer and thus prevents confusing back-scatter.
   d. Clinical considerations.
      (1) **Retinal detachment** is the result of separation between the anterior and posterior layers, re-establishing the primitive optic vesicle.
         (a) The separation usually starts anteriorly and therefore is easily missed until well advanced.
         (b) The predisposition to detachment is bilateral, so care must be taken to protect the other eye.
         (c) The layers may be reattached by coagulation procedures.
      (2) Because the optic nerve is a portion of the central nervous system (CNS) and is, as such, surrounded by layers of meninges, elevated cerebrospinal fluid (CSF) pressure produces edema of the optic disk (**papilledema**), which can be detected by ophthalmoscopic examination.
      (3) The **central artery of the retina** enters the eyeball with the optic nerve and branches over the retina.
         (a) The retinal vessels can be examined with an ophthalmoscope.
         (b) These are the only arteries in the body that can be examined directly for signs of systemic disease such as hypertension and diabetes.

4. The chambers of the eye.
   a. The **anterior chamber** lies between the cornea and the iris; the **posterior chamber** lies between the iris and the lens (see Fig. 30-7).
      (1) These chambers contains the thin, watery aqueous humor.
         (a) **Aqueous humor** is secreted by the ciliary process into the posterior chamber.
         (b) It passes through the pupil into the anterior chamber.
         (c) It drains into the venous system through **Schlemm's canal** at the angle of the anterior chamber.
      (2) If drainage is impaired, intraocular pressure increases and the retinal blood flow is thereby impaired, producing retinal ischemia (**glaucoma**) and blindness.
   b. The **vitreous body**, a transparent and semigelatinous material, fills the **vitreous chamber** behind the lens (see Fig. 30-6).

5. The **lens** separates the aqueous and vitreous chambers (see Figs. 30-6 and 30-7).
   a. It is composed of highly ordered connective tissue cells; the high degree of order confers transparency.
   b. It is enclosed in an elastic capsule, into which the zonular fibers insert.
   c. The lens itself is deformable and elastic.
   d. Its refractive index is slightly different from the aqueous and vitreous humors, providing some degree of refraction.
      (1) The tension exerted on the lens by the zonular fibers adjusts the shape of the lens and thereby alters the refracting power.
         (a) When the ciliary muscle contracts and releases tension on the zonular fibers, the lens assumes a more spherical shape and strongly refracts light rays, accommodating the eye for near vision.
         (b) When the ciliary muscle relaxes, the tension on the zonular fibers is restored, and the lens is flattened so that it weakly refracts light rays, accommodating the eye for far vision.

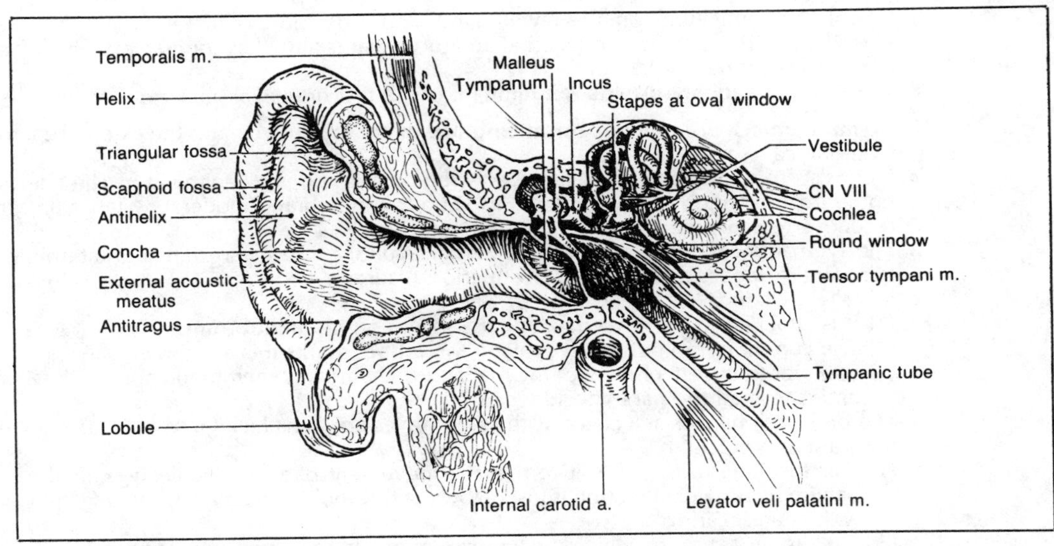

Figure 30-8. *The right ear.*

(2) With aging, the lens tends to harden and lose intrinsic elasticity. The aged lens cannot be deformed sufficiently to accommodate to the near and far extremes (**presbyopia**), both of which may require correction with bifocal lenses.

(3) The most common cause of blindness is **cataracts**, which are progressive opacities that result from degenerative changes. When of sufficient size to impair vision, the lens can be removed surgically.

III. THE EAR.
The ear is divided into external, middle, and inner portions.

A. THE EXTERNAL EAR
consists of the auricle (pinna), the external auditory meatus, and the tympanic membrane.

1. The **auricles**, lying on either side of the head and directed slightly forward, concentrate sound waves and enable stereophonic localization of the source (Fig. 30-8).
   a. The skeleton of the ear is a single convoluted plate of elastic cartilage. The skin of the auricle is firmly attached to the underlying perichondrium.
   b. Anteriorly, the auricle is attached to the calvaria; posteriorly, there is a free rolled edge that forms the **helix**.
      (1) The helix begins as the crus in the concha just superior to the external auditory meatus.
      (2) A small superior (darwinian) **tubercle** is a vestige of the point of the ear.
      (3) The helix ends as a flabby **lobule** or **ear lobe**.
   c. A second ridge, the **antihelix**, runs approximately parallel to the helix, dividing the auricle into an outer **scaphoid fossa** and the deeper **concha** (*konche, G.* shell).
      (1) The antihelix begins anterosuperiorly as two crura that define the **triangular fossa**.
      (2) Inferiorly, the antihelix terminates as the **antitragus**.
   d. The **tragus** is a projection from the anterior portion of the external ear.
      (1) Directed posteriorly, it partially covers and protects the external auditory meatus.
      (2) The tragus is separated from the antitragus by the intertragic incisure.
2. The **external auditory (acoustic) meatus** extends from the concha to the tympanic membrane (see Fig. 30-8).
   a. It is somewhat S-shaped and has two distinct portions.
      (1) The external one-third is formed by a continuation of the elastic cartilage of the concha.
         (a) It is directed upwards and backwards.
         (b) The skin lining the external auditory meatus contains sebaceous glands associated with hair follicles, as well as ceruminous glands.
         (c) Ear wax consists of the secretions of both glands in addition to desquamated cells and dust.
         (d) Excessive accumulations of ear wax can clog the external auditory meatus, especially if one tends to remove it with the fifth digit, formerly named the auricularis (*L.* ear).
      (2) The internal two-thirds of the external auditory meatus runs within a bony canal (see Fig. 30-8).
         (a) It is directed slightly downwards.

(b) The epithelium contains fewer glands and is devoid of hairs.

b. The shape of the concha and external auditory meatus amplifies sound waves by a factor of five to ten times (5–10 dB).

c. The external auditory meatus is a remnant of the first branchial (gill) groove.

3. The **tympanum** (G. drum) [**tympanic membrane** or **ear drum**] separates the external ear from the internal ear (see Fig. 30-9).

a. It consists of a sheet of fibrous tissue covered on both sides by epithelium. The fibers are both circularly and radially arranged to keep the membrane moderately tense and thus receptive to sonic vibrations.

b. The tympanum is tilted across the external auditory meatus so that the anteroinferior quadrant is deeper than the posterosuperior quadrant.

c. It is slightly concave.

(1) It is attached to the **malleus** along the length of the **manubrium**.

(2) The central portion of the concavity, the **umbo**, marks this attachment.

(3) Superior to the attachment of the malleus, the tympanic membrane appears less tense and is termed the **pars flaccida**.

d. The pressure changes associated with sonic waves are transduced into mechanical vibrations at the tympanum.

(1) This mechanism is so sensitive that the movement of air molecules against the tympanum by brownian motion is just below the threshold of hearing in a young individual with acute hearing.

(2) The distance through which the tympanum moves in response to normal (60 dB) conversation is measured in nanometers.

4. The innervation of the external ear.

a. The anterior aspect of the auricle and a variable part of the external auditory meatus is innervated by the **auriculotemporal branch** of the **mandibular nerve** (CN V_3) and by the **auricular branch** of the **facial nerve** (CN VII); the posterior aspect of the auricle is innervated by the **great auricular** and **lesser occipital nerves**, which arise from spinal levels C2 and C3.

b. The external auditory meatus is also innervated by twigs from the glossopharyngeal nerve (CN IX) and the vagus nerve (CN X). This explains why a patient may gag or cough when an insect enters the external auditory meatus or when cerumen is removed by curettage.

B. THE MIDDLE EAR consists of the tympanic cavity with its extensions and the auditory ossicles.

1. The **tympanic cavity (antrum)** is a space between the squamous and petrous portions of the temporal bone (Fig. 30-9).

a. The **lateral wall** is formed in large part by the tympanic membrane and the lateral wall of the epitympanic recess.

(1) The **chorda tympani**, a branch of the facial nerve, passes across the tympanium and manubrium of the malleus.

(a) It enters the tympanic cavity at the canaliculus of the chorda tympani (iter chordae posterius) in the posterior wall.

(b) It leaves the tympanic cavity through a canal (iter chordae anterius) in the petrotympanic fissure.

(2) This nerve conveys taste sensation from the anterior two-thirds of the tongue, as well as parasympathetic presynaptic secretomotor fibers to the submandibular ganglion.

b. The **roof** (tegmen tympani) separates the epitympanic recess from the middle cranial fossa. The epitympanic recess, a superior extension of the antrum, contains the head of the malleus and the body of the incus.

c. The **floor** of the antrum is a plate of bone separating the middle ear cavity from the jugular canal.

d. The **posterior wall** of the antrum contains numerous communications with the mastoid air cells.

(1) Infection of the middle ear may spread into these spaces and be difficult to treat.

(2) Projecting from the posterior wall is a pyramidal eminence that contains the stapedius muscle.

e. The **anterior wall** separates the antrum from the carotid canal.

(1) The **auditory (pharyngotympanic, eustachian) tube** opens into the anterior wall at the **tympanic orifice**.

(a) It connects the middle ear with the nasopharynx and serves as a means of equalizing the pressure across the tympanium. A pressure differential of 100–150 mm Hg will rupture the tympanium.

(b) It is a pathway for spread of infection from the nasopharynx to the middle ear.

(c) It is a remnant of the first branchial pouch.

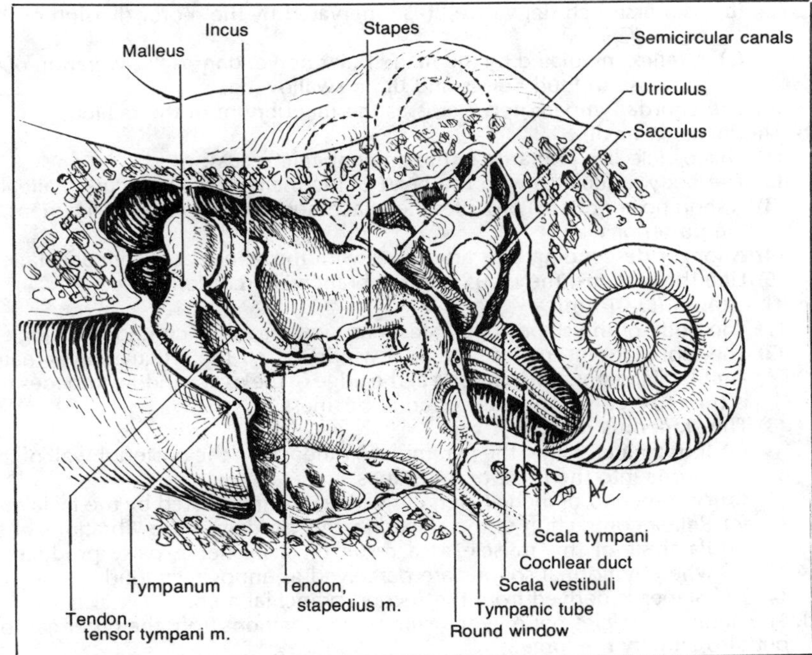

Figure 30-9. *The right middle ear and inner ear.*

(2) The canal of the tensor tympani muscle opens into the anterior wall just superior to the tympanic orifice.

f. The **medial wall** is the most complex.

(1) A **superior prominence** marks the position of the lateral semicircular canal of the inner ear.

(2) The **prominence of the facial (fallopian) canal** marks the course of the horizontal portion of the facial nerve. Occasionally there is only a layer of periostium between the middle ear and the nerve, and it is easy to see how **otitis media** may involve the facial nerve.

(3) The **oval window** (fenestra vestibuli) receives the foot-plate of the stapes and transmits the sonic vibrations of the ossicles to the perilymph of the scala vestibuli.

(4) The **tympanic bulla (promontory)** is formed by the underlying basal turn of the cochlea.

(a) The **tympanic plexus** (Jacobson's) passes across this promontory and contains sensory contributions from the glossopharyngeal and vagus nerves, which distribute to the tympanium and external auditory meatus.

(b) Anteriorly from this plexus, the **lesser superficial petrosal nerve** conveys the presynaptic parasympathetic secretomotor fibers of CN IX origin to the otic ganglion.

(5) The **round window** (fenestra tympani) is covered by an elastic membrane (the secondary tympanic membrane). It accommodates the pressure waves transmitted to the perilymph of the scala tympani.

2. The three **auditory ossicles** bridge the tympanic cavity and transmit sonic vibrations from the external ear to the inner ear (see Fig. 30-9).

a. The **malleus** (*L.* hammer).

(1) The manubrium (*L.* handle) of the malleus is attached along its length to the tympanium.

(2) The head projects into the epitympanic recess and articulates with the incus.

(3) An anterior process provides attachment for the anterior ligament, which passes through the petrotympanic fissure and seems to be developmentally continuous with the sphenomandibular ligament. Both are remnants of Meckel's cartilage and are thus first branchial arch derivatives.

(4) The head of the malleus is stabilized by a superior ligament to the tegmentum tympani.

(5) The **tensor tympani muscle** (see Fig. 30-8).

(a) It originates within the like-named canal in the anterior wall and inserts into the neck of the malleus.

(b) As a first arch derivative, it is innervated by the motor division of the mandibular nerve (CN V_3).

(c) A reflex, mediated by the mandibular nerve, damps the vibration of the malleus in response to loud noises and upon swallowing.

(6) The chorda tympani nerve crosses the manubrium of the malleus.

b. The **incus** (*L.* anvil).

(1) This ossicle lies primarily in the epitympanic recess.

(2) The body articulates with the head of the malleus at the incudomalleal joint.

(3) A short posterior crus provides the attachment for the posterior ligament, which runs to the posterior wall.

(4) A longer descending crus articulates with the stapes at the incudostapedial joint.

(5) Like the malleus, the incus is a first branchial arch derivative.

c. The **stapes** (*L.* stirrup).

(1) The body of the stapes bifurcates into two limbs, which end in a single foot-plate.

(2) The foot-plate inserts into the oval window, and the articulation is maintained by an annular ligament. **Otosclerosis** at the edge of the oval window impedes movement and is the most common cause of adult deafness.

(3) The **stapedius muscle**.

(a) It originates within the pyramidal eminence on the posterior wall of the antrum and inserts into the head of the stapes.

(b) As a second branchial arch derivative, it is innervated by the facial nerve (CN VII).

(c) Reflex contraction of the stapedius muscle damps the vibrations of the stapes.

(d) Paralysis of this muscle as a result of facial nerve palsy produces **hyperacusis**, whereby normal sounds are perceived as annoyingly loud.

(4) The stapes is derived from the second branchial arch.

d. The auditory ossicles not only transmit sonic vibrations from the outer ear to the inner ear but also amplify the force.

(1) The area ratio of the tympanium to the oval window is about 18 to 1. However, the inferior crus of the incus is not as long as the handle of the malleus, so the excursion of the stapes is only about half that of the tympanium. The result is a net mechanical advantage of about 10 times.

(2) This amplification of force compensates for the differences in impedance between the air on one side of the tympanium and the perilymph on the other side of the oval window.

C. THE INNER EAR is contained within the petrous portion of the temporal bone and consists of a vestibular portion concerned with balance and a cochlear portion concerned with audition (see Fig. 30-9).

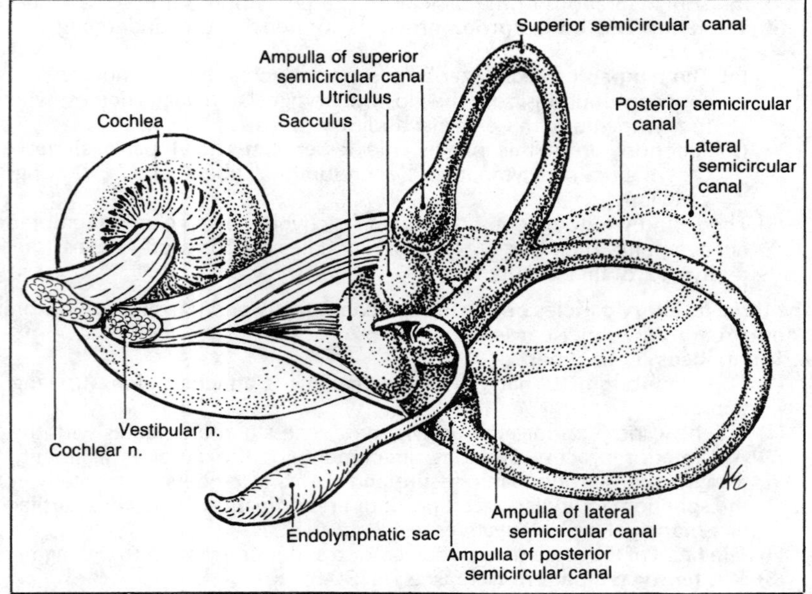

Figure 30-10. *The right membranous labyrinth.*

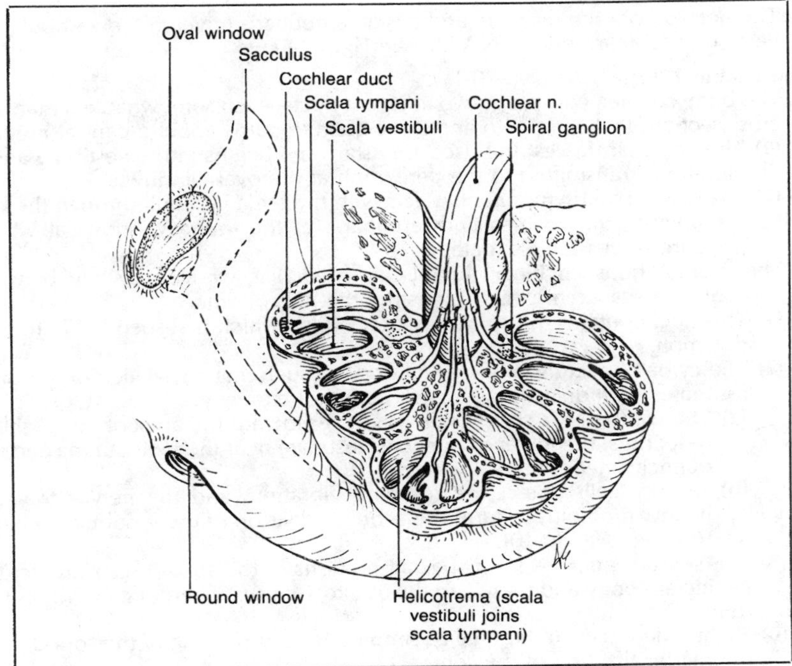

Figure 30-11. *The right cochlea.*

1. The inner ear is composed of a series of bony canals, the **osseous labyrinth**, within which is a system of continuous membranous canals, the **membranous labyrinth**.
   a. The osseous labyrinth is filled with perilymph, in which is suspended the membranous labyrinth.
   b. The membranous labyrinth is filled with endolymph and contains the sensory organs (Fig. 30-10).

2. The **vestibular apparatus**.
   a. The bony labyrinth of the vestibular portion.
      (1) The **vestibule** is a chamber in the osseous labyrinth situated behind the oval window.
         (a) From this chamber radiate three bony semicircular canals and one of the chambers (scala vestibuli) of the spiral cochlea.
         (b) Within these chambers and canals lie the comparable portions of membranous labyrinth.
      (2) The **utricle** and **saccule** are dilations of the membranous labyrinth within the vestibule.
         (a) Within each of these dilations is a sensory **macula** that projects into the endolymph.
         (b) The cytoarchitecture of the **macula** is such that sensory nerve endings are deformed in response to static gravity and inertia, as well as in response to vibrations.
   b. The three semicircular canals of the bony labyrinth.
      (1) These are arranged in mutually perpendicular planes.
         (a) The **anterior (superior) semicircular canal** projects vertically, with the long axis directed anteromedially at about 45°.
         (b) The **lateral semicircular canal** is nearly horizontal and projects slightly into the middle ear cavity.
         (c) The **posterior semicircular canal** projects vertically, with the long axis directed posterolaterally at about 45°.
         (d) The anterior semicircular canal is parallel to the contralateral posterior semicircular canal.
      (2) The membranous semicircular canals are suspended in perilymph.
         (a) A dilation of the membranous labyrinth at one end of each semicircular canal contains a **crista ampullaris**.
         (b) The cytoarchitecture of the **crista ampullaris** is such that sensory nerve endings are stimulated on structural deformation caused by rotational inertia on the enclosed endolymph.

c. The nerves from the maculae and cristae ampullaris travel in the vestibular portion of the vestibulocochlear nerve (CN VIII) [see Fig. 29-8].

3. The **cochlear apparatus** (Fig. 30-11).
   a. The bony cochlea consists of two adjacent ducts, each somewhat less than semicircular in cross section, that spiral two and three-quarters turns about a central **modiolus**.
      (1) The upper **scala vestibuli** (scala, *L.* stairway) begins in the vestibule and receives the vibrations transmitted to the perilymph at the oval window.
      (2) The lower **scala tympani** connects with the scala vestibuli through the **helicotrema** at the apex of the cochlea and terminates at the round window, at which the sound pressure waves are dissipated.
   b. The membranous **cochlear duct** (scala media) is wedged distally between the scala vestibuli and scala tympani as far as the the helicotrema.
      (1) This duct contains the **spiral organ** (Corti's), which is suspended in the contained endolymph.
      (2) The cytoarchitecture of the spiral organ is such that a specific portion of this structure resonates harmonically with each audible frequency.
         (a) The width of the spiral organ is greater toward the apex of the cochlea than at the base; thus, the lower frequencies resonate near the helicotrema and the higher frequencies near the oval window.
         (b) Sensory cells detect the resonant vibration, and the nerves travel through the hollow modiolus to form the acoustic division of the vestibulocochlear nerve (CN VIII) [see Fig. 30-10].
      (3) The sensory neurons from specific portions of the spiral organ project to the medial geniculate body and thence to the auditory cortex where the activity is perceived as a tone.
      (4) Slight differences in the quality, amplitude, and phasing of the sound result in stereophonic localization of the source.

Part VIII Somatic Neck and Neurocranium

STUDY QUESTIONS

Directions: Each question below contains five suggested answers. Choose the **one best** response to each question.

1. Which of the following groups of structures is enclosed by the prevertebral fascia?

(A) Infrahyoid (ribbon) muscles
(B) Muscles of the cervical vertebral column
(C) Sternomastoid and trapezius muscles
(D) Thyroid gland
(E) Trachea and esophagus

2. Which of the following nerves of the cervical plexus has motor function?

(A) Greater occipital
(B) Lesser occipital
(C) Posterior auricular
(D) Suboccipital
(E) Transverse cervical

3. Which of the following structures passes through the foramen lacerum?

(A) Internal jugular vein
(B) Maxillary division of the trigeminal nerve
(C) Middle meningeal artery
(D) Ophthalmic vein
(E) None of the above structures

4. The crista galli serves as an attachment for

(A) the diaphragma sella
(B) the falx cerebelli
(C) the falx cerebri
(D) the tentorium cerebelli
(E) none of the above

5. Cerebrospinal fluid enters the subarachnoid space at the

(A) arachnoid villi
(B) choroid plexuses
(C) foramina of Luschka and Magendie
(D) foramen of Monro
(E) iter

6. A subdural hematoma is usually caused by

(A) fracture of the diploic space
(B) laceration of the superficial temporal artery
(C) leakage from a cerebral vein
(D) rupture of a cerebral artery
(E) tearing of a meningeal vessel

7. A cerebral vascular accident in the primary visual area of the left cerebral hemisphere results in

(A) binasal heteronymous hemianopsia
(B) bitemporal heteronymous hemianopsia
(C) contralateral homonymous hemianopsia
(D) ipsilateral homonymous hemianopsia
(E) total blindness of the right eye

8. All of the following arteries are branches of the internal carotid artery EXCEPT the

(A) anterior cerebral
(B) middle cerebral
(C) ophthalmic
(D) posterior cerebral
(E) posterior communicating

9. All of the following statements concerning lacrimation are correct EXCEPT

(A) blinking of the eyelids expresses small amounts of lacrimal fluid from the gland
(B) lacrimal secretion is controlled by parasympathetic nerves
(C) the lacrimal canaliculi drain through an ampulla into the lacrimal sac
(D) blinking causes the lacrimal sac to aspirate lacrimal fluid from the lacus lacrimalis
(E) the nasolacrimal duct ends in the hiatus semilunaris of the middle nasal meatus

10. The cell bodies of the neurons responsible for pupillary dilation are located in the

(A) accessory oculomotor nucleus
(B) ciliary body
(C) ciliary ganglion
(D) pterygopalatine ganglion
(E) superior cervical ganglion

Directions: Each question below contains four suggested answers of which **one or more** is correct. Choose the answer

A if **1, 2, and 3** are correct
B if **1 and 3** are correct
C if **2 and 4** are correct
D if **4** is correct
E if **1, 2, 3, and 4** are correct

11. The infrahyoid (ribbon) muscles are innervated by the

(1) superior ramus of the ansa cervicalis
(2) hypoglossal nerve
(3) inferior ramus of the ansa cervicalis
(4) spinal accessory nerve

12. Nerves passing through the cavernous sinus include the

(1) oculomotor
(2) abducens
(3) trochlear
(4) ophthalmic

13. True statements concerning the precentral gyrus of the brain include which of the following?

(1) It is the primary motor area
(2) It receives input from the basal ganglia and cerebellum
(3) It projects to brain stem nuclei and the gray matter of the spinal cord
(4) It receives direct sensory input

14. Palsy of the right abducens nerve results in diplopia when the gaze is directed

(1) up and to the right
(2) laterally to the right
(3) down and to the right
(4) laterally to the left

15. Loud low-pitched sounds cause

(1) reflex activity along a branch of CN V
(2) reflex activity along a branch of CN VII
(3) a bulging of the round window membrane into the middle ear cavity
(4) maximal vibration in the apical portion of Corti's spiral organ

Directions: The groups of questions below consist of lettered choices followed by several numbered items. For each numbered item select the **one** lettered choice with which it is **most** closely associated. Each lettered choice may be used once, more than once, or not at all.

Questions 16–18

For each muscle action listed below, select the effect with which it is most commonly associated.

(A) Rotation of the head to the right
(B) Lateral flexion of the head to the right
(C) Both
(D) Neither

16. Spasmotic contracture (torticollis) of the right sternomastoid muscle

17. Normal action of the right scalene group of muscles

18. Normal action of the right splenius group of muscles

Questions 19 and 20

For each case history listed below, select the nerve damage with which it is most likely to be associated.

(A) Right CN V damage
(B) Right CN VII damage
(C) Both
(D) Neither

19. After an injury to the face, stimulation of the right cornea resulted in blinking of the left eye but not of the right eye

20. After sustaining a fracture that separated the facial skull from the somatic skull (La Fort type III), stimulation of the right cornea failed to produce blinking in either eye. Stimulation of the left cornea produced blinking in the left eye only.

ANSWERS AND EXPLANATIONS

1. The answer is B. *(Chapter 27 I B 2 d; Figure 27-1)* The prevertebral fascia encloses and invests the somatic musculature of the neck—that is, muscles of the cervical vertebral column. The sternomastoid and trapezius muscles are associated with the superficial layer of the deep cervical fascia, while the strap muscles lie between that layer and the pretracheal fascia. The visceral structures of the neck are bounded by the pretracheal fascia.

2. The answer is D. *[Chapter 27 II D 1 a–b, 2 a (1)–(2)]* The suboccipital nerve supplies the muscles of the occipital triangle and has no sensory function. All of the other named branches of the cervical plexus, with the exception of the ansa cervicalis, are sensory nerves.

3. The answer is E. *[Chapter 28 III B 5 c (3) (b) (ii), C 2 b (2) (f)]* The foramen lacerum is a defect in the floor of the middle cranial fossa and the carotid canal. It is covered by fibrocartilage and periosteum. Normally it does not transmit any structure into or out of the cranial cavity.

4. The answer is C. *[Chapter 28 III C 2 b (1) (b); IV A 1 b (1) (a)–(d)]* In the anterior cerebral fossa, the falx cerebri attaches to the crista galli, the continuation of the vertical plate of the ethmoid bone. The diaphragma sella covers the pituitary fossa; the tentorium cerebelli separates the middle from the posterior cranial fossae; and the falx cerebelli lies in the posterior cranial fossa.

5. The answer is C. *(Chapter 28 IV A 2 b (1) (b); Chapter 29 VI B)* Cerebrospinal fluid (CSF), secreted into the ventricles by the choroid plexuses, enters the subarachnoid space through the foramina of Luschka and Magendie in the roof of the fourth ventricle. CSF passes from the lateral ventricles through the foramen of Monro into the third ventricle and thence through the iter into the fourth ventricle. CSF enters the superior sagittal sinus via arachnoid villi.

6. The answer is C. *(Chapter 28 IV B 3)* Cerebral veins are most vulnerable to tearing as they pass between the arachnoid and dura mater to drain into the venous sinuses. Injury here results in extravasation of blood in the subdural space (subdural hematoma). A meningeal artery tear produces an epidural hematoma, while a ruptured cerebral artery bleeds into the subarachnoid space.

7. The answer is C. *[Chapter 29 II A 3 e (3) (a) (iii)]* Because the optic projections to the visual cortex contain fibers that convey information from the contralateral fields, loss of the primary visual area on one side produces loss of the same contralateral visual field in each eye—contralateral homonymous hemianopsia.

8. The answer is D. *[Chapter 29 VII B 2 d–f, 3 d (4)]* The posterior cerebral artery is a branch of the basilar artery, the continuation of the joined vertebral arteries. The basilar artery not only supplies the brain stem but also supplies the occipital lobe of the cerebrum.

9. The answer is E. *(Chapter 30 I B 8–9)* The nasolacrimal duct enters the inferior meatus of the nose. The hiatus semilunaris in the middle meatus receives drainage from the frontal, ethmoidal, and maxillary sinuses.

10. The answer is E. *(Chapter 30 I D 4 a)* The cell bodies of the neurons that innervate the dilator pupillae muscle are located in the superior cervical ganglion. These receive presynaptic stimulation from neurons located in the uppermost thoracic levels of the spinal cord. Generally, the sympathetic neurons follow perivascular pathways to the site of innervation.

11. The answer is B (1, 3). *[Chapter 27 II D 2 a (2) (b)]* The superior ramus of the ansa cervicalis from spinal nerve C1 runs with (but is not a part of) the hypoglossal nerve to innervate the geniohyoid, thyrohyoid, and sternohyoid muscles. The inferior ramus of the ansa cervicalis from spinal nerves C2 and C3 innervates the omohyoid and thyrohyoid muscles. The two limbs of the ansa communicate and provide cross innervation to some of these muscles such as the sternothyroid.

12. The answer is E (all). *(Chapter 28 IV A 1 b (3) (c); Figure 28-11)* The oculomotor, trochlear, abducens, and ophthalmic nerves run through the cavernous sinus en route to the orbit. The carotid artery also lies in the medial wall of this sinus.

13. The answer is A (1, 2, 3). [Chapter 29 II A 3 e (1) (a)] The precentral gyrus of the frontal lobe is the primary motor area. Its somatotopically organized pyramidal cells project to the lower motor neurons of the brain stem and spinal cord. It receives input from the cortical sensory areas as well as modulating influences from the basal ganglia and cerebellum. The motor area receives no direct sensory projections.

14. The answer is A (1, 2, 3). *(Chapter 29 V C 1 c (3); Chapter 30 I C 4 c, 5; Table 30-1)* Because the abducens nerve innervates the lateral rectus muscle, abducens palsy results in a medial strabismus. The diplopia is minimized when the gaze is directed toward the opposite side.

15. The answer is E (all). *[Chapter 30 III B 1 f (5), g (1) (e), (3) (c), C 3 b (2) (a)]* Low-pitched sounds cause harmonic resonance in the apical portion of the spiral organ. Loud sounds result in reflex contraction of the tensor tympani muscle (innervated by CN V) and the stapedius muscle (innervated by CN VII). Every action of the stapes at the oval window produces an opposite action at the round window.

16–18. The answers are: 16-B, 17-D, 18-A. *[Chapter 27 II C 1 a (3), 2 a, 2 c (1)]* The sternomastoid muscles turn the head toward the contralateral side and flex the neck toward ipsilateral side. The scalene muscles, acting as accessory respiratory muscles, elevate the first and second ribs and have no action on the head. The splenius group, acting bilaterally, extends the head; acting unilaterally, it turns the head toward the ipsilateral side.

19–20. The answers are 19-B, 20-C. *[Chapter 30 I B 5 a (4), 7 b (2)]* The trigeminal nerve (CN V) is the afferent limb of the blink reflex. Complex neural pathways in the brain stimulate both ipsilateral and contralateral facial nerve nuclei so that a bilateral (consensual) blink response is mediated by the facial nerves, forming the efferent limb of the blink reflex. The results of the tests may be deduced from this knowledge.

Part IX
Facial Cranium and Visceral Neck

31
The Facial Skeleton

I. INTRODUCTION

A. DEVELOPMENTAL CONSIDERATIONS

1. The facial cranium develops in conjunction with the sensory organs for vision and smell, the nasal passages for respiration, and the oral stoma for taste and ingestion.
2. Portions of the facial skeleton and the anterior skeletal structures of the neck are the visceral derivatives.
   a. The face and anterior neck superior to the hyoid bone have lost the usual somatic covering, and the visceral (branchiomeric) musculature has come to lie on the surface.
   b. Because these structures derive from the primitive gill-arch system, they exhibit **branchiomeric segmentation** (*branchios, G.* gill).
3. The basic innervation of the visceral portion of the head and neck.
   a. One series of cranial nerves is related to the organs of special senses: the olfactory nerve (CN I), the optic nerve (CN II), and the vestibulocochlear nerve (CN VIII).
   b. A second series of cranial nerves is related to muscles with somatic segmentation in the head.
      (1) The extraocular muscles are derived from cephalic somites and innervated by the oculomotor nerve (CN III), the trochlear nerve (CN IV), and the abducens nerve (CN VI).
      (2) In addition, many of the muscles of the tongue are derived from cephalic somites and are innervated by the hypoglossal nerve (CN XII).
   c. A third series of cranial nerves is associated with muscles of visceral segmentation and is related to the gill arches.

B. THE ORGANIZATION of the anterior face and visceral neck.

1. In the pharynx of lower vertebrates (e.g., fish), a series of gill clefts pass completely through the pharyngeal wall.
   a. Respiration occurs by passing swallowed water through the gill slits, where gaseous exchange occurs between the water and the very superficial branchial blood vessels.
   b. Both cephalad and caudal to the branchial cleft (trema) are branchial arches that are composed of bone (or cartilage), muscles, blood vessels, and nerves.
   c. In man, remnants of these gill arches remain, many being incorporated into other structures and taking on other functions (Fig. 31-1).
      (1) The first arch derivatives include:
         (a) The mandible, sphenomandibular ligament, malleus, and incus.
         (b) The muscles of mastication as well as the mylohyoid, anterior belly of the digastric, tensor tympani, and tensor veli palatini muscles.
         (c) The trigeminal nerve (CN V).
      (2) The second arch derivatives include:
         (a) The lesser horn of hyoid bone, stylohyoid ligament, styloid process, and stapes.
         (b) The muscles of facial expression as well as the stapedius, stylohyoid, and posterior belly of the digastric.
         (c) The facial nerve (CN VII).
      (3) The third arch derivatives include:
         (a) The body and greater horn of hyoid bone.
         (b) The stylopharyngeus muscle.
         (c) The glossopharyngeal nerve (CN IX).

(4) The fourth arch derivatives include:
(a) Laryngeal cartilages.
(b) Most of the pharyngeal musculature.
(c) The superior laryngeal branch of the vagus nerve (CN X).

(5) The fifth and sixth arch derivatives include:
(a) Laryngeal cartilages.
(b) Most of the laryngeal musculature.
(c) The recurrent laryngeal branch of the vagus nerve (CN X) [the cranial accessory portion of CN XI].

d. While many branchiomeric muscles maintain relationships with the bones associated with the arch, some muscle migration also occurs.
(1) The muscles of facial expression migrate cranially from the second arch to cover the facial skeleton and to overlie the muscles of the first arch.
(2) The diaphragm migrates caudally into the thoracic region.

2. The branchiomeric nerves are all mixed nerves, containing afferent and efferent fibers.
a. These mixed nerves run from the brain stem to the region of the cleft and divide in a characteristic manner forming **pretrematic** (*trema,* G. hole) and **post-trematic** branches (see Fig. 31-1).
(1) The afferent fibers pass both anterior and posterior to the remnant of the branchial cleft.
(2) The efferent fibers run in the post-trematic division only.

b. The pretrematic branch of the branchiomeric nerves tends to join the post-trematic branch of the preceding branchial arch. Thus each arch has, to some extent, a dual innervation.
(1) This dual innervation accounts for the maxillary nerve (the pretrematic branch of the trigeminal nerve for the first branchial arch) being totally sensory and the mandibular nerve (the post-trematic branch of the trigeminal nerve) being mixed.
(2) It also explains why the chorda tympani (the pretrematic branch of the facial nerve for the second branchial arch) is sensory and joins the lingual branch of the mandibular nerve.
(3) Similarly, the tympanic nerve (the pretrematic branch of the glossopharyngeal nerve for the third branchial arch) is sensory and joins the acoustic branch of the facial nerve to innervate the external auditory meatus.

II. THE FACIAL CRANIUM

A. THE FACIAL CRANIUM fills a space bounded posteriorly by the sphenoid bone and superiorly by the floor of the anterior cranial fossa.

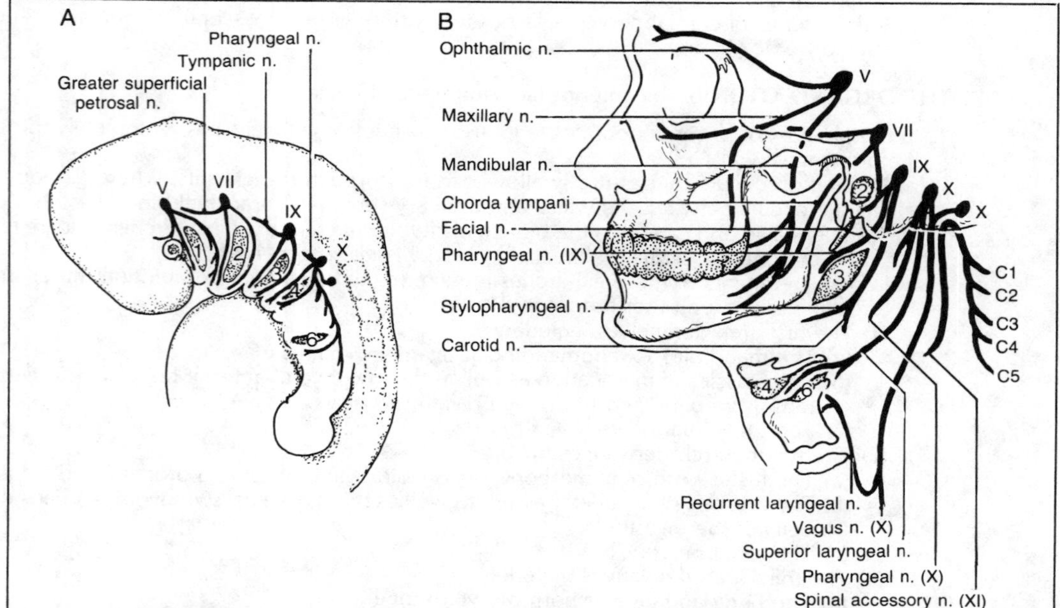

Figure 31-1. *The developmental organization of the cranial nerves. A,* Each branchiomeric nerve has a sensory pretrematic division and a mixed post-trematic division, which innervate the adjacent sides of each gill cleft. *B,* The same pattern can be discerned in the adult.

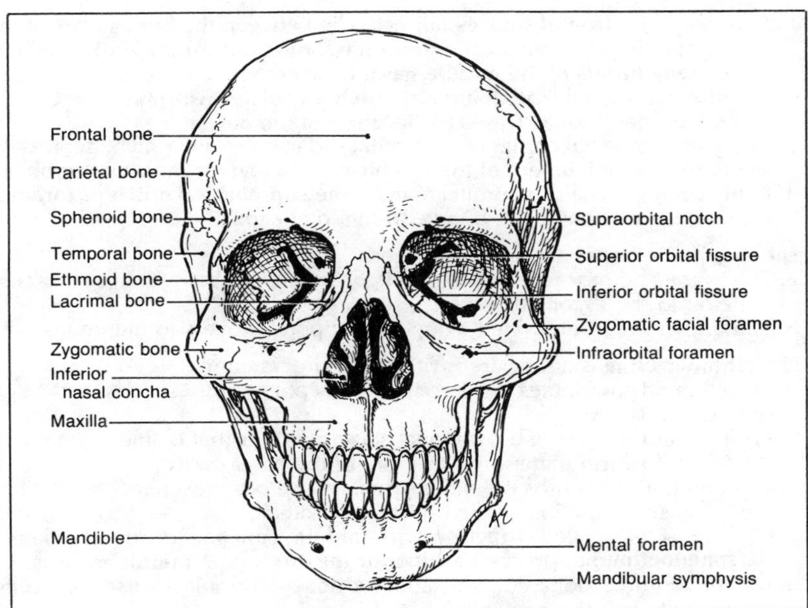

Figure 31-2. *The facial skeleton.*

1. The facial cranium is the external counterpart of the step between the anterior and middle cranial fossae.

2. It consists of 14 bones.
   a. There are two midline bones.
      (1) The **ethmoid bone.**
      (2) The **vomer.**
   b. There are two lateral bones that fuse in the midline.
      (1) The **maxilla.**
      (2) The **mandible.**
      (3) The **hyoid bone**. (Sometimes included, this bone is part of the visceral neck.)
   c. There are five pairs of separate lateral bones.
      (1) The **nasal bones.**
      (2) The **lacrimal bones.**
      (3) The **inferior nasal conchae.**
      (4) The **palatine bones.**
      (5) The **zygomatic bones.**

3. In addition, the facial skeleton shares a number of the bones of the neurocranium, including the frontal, ethmoid, sphenoid, temporal, and basioccipital bones.

4. In the upper face, the orbital margin is bounded by the frontal, zygomatic, and maxillary bones. The sphenoid, palatine, ethmoid, lacrimal, and nasal bones also contribute to the orbital walls (Fig. 31-2).

5. On either side of the sagittal plane in the midface, a pyramidal stack of hollow bones surrounds the nasal cavity (see Fig. 31-2).
   a. Its apex is the ethmoid bone; its floor (the hard palate) is formed by the maxilla and palatine bones; and in between, the vomer contributes to the nasal septum.
   b. On either side are the maxilla, palatine, inferior nasal concha, lacrimal, and nasal bones (see Fig. 31-2).

6. In the lower face, the mandible surrounds the floor of the mouth and pharynx. It articulates with the temporal bone of the neurocranium to complete the skull (see Fig. 31-2).

B. THE BONES OF THE FACIAL CRANIUM (see Fig. 31-2).

1. The **frontal bone** is shared with the neurocranium.
   a. The squamous portion of the frontal bone turns sharply toward the posterior at the orbital margins to form the **orbital plates**.
   b. Between the orbital plates is the **ethmoid notch**, in which the cribriform plate of the ethmoid bone lodges.

c. Anteriorly, the **frontal sinuses (air cells)** lie between the bony tables of the frontal bone. The frontal sinuses drain via the **frontonasal duct (infundibulum)**, which empties into the **hiatus semilunaris** of the **middle nasal meatus.**

d. Anteriorly in the midline, the **nasal notch** articulates with the nasal bone and the frontal process of the maxilla. This is the location of the **nasion**.

e. At the anteromedial corner of the orbital surface, there is a slight depression (sometimes a spine) for the attachment of the **trochlea** (*L.* pulley) for the superior oblique muscle.

f. In the center of the supraorbital margin, the **supraorbital notch** or **foramen** transmits the supraorbital neurovascular bundle to the forehead.

2. The **zygomatic bones** are laterally placed.

a. The zygomatic processes of the maxillary and frontal bones transmit forces from the facial skeleton to the zygomatic bone.

b. These forces are transmitted along the temporal process to the neurocranium.

3. The **ethmoid bone** is also shared with the neurocranium.

a. The ethmoid contributes to the roof, walls, septum, sinuses, and conchae of the nasal cavity (see Fig. 31-3A).

(1) It sits at the apex of a pyramidal stack of bones that define the nasal opening.

(2) The **cribriform plate** forms the roof of the nasal cavity.

b. The **ethmoid labyrinths** on either side form the posterosuperior walls of the nasal cavity, medially, and a portion of the orbital wall, laterally.

(1) Thin scroll-shaped **superior** and **middle conchae** divide the lateral wall into a **sphenoethmoidal recess**, a **superior meatus**, and a **middle meatus**.

(2) The ethmoid bone contains about a dozen **ethmoid sinuses** (**air cells**), which open medially into the nasal cavity.

(a) The posterior ethmoid sinuses open into the superior meatus.

(b) The middle and anterior ethmoid sinuses open into the **hiatus semilunaris** of the middle meatus (see Fig. 31-3A).

(3) The lateral plates of the ethmoid labyrinth form the **lamina papyracea** of the medial orbital wall.

c. In the midline, the vertical plate intersects the cribriform plate as the **crista galli** above, and articulates with the vomer below, to form a portion of the bony **nasal septum**.

d. On each side in the frontal bone are two tiny canals, which transmit the anterior and posterior ethmoid nerves and arteries to the superior surface of the cribriform plate and to the nasal cavity.

4. The **nasal bones** form the bridge of the nose and articulate with the frontal processes of the maxilla as well as with the ethmoid and frontal bones of the neurocranium (see Fig. 31-2).

5. The **lacrimal bones** are the smallest bones of the facial cranium (see Fig. 30-1).

a. Each lies between the frontal process of the maxilla and the ethmoid labyrinth.

b. They are very thin bones.

c. Each lacrimal bone bears a groove (lacrimal fossa) for the lacrimal sac.

6. The fused **maxillae** form the largest bone of the facial cranium (see Fig. 31-2).

a. Each maxilla lies below the frontal bone and ethmoid labyrinth.

(1) It is more or less pyramidal in shape.

(2) From it extend several processes.

(a) The lateral zygomatic process articulates with the zygomatic bone and marks the boundary of the anterior face and the infratemporal fossa.

(b) The alveolar process (the superior alveolar margin) supports the upper teeth.

(i) The alveolar process has sockets for eight teeth on each side.

(ii) The posterior end of the alveolar process continues a little way beyond the third molar as the **maxillary tubercle.**

(c) The frontal process extends around the anterior end of the ethmoid bone to articulate with the frontal bone.

(i) It forms a portion of the medial margin of the orbit.

(ii) A bony canal in the frontal process of the maxilla transmits the **nasolacrimal duct**, which drains the lacrimal sac into the inferior meatus of the nasal cavity.

(d) On its inner surface above the alveolar margin, the horizontal palatine process forms the anterior floor of the nasal cavity and the roof of the oral cavity.

(i) These are fused in the midline to form the anterior four-fifths of the **hard palate**.

(ii) The horizontal palatine process is pierced anteriorly by the **incisive foramen**, which transmits the nasopalatine neurovascular bundle.

b. The posterior end of the ethmoid bone projects superiorly beyond the maxilla. The narrow space behind the maxilla (the pterygopalatine fossa) is closed medially by the perpendicular plate of the palatine bone.

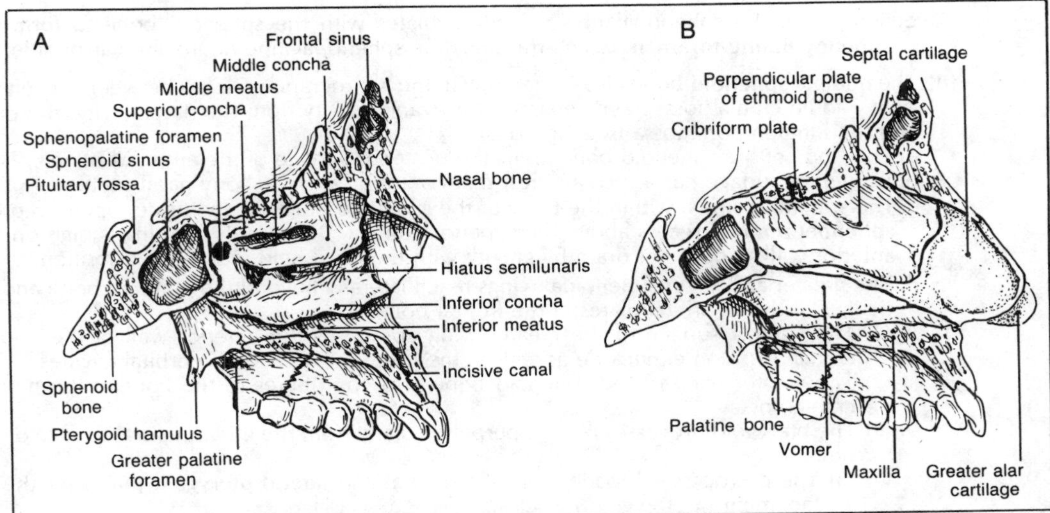

Figure 31-3. *The bones of the nasal cavity. A,* The lateral wall. *B,* The nasal septum.

- **c.** The superior orbital surface of the maxilla is grooved and tunneled by the infraorbital nerve, which passes into the face through the **infraorbital foramen**.
- **d.** The large **maxillary sinus** (Highmore's antrum) lies within the maxillary bone.
  - **(1)** The opening is high on the medial surface so that it does not drain by gravity when the head is held erect.
  - **(2)** It drains with the frontal sinus and several ethmoid sinuses into the **hiatus semilunaris**.
    - **(a)** This explains why infection so readily spreads among these nasal sinuses.
    - **(b)** Once infected it is difficult to clear because the cilia, which would normally sweep the mucus up to the opening, are disabled. A collection of stagnant and putrid mucus may have to be drained surgically.
  - **(3)** The floor of the maxillary sinus is formed by the alveolar process; only a thin layer of bone separates the sinus cavity from the roots of the teeth. Sinusitis may produce toothache.
  - **(4)** The infraorbital canal frequently projects from the root of the sinus.

7. The **inferior nasal concha.**
- **a.** This bone in the lateral wall of the nasal cavity curls above the inferior meatus (Fig. 31-3A).
- **b.** It articulates anteriorly with the perpendicular plate of the ethmoid bone.
- **c.** It forms the inferior boundary of the **hiatus semilunaris**, into which the frontal, anterior, and middle ethmoidal sinuses drain anterosuperiorly and the maxillary sinus drains posteroinferiorly.

8. The midline **vomer** is a thin plate of bone that completes the posterior portion of the bony nasal septum (Fig. 31-3B).
- **a.** Anteriorly it is grooved for articulation with the septal cartilage.
- **b.** Posteriorly it splits into two alae to accommodate the sphenoidal crest.
- **c.** It is usually deviated to one side. Simple preliminary observation of the nasal septum can save much time when a nasogastric or nasotracheal tube must be inserted.

9. The **palatine bones** have both vertical and horizontal processes.
- **a.** The perpendicular plate of the palatine bone is flush with the posteromedial edge of the maxilla and forms the lateral walls of the posterior portion of the nasal cavity (see Fig. 31-3A).
- **b.** The horizontal palatine process parallels the palatine process of the maxilla and forms the posterior one-fifth of the **hard palate** (see Fig. 31-6A).
  - **(1)** The horizontal processes of the maxilla and palatine bones articulate to form the **hard palate**.
  - **(2)** The posterior edge of the hard palate is sharp and concave on either side of the nasal spine.
- **c.** Posteriorly, nerves and vessels within the laterally situated perpendicular plate of the palatine bone pierce the adjacent hard palate through the **greater** and **lesser palatine foramina**.
- **d.** Medially, the horizontal plate of the palatine bone articulates with the vomer, completing the nasal septum.

e. Superiorly, the sphenopalatine notch articulates with the sphenoid bone to form the **sphenopalatine foramen**, which transmits the sphenopalatine neurovascular bundle.

10. The midline **sphenoid bone** also is shared with the neurocranium. It lies between the left and right middle cranial fossae and separates the cranial cavity from the nasal cavity, the orbit, and the infratemporal fossa (see Fig. 31-3).

a. The body of the sphenoid bone abuts the posterior surface of the ethmoid bone.

b. Its inferior surface has a median crest that contributes to the bony nasal septum.

c. The **sphenoid sinus**, within the body of the sphenoid bone on each side, opens into the **sphenoethmoidal recess** above the superior concha. Because its opening is high on the anterior wall, it does not drain by gravity with the head held in the erect position.

d. The greater and lesser sphenoidal wings reach laterally beyond the ethmoid bone and articulate with the orbital plates of the frontal bone.

(1) The vertical surface of the greater wing comprises the anterior wall of the middle cranial fossa on either side as well as the posterior walls of the orbital cavities.

(2) The greater wing of the sphenoid bone also contributes to the floor of the middle cranial fossa.

(3) The pterygoid processes project perpendicularly from the infratemporal surface of the skull.

(a) These processes broaden into the **medial** and **lateral pterygoid plates** for the attachment of pterygoid muscles.

(b) Between the two plates, a triangular defect (the scaphoid fossa) is filled by the pyramidal process of the palatine bone.

e. The **pterygopalatine fossa** (Fig. 31-4). The maxilla does not extend as far posteriorly as does the ethmoid bone, leaving a gap between the pterygoid process of the greater wing of the sphenoid bone and the maxilla.

(a) The pterygopalatine fossa is open laterally, communicating with the infratemporal fossa through the **pterygomaxillary fissure**.

(b) Inferiorly, the pterygomaxillary fissure is closed by the approximation of the maxillary process, the palatine bone, and the pterygoid process.

(i) As the pterygoid process passes inferiorly, it approaches the posterior wall of the maxilla, and the pterygopalatine fossa tapers.

(ii) The maxillary tubercle closes the most dependent portion of the pterygopalatine fissure laterally, forming the **greater palatine canal** between the alveolar process of the maxilla, the palatine bone, and the pterygoid process.

(iii) The **greater palatine canal** transmits the palatine neurovascular bundle to the oral cavity.

(c) The posterior wall is formed by the pterygoid process of the greater wing of the sphenoid bone. It contains three foramina.

(i) The **pterygoid (vidian) canal** traverses the root of the pterygoid process and contains the presynaptic parasympathetic neurons of the greater superficial petrosal branch of the facial nerve, which synapse in the pterygopalatine ganglion. It also contains postsynaptic sympathetic fibers from the carotid perivascular plexus via the deep petrosal nerve.

(ii) The **foramen rotundum** pierces the greater wing of the sphenoid bone just anterior to the root of the pterygoid process and transmits the maxillary division of the trigeminal nerve (CN V_2).

(iii) The **pterygovaginal canal** contains vessels that communicate with the pharynx.

(d) The medial wall is formed by the perpendicular plate of the palatine bone. The superior portion of this bone is deficient (the sphenopalatine notch) and forms the sphenopalatine foramen at the articulation with the sphenoid bone, through which nerves and vessels reach the nasal cavity.

(e) The anterior wall is formed by the posterior surface of the maxilla and the greater wing of the sphenoid bone. However, these bones fail to fuse, forming the **inferior orbital fissure**, through which the maxillary nerve reaches the orbit and face.

(f) The greater wing of the sphenoid bone is separated from the lesser wing, forming the **superior orbital fissure**, which also communicates with the orbit and transmits the oculomotor nerve (CN III), the trochlear nerve (CN IV), the ophthalmic division of the trigeminal nerve (CN V_1), and the abducens nerve (CN VI), as well as the ophthalmic vein.

C. FACIAL FRACTURES

1. Zygomatic fractures result from trauma to the cheek bone with loss of stability in the zygomatic arch. Because the zygomaticotemporal and zygomaticofacial branches of the maxillary nerve pass through foramina in the zygomatic bone, there may be associated paresthesias.

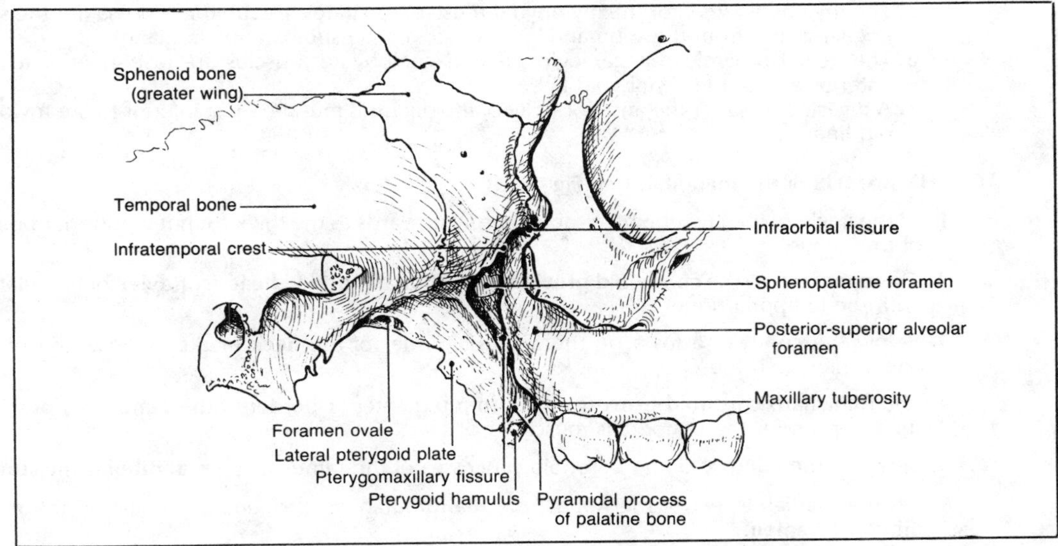

Figure 31-4. Infratemporal fossa and pterygopalatine fossa.

2. Fractures of the medial orbital wall breach separation of the nasal and orbital cavities so that blowing the nose may produce exophthalmus.
3. Ethmoidal fracture may produce rhinorrhea [discharge of cerebrospinal fluid (CSF) through the nose].
4. Maxillary fractures.
   a. Fracture of the orbital floor may result in herniation of the periorbital fat into the maxillary sinus with enophthalmus.
   b. Collapse of the thin anterior wall of the maxillary sinus may denervate the anterior maxillary teeth because the anterior-superior alveolar nerves run in tunnels through this bone.
   c. The maxilla may be fractured transversely at the level of the nasal floor (La Fort type I).
5. A severe fracture may separate the visceral face from the neurocranium (La Fort type III), allowing the face to move posteriorly and obstruct the airway.

III. THE MANDIBLE

A. **THE MANDIBLE** forms the lower jaw (Fig. 31-5).

1. This U-shaped bone bounds the floor of the mouth inferior to the alveolar margin.
2. The horizontal portion is the **body**; the two posterior vertical portions are the **rami**. The extension of the ramus into the body forms the **oblique line**.
3. An **alveolar ridge** runs along the superior border of the body.
   a. The alveolar ridge has sockets for eight teeth on each side.
   b. It formed in response to the presence of the teeth and the forces transmitted by them to the jaw.
   c. In the edentulous individual, the alveolar ridge atrophies.
4. The thick rounded inferior border of the body of the mandible is termed the **base**.

B. **THE BODY** of the mandible.

1. The **lateral surface** of the body (Fig. 31-5A).
   a. The halves of the body are joined at birth by dense fibrous tissue, the **symphysis menti**.
      (1) The bones fuse after 2 years of age.
      (2) It is marked in the adult by a faint midline ridge, which widens inferiorly to the **mental tubercles**.
   b. The **mental foramen**, located inferior to the second premolar tooth, transmits the mental branch of the inferior alveolar nerve, which arises from the mandibular division of the trigeminal nerve.
2. The **medial surface** of the body is marked by the insertions of the suprahyoid muscles (Fig. 31-5B).

a. The mylohyoid line (for the mylohyoid muscle) separates the sublingual fossa (for the sublingual gland) from the submandibular fossa (for the submandibular gland).
b. The genial tubercles for genioglossus and geniohyoid muscles are just superior to the anterior end of the mylohyoid line.
c. A digastric fossa for the anterior belly of the digastric muscle is just inferior to the mylohyoid line.

C. **THE RAMUS** of the mandible (see Fig. 31-5)

1. At the angle of the jaw, the base turns sharply upwards as the thick rounded posterior border of the ramus.
2. This continues to the **condyloid process**, which has a rounded **head (condyle)** for articulation with the temporal bone.
3. Below the condyle, a fossa on the anterior border of the **neck** marks the insertion of the lateral pterygoid muscle.
4. The thin sharp **coronoid process** arises from the anterior border of the ramus and provides the insertion for the temporalis muscle.
5. Between the condyloid and coronoid processes of the ramus is the **mandibular incisure**.
6. On the medial aspect of the ramus, the **mandibular canal** opens proximally at the **mandibular foramen**.
   a. The small triangular **lingula** guards the anterior margin of the mandibular foramen and gives attachment to the **sphenomandibular ligament**, from which the mandible swings.
   b. Just before the inferior alveolar nerve enters the mandibular foramen, it gives off its mylohyoid branch, which forms a groove along the medial surface of the mandible below the mylohyoid line.
   c. The inferior alveolar nerve enters the mandibular foramen and continues through the mandibular canal to the **mental foramen**.
   d. Beyond the mental foramen, the inferior alveolar nerve continues along the incisive canal.
7. A series of oblique ridges at the angle of the mandible corrugate the medial surface for the insertion of the medial pterygoid muscle and the lateral surface for the insertion of the masseter muscle.
   a. These muscles insert into periosteum and not into the bone of the mandible because the periosteum does not have Sharpey's fibers in this region.
   b. They form a **mandibular sling** at the angle.

D. Mandibular fracture is common, only second in frequency to fracture of the nasal bone.

1. Its U-shape makes it especially susceptible to double fractures.
2. Frequent combinations include fractures of the symphysis or canine area and the opposite angle or condylar neck.
3. A mandibular fracture may damage the inferior alveolar nerve.

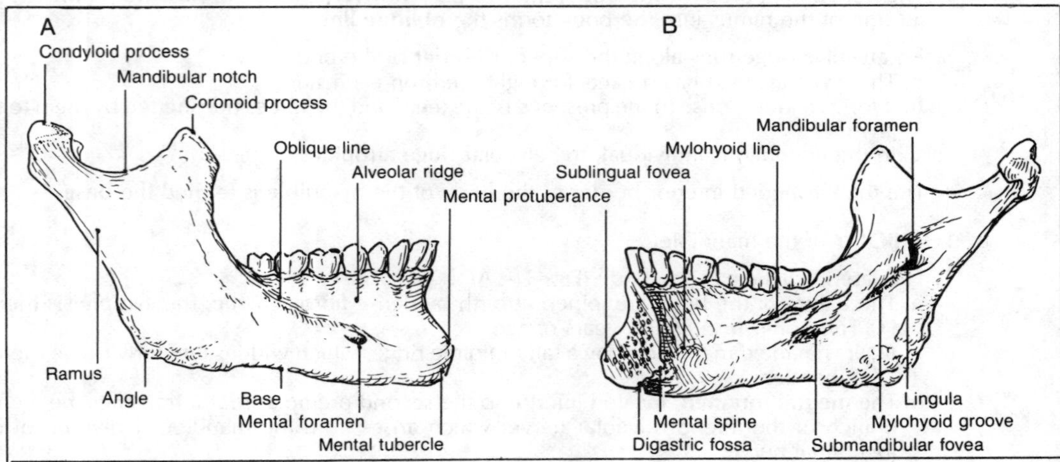

Figure 31-5. *The mandible. A,* The external aspect. *B,* The internal aspect.

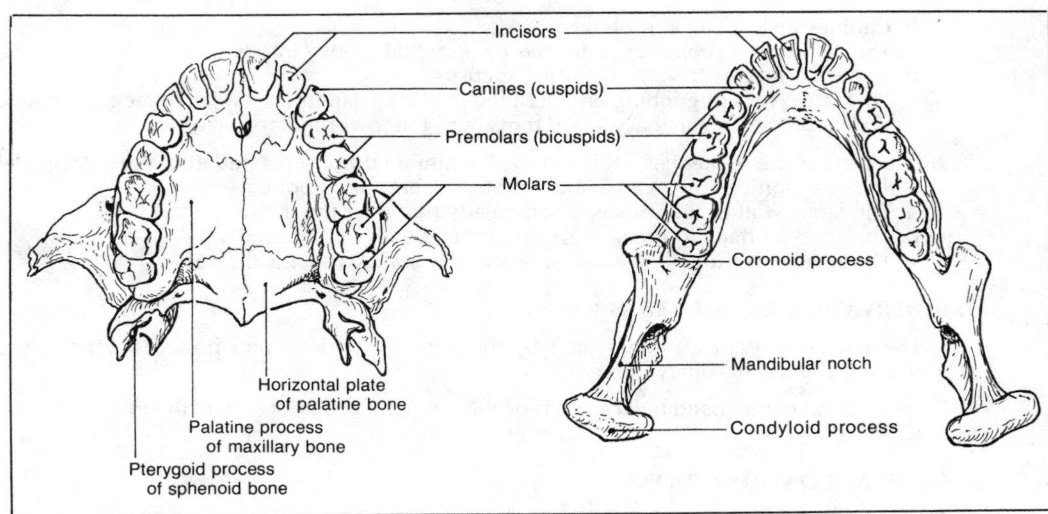

Figure 31-6. *The teeth. A,* The maxilla. *B,* The mandible.

IV. THE TEETH

A. INTRODUCTION

1. The teeth are not classified as skeletal elements.
2. In children there are 20 **deciduous** (milk) **teeth**. In each jaw on each side there are:
   a. Two incisors, which come in at approximately 6 and 8 months.
   b. One canine, which comes in at approximately 10 months.
   c. Two molars, which come in during the second year.
3. In the adult there are 32 **permanent teeth**. In each jaw on each side there are:
   a. Two incisors, which erupt in the seventh and eighth years.
   b. One canine, which erupts in the tenth year.
   c. Two premolars (bicuspid teeth), which erupt in the ninth and eleventh years.
   d. Three molars (tricuspid teeth), the first of which comes in during the sixth year, followed by the second molar in the early teen years and the "wisdom teeth" in the late teen years.

B. THE STRUCTURE OF THE TOOTH

1. Each tooth is composed of two principal parts.
   a. The root is embedded in the alveolar part of the mandible or maxilla.
   b. The crown projects into the oral cavity, composed of one or more cusps.
   c. An intermediate cervical part is related to the gingiva (gum).
2. The basic structural element is **dentine**—a yellowish substance that is nurtured through the fine dentinal tubules of odontoblasts lining the central pulp space.
3. The crown is covered with **enamel**—a hard white crystalline substance formed before the tooth erupts.
   a. After eruption this inert material changes little. Its only change is to adsorb fluoride ions, which reduce its solubility in the acid metabolites of oral bacteria.
   b. Like bone, enamel is stained by tetracycline antibiotics, but only while it is being formed. Therefore, tetracycline should not be given to children or pregnant women.
4. The root is covered with **cement**, which is connected to the bone of the socket by a layer of modified periosteum, the **periodontal ligament**.

C. THE ORGANIZATION OF THE TEETH

1. Each alveolar margin bears 16 teeth—2 incisors, 1 canine, 2 premolars, and 3 molars on either side, each classified according to the shape of the crown (Fig. 31-6).
   a. **Incisors** have a thin cutting edge.

b. **Canines** have a single prominent cone.
c. The crown of a **premolar** is divided by a sagittal groove into two cusps.
d. The adult **molars** have three or more cusps.
   (1) Many years of grinding an unrefined diet may flatten the molar surfaces completely.
   (2) The upper molars have three roots, and the lower molars have two.

2. The bone of the socket has a thin cortex, the **lamina dura**, separated from the adjacent labial and lingual cortices by a variable amount of trabeculated bone.
   a. The labial wall of the socket is particularly thin over the incisor teeth.
   b. This is the surface to break in order to remove an incisor tooth.
   c. The reverse is true of the molars, where the lingual route is the easier.

D. INNERVATION OF THE TEETH (see Fig. 32-4).

1. The maxillary teeth are innervated by the anterior, middle, and posterosuperior alveolar branches of the maxillary nerve (CN V_2).

2. Innervation of the mandibular teeth is by the inferior alveolar branch of the mandibular nerve (CN V_3).

E. CLINICAL CONSIDERATIONS

1. Advanced infection of the root involves the periodontal ligament and erodes through the lamina dura (a sign of chronic infection that is apparent on an x-ray plate).

2. Abscess involving the mandibular teeth may spread through the lower jaw to emerge on the face or in the floor of the mouth. The mandibular nerve also innervates a portion of the ear, and the pain of an infected lower tooth may be referred to the ear.

3. Abscess involving the maxillary teeth may spread through the upper jaw to emerge on the face or in the roof of the mouth. The relations of the upper teeth may influence the mode of presentation.
   a. Infection of a second incisor may spread along the palate.
   b. An infected canine may point in the face, thrombose the angular facial vein, and spread along the superior ophthalmic vein to the cavernous sinus (see Chapter 29).
   c. Because the roots of the maxillary molars project into the maxillary sinus, infection of an upper tooth may produce symptoms of sinusitis with pain referred to the distribution of the maxillary nerve.
   d. Similarly, sinus infection may irritate the nerves to these teeth, causing toothache.

32 The Face

I. FEATURES

A. The face comprises derivatives of the primitive gut (the nose and mouth) and special sensory receptors (the eye and ear).

1. At the oral fissure, the nostrils, and the palpebral fissures, the skin of the face is continuous with the mucous membrane lining these structures.
2. Around each orifice a system of subcutaneous muscles is arranged as sphincters and dilators, the muscles of facial expression. These muscles are innervated by the facial nerve (CN VII).
3. The orifices and their associated superficial structures are the features of the face. Subtle variations in the proportions of each feature determines an individual's facial identity.

B. The palpebral fissures and eyes have been discussed (see Chapter 30 I, II).

C. The external nose is a pyramidal anterior extension of the nasal cavity.

1. Its skeleton is in part bony and in part cartilaginous.
2. The frontal process of the maxilla and the nasal bones articulate with the nasal notch of the frontal bone to form the arch of the nose.
3. The septal cartilage is a quadrangular plate of fibrocartilage thickened around its margin.
   - **a.** It lies approximately in the midline sagittal plane.
   - **b.** Its posterior edge meets the septal plate of the ethmoid bone and the vomer.
   - **c.** The anterior edge emerges from behind the nasal bones as the ridge of the nose.
   - **d.** It articulates with the lateral and alar plates to complete the cartilaginous portion of the nose.

D. The oral fissure is bounded by the superior and inferior labial folds.

1. These meet laterally at the commissures.
2. The labial margin has prominent superficial vascular papillae (thelia), which give it a vermilion hue.
3. A thin median frenulum extends from the inner surface to the gum.
4. Between the epithelial surfaces are the pea-sized labial salivary glands and the orbicularis oris muscle.

E. The external ear has been discussed (see Chapter 30 III).

II. MUSCLES OF FACIAL EXPRESSION

A. INTRODUCTION

1. All of the **muscles of facial expression** are derived from a single plate of muscle, derived from the second branchial arch, which migrates over the skull.
2. The **facial nerve** (CN VII) is the nerve of the second branchial arch and therefore innervates all of the muscles of facial expression.

3. Most of these muscles comprise sphincters and dilators that guard the facial openings.
4. The muscles of facial expression are important for the display of emotions in humans.

B. ORAL MUSCULATURE (Fig. 32-1).

1. The **orbicularis oris** is the sphincter of the mouth.
   a. It lies circumferential to the margin of the lips with muscle fascicles passing in a series of loops around the mouth.
   b. Peripherally, some fascicles decussate at the commissure and pass into the face along the various dilator muscles, and other fascicles pass into the **modiolus** (Lightoller's) on either side of the mouth.
   c. It purses or puckers the lips.
2. The dilator muscles have either facial or bony origins, and their insertions blend with the muscular substance of the lip.
   a. They are numerous and small (often the name in print is longer than the muscle itself).
      (1) Many of these muscles insert into the **modiolus.**
      (2) It is in part because of this arrangement that an almost infinite range of facial expression is possible.
   b. The dilator muscles may be resolved into four layers.
      (1) The *subcutaneous* **risorius muscle** is small and variable. It arises from the parotid fascia and runs horizontally to insert at the angle of the mouth. It retracts the modiolus.
      (2) The three *superficial muscles* form two inverted V's as they diverge from their origins at the orbital margins to their insertions.
         (a) The **zygomaticus major** inserts on the angle of the mouth; the **zygomaticus minor** inserts on the lip. They act to raise the lip.
         (b) The **levator labii superioris alaeque nasi** inserts on the lip, laterally, and on the major alar cartilage, medially. It also acts to raise the lip.
      (3) The four muscles of the *middle layer* are named according to function.
         (a) The **triangular depressor anguli oris**, arising from the oblique line on the body of the mandible, and the **levator anguli oris**, arising from the incisive fossa of the maxilla, converge at the modiolus. The names are descriptive of their actions.
         (b) The quadrangular **depressor labii inferioris**, arising from the oblique line on the body of the mandible, and the **levator labii superioris**, arising from the orbital margin, insert on the lower and upper lips, respectively. The names are descriptive of their actions.
      (4) The *deepest layer* is formed by two muscles.
         (a) The **mentalis muscle** arises from the incisive fossa of the mandible, inserts on the lower lip near the midline, and depresses the lower lip.
         (b) The **buccinator muscle** (*L.* trumpeter) arises from the maxilla and the mandible, posterior to the third molar teeth.
            (i) Between these two points it interdigitates with fibers of the superior pharyngeal constrictor in the **pterygomandibular raphe**.
            (ii) The middle fibers decussate at the angle of the mouth; the peripheral fibers pass directly into the lips.
            (iii) Most importantly, the buccinator serves to draw the cheeks against the teeth in mastication.

C. NASAL MUSCULATURE (see Fig. 32-1)

1. The **compressor naris muscle**, the transverse part of the **nasalis muscle**, comprises the nasal sphincter. Its function is rudimentary in humans.
   a. It is connected with its fellow of the opposite side by an aponeurosis.
   b. It pulls down on the bridge of the nose, slightly constricting the nostrils.
2. The **dilator naris**, the alar part of the **nasalis muscle**, comprises the nasal dilator.
   a. It inserts along with the nasal part of the levator labii superioris alaeque nasi on the alar cartilage.
   b. Both act to flare the nostrils.

D. THE ORBITAL MUSCULATURE (see Fig. 32-1).

1. The **orbicularis oculi**, the orbital sphincter, has two parts.
   a. The thick **orbital portion** encircles the eyes in continuous loops.
      (1) It arises from the medial orbital margin and the medial palpebral ligament.
      (2) When it contracts, the eye is tightly closed (the wink).
   b. The **palpebral portion** crosses the eyelid beneath the skin.

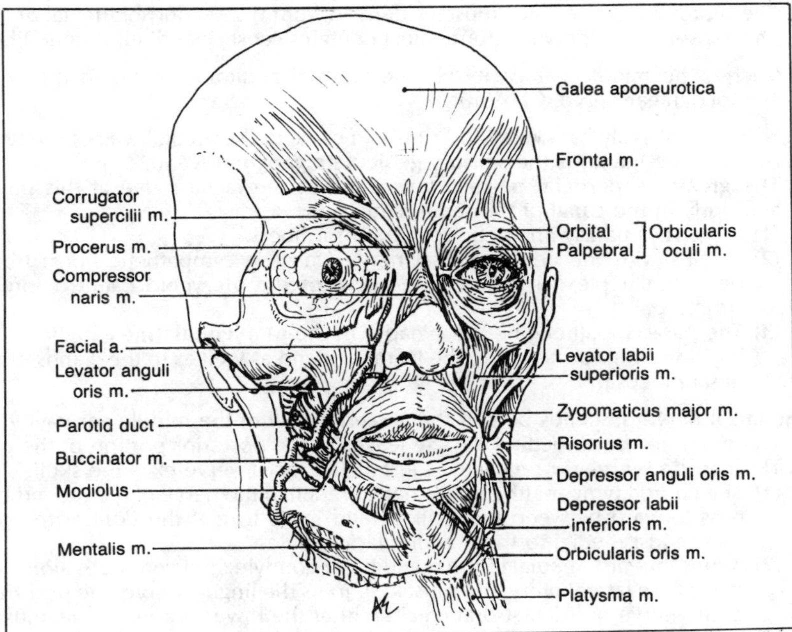

Figure 32-1. *The muscles of facial expression.* The left superficial muscles and the right deeper muscles are depicted.

(1) Contraction of the palpebral portion of this muscle produces the blink.
(2) Its fibers are so arranged that when they shorten they move toward geodesic lines—the shortest distance between two points over the surface of a sphere—and thus close the palpebral fissure.
(3) Along with the other muscles of facial expression, it is innervated by the facial nerve (CN VII).
  (a) This nerve forms the efferent limb of the **blink reflex**.
  (b) Paralysis of the palpebral portion of the orbicularis oculi results in inability to close the eyelids. This is a major problem in Bell's (facial nerve) palsy because, if the eye cannot blink, the conjunctiva dries and the cornea may ulcerate.
(4) The origin encircles the lacrimal sac, and those fibers posterior to the sac may be referred to as the lacrimal portion.
(5) At the lateral canthus, the fibers decussate in a fiberous raphe.

2. Two small muscles, the **procerus** and **corrugator supercilii**, pull the eyebrows downward and inward in an expression of concentration or to shield the eyes from bright light.

3. The **levator palpebrae superioris muscle** acts directly on the upper eyelid to open the palpebral fissure.

4. The **occipitofrontalis muscle** widens the eyes very slightly by raising the eyebrows, as in an expression of fear or surprise.

E. THE AURICULAR MUSCULATURE

1. The anterior and superior auricular muscles converge on the auricle from the epicranial aponeurosis.

2. The posterior auricular muscle arises from the mastoid process.

F. THE PLATYSMA migrates over the anterior surface of the neck and may extend onto the upper thoracic wall (see Fig. 32-1).

III. THE INNERVATION OF THE SUPERFICIAL FACE

A. THE FACIAL NERVE (CN VII) [Figs. 32-2 and 32-3].

1. It is formed by the union of two roots, which represent pretrematic and post-trematic divisions (see Fig. 31-1).

a. The motor root innervates muscles derived from the second branchial arch.
b. The sensory root (nervus intermedius) contains sensory and autonomic fibers.

2. It reaches the middle ear cavity via the internal auditory meatus in the company of the vestibulocochlear nerve (CN VIII).

3. Here it turns sharply backward and runs in a canal in the medial wall of the tympanic cavity.
   a. At the angle (genu) is the sensory **geniculate ganglion** (*genu, L.* knee).
   b. The **greater superficial petrosal nerve** leaves the facial nerve at this point and passes anteriorly in the canal of the same name.
      (1) It conveys parasympathetic secretomotor fibers.
      (2) It joins with the deep petrosal nerve (carrying sympathetic fibers from the carotid perivascular plexus) to form the **nerve of the pterygoid canal** before entering that structure.
      (3) The parasympathetic nerves synapse in the **pterygopalatine ganglion** before continuing along various nerves to the lacrimal gland as well as to the glands of the palate and nasal mucosa.

4. The facial nerve descends behind the posterior wall of the middle ear cavity.
   a. The **nerve to the stapedius muscle** leaves in the posterior portion of the canal.
   b. The **chorda tympani nerve** leaves just as the facial nerve exits the skull.
      (1) The chorda tympani (the pretrematic branch of the nerve of the second branchial arch) runs forward between the malleus and incus to exit the skull at the **petrotympanic fissure** just medial to the mandibular condyle.
      (2) It runs forward, medial to the mandibular condyle, and continues along the medial surface of the lateral pterygoid muscle. It joins the lingual nerve (the post-trematic branch of the nerve of the first branchial arch) at the lower border of this muscle.
      (3) It conveys parasympathetic secretomotor fibers, which synapse in the **submandibular ganglion** before continuing along the lingual nerve to the submandibular and sublingual salivary glands.
      (4) Sensory fibers convey **taste** from the anterior two-thirds of the tongue.

5. The facial nerve exits the skull via the **stylomastoid foramen**.
   a. As soon as it exits the skull, the facial nerve gives off digastric, stylohyoid, and posterior auricular branches.
   b. The posterior auricular nerve divides into a small auricular branch (to the posterior auricular muscle) and a larger occipital branch (to the occipital belly of the occipitofrontalis muscle).

6. The facial nerve passes anterolaterally through the parotid gland deep to the lobule of the auricle.
   a. A plexus of branches divides the parotid gland into superficial and deep lobes.

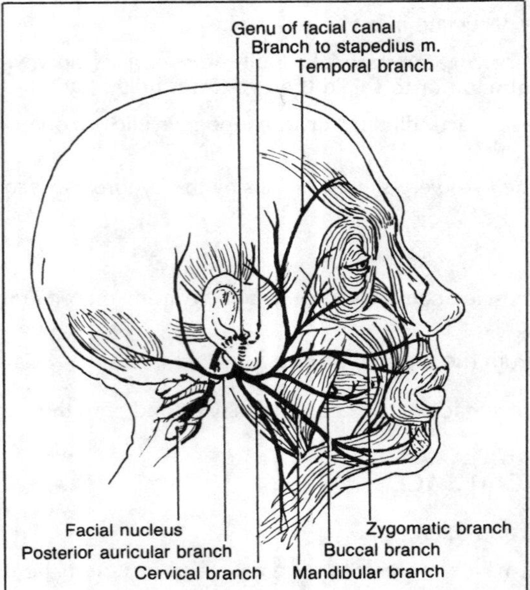

Figure 32-2. *The branchiomotor component of the facial nerve (CN VII).* This portion of the facial nerve innervates the muscles of facial expression—the posterior belly of the digastric, the stylohyoid, and the stapedius muscle.

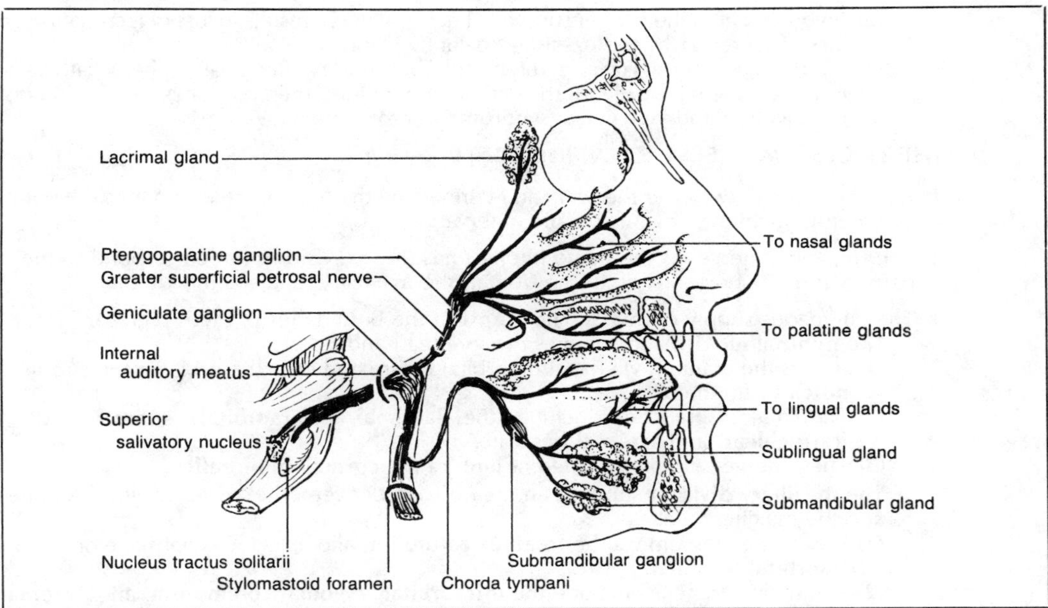

Figure 32-3. *The parasympathetic and sensory components of the facial nerve (CN VII).* The autonomic portions of the facial nerve run to the pterygopalatine and submandibular ganglia. The sensory portion conveys taste from the anterior portion of the tongue, with cell bodies of the sensory neurons located in the geniculate ganglion.

b. Each branch is separated from the parotid parenchyma by a fascial plane.

c. Five major branches emerge onto the face where they are very superficial and easily damaged by facial trauma.

(1) The **temporal branch** supplies the anterior and superior auricular muscles, the frontal belly of occipitofrontalis, the corrugator supercilii, and the orbicularis oculi (all of the muscles above the eyes). This branch conveys the efferent limb of the **corneal blink reflex.**

(a) The motor nuclei whose fibers run in the temporal nerves receive a bilateral corticomedullary innervation.

(b) Therefore, unilateral supranuclear lesions (such as the common stroke) do not affect the forehead to the degree that they affect the rest of the face.

(2) The **zygomatic branch** passes to the orbicularis oculi. This branch also mediates the efferent limb of the blink reflex.

(3) The **buccal branch** supplies the numerous muscles of the cheek, including the buccinator muscle, the muscles of the upper lip, and the muscles of the nose.

(4) The **mandibular branch** loops down over the mandible. It supplies the risorius muscle and the depressors of the lower lip.

(5) The **cervical branch** innervates the platysma muscle.

7. Symptoms and signs produced by injury to the facial nerve are related to the site of the lesion.

a. Lesions involving the terminal branches of the facial nerve produce an imbalance of the muscles of facial expression with unilateral expressionless drooping of the face.

(1) Paralysis of the orbicularis oculi can result in epiphora with conjunctival and corneal ulceration. Obviously, the corneal blink reflex is absent.

(2) Paralysis of the orbicularis oris results in drooling and difficulty in masticating food.

b. Lesions at the stylomastoid foramen produce both of the above symptoms.

c. Inflammation of the nerve within the facial canal is the most common cause of Bell's palsy. There is no effective treatment. Most cases recover spontaneously, often with little or no nerve damage.

(1) All of the above symptoms occur.

(2) Involvement of the chorda tympani results in loss of taste from the anterior two-thirds of the tongue and reduced salivary secretion on the ipsilateral side.

d. Lesions in the mid regions of the facial canal.

(1) All of the above symptoms occur.

(2) Hyperacusis results from involvement of the nerve to the stapedius muscle.

e. Lesions in the internal auditory meatus.

(1) All of the above symptoms (except epiphora) occur.

(2) Involvement of the greater superficial petrosal nerve results in loss of lacrimation and in loss of secretion by the nasal and palatine glands.

(3) Upon regeneration, if the secretomotor fibers regrow along each other's pathways and innervate the wrong gland, the anticipation of food then produces lacrimation instead of salivation and vice versa (syndrome of crocodile tears).

B. THE TRIGEMINAL NERVE (CN V) [Fig. 32-4]

1. A coronal line drawn over the scalp and connecting the tragi of the ears marks the boundary between branchiomeric and somatic innervation.

2. The trigeminal nerve (as the name implies) has three divisions that correspond to the three parts of the first branchial arch: the frontal, the maxillary, and the mandibular.

3. The cutaneous supply of the face conforms to the boundaries of these regions.
   a. The **ophthalmic division** supplies the forehead and nose.
      (1) It exits the cranium via the supraorbital fissure and exits the orbit via the **supraorbital notch** or foramen.
      (2) The major branches include the **lacrimal**, **supraorbital**, **supratrochlear**, **infratrochlear**, and **external nasal nerves**.
      (3) These nerves supply the afferent limb of the **corneal blink reflex**.
   b. The **maxillary division** supplies an area of skin between the lower eyelid, the upper lip, and the maxillary teeth.
      (1) It exits the cranium via the **foramen rotundum** and exits the facial skeleton via the **infraorbital foramen**.
      (2) The major branches include the **infraorbital**, **zygomaticotemporal**, and **zygomaticofacial nerves**.
      (3) This nerve supplies the afferent limb of the **sneeze reflex**.
   c. The **mandibular division** supplies a lateral area that curves around the side of the face (but the angle of the jaw is supplied by the great auricular nerve, C2–C3), as well as general sensation to the supratentorial meninges, all of the mucous membranes of the face, and the mandibular teeth.
      (1) It exits the cranium via the **foramen ovale** and exits the skull via the **mental foramen**.
      (2) The **auriculotemporal nerve** innervates the lateral aspect of the scalp.
      (3) The **buccal nerve** innervates the cheek (but not the buccinator muscle).
      (4) The **mental nerve** innervates the chin (but not the mentalis muscle).
      (5) The **mandibular nerve** is the only division to contain motor fibers.
         (a) These innervate the **muscles of mastication** (masseter, temporalis, medial, and lateral pterygoids), which are deep muscles of the face.
         (b) This nerve supplies both afferent and efferent limbs for the **jaw-jerk reflex**.

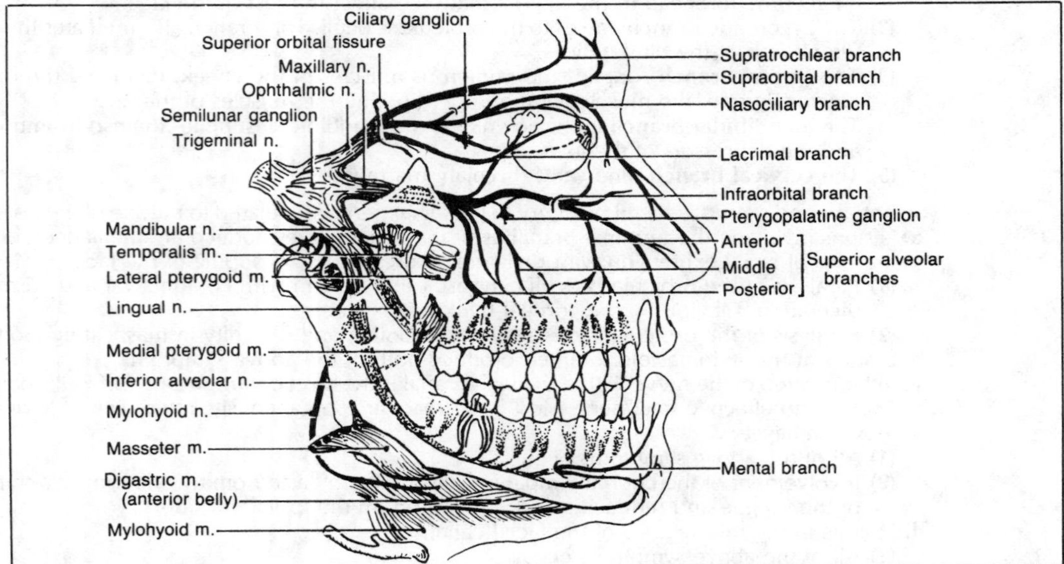

Figure 32-4. *The trigeminal nerve (CN V).* The three divisions of the trigeminal nerve provide sensory innervation to the face. The motor root of the mandibular division innervates the muscles of mastication, the tensor tympani muscle, and the tensor veli palatini muscle.

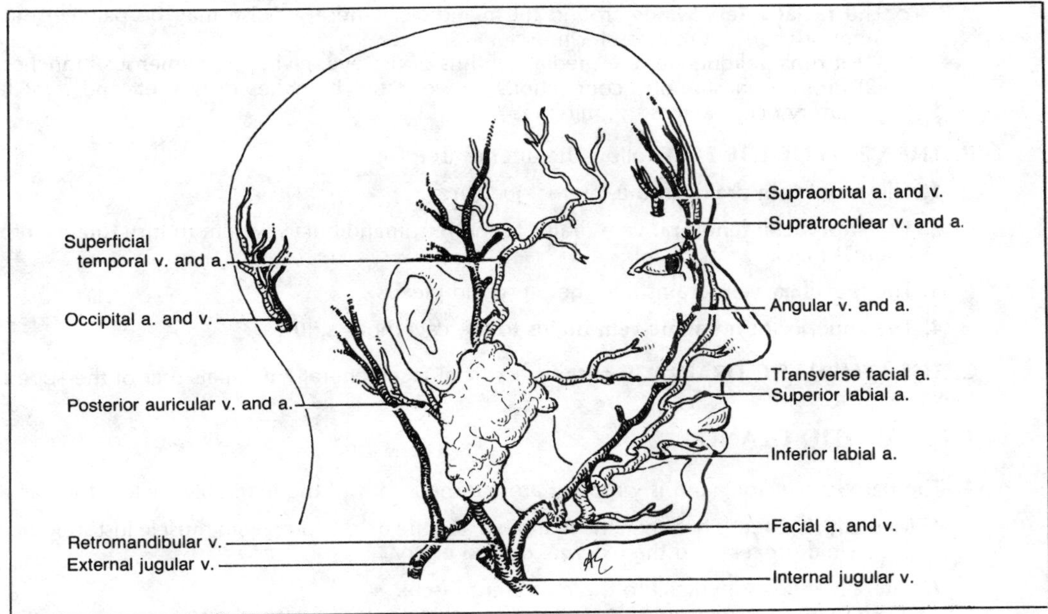

Figure 32-5. *The vasculature of the superficial face.*

(c) This nerve also innervates the mylohyoid muscle, the anterior belly of the digastric muscle, the tensor tympani muscle, and the tensor veli palatini muscle.

d. Signs and symptoms produced by lesions of the trigeminal nerve.

(1) **Trigeminal neuralgia** (tic douloureux) produces excruciating facial pain that may be initiated by touching a trigger area.

(a) The pain is often in the distribution of the mandibular nerve.

(b) The trigger area is often maxillary.

(c) So severe and unremitting is the pain that patients may be driven to suicide.

(2) Inflammation of the cavernous sinus or a lesion at the superior orbital fissure results in paresthesias of the forehead and nose with loss of the **corneal blink reflex.**

(3) A lesion at the foramen rotundum or within the infraorbital canal results in paresthesias of the midface and maxillary teeth with loss of the **sneeze reflex.**

(4) A lesion at the foramen ovale results in paresthesias along the mandible, the mandibular teeth, and the side of the face, as well as paralysis of the muscles of mastication, hearing aberrations, and loss of the **jaw-jerk reflex.**

IV. THE VASCULATURE OF THE SUPERFICIAL FACE

A. THE ARTERIES OF THE FACE are branches of the external carotid artery with a small contribution from the internal carotid artery (Fig. 32-5).

1. The **ophthalmic artery** is a branch of the **internal carotid artery** and generally follows the distribution of the ophthalmic division of the trigeminal nerve.

2. The **external carotid artery** gives rise to several major branches.

a. The **ascending pharyngeal artery.**

b. The **superior thyroid artery.**

c. The **lingual artery.**

d. The **facial (external maxillary) artery.**

e. The **occipital artery.**

f. The **superficial temporal artery.**

g. The **(internal) maxillary artery.**

3. The maxillary artery and the superficial temporal artery arise from the external carotid artery within the parotid gland.

a. The **maxillary artery** passes through the pterygomaxillary fissure, the pterygopalatine fossa, the infraorbital groove, and the infraorbital canal to emerge below the eye.

b. The **superficial temporal artery** gives off a transverse facial branch before it crosses the zygomatic arch just anterior to the ear. Where the superficial temporal artery courses along the bone, a pulse may be palpated.

c. The **facial artery** winds around the mandible (where a pulse may be palpated) at the anterior edge of the masseter muscle.
(1) It runs obliquely to the medial canthus of the eye, giving off numerous branches.
(2) Profuse anastomotic connections between the branches of the left and right facial artery occur across the midline.

B. THE VEINS OF THE FACE follow the arteries (see Fig. 32-5).

1. The **facial vein** drains to the internal jugular.

2. The **superficial temporal vein** drains to the retromandibular, which, in turn, drains into the internal jugular vein.

3. The **maxillary vein** drains into the pterygoid plexus.

4. The **superior ophthalmic vein** drains to the cavernous sinus.

C. THE LYMPHATIC DRAINAGE of the superficial face generally parallels that of the superficial veins.

V. THE PAROTID GLAND

A. The parotid salivary gland is wrapped around the ramus of the mandible, below the ear.

1. Medially, it extends between the mandible and the medial pterygoid muscle just superficial to the styloid process and the external carotid artery.

2. Laterally it lies superficial to the masseter muscle.
a. An anterior extension (or facial process) surrounds the parotid duct.
b. It may be separate from the remainder as the **accessory parotid gland**.

3. Posteriorly are the external auditory meatus and the mastoid process.

4. The superficial part of the gland is divided into a superficial and deep lobe by the plexus formed by the branches of the facial nerve.

B. The parotid gland drains into the mouth via the **parotid** (Stensen's) **duct**.

1. The firm cord-like parotid duct dives medially at the anterior edge of the masseter to enter the mouth opposite the second upper molar.

2. The oblique course of the duct between the mucous membrane and the buccinator acts as a valve. Despite this structure, retrograde spread of bacteria is fairly common in debilitated and dehydrated patients.

3. The parotid gland produces a thin serous saliva.

C. Parasympathetic secretomotor fibers reach the parotid gland by a very circuitous route (Fig. 32-6).

1. The presynaptic neurons lie in the **inferior salivatory nucleus** of the glossopharyngeal nerve (CN IX).

2. These fibers leave the **glossopharyngeal nerve** in the jugular canal as the **tympanic branch** (Jacobson's), which enters the middle ear cavity from the posterior wall and forms a plexus on the medial wall.

3. The **lesser superficial petrosal nerve** leaves this plexus through a canal of the same name in the anterior wall and gains access to the middle cranial fossa.

4. The lesser superficial petrosal nerve courses briefly across the petrous temporal bone and exits via either the foramen ovale or the innominate foramen.

5. The fibers synapse in the **otic ganglion**.

6. The postsynaptic secretomotor fibers leave the otic ganglion and are distributed by the **auriculotemporal nerve** to the parotid gland. Because the auriculotemporal nerve also contains sudomotor fibers to the sweat glands of the scalp, if the nerve is severed the fibers can regenerate along each other's pathways and innervate the wrong gland. The anticipation or taste of food then produces sweating instead of salivation (Frey's syndrome).

D. Tumors of the parotid usually arise in the superficial lobe without involvement of the facial nerve. The surgeon is presented with a challenge since iatrogenic facial nerve injury has major effects.

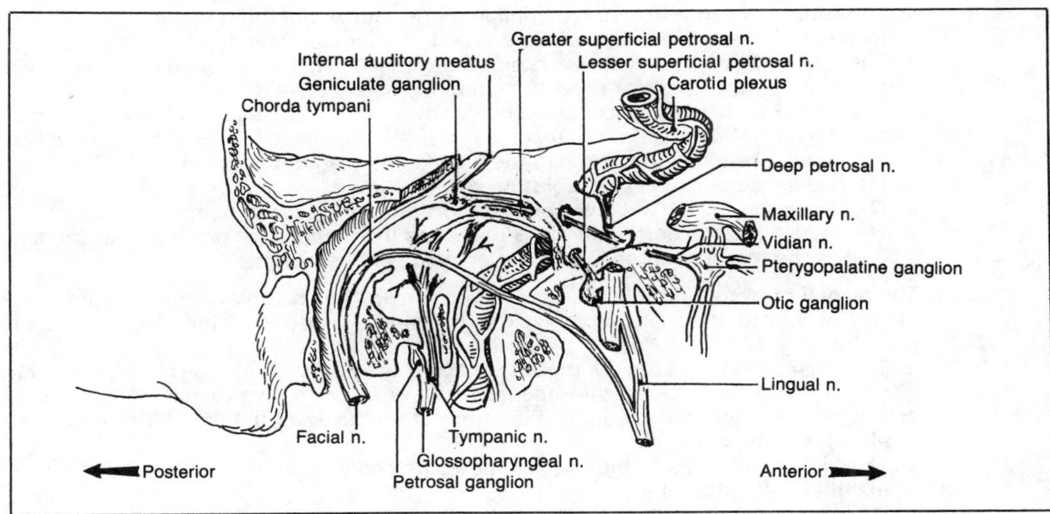

Figure 32-6. *The course of the facial and glossopharyngeal nerves in the middle ear.*

VI. THE DEEP FACE

A. THE INFRATEMPORAL FOSSA lies between the ramus of the mandible and the pharynx.

1. The parotid gland curves around the posterior border of the mandible into the infratemporal fossa. The carotid sheath passes between the parotid gland and the intrinsic pharyngeal musculature.
2. The infratemporal surface of the sphenoid bone forms the roof (see Fig. 31-4).
   a. It is pierced by the foramen ovale and foramen spinosum.
   b. It bears a pterygoid process, which expands inferiorly into medial and lateral pterygoid plates.
      (1) The medial pterygoid plate provides attachment for the superior constrictor muscle.
      (2) The lateral pterygoid plate provides attachments for the medial and lateral pterygoid muscles.
3. The anterior wall is formed by the maxilla.
4. The medial wall is formed by the temporal bone and the lateral pterygoid plate of the sphenoid bone.

B. THE MUSCLES OF MASTICATION all arise within the infratemporal fossa, insert on the ramus of the mandible, and are innervated by the mandibular nerve.

1. The **temporalis muscle** occupies the temporal fossa (Fig. 32-7A).
   a. Fibers arise beneath the temporal fascia from a broad origin on the lateral aspect of the skull inferior to the temporal line.
   b. They pass deep to the zygomatic arch and converge to insert on the coronoid process of the mandible.
   c. The anterior and medial fascicles elevate the jaw; the posterior fascicles retract the jaw.
2. The **masseter** arises from medial and lateral surfaces of the zygomatic arch (see Fig. 32-7A).
   a. It completely covers the outer surface of the ramus of the mandible and inserts into the periosteum in the region of the angle of the mandible.
   b. A **buccal fat pad** separates the masseter from the buccinator muscle.
   c. The masseteric neurovascular bundle passes through the mandibular notch to enter the deep surface of this muscle.
   d. The masseter is overlapped by the superficial lobe of the parotid gland.
   e. It is a strong elevator of the jaw.
   f. The masseter may hypertrophy as a result of **bruxism**, the nocturnal habit of clenching the jaw at times of nervous tension.
3. The **lateral pterygoid muscle** lies in a horizontal plane and has two heads that converge to their insertions (Fig. 32-7B).

a. The superior head arises from the infratemporal surface of the sphenoid bone and inserts mainly into the articular capsule and disk of the temporomandibular joint.
b. The inferior head arises from the lateral surface of the lateral pterygoid plate and inserts into a fossa on the condylar process of the mandible.
c. The lateral pterygoid muscle pulls the condyle and its articular disk forward (protusion) and downward (depression) so that the mandible effectively rotates not about the condyle but about a resultant axis through the lingula and mandibular foramen.
(1) Together, the lateral pterygoids protude the jaw.
(2) Singly, they swing the jaw from side to side to produce grinding movements.
d. The medial surface of the muscle is related to the mandibular nerve and its branches as well as the maxillary artery and the pterygoid venous plexus.

4. The **medial pterygoid muscle** has a continuous origin, which curves around the inferior edge of the lower head of the lateral pterygoid muscle and is divided by it into deep and superficial parts (see Fig. 32-7B).
a. The deep part arises from the medial surface of the pterygoid plate and the pyramidal process of the palatine bone, the superficial part arises from the tubercle of the maxilla.
b. It inserts into the periosteum of the ramus at the angle of the mandible and is a strong elevator of the mandible.
c. This thick quadrilateral muscle resembles its counterpart on the outer surface of the mandible—the masseter.
(1) Together with the masseter, it forms the **masseteric (mandibular) sling**, which supports the angle of the mandible.
(2) The masseteric sling is not attached to the bone by Sharpey's fibers.
d. The lateral surface is separated from the mandible by a lobe of the parotid gland. The lingual nerve, the chorda tympani nerve, the inferior alveolar nerve, and the inferior alveolar artery all course along the lateral surface of this muscle.
e. The medial surface is related to the tensor veli palatini muscle, the styloglossus muscle, and the stylopharyngeus muscle, which lie external to the intrinsic pharyngeal musculature.

5. While not grouped with the muscles of mastication, the digastric, mylohyoid, and geniohyoid muscles insert onto the mandible and act to produce motion at the temporomandibular joint.

C. THE TEMPOROMANDIBULAR JOINT (Fig. 32-8).

1. The temporomandibular joint is formed by the **condyle** of the mandible together with the **articular tubercle** and the **mandibular fossa** of the temporal bone.

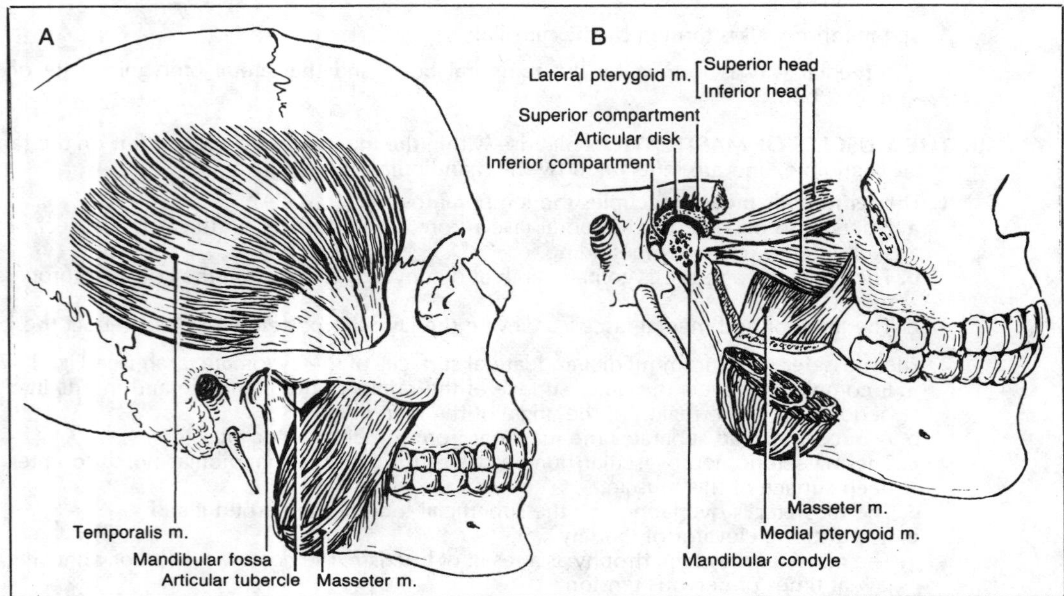

Figure 32-7. *The muscles of mastication. A,* The temporalis and masseter muscles. *B,* The medial pterygoid and lateral pterygoid muscles.

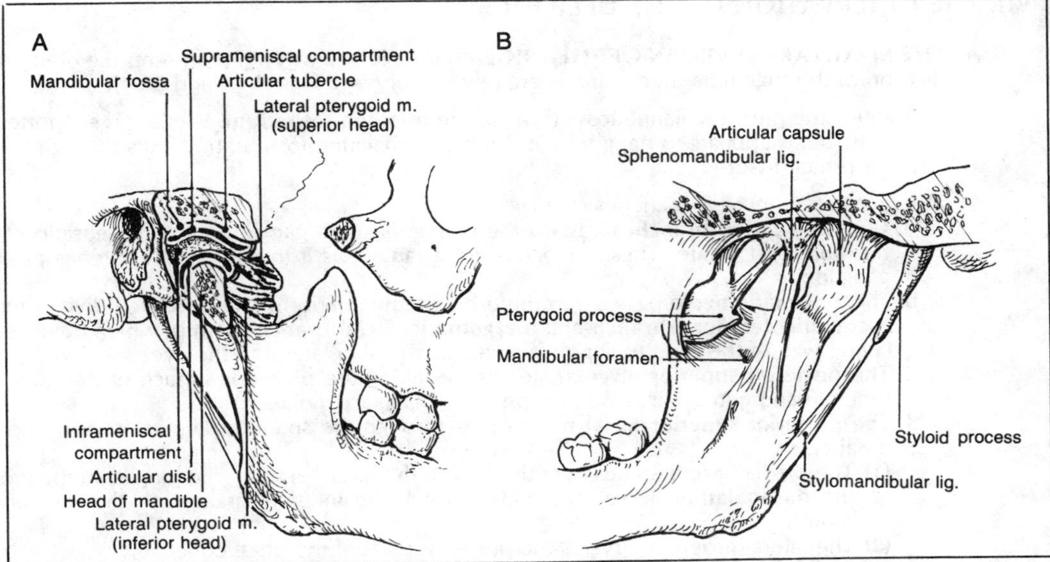

Figure 32-8. *The temporomandibular joint. A,* Lateral aspect with the joint capsule opened to show the articular disk. *B,* Medial aspect.

2. A fibrocartilage biconcave articular disk is interposed between the rounded condyle of the mandible and the corresponding sigmoid articular surface of the temporal bone, dividing the joint into suprameniscal and inframeniscal compartments.
   a. In the **inframeniscal compartment**, the condyle simply rotates on the articular disk.
   b. In the **suprameniscal compartment**, the condyle and disk slide anteriorly and posteriorly over the articular tubercle of the temporal bone.

3. During simple opening and closing of the mouth, the compound axis passes through the lingula of the ramus.
   a. This action might be expected because of the neurovascular bundle and sphenomandibular ligament, which join the mandible at this point.
   b. As the mouth is opened, the condyle slides anteriorly with the disk due to the traction on both by the two heads of the lateral pterygoid muscle.
   c. If the condyle passes beyond the apex of the articular tubercle, spasm of the temporalis muscle may jam it under the zygoma, locking the jaw.
      (1) The physician or oral surgeon faced with this problem must first pull downward on the mandible to overcome the pull of temporalis, masseter, and medial pterygoid muscles.
      (2) As the jaw snaps back into place, he or she should then beware of the forceful apposition of the molars upon sudden release.

4. The lateral pterygoid muscles can act together to protract the jaw or, by acting on one side alone, rotate the mandible about a vertical axis.

D. THE PTERYGOPALATINE FOSSA is a narrow cleft between the maxilla and the pterygoid process (*pteryx, G.* wing) of the sphenoid bone, separated from the nasal cavity by the perpendicular plate of the palatine bone (see Fig. 31-4).

1. It contains branches of the pterygopalatine ganglion, the vidian nerve, the maxillary nerve, and the maxillary artery. It has been called the neurovascular junction of the deep face.

2. The vascular structures enter this fossa laterally; the nerves enter through the posterior wall (Fig. 32-9).

3. The pterygopalatine fossa communicates with several other cavities or spaces, including:
   a. The middle cranial fossa via the foramen rotundum and the pterygoid canal.
   b. The infratemporal fossa via the pterygomaxillary fissure.
   c. The orbit via the inferior orbital fissure.
   d. The nasal cavity via the sphenopalatine foramen.
   e. The oral cavity via the palatine canal.
   f. The pharynx via the pterygovaginal canal.

VII. THE INNERVATION OF THE DEEP FACE

A. THE MAXILLARY DIVISION OF THE TRIGEMINAL NERVE (CN V_2) represents the pretrematic division of the trigeminal nerve and is entirely sensory (see Figs. 31-1 and 32-4).

1. It enters the pterygopalatine fossa through the **foramen rotundum**. While it is supported by fatty tissue, it appears to hang free in the pterygopalatine fossa until it exits through the inferior orbital fissure.

2. It gives off numerous branches (see Fig. 32-9).

a. **Communicating branches** pass to the pterygopalatine ganglion carrying sensation from the nose and palate. These branches are arranged as if to suspend the pterygopalatine ganglion.

b. The **zygomatic nerve** passes into the orbit via the inferior orbital fissure. It divides into a **zygomaticotemporal branch** and a **zygomaticofacial branch**, both of which traverse the lateral wall of the orbit to supply a small area of skin lateral to the eye.

c. The **posterior-superior alveolar nerves** run along the posterior surface of the maxillary tubercle and enter that bone to supply the maxillary molars.

d. The **posterior-superior nasal nerves** pass through the sphenopalatine foramen into the nasal cavity.

(1) The medial group arches over the roof to the nasal septum. The longest of the group, the **nasopalatine nerve**, descends through the incisive foramen to the roof of the mouth.

(2) The lateral group innervates the posterior ends of the nasal conchae.

e. The **palatine nerves** continue down the pterygopalatine fossa to the pterygopalatine canal.

(1) The **greater palatine nerve** passes through the greater palatine foramen. It then runs anteriorly along the inner aspect of the alveolar margin in a groove on the surface of the hard palate.

(2) The **lesser palatine nerves** emerge through the lesser palatine foramen. They innervate the soft palate, the tonsils, and the palatoglossal folds.

f. The main trunk of the maxillary nerve passes through the inferior orbital fissure to become the **infraorbital nerve** in the floor of the orbital cavity.

(1) It gives off **middle-superior alveolar** and **anterior-superior alveolar nerves** which run within the walls of the maxilla to supply the maxillary premolars and incisors. The anterior-superior alveolar nerve also gives off a small branch to the lateral wall of the nasal cavity.

(2) It traverses the orbital margin in a bony canal, the **infraorbital foramen**, and terminates in a spray of small branches to the nose, lip, and lower eyelid.

g. The maxillary nerve may be anesthetized at the foramen rotundum by inserting a needle through the mandibular notch and guiding it anteriorly along the pterygoid plate until it slips into the pterygopalatine fossa. Alternatively, the needle may be passed into the greater palatine foramen and upward along the pterygopalatine canal.

3. The **pterygopalatine ganglion** lies in the pterygopalatine fossa (see Fig. 32-9).

a. The **greater superficial petrosal nerve**, a branch of the facial nerve, contains preganglionic parasympathetic secretomotor fibers.

(1) It is joined at the foramen lacerum by the sympathetic fibers of the **deep petrosal nerve** to form the **(vidian) nerve of the pterygoid canal**.

(2) This nerve runs in the base of the pterygoid process to the **pterygopalatine ganglion**, where parasympathetic fibers synapse.

(3) Postsynaptic fibers follow the branches of the ganglion to join branches of both the **maxillary** and the **facial nerves**. They are distributed medially along the nasopalatine branches to the nasal cavity, inferiorly along the palatine branches to the palate, anteriorly along the infraorbital nerve to the maxillary sinus.

(4) **Secretomotor fibers** for lacrimation apparently pass to the infraorbital nerve and its zygomaticotemporal branch and thence to the lacrimal branch of the ophthalmic nerve (via a communicating branch), which carries them to the lacrimal gland.

b. Short sensory branches also pass between the maxillary nerve and the ganglion at the superior end of the fossa. Through these pass fibers conveying general sensation from the nasal cavity (**sneeze reflex**) and palate.

B. THE MANDIBULAR NERVE (CN V_3) represents the post-trematic branch of the trigeminal nerve (see Figs. 31-1 and 32-4) and is a mixed nerve.

1. It leaves the cranium through the **foramen ovale**, where the large sensory root is joined by a small motor root. The otic ganglion lies just medial to the mandibular nerve at this point.

2. The mandibular nerve gives off a recurrent meningeal branch and motor branches to the

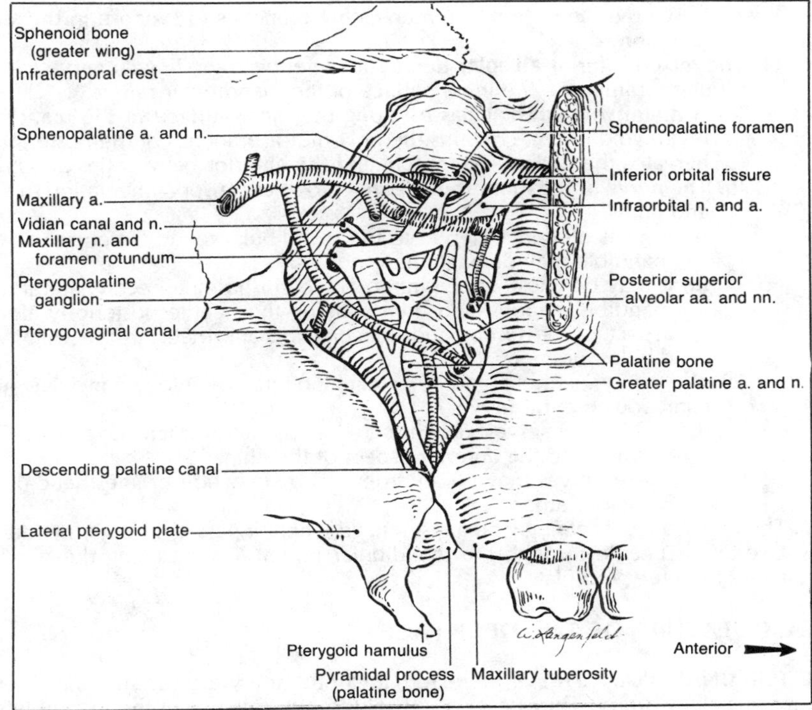

Figure 32-9. *The contents of the pterygopalatine fossa.*

muscles of mastication before dividing into anterior and posterior trunks. All of these structures lie on the medial surface of the lateral pterygoid muscle.

a. The *anterior trunk* is primarily motor and passes inferior to the lateral pterygoid muscle.

(1) The branch to the medial pterygoid muscle emerges from the medial aspect of the mandibular nerve.

(a) It gives off slender filaments to the tensor veli palatini muscle.

(b) It also supplies the tensor tympani muscle.

(2) The branches to the masseter, lateral pterygoid, and temporalis muscles emerge from the lateral aspect of the mandibular nerve.

(3) The meningeal branch enters the cranium through the foramen spinosum.

(4) The terminal sensory branch passes inferior to the lateral pterygoid muscle and courses toward the anterior aspect of the condylar process of the mandible.

b. The *posterior trunk* gives off several branches.

(1) The sensory **auriculotemporal nerve** is formed by the union of two branches of the posterior trunk. These branches pass on either side of the middle meningeal artery.

(a) This nerve runs between the neck of the mandible and the external auditory meatus to reach the temple.

(b) Parasympathetic secretomotor fibers from the otic ganglion pass along this nerve to innervate the parotid gland.

(c) The auriculotemporal nerve has a cutaneous distribution.

(2) The sensory **lingual nerve** is joined by the **chorda tympani nerve** (the pretrematic division of the facial nerve) at the inferior border of the lateral pterygoid muscle.

(a) It descends beyond the lateral pterygoid muscle and onto the lateral surface of the medial pterygoid muscle.

(b) It enters the floor of the oral cavity on the lateral surface of the styloglossus muscle below the origin of the superior constrictor muscle.

(c) At the posterior end of the inferior alveolar margin, it lies adjacent to the third molar.

(d) On the surface of the hyoglossus muscle, it passes superficial to the submandibular duct.

(e) It conveys general sensation from the anterior two-thirds of the tongue with cell bodies located in the **semilunar (gasserian) ganglion**.

(f) Taste sensation from the anterior two-thirds of the tongue is carried centrally in the chorda tympani nerve; the cell bodies are located in the **geniculate ganglion**.

(g) Parasympathetic secretomotor fibers from the chorda tympani leave the lingual nerve by a few short communicating branches to synapse in the **submandibular ganglion**.

(3) The mixed **inferior alveolar nerve** runs inferior to the lingual nerve to enter the mandibular foramen on the inner surface of the mandibular ramus.

(a) A **mylohyoid branch** passes along the inner surface of the mandible below the mylohyoid line. It contains the only motor branches of the posterior trunk and innervates the mylohyoid muscle and the anterior belly of the digastric muscle.

(b) The **nerve of the mandibular canal** provides sensory innervation to the lower teeth and gums.

(i) It gives off dental nerves to the mandibular teeth. (Fracture of the body of the mandible may damage these nerves.)

(ii) It may be anesthetized intraorally by inserting a needle lateral to the pterygomandibular raphe and then walking the needle posteriorly along the medial aspect of the ramus and injecting the anesthetic in the vicinity of the mandibular foramen.

(c) The nerve of the mandibular canal bifurcates into an **incisive nerve** and the cutaneous **mental nerve**.

(i) The mental nerve exits through the **mental foramen** and provides cutaneous innervation to the anterior aspect of the chin.

(ii) These nerves may be anesthetized by injection of anesthetic directly into the mental foramen.

c. The mandibular nerve may be anesthetized where it passes through the foramen ovale by inserting a needle through the mandibular notch and marching the needle posteriorly along the pterygoid plate.

VIII. THE VASCULATURE OF THE DEEP FACE

A. THE TERMINAL BRANCHES of the external carotid artery are the superficial temporal artery and the maxillary artery. They pass on either side of the ramus of the mandible in the parotid gland.

1. The **superficial temporal artery** arises from the external carotid artery in the parotid gland.

2. The **maxillary artery** runs anteriorly out of the parotid gland on the deep surface of the mandibular ramus, where it crosses superficial to the inferior alveolar and lingual nerves.

a. The maxillary artery gives rise to several branches within the infratemporal fossa (Fig. 32-10).

(1) Small tympanic and deep auricular branches supply the ear.

(2) The **middle meningeal artery** enters the cranium through the foramen spinosum surrounded by a plexus of postganglionic sympathetic neurons.

(3) An **accessory meningeal artery** accompanies the mandibular nerve through the foramen ovale.

(4) The **inferior alveolar artery** runs with the inferior alveolar nerve and its branches to the lower jaw.

b. The maxillary artery usually passes deep to the inferior head of the lateral pterygoid muscle. Branches of the pterygoid part of the maxillary artery correspond to the branches of the anterior division of the mandibular nerve. They are the masseteric, buccal, deep temporal, and pterygoid arteries.

c. The terminal portion of the maxillary artery passes through the pterygomaxillary fissure to enter the pterygopalatine fossa. Branches in the pterygopalatine fossa accompany the branches of the maxillary nerve into the nose, palate, and upper jaw (see Fig. 32-9).

(1) The **pharyngeal branch** courses posteriorly through the palatovaginal canal.

(2) The **posterior-superior alveolar arteries** supply the maxillary molars.

(3) The **sphenopalatine artery** has a similar distribution to the posterior nasal nerves, supplying the medial and lateral walls of the nasal cavity.

(4) The **descending palatine artery** supplies the hard and soft palates.

(5) The **infraorbital artery** accompanies the infraorbital nerve and supplies the remainder of the maxillary teeth.

B. The **pterygoid venous plexus** lies in close association with the lateral pterygoid muscle and the middle portion of the maxillary artery.

1. It receives drainage from the veins of the nasal cavities, the paranasal sinuses, and the oral cavity as well as from the structures of the infratemporal fossa.

a. The vessels that drain into the pterygoid plexus have anastomotic connections with the

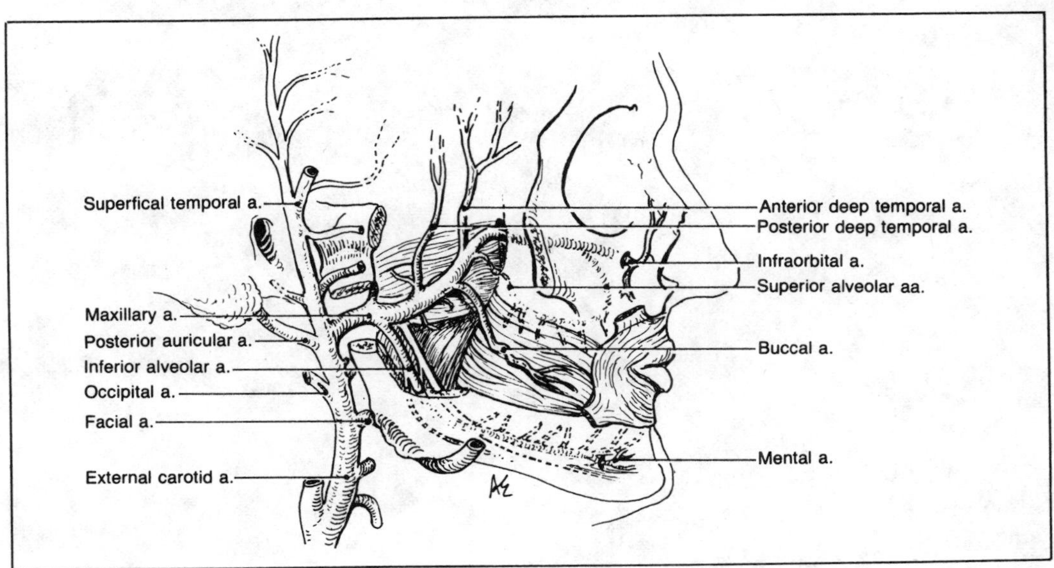

Figure 32-10. *The arteries of the deep face.*

superficial veins of the face by numerous pathways, including the **deep facial (buccal) vein**, the **inferior ophthalmic vein**, and the **pharyngeal plexus**.

b. The pterygoid plexus anastomoses with the cavernous sinus via the **sphenoid emissary vein** (of the foramen of Vesalius).

c. Thus, infection in the danger zone about the nose may spread to the pterygoid plexus and from there into the cranial, orbital, or pharyngeal regions.

2. Posteriorly, the pterygoid plexus coalesces with the maxillary vein, which drains into the external jugular vein via the retromandibular vein.

33
The Cranial and Cervical Viscera

I. INTRODUCTION

A. The cranial end of the primitive gut develops into two systems: the respiratory tract and the alimentary tract.

B. Despite this functional dichotomy, the anatomical division is complete in man from the palate to the larynx.

II. THE NASAL CAVITY

A. THE NASAL CAVITY is a pyramidal bony chamber that lies beneath the anterior cranial fossa.

1. Anteriorly, it is capped by the cartilaginous external nose and opens at the **nares**.
2. Posteriorly, it opens into the nasopharynx at the **choanae** (see Fig. 33-3).
3. The thin median cartilaginous and bony **nasal septum** extends from the roof to the floor and is usually deviated to one side of the midline.
4. The bones of the lateral walls contain sinuses (air cells) and bear three scroll-like **conchae (turbinate bones)**. The inferior concha is a separate bone.
5. The floor is formed by the palate.

B. THE SUPERIOR, MIDDLE, AND INFERIOR conchae curl over the corresponding meatus.

1. The **sphenoethmoidal recess** is the narrow space above the small superior concha. The ostium of the sphenoid sinus lies posteriorly in this recess (see Fig. 33-1).
2. The **superior meatus** below the superior concha receives several ducts from the posterior ethmoid sinuses.
3. In the **middle meatus**, the **hiatus semilunaris** curves around the inferior edge of a rounded prominence, the **ethmoid bulla**, which contains the middle ethmoid sinuses.
   a. The frontal sinus, as well as the middle and anterior ethmoid sinuses, drains along a canal, the **infundibulum,** which enters the anterosuperior end of the hiatus semilunaris.
   b. The **ostium of the maxillary sinus** opens into the posteroinferior end of the hiatus semilunaris.
4. The **inferior meatus** receives the **nasolacrimal duct** anteriorly. A small fold of mucous membrane at the nasal orifice acts as a valve.
5. Anterior to the middle meatus is a shallow depression, the **antrum,** which is limited superiorly by a small rudimentary concha (agger nasi). Inferiorly, it is continuous with the **vestibule** of the nose.

C. THE MUCOUS MEMBRANES

1. The vestibule is lined by skin that bears short thick hairs, called **vibrissae**.
2. The mucous membrane lining the apex of the nasal cavity contains olfactory cells.
   a. Approximately 20 olfactory nerves convey the sensation of smell from the olfactory cells through the cribriform plate of the ethmoid bone to synapase in the olfactory bulbs.
   b. These olfactory nerves have small meningeal sheaths.

c. They are delicate and prone to injury.

3. Elsewhere, the mucosa is of the pseudostratified columnar type, containing numerous mucus-secreting glands.
   a. The layer of mucosa, covering the middle and inferior concha, traps particulate matter and moistens the inspired air.
   b. The polluted mucus is swept backwards by cilia into the nasopharynx and is swallowed.
      (1) The cilia are affected by low temperatures and cease to beat at about 10°C, explaining in part why the nose "runs" in cold weather.
      (2) Upper respiratory tract infections and cigarette smoke also inhibit ciliary action.

D. THE VASCULATURE OF THE NASAL CAVITY

1. A profuse arterial anastomosis is fed by branches of the internal and external carotid arteries (Fig. 33-1).
   a. The **ethmoidal arteries**, branches of the ophthalmic artery, supply the anterior and superior regions.
   b. The **sphenopalatine artery**, a branch of the maxillary artery, supplies the posterior region.
   c. The **greater palatine artery**, also a branch of the maxillary artery, supplies an anteroinferior region.
   d. The **facial artery**, a branch of the external carotid, supplies the inferior and interior portions through its superior labial and ascending palatine branches.
   e. Trauma, hypertension, infection and clotting disorders are associated with epistaxis (nosebleeds).
      (1) Treatment depends upon the cause, but topical pressure, vasoconstrictors, and cautery are usually effective.
      (2) Ligation is impractical because the branches are deep, although limited success has been reported with external carotid ligation.
2. There is an extensive **submucous venous plexus**. The vasomotor control of this plexus is a delicately balanced system capable of accommodating changes in humidity, temperature, and air flow.
   a. A submucous plexus of veins functions as a heat exchanger.
   b. Vasodilation and venous engorgement is accompanied by a certain amount of mucosal edema, which may occlude the nasal airway.
      (1) Disturbances in control are sometimes the result of allergic responses (such as those to pollen and housedust) but often are of uncertain etiology.
      (2) Vasoconstrictor agents produce temporary relief because rebound vasodilation occurs and tolerance quickly develops.
   c. The veins tend to drain into the pterygoid plexus.

E. THE INNERVATION OF THE NOSE is divided, like the arterial supply, into an anterior ethmoidal (ophthalmic) area and a posterior sphenopalatine (maxillary) area.

1. The **anterior ethmoid nerve** and the **posterior ethmoid nerve**, which supply the anterior half, are branches of the nasociliary nerve, a division of CN V_1.
2. The posterior half of the nasal cavity is supplied by **posterior nasal branches** of the maxillary nerve, which pass through the sphenopalatine foramen in company with the artery of that name.
3. A small area around the anterior end of the inferior concha is supplied by the anterior-superior alveolar branch of the maxillary nerve.

F. THE PARANASAL SINUSES

1. The **frontal sinus** develops as the superior extension of an anterior ethmoid sinus.
   a. The left and right frontal sinuses are usually asymmetrical.
   b. They usually drain into the hiatus semilunaris of the middle meatus via the infundibulum.
2. The **ethmoid labyrinth** is composed of 6 to 18 thin walled air cells (sinuses).
   a. These open medially above and below the middle concha with considerable variation.
   b. Infection of these sinuses may erode through the lamina papyracea and spread into the orbit.
3. The **sphenoid sinuses** fill the body of the sphenoid bone. They are divided into the right and left by a thin septum.
   a. The septum of the sphenoid process rarely lies in the midline.
   b. The sphenoid sinus opens by a small ostium high on the anterior wall, so that drainage cannot occur by gravity with the head held erect.

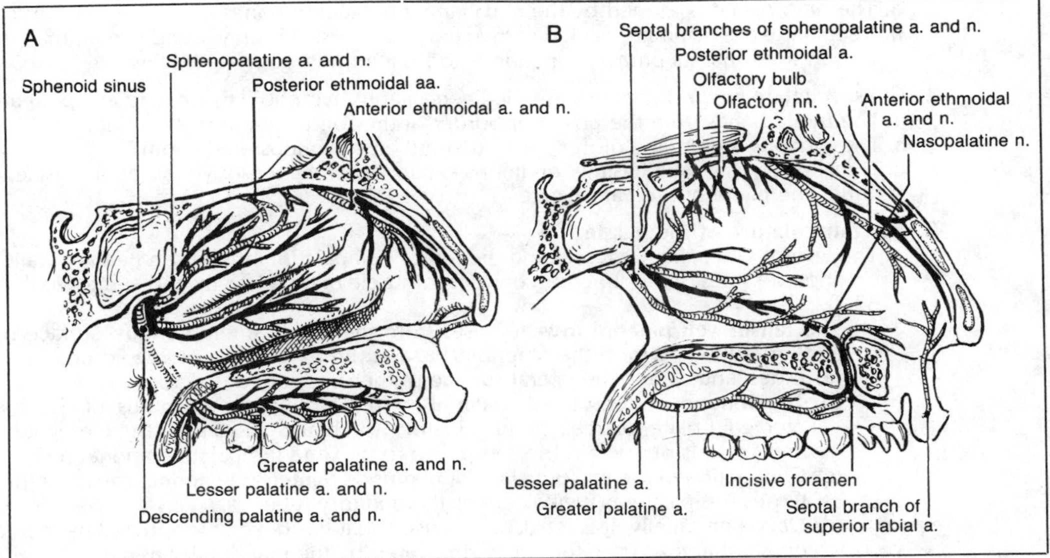

Figure 33-1. *The nasal cavity. A,* The neurovascular structures of the lateral wall. *B,* The neurovascular structures of the nasal septum.

c. Just lateral to the sphenoid sinus are the optic nerve, internal carotid artery, maxillary nerve, and pyterygoid (vidian) nerve. Infection of the sinus may erode the thin walls and involve these structures.

d. The pituitary gland may be reached through the nose.

(1) This transphenoidal approach follows the septum of the nose through the body of the sphenoid.

(2) Care must be taken to avoid the cavernous sinus and the internal carotid artery in order to avoid potentially catastrophic hemorrhage.

4. The **maxillary sinus** (antrum of Highmore) is located in the body of the maxilla.

a. Much of the medial wall of the **maxillary sinus** is composed of cartilage.

b. The slit-like osteum is high on the medial surface, so that it does not drain by gravity with the head held erect.

(1) It is closely related to the orifices of the frontal and ethmoid sinuses at the hiatus semilunaris. Infection in any one of these quickly spreads to the maxillary sinus.

(2) The maxillary sinus is particularly prone to infection, which tends to be complicated by the collection of a stagnant pool of mucus. This pool may be drained surgically by piercing the bony nasal wall of the sinus beneath the inferior concha.

c. The sinuses are lined by mucoperiosteum, which is thinner and less richly supplied with blood vessels and glands than the mucosa of the nasal cavity. Cilia sweep mucus towards the ostia. In mucosal infections the cilial action is inhibited and mucus collects.

d. Maxillary sinusitis mimics the clinical sign of maxillary tooth abscess.

(1) Approximately 89 percent of all cases of maxillary sinusitis are related to an infected tooth.

(2) Infection may also spread from the maxillary sinus to the upper teeth.

III. THE ORAL CAVITY

A. THE ORAL CAVITY extends from the oral fissure to the oropharyngeal isthmus between the palatoglossal arches (Fig. 33-2).

1. Superiorly, the palate forms the roof of the mouth.

2. Anteriorly and laterally, the oral cavity has two walls: an inner bony wall, which supports the teeth, and an outer fleshy wall.

3. Between the walls is the vestibule, and inside the alveolar ridges is the oral cavity proper.

4. Inferiorly, the mylohyoid muscle supports the floor of the oral cavity.

B. THE PALATE

1. The **hard palate** forms the anterior four-fifths of the palate.

a. The hard palate is formed by the maxillary and palatine bones.
b. It separates the respiratory and digestive tracts. Because it is only found in mammals, it can be surmised that its primary function is to isolate the mouth in suckling.

2. The **soft palate (velum)** contains muscle fibers, glands, lymphoid tissue, and an aponeurosis.
a. The **uvula** hangs from the posterior border of the soft palate in the midline.
b. The mucous membrane of the palate is stratified squamous epithelium.
c. The aponeurosis, which is the tendinous expansion of the tensor veli palatini muscle, continues the plane of the hard palate.

3. **The musculature of the palate.**
a. The levator veli palatini muscle and the tensor veli palatini muscle run nearly parallel but on opposite sides of the superior constrictor muscle of the pharynx and auditory tube (Fig. 33-3).
(1) The **tensor veli palatini muscle** arises from the lateral (membranous) surface of the auditory tube and from the scaphoid fossa at the base of the pterygoid process.
(a) It descends along the lateral surface of the superior constrictor muscle.
(b) The tendon of tensor veli palatini passes around the hamulus of the medial pterygoid plate, pierces the buccinator, which also arises from the pterygoid plate and spreads out within the soft palate to become the palatine aponeurosis.
(c) Contraction of the tensor veli palatini muscle tightens the aponeurosis and thereby tenses the palate as well as opens the auditory tube.
(d) Developmentally, it is a first arch muscle that has been appropriated by the developing pharynx. Therefore, it is innervated by the mandibular nerve.
(2) The **levator veli palatini muscle** arises from the petrous portion of the temporal bone and the medial (cartilaginous) surface of the auditory tube.
(a) It follows the auditory tube over the superior constrictor muscle and inferiorly along its inner surface.
(b) It inserts into the tensor aponeurosis of the soft palate.
(c) It acts to elevate the palate and somewhat to open the auditory tube.
(d) Since it lies within the pharynx, it is innervated by the vagus nerve (via the pharyngeal plexus).
b. The **palatoglossus muscle** arches upward from the side of the tongue to the inferior surface of the soft palate (see Fig. 33-3).
(1) It underlies the anterior faucial pillar (palatoglossal fold).
(2) It acts as a sphincter (in conjunction with the base of the tongue) between the oral cavity and the pharynx.
(3) It is innervated by the vagus nerve (CN X) through the pharyngeal plexus.
c. The **palatopharyngeus muscle** arises laterally from the palate and inserts into the pharyngeal musculature.

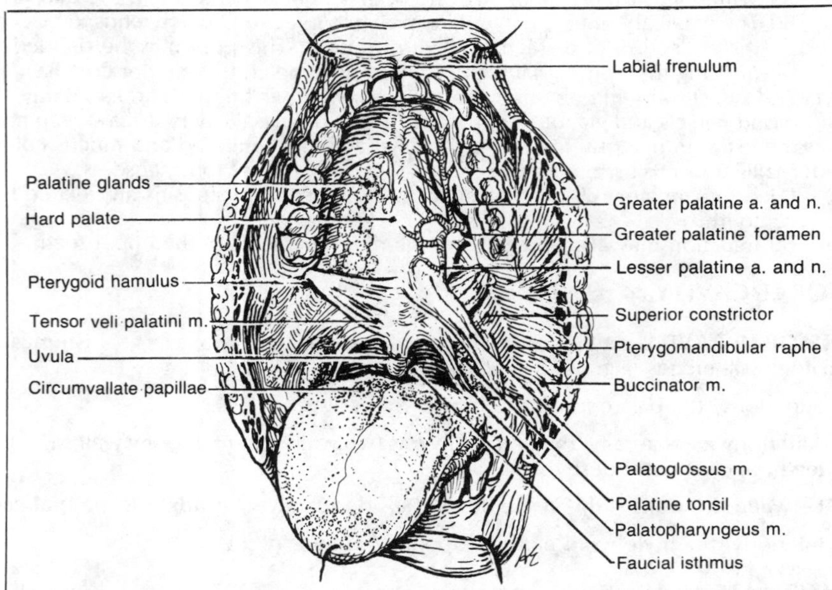

Figure 33-2. *The oral cavity.*

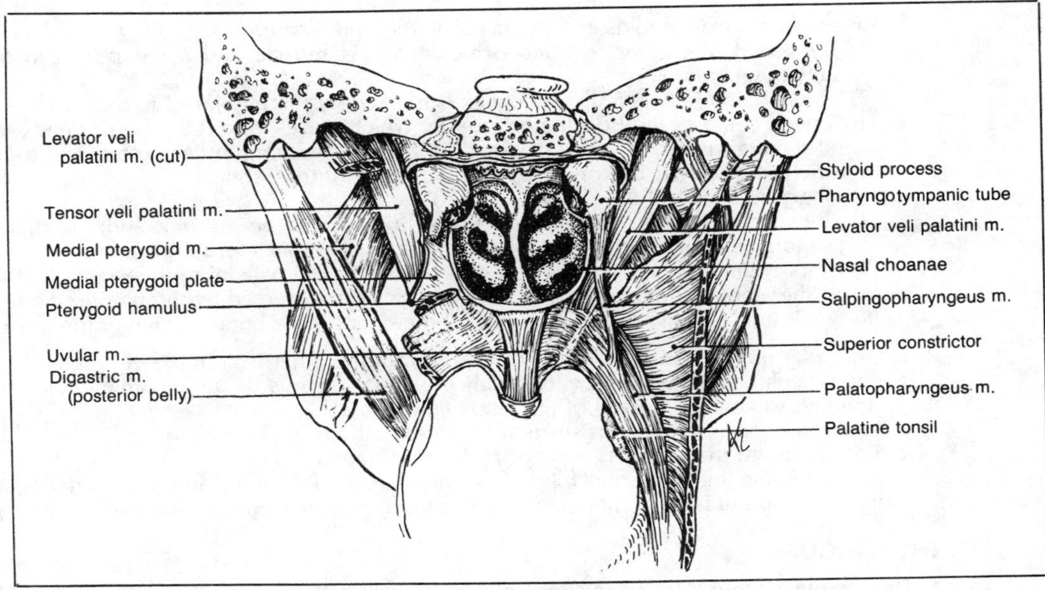

Figure 33-3. *The musculature of the soft palate.* The viewpoint is posterior.

(1) It underlies the posterior faucial pillar (palatopharyngeal fold).
(2) It also acts as a sphincter (in conjunction with the base of the tongue) between the oral cavity and the pharynx.
(3) It helps to raise the larynx in swallowing (deglutition).
(4) It is innervated by the vagus nerve (CN X) through the pharyngeal plexus.

d. The paired **uvula muscles** descend from the spine of the hard palate into the uvula on each side.
(1) They are innervated by the vagus nerve (CN X).
(2) These muscles contract when the palate is raised.
(a) When one muscle contracts alone, it draws the uvula to the same side.
(b) The integrity of the vagus nerve is observed when the examining physician has the patient say "ahhhhhh."

e. Closure of the nasopharyngeal isthmus by elevation of the palate (establishment of the velopharnygeal seal) is important in swallowing, speaking, and blowing.
(1) During quiet nasal respiration, the soft palate hangs vertically.
(2) In swallowing, the soft palate is raised to prevent food from entering the nose.
(3) In blowing or in the production of the explosive consonants, the escape of air through the nose is prevented so that the pressure can build up within the mouth.

4. The vasculâture of the palate.
a. The **arterial supply** is from several sources:
(1) The **greater palatine artery,** which gives off the lesser palatine artery, which supplies the hard and soft portions, respectively.
(2) The **ascending palatine branch** of the facial artery.
(3) The **tonsillar branch** of the facial artery.
b. Veins drain to the **pterygoid plexus** and a peritonsillar plexus.

5. Innervation of the palate.
a. General sensation for the hard palate is carried by the **greater palatine nerve** and **nasopalatine nerve**; the soft palate is innervated by the **lesser palatine nerve** and the glossopharyngeal nerve.
b. Motor innervation is provided by the mandibular division of the trigeminal nerve (CN V_3) and the vagus nerve (CN X).

6. Cleft palate, a developmental anomaly, prevents the establishment of the velopharyngeal seal. If surgically uncorrected, the individual experiences difficulty with speech and deglutition.

C. THE WALLS OF THE ORAL CAVITY.

1. The fleshy walls of the oral cavity comprise a layer of muscle between the skin of the face and the mucous membrane of the **vestibule** (see Fig. 33-2).

a. Anteriorly, the labial folds are separated by the oral fissure.

(1) The labial folds (lips) contain **orbicularis oris muscle** and some pea-sized **labial salivary glands.**

(2) The junction between lip and cheek is marked by the nasolabial folds.

b. The cheek contains the **buccinator muscle**, which arises from the posterior ends of the alveolar margins and the anterior aspect of the **pterygomandibular raphe.** The superior constrictor muscle arises from the posterior aspect of this raphe.

(1) Fibers pass around the cheeks and into the lips.

(2) The buccinator muscle's main function is to compress the vestibule and, together with the tongue, to keep the food on the cusps of the molars for mastication.

(3) At its origin it is pierced by the tendon of tensor veli palatini muscle.

(4) At the anterior border of the masseter muscle, it is pierced by the parotid duct.

(5) Small **buccal salivary glands** surround the duct on the outer surface of the muscle.

2. The alveolar margins and associated dentition form the inner wall of the vestibule.

a. The alveolar margins are covered with the dense fibrous tissue and thinly keratinized stratified squamous mucosa of the **gums (gingivae)**.

b. In a young person the gums surround the cervical portion of the teeth.

c. The teeth and their sockets were discussed in Chapter 31.

d. In the midline there are small folds of mucosa related to the inner and outer surfaces of the gums: **frenulum labia** of the upper and lower lips and the **frenulum linguae** of the tongue.

D. THE TONGUE

1. The **tongue** is a muscular organ concerned with swallowing, speech, and taste (see Fig. 33-4).

a. The oral and pharyngeal portions of the tongue have different origins, different mucosa, and different innervation.

b. A V-shaped sulcus lies between the second and third arch derivatives, which comprise the oral and lingual parts.

(1) The apex of the V points posteriorly.

(2) It has a **foramen caecum** from which the **thyroglossal duct** originated.

2. The oral portion of the tongue has a tip, a dorsal surface, a ventral surface, and a root.

a. Small projections of the lamina propria over the dorsum of the oral (presulcal) tongue create papillae, which increase the surface area of mucosa available for taste receptors.

(1) They vary in size from the large flat-topped **vallate papillae** through the **fungiform** to the conical **filiform.**

(2) The vallate papillae are usually confined to the area immediately adjacent to the sulcus.

(3) Small transverse mucosal folds at the lateral edges of the tongue create **foliate papillae**.

b. The mucosa of the ventral surface of the tongue is smooth.

(1) Two fringed folds of mucosa, the **plica fimbriata**, extend from the floor of the mouth almost to the tip of the tongue.

(2) A median frenulum also extends the length of the ventral surface.

(3) Between these mucosal folds the lingual veins are visible.

(4) At the root of the tongue the frenulum is related to the paired **sublingual papillae** for the ducts (Wharton's) of the submandibular salivary glands.

(5) The associated sublingual glands create sublingual folds around the root of the tongue.

(6) The ventral surface is highly vascular, and the vessels are close to the surface. This is a route for sublingual absorption of substances such as nitroglycerin and nicotine.

3. The pharyngeal portion of the tongue has no papillae.

a. The boundaries between the oral and pharyngeal tongue are blurred a little by the extension of third arch primordia superfically across the sulcus, so that the area marked by the vallate papillae (in front of the sulcus) is also innervated by the glossopharyngeal nerve.

b. The lumpy surface of the mucosa covers the lymphoid tissue of the **lingual tonsil**.

4. The **root** of the tongue lies deep to the mandible and above the hyoid bone.

a. Extrinsic muscles course through the root of the tongue to the hyoid bone, the soft palate, the styloid process, and the genial tubercle of the mandible (Fig. 33-4A).

(1) The **hyoglossus muscle** originates along the ramus of the hyoid bone and interdigitates with the styloglossus muscle and the lateral intrinsic musculature of the tongue.

(a) Superficial to the hyoglossus muscle are:

(i) The hypoglossal nerve (CN XII), which passes near the inferior border of the muscle.

(ii) The submandibular duct, which passes superior to the muscle to empty into the oral vestibule.

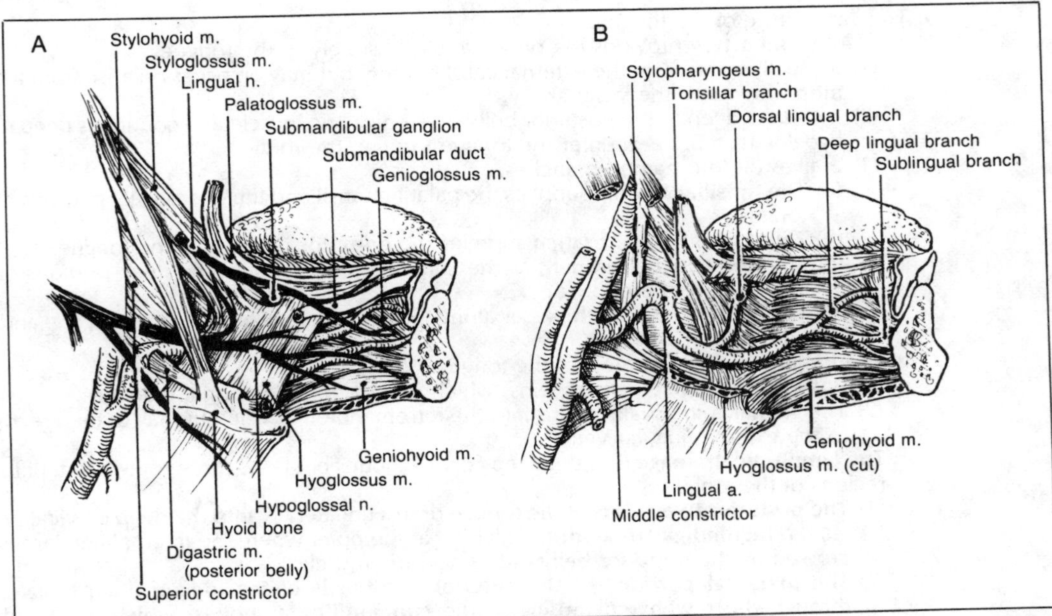

Figure 33-4. *The tongue. A,* The lingual musculature. *B,* The hyoglossus muscle has been partially removed to show the deep neurovascular structures.

(iii) The lingual nerve, which winds around the submandibular duct, crossing it first on its lateral aspect and then on its medial aspect.

(iv) A portion of the submandibular gland.

(b) Deep to the hyoglossus muscle are:

(i) The lingual artery, which runs its tortuous course deep to the hyoglossus muscle and superficial to the genioglossus muscle and giving off dorsal and deep branches and the sublingual artery.

(ii) The terminal portion of the deep branch of the glossopharngeal nerve passes posterior to this muscle to enter the posterior portion of the tongue.

(c) The hyoglossus draws the side of the tongue inferiorly.

(d) It is innervated by the hypoglossal nerve.

(2) The **genioglossus muscle** arises from the superior genial tubercle, just superior to the geniohyoid.

(a) It passes posteriorly deep to the hyoglossus muscle.

(b) Some fibers continue beyond the tongue into the middle constrictor muscle.

(c) The genioglossus muscle pulls the base of the tongue anteriorly in protrusion.

(i) Acting singly they push the protruded tongue to the opposite side.

(ii) It is innervated by the hypoglossal nerve, the integrity of which is checked by the examining physician when he or she has the patient stick out the tongue.

(3) The **styloglossus muscle** arises from the styloid process and interdigitates with the hyoglossus muscle and the lateral intrinsic musculature of the tongue.

(a) It draws the posterior portion of the tongue superiorly and posteriorly and is the principal retractor of the tongue.

(b) It is innervated by the hypoglossal nerve.

(4) The **palatoglossus muscle** lies within the palatoglossal arch at the oropharnygeal isthmus.

(a) It acts more on the palate than on the tongue, establishing the palatoglossal sphincter.

(b) Unlike the other muscles of the tongue, it is innervated by the vagus nerve (CN X).

b. The **intrinsic muscles** of the tongue are arranged in longitudinal, transverse, and vertical groups (see Fig. 33-4A).

(1) They arise and insert within the substance of the tongue.

(2) The two halves of the tongue are separated by a fibrous septum into which some of these muscles insert.

(3) They act to change the shape of the tongue (Fig. 33-4B).

(4) All are striated muscles of somatic origin and are innervated by the hypoglossal nerve (CN XII).

5. The vasculature of the tongue.

a. The **lingual artery** provides the principal blood supply to the tongue.

(1) It usually arises from the external carotid artery but may on occasion arise from a common trunk with the facial artery.

(2) It courses deep to the posterior belly of the digastric muscle and continues deep to the hyoglossus muscle to enter the tongue through the root.

(3) It gives off four named branches.

(a) The **tonsillar branch** supplies the palatine tonsil and the pharyngeal portions of the tongue.

(b) The **dorsal lingual branch** supplies the posterior portions of the tongue.

(c) The **deep lingual branch**, as the name implies, supplies the deep portions of the central part of the tongue.

(d) The **sublingual branch** passes along the genioglossus muscle to supply the anterior portion of the tongue.

b. The **lingual vein** drains most of the tongue.

(1) It is formed by like-named veins, which drain along the arterial branches.

(2) The lingual vein usually drains into the retromandibular vein but may drain directly into the external jugular vein.

c. The **lymphatic drainage** is to the deep cervical nodes but takes three routes from different regions of the tongue.

(1) The **posterior two-thirds** of the tongue drains unilaterally into the deep cervical nodes via the **jugulodigastric nodes**, which lie at the point where the internal jugular vein is crossed by the posterior belly of the digastric muscle.

(2) The **marginal portions of the anterior two-thirds** of the tongue drain unilaterally through the mylohyoid muscle to the **submandibular nodes**, which in turn drain through the **juguloomohyoid nodes** into the deep cervical nodes.

(3) The **central portions of the anterior two-thirds** of the tongue drain **bilaterally** through the mylohyoid muscle to the **submental nodes** and the **juguloomohyoid nodes**, which drain into the deep cervical nodes.

(4) The lymphatic drainage of the lips and tongue is an important consideration in the spread of tobacco-associated carcinoma of these structures.

6. Innervation of the tongue is by branches of the trigeminal, glossopharyngeal, and hypoglossal nerves (see Fig. 33-4B).

a. The **lingual nerve** (CN V_3), a portion of the post-trematic nerve of the first arch, supplies branchiomeric sensation for the anterior two-thirds of the tongue.

b. The **chorda tympani**, the pretrematic division of the **facial nerve** (CN VII), the nerve of the second branchial arch, carries taste and secretomotor fibers for the submandibular gland, the sublingual gland, and the intrinsic glands (Nuhn's) of the anterior two-thirds of the tongue.

c. The **glossopharyngeal nerve** (CN IX) supplies branchiomeric sensation and taste for the posterior third of the tongue.

d. The **hypoglossal nerve** (CN XII) supplies general somatic motor innervation to the tongue. This nerve innervates the hyoglossus, genioglossus, and styloglossus muscles, as well as the intrinsic muscles of the tongue.

E. THE SUBMANDIBULAR AND SUBLINGUAL SALIVARY GLANDS

1. The **submandibular gland** is folded around the posterior edge of the mylohyoid muscle, so that a portion lies within the floor of the oral cavity and a portion is external to the oral cavity.

a. The submandibular gland is ensheathed by the investing layer of deep cervical fascia.

b. The **submandibular duct** (Wharton's) runs from the anterior end of the deep portion, superficial to the hyoglossus muscle.

c. It empties into the oral cavity at a small papilla at the side of the frenulum of the tongue.

d. The lingual nerve winds around the duct from lateral to medial.

2. The **sublingual gland** surrounds the terminal portion of the submandibular duct

a. It empties directly into the floor of the mouth by 10 to 20 short ductules.

b. A few ductules also drain via the submandibular duct.

3. These glands are innervated by parasympathetic secretomotor fibers from the facial nerve (CN VII).

a. Parasympathetic fibers from the facial nerve join the lingual nerve via the chorda tympani

b. The submandibular ganglion is attached to the lingual nerve by a few short connecting branches.

c. The presynaptic fibers synapse in the submandibular ganglion and return to the lingual

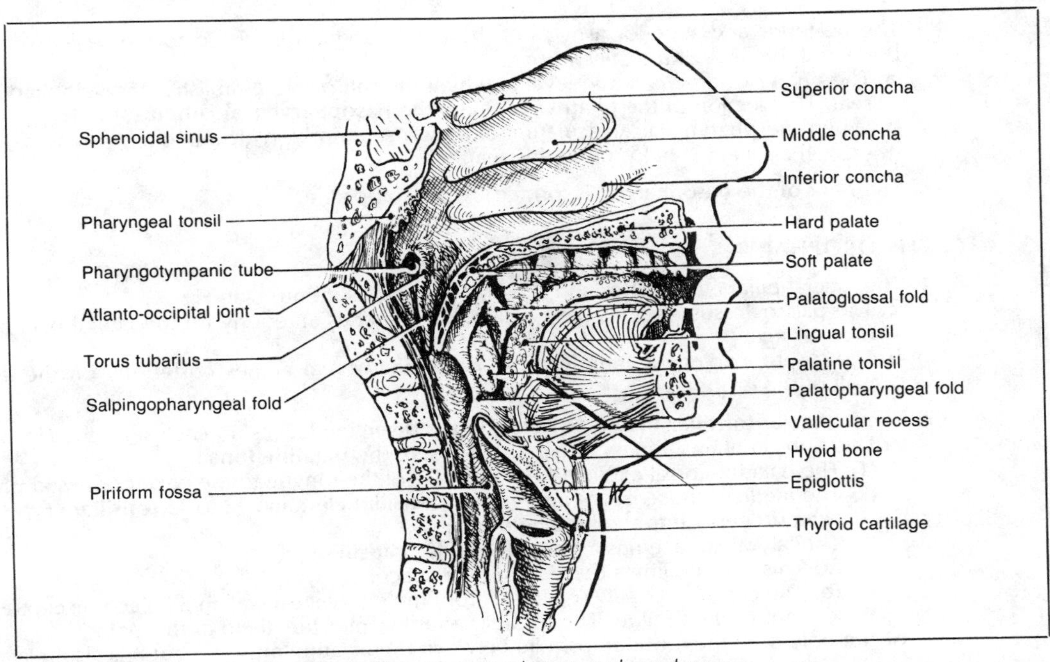

Figure 33-5. *The nasopharynx and oropharnyx.*

nerve to be distributed to the submandibular, sublingual glands as well as the intrinsic glands of the anterior tongue.

IV. THE PHARYNX

A. THE PHARYNX extends between the rami of the mandible and the base of the cranium.

1. The upper end of the pharynx may be divided in the plane of the palate into the nasopharynx and the oropharynx.
2. It opens anteriorly into the nasal and oral cavities. It continues inferiorly into the larynx and esophagus.
3. The walls consist of mucosa and striated muscle.
4. On either side of the pharynx, the muscles of mastication pass in the infratemporal fossa from the cranium to the mandible.
5. Posteriorly, the pharynx is apposed to the prevertebral fascia of the somatic neck.

B. THE NASOPHARYNX

1. The nasopharynx begins at the nasal choanae (Fig. 33-5).
2. The **auditory (pharyngotympanic) tube** opens on the lateral wall of the nasopharynx at the level of the inferior meatus of the nasal cavity.
   a. The cartilaginous wall of the tube raises a **tubal elevation** (torus tubarius).
   b. A small aggregation of lymphoid tissue forms the **tubal tonsil** in this region.
   c. Infection in the nasopharynx may track along the auditory tube to produce **otitis media**. Hypertrophy or edema of the tubal tonsil may occlude the auditory tube with accumulation of secretions .
   d. The auditory tube is a remnant of the first branchial pouch.
3. The **salpingopharyngeus muscle** (*salpynx, G.* horn) originates from the end of the auditory tube and inserts into the musculature of the pharynx (see Fig. 33-6).
   a. It acts to raise the pharynx during swallowing.
   b. The **salpingopharyngeal fold** overlies this muscle.
4. The pharyngeal tonsil is formed by lymphoid tissue embedded in the posterior wall of the nasopharynx (see Fig. 33-5). Hypertrophy of the pharyngeal tonsil (adenoids) may interfere with nasal respiration and phonation.

5. The posterior and superior aspects of the nasopharynx are related to the basilar occipital bone and the arch of the atlas.
   a. The pharynx narrows at the level at which the soft palate establishes the velopharyngeal seal. This section of the pharnyx is called the **nasopharyngeal isthmus**.
   b. When the pharyngeal musculature is drawn superiorly in deglutition, a slight posterior ridge (Passavant's) helps close this seal.

6. The floor of the nasopharynx is formed by the soft palate.

C. THE OROPHARYNX

1. The faucial pillars (arches) bound the oropharyngeal isthmus (see Fig. 33-5).
   a. The palatoglossus muscles raise palatoglossal arches anteriorly on the lateral wall of the oropharynx.
   b. The palatopharyngeus muscles raise palatopharyngeal arches posteriorly on the lateral wall of the oropharynx.

2. The triangular **tonsillar fossae** lie between the diverging fauces on each side (see Fig. 33-5).
   a. This sinus contains a mass of lymphoid tissue, the **palatine tonsil**.
      (1) The palatine tonsil extends from the base of the tongue to the edge of the soft palate.
      (2) The medial surface has a superior **intratonsillar cleft** and 12 to 15 **tonsillar crypts** that extend deeply into the lymphoid tissue.
      (3) The lateral surface has a fibrous **tonsillar capsule**.
         (a) This capsule limits the peritonsillar space.
         (b) The capsule is easily separated from the pharyngeal wall in all places except at the root of the tongue. The tonsillar arteries enter the gland at this point.
   b. Not only does the tonsillar fossa lie between the palatine muscles, but it is also related to the space between the superior and middle pharyngeal constrictors (Fig. 33-6).
      (1) Posteriorly, the gap is limited by the origin of the middle constrictor and the stylohyoid ligaments from which it arises.
      (2) Anteriorly, the hyoglossus muscle ascends to the lateral surface of the tongue.
      (3) The stylopharyngeus muscle carries the glossopharyngeal nerve between the superior and middle constrictors.
         (a) Therefore, both structures are laterally related to the tonsil. The nerve is at risk during tonsillectomy.
         (b) If the styloid process is unusually long (4 percent of individuals), a swollen tonsil may compress the glossopharyngeal nerve against the bone with pain referred to the pharynx and ear.
   c. The tonsil is supplied by tonsillar branches from several sources (see Fig. 33-6), including:
      (1) The lesser palatine branch of the maxillary artery.
      (2) The ascending palatine branch of the facial artery.
      (3) The tonsillar branch of the lingual artery.
      (4) The ascending pharyngeal artery.
      (5) The tonsillar branch of the facial artery is the primary source.
   d. A **peritonsillar venous plexus** is drained by veins that parallel the arterial branches.
      (1) The principal drainage is by the tonsillar branch of the lingual vein.
      (2) However, any of the veins that drain the tonsil may be quite large and be a source of a profuse venous hemorrhage after tonsillectomy.
   e. Currently, the treatment of choice for tonsillitis is conservative (antibiotic) therapy. Tonsillectomy is sometimes indicated in persistent infection, particularly if complicated by sinusitis, otitis media, or direct spread into the loose tissue of the pharynx (quinsy).
   f. The tonsillar fossa represents the second branchial pouch.

3. A ring of lymphoid tissue (Waldeyer's ring) extends around the pharynx, including the pharyngeal, tubal, palatine, and lingual tonsils.
   a. The adenoids hypertrophy and regress by 8 years.
   b. The palatine tonsils hypertrophy somewhat later and regress by puberty.
   c. The lingual tonsils enlarge at about the time of puberty and regress very little during adult life.

D. THE HYPOPHARYNX (see Fig. 33-9).

1. The anterior wall is formed by the base of the tongue.

2. In the midline, the **epiglottis** marks the upper surface of the larynx and guards the opening into the larynx.
   a. There is a **median glossoepiglottic fold** from the base of the tongue to the epiglottis, with a **vallecular recess** to each side.

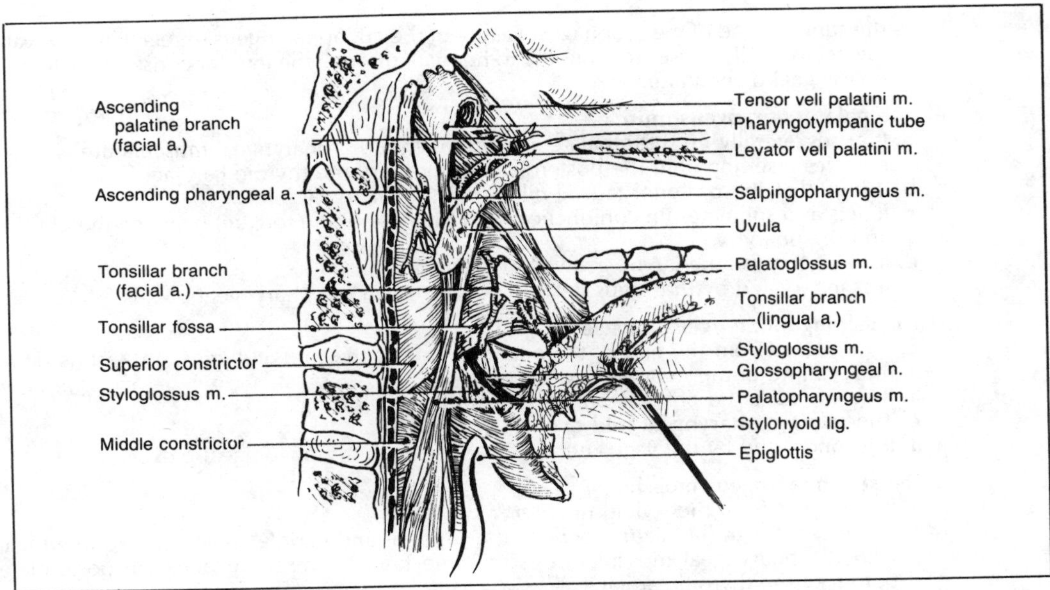

Figure 33-6. *The internal pharyngeal musculature.*

b. Two **lateral glossoepiglottic (pharyngeoepiglottic) folds** run from the anterolateral walls of the hypopharynx to the base of the epiglottis and define the inferolateral boundaries of the vallecullar recesses.

c. The vallecular recesses represent the remnants of the third branchial pouches.

3. The **piriform recesses** of the hypopharynx extend inferior to the lateral glossoepiglottic folds on either side of the larynx.
   a. These dilate considerably. Food and liquid are diverted to either side of the larynx into these recesses upon deglutition.
   b. Swallowed foreign bodies may lodge in these recesses.

E. THE PHARYNGEAL MUSCULATURE consists of three overlapping constrictors and three diagonal muscles (see Fig. 33-6; Fig. 33-7).

1. The **superior constrictor** lies deep to the ramus of the mandible in the infratemporal fossa.
   a. Anteriorly, it is attached to the pterygoid plate, the pterygomandibular raphe, which it shares with the buccinator muscle, and the posterior portions of the maxillary and mandibular alveolar processes.
   b. Posteriorly, it attaches to the pharyngeal tubercle of the basioccipital bone, as well as to the posterior pharyngeal raphe.
   c. The lateral superior edge is free and does not meet the cranium.
      (1) The auditory tube passes through this hiatus to open into the nasopharynx.
      (2) The tensor veli palatini muscle descends vertically from the auditory tube, lateral to this hiatus, to reach the hamulus; it then turns medially through the hiatus to insert into the soft palate.
      (3) The levator veli palatini muscle remains deep to the superior constrictor.

2. The **middle constrictor** is fan shaped.
   a. Anteriorly, it arises from the stylohyoid ligament and the greater and lesser horns of the hyoid bone.
   b. Its fibers pass posteriorly, external to those of the superior constrictor, to insert into the pharyngeal raphe.
   c. A gap exists between the lowest origin of the superior constrictor and the uppermost origin of the middle constrictor. The stylopharyngeus muscle, the pharyngeal branch of the glossopharyngeal nerve, and the tonsillar branch of the facial artery pass through this gap.

3. The **inferior constrictor** is fan shaped superiorly but tubular inferiorly.
   a. It arises from the oblique line on the lamina of the thyroid cartilage and from the cricoid cartilage.
   b. The fibers pass posteriorly, external to the middle constrictor, to insert into the pharyngeal raphe.
   c. The cricoid portion of this muscle (the **cricopharyngeus muscle**) functions as a sphincter at

the superior end of the esophagus. Failure of the cricopharyngeus to relax during swallowing occasionally causes the mucosa to herniate through the inferior constrictors, forming a pharyngeal diverticulum.

4. The **palatopharyngeus muscle.**
   - **a.** It arises laterally from the palate and inserts into the pharyngeal musculature with some fascicles inserting onto the posterolateral border of the thyroid cartilage.
   - **b.** It underlies the posterior faucial pillar (palatopharyngeal fold).
   - **c.** It acts as a sphincter (in conjunction with the base of the tongue) between the oral cavity and the pharynx.
   - **d.** It helps to raise the larynx in swallowing (deglutition).
   - **e.** It is innervated by the vagus nerve (CN X) through the pharyngeal plexus.

5. The **salpingopharyngeus muscle.**
   - **a.** It originates from the end of the auditory tube at the torus tubarius and inserts into the musculature of the midpharynx.
   - **b.** It acts to raise the pharynx and larynx during swallowing.
   - **c.** The **salpingopharyngeal fold** overlies this muscle.
   - **d.** It is innervated by the vagus nerve (CN X) through the pharyngeal plexus.

6. The **stylopharyngeus muscle.**
   - **a.** It originates from the styloid process.
   - **b.** It passes through the hiatus between the superior and middle constrictors to interdigitate with the pharyngeal musculature with some fascicles inserting onto the posterolateral border of the thyroid cartilage.
   - **c.** It raises the pharynx during deglutition.
   - **d.** It is the only muscle innervated by the glossopharyngeal nerve (CN IX).

F. THE INNERVATION OF THE PHARYNX is by the glossopharyngeal and vagus nerves.

1. The **glossopharyngeal nerve** (CN IX) comprises the pretrematic and post-trematic portions of the nerve of the third branchial arch. It has fibers belonging to all of the functional categories except somatic efferent (Fig. 33-8).
   - **a.** The glossopharyngeal nerve passes out the cranium in a deep groove on the medial margin of the **jugular foramen.**
     - **(1)** The pretrematic **tympanic branch** conveys sensory neurons to the ear and parasympathetic neurons continue along the **lesser superficial petrosal nerve** to the **otic ganglion** for parotid gland secretion.
     - **(2)** It has a small **superior (jugular) ganglion** and **inferior (petrosal) ganglion** just below the jugular foramen. These contain the cell bodies of the sensory neurons.

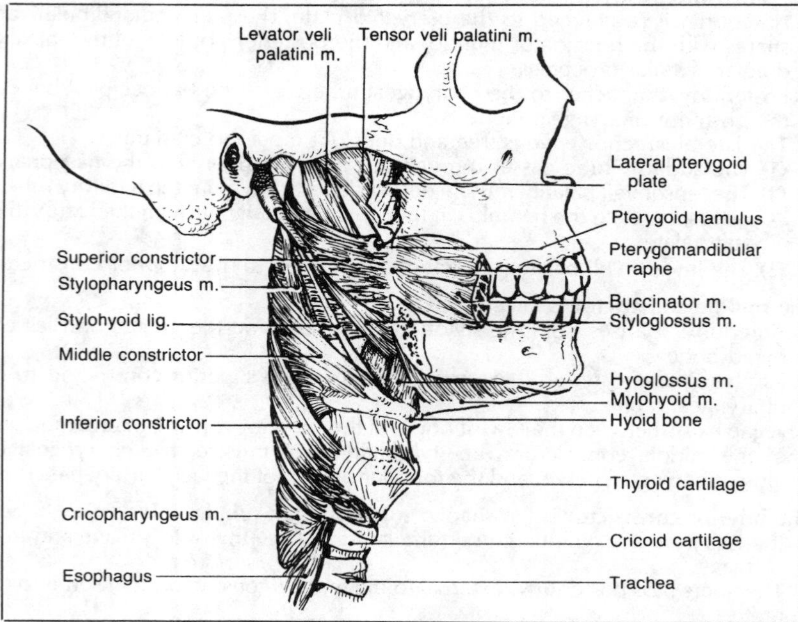

Figure 33-7. *The external pharyngeal musculature.*

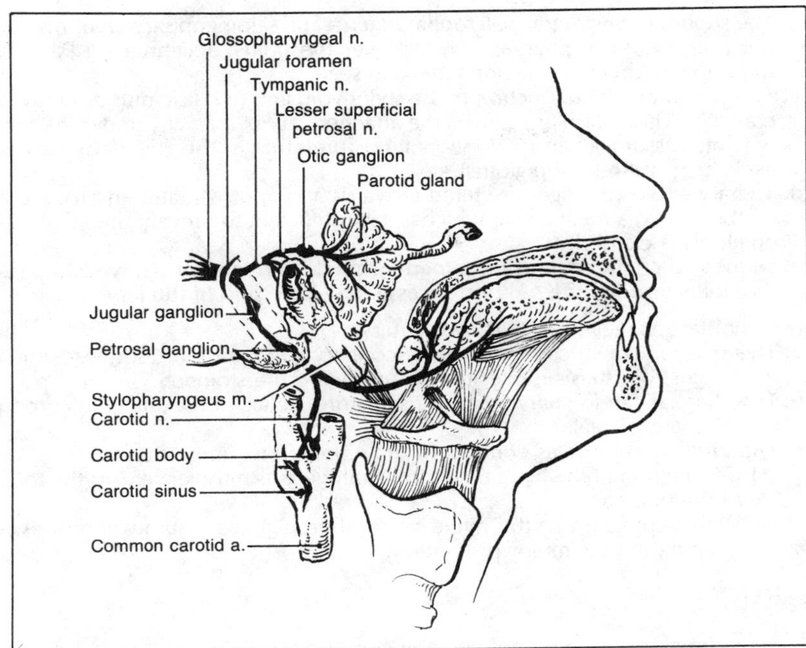

Figure 33-8. *The glossopharyngeal nerve.*

b. It passes laterally between the internal jugular vein and internal carotid artery; it then winds around the stylopharyngeus to enter the pharynx between the superior and middle constrictors.

(1) The **pharyngeal branch** (also pretrematic) supplies branchiomeric sensation from the posterior third of the tongue and the walls of the pharynx.

(a) This is the afferent limb of the **gag reflex**.

(i) The integrity of the glossopharyngeal is tested by the examining physician with a wooden tongue depressor.

(ii) Stimulation in the external auditory meatus may initiate a referred gag reflex.

(b) The pharyngeal branch also supplies taste sensation from the posterior third of the tongue.

(2) The post-trematic motor branch supplies the stylopharyngeus muscle.

(a) This seems to be the only branchiomeric muscle supplied by the glossopharyngeal nerve, although some sources also claim the middle constrictor.

(b) The post-trematic sensory branch is associated with the carotid body and sinus and mediates the **carotid reflex**.

2. The **vagus nerve** (CN X) supplies all of the branchiomeric musculature of the pharynx (except the stylopharyngeus muscle) through its pharyngeal branch (see Fig. 33-13).

G. DEGLUTITION (swallowing) may be divided into three phases: oral, pharyngeal, and esophageal.

1. The first (oral) phase is voluntary.

a. Elevation of hyoid bone by the digastric and mylohyoid muscles raises the tongue to the roof of the mouth.

b. The intrinsic muscles press the tip of the tongue against the maxillary incisors and the hard palate to squeeze the food toward the pharynx.

c. The palatoglossus elevates the base of the tongue to squeeze the food through the fauces into the pharynx.

d. The styloglossus draws the base of the tongue posteriorly, propelling the bolus into the oropharynx.

e. Together with the elevated and retracted tongue, the palatoglossus and palatopharyngeus muscles close the oropharyngeal isthmus behind the bolus.

2. The start of the pharyngeal phase is also voluntary, but once started, cannot be comfortably interrupted.

a. The soft palate is elevated by the action of the levator veli palatini and tensor veli palatini muscles to seal off the nasopharyngeal isthmus.

b. The superior constrictor, palatopharyngeus, and salpingopharyngeus muscles draw the upper portion of the pharynx upward over the bolus, accentuating Passavant's ridge and thereby reinforcing the velopharyngeal seal.
c. The concomitant contraction of the stylohyoid and digastric muscles draw the hyoid bone cranially. This action also draws the attached larynx cranially under the tongue so that the epiglottis assumes a more horizontal posture—that is, the voice box (larynx) rises to close itself against the lid (epiglottis).
d. The arytenoid cartilages are tilted forwards and approximated in order to assist in closing off the larynx, a mechanism which is itself sufficient to prevent the inhalation of food if the epiglottis is excised.
e. Sequential contraction of the superior middle and inferior pharyngeal constrictors propel the bolus through the piriform recesses to either side of the larynx.

3. The final (esophageal) phase is involuntary.
a. The inferior constrictor and the cricopharyngeus muscles squeeze the bolus into the esophagus; peristalsis moves the bolus toward the stomach.
b. The lingual and pharyngeal musculature relax, breaking the oropharyngeal and velopharyngeal seals.
c. The infrahyoid muscles contract to draw the larynx inferiorly.
(1) Contraction of the hyoglossus and genioglossus muscles returns the tongue to the floor of the oral cavity.
(2) With depression of the hyoid bone, the epiglottis assumes a more vertical position, opening the larynx for respiration.

V. THE LARYNX

A. The larynx functions as a compound sphincter (Fig. 33-9).

1. It closes the airway during swallowing and during Valsalva's maneuver, (as in coughing, urination, and defecation).

2. With fine motor control, it constricts the airway for phonation.

3. The larynx is formed by a rigid framework of bones, cartilages, and ligaments.
a. The hyoid, thyroid, cricoid, and epiglottis are single symmetrical structures.
b. The arytenoid, corniculate, and cuneiform cartilages are paired.

B. THE LARYNGEAL OSSICLES (Fig. 33-10).

1. The **hyoid bone** is U-shaped with a median **body**, paired **lesser cornua (horns)**, laterally, and paired **greater cornua (horns)**, posteriorly.
a. The **stylohyoid ligament** runs between the styloid process and the lesser horn.
b. Numerous muscles are attached along the outer surface of the greater horn and body.
c. The **thyrohyoid membrane** runs from the smooth medial surface of the hyoid bone to attach along the upper border of the thyroid cartilage (see Fig. 33-10).

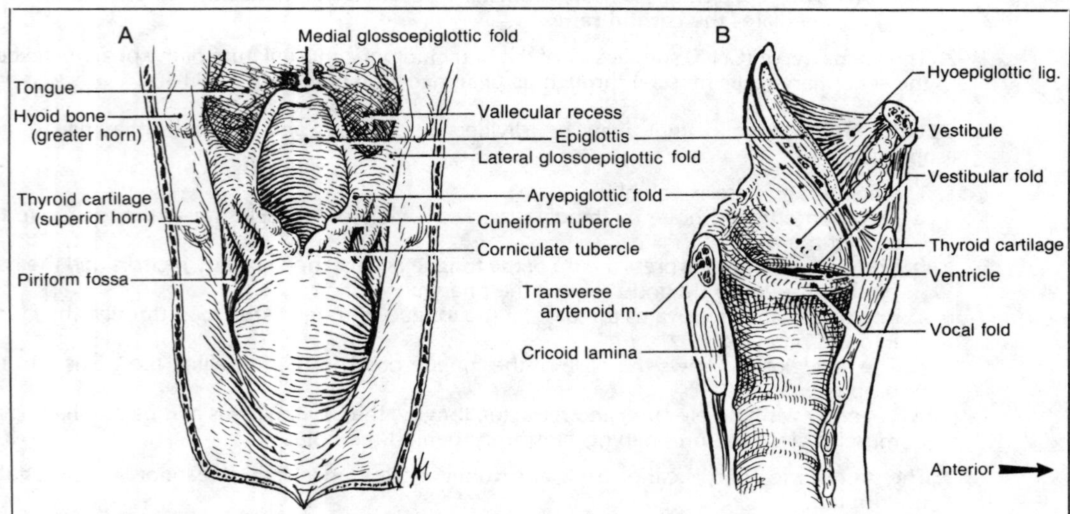

Figure 33-9. *The hypopharynx and larynx. A,* Posterior aspect. *B,* Lateral aspect.

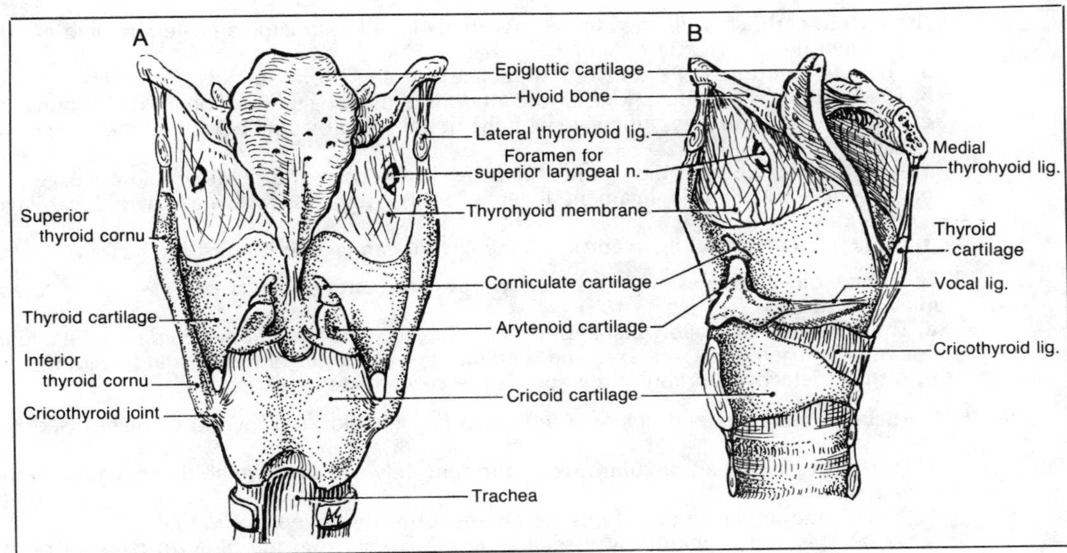

Figure 33-10. *The laryngeal ossicles. A,* Posterior aspect. *B,* Lateral aspect.

(1) The thyrohyoid membrane is thickened anteriorly to form the **medial thyrohyoid ligament.**
(2) It is thickened laterally to form the **lateral thyrohyoid ligament**, which frequently contains a small cartilaginous nodule.
(3) It is pierced superolaterally by the internal branch of the superior laryngeal neurovascular bundle.

2. The **epiglottis** is a leaf-shaped stalk that projects at an angle posterosuperiorly into the hypopharynx (see Fig. 33-9).
   a. It attaches to the anterior inner surface of the thyroid cartilage.
   b. The anterior surface of the epiglottis is joined to the hyoid by the **hyoepiglottic ligament**.
   c. The superior surface of the epiglottis has a **medial glossoepiglottic fold** between two **vallecular recesses**, which are limited inferolaterally by the **lateral glossoepiglottic folds**.
   d. On either side of the laryngeal opening inferior to the lateral glossoepiglottic folds are the **piriform fossae** of the hypopharynx.
   e. The innervation of the epiglottis is divided.
      (1) Branchiomeric sensation from the upper epiglottic surface is carried by the glossopharyngeal nerve.
      (2) The taste buds of the epiglottis are innervated by the internal branch of the superior laryngeal nerve, which arises from the vagus nerve and represents the pretrematic division of the nerve of the fourth branchial arch.
      (3) Branchiomeric sensation from the lower surface of the epiglottis and supraglottic larynx also is carried by the internal branch of the superior laryngeal branch of the vagus nerve. This provides the afferent limb of the **cough reflex**.

3. The **thyroid cartilage** comprises two laminae fused in the anterior midline (see Fig. 33-10A).
   a. The *upper border* has a median **thyroid notch**.
      (1) Below the notch is a **laryngeal prominence**.
      (2) In men, the angle between the laminae is more acute; the thyroid notch and laryngeal prominence (Adam's apple) are more apparent than in women. This results in longer vocal cords that vibrate at a lower frequency.
   b. The smooth *outer surface* ends posteriorly on either side at an **oblique line** between the superior and inferior tubercles. The linear origins of the thyrohyoid muscle and inferior constrictor are separated by the insertion of the sternothyroid muscle along the crest of the **oblique line**.
   c. The *posterolateral border* is thickened for the insertion of the stylopharyngeus and palatopharyngeus muscles.
      (1) Its ends are drawn out into the **superior cornu** and the **inferior cornu**.
      (2) The short inferior cornu has a synovial articulation with the lamina of the cricoid cartilage. The joint allows a hinge-like rotation about a single axis between the thyroid and cricoid cartilages.

4. The **cricoid cartilage** is likened to a signet ring with a broad lamina posteriorly and narrow arch anteriorly (see Fig. 33-10A).
   a. The inferior cornua of the thyroid articulate with the lateral surface.
   b. The **lamina** has a median ridge for the tendons of the longitudinal fibers of the esophagus.
   c. The arytenoid cartilages articulate with the upper posterolateral borders of the cricoid cartilage.
   d. A posterior cricoarytenoid ligament attaches to the base of each arytenoid cartilage.
   e. A medial cricothyroid ligament extends between the cricoid and thyroid cartilages anteriorly.
   f. A thick **cricotracheal ligament** binds the cricoid cartilage to the first tracheal ring.

5. The **arytenoid cartilages** have a **base**, an **apex**, an **anterior vocal process**, and a **lateral muscular process** (see Fig 33-10B).
   a. The base forms a shallow ball-and-socket articulation with the upper border of the cricoid cartilage. There are three axes and therefore three degrees of freedom at this joint.
   b. A small detached portion of the apex is the **corniculate cartilage**.

6. **Fibroelastic plates** extend between the arytenoid, cricoid, and thyroid cartilages (see Fig. 33-10B).
   a. The fibroelastic **quadrangular membrane** runs between the arytenoid cartilages and the epiglottis.
      (1) The free upper border of this membrane forms the **aryepiglottic fold**.
         (a) This fold forms the wall separating the larynx from the piriform recesses of the hypopharynx.
         (b) A small rod-like **cuneiform cartilage** lies within the free edge of each fold.
      (2) The free lower border of this membrane lies within and forms the horizontal **vestibular fold** (false vocal cord).
      (3) The laryngeal **vestibule** lies between the aryepiglottic folds and the vestibular folds.
   b. Inferior to the **vestibular folds** is a fusiform **laryngeal sinus [ventricle]** (Figs. 33-9B and 33-12A).
      (1) The pouch of mucous membrane, the **saccule**, extends superiorly from the anterior end of the sinus.
         (a) It lies between the quadrangular membrane and the small thyroepiglottic muscle.
         (b) Numerous mucous glands open into the saccule, which empty their secretion when the thyroepiglotticus contracts.
      (2) The laryngeal sinus seems to represent the fourth branchial pouch.

C. THE INTRINSIC MUSCULATURE OF THE LARYNX (Fig. 33-11).

1. The movements of the larynx occur at the cricoarytenoid and cricothyroid joints.
   a. The arytenoid cartilage can slide along the cricoid in abduction and adduction, rotate about a vertical axis, and tilt forward and backward slightly.

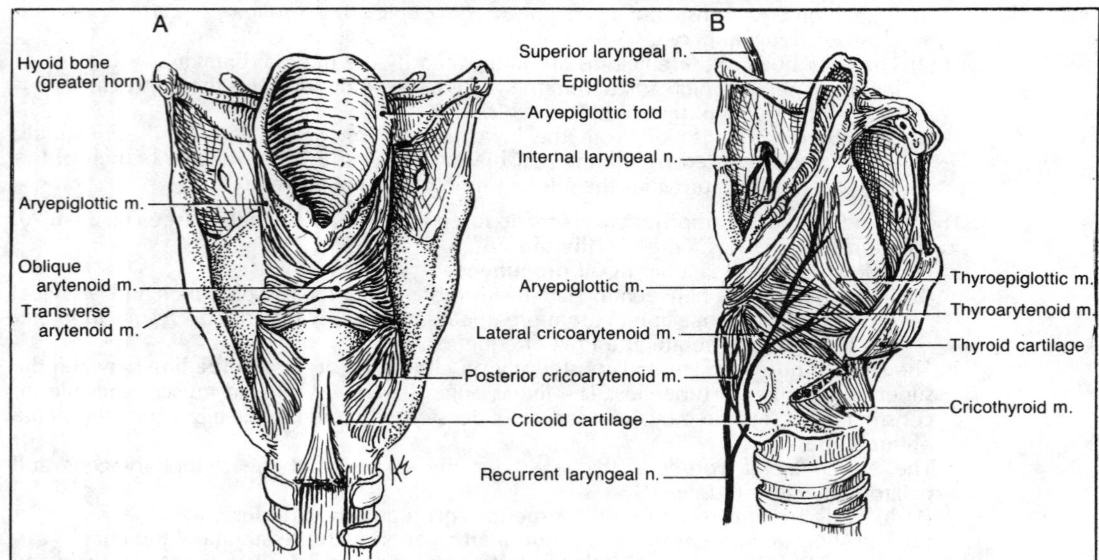

Figure 33-11. *The laryngeal musculature. A,* Posterior aspect, *B,* Lateral aspect.

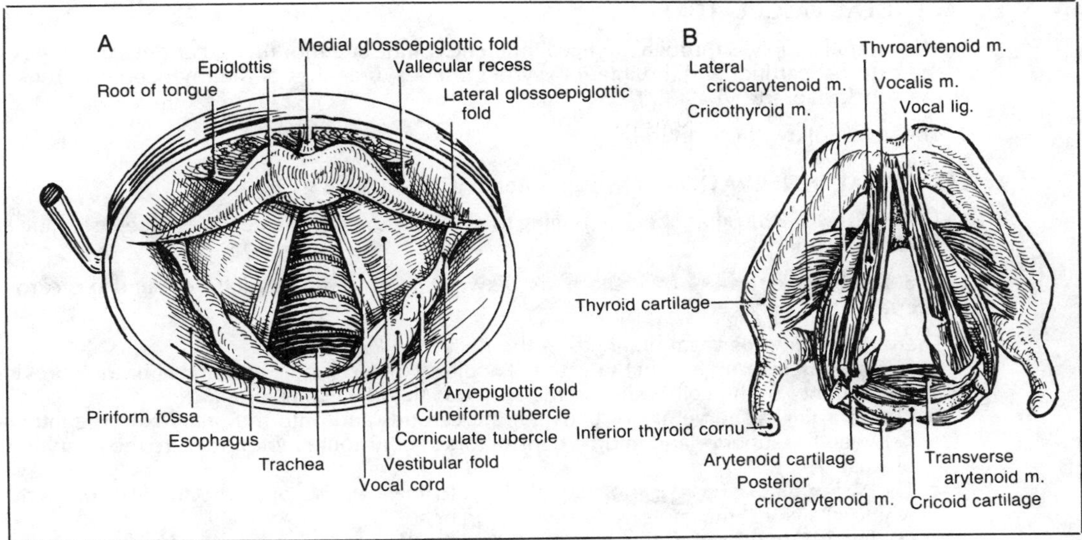

Figure 33-12. *The vocal cords. A,* As seen through a laryngoscope. *B,* The vocalis muscle.

b. These movements are conjoined so that the abduction always occurs with external rotation, and adduction occurs with internal rotation.

2. Muscles radiate from the muscular process and posterior surface of the arytenoid (see Fig. 33-11).

a. The **cricothyroid muscle.**
- **(1)** This muscle runs between the external surfaces of the thyroid and cricoid arches.
- **(2)** In contraction it approximates the anterior edges of the cricoid and thyroid cartilages, thereby stretching and tensing the vocal folds.
- **(3)** This is the only muscle innervated by the external branch of the superior laryngeal branch of the vagus nerve.

b. The **posterior cricoarytenoid muscle.**
- **(1)** Fibers converge to the muscular process of the arytenoid cartilage from a broad origin on the posterior lamina of the cricoid cartilage.
- **(2)** The horizontal fibers rotate the arytenoid laterally, thereby swinging the vocal process outward to abduct the vocal folds.
- **(3)** The vertical fibers tilt the arytenoid backwards and thereby tense the vocal cords.
- **(4)** This is the only dilator muscle of the rima glottidis.

c. The **transverse arytenoid muscle** and the **oblique arytenoid muscle**.
- **(1)** These fibers run between the medial surfaces of the arytenoid cartilages and fibers from the oblique arytenoid muscles, which continue to the epiglottis as the **aryepiglotticus muscle**.
- **(2)** These adduct the arytenoid cartilages, thereby adducting the vocal cords.

d. The **lateral cricoarytenoid muscle**.
- **(1)** It arises from the outer surface of the cricoid arch and inserts onto the muscular process of the arytenoid cartilage.
- **(2)** It rotates the arytenoid cartilage medially and tilts it forward, thereby adducting the vocal cords and reducing the tension on the cords.

e. The **superior thyroarytenoid** muscle.
- **(1)** This muscle passes obliquely from the thyroid cartilage to the aryepiglottic folds and arytenoid cartilage superficial to the quadrangular membrane.
- **(2)** Superiorly it is the **thyroepiglotticus muscle** that acts to widen the laryngeal inlet.
- **(3)** Inferiorly it is the **thyroarytenoid muscle** (see Fig. 33-11B; Fig. 33-12).
  - **(a)** The thyroarytenoid is another internal rotator of the arytenoid cartilage and thereby adducts the vocal folds.
  - **(b)** At the same time it draws the arytenoids forwards, slackening the vocal cords.
  - **(c)** The most medial portion of the thyroarytenoid is termed the **vocalis muscle**.
    - **(i)** Although the vocalis slackens the posterior portion of the cord, it tenses the anterior portion and thus raises the pitch of the voice.
    - **(ii)** It also contracts during deglutition to close the rima glottidis, the space between the vocal cords.

D. LARYNGEAL VASCULATURE

1. The arterial supply is through laryngeal branches of the superior thyroid arteries (branches of the external carotid) and the inferior thyroid arteries (branches of the thyrocervical trunk of the subclavian artery).

2. The veins follow these arteries.

E. LARYNGEAL INNERVATION is by vagus nerve (CN X) [Fig. 33-13].

1. The **vagus nerve** also has fibers belonging to all of the functional categories except somatic efferent.

2. The vagal trunk passes down the neck between the internal jugular vein and the carotid arteries.

3. There are numerous vagal branches in the neck (see Fig. 33-13).
   a. The **auricular branch** from the external auditory meatus and a **meningeal branch** provide the majority of the cell bodies in the **superior (jugular) ganglion**.
   b. The **pharyngeal nerve** passes between the carotids and onto the pharynx in the interval between the superior and middle constrictors to contribute motor fibers to the pharyngeal plexus.
   c. The **superior laryngeal nerve** passes deep to the carotids onto the middle constrictor, where it divides into motor and sensory branches.
      (1) The **internal branch** pierces the thyroid membrane.
         (a) It is sensory between the inferior surface of the epiglottis and the vocal folds and also conveys taste fibers from the epiglottis.
         (b) It provides the afferent **limb** of the **cough reflex**.
         (c) Its cell bodies lie in the **inferior (nodose) ganglion**.
         (d) It seems to represent the pretrematic division of the nerve of the fourth branchiomeric arch.
      (2) The **external branch** descends on the surface of the inferior constrictor to the cricothyroid muscle.
         (a) It has a posterior relation to the upper pole of the thyroid gland and must be avoided when clamping the upper vascular peduncle in thyroidectomy.
         (b) This seems to represent the post-trematic division of the nerve of the fourth branchial arch.
   d. The **inferior (recurrent) laryngeal nerve** takes a different course on each side of the neck.
      (1) During development, the recurrent laryngeal is associated with the sixth branchial arch.
         (a) On the right side, the fifth and sixth arches degenerate, allowing the recurrent laryngeal nerve to pass around the subclavian artery which is associated with the fourth branchial arch.
         (b) On the left side, the recurrent laryngeal is trapped below the ligamentum arteriosum, the remnant of the artery of the sixth branchial arch.
      (2) Each recurrent laryngeal nerve ascends the neck in the groove between the esophagus and the trachea.
      (3) It is intimately related to the posterior surface of the thyroid gland, lying among the terminal branches of the inferior thyroid artery between two layers of pretracheal fascia.
      (4) At the level of the inferior border of the inferior constrictor (cricopharyngeus muscle) it dives into the larynx and ascends posterior to the articulation between the thyroid and cricoid cartilages.
      (5) It is motor to all of the muscles of the larynx except the cricothyroid muscle and sensory to the mucosa below the vocal folds.
         (a) The recurrent nerve is occasionally injured during thyroidectomy.
         (b) When paralyzed, the vocal fold adopts a middle position. This lack of tension produces hoarseness.
         (c) Because the vocal folds have several adductors and only one abductor, laryngospasm apposes the vocal folds. This can be fatal if the patient is not intubated immediately.

4. The majority of fibers in the vagus are general visceral afferents and efferents associated with the thoracic and abdominal viscera.

F. PHONATION

1. Phonation is accomplished by forcing air between the closely apposed vocal cords (lateral cricoarytenoid and transverse arytenoic muscles) and thereby causing the column of air in the larynx to vibrate.

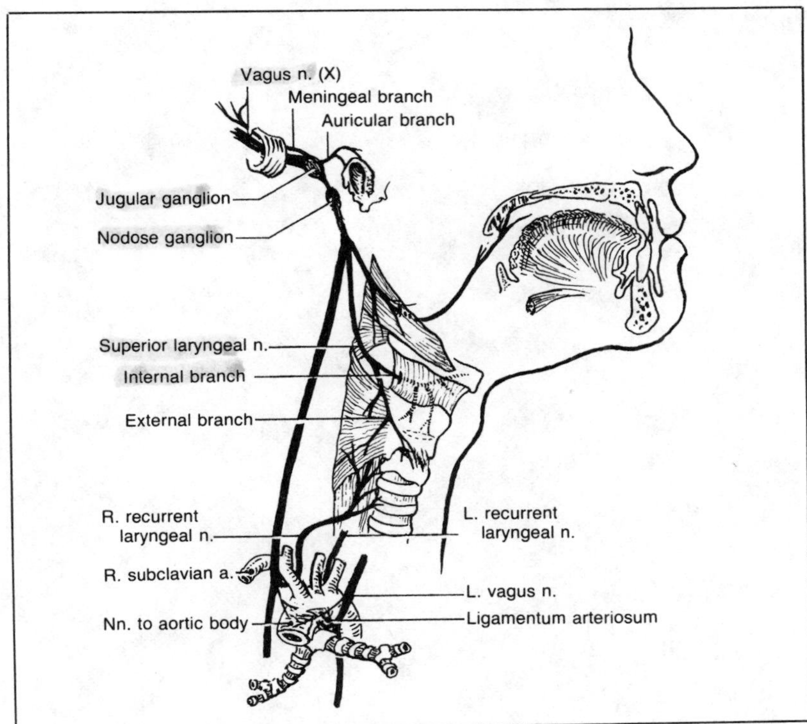

Figure 33-13. *The vagus nerve.*

a. By tensing the cords (cricothyroid muscle), the vibrations become strong and the pitch rises.
b. By effectively shortening the cords (vocalis muscle), the pitch also rises.
c. The vibrations are transmitted to the pharyngeal, oral, and nasal passages, as well as to the paranasal, thereby producing resonance.

2. Speech is accomplished by varying the resonance characteristics of the nasopharynx, oropharnyx, and oral cavity to produce the vowel sounds; interrupting the resonance generally produces the consonants.
a. The posterior tongue and the palatine musculature change the shape of the oropharynx and intermittently limit the access to the nasopharynx.
b. The mandibular musculature, the tongue, and the facial musculature about the lips change the shape of the oral cavity and thereby vary the resonance characteristics.
c. Finally, the tongue, in combination with the palate, teeth, and lips interrupt the resonance.

34
The Anterior Cervical Triangle

I. INTRODUCTION

A. The anterior triangle is bounded by the anterior border of the sternomastoid, the base of the mandible, and the ventral midline.

B. The cleavage lines (Langer's lines) run horizontally around the neck. A transverse incisional scar is almost invisible (see Fig. 2-1).

C. What little subcutaneous tissue there is contains the fibers of platysma.

1. This cutaneous muscle acts to flatten the skin of the neck, which can be useful in shaving.
2. It is innervated by the cervical branch of the facial nerve.

D. The deep cervical fascia was discussed in Chapter 27.

E. The muscles of the anterior triangle surround and support the cranial and cervical portions of the gut and are best understood by reference to the basic structure of the oral cavity, the pharynx, and the larynx (see Fig. 34-1).

1. A wall of musculature surrounds the floor of the mouth, the pharynx, the larynx, the esophagus, the trachea, the salivary glands, and the thyroid gland. These branchiomeric muscles are derived from primitive gill structures.
2. The U-shaped hyoid bone, the thyroid cartilage, and the cricoid cartilage support the anterior aspect of the pharynx and larynx.
3. Posteriorly the flat anterior wall of the vertebral column lies in a coronal plane between the somatic and visceral necks.
4. A neurovascular bundle, enclosed by the carotid sheath, runs cranially anterior to the cervical transverse processes on either side of the visceral structures (see Fig. 27-1).

II. THE HYOID BONE

A. This U-shaped bone consists of a median **body**, paired **lesser cornua (horns)** laterally, and paired **greater cornua (horns)** posteriorly.

B. The **stylohyoid ligament** runs between the styloid process and the lesser horn.

C. Muscles are attached along the outer surface of the greater horn and body.

1. Along the greater cornu, the middle constrictor attaches posteriorly; the hyoglossus, digastric and stylohyoid attach superiorly; the thyrohyoid attaches inferiorly.
2. Along the body, the geniohyoid and mylohyoid attach superiorly; the omohyoid and sternohyoid attach inferiorly.

III. THE MUSCULATURE OF THE ANTERIOR TRIANGLE

A. THE STERNOMASTOID MUSCLE passes anteriorly and medially from the mastoid process of the cranium to its sternal and clavicular attachments (Fig. 34-1).

1. It defines the anterior triangle.
2. It is innervated by the spinal accessory nerve and fibers from the cervical plexus (C2, C3).
3. It acts to rotate the head to the opposite side and flex the head to the same side.

B. Two superficial muscles, the digastric and the omohyoid, insert onto the hyoid bone. They divide the anterior triangle into digastric, mylohyoid, carotid, and muscular triangles.

1. The two bellies of digastric muscle radiate towards the superior corners of the anterior triangle from a central attachment on the hyoid bone and thus define the **digastric triangle** superiorly and the **mylohyoid triangle** anteriorly (see Fig. 33-4).
2. The omohyoid muscle divides the rest of the anterior triangle into a **carotid triangle** posteriorly and a **muscular triangle** anteriorly (see Fig. 27-8).

C. THE INFRAHYOID (STRAP OR RIBBON) MUSCLES (see Fig. 34-1)

1. These muscles are named for their appearance.
   a. The **sternohyoid muscle**, running between the hyoid bone and the sternum, is the longest and most superficial strap muscle.
   b. The short **thyrohyoid muscle** runs between the hyoid bone and the oblique line of the thyroid cartilage.
   c. The **sternothyroid muscle** continues the path of the thyrohyoid muscle between the oblique line on the lamina of the thyroid cartilage and the sternum.
   d. The **omohyoid muscle** (*omos*, *G.* shoulder) runs between the hyoid bone and the superior transverse ligament of the scapular notch.
      (1) The anterior belly runs upwards across the anterior triangle to the hyoid bone.
      (2) The posterior belly runs along the base of the posterior triangle, where it is bound to the clavicle by the investing layer of deep cervical fascia.
      (3) A tendinous intersection between the two bellies is attached to the investing fascia deep to the sternomastoid muscle and allows a directional change from vertical to nearly horizontal.
2. The infrahyoid muscles are innervated by the **ansa cervicalis** (C1–C3).
3. These raise, depress, and stabilize the hyoid bone and larynx during swallowing and phonation.
4. The sternohyoid and sternothyroid muscles bind the thyroid gland to the anterior surface of the neck.

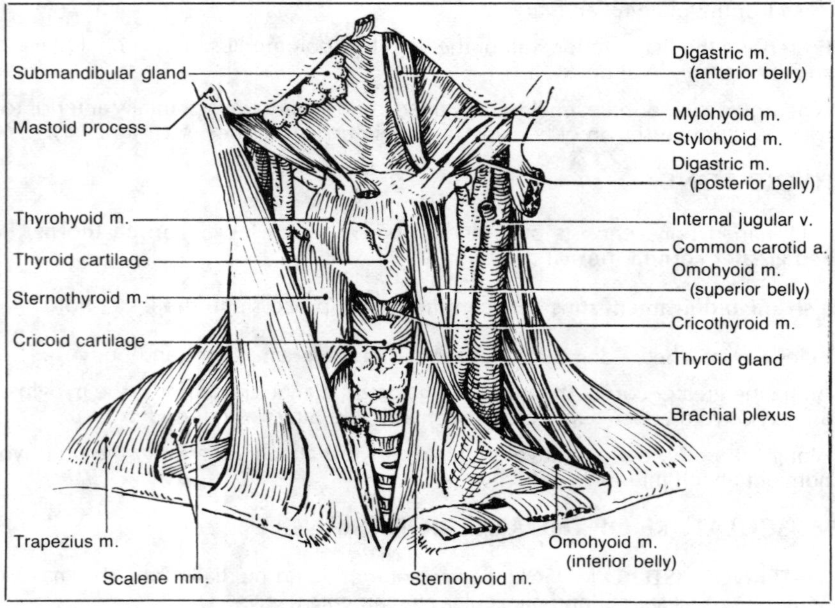

Figure 34-1. *The musculature of the anterior cervical triangle.*

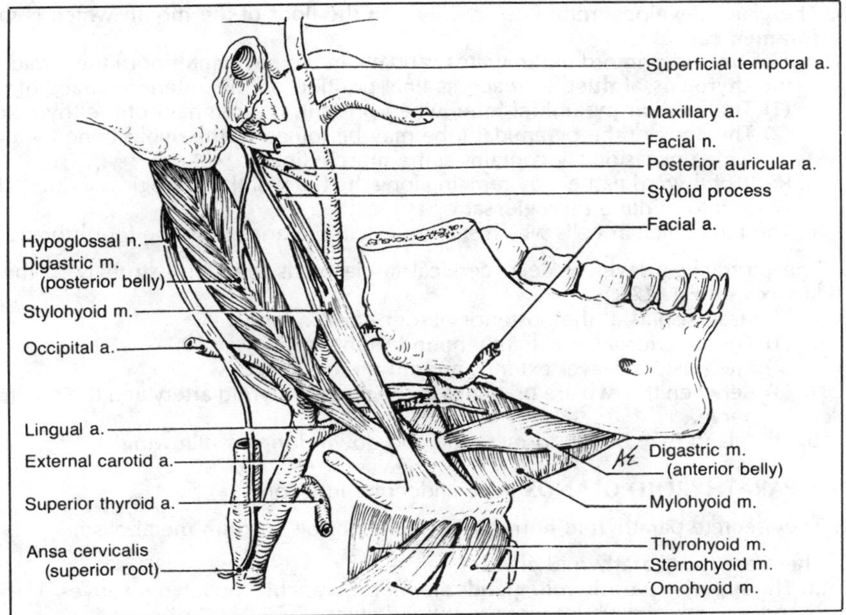

Figure 34-2. *The nerves and vessels of the submandibular triangle.*

D. THE SUPRAHYOID MUSCLES surround the floor of the mouth (Fig. 34-2).

1. The **digastric muscle**, as the name implies, consists of two bellies with an intervening tendon.
   a. The **posterior belly of the digastric muscle** originates from the mastoid process and passes through a trochlea on the superior surface of the hyoid bone near the lesser horn. This portion is innervated by the digastric branch of the facial nerve.
   b. The **anterior belly** continues from the trochlea to the digastric fossa of the mandible. This portion is innervated by the mylohyoid branch of the inferior alveolar nerve.

2. The **stylohyoid muscle** runs from the styloid process to the lesser horn of the hyoid bone, where its tendon of insertions splits for the passage of the digastric muscle. This muscle is innervated by the facial nerve.

3. The **mylohyoid muscle** arises from the mylohyoid line on the medial surface of the body of the mandible.
   a. Fibers run anteroinferiorly to meet the corresponding fibers from the opposite side in a midline fibrous raphe.
   b. This raphe is tethered to the hyoid bone.
   c. This muscle forms a sling beneath the floor of the mouth.

4. Three muscles pass through the interval between the superior and middle constrictors—the **styloglossus muscle**, the **stylopharyngeus muscle**, and the **hyoglossus muscle** (see Fig. 33-7). These muscles were discussed in Chapter 33 IV E.

5. The three **pharyngeal constrictors** that comprise the innermost layer were discussed in Chapter 33.

IV. THE THYROID GLAND AND PARATHYROID GLANDS

A. THE THYROID GLAND has endocrine functions.

1. It produces thyroxine and thyrocalcitonin, which regulate metabolism and calcium balance, respectively.

2. Each pear-shaped **lobe** extends around the trachea and esophagus to the carotid sheath (see Fig. 34-1).
   a. Superiorly it is limited by the insertion of the sternothyroid muscle onto the thyroid cartilage.
   b. Its base reaches the level of the fifth tracheal ring.

3. An **isthmus** connects the left and right lobes and lies anterior to the upper tracheal rings.

4. The gland develops from a diverticulum in the floor of the mouth which is marked by the **foramen caecum**.
   a. The thyroid primordium migrates caudally in the neck anterior to the hyoid bone, trailing the **thyroglossal duct**, to reach its final position on the anterior surface of the trachea.
      (1) The **median pyramidal lobe**, when present, is a remnant of the thyroglossal duct.
      (2) The apex of the pyramidal lobe may be joined to the hyoid bone by a fibrous band, which occasionally contains some muscle fibers.
   b. Rests of thyroid tissue may remain along the track of the thyroglossal duct. These may develop into midline **thyroglossal cysts**.
   c. The parafollicular cells which secrete calcitonin arise from the fourth branchial pouches.
5. The pretracheal layer of deep cervical fascia binds the gland strongly to the trachea and larynx (see Fig. 27-1).
   a. The fascia splits at the posterior border of each lobe.
      (1) The anterior layer is tightly bound to the larynx.
      (2) The posterior layer extends around the esophagus.
      (3) Between the two are branches of the inferior thyroid artery and the recurrent laryngeal nerve.
   b. The gland may be felt to move up and down during swallowing.

B. THE PARATHYROID GLANDS have endocrine function.

1. They secrete parathyroid hormone, which controls calcium metabolism.
2. There are four parathyroid glands.
   a. The **superior parathyroid glands** are fourth branchial pouch derivatives. They lie posterior to the apex of each lobe of the thyroid gland.
   b. The **inferior parathyroid glands** are third branchial pouch derivatives. They are related to the base of each lobe. However, they occasionally may follow the thymus (also a third pouch derivative) into the thorax, which is frustrating if the gland develops a tumor requiring excision.
   c. These glands are small and brownish-pink due to their vasularity. They can be difficult to find at operation.
      (1) Inadvertent total parathyroidectomy is followed by a gradual fall in blood calcium levels, which can cause fatal tetany.
      (2) These glands are extremely hardy and will continue to function if transplanted from an excised thyroid into the sternomastoid muscle.

C. THE THYROID VASCULATURE

1. The arterial supply to the thyroid and parathyroid glands is by the superior thyroid artery (a branch of the external carotid artery), the inferior thyroid artery (a branch of the thyrocervical trunk of the subclavian artery), and occasionally a median thyroidea ima artery from the brachiocephalic artery.
2. A venous plexus on the surface of the gland and over the trachea drains into superior, middle, and inferior thyroid veins.

D. CLINICAL CONSIDERATIONS

1. An enlarged thyroid gland may extend superiorly in the carotid triangle and down into the thoracic inlet. Pressure of the trachea at the thoracic inlet may cause shortness of breath and a wheeze, especially when the patient is asked to raise his arms above his head.
2. For a tracheostomy the surgeon may identify and ligate bleeding vessels or even divide the thyroid isthmus, placing the tracheal stoma just above the jugular notch. In an emergency tracheostomy is frequently accomplished through the cricothyroid membrane, a relatively avascular region.

V. THE VASCULATURE OF THE ANTERIOR NECK

A. THE CAROTID SHEATH contains the neurovascular bundle of the neck (see Fig. 27-1).

1. For the most part, the carotid sheath is under cover of the sternomastoid muscle.
2. It lies anterior to the transverse processes of the cervical vertebrae.
   a. Arterial pulsations can be felt by gently pressing the vessels against the unyielding bone. Further pressure occludes the artery. It is therefore unwise to feel for the carotid pulses bilaterally, as one would do anywhere else in the body when evaluating blood flow.

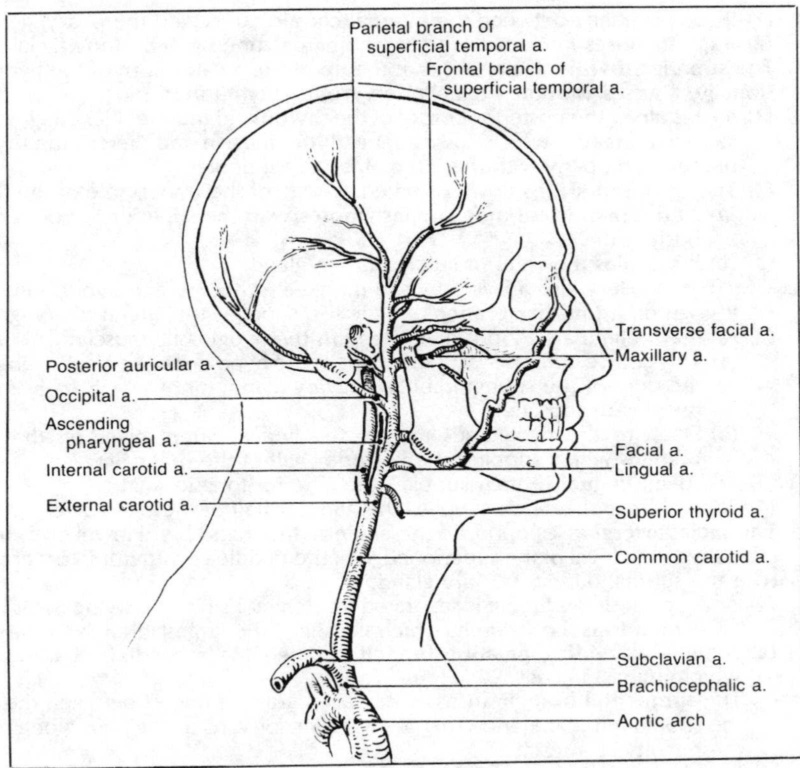

Figure 34-3. *The branches of the carotid arteries in the neck.*

b. In the elderly, atheromatous plaques may be dislodged upon carotid palpation. It is wise, therefore, to take the carotid pulse on the right side since a stroke induced in the non-dominant cerebral hemisphere would be less devastating.

B. THE ARTERIAL SUPPLY (Fig. 34-3)

1. The **carotid arteries** supply most of the head and neck.

a. The **left common carotid artery** arises from the aortic arch.

b. The **right common carotid artery** arises with the right subclavian artery from the brachiocephalic artery, which is the remnant of the artery of the right fourth branchial arch.

c. Each common carotid artery bifurcates at the level of the superior edge of the thyroid cartilage (C4).

d. Two receptors at the bifurcation provide feedback to the vasomotor and respiratory centers concerning the pressure and oxygenation of the blood en route to the brain.

(1) The **carotid sinus** is a fusiform dilatation that functions as a baroreceptor.

(a) The afferent limb of the inhibitory **carotid reflex** is mediated by the carotid branch of the glossopharyngeal nerve (CN IX). This branch is part of the post-trematic division of the nerve of the third branchial arch (see Fig. 31-1).

(b) The efferent limb of the inhibitory **carotid reflex** is mediated by the vagus nerve (CN X).

(c) Supraventricular tachycardia, is susceptible to vagal effects and may be converted to normal sinus rhythm by carotid sinus massage.

(2) The **carotid body** is very vascular and lies between the bifurcation.

(a) It is a chemoreceptor sensitive to blood oxygenation and to a lesser extent to pH and carbon dioxide partial pressure (P_{CO_2}).

(b) It occasionally gives rise to tumors that are difficult to excise because of their location.

2. The **external carotid artery** supplies most of the face and visceral neck with eight major branches (see Fig. 34-3).

a. The **ascending pharyngeal artery** is a long slender branch, which arises close to the bifurcation of the common carotid artery.

(1) It runs cranially between the internal carotid artery and the pharyngeal constrictors.
(2) It anastamoses freely with the ascending palatine branch of the facial artery.

b. The **superior thyroid artery** arises at the tip of the greater horn of the hyoid bone. Occasionally, it arises with the lingual artery from a common trunk.
(1) It runs along the posterior border of the thyrohyoid muscle. Here it gives off an internal laryngeal branch, which passes beneath the muscle and pierces the thyrohyoid membrane in company with the internal laryngeal nerve.
(2) The main trunk runs down the medial edge of the upper pole of the thyroid gland.
(a) At the thyroid isthmus, it anastamoses with the superior thyroid artery of the opposite side.
(b) It supplies the anterior surface of the gland.

c. The **lingual artery** also arises opposite the greater horn of the hyoid bone.
(1) It gives off a **tonsillar branch**, which also supplies the lateral pharyngeal walls.
(2) After a preliminary loop it dives deep to the hyoglossus muscle.
(a) This course contrasts with the paths of the hypoglossal nerve, the lingual nerve, and the duct of the submandibular salivary gland, all of which pass superficial to the hyoglossus muscle.
(b) Deep to the hyoglossus muscle the lingual artery gives off the **dorsal lingual branch**, which supplies the posterior half of the oral cavity.
(3) The **deep lingual branch** supplies the anterior tongue.
(4) The **sublingual branch** supplies the anterior half of the oral cavity.

d. The **facial artery** arises opposite the angle of the mandible. However, it has to continue cranially to the level of the superior edge of the middle constrictor in order to pass laterally over the submandibular salivary gland.
(1) At the apex of this loop it is separated from the tonsil by the styloglossus muscle. Here it gives off a **tonsillar branch**, which is usually the largest branch to the tonsil.
(2) A small **ascending palatine branch** parallels the ascending pharyngeal artery with which it anastomoses.
(3) The **submental branch** arises as the facial artery emerges between the mandible and the submandibular gland. This branch runs forward to the chin along the insertion of the mylohyoid muscle.
(4) In a sinuous course the facial artery continues over the ramus of the mandible to enter the face.
(a) Here the facial pulse is palpable.
(b) It then gives off **inferior** and **superior labial branches.**
(5) It continues lateral to the nose as the **angular artery.**

e. The **occipital artery** arises from the posterior aspect of the external carotid artery.
(1) Deep to the origin of the digastric muscle, the occipital artery continues posteriorly, grooving the medial surface of the mastoid process, to emerge at the apex of the posterior cervical triangle.
(2) It gives off two branches to the sternomastoid. The upper branch is associated with the spinal accessory nerve; the lower with the hypoglossal nerve.

f. The **posterior auricular artery** also arises posteriorly from the external carotid artery.
(1) It ascends between the superior surface of the posterior belly of the digastric muscle and the parotid gland.
(2) It also grooves the base of the skull as it passes between the mastoid process and the auricle.

g. The **superficial temporal artery** ascends anterior to the external auditory meatus and supplies the scalp.

h. The **(internal) maxillary artery** supplies the deep face as well as portions of the nasal and oral cavities. It was discussed in Chapter 32.

3. The **internal carotid artery** ascends to the **carotid canal** in the base of the cranium without giving off any branches in the neck.
a. The styloid process and its associated muscles intervene between the internal and external carotid arteries.
b. Its **ophthalmic branch** supplies the upper portion of the face.

C. THE VENOUS RETURN (Fig. 34-4)

1. The **internal jugular vein** exits the skull in company with the glossopharyngeal, vagus, and spinal accessory nerves.
a. The **jugular foramen** is posterior to the carotid canal.
b. As it descends through the neck in the carotid sheath, the internal jugular vein winds laterally.
c. It joins the **subclavian vein** by an acute angle behind the sternoclavicular joint to form the **brachiocephalic vein**.

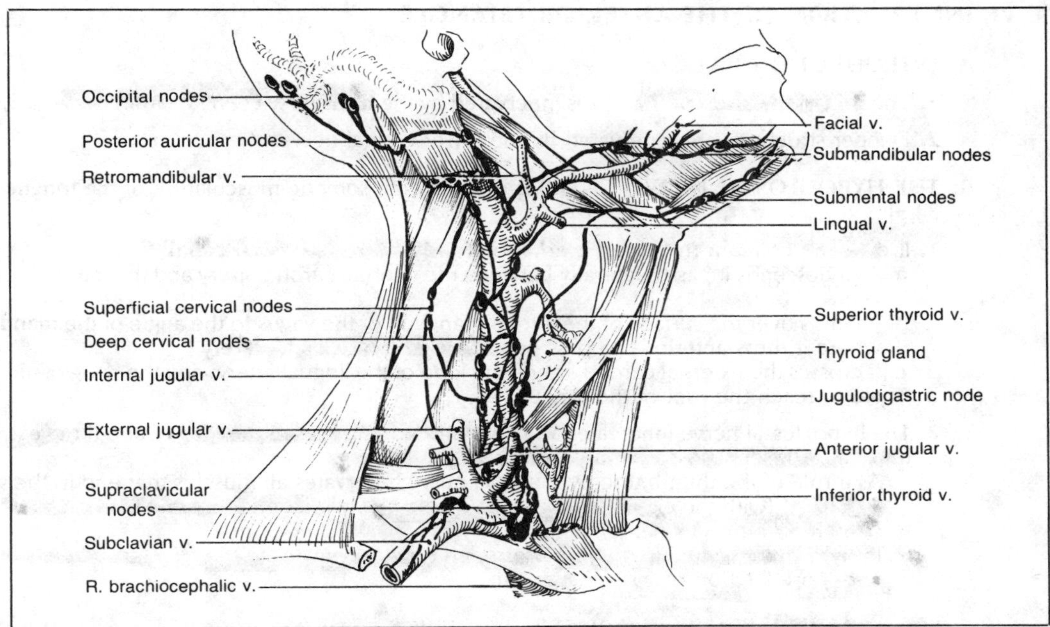

Figure 34-4. *The venous and lymphatic drainage of the head and neck.*

d. In the root of the neck, the subclavian and jugular veins lie anterior to the corresponding arteries, facilitating emergency venous access.

2. The **external jugular vein** is formed at the angle of the jaw by the union of the posterior division of the **retromandibular vein** and the **posterior auricular vein** (see Fig. 28-8).

a. It tends to be superficial and variable.

b. It receives four tributaries.

(1) The suprascapular vein accompanies the like-named artery.

(2) The superficial cervical vein accompanies the like-named artery.

(3) The anterior jugular vein comes from the anterior triangle.

(4) The posterior jugular vein comes from the apex of the posterior triangle.

c. It passes in the superficial fascia of the posterior triangle to the middle of the clavicle, where it joins the subclavian vein.

d. The external jugular vein may be demonstrated by obstructing its drainage with light finger pressure or by having the patient perform Valsalva's maneuver.

e. Careful observation of the external jugular will usually reveal venous pressure waves. The level of the column of blood in the external jugular veins is a useful indication of right atrial pressure.

D. LYMPHATIC DRAINAGE OF THE ANTERIOR NECK (see Fig. 34-4)

1. The lymph from the face and anterior neck drains by superficial cervical nodes that parallel the external jugular vein and by deep cervical nodes that parallel the internal jugular vein.

a. The **occipital nodes**, the **retroauricular nodes**, the **parotid nodes**, and some **anterior cervical nodes** drain into the **external jugular (superficial) nodes**.

b. The lymphatics of the superficial face and deep structures generally drain via the **submandibular**, **submental**, and **jugulodigastric nodes** into the **internal jugular (deep) nodes**.

(1) Just below the posterior belly of the digastric muscle, the jugulodigastric node drains the tonsillar region and is valuable in the diagnosis of pharyngeal inflammation.

(2) Some anterior cervical nodes also drain into the deep lymphatic chain.

c. In the root of the neck, the deep lymphatics form the **jugular trunk**.

(1) On the right side the jugular trunk ends at the junction of the internal jugular vein and the subclavian vein.

(2) On the left side it joins the thoracic duct.

2. The **thoracic duct** enters the neck behind the left subclavian artery.

a. From the thorax, it ascends into the neck to the level of C6, where it crosses anterior to the vertebral artery and the thyrocervical trunk.

b. It drains into either the jugular or the subclavian vein.

VI. INNERVATION OF THE ANTERIOR TRIANGLE

A. INTRODUCTION

1. The skin of the anterior triangle is innervated by the **transverse cervical (colli) nerve** (C2, C3).
2. Deeper structures are innervated by the lower cranial nerves.

B. THE HYPOGLOSSAL NERVE (CN XII) is motor to the somatic musculature of the tongue (Fig. 34-5)

1. It exits the cranium through the **anterior condylar (hypoglossal) canal.**
   a. As it descends it passes laterally between the internal carotid artery and the internal jugular vein.
   b. It winds over the surface of the inferior ganglion of the vagus to the angle of the mandible, where it turns anteriorly to pass superficial to the occipital artery.
   c. It crosses the external carotid artery, the loop of the lingual artery, and the hyoglossus muscle to reach the root of the tongue.
2. The hypoglossal nerve innervates all of the intrinsic muscles of the tongue and three extrinsic muscles—the styloglossus, hyoglossus, and genioglossus.
   a. As a rule of the thumb, the hypoglossal nerve innervates all muscles that end in the suffix "glossus" (with the exception of the palatoglossus, which is innervated by the pharyngeal branch of the vagus nerve).
   b. In hypoglossal nerve injury, the protruded tongue deviates to the affected side due to the unopposed action of the genioglossus.

C. THE ANSA CERVICALIS innervates the strap muscles.

1. Fibers from the first cevical nerve (C1) run to the hypoglossal nerve via a communicating branch (see Fig. 34-5).
   a. Some of these fibers accompany the hypoglossal nerve to be distributed to the thyrohyoid muscle and the geniohyoid muscle.
   b. Other fibers leave the company of the hypoglossal nerve as the **superior ramus** of the **ansa**

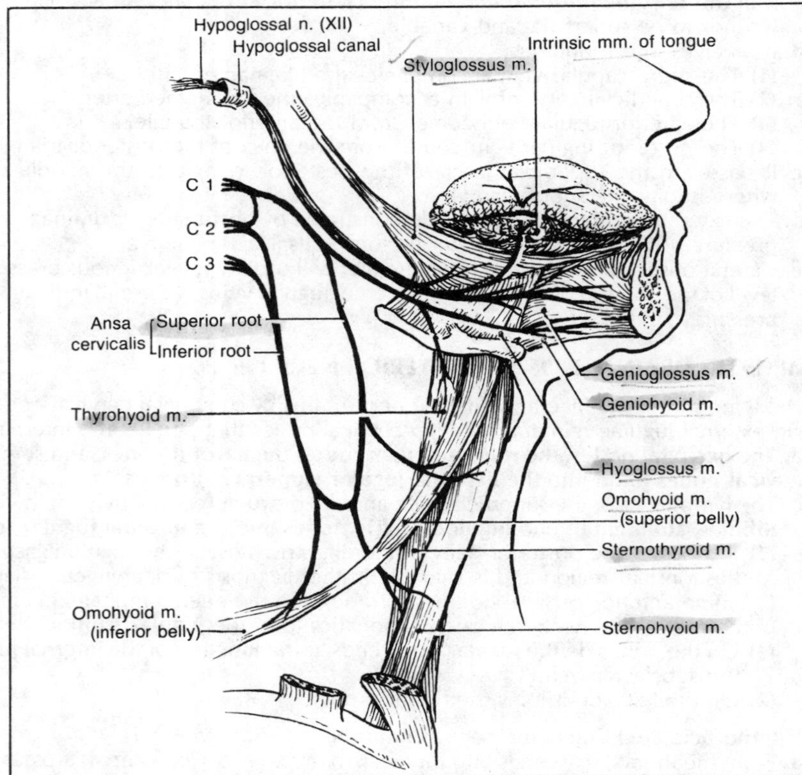

Figure 34-5. *The hypoglossal nerve.*

cervicalis (whence the former name **descendens hypoglossi**) to innervate the sternohyoid and sternothyroid muscles.

2. The **inferior ramus** of the **ansa cervicalis** is formed by the union of branches from C2–C3 (whence the former name **descendens cervicalis**) and innervates the omohyoid muscle and a portion of the sternothyroid muscle.
3. The superior and inferior roots anastomose to form a dependent loop on the surface of the carotid sheath.

D. CERVICAL SYMPATHETIC NERVES

1. The cervical sympathetic chain runs cranially in the neck posterior to the carotid sheath (see Fig. 27-1).
   a. It is thinner and more medially placed than the vagus nerve.
   b. Preganglionic fibers run up the chain from the upper thoracic levels to synapse in the superior, middle, and inferior cervical ganglia.
2. The large **superior cervical ganglion** sits immediately below the skull.
   a. **Gray rami communicantes** pass postganglionic fibers to the upper four cervical nerves carrying secretomotor, vasomotor, and pilomotor fibers.
   b. Fibers to the head form a perivascular plexus around the internal carotid artery.
      (1) They innervate, among other things, the pupillodilators of the eyes, the palpebral muscle (Muller's), and the orbitalis muscle (Muller's), as well as the sweat glands and the vasoconstrictors.
      (2) Lesions of the cervical sympathetic chain therefore produce meiosis, ptosis, enophthalamos, anhydrosis, and flushing (Horner's syndrome).

VII. SUMMARY OF THE CRANIAL NERVES (Table 34-1)

A. SPECIAL SOMATIC AFFERENT (SSA)

1. The **olfactory nerve** (CN I):
   a. Mediates smell.
   b. Reaches the nasal olfactory mucosa through the cribriform plate of the ethmoid bone.
2. The **optic nerve** (CN II):
   a. Mediates vision.
   b. Reaches the eyeball through the optic foramen.
3. The **vestibulocochlear nerve** (CN VIII):
   a. Mediates balance and audition.
   b. Reaches the inner ear through the internal auditory meatus.

B. GENERAL SOMATIC EFFERENT (GSE)

1. The **oculomotor nerve** (CN III):
   a. This nerve reaches the orbit through the superior orbital fissure.
   b. The somatic motor component from the oculomotor nucleus innervates:
      (1) The levator palpebrae superioris muscle.
      (2) The superior rectus muscle.
      (3) The medial rectus muscle.
      (4) The inferior rectus muscle.
      (5) The inferior oblique muscle.
   c. The parasympathetic component originates in the accessory oculomotor nucleus.
      (1) Presynaptic fibers leave the ciliary root of the inferior division.
      (2) Synapse occurs in the ciliary ganglion.
      (3) Postsynaptic fibers course in the short ciliary nerves to the eyeball.
      (4) The parasympathetic component mediates pupillary constriction and accommodation.
2. The **trochlear nerve** (CN IV):
   a. Originates in the trochlear nucleus.
   b. Enters the orbit through the superior orbital fissure.
   c. Innervates the superior oblique muscle only.
3. The **abducens nerve** (CN VI):
   a. Originates in the abducens nucleus.
   b. Enters the orbit through the superior orbital fissure.
   c. Innervates the lateral rectus muscle only.

Table 34-1. Summary of the Cranial Nerves

Functional Component	Branches	Cell Bodies	Course and Cranial Foramen	Distribution	Associated Nucleus
		The Olfactory Nerve (CN I)			
Afferent					
SSA	. . .	Olfactory epithelium	Cribriform plate	Olfactory epithelium	Olfactory bulb and tract
		The Optic Nerve (CN II)			
Afferent					
SSA	. . .	Bipolar cells of retina	Optic foramen	Rods and cones of nasal and temporal retina	Ganglion cells of retina to lateral geniculate body to occipital lobe
		The Oculomotor Nerve (CN III)			
Efferent					
GSE	Superior	Oculomotor nucleus	Superior orbital fissure	Levator palpebrae, superioris muscle, superior rectus muscle	. . .
	Inferior	Oculomotor nucleus	Superior orbital fissure	Medial rectus muscle, inferior rectus muscle, inferior oblique muscle	. . .
GVE	Ciliary	Accessory oculomotor nucleus, ciliary ganglion	Superior orbital fissure, motor root, ciliary ganglion, short ciliary nerves	Ciliary muscle (accommodation), iris (constriction)	. . .
		The Trochlear Nerve (CN IV)			
Efferent					
GSE	. . .	Trochlear nucleus	Superior orbital fissure	Superior oblique muscle	. . .
		The Trigeminal Nerve (CN V)			
Afferent					
GSA	Ophthalmic: Lacrimal, Nasociliary, Supratrochlear, Supraorbital	Semilunar ganglion	Superior orbital fissure	Forehead, mucous membranes of nasal cavity and sinuses, conjunctiva (corneal blink reflex)	Trigeminal sensory nucleus
	Maxillary: Infraorbital, Nasopalatine, Descending palatine, Superior alveolar, Zygomaticofacial, Zygomaticotemporal	Semilunar ganglion	Foramen rotundum	Midface, upper teeth, mucous membranes of nasal cavity, maxillary sinus and hard palate (sneeze reflex)	Trigeminal sensory nucleus

Table 34-1. Continued

Functional Component	Branches	Cell Bodies	Course and Cranial Foramen	Distribution	Associated Nucleus
	Mandibular: Inferior alveolar lingual Mental Mylohyoid Buccal Auriculo-temporal	Semilunar ganglion	Foramen ovale	Mandible, lower teeth, cheeks, anterior two-thirds of tongue (jaw-jerk reflex)	Trigeminal sensory nucleus
Efferent SVE	Motor root of mandibular n.: Muscular Mylohyoid	Motor nucleus of CN V	Foramen ovale	Muscles of mastication, tensor tympani muscle, tensor veli palatini muscle, mylohyoid muscle, anterior belly of the digastric muscle (jaw-jerk reflex)	. . .
			The Abducens Nerve (CN VI)		
Efferent GSE	. . .	Abducens nucleus	Superior orbital fissure	Lateral rectus muscle	. . .
			The Facial Nerve (CN VII)		
Afferent GSA	Auricular	Geniculate ganglion	Facial canal	External ear	Spinal nucleus of CN V
GVA	Greater superficial petrosal	Geniculate ganglion	Hiatus of the facial canal	Nasal mucosa	Nucleus solitarius
SVA	Chorda tympani via lingual (V)	Geniculate ganglion	Petrotympanic fissure, middle ear, facial canal, internal auditory meatus	Taste, anterior two-thirds of tongue	Nucleus solitarius
Efferent GVE	Greater superficial petrosal to nerve of pterygoid canal	Superior salivatory nucleus, pterygopalatine ganglion	Hiatus of the facial canal, pterygoid canal	Secretomotor, lacrimal, nasal, and palatine glands	. . .
	Chorda tympani	Superior salivatory nucleus, submandibular ganglion	Facial canal, middle ear, petrotympanic fissure, lingual nerve (V)	Secretomotor, submandibular, and intrinstic lingual glands	. . .
SVE	Temporal Zygomatic Buccal Mandibular Cervical	Facial nucleus	Stylomastoid foramen	Muscles of facial expression, stapedius muscle, posterior belly of digastric muscle, stylohyoid muscle (corneal blink reflex)	. . .

Table 34-1. Continued

Functional Component	Branches	Cell Bodies	Course and Cranial Foramen	Distribution	Associated Nucleus
			The Vestibulocochlear Nerve (CN VIII)		
Afferent					
SSA	Vestibular	Vestibular ganglion thelium	Internal auditory meatus	Utricle, saccule, and semicircular canals	Vestibular nuclei
	Cochlear	Spiral ganglion	Internal auditory meatus	Spiral organ	Cochlea nuclei to medial geniculate body to temporal lobe
			The Glossopharyngeal Nerve (CN IX)		
Afferent					
GSA	Tympanic	Jugular ganglion (IX)	Tympanic canal	External auditory meatus and middle ear	Spinal nucleus of CN V
GVA	Lingual	Petrosal ganglion	Jugular foramen	Sensation, posterior third of tongue	Nucleus solitarius
	Pharyngeal	Petrosal ganglion	Jugular foramen	Nasopharynx, oropharynx (gag reflex)	Nucleus solitarius
	Carotid	Petrosal ganglion	Jugular foramen	Carotid sinus and body (baroreflex and chemoreflex)	Nucleus solitarius
SVA	Lingual	Petrosal ganglion	Jugular foramen	Taste, posterior third of tongue	Nucleus solitarius
Efferent					
GVE	Tympanic and lesser superficial petrosal	Inferior salivatory nucleus, otic ganglion	Tympanic canal, middle ear, lesser superficial petrosal, and foramen ovale	Secretomotor, parotid gland	. . .
SVE	Stylopharyngeal	Nucleus ambiguus	Jugular foramen	Stylopharyngeus muscle	. . .
			The Vagus Nerve (CN X)		
Afferent					
GSA	Auricular	Jugular ganglion (X)	Tympanic canal	External auditory meatus	Spinal nucleus of CN V
GVA	Aortic plexus	Nodose ganglion	Jugular foramen	Aortic body (chemoreceptor reflex)	Nucleus solitarius
	Superior laryngeal: Internal	Nodose ganglion	Jugular foramen	Superior larynx (cough reflex)	Nucleus solitarius
	Inferior laryngeal	Nodose ganglion	Jugular foramen	Inferior larynx	Nucleus solitarius
	Vagal	Nodose ganglion	Jugular foramen	Thoracic and upper abdominal viscera	Nucleus solitatius

Table 34-1. Continued

Functional Component	Branches	Cell Bodies	Course and Cranial Foramen	Distribution	Associated Nucleus
SVA	Superior laryngeal: Internal	Nodose ganglion	Jugular foramen	Taste from epiglottis	Nucleus solitarius
Efferent					
GVE	Vagal	Dorsal motor nucleus of CN X, distal ganglion	Jugular foramen	Visceral smooth muscle and gland control	. . .
SVE	Pharyngeal	Nucleus ambiguus	Jugular foramen	Palate muscles (except tensor veli palatini) and pharyngeal muscles (except stylopharyngeus)	. . .
	Superior laryngeal: External	Nucleus ambiguus	Jugular foramen	Cricothyroid muscle	. . .
	Contributes to inferior laryngeal	Nucleus ambiguus	Jugular foramen	Laryngeal muscles (except cricothyroid muscle)	. . .
The Spinal Accessory Nerve (CN XI)					
Efferent					
SVE	Contributes to inferior laryngeal	Nucleus ambiguus	Jugular foramen	Laryngeal muscles (except cricothyroid muscle)	. . .
SSE	Sternomastoid and trapezius	Lateral column of upper cervical cord	Vertebral canal, foramen magnum, jugular foramen	Sternomastoid muscle and trapezius muscle	. . .
The Hypoglossal Nerve (CN XII)					
Efferent					
GSE	. . .	Hypoglossal nucleus	Hypoglossal (anterior condylar) canal	Intrinsic tongue muscles, genioglossus muscle, hyoglossus muscle, styloglossus muscle	. . .

4. The **hypoglossal nerve** (CN XII):
   a. Originates in the hypoglossal nucleus.
   b. Leaves the cranium through the anterior condylar (hypoglossal) canal.
   c. Innervates:
      (1) The intrinsic muscles of the tongue.
      (2) The styloglossus muscle.
      (3) The hyoglossus muscle.
      (4) The genioglossus muscle.
   d. Causes protrusion of the tongue toward the injured side in the presence of a lesion.

C. MIXED BRANCHIOMERIC NERVES

1. The **trigeminal nerve** (CN V).

a. The **ophthalmic division** (CN V_1):
  (1) Enters the orbit through the superior orbital fissure.
  (2) Is sensory to the forehead and orbital area with cell bodies in the trigeminal (semilunar, gasserian) ganglion.
  (3) Has the following major branches.
    (a) Lacrimal.
    (b) Frontal (which bifurcates into supraorbital and supratrochlear).
    (c) Nasociliary.
  (4) Comprises the afferent limb of the corneal blink reflex.

b. The **maxillary division** (CN V_2):
  (1) Exits the cranium through the foramen rotundum and passes through the pterygopalatine fossa.
  (2) Is sensory to the mid-face, nose, palate, and maxillary teeth with cell bodies in the trigeminal (semilunar, gasserian) ganglion.
  (3) Has the following major branches.
    (a) Infraorbital.
    (b) Greater palatine.
    (c) Nasal.
    (d) Anterior, middle, and posterior superior alveolar.
  (4) Mediates the afferent limb of the sneeze reflex.

c. The **mandibular division** (CN V_3):
  (1) Leaves the cranium through the foramen ovale.
  (2) Is branchiomotor to the following muscles.
    (a) Masseter.
    (b) Temporalis.
    (c) Lateral pterygoid.
    (d) Medial pterygoid.
    (e) Tensor tympani.
    (f) Tensor veli palatini.
    (g) Mylohyoid.
    (h) Anterior belly of the digastric.
  (3) Is sensory to the jaw, mandibular teeth, and anterior two-thirds of the tongue with cell bodies in the trigeminal ganglion.
  (4) Mediates the jaw-jerk reflex.

2. The **facial nerve** (CN VII):

a. Leaves the cranium through the stylomastoid foramen.

b. Is branchiomotor to the following muscles.
  (1) The muscles of facial expression, mediating the efferent limb of the corneal blink reflex.
  (2) The stapedius.
  (3) The stylohyoid.
  (4) The posterior belly of the digastric.

c. Has a parasympathetic component that arises from the superior salivatory nucleus and contributes to the nervus intermedius.
  (1) Lacrimation and nasopalatine secretion.
    (a) Presynaptic fibers leave the facial nerve at the genu to course in the greater superficial petrosal nerve and then the nerve of the pterygoid canal.
    (b) Synapse occurs in the pterygopalatine ganglion.
    (c) Postsynaptic fibers course along the maxillary nerve and the zygomaticofacial branch to reach the lacrimal branch of the ophthalmic nerve.
    (d) Other postsynaptic fibers join the nasopalatine branch of the maxillary nerve and enter the sphenopalatine foramen to innervate the glands of the nose.
    (e) Still other postsynaptic fibers join the descending palatine branch of the maxillary nerve and pass through the palatine canal to innervate the glands of the palate.
  (2) Salivation.
    (a) Presynaptic fibers leave the facial nerve in the vicinity of the stylomastoid foramen as the chorda tympani, which passes through the middle ear and joins the lingual nerve.
    (b) Small rami communicantes convey these fibers to the submandibular ganglion, where they synapse.
    (c) Postsynaptic fibers rejoin the lingual nerve to reach the submandibular, sublingual, and anterior lingual glands.

d. Has sensory fibers that mediate taste from the anterior two-thirds of the tongue via the chorda tympani. The cell bodies of these fibers are located in the geniculate ganglion.

3. The **glossopharyngeal nerve** (CN IX):
   a. Leaves the cranium through the jugular foramen.
   b. Is branchiomotor to the stylopharyngeus muscle only.
   c. Conveys parasympathetic secretomotor fibers to the parotid gland.
      (1) Fibers from the inferior salivatory nucleus, as well as general sensory fibers, leave the nerve in the jugular canal to form the tympanic nerve.
      (2) From the tympanic nerve, fibers run in the lesser superficial petrosal nerve to the otic ganglion.
      (3) The postsynaptic fibers pass along the auriculotemporal branch of the mandibular nerve to the parotid gland.
   d. Has sensory cell bodies in the superior and inferior glossopharyngeal ganglia.
      (1) General sensation from the external auditory meatus.
      (2) Visceral sensation from the posterior third of the tongue and pharynx, mediating the afferent limb of the gag reflex.
      (3) Taste from the posterior third of the tongue.
      (4) Reflex afferents from the carotid body and sinus.

4. The **vagus nerve** (CN X):
   a. Leaves the cranium via the jugular foramen.
   b. Is branchiomotor.
      (1) The pharyngeal branch is motor to the following muscles.
         (a) Levator veli palatini.
         (b) Uvulae.
         (c) Palatopharyngeus.
         (d) Palatoglossus.
         (e) Salpingopharyngeus.
         (f) Superior, middle, and inferior pharyngeal constrictors.
      (2) The superior laryngeal branch is motor to the cricothyroid muscle.
      (3) The inferior laryngeal branch is motor to all of the remaining laryngeal muscles.
   c. Provides parasympathetic innervation to the thoracic and much of the abdominal viscera.
   d. Has sensory cell bodies in the superior and inferior vagal ganglia.
      (1) General sensation to the external auditory meatus.
      (2) Visceral sensation to the larynx and trachea. Mediates the afferent limb of the cough reflex.
      (3) Taste fibers from the epiglottis.

5. The **spinal accessory nerve** (CN XI):
   a. Leaves the cranium through the jugular foramen.
   b. Is branchiomotor to the sternomastoid and trapezius.

Part IX Facial Cranium and Visceral Neck

STUDY QUESTIONS

Directions: Each question below contains five suggested answers. Choose the **one best** response to each question.

1. The buccal area has a dual nerve supply. Which of the following statements concerning the two buccal nerves is true?

(A) The fibers of one nerve are associated with the geniculate ganglion, while those of the other are not associated with any ganglion
(B) The fibers of one nerve are associated with the submandibular ganglion and those of the other with the pterygopalatine ganglion
(C) The fibers of one nerve are associated with the trigeminal ganglion and those of the other with the geniculate ganglion
(D) The fibers of one nerve are associated with trigeminal ganglion and those of the other with the pterygopalatine ganglion
(E) The fibers of one nerve are associated with the trigeminal ganglion, while those of the other are not associated with a ganglion

2. The epiglottis is described by all of the following statements EXCEPT that

(A) during swallowing, the epiglottis becomes horizontal to close the laryngeal aditus
(B) it contains taste buds innervated by the vagus nerve
(C) it is connected to the root of the tongue
(D) the piriform recesses lie on either side of it in the hypopharynx
(E) two lateral aryepiglottic folds connect it to the laryngeal cartilages

3. The tissues of the hard and soft palates receive an autonomic neural innervation that is described by all of the following statements EXCEPT

(A) preganglionic parasympathetic fibers travel along the greater superficial petrosal nerve, a branch of VII
(B) postganglionic sympathetic fibers arrive via the deep petrosal nerve
(C) the lesser superficial petrosal nerve contributes to the nerve of the pterygoid canal
(D) the greater and lesser palatine nerves pass through the pterygopalatine canal
(E) the anterior portion of the hard palate is supplied by the nasopalatine nerve, which enters the nose through the sphenopalatine foramen

4. Anesthesia of the maxillary premolar teeth can be effected by infiltrating the nerve as it leaves the

(A) foramen rotundum
(B) greater palatine foramen
(C) incisive canal
(D) infraorbital foramen
(E) lesser palatine foramen

Directions: Each question below contains four suggested answers of which **one or more** is correct. Choose the answer

A if **1, 2, and 3** are correct
B if **1 and 3** are correct
C if **2 and 4** are correct
D if **4** is correct
E if **1, 2, 3, and 4** are correct

5. When cell bodies in the left semilunar ganglion are damaged by a viral infection, the signs and symptoms include

(1) loss of sensation of pain from the anterior two-thirds of the tongue
(2) loss of taste from the anterior two-thirds of the tongue
(3) inability to elicit a corneal blink reflex from the left side
(4) left facial paralysis

6. Which of the following signs and symptoms are produced by a fracture passing through the left stylomastoid foramen and injuring the contained nerve?

(1) Hyperacusis in the left ear
(2) Loss of lacrimation on the left side
(3) Loss of left parotid gland secretion
(4) Facial palsy

7. The articular disk, or meniscus, of the temporomandibular joint is characterized by

(1) a fibrocartilage composition
(2) an attachment from the medial pterygoid muscle
(3) a downward and forward movement during protrusion of the mandible
(4) separation of the medial and lateral joint compartments

8. An infection that spreads into the pterygopalatine fossa from the pterygoid venous plexus may subsequently track into which of the following cavities?

(1) Nasal cavity
(2) Middle cranial fossa
(3) Oral cavity
(4) Orbital cavity

9. Symptoms that result from the administration of local anesthetic into the greater palatine canal far enough to reach the ganglion situated superior to the canal include

(1) dry nasal mucosa from the loss of secretion of the nasal glands
(2) dry mouth from loss of secretion of the parotid gland
(3) dry eyes from loss of secretion of the lacrimal glands
(4) dry mouth from loss of secretion of submandibular and sublingual glands

10. Which of the following structures drain into the middle nasal meatus?

(1) Frontal sinus
(2) Posterior ethmoid sinuses
(3) Maxillary sinus
(4) Nasolacrimal duct

11. Sinuses that drain by gravity when the head is erect include the

(1) frontal sinus
(2) sphenoid sinus
(3) ethmoid sinuses
(4) maxillary sinus

12. Muscles that remain functional when the hypoglossal nerve deteriorates from a tumor in the brain stem include

(1) the genioglossus
(2) the palatoglossus
(3) the hyoglossus
(4) the geniohyoid

13. The palatine tonsil receives blood supply from the

(1) facial artery
(2) ascending pharyngeal artery
(3) maxillary artery
(4) lingual artery

14. Which of the following statements about the carotid sinus and carotid body are correct?

(1) The body functions to regulate cardiac and respiratory rates
(2) They are found in the bifurcation of the common carotid artery
(3) The sinus functions to regulate cardiac and respiratory rates
(4) The carotid sinus is a pressure receptor and the carotid body a chemoreceptor

15. Correct statements concerning the inferior laryngeal nerve include which of the following?

(1) It has an internal branch that conveys sensation from the larynx superior to the vocal cords
(2) It produces muscle contraction that lengthens the (true) vocal folds
(3) It conveys taste from the epiglottis
(4) It innervates all of the laryngeal musculature by an external branch except the cricothyroid muscle

16. Muscles that are innervated by the ansa cervicalis include the

(1) omohyoid
(2) thyrohyoid
(3) sternohyoid
(4) geniohyoid

17. Structures that drain into the deep cervical lymph nodes include the

(1) occipital region
(2) nasal sinuses
(3) parotid gland
(4) teeth and gingivae

Directions: The group of questions below consists of lettered choices followed by several numbered items. For each numbered item select the **one** lettered choice with which it is **most** closely associated. Each lettered choice may be used once, more than once, or not at all.

Questions 18–23

For each reflex below, select the nerve that mediates the afferent limb of that reflex.

(A) CN V_1
(B) CN V_2
(C) CN V_3
(D) CN IX
(E) CN X

18. Carotid reflex
19. Corneal blink reflex
20. Cough reflex
21. Gag reflex
22. Jaw-jerk reflex
23. Sneeze reflex

ANSWERS AND EXPLANATIONS

1. The answer is E. *(Chapter 32 III A 6 c (3), B 3 c (3); Chapter 34 Table 34-1)* The buccal branch of the facial nerve is motor to the muscles of facial expression, including the buccinator muscle. These fibers originate in the facial nucleus of the brain stem and are not associated with a ganglion. The buccal branch of the trigeminal nerve is sensory to the cheek and has its cell bodies located in the trigeminal (semilunar) ganglion.

2. The answer is D. *(Chapter 33 IV D 2–3, V B 2 a–e)* The vallecular recesses lie to either side of the epiglottis. The piriform recesses lie inferior to the epiglottis and caudal to the lateral glossoepiglottic folds, which connect the epiglottis to the tongue.

3. The answer is C. *(Chapter 32 V C 4; Chapter 32 VII A 3 a)* The lesser superficial petrosal nerve, a continuation of the tympanic branch of the glossopharyngeal nerve, conveys parasympathetic preganglionic fibers to the otic ganglion for parotid secretion. The greater superficial petrosal nerve, a branch of the facial nerve, and the deep petrosal nerve convey the parasympathetic parasynaptic and sympathetic postsynaptic fibers, respectively, to the pterygopalatine fossa.

4. The answer is A. *(Chapter 32 VII A 2 f– g)* The maxillary premolar teeth are innervated by the middle-superior alveolar nerves. These teeth can be anesthetized effectively by infiltrating the maxillary nerve as it exits through the foramen rotundum into the pterygopalatine fossa.

5. The answer is B (1, 3). *[Chapter 29 V H 1 a (1); Chapter 32 III A 4 b (4), 7 a, B 3]* The semilunar or trigeminal ganglion, the equivalent of a dorsal root ganglion, contains the cell bodies of the sensory neurons of the trigeminal nerve (CN V). Selective damage to these neurons results in loss of sensation from the face and anterior two-thirds of the tongue. The motor division of the facial nerve supplies the muscles of facial expression and taste from the anterior two-thirds of the tongue.

6. The answer is D (4). *(Chapter 32 III A 5, 7 a–b)* That portion of the facial nerve that leaves the stylomastoid foramen is motor to the muscles of facial expression, so that injury here will cause facial paralysis only. The branch of the facial nerve to the stapedius muscle comes off higher in the facial canal, and the nerve to the lacrimal gland arises at the genu. The parotid gland is innervated by the tympanic branch of the glossopharyngeal nerve.

7. The answer is B (1, 3). *(Chapter 32 VI C 2–3)* The meniscus of the temporomandibular joint separates the superior and inferior compartments and receives the superior head of the lateral pterygoid muscle—the jaw protruder muscle. The inferior head inserts primarily onto the mandibular condyle and neck. The medial pterygoid muscle inserts onto the mandibular at the angle.

8. The answer is E (all). *(Chapter 32 VI D 3)* The branches of the nerves and vessels of the pterygopalatine fossa reach the nose, eye, and mouth through several foramina. The pterygopalatine fossa communicates with the middle cranial fossa via the foramen rotundum and by the vidian canal, with the nasal cavity via the sphenopalatine foramen, with the orbital cavity via the inferior orbital fissure, and with the oral cavity via the palatine canal.

9. The answer is B (1, 3). *[Chapter 32 VII A 2 g, 3 a (4)–(5)]* The postsynaptic parasympathetic neurons of the pterygopalatine ganglion control secretion of the nasal and oral mucosa as well as lacrimation. Injection of anesthetic into the pterygopalatine fossa, either by the inferior approach through the palatine canal or by the lateral approach through the pterygomaxillary fissure, anesthetizes the ganglion as well.

10. The answer is B (1, 3). *(Chapter 33 II B 3)* The maxillary, frontal, and anterior sinuses, as well as the middle ethmoid sinus, drain into the hiatus semilunaris within the middle nasal meatus. The sphenoid sinus and posterior ethmoid sinuses drain into the superior meatus. The nasolacrimal duct drains into the inferior meatus.

11. The answer is B (1, 3). *(Chapter 33 II F 3 b, 4 b)* The frontal and ethmoid sinuses drain by gravity with the head erect. The maxillary and sphenoid sinuses drain when the head is flexed forward or when the body is prone.

12. The answer is C (2, 4). *(Chapter 33 III D 6 d; Chapter 34 VI B 2)* The hypoglossal nerve (CN XII) innervates the somatic muscles of the tongue. The geniohyoid muscle is innervated by nerves that arise from C1 of the cervical plexus and along the distal portion of the hyoglossal nerve. Thus a brain stem lesion in the vicinity of the hypoglossal nucleus would not affect the geniohyoid innervation. The palatoglossus muscle is innervated by the branchiomeric motor division of the vagus nerve (CN X).

13. The answer is E (all). *(Chapter 33 IV C 2 c)* The blood supply to the palatine tonsil is derived from several sources, including two branches of the facial artery (tonsillar and ascending palatine), the tonsillar branch of the lingual artery, the ascending pharyngeal artery, and the lesser palatine branch of the maxillary artery.

14. The answer is E (all). *[Chapter 33 IV F 1 b (2) (b); Chapter 34 VII C 3 d (4)]* The carotid body and sinus, located at the bifurcation of the common carotid artery, monitor the partial pressure of the dissolved oxygen in the blood and the blood pressure, respectively. They are innervated by the carotid branch of the glossopharyngeal nerve which functions as the afferent limb of a reflex that controls the heart and respiratory rates.

15. The answer is D (4). *(Chapter 33 V E 4 e, F 1)* The recurrent (inferior) laryngeal nerve, a branch of the vagus nerve (CN X), innervates all of the muscles of the larynx except the cricothyroid muscle, which is innervated by the external branch of the superior laryngeal nerve and functions to lengthen the vocal folds. Taste from the epiglottic taste buds is conveyed by the superior laryngeal branch.

16. The answer is E (all). *(Chapter 34 VI C 1 a–b, 2)* The superior ramus of the ansa cervicalis, arising from C1 and accompanying the hyoglossal nerve, innervates the geniohyoid and thyrohyoid muscles. The inferior ramus, arising from C1 and C3, innervates the omohyoid and sternothyroid muscles.

17. The answer is C (2, 4). *(Chapter 34 V D 1 a–b)* The lymphatic drainage of the anteroinferior portion of the face, the nasal cavities, and the anterior portion of the oral cavity, including the anterior margin of the tongue, gingivae, and teeth, is through the submandibular lymph nodes to the deep cervical nodes. The lympathic drainage of the occipital region, the external ear, the parotid gland, and the anterosuperior portion of the face is toward the superficial cervical lymph nodes.

18–23. The answers are: 18-D, 19-A, 20-E, 21-D, 22-C, 23-B. *[Chapter 32 III B 3 a (3), c (5) (b); Chapter 33 IV F 1 b (1) (a), (2) (b); Chapter 33 V E 3 c (1) (b); Chapter 34 V B 1 d (1) (a)]* The trigeminal nerve mediates three reflexes: the ophthalmic division provides the afferent limb of the corneal blink reflex, the maxillary division the afferent limb of the sneeze reflex, and the mandibular division of both limbs of the jaw-jerk reflex. The glossopharyngeal nerve provides the afferent limbs of the gag and carotid reflexes. The vagus provides the afferent limb of the cough reflex.

Post-test

QUESTIONS

Directions: Each question below contains five suggested answers. Choose the **one best** response to each question.

1. A posterolateral herniation of the L4 to L5 intervertebral disk may affect

(A) the fourth lumbar nerve
(B) the fifth lumbar nerve
(C) the fourth and fifth lumbar nerves only
(D) the fourth and fifth lumbar and first and second sacral nerves
(E) none of the above

2. All of the following joints have more than one degree of freedom EXCEPT the

(A) humeroulnar joint
(B) metacarpophalangeal joints
(C) radiocarpal joint
(D) radiohumeral joint
(E) strernoclavicular joint

3. Epiphora (constant tearing down the face) may be caused by

(A) blockage of the nasolacrimal duct
(B) excessive activity of the parasympathetic components of CN III
(C) excessive activity of the parasympathetic components of CN IX
(D) injury to the cervical sympathetic chain
(E) lesion of the greater superficial petrosal nerve

4. Paralysis of the superior gluteal nerve results in an inability to

(A) fully extend the hip joint on the affected side
(B) fully extend the knee joint on the affected side
(C) maintain the pelvis level, so that it tilts toward the affected side with each step
(D) maintain the pelvis level, so that it tilts toward the unaffected side with each step
(E) perceive cutaneous sensation over the posterior aspect of the thigh

5. Characteristics of an indirect inguinal hernia include all of the following EXCEPT

(A) it is located adjacent to the spermatic cord
(B) it is located lateral to the inferior epigastric artery
(C) it originates lateral to the inguinal triangle
(D) it passes through the deep inguinal ring
(E) it passes through the superficial inguinal ring

6. Which of the following statements concerning internal (medial) rotation of the shoulder is true?

(A) It cannot occur if the pectoralis major muscle is paralyzed
(B) It is accomplished in part by a muscle that inserts onto the lesser tubercle
(C) It is accomplished in part by muscles innervated by the long thoracic nerve
(D) It is weakened by interruption of the suprascapular nerve
(E) It occurs around the anteroposterior axis of the glenohumeral joint

7. The femoral artery becomes the popliteal artery as it passes

(A) between the medial and lateral heads of the gastrocnemius muscle
(B) deep to the popliteus muscle
(C) deep to the soleus muscle
(D) into the posterior compartment of the leg
(E) through the adductor magnus muscle

8. All of the following problems usually accompany carpal tunnel syndrome EXCEPT

(A) paralysis of the opponens pollicis muscle
(B) paralysis of the abductor pollicis brevis muscle
(C) paralysis of lumbricales 1 and 2
(D) paralysis of the oblique head of the adductor pollicis muscle
(E) sensory disturbances on the radial half of the distal palm and first three digits

9. One means of alleviating portal hypertension is by a side-to-side anastomosis between the hepatic portal vein and the inferior vena cava. The location of the inferior vena cava is

(A) across the epiploic foramen from the portal vein
(B) anterior and to the left of the portal vein
(C) immediately to the right of the portal vein
(D) posterior and to the right of the portal vein
(E) in none of the above places

10. The posterior muscles of the back receive motor innervation from

(A) dorsal primary rami
(B) dorsal roots
(C) posterior branches of the lateral perforating nerves
(D) ventral primary rami
(E) none of the above

11. An expanding pituitary adenoma eroding the pituitary fossa and compressing the optic chiasr produces which of the following symptoms?

(A) Binasal heteronymous hemianopsia
(B) Bitemporal heteronymous hemianopsia
(C) Contralateral homonymous hemianopsia
(D) Ipsilateral homonymous hemianopsia
(E) Total blindness of the right eye

12. All of the following areas in the male are drained by inguinal lymph nodes EXCEPT the

(A) leg and foot
(B) lower abdominal wall
(C) lower part of the anal canal
(D) penis
(E) testes

13. The cell bodies that contribute to the deep petrosal nerve and dilate the pupil are located in the

(A) ciliary ganglion
(B) geniculate ganglion
(C) otic ganglion
(D) superior cervical ganglion
(E) superior glossopharyngeal ganglion

14. The annular ligament prevents dislocation of the

(A) distal end of the clavicle from the acromion process
(B) head of the humerus from the glenoid cavity
(C) head of the ulna from the triquetrum
(D) head of the radius from the radial notch of the ulna
(E) head of the third metacarpal from the base of the proximal phalanx

15. A drug addict ("mainliner") is diagnosed as having tricuspid insufficiency secondary to bacterial endocarditis (infection of the endocardium and valve leaflets). All of the following findings might be noted EXCEPT

(A) abdominal ascites
(B) distended neck veins
(C) pulmonary edema
(D) swelling of the ankles
(E) systolic murmur in the fifth right intercostal space

16. An epidural hematoma is usually caused by

(A) fracture of the diploic space
(B) laceration of the superficial temporal artery
(C) leakage from a cerebral vein
(D) rupture of a cerebral artery
(E) tearing of a meningeal vessel

17. In portal hypertension where flow of blood to or through the liver is impeded, blood will return to the systemic circulation by all of the following veins EXCEPT the

(A) esophageal veins
(B) gonadal veins
(C) hemorrhoidal veins
(D) renal veins
(E) veins of Retzius associated with the ascending colon

18. The cell bodies for the sensory fibers that convey pain, touch, and temperature from the anterior two-thirds of the tongue are located in the

(A) geniculate ganglion
(B) pterygopalatine ganglion
(C) semilunar ganglion
(D) submandibular ganglion
(E) petrosal ganglion

19. Which of the following statements concerning the transverse vertebral foramen is correct?

(A) It is associated with the transverse processes of the thoracic vertebrae
(B) It is not patent in the adult
(C) It is not present in the atlas
(D) It provides a pathway for the vertebral artery, except for C7
(E) It transmits the spinal nerve

20. All of the following muscles are involved in dorsiflexion of the ankle joint EXCEPT the

(A) extensor digitorum longus
(B) extensor hallucis longus
(C) peroneus longus
(D) peroneus tertius
(E) tibialis anterior

21. The apex of the heart is normally located

(A) at the level of the fifth thoracic vertebra in the midclavicular line
(B) deep to the left third intercostal space in the midclavicular line
(C) deep to the left fifth intercostal space in the midclavicular line
(D) directly in the midline beneath the xiphoid process
(E) in none of the above places

22. The neural supply to the maxillary incisor teeth is described by all of the following statements EXCEPT

(A) These fibers are accompanied by motor fibers that exit via the infraorbital foramen
(B) These fibers are at risk if the anterior maxillary wall is fractured
(C) These fibers exit from the skull via the foramen rotundum
(D) These fibers cross the pterygopalatine fossa
(E) These fibers run in the inferior orbital fissure

23. Which of the following muscles is most efficient at extending the knee when the hip joint is extended?

(A) Biceps femoris
(B) Rectus femoris
(C) Vastus intermedius
(D) Vastus lateralis
(E) Vastus medialis

24. Tapping the patellar ligament activates afferent impulses from muscle spindles and elicits a knee-jerk reflex, testing which of the following spinal nerves or spinal cord levels?

(A) T12–L2
(B) L2–L4
(C) L4–S1
(D) S1–S3
(E) S3–S5

25. The secretory function of the parotid gland is likely to be effected by

(A) a severe or prolonged middle ear infection
(B) anesthesia of structures exiting the stylomastoid foramen
(C) facial nerve palsy
(D) severing the glossopharyngeal nerve as it passes between the superior and middle constrictors
(E) severance of the cervical sympathetic chain

26. A diastolic murmur heard most distinctly in the second intercostal space to the right of the sternum is most attributable to

(A) aortic insufficiency
(B) mitral stenosis
(C) patent ductus arteriosus
(D) pulmonary stenosis
(E) tricuspid insufficiency

27. Which of the following nerves is jeopardized by a fracture of the proximal neck of the fibula?

(A) Common peroneal
(B) Saphenous
(C) Sural
(D) Tibial
(E) None of the above nerves

Directions: Each question below contains four suggested answers of which **one or more** is correct. Choose the answer

A if **1, 2, and 3** are correct
B if **1 and 3** are correct
C if **2 and 4** are correct
D if **4** is correct
E if **1, 2, 3, and 4** are correct

28. A 64-year-old man with severe angina pectoris is found by coronary angiography to have a 90 percent left coronary artery occlusion. A venous homograph is not possible because of severe venous varicosities; thus, the patient is treated surgically by diverting his left internal thoracic artery into the left coronary artery distal to the occlusion. Postoperatively the chest wall region originally supplied by the left internal thoracic artery receives blood flow from the

(1) left posterior intercostal arteries
(2) left pericardiophrenic artery
(3) left inferior epigastric artery
(4) posterior descending coronary artery

29. Cranial nerves with parasympathetic components include

(1) CN III
(2) CN VII
(3) CN IX
(4) CN X

30. A surgeon identifies a segment of gut as jejunum. He forms his decision based upon which of the following external characteristics of the jejunum?

(1) An arterial supply with few arcades
(2) A relatively thick wall
(3) Long arteriae rectae
(4) A mesentery that is thicker due to fat deposits

31. Which of the following muscles are flexors of the thigh?

(1) Iliopsoas
(2) Rectus femoris
(3) Sartorius
(4) Pectineus

32. An occlusion of the axillary artery does not seriously interfere with the blood supply to the upper limb because of circumscapular anastomoses. Branches of the axillary and subclavian arteries that contribute to this anastomosis include the

(1) transverse cervical artery
(2) posterior humeral circumflex artery
(3) suprascapular artery
(4) subscapular artery

33. The ethmoid bone, an endochondral bone, contributes to the

(1) cribriform plate
(2) superior concha
(3) lamina papyracea of the orbit
(4) inferior concha

34. True statements about the prostate gland include which of the following?

(1) If its median lobe is enlarged, it will obstruct the urethra
(2) It empties into the prostatic urethra on either side of the urethral crest
(3) The dorsal vein of the penis empties into a plexus of veins around its base
(4) It contributes fructose and choline to seminal fluid

35. A fall on the outstretched hand can result in a broken clavicle rather than a clavicular dislocation. Resistance to dislocation of the clavicle is provided by the

(1) coracoclavicular ligament
(2) combined contractions of the subclavius and pectoralis muscles
(3) articular disk and its attachments at the sternoclavicular joint
(4) actions of the rotator cuff muscles

36. A horizontal incision that transects the anterior wall of the rectus sheath at the level of the umbilicus will cut through the

(1) aponeurosis of the internal oblique muscle
(2) aponeurosis of the transverse abdominis muscle
(3) aponeurosis of the external oblique muscle
(4) transversalis fascia

37. In the adult, the blood supply to the head of the femur is by the

(1) lateral femoral circumflex artery
(2) obturator artery
(3) medial femoral circumflex artery
(4) artery of the ligamentum teres

38. A fracture that passes through the right occipital condyle from the anterior condylar canal to the posterior condylar canal may involve

(1) loss of the gag reflex on the ipsilateral side
(2) inability to direct the tongue to the opposite side
(3) the vertebral artery
(4) an emissary vein

39. Which of the following structures are both medial to the biceps tendon and deep to the bicipital aponeurosis?

(1) Branchial artery
(2) Deep brachial artery
(3) Median nerve
(4) Median cubital vein

40. A fracture of the sphenoid bone that passes through the foramen ovale may result in paralysis to which of the following muscles?

(1) Tensor tympani
(2) Posterior belly of the digastric
(3) Masseter
(4) Buccinator

41. A 24-year-old woman is taken to the emergency room following 12 hours of nausea, vomiting, and right lower quadrant pain. She has a temperature of 101°F, white blood count (WBC) of 13,700/mm^3 (normal 5,000–10,000), and evidence of peritonitis in the lower right quadrant. She is brought to surgery for an emergency appendectomy. At operation, the appendix is normal. Alternative causes of this person's illness include

(1) ovarian tortion
(2) Meckel's diverticulitis
(3) salpingitis
(4) urinary tract infection

42. An elderly man is taken to the emergency room because he has an object lodged in his throat. A small steak bone is removed by esophagoscopy and a small posterior tear of the esophagus is noted. The patient is admitted and placed on broad-spectum antibiotic therapy to prevent infection from developing in the

(1) anterior mediastinum
(2) posterior mediastinum
(3) pretracheal space
(4) retrovisceral space

43. Extension of a malignant tumor in the apical segment of the upper lobe of the right lung may involve

(1) diminshed right axillary pulse
(2) vocal hoarseness
(3) bilateral facial venous engorgement
(4) paradoxical movement of the right hemidiaphragm

SUMMARY OF DIRECTIONS

A	B	C	D	E
1, 2, 3 only	1, 3 only	2, 4 only	4 only	All are correct

44. A surgeon making a deep lumbar incision along the twelfth rib to perform a right adrenalectomy notes the bright red blood in the field grow suddenly dark. He knows that he has probably

(1) incised the spleen
(2) opened the parietal peritoneum
(3) punctured the renal capsule
(4) entered the costodiaphragmatic recess

45. True statements about the postcentral gyrus of the parietal lobe include which of the following?

(1) It is somatotopically organized
(2) It receives sensory input relayed from the thalamus
(3) It is concerned with the opposite side of the body
(4) It is the site of interpretation and recognition of sensory information

46. Which of the following statements concerning the scaphoid bone are true?

(1) It participates in the midcarpal joint
(2) It is the most susceptible of the carpal bones to fracture
(3) It receives an attachment for the transverse carpal ligament
(4) It articulates maximally with the radius in adduction

47. Structures found in the deep perineal pouch include the

(1) external urethral sphincter muscle in the female
(2) bulbourethral glands of the male
(3) deep transverse perineal muscle
(4) vestibular glands of the female

48. A young man suffers a knee injury in a rugby scrum. The physician diagnoses tears of the medial collateral ligament, the medial meniscus, and the anterior cruciate ligament. The signs that lead to this diagnosis include

(1) abnormal abduction of the knee
(2) abnormal adduction of the knee
(3) locking of the knee in partial flexion
(4) abnormal posterior movement of the tibia on the femur

49. The movement of the ribs during thoracic expiration involves

(1) decrease of the transverse thoracic diameter
(2) movement at the costovertebral joints
(3) release of stored energy within the costal cartilages
(4) inward rotation of the ribs

50. The right adrenal gland receives its blood supply from the

(1) renal artery
(2) aorta, direct
(3) inferior phrenic artery
(4) superior mesenteric artery

51. In addition to the common carotid artery, the carotid sheath contains the

(1) external carotid artery
(2) sympathetic chain
(3) vagus nerve
(4) external jugular vein

52. A young man with no prior medical problems is confined in bed for 6 weeks after breaking his leg. The day after sitting in a chair for the first time since the accident, the patient develops mild weakness in the right arm and facial muscles and slurring of speech. A cerebral embolus from the legs is suspected, but a pulmonary embolus is expected. The anatomic explanation for this "paradoxical embolus" might be

(1) a portacaval anastomosis
(2) a patent ductus arteriosus
(3) a thrombus in the right atrial appendage
(4) an atrioseptal defect

Directions: The groups of questions below consist of lettered choices followed by several numbered items. For each numbered item select the **one** lettered choice with which it is **most** closely associated. Each lettered choice may be used once, more than once, or not at all.

Questions 53–56

For each organ listed below, select the region to which pain in that organ is usually referred.

(A) Inguinal and pubic regions
(B) Perineum, posterior thigh, and leg
(C) Both
(D) Neither

53. The ovary
54. The uterus
55. The epididymis
56. The testis

Questions 57–60

For each functional event listed below, select the stage of the cardiac cycle in which it is most likely to occur

(A) Atrial systole
(B) Ventricular diastole
(C) Both
(D) Neither

57. The atrioventricular valves close
58. The semilunar valves close
59. Blood flow through the myocardium is maximal
60. Blood flows into the ventricles

ANSWERS AND EXPLANATIONS

1. The answer is B. *[Chapter 19 I E 4 b (5) (b)]* Because the inferior intervertebral notch of the lumbar vertebrae is so deep, a posterolateral herniation of the L4–L5 intervetebral disk would not affect the fourth lumbar nerve. However, the herniation will place pressure on the subsequent nerve (L5) as it passes toward its intervertebral foramen. If the herniation is large, it may also involve subsequent roots and even involve roots across the midline.

2. The answer is A. *(Chapter 7 III A 1 a)* The particular shape of the trochlea of the humerus and the trochlear notch of the ulna results in only one degree of freedom at the humeroulnar joint, permitting only flexion and extension of the elbow. All of the other joints listed have two degrees of freedom.

3. The answer is A. *[Chapter 30 I B 9 d (4) (b)]* Blockage of the nasolacrimal duct results in epiphora as does excessive stimulation of the parasympathetic component of the facial nerve (CN VII). Conversely, a lesion of the greater superficial petrosal nerve results in a loss of lacrimation. Sympathetic stimulation, or lack thereof, appears to have little effect on lacrimation.

4. The answer is D. *(Chapter 23 II E 1 d)* The superior gluteal nerve innervates the gluteus medius and gluteus minimus muscles, which, as strong abductors of the thigh, act to maintain the pelvis level when the opposite leg is lifted from the ground. A superior gluteal nerve palsy results in a downward tilt of the pelvis when the leg of the unaffected side is lifted off the ground (abductor lurch).

5. The answer is A. *[Chapter 15 X E 3 c (1)]* An indirect inguinal hernia is located *within* not adjacent to the spermatic cord. The direct inguinal hernia is located adjacent to the spermatic cord, medially. The indirect inguinal hernia passes through the deep and the superficial inguinal rings. It must also pass lateral to the inferior epigastric artery to enter the inguinal canal. This places its origin lateral to the inguinal triangle.

6. The answer is B. *(Chapter 6 II D 4 g (3); Table 6-2)* The prime internal rotator of the glenohumeral joint is the subscapularis muscle, which inserts on the lesser tubercle of the humerus. It is innervated by the upper subscapular nerve, as well as in small part by the lower subscapular nerve. Rotation of the arm occurs about a vertical axis through the glenohumeral joint.

7. The answer is E. *(Chapter 24 V A 2 f, 3)* Upon passing through the adductor canal (Hunter's), which is between the hamstring and anterior portions of the adductor magnus muscle, the femoral artery becomes the popliteal artery.

8. The answer is D. *[Chapter 10 VII B 3 (b), C 1 (b)]* Both the oblique and transverse heads of the adductor pollicis muscle, as well as the deep head of the flexor pollicis and lumbricales 3 and 4, are innervated by the ulnar nerve. Because the ulnar nerve does not pass beneath the transverse carpal ligament, the muscles it innervates are unaffected by carpal tunnel syndrome.

9. The answer is A. *(Chapter 16 V E 3; Chapter 17 VI B 3 a (1); XIII D 6 a)* The inferior vena cava lies posterior to the epiploic foramen. The hepatic portal vein lies anterior to the epiploic foramen in the hepatoduodenal ligament, where it is posterior to the hepatic artery and common bile duct.

10. The answer is A. *(Chapter 19 II B 5)* The intrinsic musculature of the back receives innervation from the dorsal primary rami of the spinal nerves. The ventral primary rami innervates the superficial muscles of the back as well as those of the anterior and lateral body wall and extremities.

11. The answer is B. *[Chapter 29 V A 2 c (3) (b)]* Compression of the optic chiasm involves the decussating nerve fibers from the nasal portions of each retina. Because the lateral visual fields are projected onto the nasal portion of the retina, the result is bitemporal heteronymous hemianopsia (tunnel vision).

12. The answer is E. *(Chapter 22 V A 5 c)* The lymphatic drainage of the testes is primarily along the testicular neurovascular bundle to the para-aortic lymph nodes. Portions of the penis, lower abdominal wall, leg, and foot drain to the inguinal nodes. Portions of the penis and anal canal drain along the internal pudendal vessels to the pelvic nodes.

13. The answer is D. *[Chapter 30 I D 4 a (2); Chapter 34 VI D 2 b (1)]* Postsynaptic sympathetic neurons that innervate the head are located in the superior cervical ganglion. Some of these form the deep petrosal nerve, which bring sympathetic fibers into the pterygopalatine fossa. Other neurons innvervate the smooth muscle (Müller's) in the eyelid, which prevents ptosis, and also innervate the iridial dilator muscle of the eye.

14. The answer is D. *[Chapter 7 III A 3 b (1)]* The annular ligament stabilizes the proximal radioulnar joint, allowing the radius to rotate relative to the ulnar in pronation and supination.

15. The answer is C. *(Chapter 13 XI A 1 a)* Mitral insufficiency produces pulmonary edema. Tricuspid insufficiency, evident as a blowing systolic murmur in the right fifth intercostal space, produces systemic venous congestion, which in turn results in pitting edema and ascites.

16. The answer is E. *(Chapter 28 IV B 2, 3)* A meningeal artery tear produces an epidural hematoma. Leakage from cerebral veins results in extravasation of blood in the subdural space (subdural hematoma). A ruptured cerebral artery bleeds into the subarachnoid space. Rupture of a superficial vessel, such as the superficial temporal artery, produces an epicranial hematoma.

17. The answer is B. *(Chapter 17 XIII D 2–5)* The hepatic portal anastomoses with the systemic system about the esophagus, the rectum, the umbilicus, and between secondarily retroperitoneal sections of the gut and dorsal body wall. There are no anastomoses between the gondal veins and those of the gut.

18. The answer is C. *(Chapter 29 V H 1 a (1); Chapter 33 III D 6 a, b, E 3)* The semilunar (trigeminal) ganglion, the equivalent of a dorsal root ganglion, contains the cell bodies of the sensory neurons of the trigeminal nerve (CN V). The lingual branch of the mandibular division of the trigeminal nerve conveys pain, touch, and temperature sensation from the anterior two-thirds of the tongue.

19. The answer is D. *[Chapter 19 I D 1 c (2) (b)]* The costal and transverse processes of the developing cervical vertebrae fuse about the vertebral artery to form the transverse foramina. While each cervical vertebrae has a pair of transverse foramina, the vertebral artery does not pass through those of C7.

20. The answer is C. *(Chapter 25 III C 1 b (1); Figure 25-4)* The peroneus longus muscle, passing posterior to the lateral malleolus, is posterior to the transverse axis of the ankle joint and is therefore a plantar flexor of the foot.

21. The answer is C. *(Chapter 13 III B 2 c)* The apex of the heart, formed by the tip of the left ventricle, lies deep to the fifth intercostal space in the midclavicular line. Here the apical pulse may be palpated.

22. The answer is A. *[Chapter 31 III F 1; Chapter 32 VII A 2 c, f (1)]* The innervation of the maxillary incisor teeth is by the anterior, middle, and posterosuperior alveolar branches of the maxillary division of the trigeminal nerve (CN V). The maxillary division of the trigeminal is purely sensory.

23. The answer is B. *[Chapter 3 III C 2; Chapter 24 III B 2 a–e).* The rectus femoris, spanning both hip and knee joints, is stretched when the hip is extended and therefore has greater intrinsic force-generating capacity. The three vasti do not span the hip joint, and the biceps femoris, also a two-joint muscle, is a knee flexor.

24. The answer is B. *(Chapter 24 VI B 1 c)* The afferent and efferent limbs of the knee-jerk reflex are carried by the femoral nerve, which arises mainly from spinal segments L2–L4.

25. The answer is A. *(Chapter 30 III B 1 f (4) (a); Chapter 32 V C 1–3; Chapter 33 III E)* The parotid gland is innervated by the tympanic branch of the glossopharyngeal nerve. This nerve passes through the temporal bone and enters enters the middle ear cavity where it forms the tympanic plexus on the tympanic bulla. It is at risk during otitis media.

26. The answer is A. *(Chapter 13 IX C 2 c)* The closure of the aortic valve is heard most distinctly in the second intercostal space to the right of the sternum. An insufficient aortic valve produces a diastolic murmur. The pulmonic valve is best heard in the second intercostal space to the left of the sternum, and a patent ductus arteriosus produces a constant rumbling murmur to the left of the pulmonic area.

27. The answer is A. *(Chapter 25 VI B)* The common peroneal nerve passes lateral to the neck of the fibula, where it is subject to blunt trauma or involvement in a fibular fracture. Common peroneal nerve palsy results in "foot drop."

28. The answer is B (1, 3). *(Chapter 11 V B 2 a–c)* The internal thoracic artery and its musculophrenic branch anatomose with the posterior intercostal arteries. The superior epigastric branch anastomoses with the inferior epigastric artery.

29. The answer is E (all). *(Chapter 29 V E 1–4)* The parasympathetic neurons for pupillary reflexes and accommodation run in the oculomotor nerve (CN III). Those for lacrimation, as well as those for nasal, submandibular, and sublingual secretion, initially run with the facial nerve (CN VII). Those for parotic secretion run initially with the glossopharyngeal nerve (CN IX). The vagus nerve contains many parasympathetic functions associated with the thoracic and abdominal viscera.

30. The answer is A (1, 2, 3). *(Chapter 17 Table 17-1, Figure 17-12)* The mesentery of the jejunum is thin and transparent because there are few fat deposits in this portion. In addition, the jejunum has a greater diameter than the ileum.

31. The answer is E (all). *(Chapter 23 II C 3 a (1); Table 23-1)* The iliopsoas, rectus femoris, sartorius, and pectineus muscles are all flexors of the thigh and, as such, oppose the actions of the hamstring group and gluteus maximus muscle.

32. The answer is E (all). *[Chapter 6 III C 1 c (3) (b)–(c), 2 a (3)]* The axillary artery anatomoses in the circumscapular area with all of the arteries listed in the question. Because of the profuse anastomotic vascular circulation, blockage or ligation of the axillary artery proximal to the posterior humeral circumflex artery has little effect on the circulation to the upper extremity.

33. The answer is A (1, 2, 3). *[Chapter 28 III C 2 b (1) (b) (iii); Chapter 30 I A 1 b (1)]* The ethmoid bone is shared between the neurocranium and the facial skeleton. Medially, it forms the cribriform plate, a portion of the nasal septum. Laterally, it forms the lamina papyracea portion of the medial orbital walls as well as the superior and middle conchae.

34. The answer is A (1, 2, 3). *(Chapter 21 IV F 2; Chapter 22 II C 1 a; V E 1–4)* The prostate gland discharges on either side of the urethral crest of the prostatic urethra. It contributes phosphatase, citric acid, and fibrinogen to the seminal fluid. The prostatic venous plexus receives the dorsal vein of the penis. An enlarged median lobe (benign prostatic hypertrophy) projects into the urinary bladder and prevents complete emptying.

35. The answer is B (1, 3). *(Chapter 6 II D 1, 2)* The articular disk and strong sternoclavicular ligaments stabilize the clavicle proximally; the coracoclavicular ligament stabilizes the clavicle distally. This stability, along with the pronounced sigmoid shape of the clavicle, predisposes the clavicle to fracture rather than dislocation.

36. The answer is B (1, 3). *(Chapter 15 IV C 2 a; IX A 1)* The anterior leaf of the rectus sheath superior to the umbilicus is formed by the fused aponeuroses of the external oblique and internal oblique muscles.

37. The answer is B (1, 3). *(Chapter 23 II D 2)* The retinacular of the anterior and posterior femoral circumflex arteries supply the neck and head of the adult femur. The artery of the ligamentum teres, a branch of the obturator artery, degenerates in the adult and becomes of marginal significance at most. This circumstance is responsible for the high incidence of femoral head necrosis accompanying fracture of the femoral neck.

38. The answer is C (2, 4). *(Chapter 28 III B 5 b (3)–(4); Chapter 29 V C 2)* The anterior condylar (hypoglossal) canal transmits the hypoglossal nerve (CN XII), which innervates the muscle of the tongue. The posterior condylar canal transmits an emissary vein, and hemorrhage may produce a hematoma in the vicinity of the mastoid process. The gag reflex is associated with the glossopharyngeal nerve (CN IX), which transits the jugular foramen. The vertebral artery enters the cranial cavity through the foramen magnum.

39. The answer is B (1, 3). *(Chapter 7 IV A 3, C 3 b; V B 2)* In the cubital fossa the median nerve and the brachial artery lie beneath the bicipital aponeurosis and just medial to the biceps tendon. The median cubital vein is superficial to the bicipital aponeurosis.

40. The answer is B (1, 3). *[Chapter 32 III B 3 c (5) (a)–(c)]* In addition to the muscles of mastication. the motor division of the mandibular nerve innervates the tensor veli palatini, the tensor tympani, and the mylohyoid muscles as well as the anterior belly of the digastric muscle. The posterior belly of the digastric and the buccinator muscle are both innervated by the facial nerve.

41. The answer is E (all). *[Chapter 17 IX F 3; X B 2 h (1) (d)–(e); XV B; Table 17-2; Chapter 22 II B 3; III; VI C 5 b; VI C 7 a; VI D 2 g (4)]* The urinary tract, the ovary, and the oviduct refer pain to the right lower quadrant, while the appendix and Meckl's diverticulum initially refer to the umbilical region. Peritonitis secondary to inflammation of any of these structures will produce peritoneal pain in the right lower quadrant.

42. The answer is C (2, 4). *[Chapter 27 I B 2 e (1) (a), (b)]* The retrovisceral space lies posterior to the esophagus, and infection within this space may spread inferiorly to produce posterior mediastinitis. The pretracheal space lies anterior to the trachea and is continuous with the superior mediastinum and anterior mediastinum.

43. The answer is E (all). *(Chapter 12 III B 8 a; Chapter 14 II H 5)* Because of the proximity of the apical bronchopulmonary segment of the right lung to the right subclavian artery and right brachiocephalic vein, a tumor in this segment could compress these vascular structures, diminishing axillary pulse and restricting venous return from the head and extremities. In addition, involvement of the recurrent laryngeal nerve, which passes around the subclavian artery, may produce hoarseness, whereas involvement of the phrenic nerve may produce paralysis of a hemidiaphragm.

44. The answer is D (4). *(Chapter 12 II D 2 d (2); Chapter 18 IV E 4 d)* The pleural cavity extends inferior to the twelfth rib, posteriorly. Extension of an abdominal incision into the costodiaphragmatic recess of the pleural cavity will result in pneumothorax.

45. The answer is A (1, 2, 3). *[Chapter 29 II A 3 e (2)]* The postcentral gyrus of the parietal lobe is the primary sensory area, receiving information from the opposite side of the body relayed through the thalamus. This region of the cortex is somatotopically organized so that specific regions correspond to various regions of the body. The adjacent sensory association areas interpret and integrate sensory stimuli and enable recognition.

46. The answer is A (1, 2, 3). *(Chapter 8 II B 1 a; III A 2 f)* The scaphoid bone, articulating maximally with the radius in abduction, transmits the major forces from the hand to the forearm and is therefore the carpal bone most prone to fracture. The scaphoid provides a strong attachment for the transverse carpal ligament and participates in both the radiocarpal and midcarpal joints.

47. The answer is A (1, 2, 3). *(Chapter 21 IV D 1; V C 4 f; D 4)* The deep perineal pouch contains the deep transverse perineal muscle and its specialized central portion, the external urethral sphincter, and the bulbourethral glands in the male. The vestibular glands of the female are in the superficial pouch.

48. The answer is B (1, 3). *(Chapter 24 II C 3 b (5), 4 b (2) (d); IV B 3 c)* The "unhappy triad" described in the case is the result of rotation injury of the partially flexed knee. A torn medial collateral ligament permits abnormal abduction and a torn anterior cruciate permits *anterior* displacement of the tibia on the femur (anterior drawer sign). A torn medial meniscus may get caught between the articular surfaces, locking the knee in partial flexion.

49. The answer is E (all). *(Chapter 11 VII C 3; Figure 11-7A)* The expiratory action of the internal intercostal muscles and the release of stored energy within the costal cartilages (extrinsic elastic recoil) result in *downward* movement of the ribs about an axis described by the costovertebral joints as well as an *inward* rotation of the ribs. The transverse diameter of the thoracic cage is thereby diminished.

50. The answer is A (1, 2, 3). *(Chapter 18 III C 1)* Each adrenal gland receives arterial branches from the inferior phrenic arteries, the aorta, and the renal arteries. There normally are no arterial anastomoses between systemic arteries and the superior mesenteric artery.

51. The answer is B (1, 3). *(Chapter 27 I B 2 c)* The carotid sheath encloses the common carotid artery and, after its bifurcation, the internal and external carotid arteries. In addition, it contains the internal jugular vein, the vagus nerve, and the carotid branch of the glossopharyngeal nerve.

52. The answer is D (4). *(Chapter 13 XI B 2 a; Chapter 5 Figure 5-1)* An atrioseptal defect can be asymptomatic for most of ones life because there is normally a left to right shunt in these individuals. Problems only arise if a right to left shunt should occur due to a pathophysiologic disturbance or, as in this case, an embolus traverses the atrioseptal defect. The ductus arteriosus enters the aorta after the carotids are given off.

53–56. The answers are: 53-A, 54-C, 55-B, 56-A. *[Chapter 22 II A 3 b–c, V A 6 b, B 4 b; VI C 5 b; D 2 g (4)]* Afferent nerves from the pelvic viscera travel along autonomic pathways. Afferents from the ovary, testis, upper to middle ureter, uterine tubes, urinary bladder, and uterine body travel along the least splanchnic nerve to the lower thoracic segment and along the lumbar splanchnic nerves to the upper lumbar segments of the spinal cord; thus, pain is referred to the inguinal and pubic regions as well as the lateral and anterior aspects of the thigh. Afferents from the epididymis, uterine cervix, and distal ureter travel along the pelvic splanchnic nerves to the midsacral spinal segments; thus, pain is referred to the perineum, posterior thigh, and leg.

57–60. The answers are: 57-D, 58-B, 59-B, 60-C. *(Chapter 13 IV A 4 b; IX A 1–4)* The ventricles fill during ventricular diastole and atrial systole. The atrioventricular valves close in early ventricular diastole. The blood flow through the coronary circulation is maximal during ventricular diastole when the pressure differential between the aorta and myocardium is maximal.

Index

Individual arteries, bones, muscles, nerves, and veins are listed under those headings (e.g., Bones, hyoid; Arteries, carotid). Other structures are listed alphabetically throughout the index (e.g., Knee joint).

Page numbers in italics designate figures; numbers followed by (t) designate tables.

A

B

C

D

E

F

J

K

L

N

O

P

Q

R

T

U

W

X - Y - Z

anatomy

The National Medical Series for Independent Study

YOU CAN USE THIS BOOK FOR

- Course study
- Review
- Exam preparation

The National Medical Series for Independent Study presents a new system for effective learning and review. It includes outlines of the basic and clinical sciences, board-type questions, and annotated answers. The books in the Basic Science Group have been designed specifically for the medical student who wants to:

- Master large amounts of information in a limited amount of time
- Prepare for FLEX and Part I of the National Boards

LOOK FOR THE OTHER TITLES IN THE BASIC SCIENCE GROUP

- biochemistry
- histology and embryology
- microbiology
- pathology
- pharmacology
- physiology

ALSO LOOK FOR THE NMS™ *CLINICAL SCIENCE GROUP*

- medicine
- obstetrics and gynecology
- pediatrics
- preventive medicine and public health
- psychiatry
- surgery

0-471-09624-5